Organic Photovoltaics

OPTICAL ENGINEERING

Founding Editor
Brian J. Thompson
University of Rochester
Rochester, New York

Organic Photovoltaics: Mechanisms, Materials, and Devices

Edited by Sam-Shajing Sun and Niyazi Serdar Sariciftci

CRC Press
Taylor & Francis Group
Boca Raton London New York

CRC Press is an imprint of the
Taylor & Francis Group, an **informa** business

Back cover illustration by Marcia Minwen Sun. 6-year-old daughter of the editor Dr. Sam-Shajing Sun.

CRC Press
Taylor & Francis Group
6000 Broken Sound Parkway NW, Suite 300
Boca Raton, FL 33487-2742

© 2005 by Taylor & Francis Group, LLC
CRC Press is an imprint of Taylor & Francis Group, an Informa business

First issued in paperback 2019

No claim to original U.S. Government works

ISBN 13: 978-0-367-44649-9 (pbk)
ISBN 13: 978-0-8247-5963-6 (hbk)

Library of Congress Cataloging-in-Publication Data

Organic photovoltaics : mechanism, materials, and devices / [edited by] Sam-Shajing Sun, Niyazi Serdar Sariciftci.
    p. cm. -- (Optical engineering)
    Includes bibliographical references and index.
    ISBN 0-8247-5963-X (alk. paper)
    1. Photovoltaic cells. 2. Organic semiconductors. I. Sun, Sam-Shajing. II. Sariciftci, Niyazi Serdar. III. Optical engineering (Marcel Dekker, Inc.)

TK8322.O73 2004
621.3815'42--dc22
    2004059378

Visit the Taylor & Francis Web site at
http://www.taylorandfrancis.com

and the CRC Press Web site at
http://www.crcpress.com

This book is dedicated to the 50th anniversary of the invention of silicon p/n junction type solar cells (G. Pearson, D. Chaplin, and C. Fuller at Bell Labs), the 20th anniversary of the invention of organic donor/acceptor binary type solar cells (C.W. Tang at Eastman Kodak), and the 50th anniversary of the founding of the International Solar Energy Society (ISES).

Foreword 1

Photovoltaic materials can be used to fabricate solar cells and photo detectors. Inorganic photovoltaic devices were first demonstrated at the Bell Laboratories more than 50 years ago. Today, silicon solar cells are "big business". Their initial applications were in earth satellites. A wider range of applications quickly emerged. Because solar energy is perhaps the most obvious renewable energy source, large-scale application of solar cell technology for the production of energy for our future civilization is, and must be, a high priority — a priority that becomes ever more important as oil prices continue to increase and fossil fuel burning continues to degrade the global environment. The silicon solar cell technology suffers, however, from two serious disadvantages: The production cost is relatively high, and the rate at which new solar cell area can be produced is limited by the basic high-temperature processing of silicon.

Semiconducting polymers (and organic materials, more generally) provide an alternative route to solar cell technology. Since these plastic materials can be processed from solution and printed onto plastic substrates, they offer the promise of being lightweight, flexible, and inexpensive. Moreover, web-based printing technology is capable of generating large areas with rapid throughput. Although the overall power conversion efficiency of current organic solar cell is relatively low compared to that available from silicon technology, the efficiency can be improved through systematic molecular engineering and the development of device architecture that is optimally matched to the properties of these new photovoltaic materials. *Organic Photovoltaics: Mechanisms, Materials, and Devices*, edited by Sun and Sariciftci, is a very timely and comprehensive volume on the subject. It is certainly an important starting point and reference for all those involved in research and education in this subject and related fields.

Alan J. Heeger, Nobel Laureate
University of California at Santa Barbara

Foreword 2

The first true inorganic solar cell had its roots in a two-phase process presently called photoelectrochemistry. In 1839 Becquerel observed that a photovoltage resulted from the action of light on an electrode in an electrolytic solution. In the 1870s it was discovered that the solid material selenium demonstrated the same effect and by the early 1900s selenium photovoltaic cells were widely used in photographic exposure meters. By 1914 these cells were still less than 1% efficient. In 1954, Chapin reported a solar conversion efficiency of 6% for a single-crystal silicon cell, marking the beginning of modern-day photovoltaics. By 1958, small-area silicon solar cells had reached an efficiency of 14% under terrestrial sunlight.

On March 17, 1958, the world's first solar powered satellite Vanguard 1 was launched. It carried two separate radios: the battery-powered transmitter operated for 20 days and the solar cell powered transmitter operated until 1964, when it is believed that the transmitter circuitry failed. Setting a record for satellite longevity at that time, Vanguard 1 proved the merit of (inorganic) space solar cell power. Current world-class space solar cells are based on multijunction GaAs and related materials with efficiencies over 30%. But these cells are expensive and relatively brittle. Terrestrial power generation is dominated by silicon-based photovoltaics in three forms: single crystal, polycrystalline, and thin-film. While these photovoltaics are much cheaper than GaAs, turning essentially sand into pure silicon is very energy intensive. Despite such shortcomings of inorganic solar cells, the world's annual generating capacity of solar cells in the past 15 years has increased tenfold up to nearly 600 MWp per annum.

Organic solar cells were only recently produced; the first organic solar cell device was described by Tang in 1986. After less than two decades of research, these cells are approaching a world-record efficiency of 3%. However, solar cell technology does not fit neatly into an organic or inorganic paradigm. A hybrid device employing dye-sensitized, nanocrystalline inorganic materials based on a photoelectrochemical process was first developed by Grätzel in 1991. These cells have achieved efficiencies surpassing the 10% limit. However, the predominant mass in these systems consists of inorganic materials. And of course, research is quite dynamic in the area of nanoparticle(-enhanced) solar cells.

As we enter into the third century of photovoltaics, Zweibel and Green in a recent editorial observed that

> ... photovoltaics is poised to progress from the specialized applications of the past to make a broader impact on the public consciousness, as well as on the well-being of humanity ....

It is important to keep in mind the fact that solar cells are made to generate electricity. Each solar power application results in its own unique set of challenges.

These challenges can be addressed by a variety of technologies that overcome specific issues involving available area, efficiency, reliability, and specific power at an optimal cost. The particular application may be in aerospace, defense, utility, consumer, grid-based, off-grid (housing), recreational, or industrial settings.

Organic photovoltaics will most likely provide solutions in applications where price or large area challenges, or both, dominate, such as consumer or recreational products. Inorganic materials will most likely predominate in aerospace and defense applications: satellites, non-terrestrial surface power, and planetary exploration. However, as efficiencies, environmental durability, and reliability improve for organic photovoltaics, other applications such as off-grid solar villages and utility-scale power generation may well be within reach. Technology developments based upon the work described in this monograph will likely revolutionize the way we exploit the sun's energy to generate electricity on the Earth and beyond.

Aloysius F. Hepp and Sheila G. Bailey
Photovoltaic and Space Environments Branch
NASA Glenn Research Center
Cleveland, Ohio, U.S.A.

Preface

Energy and environment have become two of the most critical subjects of wide concern nowadays, and these two topics are also correlated to each other. An estimated 80% or more of today's world energy supplies are from the burning of fossil fuels such as coal, gas, or oil. However, carbon dioxide and toxic gases released from fossil fuel burning contribute significantly to environmental degradation, such as global warming, acid rains, smog, etc. In addition, fossil fuel deposits on Earth is not unlimited. Due to today's increased demands for energy supplies coupled with increased concerns of environmental pollution, alternative renewable, environmentally friendly as well as sustainable energy sources become desirable.

Sunlight is an unlimited (renewable and thus sustainable), clean (non-polluting), and readily available energy source, which can be exploited even at remote sites where the generation and distribution of electric power present a challenge. Today, any crude oil supply crisis or environmental degradation concern resulting from fossil fuel burning has prompted both people and government to consider solar energy resource more seriously. The technique of converting sunlight directly into electric power by means of photovoltaic (PV) materials has already been widely used in spacecraft power supply systems, and is increasingly extended for terrestrial applications to supply autonomous customers (portable apparatus, houses, automatic meteo stations, etc.) with electric power. According to U.S. DOE EIA, NREL U.S. PV Industry Technology Roadmap 1999 Workshop and Strategies Unlimited, photovoltaics is becoming a billion dollar per annum industry and is expected to grow at a rate of 15 to 20% per year over the next few decades. Nevertheless, one major challenge for large-scale application of the photovoltaic technique at present is the high cost of the commercially available inorganic semiconductor-based solar cells. In contrast, recently developed organic and polymeric conjugated semiconducting materials appear very promising for photovoltaic applications due to several reasons:

1. Ultrafast optoelectronic response and charge carrier generation at organic donor–acceptor interface (this makes organic photovoltaic materials also attractive for developing potential fast photo detectors)
2. Continuous tunability of optical (energy) band gaps of materials via molecular design, synthesis, and processing
3. Possibility of lightweight, flexible shape, versatile device fabrication schemes, and low cost on large-scale industrial production
4. Integrability of plastic devices into other products such as textiles, packaging systems, consumption goods, etc.

While there exists a number of books covering general concepts of photovoltaic and inorganic photovoltaic materials and devices, there are very few comprehensive and dedicated books covering the recent fast-developing organic and polymeric photovoltaic materials and devices. It is therefore the mission of this book to present an overview and brief summary of the current status of organic and polymeric photovoltaic materials, devices, concepts, and ideas. This book is well suited for people who are involved in the research, development and education in the areas of photovoltaics, photo detectors, organic optoelectronic materials and devices, etc. This book would also be helpful for all those who are interested in the issues related to renewable and clean energy, particularly solar energy technologies.

Sam-Shajing Sun, Ph.D.
Niyazi Serdar Sariciftci, Ph.D.

Acknowledgments

I would like to acknowledge the following people for their important or special contributions/roles to this book project:

(1) Taisuke Soda at CRC Press for his enthusiasm, patience, and professional assistance during the entire project period. I also thank Robert Sims, Helena Redshaw at CRC Press, Balaji Krishnasamy at SPI-Kolam, and others for their professional and efficient assistance.

(2) All authors who contributed to this book for their high quality and timely contributions.

(3) Professor Serdar Sariciftci at Linz Organic Solar Cell Institute in Austria. Professor Sariciftci accepted my invitation to serve as a co-editor of this important book. He was very helpful and always quickly responded to my queries and concerns during the project.

(4) Drs. Aloysius Hepp and Sheila Bailey at NASA Glenn Research Center. Both Drs. Hepp and Bailey assisted our polymer photovoltaic research projects during the past several years. I remember when Dr. Hepp encouraged me during my first polymer photovoltaic project in 1999; he drove all the way from Cleveland, Ohio, to my office at Norfolk, Virginia, and showed me some of the great potential applications of organic and polymer photovoltaic materials in the future. It was those sponsored research efforts that eventually lead me to initiate this book project. Drs. Hepp and Bailey also advocated and contributed to this book with their valuable review chapter and a Foreword.

(5) Dr. Charles Lee at Air Force Office of Scientific Research. Dr. Lee also advocated and effectively assisted our photovoltaic polymer research and educational projects.

(6) Dr. Zakya Kafafi at Naval Research Laboratory. Dr. Kafafi shared with me her valuable insights into editing other books and also contributed (with her coauthor) an excellent review chapter to this book.

(7) Professor Alan Heeger at UC Santa Barbara. Professor Heeger constantly showed his strong interest and advocacy to organic and polymer related photovoltaic projects. He also contributed an important Foreword to this book.

(8) Dr. Aleksandra Djurišić at the University of Hong Kong. Many chapters of this book were reviewed and commented by Dr. Djurišić. Many authors expressed their appreciation for the excellent reviews and suggestions, and most of those credits should go to Dr. Djurišić.

(9) Last but not the least, my parents (D. X. Sun & W. R. Sha) who are taking care of my little daughter (Marcia Minwen Sun), and my wife (Li Sun) who allows me to work overtime so often.

Sam-Shajing Sun, Ph.D.
Norfolk State University
Norfolk, Virginia, USA
August 2004

Editors

Sam-Shajing Sun (孙沙京), Ph.D., Editor

Dr. Sun obtained his B.S. in physical chemistry from Peking (Beijing) University in 1984, his M.S. degree in inorganic/analytical chemistry from California State University at Northridge in 1991, and his Ph.D. in polymer/materials chemistry from University of Southern California in 1996. Dr. Sun's Ph.D. dissertation (under the direction of Professor Larry R. Dalton) was titled "Design, Synthesis, and Characterization of Novel Organic Photonic Materials". After a postdoctoral experience at the Loker Hydrocarbon Institute, Dr. Sun joined faculty team at Norfolk State University in 1998. Since then, Dr. Sun has won a number of US government research/educational grant awards in the area of optoelectronic polymers. He also founded/co-founded, and is currently leading a couple of research/educational centers on advanced optoelectronic and nano materials. Dr. Sun teaches organic and polymer chemistry in both undergraduate and graduate programs in chemistry and materials sciences. Dr. Sun's expertise and main research interests are in the design, synthesis, processing, characterization, and modeling of novel polymeric solid-state materials and thin film devices for optoelectronic applications. Dr. Sun is particularly interested in developing self-assembled macromolecules (SAM) that can efficiently convert Sun light into electricity.

Niyazi Serdar Sariciftci, Ph.D., Co-editor

Dr. Sariciftci obtained his M.S. in physics at University of Vienna in 1986, and his doctorate in physics at University of Vienna in 1989. Dr. Sariciftci's Ph.D. dissertation (under the direction of Prof. Dr. Hans KUZMANY and Prof. Dr. Adolf NECKEL) was titled "Spectroscopic investigations on the electrochemically induced metal to insulator transitions in polyaniline" with specialization on *in situ* Optical, Raman and FTIR spectroscopy during doping processes. From 1989-1991, Dr. Sariciftci joined Physics Institute of University of Stuttgart, Fed. Rep. of Germany, in the project "Molecular Electronics" SFB 329 of "Deutsche Forschungsgemeinschaft", c/o Prof. Michael MEHRING, with specialization on electron spin resonance (ESR), electron nuclear double resonance (ENDOR) and photoinduced electron transfer on supramolecular structures. From 1992-1996, Dr. Sariciftci was a senior research associate at the Institute for Polymers & Organic Solids at the University of California, Santa Barbara (directed by Prof. Alan J. Heeger, Nobel Laureate 2000 for Chemistry) working in the fields of photoinduced optical, magnetic resonance and transport phenomena in conducting polymers. In 1996, Dr. Sariciftci was appointed as the Ordinarius Professor (Chair) of the Institute for Physical Chemistry at the Johannes Kepler University in Linz/Austria. In 2000, Dr. Sariciftci founded the Linz Institute for Organic Solar Cells (Linzer Institut für organische Solarzellen or LIOS).

Contributors

Maher Al-lbrahim,
Thuringian Institute for Textile and
Plastics Research, Rudolstadt, Germany

Gehan A.J. Amaratunga,
Cambridge University, Cambridge, UK

Mats R. Andersson,
Chalmers University of Technology,
Gothenburg, Sweden

Neal R. Armstrong,
University of Arizona, Tucson,
AZ, USA

Sheila G. Bailey,
National Aeronautics and Space
Administration Glenn Research Center,
Cleveland, OH, USA

Stephen Barlow,
Georgia Institute of Technology,
Atlanta, GA, USA

David Beljonne,
University of Mons-Hainaut,
Mons, Belgium, and Georgia Institute
of Technology, Atlanta, GA, USA

Robert E. Blankenship,
Arizona State University, Tempe,
AZ, USA

Carl E. Bonner,
Norfolk State University, Norfolk, VA,
USA

Jean-Luc Brédas,
University of Mons-Hainaut,
Mons, Belgium, and Georgia Institute
of Technology, Atlanta, GA, USA

Cyril Brochon,
Université Louis Pasteur ECPM,
Strasbourg, France

Annick Burquel,
University of Mons-Hainaut,
Mons, Belgium

Kevin M. Coakley,
Stanford University, Stanford,
CA, USA

Jérôme Cornil,
University of Mons-Hainaut,
Mons, Belgium, and Georgia Institute
of Technology, Atlanta, GA, USA

Liming Dai,
University of Dayton, Dayton,
OH, USA

Richey M. Davis,
Virginia Polytechnique Institute and
State University, Blacksburg, VA, USA

Aleksandra Djurišić,
University of Hong Kong,
Hong Kong, China

Benoit Domercq,
Georgia Institute of Technology,
Atlanta, GA, USA

Martin Drees,
Virginia Polytechnic Institute and
State University, Blacksburg, VA, USA

Hélène Dupin,
University of Mons-Hainaut,
Mons, Belgium

Abay Gadisa,
Linköping University,
Linköping, Sweden

Yongli Gao,
University of Rochester, Rochester, NY,
USA

Brian A. Gregg,
National Renewable Energy
Laboratory, Golden, CO, USA

Joshua A. Haddock,
Georgia Institute of Technology,
Atlanta, GA, USA

Georges Hadziioannou,
Université Louis Pasteur ECPM,
Strasbourg, France

Randy Heflin,
Virginia Polytechnique Institute and
State University, Blacksburg, VA, USA

Aloysius F. Hepp,
National Aeronautics and
Space Administration Glenn Research
Center, Cleveland, OH, USA

Masahiro Hiramoto,
Osaka University, Osaka, Japan

Harald Hoppe,
Johannes Kepler University of Linz,
Linz, Austria

Olle Inganäs,
Linköping University, Linköping,
Sweden

Michael H.-C. Jin
Ohio Aerospace Institute, Brook Park,
OH, USA and
National Aeronautics and Space
Administration Glenn Research Center,
Cleveland, OH, USA

Zakya H. Kafafi,
Naval Research Laboratory,
Washington, D.C., USA

Bernard Kippelen,
Georgia Institute of Technology,
Atlanta, GA, USA

Chung Yin Kwong,
University of Hong Kong,
Hong Kong, China

Emmanuel Kymakis,
Cambridge University, Cambridge, UK.
Current Address: Technological
Education Institute, Crete, Greece

Paul A. Lane,
Naval Research Laboratory,
Washington, D.C., USA

Vincent Lemaur,
University of Mons-Hainaut,
Mons, Belgium

Wendimagegn Mammo,
Chalmers University of Technology,
Gothenburg, Sweden

Seth R. Marder,
Georgia Institute of Technology,
Atlanta, GA, USA

Michael D. McGehee,
Stanford University, Stanford,
CA, USA

Fanshun Meng,
East China University of Science &
Technology, Shanghai, China

Britt Minch,
University of Arizona, Tucson, AZ, USA

Richard Mu,
Fisk University, Nashville, TN, USA

John Perlin,
Santa Barbara, CA, USA

Nils-Krister Persson,
Linköping University, Linköping,
Sweden

Erik Perzon,
Linköping University, Linköping,
Sweden

Ryne P. Raffaelle,
Rochester Institute of Technology,
Rochester, NY, USA

L.S. Roman,
Federal University of Paraná,
Curitiba-PR, Brazil

Niyazi Serdar Sariciftci,
Johannes Kepler University of Linz,
Linz, Austria

Rachel A. Segalman,
Université Louis Pasteur ECPM,
Strasbourg, France

Steffi Sensfuss,
Thuringian Institute for Textile and
Plastics Research, Rudolstadt,
Germany

Venkataramanan Seshadri,
University of Connecticut, Storrs,
CT, USA

Gregory A. Sotzing,
University of Connecticut, Storrs, CT,
USA

Michelle C. Steel,
University of Mons-Hainaut,
Mons, Belgium

Sam-Shajing Sun,
Norfolk State University, Norfolk, VA,
USA

Mattias Svensson,
Chalmers University of Technology,
Gothenburg, Sweden

He Tian,
East China University of Science &
Technology, Shanghai, China

Akira Ueda,
Fisk University, Nashville, TN, USA

Xiangjun Wang,
Linköping University, Linköping,
Sweden

Marvin H. Wu,
Fisk University, Nashville, TN, USA

Wei Xia,
University of Arizona, Tucson,
AZ, USA

Seunghyup Yoo,
Georgia Institute of Technology,
Atlanta, GA, USA

Fengling Zhang,
Linköping University, Linköping,
Sweden

Contents

Contents

Section 1
General Overviews

1

The Story of Solar Cells

John Perlin
Santa Barbara, CA, USA

Contents

Abstract This article tracks the development of early inorganic materials to directly convert sunlight into electricity. It describes the technologies as well as their applications. The role of individuals in developing solar cells and their many usages is also emphasized.

Keywords photovoltaics, solar cells, selenium, silicon, history

1.1. THE FIRST SOLID-STATE SOLAR CELL

The direct ancestor of solar cells currently in use originated in the last half of the 19th century, with the construction of the world's first seamless communication network through transoceanic telegraph cables. While laying them under sea to permit instantaneous communications between continents, engineers experimented with selenium for detecting flaws in the wires as they were submerged. Early researchers working with selenium discovered that the material's performance depended upon the amount of sunlight falling on it. The influence of sunlight on selenium aroused the interest of scientists throughout Europe, including William Grylls Adams and his student Richard Evans Day. During one of their experiments with selenium, they observed that light could cause a solid material to generate electricity, which was something completely new.

But the science of the day, not yet sure whether atoms were real, could not explain why selenium produced electricity when exposed to light. Therefore, most of the scientists scoffed at Adams and Day's work. It took the discovery and acceptance of electrons and that light contains packets of energy called photons for the field

3

of photovoltaics to gain credibility in the scientific community. By the mid-1920s scientists theorized that when light hits materials like selenium, the more powerful photons pack enough energy to knock poorly bound electrons from their orbits. When wires are attached, the liberated electrons flow through them as electricity. Many researchers envisioned the day when banks of selenium solar cells would power factories and light homes. However, no one could build selenium solar cells efficient enough to convert more than half a percent of the sun's energy into electricity, which was hardly sufficient to justify their use as a power source.

1.2.   THE DISCOVERY OF THE SILICON SOLAR CELL

An accidental discovery by scientists at Bell Laboratories in 1953 revolutionized solar cell technology. Gerald Pearson and Calvin Fuller led the pioneering effort that took the silicon transistor, now the principal electronic component used in all electrical equipment, from theory to working device. Fuller had devised a way to control the introduction of impurities necessary to transform silicon from a poor to a superior conductor of electricity. He gave Pearson a piece of his intentionally contaminated silicon. Among the experiments he did with the specially treated silicon included exposing it to the sun while hooked to a device that measured the electrical flow. To Pearson's surprise, he observed an electric output almost five times greater than the best selenium produced. He immediately ran down the hall to tell his good friend, Daryl Chapin, who had been trying to improve selenium to provide small amounts of intermittent power for remote locations, "Don't waste another moment on selenium!" and handed him the piece of silicon he had just tested. After a year of painstaking trial-and-error research and development, Bell Laboratories showed the world the first solar cells capable of producing useful amounts of power. Reporting the event on page one, the *New York Times* stated that the work of Chapin, Fuller, and Pearson "may mark the beginning of a new era, leading eventually to the realization of one of mankind's most cherished dreams — the harnessing of the almost limitless energy of the sun for the uses of civilization."

Few inventions in the history of Bell Laboratories evoked as much media attention and public excitement as their unveiling of the silicon solar cell (Figure 1.1). Commercial success, however, failed to materialize due to the prohibitive costs of solar cells. Desperate to find marketable products run by solar cells, manufacturers used them to power novelty items such as toys and the newly developed transistor radio. With solar cells powering nothing but playthings, one of the inventors of the solar cell, Daryl Chapin could not hide his disappointment, wondering "What to do with our new baby?"

1.3.   THE FIRST PRACTICAL APPLICATION OF SILICON SOLAR
CELLS

Unknown to Chapin at that time, powerful backing of the silicon solar cell was developing at the Pentagon. In 1955, the American government announced its intention of launching a satellite. The prototype had silicon solar cells for its power plant. Since power lines could not be strung out to space, satellites needed a reliable,

Figure 1.1. A public display of one of the first modules of the Bell silicon solar cell. It appeared at the first international solar energy conference held in Tucson, Arizona [1].

long-lasting, autonomous power source. Solar cells proved to be the perfect answer (Figure 1.2).

The launching of the Vanguard, the first satellite equipped with solar cells, demonstrated their value (Figure 1.3). Preceding satellites, run by batteries, lost power in a week's time, rendering equipment worth millions of dollars useless. In contrast, the solar-powered Vanguard continued to communicate with Earth for many years, allowing the completion of many valuable experiments.

The success of the Vanguard's solar power pack broke down the existing prejudice at that time toward the use of solar cells in space. As the space race between the Americans and Russians intensified, both adversaries urgently needed solar cells. The demand opened a relatively large business for companies manufacturing them. More importantly, for the first time in the history of solar power, the sun's energy proved indispensable to society; without the secure, reliable electricity provided by photovoltaics, the vast majority of space applications so vital to our everyday lives would never have been realized.

1.4. TERRESTRIAL APPLICATIONS

While prospects were looking up for solar cells in space in the 1960s and early 1970s, their astronomical price kept them distant from Earth. In 1968, Dr. Elliot Berman decided to quit his job as an industrial chemist to develop inexpensive solar cells to bring the technology from space to Earth. Berman prophetically envisioned that with a large drop in price, photovoltaics could play a significant role in supplying electrical power to locations on Earth where it is difficult to run a power line. After 18 months of searching for venture capital, Exxon executives liked Berman's approach,

Figure 1.2. A technician attaches a nose cone embedded with two clusters of solar cells for the first trip of solar cells into space. The subsequent launch proved the efficacy of solar cells to power satellites [1].

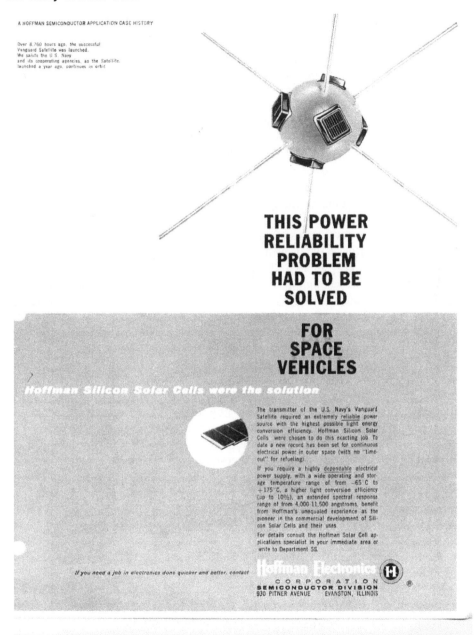

A HOFFMAN SEMICONDUCTOR APPLICATION CASE HISTORY

Over 8,760 hours ago, the successful
Vanguard Satellite was launched.
We salute the U.S. Navy
and its cooperating agencies, as the Satellite,
launched a year ago, continues in orbit

THIS POWER
RELIABILITY
PROBLEM
HAD TO BE
SOLVED

FOR
SPACE
VEHICLES

Hoffman Silicon Solar Cells were the solution

The transmitter of the U.S. Navy's Vanguard Satellite required an extremely reliable power source with the highest possible light energy conversion efficiency. Hoffman Silicon Solar Cells were chosen to do this exacting job. To date a new record has been set for continuous electrical power in outer space (with no "time-out" for refueling).

If you require a highly dependable electrical power supply, with a wide operating and storage temperature range of from −65 C to +175 C, a higher light conversion efficiency (up to 10%), an extended spectral response range of from 4,000-11,500 angstroms, benefit from Hoffman's unequaled experience as the pioneer in the commercial development of Silicon Solar Cells and their uses.

For details consult the Hoffman Solar Cell applications specialist in your immediate area or write to Department SS.

If you need a job in electronics done quicker and better, contact **Hoffman Electronics** (H)

CORPORATION
SEMICONDUCTOR DIVISION
930 PITNER AVENUE   EVANSTON, ILLINOIS

Figure 1.3. Hoffman Electronics shows its success in installing solar cells on the Vanguard, the first satellite to use solar cells and the pioneer that paved their way as the primary power source in space [1].

inviting him to join their laboratory in late 1969. Dismissing other solar scientists' obsession with efficiency, Berman concentrated on lowering costs by starting with lower grade and therefore cheaper silicon and finishing up with less expensive materials for packaging the modules. By decreasing the price from $200 per watt to $20 per watt, solar cells could compete with power equipment needed to generate electricity distant from utility poles.

The oil companies became the first major customers of solar modules. They had both the need and the money. Oil rigs in the Gulf of Mexico had to have warning lights and horns. Most of the oil rigs relied on huge flashlight-like batteries to run warning lights and horns. The batteries needed maintenance and also had to be replaced approximately every 9 months. The replacement required a large boat with an onboard crane, or a helicopter. In contrast, a small skiff could transport the much lighter solar module and accompanying rechargeable battery, resulting in tremendous savings. By 1980, photovoltaics had become the standard power source for warning lights and horns on rigs in the Gulf of Mexico and throughout the world (Figure 1.4).

Oil and gas companies also need small amounts of electricity to protect well casings and pipelines from corroding. Sending current into the ground electrochemically destroys the corroding molecules that cause this problem. Yet many oil and gas fields both in America and in other places of the world like the Middle East and North Africa are in areas far away from power lines but have plenty of sunshine. In such cases, solar modules have proven to be the most cost-effective way of providing the needed current to keep pipes and casings corrosion-free.

Money spent on changing non-rechargeable batteries on the U.S. Coast Guard's buoys exceeded the buoys' original cost. Hence, Lloyd Lomer, a lieutenant commander in the Guard, felt that switching to photovoltaics as a power source for buoys made economic sense. But his superiors, insulated from competition, balked at the proposed change. Lomer continued his crusade, winning approval to put up a test system in the most challenging of environments of Ketchikan, Alaska for solar devices. The success of the photovoltaic-powered buoy in Alaska proved Lomer's point. Still, his boss refused to budge. Going to higher authorities in government, Lomer eventually won approval to convert all buoys to photovoltaics. Almost every coast guard service in the world has followed suit.

Solar pioneer Elliot Berman saw the railroads as another natural area for the application of photovoltaics. One of his salesmen convinced the Southern Railway (now Norfolk Southern) to try photovoltaics for powering a crossing signal at Rex, Georgia. These seemingly fragile cells did not impress veteran railroad workers as capable of powering much of anything. They therefore put in a utility-tied back up. But a funny thing happened in Rex, Georgia that turned quite a few heads. That winter the lines went down on several occasions due to heavy ice build up on the wires. And the only electricity for miles around came from the solar array. As one skeptic remarked, "Rex, Georgia taught the Southern that solar worked!"

With the success of the experiment at Rex, Georgia, the Southern decided to put photovoltaics to work for the track circuitry, the railroad's equivalent of air traffic control, keeping trains at a reasonable distance from one another to prevent head-on or back-ender collisions. The presence of a train changes the rate of flow of electricity running through the track. A decoder translates that change to throw signals and switches up and down the track to ensure safe passage for all trains in the vicinity. At remote spots along the line, photovoltaics provided the needed electricity.

Other railroads followed the Southern's lead. In the old days, telegraph and then telephone poles ran parallel to most of the railroad tracks. Messages sent through these wires kept railroad stations abreast of matters paramount to the safe and smooth functioning of a rail line. But by the mid-1970s, wireless communications could do the same tasks. The poles then became a liability and maintenance an expense. Railroads started to dismantle their poles. Whenever they found their track circuitry devices too far away from utility lines, they began relying on photovoltaics.

Figure 1.4. Installing a solar panel on an oil rig [1].

The U.S. Army Signal Corps, which pioneered the use of solar power in space by equipping the solar pack on the Vanguard, were the first to bring photovoltaics down to Earth. In June 1960, the Corps sponsored the first transcontinental radio broadcast generated by the sun's energy to celebrate its 100th anniversary. The

Corps, probably the strongest supporter of photovoltaics in the late 1950s and the early 1960s, envisioned that solar broadcast would lead others to use photovoltaics to help provide power to run radio and telephone networks in remote locations.

Fourteen years later GTE John Oades realized the Corps' dream by putting up the first photovoltaic-run microwave repeater in the rugged mountains above Monument Valley, Utah (Figure 1.5). By minimizing the power needs of microwave repeater, Oades could power the repeater with a small photovoltaic panel instead, saving all the expenses formerly attributed to them.

His invention allowed people living in towns in the rugged American west, hemmed in by mountains, to enjoy the luxury of long-distance phone service that most other Americans took for granted. Prior to the solar-powered repeater, the cost incurred by phone companies to bring in cable or lines was high, forcing people to drive for hours, sometimes through blizzards on windy roads, just to make long-distance calls.

Australia had an even more daunting challenge to bring modern telecommunication services to its rural customers. Though about the same size as the United States, only 22 million people lived in Australia in the early 1970s, the time when its

Figure 1.5.   John Oades, on left, stands on top of solar-powered repeater while inspecting the installation of his solar-powered microwave repeater [1].

government mandated Telecom Australia to provide every citizen, no matter how remotely situated, with the same radio, telephone, and television service as its urban customers living in Sydney or Melbourne enjoyed.

Telecom Australia tried, but without success, traditional stand-alone power systems like generators, wind machines, and non-rechargeable batteries to run autonomous telephone receivers and transmitters for its rural customers. Fortunately, by 1974, it had another option, relatively cheap solar cells manufactured by Elliot Berman's firm, Solar Power Corporation. The first photovoltaic-run telephone was installed in rural Victoria. Then hundreds of others followed. By 1976, Telecom Australia judged photovoltaics as the preferred power source for remote telephones. In fact, the solar-powered telephones proved to be so successful that engineers at Telecom Australia felt ready to develop large photovoltaic-run telecommunication networks linking towns with colorful names like Devil's Marbles, Tea Tree, and Bullocky Bone to Australia's national telephone and television service. Thanks to photovoltaics, people in these and neighboring towns could dial long distance directly instead of having to call the operator and shout into the phone to be understood. Also, they did not have to wait for newsreels to be flown to their local station to view news already hours, if not days old. Totally, 70 solar-powered microwave repeater networks were created by the early 1980s, the longest spanning 1500 miles. The American and Australian successes showed the world in grand fashion that solar power worked and benefited thousands of people. In fact, by 1985, a consensus in the telecommunications field found photovoltaics to be the power system of choice for remote communications.

A few years earlier, in Mali, a country right below the Sahara, suffered, along with its neighbors, from drought so devastating that thousands of people and livestock dropped like flies. The Malian government knew it could not save its population without the help of people like Father Verspieren, a French priest who lived in Mali for decades and had successfully run several agricultural schools. The government asked Verspieren to form a private company to tap the vast aquifers that run underneath the Malian desert. Verspieren saw drilling as the easy part. His challenge was pumping. No power lines ran nearby and generators lay idle for lack of repairs or fuel. Then he heard about a water pump in Corsica that ran without moving parts, without fuel, without a generating plant, just on energy from the sun. Verspieren rushed to visit the pioneering installation. When he saw the photovoltaic pump, he knew that only this technology could save the people of Mali. By the late 1970s, Father Verspieren dedicated Mali's first photovoltaic-powered water pump with these words, "What joy, what hope we experience when we see that sun which once dried up our pools now replenishes them with water!" (Figure 1.6).

By 1980, Mali, one of the poorest countries in the world, had more photovoltaic water pumps per capita in the world than any other country thanks to Verspieren's efforts. The priest demonstrated that success required the best equipment, a highly skilled and well-equipped maintenance service, and financial participation by consumers. Consumers invested their money for buying the equipment and ongoing maintenance but more importantly, they came to regard the panels and pumps as valued items, which they would help care for. Most successful photovoltaic water-pump project in the developing world have followed Versperien's example. When Father Verspieren initiated his photovoltaic water-pumping program, less than ten photovoltaic pumps existed throughout the world. Now multitudes of them provide water to people, livestock, and crops.

Despite the success of photovoltaic applications throughout the world in the 1970s and early 1980s, institutions responsible for rural electrification programs in

Figure 1.6. Young Malian boys watch a photovoltaic-driven pump fill their village's formerly dry cistern with water [1].

developing countries did not consider installing solar electric panels to power villages distant from urban areas. Working from offices either in the west or in large cities in poorer countries, the developing countries only considered central power stations run by nuclear energy, oil, or coal. Not having lived "out in the bush," these "experts" did not realize the gargantuan investment required to string wires from power plants to the multitudes residing in small villages miles away. As a consequence, only largely populated areas received electricity, leaving billions in the countryside without it. The disparity in energy distribution helped cause migration to the cities in Africa, Asia, and Latin America. As a consequence, megalopolises like Mexico City, Lagos, and Mumbai faced the ensuing problems such as crime, diseases like acquired immuno-deficiency syndrome (AIDS), pollution, and poverty. The vast majority of people, still living in the countryside, do not have electricity. To have some modicum of lighting, they have had to rely on *ad hoc* solutions like kerosene lamps. People buying radios, tape recorders, and televisions also must purchase batteries. It has become apparent that mimicking the western approach to electrification has not worked in the developing world. Instead of trying to put up wires and poles, which no utility in these regions can afford, bringing a 20- to 30-year supply of electricity contained in solar modules to consumers by animal or by vehicle makes greater sense (Figure 1.7). Instead of having to wait for years to construct a centralized power plant and if ever constructed, waiting for power lines to be connected to deliver electricity, it takes less than a day to install an individual photovoltaic system.

Ironically, the French Atomic Energy Commission pioneered electrifying remote homes in the outlying Tahitian Islands with photovoltaics. The Atomic

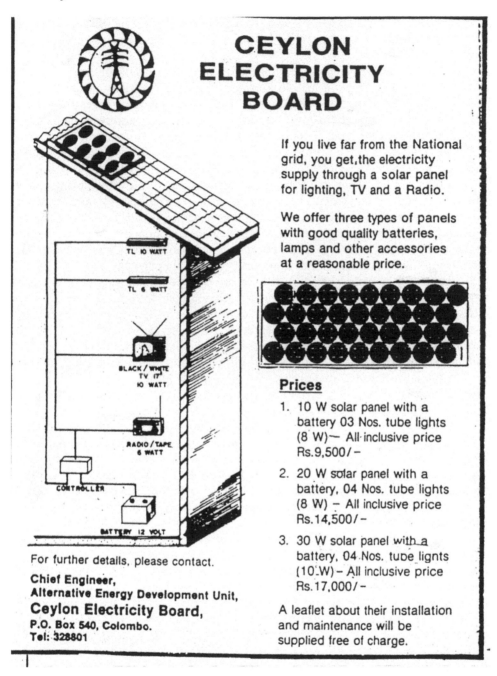

Figure 1.7. In 1984, advertisements like this one appeared in major Sri Lankan newspapers to inform villagers living far away from utility lines that the Ceylon Electricity Board, the national utility, offered them access to electricity from solar cells installed on-site [1].

Energy Commission believed that electrifying Tahiti would mollify the ill feeling in the region created by its testing of nuclear bombs in the South Pacific. The Commission considered all stand-alone possibilities including generators, wind machines, and biogas from the husks of coconut shells before choosing photovoltaics. In 1983, 20% of the world's production of solar cells found its way to French Polynesia. By 1987, half of all the homes on these islands received their electricity from the sun.

Rural residents in Kenya have felt that if they have to wait for electricity from the national utility, they will be old and gray. To have electricity instantly, many Kenyans have bought photovoltaic units. The typical module ranges from 12 to 25 W, enough to charge a battery to run three low-wattage fluorescent lights and a television for 3 h after dark. Electric lighting provided by photovoltaics does away with the noxious fumes and threat of fire people had to contend with when lighting their homes with kerosene lamps. Thanks to photovoltaics, children can now do their lessons free from eye strain and tearing and from coming down with hacking coughs they experienced when studying using kerosene lamps. Their performance in school has soared with better lighting. Electric lighting also allows women to sew and weave products, which bring in cash, at night. Newly gained economic power gives women more say in matters such as contraception. In fact, solar electricity has proven more effective in lowering fertility rates than birth control campaigns.

Electricity provided by photovoltaics allows people living in rural areas to enjoy the amenities of urban areas without leaving their traditional homes for the cities. The government of Mongolia, for example, believed that to improve the lot of its nomadic citizens would require herding them into communities so that the government could connect them to electricity generated by centralized power. The nomads, however, balked. Photovoltaics allowed the nomadic Mongolians to take part in the government's rural electrification program without giving up their traditional lifestyle. Whenever they moved, the solar panels also came along, with yurt and yaks and whatever else they valued.

With the growing popularity of photovoltaics in the developing world, solar thievery has become common. Perhaps nothing better demonstrates the high value people living in the developing countries place on photovoltaics than the drastic increase in solar module thefts over the last few years. Before the market for photovoltaics took off in the developing world, farmers would build an adobe fence around their photovoltaic panels to keep the livestock out. Now razor wire is used to cover these enclosures to bar solar outlaws.

As the price of solar cells continues to drop, devices run by solar cells have seeped into the suburban and urban landscapes of the developed world. For example, construction crews have to excavate a location to place underground electrical transmission lines every time; instead installing photovoltaics makes more economic sense. Economics, for example, has guided highway departments throughout the world to use solar cells for emergency call boxes along roads. In California alone, there are more than 30,000 call boxes. The savings have been immediate. Anaheim, California would have had to spend almost $11 million to connect its 11,000 call boxes to the power grid. Choosing photovoltaics instead, the city had to pay only $4.5 million. The city of Las Vegas found it cheaper to install photovoltaics to illuminate its new bus shelters than dig up the adjacent street and sidewalk to place new wires for the job.

In the mid-1970s and early 1980s, when governments of developing countries began to fund photovoltaic projects, they copied the way electricity had been traditionally produced and delivered, favoring the construction of large fields of photovoltaic panels far away from where the electricity would be used and delivering the electricity by building miles of transmission lines. Others, however, have questioned this approach. When Charles Fritts built the first selenium photovoltaic module in the 1880s and boldly predicted that it would soon compete with Thomas Edison's generators, he envisioned "each building to have its own power plant." Since the late 1980s, most people in the photovoltaic business have come around to Fritts' view. Instead of having to buy a huge amount of land, something unthinkable in densely populated Europe and Japan, to place a field of panels, many people began to ask why not turn each building into its own power station (Figure 1.8).

Using photovoltaics as building materials allows them to double as windows, roofing, skylights, facades, or any type of covering needed on a home or building, and makes sense since the owner gets both building material and electrical generator rolled into one package. Having the electrical production situated where the electricity will be used offers many advantages like easing the burden on the electric grid, eliminating losses that occur in transmission between power stations and end users, helping prevent brownouts and blackouts, reducing vulnerability to terrorism, and generally simplifying power production by eliminating many steps from extraction, processing, and transporting fuel to a power station and then sending that power over great distances to electrify a home. Other advantages of photovoltaics include:

Completely renewable nature
Environmentally benign construction
Absence of moving parts
Modularity to meet any need from milliwatts to gigawatts
Readily available and permanent fuel source

Figure 1.8. (Color figure follows page 348). Solar panels cover the rooftops of a Bremen, Germany housing complex [1].

1.5. THE FUTURE OF PHOTOVOLTAICS

Though the photovoltaics industry has experienced a phenomenal annual 20% growth rate over the last decade, it has just started to realize its potential. While over a million households in India alone get their electricity from solar cells, more than two billion still have no electrical service. The continuing revolution in telecommunications is bringing a greater emphasis on the use of photovoltaics. As with electrical service, the expense of stringing telephone wires keeps most of the developing world without communication services that people living in the more developed countries take for granted. Photovoltaic-run satellites and cellular sites, and a combination of the two, offer the only hope to bridge the digital divide. Photovoltaics could allow everyone the freedom to dial up at or near home and of course, hook up to the Internet.

Opportunities for photovoltaics in the developed world also continue to grow. In the U.S. and Western Europe, thousands of permanent or vacation homes are too distant for utility electric service. If people live in a vacation home that is more than 250 yards from a utility pole, paying the utility to string wires to their place costs more than supplying their power needs with photovoltaics. Fourteen thousand Swiss Alpine chalets and thousands of others from Finland to Spain to Colorado get their electricity from solar energy.

Many campgrounds now prohibit recreational vehicle (RV) owners from running their engines to power generators that run appliances inside the campground. The exhaust gases pollute, and the noise irritates other campers, especially at night. Photovoltaic panels, mounted on the roofs of RVs, provides the electricity needed without bothering others.

Restricting carbon dioxide emissions to help moderate global warming could start money flowing from burning fossil fuels to photovoltaic projects elsewhere. The damage wrought by the 1997–1998 El Nino gives us a taste of the harsher weather expected as the Earth warms. The anticipated increase in natural disasters, brought about by a more disastrous future climate, as well as the growing number of people living in catastrophe-prone regions, make early warning systems essential.

The ultimate early warning device may consist of pilotless photovoltaic-powered weather surveillance airplanes, the prototype of which is the *Helios*. The *Helios* has flown higher than any other aircraft. Solar cells make up the entire top of the aircraft, which consists of only a wing and propellers. Successors to the *Helios* will have fuel cells on the underside of the wing. They will get their power from the photovoltaic panels throughout the day, extracting hydrogen and oxygen from the water discharged by the fuel cells the night before. When the sun sets, the hydrogen and oxygen will power the fuel cells, generating enough electricity at night to run the aircraft. Water discharged in the process will allow the diurnal cycle to begin the next morning. The tandem use of solar cells and fuel cells will allow the aircraft to stay aloft forever, far above the turbulence, watching for and tracking hurricanes, and other potentially dangerous weather and natural catastrophes.

Revolutionary lighting elements called light-emitting diodes (LEDs) produce the same quality of illumination as their predecessors with only a fraction of energy. LEDs therefore significantly reduce the amount of panels and batteries necessary for running lights, making a photovoltaic system less costly and less cumbersome. They have enabled photovoltaics to take over from gasoline generators the mobile warning signs used on roadways to alert motorists about lane closures and other temporary problems that drivers should know about. The eventual replacement of household lighting by LEDs will do the same for photovoltaics in homes.

To bring photovoltaics to mainstream will require further reductions in their cost. Many researchers believe that greater demand could do the trick because for every doubling of production, the price drops 20%. Others believe that new methods of producing silicon solar cells will drop the price significantly. Some researchers are exploring the development of producing less expensive silicon feed stock. At present, most photovoltaic material is made from silicon grown as large cylindrical single crystals or cast in multiple-crystal blocks. Cutting cells only 300 or 400 μm thick from such bulky materials demands excessive cutting, and half of the very expensive starting material ends up on the floor as dust.

New less costly and wasteful ways of manufacturing solar cells promise much lower prices. A number of companies, for example, have begun producing cells directly from molten silicon; the hardened material, only about 100 μm thick, is then fitted into modules. Other companies have developed processes to spray photovoltaic material onto supporting material. All these new techniques have potential for mass production.

There are skeptics, however, who believe that today's techniques will never reach a low enough price for mass use. Some optimists see emerging nanotechnology as the answer. Authors contributing to this book are working with organic compounds that can absorb light and change it into electricity. They envision depositing these compounds on film-like material, which would cost very little to produce, and could be easily adhered to building surfaces. Commercialization is yet to begin.

In truth, the number of potentially inexpensive ways to make solar cells being pursued is dazzling. When Bell Laboratories first unveiled the silicon solar cell, their publicist made a bold prediction:

> The ability of transistors to operate on very low power gives solar cells great
> potential and it seems inevitable that the two Bell inventions will be closely linked
> in many important future developments that will influence the art of living.

Already, the tandem use of transistors and solar cells for running satellites, navigation aids, microwave repeaters, televisions, radios, and cassette players in the developing world and a myriad of other devices has fulfilled the Bell prediction. It takes no great leap of the imagination to expect the transistor and solar cell revolution to continue until it encompasses every electrical need from space to Earth.

REFERENCES

1. J. Perlin, *From Space to Earth: The Story of Solar Electricity*. Harvard University Press: Cambridge, MA, 2002.
2. www.californiasolarcenter.org/history.html.

2

Inorganic Photovoltaic Materials and Devices: Past, Present, and Future

Aloysius F. Hepp and Sheila G. Bailey
Photovoltaic and Space Environments Branch, NASA Glenn Research Center, Cleveland, OH, USA

Ryne P. Raffaelle
Department of Physics and Microsystems Engineering, Rochester Institute of Technology, Rochester, NY, USA

Contents

Abstract This chapter describes recent aspects of advanced inorganic materials used in photovoltaics or solar cell applications. Specific materials examined will be high-efficiency silicon, gallium arsenide and related materials, and thin-film materials, particularly amorphous silicon and (polycrystalline) copper indium selenide. Some of the advanced concepts discussed include multijunction III–V (and thin-film) devices, utilization of nanotechnology, specifically quantum dots, low-temperature chemical processing, polymer substrates for lightweight and low-cost solar arrays, concentrator cells, and integrated power devices. While many of these technologies will eventually be used for utility and consumer applications, their genesis can be traced back to challenging problems related to power generation for aerospace and defense applications. Because this overview of inorganic materials is included in a monogram focused on organic photovoltaics, fundamental issues common to all solar cell devices (and arrays) will be addressed.

Keywords copper indium selenide, (poly)crystalline, gallium arsenide, high-efficiency solar cells, inorganic photovoltaic materials, multijunction devices, quantum dots, silicon solar cells, amorphous silicon, thin films

2.1. INTRODUCTION

2.1.1. Recent Aspects of Advanced Inorganic Materials

A variety of different cell types are currently under development for future utility, consumer, military, and space solar power (SSP) applications. The vast majority of these needs are still currently met by the use of single-crystal silicon [1]. However, a variety of other materials and even solar cell types are vying for the opportunity to replace silicon, if not for all applications at least in certain specific ones. Close relatives to single-crystal silicon such as polycrystalline silicon and amorphous silicon are receiving much attention [2]. More exotic approaches such as polycrystalline thin-film $CuInGaSe_2$ [3] or CdTe [4], dye-sensitized or titania solar cells [5], and even nanomaterial-enhanced conjugated polymer cells [6] are also garnering much interest. However when it comes to highest efficiency cells, multijuction III–V devices are currently the leader. Of course, it is expected that current triple-junction gallium arsenide (GaAs) cell technology will give way to quadruple junctions or possible nanostructured approaches. However, ordinary Ge substrates may well give way to Ge on Si substrates. These have the advantages of lower cost, higher strength, and lighter weight. In addition to traditional lattice-matched multijunction GaAs cells, we may also see the emergence of lattice mismatched or polymorphic cells [7].

2.1.2. Focus: Advanced Materials and Processing

There is currently a tremendous amount of research directed towards a thin film alternative to traditional crystalline cells. Thin-film cells are quite attractive due to the fact that many of the proposed fabrication methods are inexpensive and lend themselves well to mass production [8]. These cells can be made to be extremely lightweight and flexible, especially if produced on polymeric substrates. Thin-film cells have been investigated for quite some time now. Cu_2S–CdS cells were developed as far back as the mid-1970s. However, even the best results to date have been plagued by low efficiencies and poor stability. The inorganic materials that have received the most attention are amorphous silicon, CdTe, and $CuInGaSe_2$ [2–4]. Recently thin-film polymeric cells that incorporate inorganic components such as CdSe or $CuInS_2$ quantum dots have garnered much attention [6,9]. Many researchers believe that these materials will hold the key to inexpensive, easily deployed, large area, highly specific power arrays. This is due in part to the possibility of roll-to-roll processing using low-cost spray chemical deposition or direct-write approaches to producing thin-film solar cells on inexpensive lightweight substrates with these materials [3,8,10].

2.1.3. Increased Efficiency

Multijunction technology has provided the most dramatic increase in inorganic solar cell technology. Many groups have reported laboratory efficiencies well in excess of 30% AM1.5. Industrial lot averages are already above a 28% threshold [11]. The move towards a quadruple junction solar cell has been problematic due to lattice constant constraints. However, there are several new programs looking at lattice mismatch approaches to multijunction III–V devices, which can give a better match to the solar spectrum than what is available in a lattice-matched triple-junction cell. In addition, there is now a considerable amount of attention paid to the use of nanostructures (i.e., quantum dots, wires, and wells) to improve the efficiencies of III–V devices. Recent theoretical results and experimental advances have shown that nanostructures may afford dramatic improvement in cell efficiency [12,13].

2.1.4. Increased Specific Power

When considering specific power, or the power per mass of a solar cell or array, it is clear that crystalline technology will be severely limited by the mass of the substrate on which it is made. In order to achieve the types of specific power required to meet portable power needs of consumer or military markets, as well as many large-scale space power needs, the development of efficient thin-film photovoltaics on polymeric substrates. When considering array-specific power, it is important to note that a cell-specific power considerably higher than the array-specific power will be necessary. The difference between the cell- and array-specific power must be sufficient to make up for the interconnects, diodes, and wiring harnesses. Roughly speaking, the cell mass usually accounts for approximately half the mass of the total balance of the systems [14,15].

In addition to improving the cell efficiency or making the cells lighter, gains in array-specific power may be made by an increase in the operating voltage. Higher array operating voltages can be used to reduce the conductor mass. A typical array designed for 28 V operation at several kilowatts output, with the wiring harness comprising ~10% of the total array mass, yields a specific mass of ~0.7 kg/kW. If this array was designed for 300 V operation, it could easily result in the reduction of the harness-specific mass by at least 50%. This alone would increase the specific power by 5% or more without any other modification.

The extremely high-specific power arrays that need to be developed for several proposed space programs like solar electric propulsion (SEP) or space solar power (SSP) will require lightweight solar arrays capable of high-voltage operation in the space plasma environment. SEP missions alone will require 1000 to 1500 V to drive electric propulsion spacecraft (i.e., no voltage step-up is required to operate the thrusters). National Aeronautics and Space Administration (NASA) has benchmarked a thin-film stand-alone array-specific power 15 times the state-of-the-art (SOA) III–V arrays, area power density 1.5 times that of the SOA III–V arrays, and specific costs 15 times lower than the SOA III–V arrays [15,16].

2.2. OVERVIEW OF SPECIFIC MATERIALS

2.2.1. High-Efficiency Silicon

Silicon solar cells are the most mature of all solar cell technologies and have been used on the vast majority of consumer, military, and space applications. In the early 1960s, silicon solar cells were ~11% efficient, relatively inexpensive, and well suited for the low power and limited lifetime application. The conversion efficiency of current "standard technology" silicon ranges from around 12% to 15% under standard test conditions. The lower efficiency cells are generally more resistant to radiation [7,14,17].

There have been many enhancements to silicon cells over the years to improve their efficiency. Textured front surfaces for better light absorption, extremely thin cells with back-surface reflectors for internal light trapping, and passivated cell surfaces to reduce losses due to recombination effects are just a few examples. The highest measured efficiency for a large-area (i.e., 5 in.^{2}) crystalline silicon solar cell stands at 21.5%. Currently, high-efficiency silicon cells approaching 17% AM0 efficiency in production lots are available from Japanese and German producers. Tecstar and Spectrolab, the two large U.S. companies, which have produced virtually all of the domestic silicon space solar cells, only offer the conventional type of silicon cells with efficiencies around 14.8% because they have concentrated their developmental efforts on GaAs-based multijunction cells. The advantage of high-efficiency silicon cells over III–V cells lies in their relatively lower cost, lower material density, and higher strength [1,7,17].

2.2.2. Polycrystalline Silicon

A lower cost alternative, although less efficient, to standard crystalline silicon technology is the use of polycrystalline silicon [18]. This material is manufactured by pouring liquid silicon into a mold. Upon solidification, multicrystallites form with associated grain boundaries. The resulting blocks of material are sliced into suitable wafers. Polycrystalline silicon (p-Si) has been extensively studied over the past decade due to its applications in optoelectronics. The light absorption properties along with its simple manufacture and broadly tunable morphology have made it an attractive photovoltaic material for some time now. Due to the defects associated with the grain boundaries, the best p-Si solar cell efficiencies stand at 19.8%, less than its monocrystalline silicon counterpart [19].

2.2.3. Amorphous Silicon

Silicon-based solar cells can also be produced using amorphous thin films that are evaporated onto glass or even polymeric substrates [2]. This approach requires very little active material and thus can relate to tremendous savings in production costs. These types of cells are well suited to large-area roll-to-roll processing [8]. Unfortunately this method of manufacture results in high defect densities and thus is even less efficient than PSi solar cells. However, this approach does lend itself to multilayer processing and thus multijunction solar cells are possible. This approach has resulted in dramatic increases in overall amorphous silicon solar cell efficiencies. Multijunction amorphous silicon solar cell efficiencies as high as 13% AM1.5 have been reported [2].

2.2.4. Gallium Arsenide and Related III–V Materials

GaAs has been a material of interest to the solar cell community for many years [20]. In 1955, an RCA group was funded by the U.S. Army Signal Corps, and later by the Air Force, to work on the development of GaAs-based cells. GaAs has a nearly ideal direct bandgap of 1.42 eV for operation in our solar spectrum. It also has favorable thermal stability and radiation resistance when compared to silicon. However, it took nearly 30 years of efficiency improvements until the use of GaAs-based cells could be argued because of its much higher costs. The arena in which this premium could most easily be justified was in space utilization. In fact, now the vast majority of cells launched for SSP are multijunction III–V cells [14,15].

The lattice-matched heterojunction cells developed in the early 1970s led to the acceptance of GaAs as a viable photovoltaic material. The Air Force launched the Manufacturing Technology for GaAs Solar Cells (MANTECH) program in 1982. This program was designed to develop metal organic chemical vapor deposition (MOCVD) techniques necessary for the large-scale production of GaAs solar cells [21].

The other significant development in GaAs technology was in the use of alternative substrates. In 1986, the Air Force supported work by the Applied Solar Energy Corp. (ASEC) in which they developed GaAs cells grown on Ge substrates. This was possible due to the similarity in the lattice constants and thermal expansion coefficients of the two materials. This resulted in the improvement in the mechanical stability of the cells and a lowering of the production costs.

The development of GaAs-based cells continued throughout the late 1980s with the primary focus on the development of multijunction approaches to photovoltaic conversion. The initial work was mechanical stacked cells; however, this quickly transitioned into epitaxially grown dual junctions with tunnel junctions in between. Overall cell efficiencies increased dramatically during this time (Figure 2.1) [22].

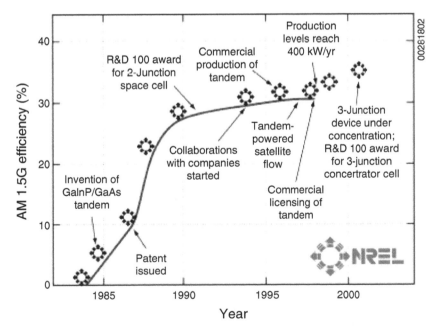

Figure 2.1. GaAs-based solar cell development (courtesy of the National Renewable Energy Laboratory).

2.2.5. Thin-Film Materials

Thin-film photovoltaics is an intriguing technology due to flexible lightweight construction, permitting arrays to be "molded" onto nonrigid or uniform structures for innovative power systems. Photovoltaic modules based on ternary chalcopyrite absorber materials (i.e., I–III–VI$_2$ Cu(In:Ga)(S:Se)$_2$) have been the focus of intense investigation for over two decades [3]. The use of chalcopyrite absorbers is attractive since their bandgaps correlate well with the maximum photon power density in the solar spectrum for both terrestrial (AM1.5) [8] and space applications (AM0) (Figure 2.2) [14–16]. Additionally, by adjusting the percent atomic composition of either Ga for In or S for Se, or both, the bandgap can be tuned from 1.0 to 2.4 eV, thus permitting fabrication of high or graded bandgaps [3,8].

One of the first thin-film cells, Cu$_2$S/CdS, was developed for space applications. Reliability issues eliminated work on this particular cell type for both space and terrestrial considerations even though AM1.5 efficiencies in excess of 10% were achieved. Thin-film cells require substantially less material and thus lower mass and promise the advantage of large area and low-cost manufacturing [2–4,8].

The development of other wide bandgap thin-film materials that can be used in conjunction with CuInSe$_2$ (CIS) to produce a dual junction device is underway. As already demonstrated in III–V cells for space use, a substantial increase over single-junction device efficiency is possible with a dual-junction device. NASA and National Renewable Energy Laboratory (NREL) have both initiated a dual-junction CIS-based thin-film device programs [3,16]. The use of Ga to widen the bandgap of CIS and thus improve the efficiency is already well known. The substitution of S for Se also appears to be an attractive top cell material. The majority of thin-film devices developed thus far have been on heavy substrates such as glass. However, progress is continuing to reduce substrate mass through the use of thin metal foils and light-

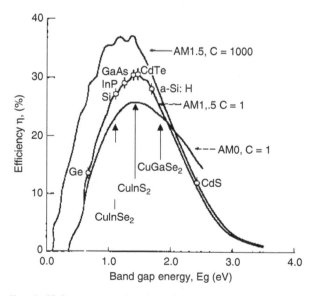

Figure 2.2. Predicted efficiency versus bandgap for thin-film photovoltaic materials for solar spectra in space (AM0) and on the surface of the Earth (AM1.5) at 300 K compared with bandgaps of other photovoltaic materials with unconcentrated ($C = 1$) and high concentration ($C = 1000$) sunlight.

weight flexible polyimide or plastic substrates. A major problem with the use of plastic substrates is the incompatibility with many of the deposition processes. The most efficient thin-film cells to date are made by a combination of co-evaporation of the elements and subsequent annealing. AM0 cell efficiencies as high as 7% have been measured for $CuIn_{0.7}Ga_{0.3}S_2$ (e.g., 1.55 eV) thin-film devices on flexible substrates (R. Birkmire, personal communication, 2003). The use of plastic substrates such as PBO, Upilex or Kapton puts an unacceptable restriction on the processing temperatures. The current world records for thin-film CdTe and $CuInGaSe_2$ solar cells AM1.5 efficiencies stand at 16.5% and 19.2%, respectively [3,4,7].

2.3. ADVANCED CONCEPTS

2.3.1. Multijunction III–V Devices

Investigations into further efficiency improvement toward the end of the 20th century turned to the development of multiple junction cells and concentrator cells. Much of the development of multijunction GaAs-based photovoltaics was supported by a cooperative program funded by the Air Force (Manufacturing Technology program, Space Vehicles Directorate, and Space Missile Center), and NASA. This work resulted in the development of a "dual-junction" cell, which incorporates a high-bandgap GaInP cell grown on a GaAs low-bandgap cell. The 1.85 eV GaInP converts higher energy photons and the GaAs converts the lower energy photons. Commercially available dual-junction GaA/GaInP cells have an AM0 efficiency of 22% with a V_{oc} of 2.06 V [20–22].

The highest efficiency solar cells currently available are triple-junction cells consisting of GaInP, GaAs, and Ge (Figure 2.3). They are grown in series connected layers and have been produced with an efficiency of 26.8% and V_{oc} of

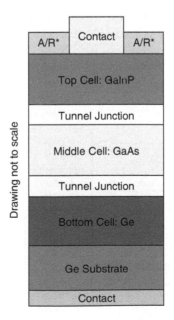

*A/R Anti-Reflective Coating

Figure 2.3. Multijunction GaInP/GaAs solar cell structure (courtesy of Spectrolab, Inc.).

2.26 V in production lots, and with 29% efficiency in the laboratory. Emcore, Inc. and Spectrolab Inc., currently produce cells that are commercially available in the 25% to 27% efficiency range. Their high efficiencies are due to their ability to convert a larger portion of the available sunlight. In fact, it is expected that as better materials for a triple junction are developed and eventually a quadruple junction is developed, the cell efficiencies will eclipse the 40% efficiency barrier (Figure 2.4). Hughes Space and Communications Company's HS601 and HS702 spacecrafts currently use multi-junction technology as do most other contractors for their high-performance space-crafts [14,15,21,22].

NREL recently announced a new world record conversion efficiency for a multjunction GaAs solar cell at 32.5% [23]. Spectrolab, the supplier of more than half of the world's spacecraft solar cells, has reached a milestone of 25,000 triple-junction GaAs solar cells, with an average conversion efficiency of 24.5%. They are currently producing 1 MW of cells for commercial sales to a variety of array manufacturers (e.g., Hughes Space and Communication Company, Ball Aerospace & Technologies Group, Lockheed Martin, and Boeing). Their high-efficiency cells retain 86% of their original power after 15 years of operation. There is currently more than 50 kW of Spectrolab dual-junction solar cells in operation [14].

The new multijunction III–V cells have reduced solar array size and mass over the previously used silicon cells to achieve comparable power levels. This is especially important in space applications. Scientists expect the majority of the 800 commercial and military spacecrafts to be launched in the next 5 years to use multijunction technology. This should result in the lower costs for telecommunications, internet, television, and other wireless services. The Air Force Research Laboratory (AFRL) recently initiated a 35% efficient four-junction solar cell program.

Multijunction III–V cells are expensive to produce. The development of large-area arrays using these cells can become prohibitively costly. One option to reduce the overall cost is to use the cells in solar concentrators, where a lens or a mirror is used to decrease the required cell area [24, see section 2.3.4].

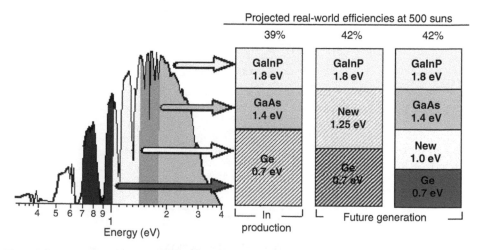

Figure 2.4. (**Color figure follows page 348**). Diagram showing the potential photoconversion of sunlight using multijunction III–V solar cells (courtesy of Spectrolab, Inc.).

2.3.2. Nanotechnology — Specifically Quantum Dots

A recent approach to increasing the efficiency of thin-film photovoltaic solar cells involves the incorporation of quantum dots [6,9]. Semiconductor quantum dots are currently a subject of great interest mainly due to their size-dependent electronic structures, in particular the increased bandgap and therefore tunable optoelectronic properties. To date, these nanostructures have been primarily limited to sensors, lasers, light-emitting diodes (LEDs), and other optoelectronic devices. However the unique properties of the size-dependent increase in oscillator strength due to the strong confinement exhibited in quantum dots and the blue shift in the bandgap energy of quantum dots are properties that can be exploited for developing photovoltaic devices and these devices offer advantages over conventional photovoltaics. Theoretical studies predict a potential efficiency of 63.2%, for a single size quantum dot, which is approximately a factor of 2 better than any SOA device available today. For the most general case, a system with an infinite number of sizes of quantum dots has the same theoretical efficiency as an infinite number of bandgaps or 86.5% [12,13].

A collection of different sizes of quantum dots can be regarded as an array of semiconductors that are individually size-tuned for optimal absorption at their bandgaps throughout the solar energy emission spectrum. This is in contrast with a bulk material in which photons are absorbed at the bandgap; result in less efficient energies above the bandgap photogeneration of carriers. In addition, bulk materials used in photovoltaic cells suffer from reflective losses for convertible photons near the bandgap, whereas for individual quantum dots, reflective losses are minimized. Some recent work has shown that quantum dots may also offer some additional radiation resistance and favorable temperature coefficients.

2.3.3. Advanced Processing for Low-Temperature Substrates

A key technical issue outlined in the 2001 U.S. Photovoltaic roadmap [25] is the need to develop low-cost, high-throughput manufacturing for high-efficiency thin-film solar cells. Thus, a key step for device fabrication for thin-film solar cells is the deposition onto flexible, lightweight substrates such as polyimides. Current methods for depositing ternary crystallite compounds often include high-temperature processes, which are followed by toxic sulfurization or selenization steps [3,4,8]. The high-temperature requirements make this protocol incompatible with all presently known flexible polyimides, or other polymer substrates. In addition, the use of toxic reagents is a limiting factor. The use of multisource inorganic and organometallic precursors in a chemical vapor deposition (CVD)-type process is more appealing due to milder process parameters. However, stoichiometric control of deposited films can be difficult to achieve and film contamination has been reported [26]. A novel alternative approach is the use of ternary single-source precursors (SSPs), which have the I–III–VI$_2$ stoichiometry "built-in" and are suitable for low-temperature deposition (Figure 2.5). Although a rich and diverse array of binary SSPs are known, characterized, reviewed, and tested, the number of known ternary SSPs and their use in deposition processes is limited [27]. We briefly summarize a highly promising technique for thin-film growth: molecular design of SSPs for use in a chemical (vapor or spray) deposition process.

Spray CVD has become an often-studied deposition technique. In this process, a precursor solution is ultrasonically nebulized and swept into a two-zone, hot-wall

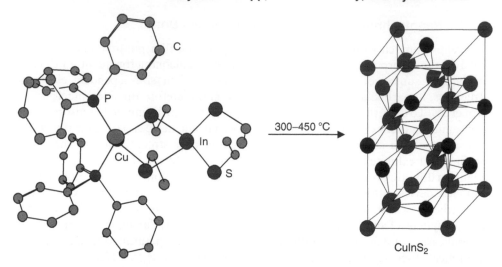

Figure 2.5. (Color figure follows page 348). Pyrolysis of a single-source precursor [{PPh$_3$}$_2$Cu(SEt)$_2$In(SEt)$_2$] to produce a semiconductor material, CuInS$_2$.

reactor (Figure 2.6). The carrier-solvent is evaporated in the warm zone, and the gaseous precursors are decomposed in the hot zone, where film growth occurs as in conventional CVD. Spray CVD maintains the most desirable features of MOCVD and spray pyrolysis, such as film growth in inert atmospheres, large-area deposition, laminar flow over the substrate, and low-temperature solution reservoir, while avoiding their major difficulties [28]. It minimizes the high volatility and temperature requirements for the precursor, which are essential in MOCVD, by delivering the precursor to the furnace as an aerosol propelled by a fast-flowing carrier gas from a low-temperature precursor reservoir analogous to that employed in spray pyrolysis. The latter feature is an important benefit that can prevent premature precursor decomposition when using thermally labile precursors.

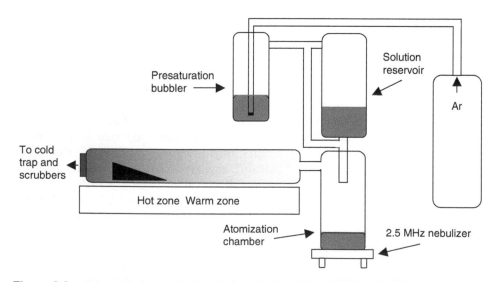

Figure 2.6. Schematic for spray chemical vapor deposition (CVD) apparatus.

In 1990 Kanatzidis et al. [29] reported the preparation of hetero-binuclear complexes consisting of tetrahedrally arranged Cu and In centers, with two bridging thiolato and selenolato groups (see Figure 2.5 for similar structure). Pyrolysis studies revealed that the Se derivative could be converted into $CuInSe_2$ at 400–450°C, but none of the precursors had been evaluated in a thin-film deposition study. Buhro and Hepp [10,30] were able to demonstrate that [{PPh_3}$_2$Cu(SEt)$_2$In (SEt)$_2$] could be utilized in a spray CVD process for depositing thin-film $CuInS_2$ below 400°C. Thin films were deposited using a dual solvent system of toluene and dichloromethane (CH_2Cl_2) as the carrier solvent. Single-phase 112 oriented $CuInS_2$ thin films were successfully deposited at a range of temperatures from 300 to 400°C; at elevated temperatures (>500°C), $CuIn_5S_8$ phase thin films were deposited. Analysis of materials showed that the films were free from any detectable impurities and highly crystalline, thus concluding that the precursor decomposes cleanly.

In the course of our investigations for improved SSPs for the spray CVD of chalcopyrite thin films to the ternary semiconductor Cu(Ga:In)(Se:S)$_2$, we have continued to expand the molecular design of SSPs based on the [{LR_3}$_2$Cu(ER′)$_2$M(ER′)$_2$] architecture [10]. Furthermore, the number of "tunable" sites within the complex serves as a utility for preparing a number of ternary chalcopyrites of varying composition, in addition to engineering the SSP to match a given spray CVD process.

Spray CVD in conjunction with SSP design provides a proof-of-concept for a reproducible, high manufacturable process. An outlook for further investigation that needs to be undertaken is as below:

(1) *Precursor design*: development of more volatile and thermally labile systems. This can be achieved by the incorporation of fluorinated or silylated functional groups. Importantly, due to the well-known propensity of fluorine to react with silyl moieties, incorporation of both elements in the molecule can serve to increase volatility, and also as a "self-cleaning" mechanism so that precursor does not decompose in undesired pathway [31].

(2) *Processing parameters*: Spray CVD has a number of tunable variables, such as droplet size, flow rate, concentration, solvent polarity, which are advantageous to achieve the desired film characteristics. Thus an in-depth study needs to address these parameters to film composition [32].

(3) *Device fabrication*: working devices from deposited films need to be tested to aid SSP design and spray CVD process parameters; this is the current focus of work on-going at NASA Glenn Research Center [33].

The work summarized here on the molecular design of SSPs for their use in a spray CVD process although still in its infancy, undoubtedly shows it as a mass producible, cost-effective method for fabricating commercial thin-film photovoltaic devices. Furthermore, we have discovered that SSPs are a valuable route to semiconductor nanoparticles [9,31].

2.3.4. Concentrator Cells

Concentrating solar collectors use devices such as Fresnel lenses and parabolic mirrors to concentrate light onto solar cells [24]. This reduces the area of cells needed and also raises the efficiency of the cells by operating them at higher light concentration. Normally these collectors are mounted onto an axis tracking system to follow the track of the sun across the sky. Figure 2.7 shows a point-focus Stirling engine concentrator system. Figure 2.8 shows a 100-kW concentrator system in Fort Davis, Texas.

2.3.5. Integrated Power Devices

NASA has been working to develop lightweight, integrated space power systems on small or flexible substrates. These systems generally consist of a high-efficiency thin-

Figure 2.7. A 5-kW point-focus Stirling engine concentrator system (courtesy of National Renewable Energy Laboratory).

Figure 2.8. A 100-kW utility-scale concentrator system (courtesy of National Renewable Energy Laboratory).

film solar cell, a high energy density solid-state Li-ion battery, and the associated control electronics in a single monolithic package. These devices can be directly integrated into microelectronic or microelectromechanical system (MEMS) devices and are ideal for distributed power systems on satellites or even the main power supply on a nanosatellite. These systems have the ability to produce constant power output throughout a varying or intermittent illumination schedule experienced by a rotating satellite or "spinner" and by satellites in a low earth orbit (LEO) by combining both generation and storage [34].

An integrated thin-film power system has the potential to provide a low mass and cost alternative to the current SOA power systems for small spacecraft. Integrated thin-film power supplies simplify spacecraft bus design and reduce losses incurred through energy transfer to and from conversion and storage devices. Researchers hope that this simplification will also result in improved reliability (Figure 2.9).

The NASA Glenn Research Center has recently developed a microelectronic power supply for a space flight experiment in conjunction with the Project Starshine atmospheric research satellite (Figure 2.10) [35]. This device integrates a seven-junction small-area GaAs monolithically integrated photovoltaic module (MIM) with an all-polymer $LiNi_{0.8}Co_{0.2}O_2$ Li-ion thin-film battery. The array output is matched to provide the necessary 4.2 V charging voltage and minimized the associated control electronic components. The use of the matched MIM and thin-film Li-ion battery storage maximizes the specific power and minimizes the necessary area and thickness of this microelectronic device. This power supply was designed to be surface mounted to the Starshine 3 satellite, which was ejected into LEO with a fixed rotational velocity of 5°/s. The supply is designed to provide continuous power even with the intermittent illumination due to the satellite rotation and LEO.

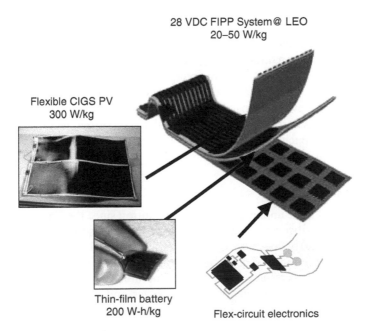

Figure 2.9. (Color figure follows page 348). Flexible integrated power pack (FIPP) (courtesy of ITN Energy Systems).

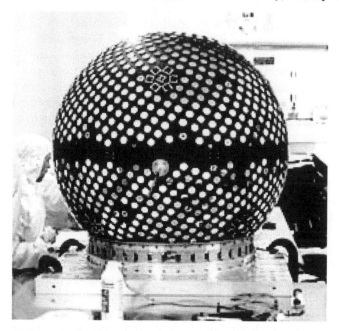

Figure 2.10.   Starshine 3 satellite (an IPS surrounded by six Emcore triple-junction III–V solar cells is shown in the middle of the top hemisphere at the satellite).

2.4.   APPLICATIONS

2.4.1.   Terrestrial

The use of photovoltaic power continues to increase steadily in the U.S. and world-wide. Programs to stimulate the use of solar power exist to support homeowners, small businesses, and industry. The government has also expanded programs to support local utility companies, state agencies, and even international developmental projects. Many other nations are supporting similar programs. The U.K. Research Council recently funded their largest solar power program to date that will support six universities and seven companies in the U.K. [37]. Japan's largest solar power generator was also recently completed in Tsukuba. The system uses 5600 solar panels and will have the capacity to generate 1 MW of power. Figure 2.11 shows a 1 MW solar power station installed by BP Solar in Toledo, Spain.

2.4.2.   Aerospace

Solar array designs have undergone a steady evolution since the Vanguard 1 satellite. Early satellites used silicon solar cells on honeycomb panels that were body mounted to the spacecraft. Early space solar arrays only produced a few hundred watts of power. However, satellites today require low-mass solar arrays that produce several kilowatts of power. Several new solar array structures have been developed over the past 40 years to improve the array-specific power and reduce the stowed volume during launch. The solar arrays presently in use can be classified into six categories:

1.   Body-mounted arrays
2.   Rigid panel planar arrays

Figure 2.11. Solar power station in Toledo, Spain (courtesy of BP Solar, Inc.).

3. Flexible panel array
4. Flexible roll-out arrays
5. Concentrator arrays
6. High-temperature and high-intensity arrays.

In addition, several proposed space missions have put other constraints on the solar arrays. Many of the proposed Earth-orbiting missions designed to study the sun require "electrostatically clean" arrays. Inner planetary missions and mission to study the sun within a few solar radii require solar arrays capable of withstanding temperatures above 450°C and functioning at high solar intensities. Outer planetary missions require solar arrays that can function at low solar intensities and low temperatures (LILT). In addition to the near-sun missions, missions to Jupiter and its moons also require solar arrays that can withstand high radiation levels.

The International Space Station (ISS) will have the largest photovoltaic power system ever present in space (Figure 2.12). It will be powered by 262,400 (8 cm × 8 cm) silicon solar cells with an average efficiency of 14.2% on eight U.S. solar arrays (each ~ 34 m × 12 m). This will generate about 110 kW of average power, which after battery charging, life support, and distribution, will supply 46 kW of continuous power for research experiments. The Russians also supply an additional 20 kW of power to ISS.

2.5. SUMMARY AND CONCLUSIONS

Zweibel and Green in a recent editorial [38] observed that:

> ... photovoltaics is poised to progress from the specialized applications of the past to make a broader impact on the public consciousness, as well as on the well-being of humanity ....

Figure 2.12.   (Color figure follows page 348). Solar arrays of the International Space Station (graphic courtesy of NASA).

We and Perlin (see Chapter 1) have endeavored to provide a context for the development of organic photovoltaics. While highlighting important applications involving uses on Earth, where price per watt will be a critical driver; the challenges of aerospace and defense applications have spurred the development of new technologies. We have highlighted a number of these advances in materials, devices, and processing in this overview. It is important to keep in mind the fact that solar cells regardless the materials or device structure are made to generate electricity. Also, an important lesson to be learned from a more practical consideration of solar cells is that a wide array of power generation applications exist, each with its own set of challenges.

These challenges can be addressed by a variety of technologies that overcome specific issues involving available area, efficiency, reliability, and specific power at an optimal cost. The particular application may be in aerospace, defense, utility, consumer, grid-based, off-grid (housing), recreational, or industrial settings. Organic photovoltaics will most likely provide solutions in applications where price or large area challenges, or both, dominate, such as consumer or recreational products. Inorganic materials will most likely predominate in aerospace and defense applications: satellites, nonterrestrial surface power, and planetary exploration. However, as efficiencies, environmental durability, and reliability improve for organic photovoltaics, other applications such as off-grid solar villages and utility-scale power generation may well be within reach.

REFERENCES

1. M. A. Green, *Silicon Solar Cells: Advanced Principles and Practice*, Bridge Printery, Sydney, 1995.

2. S. Guha, J. Yang, and A. Banerjee, Amorphous silicon alloy photovoltaic research — present and future, *Prog. Photovolt. Res. Appl.* **8**, 141–150 (2000).

3. H.-W. Schock and R. Noufi, CIGS-based solar cells for the next millennium, *Prog. Photovolt. Res. Appl.* **8**, 151–160 (2000).

4. D. Bonnet and P. V. Meyers, Cadmium telluride — material for thin film solar cells, *J. Mater. Res.* **13**, 2740–2753 (1998).

5. M. Grätzel, Perspectives for dye-sensitized nanocrystalline solar cells, *Prog. Photovolt. Res. Appl.* **8**, 171–185 (2000).

6. W. Huynh, J. Dittmer, and A. P. Alivisatos, Hybrid nanorod-polymer solar cells, *Science* **295**, 2425–2428 (2002).

7. M. A. Green, K. Emery, D. L. King, S. Igari, and W. Warta, Solar cell efficiency tables (version 20), *Prog. Photovolt. Res. Appl.* **10**, 355–363 (2002).

8. P. Sheldon, Process integration issues in thin-film photovoltaics and their impact on future research directions, *Prog. Photovolt. Res. Appl.* **8**, 77–91 (2000).

9. S. L. Castro, S. G. Bailey, R. P. Raffaelle, K. K. Banger, and A. F. Hepp, Nanocrystalline chalcopyrite materials ($CuInS_2$ and $CuInSe_2$) via low-temperature pyrolysis of molecular single-source precursors, *Chem. Mater.* **15**, 3142–3147 (2003).

10. K. K. Banger, J. A. Hollingsworth, J. D. Harris, J. Cowen, W. E. Buhro, and A. F. Hepp, Ternary single-source precursors for polycrystalline thin-film solar cells. *Appl. Organomet. Chem.* **16**, 617–627 (2002).

11. D. J. Friedman, S. R. Kurtz, K. A. Bertness, A. E. Kibbler, C. Kramer, J. M. Olson, D. L. King, B. R. Hansen, and J. K. Snyder, 30.2% efficient GaInP/GaAs monolithic two-terminal tandem concentrator cells, *Prog. Photovolt.* **3**, 47–50 (1995).

12. R. P. Raffaelle, S. L. Castro, A. F. Hepp, and S. G. Bailey, Quantum dot solar cells, *Prog. Photovolt. Res. Appl.* **10**, 433–439 (2002).

13. A. Martí, L. Cuadra, and A. Luque, Quantum Dot Intermediate Band Solar Cell, in 28th IEEE Proceedings of the Photovoltaics Specialists Conference, Anchorage, AK, September 2000, IEEE, New York, 2000, pp. 940–943.

14. P. A. Iles, Future of photovoltaics for space applications, *Prog. Photovolt. Res. Appl.* **8**, 39–51 (2000).

15. D. J. Hoffman, T. W. Kerslake, A. F. Hepp, M. K. Jacobs, and D. Ponnusamy, Thin-Film Photovoltaic Solar Array Parametric Assessment, in Proceedings 35th IECEC Conference, Vol. 1, AIAA (AIAA-00-2919), Washington, D.C., 2000, pp. 670–680.

16. A. F. Hepp, M. Smith, J. H. Scofield, J. E. Dickman, G. B. Lush, D. Morel, C. Ferekides, and N. G. Dhere, Multi-Junction Thin-Film Solar Cells on Flexible Substrates for Space Power, in Proceedings, 37th IECEC Conference, EDS (IECEC-2002-20155), Washington, D.C., NASA/TM 2002-211834, 12 pp., October 2002.

17. M. A. Green, The future of crystalline silicon solar cells, *Prog. Photovolt. Res. Appl.* **8**, 127–139 (2000).

18. J. C. Zolper, S. Narayanan, S. R. Wenham, and M. A. Green, 16.7% efficiency, laser textured, buried contact polycrystalline silicon solar cell, *Appl. Phys. Lett.* **55**, 2363–2365 (1989).

19. J. Zhao, A. Wang, and M. A. Green, 19.8% efficient, 'honeycomb' textured, multicrystalline and 24.4% monocrystalline silicon solar cells, *Appl. Phys. Lett.* **73**, 1991–1993 (1998).

20. H. J. Hovel, *Solar Cells*, Vol. 2, Academic Press, New York, 1978.

21. D. Keener, D. Marvin, D. Brinker, H. Curtis, and M. Price, Progress Toward Technology Transition of GaInP/GaAs/Ge Multijunction Solar Cells, in 26th Proceedings of the IEEE Photovoltaics Specialists Conference, Anaheim, CA, August 1997, IEEE, New York, 1997, pp. 787–781.

22. M. Yamaguchi, Multi-Junction Solar Cells: Present and Future, in Proceedings of 12th International Photovoltaics Science and Engineering Conference, Jeju, Korea, June 11–15, 2001, Kyung Hee Information Printing Co. Ltd., Seoul, Korea, 2001, pp. 291–295.

23. http://www.eere.energy.gov/aro/r_and_d.html#2

24. R. M. Swanson, The promise of concentrators, *Prog. Photovolt. Res. Appl.* **8**, 93–111 (2000).

25. http://www.nrel.gov/ncpv/intro_roadmap.html

26. M. C. Artaud, F. Ouchen, L. Martin, and S. Duchemin, $CuInSe_2$ thin films grown by MOCVD: characterization, first devices, *Thin Solid Films*, **324**, 115–123 (1998).

27. A. C. Jones and P. O'Brien, *CVD of Compound Semiconductors: Precursors Synthesis, Development & Application*, Wiley-VCH, Berlin, 1997.

28. H. Miyake, T. Hayashi, and K. Sugiyama, Preparation of $CuGa_xIn_{1-x}S_2$ alloys from In solutions, *J. Cryst. Growth* **134**, 174–180 (1993).

29. W. Hirpo, S. Dhingra, A. C. Sutorik, and M. G. Kanatzidis, Synthesis of mixed copper-indium chalcogenolates. Single source precursors for the photovoltaic material $CuInQ_2$ (Q = S, Se), *J. Am. Chem. Soc.* **115**, 597–599 (1993).

30. J. A. Hollingsworth, A. F. Hepp, and W. E. Buhro, Spray CVD of copper indium disulfide films: control of microstructure and crystallographic orientation. *Chem. Vap. Deposit.* **5**, 105–108 (1999).

31. K. K. Banger, M. H.-C. Jin, J. D. Harris, P. E. Fanwick, and A. F. Hepp, A new facile route for the preparation of single-source precursors for bulk, thin-film, and nanocrystallite I–III–VI semiconductors, *Inorg. Chem.* **42**, 7713–7715 (2003).

32. K. K. Banger, J. A. Hollingsworth, M. H.-C. Jin, J. D. Harris, E. W. Bohannan, J. A. Switzer, W. E. Buhro, and A. F. Hepp, Ternary precursors for depositing I–III–VI_2 thin films for solar cells via spray CVD. *Thin Solid Films* **431–432**, 63–67 (2003).

33. J. D. Harris, K. K. Banger, D. A. Scheiman, M. A. Smith, and A. F. Hepp, Characterization of $CuInS_2$ films prepared by atmospheric pressure spray chemical vapor deposition, *Mater. Sci. Eng.* **B98**, 150–155 (2003).

34. R. P. Raffaelle, A. F. Hepp, G. A. Landis, and D. J. Hoffman, Mission applicability assessment of integrated power components and systems. *Prog. Photovolt. Res. Appl.* **10**, 391–397 (2002).

35. http://www.azinet.com/starshine/

36. http://www.eere.energy.gov/financing/

37. http://www.solarbuzz.com

38. K. Zweibel and M. A. Green, Millennium Special Issue: PV 2000 — and beyond [editorial], *Prog. Photovolt. Res. Appl.* **8**, 1 (2000).

3

Natural Organic Photosynthetic Solar Energy Transduction

Robert E. Blankenship

Department of Chemistry and Biochemistry, Arizona State University, Tempe, AZ, USA

Contents

Abstract Most natural photosynthetic systems utilize chlorophylls to absorb light energy and carry out photochemical charge separation that stores energy in the form of chemical bonds. All photosynthetic organisms contain antenna systems, which increase the photon collection ability, and reaction center complexes, which carry out the actual photochemistry. All these components are bound to proteins in well-defined sites with unique positions. The reaction center complexes are always integral membrane proteins that are embedded in biological membranes. Antenna complexes are always attached to the membrane and may also be integral membrane proteins. The antenna complexes transfer energy to the reaction center using a Förster resonance energy transfer mechanism. The reaction center contains a special pair of chlorophylls, which act as a primary electron donor. When excited, they transfer an electron from the excited state to a nearby electron acceptor molecule. After the primary photochemistry, the electron and the hole are separated by an ultrafast series of chemical reactions that separate the charges and stabilize them against recombination. Long-term storage of energy takes place with the formation of adenosine triphosphate (ATP), reduced form of nicotinamide adenine dinucleotide phosphate (NADPH), and ultimately sugars.

Keywords photosynthesis, chlorophyll, antenna, reaction center, excited state, energy transfer, electron transfer

3.1. INTRODUCTION

Photosynthesis is a biological process in which the sun's energy is captured and stored by a series of events that convert the pure energy of light into the biochemical energy needed to power life [1]. This remarkable process provides the foundation for almost all life, and has over geologic time altered the Earth itself in profound ways. It provides all of our food and most of our energy resources.

Photosynthesis literally means "synthesis with light." *Photosynthesis is a process in which light energy is captured and stored by an organism, and the stored energy is used to drive cellular processes.* This definition is relatively broad, and will include the familiar chlorophyll-based form of photosynthesis that is the subject of this chapter, but will also include the very different form of photosynthesis carried out by some bacteria using the protein bacteriorhodopsin [2]. The bacterial rhodopsin-based form of photosynthesis, while qualifying under the general definition, is mechanistically very different from that of chlorophyll-based photosynthesis, and will not be further discussed. It operates using *cis–trans* isomerization that is directly coupled to ion transport across a membrane. No light-driven electron transfer processes are known in these systems.

The most common form of photosynthesis involves chlorophyll-type pigments, and operates using light-driven electron transfer processes. Plants, algae, cyanobacteria, and several types of more primitive anoxygenic (nonoxygen-evolving) bacteria all work in this same basic manner. All of these organisms carry out what is called "chlorophyll-based photosynthesis." This process works using light-sensitized electron transfer reactions.

3.2. PHOTOSYNTHESIS IS A SOLAR ENERGY STORAGE PROCESS

Photosynthesis uses light from the sun to drive a series of chemical reactions. The sun produces a broad spectrum of light output that ranges from gamma rays to radio waves. The entire visible range of light (400 to 700 nm), and some wavelengths in the near infrared (IR; 700 to 1000 nm), are highly active in driving photosynthesis in certain organisms, although the most familiar chlorophyll *a*-containing organisms cannot use light longer than 700 nm. Figure 3.1 shows solar spectra incident on Earth and absorption spectra for two types of photosynthetic organisms. One is a cyanobacterium that uses chlorophyll *a* and the the other is an anoxygenic bacterium that uses bacteriochlorophyll *a* as the principal photopigment. No organism is known that can utilize light of wavelength longer than about 1000 nm for photosynthesis (1000 nm and longer wavelength light comprises 30% of the solar irradiance). Infrared light has a very low-energy content in each photon, so that large numbers of these low-energy photons would have to be used to drive the chemical reactions of photosynthesis. This is thermodynamically possible but would require a fundamentally different molecular mechanism that is more akin to a heat engine than to photochemistry.

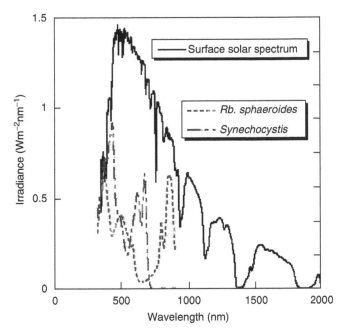

Figure 3.1. Solar irradiance spectra with inset of absorption spectra of photosynthetic organisms. Solid curve, intensity profile of sunlight at the surface of the Earth; dashed line, absorption spectrum of *Rhodobacter sphaeroides*, an anoxygenic purple photosynthetic bacterium and dot-dashed line, *Synechocystis* PCC 6803, an oxygenic cyanobacterium. The spectra of the organisms are in absorbance units (scale not shown).

3.3. WHERE PHOTOSYNTHESIS TAKES PLACE

Photosynthesis is carried out by a wide variety of organisms. In all cases, lipid bilayer membranes are critical to the energy storage, so that photosynthesis is a membrane-based process. The early reactions of photosynthesis are carried out by pigment-containing proteins, which are integrally associated with the membrane. These membranes are either the cytoplasmic membrane of bacteria or the invaginated thylakoid membranes of cyanobacteria or chloroplasts in more advanced photosynthetic organisms such as plants or algae. Chloroplasts are subcellular organelles that were originally derived from cyanobacteria, in a process called endosymbiosis [1]. Today, they retain some traces of their bacterial heritage, including their own DNA, although much of the genetic information needed to build the photosynthetic apparatus now resides in DNA located in the nucleus. Some proteins are synthesized within the chloroplast, while others are made in the cytoplasm and imported into the chloroplast. The genetics, assembly, and evolution of the photosynthetic apparatus are complex processes that are beyond the scope of this chapter.

3.4. PHOTOSYNTHETIC PIGMENTS

For light energy to be stored by photosynthesis it must first be absorbed by pigments associated with the photosynthetic apparatus. Different types of photosynthetic organisms contain different types of pigments. A representative group of pigments active in photosynthesis are shown in Figure 3.2. All chlorophyll-based

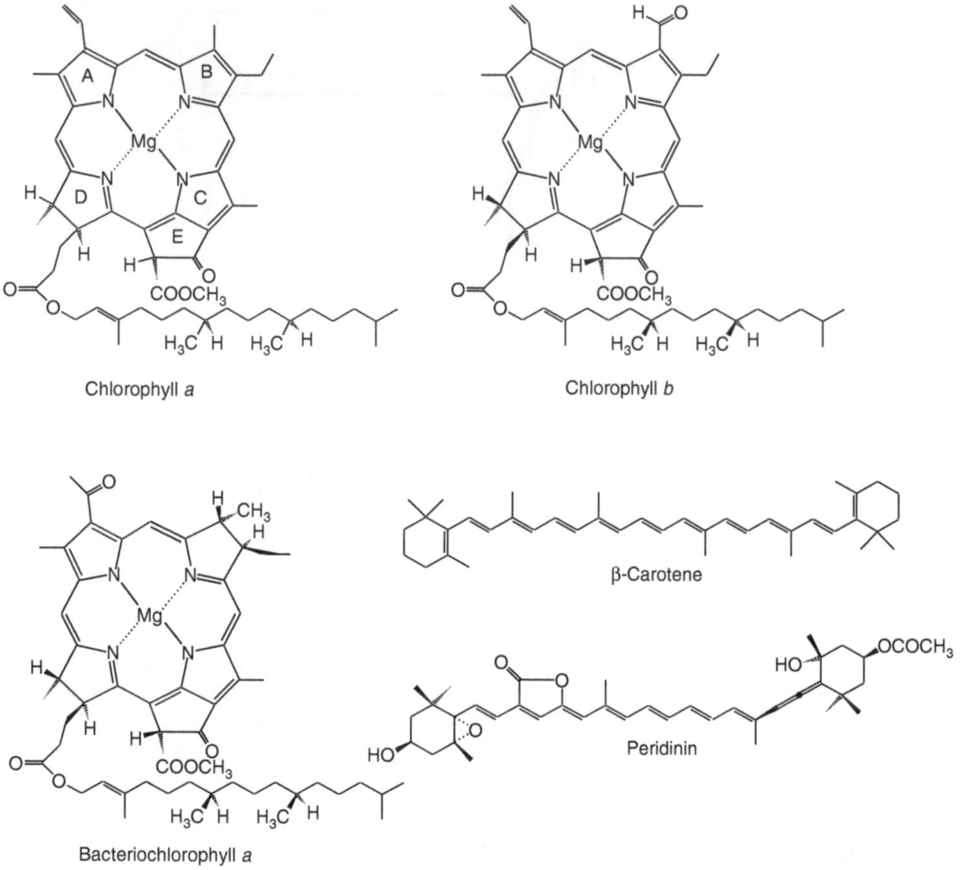

Chlorophyll *a*

Chlorophyll *b*

Bacteriochlorophyll *a*

β-Carotene

Peridinin

Figure 3.2. Chemical structures of representative photosynthetic pigments, including chlorophyll *a* and *b*, bacteriochlorophyll *a* and two carotenoids, β-carotene and peridinin. The rings are designated by letters, which apply to all the chlorophyll-type pigments. Chlorophyll *a* and bacteriochlorophyll *a* are found in both antenna and reaction center complexes, while chlorophyll *b* is only found in antenna complexes. β-Carotene is found in both antenna and reaction center complexes, while peridinin is found only in one class of antenna complex.

photosynthetic organisms contain at least one and often several types of chlorophyll-type pigments. These pigments are chlorins, similar to porphyrins but with ring D (in the lower left) reduced. In addition, an additional ring (E) called the isocyclic ring is found in all chlorophyll-type pigments. The reduction of one or more of the rings of the macrocycle reduces the symmetry and increases the absorption strength of the pigment in the visible and the near IR region where the solar spectrum is most intense (Figure 3.1). Bacteriochlorophyll *a* has ring B reduced in addition to ring D, shifting the absorption even further to the red.

In almost all cases, the central metal in chlorophyll-type pigments is Mg, although a few cases are known where Zn replaces Mg. The excited states of the pigments with these metals are relatively long, typically several nanoseconds. If transition metals such as Fe, Mn, or Cu are inserted into chlorophylls, the excited state lifetimes are shortened by many orders of magnitude due to ultrafast internal conversion via the unfilled orbitals of the metal and the pigments become useless as photosensitizers. The nature of the central metal is thus critical for photosynthesis, and pigments such as cytochromes, while important in some of the later electron transfer processes, are not capable of directly sensitizing photosynthetic energy storage. The metal-free pigments, known as pheophytins, are present in small quantities in most photosynthetic organisms, and also fulfill important functional roles in the electron transfer processes.

In almost all chlorophylls a long hydrocarbon tail is found, which is esterified to ring D. This tail serves to anchor the pigment in the protein environment. It is in most cases a polyisoprenoid group, and the most common is the phytyl group. A complete listing of photosynthetic pigments including structures and spectra can be found in Ref. [1].

The chlorophylls found in modern-day photosynthetic organisms are clearly the result of a long process of evolutionary development, which has optimized their light absorption and redox properties. Early photosynthetic organisms probably used simpler pigments such as porphyrins.

In addition to the chlorophyll-type pigments, all photosynthetic organisms contain carotenoid pigments (Figure 3.2). These pigments have several distinct functions, the most important is to provide photoprotection to the organisms. A variety of processes can occur, resulting in undesired electronic excited states such as chlorophyll triplet states and singlet oxygen. Carotenoids are remarkably efficient at quenching these excited states and protecting the pigments from photooxidative damage. In addition, carotenoids function as accessory antenna pigments (see below), absorbing light and transferring it to chlorophylls before it is used for photochemistry. Finally, carotenoids are important in the regulation of energy flow in photosynthetic systems [3].

3.5. THE FOUR PHASES OF ENERGY STORAGE IN PHOTOSYNTHETIC ORGANISMS

It is convenient to divide photosynthesis into four distinct temporal phases, beginning with photon absorption and ending with the production of stable carbon products. The four phases are:

1. Light absorption and energy delivery by antenna systems
2. Primary electron transfer in reaction centers
3. Energy stabilization by secondary processes
4. Synthesis and export of stable products.

In this chapter, we will focus on the first two of these phases.

3.6. ANTENNAS AND ENERGY TRANSFER PROCESSES

Photon absorption creates an excited state, leading to charge separation in the reaction center. Not every pigment carries out photochemistry. Most pigments function as antennas [3], collecting light and then delivering energy to the reaction center where the photochemistry takes place. The antenna system is in concept similar to a satellite dish, collecting energy and concentrating it in a receiver where the signal is converted to a different form. This is shown schematically in Figure 3.3.

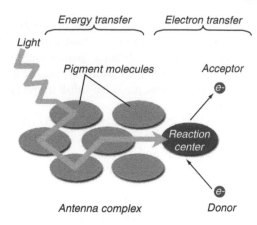

Figure 3.3. Schematic concept of antennas and reaction centers in photosynthetic systems. Antenna pigments absorb light and energy transfer processes deliver the excited state to the reaction center where electron transfer reactions store the energy. Figure courtesy of Dr. Alexander Melkozernov.

A remarkable variety of antenna complexes are known, many of which have little or no apparent structural relationship to each other either in terms of protein structure and even in pigment composition. The various types of antennas clearly result from multiple evolutionary innovations. In addition to chlorophylls, antenna pigments include carotenoids. Figure 3.4 shows structures of some of the main classes of photosynthetic antenna complexes. Most of these are distinct molecular complexes that can be separated from the reaction center complexes involved in electron transfer. They can be classified as either integral membrane proteins or peripheral membrane proteins. In some cases, such as Photosystem I from oxygenic photosynthetic organisms, the antenna pigments are an integral part of the core reaction center complex and cannot be separated without destroying the complex.

The antenna system does not do any chemistry; it works by an energy transfer process that involves the migration of electronic excited states from one molecule to another. This is a purely physical process, which depends on an energetic coupling of the antenna pigments. In almost all cases, the pigments are bound to proteins in highly specific associations as shown in Figure 3.4.

The mechanism of energy transfer in most photosynthetic antennas is the Förster mechanism, in which the energy transfer rate depends on the sixth power of the distance between the pigments, the relative orientation of the transition dipole moments of the energy donor and acceptor, and the spectral overlap of the two pigments. Efficient energy transfer between chlorophylls can take place over a distance of several tens of angstroms in times of picoseconds or less. The energy transfer process is sufficiently fast to deliver the energy to the reaction center complex in a time much less than the nanoseconds excited state lifetime of the pigment. In some types of antenna complexes, such as the LH1 and LH2 complexes of purple photosynthetic bacteria and the chlorosome antennas of green photosynthetic bacteria, the pigments are very close to each other and are strongly energetically coupled. In this case the excited state is viewed as delocalized over several molecules as an exciton [4].

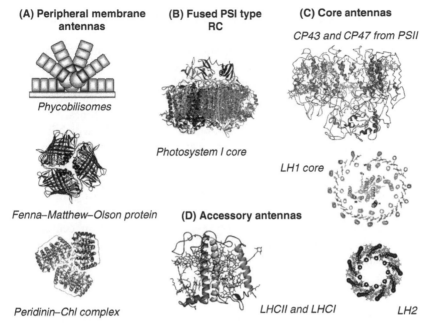

(A) Peripheral membrane antennas

Phycobilisomes

Fenna–Matthew–Olson protein

Peridinin–Chl complex

(B) Fused PSI type RC

Photosystem I core

(D) Accessory antennas

LHCII and LHCI

(C) Core antennas

CP43 and CP47 from PSII

LH1 core

LH2

Figure 3.4. (Color figure follows page 348). Examples of photosynthetic light-harvesting antenna complexes. (A) Peripheral antennas: phycobilisomes from cyanobacteria and red algae (schematic); Fenna–Matthew–Olson protein from a green sulfur bacterium *Prostheco-chloris aestuarii* (pdb code 4BCL, [9]); peridinin–Chl complex from a dinoflagellate *Amphidi-nium carterae* (pdb code 1PPR, [10]). (B) Fused PSI type RC from a cyanobacterium *Synechococcus elongatus* (pdb code 1JB0, [8]). (C) Core antennas: CP43 and CP47 from the PSII of cyanobacterium *Synechococcus elongatus* (pdb code 1FE1, [11]) and LH1 core from purple bacterium *Rhodopseudomonas palustris* (1PYH, [12]). (D) Peripheral antennas: LHCI and LHCII from algae and higher plants (pdb code 1RWT for LHCII, [13]) and LH2 from *Rhodopseudomonas acidophila* (pdb code 1KZU, [14]). Molecular graphics rendered using Web Lab Viewer from Molecular Simulations, Inc. (Figure courtesy of Dr. Alexander Melkozernov.)

Antenna systems usually incorporate an energetic and spatial funneling mechanism, in which pigments on the periphery of the complex have absorption at shorter wavelengths and therefore higher excitation energies compared to those at the core. As energy transfer takes place, the excitation energy moves from higher to lower energy pigments, at the same time moving physically towards the reaction center, where it is eventually trapped by electron transfer reactions.

Antennas greatly increase the amount of energy that can be absorbed compared to a single pigment. Under most conditions this is an advantage, because sunlight is a relatively dilute energy source. Even in full sunlight, each chlorophyll only absorbs about ten photons per second [1]. Under some conditions, however, especially if some other form of stress is present on the organism, more light energy can be absorbed, which can be productively used by the system. If unchecked, this can lead to severe damage in short order. Even under normal conditions, the system is rapidly inactivated if some sort of photoprotection mechanism is not present. Antenna systems as well as reaction centers, therefore, have extensive multifunctional regulation, protection, and repair mechanisms [3].

3.7. PRIMARY ELECTRON TRANSFER IN REACTION CENTERS

The conversion of pure energy of excited states to chemical bonds in molecules takes place in the reaction center. The reaction center is a multisubunit integral–membrane pigment–protein complex, incorporating both chlorophylls and other electron transfer cofactors such as quinones or iron–sulfur centers, along with hydrophobic peptides that thread back and forth across the membrane several times.

The reaction center contains a special dimer of pigments, which is the primary electron donor for the electron transfer cascade. These pigments are chemically identical (or nearly so) to the chlorophylls (antenna pigments), but their environment in the reaction center protein gives them unique properties. The final step in the antenna system is the transfer of energy into this dimer, creating an electronically excited dimer.

High-resolution structures are now available for reaction center complexes from anoxygenic photosynthetic bacteria, and for both Photosystems I and II from oxygenic organisms. Figure 3.5 shows representative structures of these complexes.

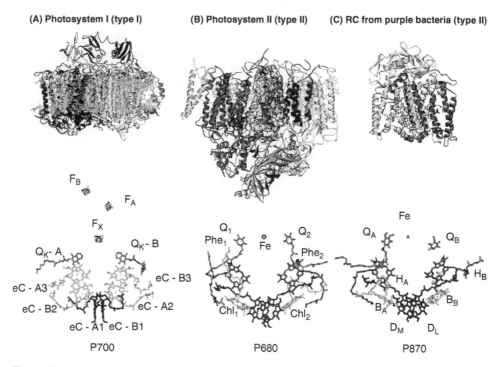

Figure 3.5. **(Color figure follows page 348)**. Structure of the photosynthetic reaction centers of Photosystem I (A), Photosystem II (B) and RC from purple bacteria (C). Upper panel: Side views of the pigment–protein complexes in the reaction centers. Lower panel: cofactors of electron transfer in the reaction centers. Molecular graphics rendered using Web Lab Viewer from Molecular Simulations, Inc. using atomic coordinates of the molecules in Protein Data Bank (code 1JB0 [8] for PSI from the cyanobacterium *Thermosynechococcus elongatus*, code 1S5L [9] for PSII from *T. elongatus*, and code 1M3X [15] for the RC from the purple bacterium *Rhodobacter sphaeroides*). Both models are shown at the same scale. (Figure courtesy of Dr. Alexander Melkozernov.)

The complete complex including all pigments and the protein part is shown in Figure 3.5A, and the pigments and other cofactors that make up the electron transfer chain are shown in Figure 3.5B.

The basic process that takes place in all reaction centers is described schematically in Figure 3.6. A chlorophyll-like pigment (P) is promoted to an excited electronic state, either by direct photon absorption or more commonly by energy transfer from the antenna system. The excited state of the pigment is an extremely strong reducing species. It rapidly loses an electron to a nearby electron acceptor molecule (A), generating an ion-pair state P^+A^-. This is the primary photochemical energy storing reaction of photosynthesis. After this step, the energy is transformed from electronic excitation to chemical redox energy.

The system is now in a very vulnerable position with respect to losing the stored energy. If the electron is transferred back to P^+ from A^-, the energy is converted to heat and lost without doing any work. The system is vulnerable to recombination and energy loss because the highly oxidizing P^+ species is physically positioned adjacent to the highly reducing A^- species so that the electron transfer can be rapid. The system is therefore both thermodynamically and kinetically vulnerable. The avoidance of these wasteful processes is achieved by a remarkable favorable combination of structural positioning of the cofactors by the protein environment and the electronic states and redox energies of the complex. Part of the efficiency derives from the fact that the recombination reaction is in the Marcus inverted region and therefore slow compared to the other, more productive

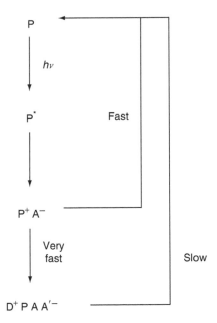

Figure 3.6. General scheme of electron transfer in photosynthetic reaction centers. Light absorption creates an excited state of the reaction center pigment (P). Electron transfer creates an ion-pair state (P^+A^-). This is unstable against recombination to the ground state. Secondary electron transfer processes spatially separate the positively and negatively charged species and the increased distance between them greatly stabilizes the system against recombination.

processes, which are in many cases poised near the top of the Marcus rate vs. free energy curve [5].

A series of extremely rapid secondary reactions therefore successfully compete with recombination and ensure that the energy is stored. These reactions spatially separate the positive and negative charges, and reduce the intrinsic recombination rate by orders of magnitude. The final result is that in less than a nanosecond, the oxidized and reduced species are separated by the thickness of the biological membrane ($\sim 30\,\text{Å}$). Slower processes further stabilize the energy that has been stored and convert it into more easily utilized forms, as described below. The system is so finely tuned that the photochemical quantum yield is nearly 1.0, so that nearly every photon that is absorbed gives rise to stable products. The energy yield is significantly less, because some energy is sacrificed to ensure a high quantum yield. The energy efficiency of photosynthesis depends on exactly what point in the process is considered, but a representative value of the efficiency of energy conversion from photons to the chemical energy that can be recovered by metabolizing or burning sugars is about 20–25% [1].

A diagram that shows the redox potentials of the various components of the reaction center complex, as well as some of the secondary electron carriers, is very informative to help visualize both the energetics and electron transfer pathway of the photosynthetic system. These diagrams are shown in Figure 3.7 for various classes of photosynthetic organisms. The electron transfer components are placed vertically at their midpoint redox potentials, which have in many cases been determined either in situ or on isolated cofactors using electrochemical techniques. A vertical arrow is

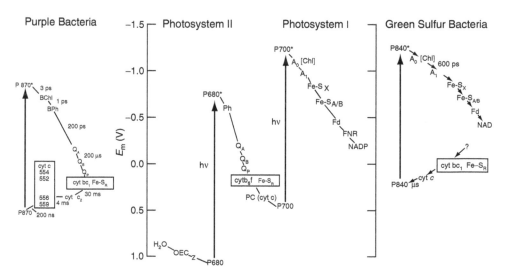

Figure 3.7. Schematic diagram of electron transfer pathway found in photosynthetic organisms. The left diagram shows the cyclic electron transfer system found in the anoxygenic purple bacteria, while the right diagram shows the system found in the anoxygenic green sulfur bacteria. The central diagram shows the electron transfer pattern, called the Z scheme, which takes place in oxygen-evolving photosynthetic organisms such as cyanobacteria and chloroplasts. The vertical axis is midpoint redox potential of the electron carriers. The excited-state redox potentials are calculated from the measured ground state potentials and the excitation energy as described in Ref. [6].

used to represent photon absorption; the length of the arrow is proportional to the energy of excitation of the primary electron donor pigment, which is determined from the absorption spectrum. The pigment is designated PXXX with the Xs giving the wavelength of the longest wavelength absorption of the pigment, such as P700 or P870. The upper end of the arrow gives the redox potential of the chlorophyll excited state, which is calculated by subtracting the excitation energy from the ground state redox potential [6]. The excited state is thus an extremely strong reducing agent and the primary photochemical process is in all known cases a light-dependent oxidation of the chlorophyll pigment with concomitant reduction of the primary acceptor.

3.8. STABILIZATION BY SECONDARY REACTIONS

The primary electron transfer event is followed by separation of the positive and negative charges by a very rapid series of secondary chemical reactions. This basic principle applies to all photosynthetic reaction centers, although the details of the process vary from one system to the next.

In some organisms, one light-driven electron transfer and stabilization is sufficient to complete a cyclic electron transfer chain. This cyclic electron transfer process is not in itself productive unless some of the energy of the photon can be stored. This takes place by the coupling of proton movement across the membrane with the electron transfer, so that the net result is a light-driven pH difference, or electrochemical gradient, collectively known as a proton motive force. This proton motive force is used to drive the synthesis of adenosine triphosphate (ATP).

The more familiar oxygen-evolving photosynthetic organisms have a different pattern of electron transfer. They have two photochemical reaction center complexes that work together in a noncyclic electron transfer chain, as shown in Figure 3.7. The two reaction center complexes are known as Photosystems I and II. Electrons are removed from water by Photosystem II, oxidizing it to molecular oxygen, which is released as a waste product. The center that oxidizes water to form molecular oxygen involves a complex of four Mn ions that transfers electrons via a tyrosine amino acid side chain to the chlorophyll that is the primary electron donor in Photosystem II [7].

The electrons extracted from water are donated to Photosystem II, and after a second light-driven electron transfer step by Photosystem I [8], eventually reduce an intermediate electron acceptor, the oxidized form of nicotinamide adenine dinucleotide phosphate ($NADP^+$). Protons are also transported across the membrane and into the thylakoid lumen during the process of the noncyclic electron transfer, creating a pH difference, which contributes to the proton motive force. The energy in this proton motive force is used to make ATP. The NADPH and ATP that are formed in the light-driven steps of photosynthesis are used to fix CO_2 to form sugars and other organic products that give energy for metablolism and growth of the organism. Excess stored energy is available to us in the form of food, fiber, and biomass.

3.9. CONCLUSIONS

Chlorophyll-based photosynthesis is an extraordinarily complex biological process. It is remarkable in its efficiency and robustness. Photosynthetic organisms occupy essentially all environments on Earth where light energy is available, even in environments where the photon flux is so weak that the chlorophylls absorb a photon only once every several hours. Challenges for the future include the construction of artificial systems that even remotely approach the extraordinary properties of the natural photosynthetic systems. Additionally, if life is present on other worlds, it will almost certainly involve the utilization of some form of photosynthetic energy storage system. While the basic physical principles of photochemical energy storage involving light absorption and excited state processes will necessarily be followed by any such system, the details will almost certainly be very different.

REFERENCES

1. R. E. Blankenship, *Molecular Mechanisms of Photosynthesis*, Blackwell Science, Oxford, UK, 2002.
2. B. W. Edmonds and H. Luecke, Atomic resolution structures and the mechanism of ion pumping in bacteriorhodopsin, *Frontiers Biosci.*, **9**, 1556–1566, 2004.
3. B. R. Green and W. W. Parson (eds), *Light-Harvesting Antennas in Photosynthesis*, Kluwer Academic Publishers, Dordrecht, The Netherlands, 2003.
4. H. van Amerongen, L. Valkunas, and R. van Grondelle. *Photosynthetic Excitons*, World Scientific, Singapore, 2000.
5. C. C. Moser, J. M. Keske, K. Warncke, R. S. Farid, and P. L. Dutton, Nature of biological electron-transfer. *Nature*, **355**, 796–802, 1992.
6. R. E. Blankenship and R. C. Prince, Excited-state redox potentials and the Z-scheme of photosynthesis. *Trends Biochem. Sci.*, **10**, 382–383 1985.
7. K. N. Ferreira, T. M. Iverson, K. Maghlaoui, J. Barber, and S. Iwata, Architecture of the photosynthetic oxygen-evolving center. *Science*, **303**, 1831–1838, 2004.
8. P. Jordan, P. Fromme, H.-T. Witt, O. Klukas, W. Saenger, and N. Krauss, Three-dimensional structure of cyanobacterial photosystem I at 2.5 Å resolution. *Nature*, **411**, 909–917, 2001.
9. D. E. Tronrud, M. F. Schmid, and B. W. Matthews, Structure and X-ray amino acid sequence of a bacteriochlorophyll *a* protein from *Prosthecochloris aestuarii* refined at 1.9 Å resolution. *J. Mol. Biol.*, **188**, 443–454, 1986.
10. E. Hofmann, P. M. Wrench, F. P. Sharples, R. G. Hiller, W. Welte, and K. Diederichs, Structural basis of light harvesting by carotenoids: peridinin–chlorophyll–protein from *Amphidinium carterae. Science*, **272**, 1788–1791, 1996.
11. A. Zouni, H.-T. Witt, J. Kern, P. Fromme, N. Krauss, W. Saenger, and P. Orth, Crystal structure of photosystem II from *Synechococcus elongatus* at 3.8 Å resolution. *Nature*, **409**, 739–743, 2001.
12. A. W. Roszak, T. D. Howard, J. Southall, A. T. Gardiner, C. J. Law, N. W. Isaacs, and R. J. Cogdell, Crystal structure of the RC-LH1 core complex from *Rhodopseudomonas palustris. Science*, **302**, 1969–1972, 2003.
13. Z. Liu, H. Yan, K. Wang, T. Kuang, J. Zhang, L. Gui, X. An, and W. Chang, Crystal structure of spinach major light-harvesting complex at 2.72 Å resolution. *Nature*, **428**, 287–292, 2004.
14. S. M. Prince, M. Z. Papiz, A. A. Freer, G. McDermott, A. M. Hawthornthwaite-Lawless, R. J. Cogdell, and N. W. Isaacs, Apoprotein structure in the LH2 complex from *Rhodopseudomonas acidophila* strain 10050: modular assembly and protein pigment interactions. *J. Mol. Biol.*, **268**, 412–423, 1997.
15. A. Camara-Artigas, D. Brune, and J. P. Allen, Interactions between lipids and bacterial reaction centers determined by protein crystallography. *Proc. Natl Acad. Sci. USA*, **99**, 11055–11060, 2002.

4

Solid-State Organic Photovoltaics: A Review of Molecular and Polymeric Devices

Paul A. Lane and Zakya H. Kafafi
Naval Research Laboratory, Washington, D.C., USA

Contents

Abstract This chapter presents an overall review of progress in organic photovoltaics devices (OPVs). Individual sections trace the development of molecular, polymer, and inorganic–organic hybrid OPVs from the earliest devices to the current state of the art. Within each class of device, a different approach has been taken with multilayer cells having the best performance for molecular OPVs and blends of polymers and fullerene derivatives having the best performance of polymer OPVs. The chapter concludes with a discussion of solid-state dye-sensitized solar cells and a comparison of OPV performance to inorganic solar cells.

Keywords organic photovoltaic, molecular, polymer, hybrid, dye-sensitized, bulk heterojunction.

4.1. INTRODUCTION

4.1.1. Overview

Organic semiconductors are a remarkable class of materials with present and potential applications in various optoelectronic devices. In principle, devices using organic materials should be cheaper and simpler to manufacture than corresponding ones using inorganic semiconductors. While the performance of organic light-emitting diodes (LEDs) has steadily improved over the past 15 years, the power conversion efficiency of organic photovoltaic devices (OPVs) remained stubbornly below 1% until recently. The performance of solid-state organic solar cells, however, has increased substantially in the last several years, with power conversion efficiencies under white light illumination (AM1.5) $\geq 3\%$ for the best molecular, polymer, and organic–inorganic hybrid solar cells. This progress is encouraging, though it should be noted that performance of organic solar cells still lags behind the performance of inorganic solar cells [1] (as high as 24% [2] for single crystal silicon).

    This chapter takes a historical approach to the field of organic photovoltaics with an emphasis on power generation from solar cells. Organic photovoltaic cells based on small molecules will first be discussed, followed by polymer devices, and finally hybrid devices combining organic and inorganic materials will be highlighted. Starting with the earliest devices and their limitations, we will review progress within each class of device and discuss approaches taken by various groups to improve device performance.

    Molecular and polymer photovoltaics are distinguished by the type of material and fabrication methods. Molecular photovoltaic devices are generally fabricated by sublimation under vacuum of successive layers of electron- and hole-transporting materials. Charge photogeneration occurs at the interface between the two layers, also known as the organic heterojunction. Figure 4.1 shows the chemical structures of some small molecules commonly used in OPVs. These include materials primarily used as dopants in polymer OPVs as well as charge transport layers in hybrid organic–inorganic devices. The first organic solar cells exhibiting reasonable power conversion efficiencies were fabricated by Tang [3] using phthalocyanine and perylene derivatives; these families of materials have accordingly been heavily investigated. Buckminsterfullerene (C_{60}) and its derivatives have also been widely investigated, both as separate layers in molecular OPVs and in bulk heterojunction polymer OPVs.

    In contrast, polymer OPVs are generally prepared by solution processing. The most efficient devices consist of blends of a conjugated polymer and a molecular sensitizer, or blends of two different conjugated polymers. Such blends have interfaces throughout the active layer, known as a bulk heterojunction. Multilayer devices consisting of two different materials, most commonly two different polymers, have also been fabricated and are discussed. Figure 4.2 shows repeat units of various conjugated polymers used in active layers of organic solar cells. Poly(p-phenylene vinylene) (PPV) and various derivatives with different side-chain substitution have been most widely used in the active layer of polymer OPVs. Derivatives of polythiophene (PT) and, more recently, polyfluorene have also been investigated by a number of research groups.

4.1.2. Device Characterization

We describe briefly the parameters used to characterize the performance of photovoltaic devices. Figure 4.3 shows a schematic diagram of the current–voltage curve of a photodiode under illumination. Devices are generally characterized by

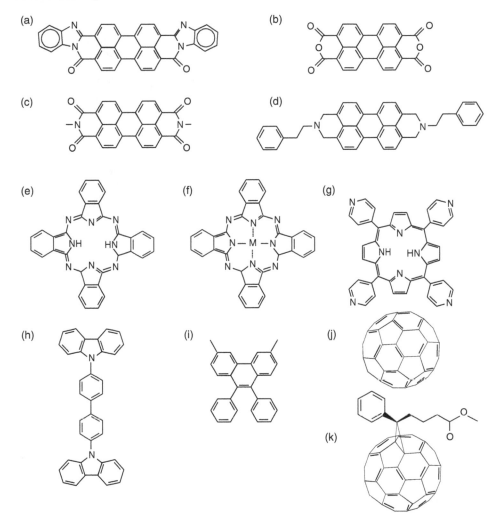

Figure 4.1. Chemical structures of molecular semiconductors used in OPVs. (a) PTCBI, (b) PTCDA, (c) Me-PTCDI, (d) Pe-PTCDI, (e) H$_2$Pc, (f) MPc (M = Zn, Cu), (g) TPyP, (h) TPD, (i) CBP, (j) C60, (k) 5,6-PCBM.

the short-circuit current (J_{sc}), the open-circuit voltage (V_{oc}), and the fill factor (ff). The fill factor of a device is defined as the ratio between the maximum power delivered to an external circuit and the potential power:

$$ff = P_m/I_{sc}V_{oc} = I_m V_m/I_{sc}V_{oc} \qquad (1)$$

The fill factor is the ratio of the darkly shaded to lightly shaded regions in Figure 4.3. The power conversion efficiency of a device is defined as the ratio between the maximum electrical power generated (P_m) and the incident optical power P_o:

$$\eta_e = I_m V_m/P_o.$$

The spectral response of OPVs is an important way to characterize such devices and optimize their performance. The device is illuminated by a monochromatic light source, generally consisting of a broadband illuminator dispersed through a monochromator. The photocurrent is measured as the function of wavelength and

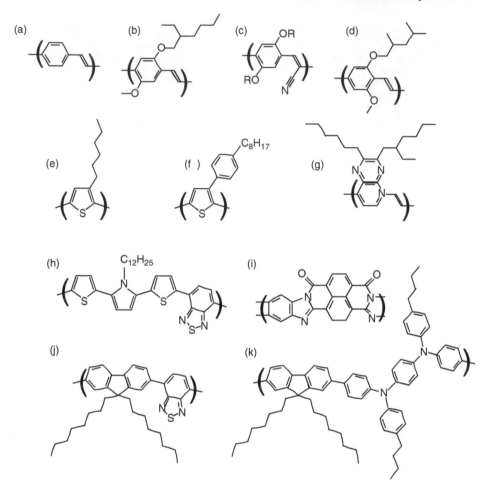

Figure 4.2. Chemical structures of conjugated polymers used in OPVs: (a) PPV, (b) MEH-PPV, (c) CN–PPV (various alkoxy cyano-derivatives have been prepared), (d) MDMO-PPV, (e) P3HT, (f) POPT, (g) EHH–PpyPz, (h) PTPTB, (i) BBL, (j) F8BT, (k) PFMO.

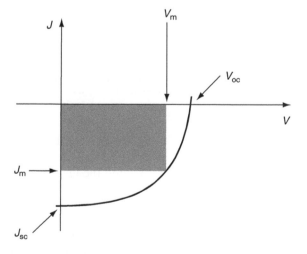

Figure 4.3. Current–voltage response of photovoltaic device under illumination. The open-circuit voltage (V_{oc}), short-circuit current (I_{sc}), and current and voltage at maximum power output (V_m and I_m, respectively) are defined.

then compared to the light intensity or photon flux. The external quantum efficiency (EQE), also known as the incident photon conversion efficiency, is given by the number of electrons generated per incident photon:

$$\text{EQE} = \frac{n_e}{n_{ph}} = \frac{I_{sc}}{P_o} \frac{hc}{\lambda e} \tag{2}$$

where P_o is the incident optical power, h is Planck's constant, c is the speed of light, λ is the wavelength of light, and e is the electrical charge. This generally follows the absorption spectrum of the materials constituting the OPV. The internal quantum efficiency, also known as the photocurrent action spectrum, is given by the ratio of the photocurrent to the absorbed photon flux.

In theory, the power efficiency of a solar cell can be calculated by integrating the EQE over the solar spectrum and multiplying by the monochromatic power efficiency. Such an approach fails because the fill factor depends on the wavelength, and device efficiency often decreases with increasing light intensity. The best approach is to use a solar simulator that replicates the solar spectrum. The AM1.5 spectrum is the solar spectrum through atmosphere, 48.2° from zenith and the total intensity is approximately 80 mW/cm². The visible and near-infrared portions of the AM1.5 spectrum [4] are shown in Figure 4.4. In many cases, solar simulator is not available and a white light source is used instead, with the spectral mismatch taken into account in the calculations of the power conversion efficiencies of OPVs. Several groups have utilized solar simulators at an independent laboratory such as the test facilities available at the U.S. National Renewable Energy Laboratory. This approach provides an independent means to directly compare solar cells from different institutions and is noted within the text where applicable.

4.2. MOLECULAR OPVs

4.2.1. Organic Heterojunction Solar Cells

Photovoltaic effects have long been observed in organic semiconductors (see Chamberlain [5] for a review of early work). Early work was inspired by photosynthesis in which light is absorbed by chlorophyll, a member of the porphyrin family.

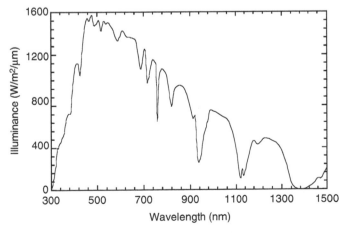

Figure 4.4. Solar irradiation spectrum for AM0 and AM1.5 illumination. From Ref. [4].

Macrocyclic molecules such as porphyrins [6] and phthalocyanines [7–9] have accordingly drawn much interest. Phthalocyanines and perylenes are widely used for xerographic applications, resulting in widespread availability and high purity at relatively low cost. These materials also have very high linear absorption coefficients (exceeding $10^5\,cm^{-1}$), permitting the use of thin films to absorb light. The earliest OPVs were Schottky-type devices (Figure 4.5(a)) in which a rectifying contact is formed at one of the organic–electrode interfaces in a metal–organic–metal sandwich [10–12] –. Such devices are intrinsically inefficient as charge photogeneration takes place only in a thin layer near the metal–organic interface, limiting the quantum yield of charge photogeneration. Exciton quenching at metal–organic interfaces [13] can also reduce photocurrent yields.

The most successful organic solar cells are based on charge generation at an interface between two different organic semiconductors, also known as an organic heterojunction (Figure 4.5(b)). Organic semiconductors have long been known to preferentially transport holes (p-type) or electrons (n-type). Early bilayer organic solar cells were made from combinations of electron-transporting dyes such as rhodamines or triphenylmethane dyes with hole-transporting dyes such as phthalocyanines or merocyanines [14]. Such organic heterojunctions generated photovoltages up to 200 mV and photocurrents on the order of 10^{-8} A in low-intensity light. Harima et al. [15] reported efficient charge photogeneration at an organic heterojunction between zinc phthalocyanine (ZnPc) and 5,10,15,20-tetra(3-pyridyl)porphyrin (TPyP) in 1984. Devices were fabricated by evaporating a 7-nm thick layer of TPyP on top of a semitransparent aluminum film, followed by a 50 nm film of ZnPc to create an organic heterojunction, and a top electrode of gold. Aluminum and gold had previously been shown to form a good electrical contact with TPyP and ZnPc, respectively [16]. The spectral response of single-layer ZnPc and bilayer TPyP/ZnPc devices was compared when illuminated through the aluminum electrode. Very strong sensitization of charge photogeneration within TPyP can be seen; the photocurrent within the Soret band of TPyP ($\lambda = 430$ nm) is roughly 30 times greater than that of the ZnPc device. The internal quantum efficiency was calculated to be as high as 17% and open-circuit voltages of ~1.0 V could be obtained. Under weak,

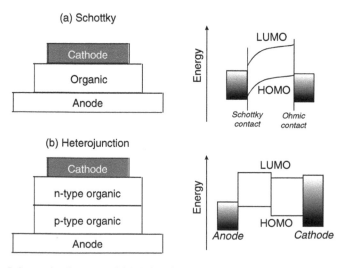

Figure 4.5. Schematic diagram of (a) Schottky-type and (b) heterojunction OPVs.

monochromatic illumination ($10\,\mu\text{W/cm}^2$ at 430 nm), the power conversion efficiency was 2%. The photoresponse decreased with thicker TPyP films due to the limited diffusion length of excitons within the TPyP film.

Tang reported a breakthrough in 1985 in organic photovoltaic performance, achieving power efficiency of nearly 1% under simulated solar illumination [3]. The device is conceptually similar to that of Harima et al., but used a perylene derivative as the n-type layer. The solar cell was fabricated by evaporating a 25-nm thick layer of copper phthalocyanine (CuPc) onto an indium tin oxide (ITO)-coated glass substrate, followed by a 45-nm thick layer of 3,4,9,10-perylene tetracarboxylic-bis-benzimidazole (PTCBI). A silver cathode was then evaporated on top of the structure. Figure 4.6 shows the current–voltage characteristics of the CuPc/PTCBI OPV [3] in the dark and under simulated AM2 solar illumination ($75\,\text{mW/cm}^2$; spectrum measured with sun 60.2° from zenith). A short-circuit current density $J_{sc} = 2.3 \pm 0.1\,\text{mA/cm}^2$, an open-circuit voltage $V_{oc} = 0.45 \pm 0.02\,\text{V}$, and fill factor ff $= 65 \pm 3\%$ were reported. The power conversion efficiency ($\eta_e = 0.95\%$), calculated from these parameters, is roughly an order of magnitude better than previous OPVs. The photovoltaic spectral response of the CuPc/PTCBI solar cell was found to roughly follow the absorption spectrum of the bilayer, reaching a peak collection efficiency (EQE) of about 15% at $\lambda = 620$ nm. Unlike the phthalcyanine–porphyrin device fabricated by Harima et al. charge photogeneration occurs in both layers. The optical density of the CuPc–PTCBI film at this wavelength is approximately 0.4 and the corresponding internal quantum efficiency is roughly 25%.

The organic heterojunction is the key to the properties of organic solar cells. In a conventional *pin* photocell, photocarriers are directly created by absorption of light within the bulk of the undoped, intrinsic layer and are swept to the electrodes by a built-in electric field due to the difference in the Fermi energy relative to the band edges of the p-type and n-type layers. Light absorption in an organic semiconductor, however, generates a neutral excited state akin to a tightly bound, Frenkel exciton. The built-in electric field arising from the different work functions of

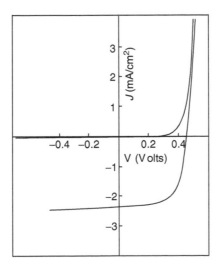

Figure 4.6. *J–V* characteristics under simulated AM2 illumination of a bilayer OPV [ITO/ (25 nm)CuPc/(45 nm)PTCBI/Ag]. (From Tang, C. W. *Appl. Phys. Lett.* **1986**, *48*(2), 183–185. Copyright American Institute of Physics, 1986. With permission.)

the cathode and anode is insufficient to ionize such a state. Rather, the photocurrent is generated by a photoinduced charge transfer reaction between a donor and an acceptor molecule and is illustrated in Figure 4.7. A difference in the electron affinity or ionization potential, or both, will drive a charge transfer reaction. Also important for photocurrent generation is a relatively slow back-transfer (or recombination) rate. Exciton dissociation can also occur at charge traps or impurities, but such films are likely to have poor charge transport.

The performance of a bilayer OPV is ultimately determined by the efficiency of charge photogeneration and charge transport. As charge transfer takes place at the organic heterojunction, absorption must take place at the interface or within the exciton diffusion length in the respective materials. In principle, the exciton diffusion length can be calculated by analyzing the spectral response of a device [17]. Typical exciton diffusion lengths have been reported in the range of 5–10 nm [18–20]. Photoluminescence (PL) quenching is a conceptually simple experiment that can be used to estimate diffusion lengths. Figure 4.8 shows the relative quenching of a polymer–C_{60} heterojunction as a function of the polymer thickness, defined as the fractional difference in emission between a neat film and polymer–C_{60} film [21]. Poly(3-[4′-(1′,4″,7‴-trioxaoctyl)phenyl]thiophene (PEOPT) was used for this experiment. Ellipsometry was used to show that the interface between the polymer and C_{60} layers was sharp. For a polymer thickness of less than 20 nm, more than 50% of PL emission was quenched. An exciton diffusion length of 5.3 nm was estimated based on these measurements.

4.2.2. Molecular OPVs with Bulk Heterojunctions

One of the primary limitations of a heterojunction solar cell is that charge photogeneration takes place only in a thin layer near the organic heterojunction. Only a fraction of incident light will be absorbed in this region, limiting the quantum efficiency of devices. One way to increase the width of the photocarrier generation region is to co-evaporate two organic pigments to create a bulk heterojunction. Ideally, electrons and holes are photogenerated in the mixed ("intrinsic") layer and are swept to the transport layers by built-in chemical and electric potentials.

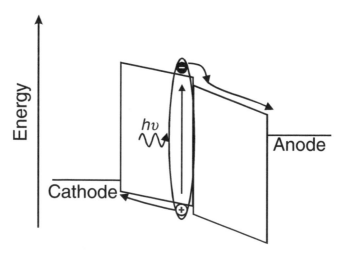

Figure 4.7.   Illustration of photoinduced charge transfer.

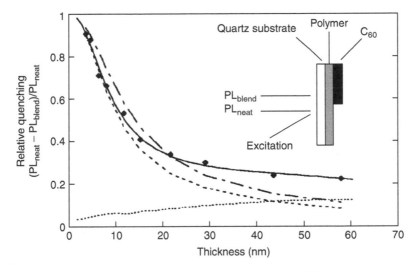

Figure 4.8. Relative quenching polymer fluorescence by C_{60} for different polymer layer thicknesses (diamonds). For comparison, the relative quenching calculated for diffusion lengths of 5.3 nm (dashed line) and 7 nm (broken-dashed line) are shown. The dotted line shows a simulation of relative quenching due to interference of absorbed and emitted light. The inset shows the experimental configuration. (Reprinted figure with permission from Theander, M.; Yartsev, A.; Zigmantas, D. Sundström, V.; Mammo, W.; Andersson, M. R.; Inganäs, O. *Phys. Rev. B*, **2000**, *61*(19), 12957–12963. Copyright American Physical Society, 2000. With permission)

The first trilayer cells were reported by Hiramoto et al. [22] who compared devices made from perylene and phthalocyanine derivatives with and without a co-deposited film between the electron-transporting layer (ETL) and the hole-transporting layer (HTL). Bilayer devices were fabricated by successive vacuum depositions of perylene and phthalocyanine on ITO-coated glass substrates. Trilayer cells were prepared by co-depositing the electron- and hole-transporting materials in the middle layer. The first set of devices used N,N'-dimethyl-3,4,9,10-perylenetetra-carboxylic acid diimide (Me-PTCDI) for the electron-transporting layer and a free-base phthalocyanine (H_2Pc) as the hole-transporting layer. Devices were also prepared from PTCBI and CuPc, the same materials used by Tang. Device structures were ITO/(40 nm Me-PTCDI)/(80 nm Pc/Au or ITO/(40 nm Me-PTCDI)/(40 nm Me-PTCDI–Pc)/(40 nm Pc)/Au.

Table 4.1 summarizes the properties of all four types of devices, measured under 100 mW/cm^2 white light illumination from a metal halide lamp. The quantum efficiency of charge photogeneration was significantly improved for both material combinations; J_{sc} improved by 60% for the PTCBI/CuPc device and by 125% for the PTCDI/H_2Pc device. The power conversion efficiency of a PTCDI/H_2Pc was doubled, though surprisingly little change was seen for the PTCBI/CuPc device. The poor fill factor of this device can be attributed to a reduced shunt resistance which in turn was due to poor film quality of the co-deposited layer. Device efficiencies were not as high as those reported by Tang, but are still significantly better than previous systems. It must be noted that a solar simulator was not used and so the values are not directly comparable with those reported by Tang.

Further improvements in device performance were reported by Meissner and co-workers, using co-deposited layer containing equal weights of C_{60} and ZnPc as the charge photogeneration layer in an OPV based on perylene and phthalocyanine

Table 4.1. Photovoltaic properties of bi- and trilayer perylene–phthalocyanine devices (from Ref. [22])

Materials	Device	J_{sc} (mA/cm^2)	V_{oc} (V)	Fill factor (%)	η_e (%)
Me-PTCDI/H$_2$Pc	Bilayer	0.94	0.54	48	0.29
Me-PTCDI/H$_2$Pc	Trilayer	2.14	0.51	48	0.63
PTCBI/CuPc	Bilayer	1.61	0.53	42	0.43
PTCBI/CuPc	Trilayer	2.56	0.57	25	0.44

derivatives [23]. The device structure was ITO/(20 nm)Me-PTCDI/(30 nm) C_{60}–ZnPC/(50 nm)ZnPc/Au. The device parameters were $J_{sc} = 5.26$ mA/cm^2 and $V_{oc} = 0.39$ V with a fill factor of 45%, yielding a power conversion efficiency of 1.05%. Devices made using TiOPc instead of ZnPc had a power conversion of 0.63% [24]. The photocurrent action spectrum, had a maximum quantum efficiency of 37.5%, more than twice the efficiency of the Tang device, however, the fill factor is somewhat lower due to reduced shunt resistance. Gebeyehu et al. fabricated a similar structure with 4,4′,4″-tris(N-3-methylphenyl-N-phenylamino)triphenylamine (m-MTDATA) as the HTL [25]. Power conversion efficiencies of 3.37% under 10 mW/cm^2 and 1.04% under 100 mW/cm^2 white light illumination were reported.

Both Harimoto et al. [22] and Rostalski and Meissner [23] suggested that the morphology of the bulk heterojunction could limit the power efficiency of molecular OPVs. A partially phase-segregated structure would be ideal as excitons would be able to diffuse to interfaces, followed by charge transport of electrons and holes within separate domains. The crystallinity and phase segregation of a film can be enhanced by evaporating onto a substrate at an elevated temperature. This, however, comes at the cost of increased surface roughness. For example, Geens et al. [26] studied the performance of OPVs based on co-evaporated films of C_{60} and a five-ring phenylene–vinylene oligomer, 2-methoxy-5-(2′-ethylhexyloxy)-1,4-bis((4′,4″-bis- styryl)styrylbenzene). Low substrate temperatures result in amorphous films with a surface roughness of only 5 nm. Increasing the substrate temperature to 100°C resulted in the formation of crystals and increased the surface roughness up to 200 nm. The "thickness" of the layer, calculated from a deposition monitor, was 100 nm. Such a rough film is unsuitable for device applications as holes reaching down to the substrate will short the device. Post-deposition annealing results in similar problems with increased surface roughness [27].

Peumans et al. [28] have shown that film morphology can be improved without the cost of increased roughness by annealing devices after a metal electrode has been evaporated on top of the organic structure. Films were prepared by evaporating 10 nm of CuPc, followed by co-evaporation of CuPc and PTCBI (3:4 ratio), and 10 nm of PTCBI. Films were capped with 100 nm of silver or left bare and annealed at 287°C for 2 min. Whereas uncapped films developed a rough surface with numerous pinholes, the roughness of an annealed, capped film was not significantly different from that of an unannealed film. The authors concluded that the metal cathode stressed the organic film during annealing, preventing the formation of pinholes while allowing phase segregation to occur in the bulk of the organic film. Although the film morphology allowed improved charge photogeneration under low intensity, monochromatic light, bilayer devices showed a higher power conversion (0.75%) under 7.8 mW/cm^2 white light illumination than annealed trilayer devices (0.065%).

Self-organization is an especially promising means to improve the morphology of molecular bulk heterojunctions. Gregg et al. prepared OPVs using liquid crystalline (LC) porphyrins and demonstrated unusual photovoltaic effects in large area ordered arrays [29]. The material used in these devices was zinc octakis(β-octyloxyethyl) porphyrin (ZnOOEP) [30]. Cells were prepared by spin-coating a spacer film onto ITO-coated glass plates and pressing plates together, using optical fringes to insure that the plates were parallel. The LC porphyrin was capillary-filled into the cell by placing 1 mg at the cell opening and heating to above the isotropic liquid phase until the cell was full and then cooling slowly through the LC phase to the solid. Devices were tested with porphyrin in the solid phase; the LC phase was used solely to order the film. Figure 4.9 shows the $J–V$ characteristics of an LC device under white light illumination at $150 \, mW/cm^2$. Even though the devices were relatively thick (1–6 μm) and used symmetric electrodes with no built-in electric field, short-circuit photocurrents up to $0.4 \, mA/cm^2$ and open-circuit voltages up to 200 mV were observed. Typical device characteristics are $J_{sc} = 250 \, mA/cm^2$ for a 1.5-μm-thick cell, falling to $15 \, \mu A/cm^2$ for a 5-μm-thick cell.

More recently, the properties of discotic LC organic semiconductors have been used to form a bulk heterojunction with superior morphology (Figure 4.10) [31]. Discotic LCs consist of disc-shaped molecules that can stack into columns. There is a high degree of order within columns and poor ordering between columns. This kind of a structure opens up the possibility of efficient charge transfer parallel to the plane of the device and charge transport within columns in the vertical direction. The ideal morphology of a photocell would consist of separate columns of electron- and hole-transporting materials in close contact with one another. Absorption by a chromophore within one column would result in a photoinduced charge transfer reaction with a molecule in a neighboring column. Charge transport of electrons and holes takes place within individual columns.

OPVs were constructed from a discotic, LC, hexaphenyl-substituted hexabenzocoronene (HBC-PhC$_{12}$) as a hole-transporting material and a soluble perylene

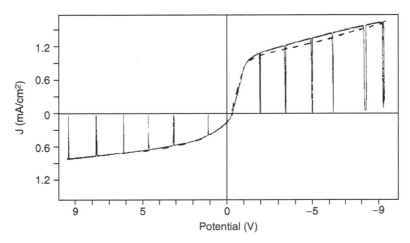

Figure 4.9. $J–V$ characteristics of liquid crystalline OPV. The voltage was scanned from –9.5 to +9.5 V (solid line) and back (dashed line) under white light illumination of $150 \, mW/cm^2$. Reprinted with permission from Gregg, B. A.; Fox, M. A.; Bard, A. J. Photovoltaic effect in symmetrical cells of a liquid crystal porphyrin. *J. Phys. Chem.* **1990**, *94*(4), 1586–1598. Copyright American Chemical Society, 1990. With permission.)

Figure 4.10. Conceptual drawing of the morphology for an idealized liquid crystalline OPV with separated electron- and hole-conducting columns.

derivative, N,N'-bis(1-ethylpropyl)-3,4,9,10-perylene tetracarboxylic diimide (EP-PTCDI), as an electron-transporting material. Intra-column hole mobilities as high as $0.22\,cm^2/V\,s$ have been measured for HBC-PhC$_{12}$ by time-resolved micro-wave conductivity (TRMC) [32]. LC ordering in spin-cast films of HBC-PhC$_{12}$ was observed by polarized light microscopy and stratification was observed in atomic force microscopy (AFM) images of EP-PTCDI/HBC-PhC$_{12}$ films. Devices were made by spin-casting a blend of 40% HBC-PhC$_{12}$ and 60% EP-PTCDI (15 mg/ml) onto an ITO-coated substrate and then evaporating an aluminum electrode on top.

The photocurrent action spectrum of the columnar LC OPV is shown in Figure 4.11(a). The EQE spectrum shows that both materials contribute to the photocurrent, with the EQE of EP-PTCDI somewhat higher than EQE of HBC-PhC$_{12}$. Double-layer devices made by evaporating EP-PTCDI onto a spin-cast HBC-PhC$_{12}$ film have the same EQE spectrum, but absolute efficiencies were one order of magnitude lower. This suggests that the goal of achieving a larger interfacial area between the electron- and hole-transporting materials has been achieved. J–V characteristics of the device under $0.47\,mW/cm^2$ illumination at 490 nm are shown in Figure 4.11(b). The device parameters are $J_{sc} = 33.5\,\mu A/cm^2$ and $V_{oc} = 0.69\,V$ with a fill factor of 40%; the maximum power conversion efficiency is 1.95% at 490 nm. Although device performance is not competitive with the best available devices, improved ordering in LC materials could realize the possibility of high-efficiency, solution-processed OPVs.

4.2.3. High-Efficiency Molecular OPVs with Exciton Blocking Layers

The Princeton group finally broke through the ~1% limit to the power conversion efficiency of molecular OPVs [33] with their best devices reaching a power efficiency of 3.6% under $150\,mW/cm^2$ AM1.5 illumination [34]. Peumans et al. fabricated trilayer devices in which an exciton blocking layer (EBL) of bathocuproine (BCP) was incorporated. BCP, a transparent wide-gap electron transporter, prevents excitons generated within the PTCBI layer from quenching at the cathode. The inset to Figure 4.12 shows the energy level diagram for this OPV structure. Excitons are excited in PTCBI and CuPc, dissociate at the organic heterojunction, and then holes

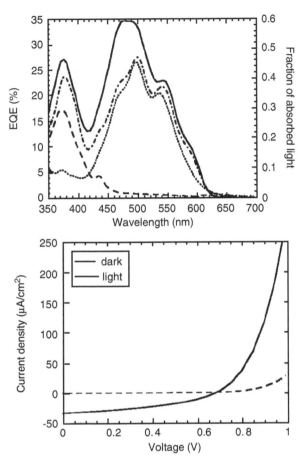

Figure 4.11. (a) EQE action spectra for a 40:60 HBC-PhC$_{12}$:perylene blend (solid line) and the fraction of absorbed light in an HBC-PhC$_{12}$ film (dashed line), a perylene film (dotted line), and a 40:60 blend film (dash-dotted line). (b) $J–V$ characteristics for this device in the dark and under illumination at 490 nm. (Reprinted with permission from Schmidt-Mende, L.; Fechtenkötter, A.; Müllen, K.; Moons, E.; Friend, R. H.; MacKenzie, J. D. *Science* **2001**, *293*, 1119–1122. Copyright AAAS, 2001. With permission.)

are transported through CuPc and electrons though PTCBI and BCP. Although the energy of the lowest unoccupied molecular orbital (LUMO) of BCP is well above the corresponding level of PTCBI, the authors suggest that the transport level of BCP, measured by inverse photoemission spectroscopy, is close in energy to that of PTCBI. Hence, BCP blocks excitons created in the PTCBI layer from quenching at the Ag electrode, but permits photocarriers to exit the structure.

Figure 4.12 shows the EQE of trilayer devices as a function of the thickness of one of the PTCBI layers. Devices were fabricated by successively evaporating layers of CuPc, PTCBI, and BCP onto an ITO-coated substrate; a silver cathode was used for these devices. The EQE of devices was measured at the peak absorbance of the layer with variable thickness; $\lambda = 620$ nm for PTCBI. The EQE rises gradually with decreasing layer thickness as the further away an exciton is created from the hetero-junction, the less likely it is to diffuse to the interface and dissociate. The EQE levels off for layers less than 20 nm thick due to the decreased absorbance in the active

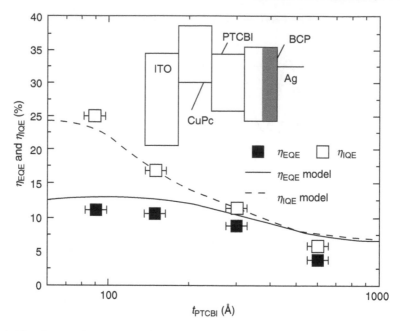

Figure 4.12. External quantum efficiency η_{EQE} (filled squares) and internal quantum efficiency η_{IQE} (open squares) for ITO/CuPc/PTCBI/BCP/Ag devices with a CuPc layer thickness of 300 Å, BCP thickness of 100 Å, and varying PTCBI thickness. The vertical error bars are smaller than the markers. Also shown are model calculations of η_{EQE} and η_{IQE}. The model assumes no quenching at the ITO–CuPc and PTCBI–BCP interfaces. Inset: Schematic energy level diagram of an ITO/CuPc/PTCBI/BCP/Ag device. The damage induced by the deposition of the Ag cathode on the BCP film is indicated by the shaded region in the diagram. (From Peumans, P.; Yakimov, A.; Forrest, S. R. *J. Appl. Phys.* **2003**, *93*(7), 3693–3723. Copyright American Institute of Physics, 2003. With permission.)

layer. The corresponding internal quantum efficiency, however, continues to increase. Comparison of the devices with and without BCP shows that the EBL does not interfere with charge transport.

Figure 4.13 shows the *J–V* characteristics of trilayer devices as a function of the illumination intensity under a simulated AM1.5 spectrum. The organic layers consisted of 15 nm of CuPc, 6 nm of PTCBI, and 15 nm of BCP–PTCBI. BCP was doped with 10% PTCBI to inhibit film recrystallization under high illumination intensities. Even though the absorbing layer is relatively thin, devices show a power conversion efficiency of ~1% under a wide range of illumination intensities. The open-circuit voltage reached 0.54 V for >10 sun illumination and device fill factors were 57% at low intensities, falling to 35% at the highest intensities (~17 suns). A light-trapping structure was fabricated in order to determine the potential collection efficiency of an OPV with an EBL. A reflective silver layer was evaporated onto the back of the substrate with a small aperture for incident light. The power conversion efficiency reached 2.4% for a cell with 6-nm thick layers of CuPc and PTCBI.

The foregoing results illustrate the utility of exciton confinement within OPVs, but the problem of optimizing the heterojunction itself still remains. The limited exciton diffusion length in PTCBI [35] (3.0 ± 0.3 nm) forces the layers to be thin in order for excitons to diffuse to the PTCBI–Pc heterojunction. The photophysics of C_{60},

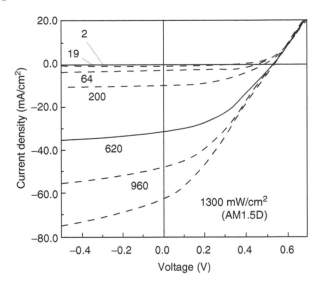

Figure 4.13. Current density vs. voltage characteristics of an ITO/(15 nm)CuPc/(6 nm) PTCBI/(15 nm) BCP/(80 nm)Ag OPV under varying AM1.5 simulated solar illumination intensities of up to 13 suns. (From Peumans, P.; Yakimov, A.; Forrest, S. R. *J. Appl. Phys.* **2003**, *93*(7), 3693–3723. Copyright American Institute of Physics, 2003.)

on the other hand, favors long exciton diffusion lengths. The triplet yield approaches unity in C_{60} as a consequence of rapid intersystem crossing (ISC) to the triplet manifold, resulting in excited state lifetimes $>10\,\mu s$ [36] and long diffusion lengths (7.7 ± 1.0 nm) [19]. C_{60} is also a strong electron acceptor, resulting in widespread use in both molecular and polymer OPVs. The attractive photophysics of C_{60} led Peumans and Forrest to use C_{60} in OPVs with EBLs [34]. Such devices reach power conversion efficiencies of 3.6% under AM1.5 spectral illumination at 150 mW/cm².

These devices also have a layer of poly(3,4-ethylenedioxythiophene)–poly(styrene-sulfonate) (PEDOT–PSS) spin-coated onto the ITO electrode prior to deposition of the photoactive layers. PEDOT–PSS has been shown to improve the performance of polymer LEDs and was first used in polymer OPVs by Roman et al. [37] and Arias et al. [38]. Peumans and Forrest found that introduction of PEDOT–PSS improves the fabrication (product) yield of OPVs to nearly 100%. Figure 4.14 shows the *J–V* characteristics of an OPV with the structure ITO/PEDOT–PSS/(5 nm CuPc)/(20 nm C_{60})/(10 nm BCP)/Al, measured under 100 mW/cm² AM1.5 illumination.* Plasma treatment of the PEDOT–PSS layer was found to improve the fill factor of these molecular OPVs. The fill factor increases from 36% for untreated PEDOT–PSS to 41% for O_2- and 49% for Ar-treated films. The short-circuit current density (11.1 mA/cm²) and open-circuit voltage (0.51–0.55 V) are insensitive to plasma treatment, with the power conversion efficiency rising from ~2% to ~3%. The effect of inserting an EBL layer of BCP is shown in the inset of Figure 4.14. J_{sc} rises moderately with BCP layer thickness up to 12 nm and then falls off due to increasing series resistance. The optimized device structure [ITO/PEDOT–PSS/(20 nm CuPc)/(40 nm C_{60})/(12 nm BCP)/Al] achieved a power conversion efficiency of $3.7 \pm 0.2\%$ at 44 mW/cm² and $3.6 \pm 0.2\%$ at 150 mW/cm².

---

* The text within the paper states an intensity of 400 mW/cm², though the figure caption states 100 mW/cm². The lower intensity figure is consistent with the reported power efficiency.

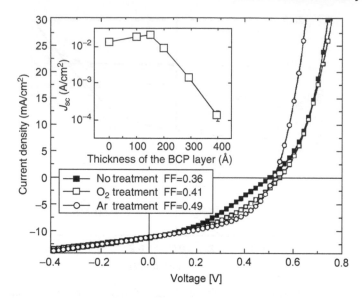

Figure 4.14. *J–V* characteristics of ITO/PEDOT:PSS/(5 nm CuPc)/(20 nm)C_{60}/(10 nm) BCP/ Al devices under AM1.5 illumination at 100 mW/cm^2. Filled squares: untreated PEDOT:PSS film; open squares: PEDOT:PSS film treated by oxygen plasma; open circles: PEDOT:PSS film treated by Ar plasma. Inset: Dependence of J_{sc} on thickness of BCP layer. (From Peumans, P.; Forrest, S.R. *Appl. Phys. Lett.* **2001**, *79*(1), 126–128. Copyright American Institute of Physics, 2001. With permission.)

We finally present a current report from the Princeton group of a molecular OPV with a power conversion efficiency of 4.2% and low series resistance [39]. Device structures were (150 nm)ITO/(20 nm)CuPc/(40 nm)C_{60}/(10 nm)BCP/(100 nm)Ag. The silver cathode was evaporated through a shadow mask, resulting in devices with areas ranging from 0.7 to 6 mm^2. The power conversion efficiency increases with incident power density, reaching a maximum power conversion efficiency of 4.2% under 4 to 12 suns simulated AM1.5 illumination. The high efficiency of this device was attributed to a very low series resistance of ~0.1 Ω/cm^2, resulting in devices with fill factors as high as 0.6. One possible explanation for the superior performance of these devices is that the nonactive regions of the substrate can serve as a thermal sink for the active area. For example, Wilkinson et al. [40] showed that extremely high current densities can be achieved in polymer LEDs with diameters of 50 μm. Their results were attributed to improved thermal management in small area devices.

4.2.4. Open-Circuit Voltage and Tandem Solar Cells

Finally, we comment on limits to the photovoltage of OPVs and, consequently, their power conversion efficiency. The upper limit on the power efficiency of solar cells using a single absorber was calculated by Shockley and Queissar [41] in 1961 to be about 31%. The maximum photovoltage is limited by the band gap of the absorbing medium; excess energy is lost during thermalization of photogenerated electron–hole pairs. The output voltage can be increased by using a higher-gap material, but without the ability to absorb a larger fraction of solar energy.

The photovoltage of inorganic photocells is determined by a built-in potential (ϕ_{bi}) equal to the difference in the Fermi energy levels of the n-doped and p-doped layers. The photovoltage of OPVs is governed by different mechanisms, resulting in open-circuit voltages that are only a fraction of the limit imposed by the optical gap. For example, the open-circuit voltage of the CuPc/PTCBI OPV (0.45 V) is only about a quarter of its potential value, calculated from the optical gap of CuPc (1.7 eV) [3]. Single component devices can produce output voltages that approach this limit, but at the cost of extremely inefficient photocurrent generation.

The built-in electrical potential in organic semiconductor devices arises from the work function difference of the anode and the cathode [42]. However, ϕ_{bi} and V_{oc} in OPVs are not closely correlated. The effects of photocarrier distributions and energy level offsets at interfaces (the chemical potential difference) also require consideration. For example, Tang [3] found that the magnitude of the photovoltage was relatively insensitive to the work function difference of the electrodes and that its polarity was determined by the organic heterojunction. The origin of V_{oc} in OPVs has been studied by a number of groups [43–46]. We draw the reader's attention to a recent paper by Gregg and Hanna [47], whose work illustrates the significance of the chemical potential difference between the donor and acceptor layers. Photocarrier pairs separate at the heterojunction, resulting in high concentrations of electrons and holes on either side of the interface. This generates a large chemical potential difference *under illumination* which, in combination with the built-in potential, drives carrier transport in opposite directions through the respective electron- and hole-transporting layers. Ultimately, the need to separate bound excitons limits both the photovoltage and the output power of OPVs.

The limits to power conversion efficiency and photovoltage can be breached through fabrication of tandem solar cells. A tandem solar cell consists of two stacked solar cells made from materials with different optical gaps. Light is first absorbed by the higher-gap cell; lower energy photons pass through the higher gap device and are absorbed by the second cell. Efficiencies of inorganic tandem solar cells have reached 30.28% under AM1.5 illumination from a double-layer InGaP/GaAs device [48] and 36.9% under concentrated sunlight (>100 suns) from a gallium indium phosphide/gallium arsenide/germanium triple-junction solar cell [49].

Tandem OPVs were first proposed by Hiramoto et al. [50] in order to address limited absorption of light at the organic heterojunction. Tandem OPVs were constructed by successively depositing a H_2Pc and Me-PTCDI onto an ITO-coated glass substrate. An intermediate gold film (<3 nm) was then deposited, followed by additional layers of H_2Pc and Me-PTCDI and, finally, a gold cathode. The Me-PTCDI layers were 70-nm thick and the H_2Pc layers were 50-nm thick; the transmittance of the bilayer was about 40%.

The performance of a bilayer cell was compared with that of tandem OPVs with and without an intermediate gold layer under 78 mW/cm^2 of white light irradiation. The open-circuit voltage of a tandem cell with a gold layer was nearly twice that of the bilayer cell, increasing from 0.44 to 0.78 V, but the short-circuit photocurrent of the tandem cell was only one-third of the photocurrent generated by a bilayer cell. The intermediate gold electrode was necessary for increased photovoltage as the inverse photovoltage generated between the stacked cells cancels out any performance benefits. The optimum gold thickness was found to be approximately 2 nm. At this thickness, the gold layer is likely to be not uniform and it was proposed that small islands of gold serve as recombination centers for electrons photogenerated in Me-PTCDI and holes generated in H_2Pc.

Recently, Yakimov and Forrest fabricated stacked solar cells with power efficiencies of 2.3–2.5% using CuPc and PTCBI [51]. The concept is similar to that of

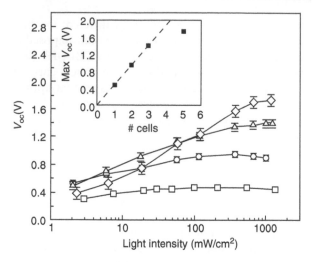

Figure 4.15. Dependence of the open-circuit voltage on light intensity for single (squares), double (circles), triple (triangles), and five-layer (diamonds) tandem OPVs. The inset shows the maximum open-circuit voltage vs. the number of heterojunctions. (From Peumans, P.; Yakimov, A.; Forrest, S. R. *J. Appl. Phys.* **2003**, *93*(7), 3693–3723. Copyright American Institute of Physics, 2003. With permission.)

Hiramoto et al. but much thinner layers were used and each tandem device consisted of up to five phthalocyanine–perylene bilayers. Silver interlayers were used rather than gold and the optimum layer thickness was found to be 0.5 nm. As with Hiramoto, the metal forms small clusters on the organic surface. Transmission electron microeoscopy indicated that the average cluster diameter is 1–5 nm and the average separation distance is 10 nm. Figure 4.15 shows the dependence of the open-circuit voltage on the light intensity for tandem solar cells with one, two, three, and five organic hetero-junctions. The inset shows the maximum photovoltage as a function of the number of layers. V_{oc} increases linearly up to three bilayers and saturates; the five-heterojunction device also requires much higher light intensity to reach the maximum V_{oc}. The power conversion efficiency was also found to be higher in tandem OPVs. The power conversion efficiency of the dual-junction device reaches a maximum value $\eta_e = 2.5 \pm 0.1\%$ at 0.6–1 sun, more than twice that of the bilayer device ($\eta_e = 1.1 \pm 0.1\%$ at 0.5 sun). A higher maximum efficiency was reached for the triple-junction device, but much high light intensities were required (10 suns). These efficien-cies are close to that achieved in solar cells with EBLs. One intriguing possibility as yet unrealized would be to stack tandem solar cells absorbing different portions of the solar spectrum to achieve higher power conversion efficiencies.

4.3. POLYMER OPVs

4.3.1. Single-Layer Polymer Devices

Reports of photovoltaic effects in π-conjugated polymers date back to 1981 [52–55]. Conjugated polymers have a number of attractive properties that have led to their use in LEDs, photovoltaic devices, field-effect transistors, and sensors. Solution processing is possible, promising inexpensive preparation of large area devices. The earliest devices were based on *trans*-poly(acetylene) (t-CH$_x$), the absorption spectrum of which is well-matched to the solar spectrum. While thin film sandwich structure devices with active layers of t-CH$_x$ exhibited photovoltaic effects, both the photo-current and fill factor were quite low. For example, Schottky-diode structures with an active layer of t-CH$_x$ prepared by Kanicki and Fedorko [53] had a power efficiency of 0.1% under white light illumination of 50 mW/cm^2 from a xenon lamp. Open-circuit voltages as high as 0.65 V were obtained, but the devices had low short-circuit currents and fill factors.

Other polymers studied over the following ten years did not show significantly improved performance. There have been several reports of photovoltaic devices based on polythiophene [54,56,57]. Power conversion efficiencies reported by Fang et al. were on the order of 10^{-2} to 10^{-3} percent [56]. PPV-based devices were studied by parallel groups at the University of Cambridge (UK) [58] and the University of California, Santa Barbara [59]. Even with illumination above the absorption edge, the power efficiency of PPV-based devices was below 10^{-2} percent.

Devices based on a single polymer face two insurmountable challenges to produce a viable solar cell. First, absorption of light does not directly generate charge carriers (electron–hole pairs or oppositely charged polarons). The primary photogenerated state is a neutral exciton with a binding energy of several tenths of an electron volt [42,60]. The electric field generated by using electrodes with differing work functions is insufficient to ionize an exciton and generate a photocurrent. The low photocurrents in single-layer polymer devices can be attributed to the low quantum yield for charge photogeneration. Studies of the effect of oxygen exposure on PPV-based devices led Harrison et al. [61] to conclude that charge photogeneration is extrinsic in nature due to quenching of excitons with defects such as interfaces, impurities, and oxygen.

The second major problem faced by any single component device is charge transport. Significant series resistance for charge carriers will reduce the fill factor of a given device and trapping will reduce the photocurrent. Many early devices had high series resistances, resulting in fill factors of 0.25 or less. Studies of a variety of conjugated polymers have shown that polymers are primarily hole transporters. For example, PPV has a much higher hole than electron mobility and the electron current is strongly reduced by trapping [62]. A time-of-flight study of PPV [63] has shown that the mobility–lifetime product and, hence, the range of holes is three orders of magnitude larger than electrons. Thus, both charge photogeneration and bipolar charge transport are problematic for any single-layer polymer OPV.

The dual problems of charge photogeneration and transport have been ad-dressed by doping polymers with a material having a higher electron affinity and better electron transport. Both conjugated polymers and molecules have been used to sensitize charge photogeneration, with fullerene-doped polymer solar cells showing particular promise. In a sensitized device, photoexcitation of the donor polymer results in electron transfer to the acceptor. Photoexcitation of the acceptor can also result in photocarrier generation by transfer of a hole to the donor. Efficient photo-induced charge transfer has resulted in a charge photogeneration yield of nearly 100% for polymer–fullerene mixtures [64]. Once the exciton has been ionized, the electric field is generated by the built-in potential.

Two approaches have been undertaken to fabricate sensitized polymer OPVs. The organic heterojunction concept, described in the previous section, has been used by a number of groups. A hole-transporting polymer film is spin-coated onto the anode, followed by evaporating a molecular sensitizer or spin-coating a second polymer. It is possible to avoid dissolving the first layer by thermal conversion of the first film to an insoluble form [65], or using a solvent for the second layer in which the already spun polymer is insolvent. Charge photogeneration occurs at the interface between the two layers, followed by charge transport within the hole- and electron-transporting layers. The primary limitation of such an approach is that charge photogeneration only occurs near the organic heterojunction. The active region is roughly the width of the interface plus the sum of the exciton diffusion length within the donor and acceptor layers. The efficiency of such devices can be improved by increasing the width of the active region.

The bulk heterojunction concept is the primary alternative to multilayer devices. Interpenetrating networks of electron donors and acceptors are potentially an ideal combination for charge photogeneration and transport. Devices have been fabricated by mixing electron-donating polymers with electron acceptors. The highest efficiencies have been achieved with blends of conjugated polymers and derivatives of buckminsterfullerene (C_{60}), though other dyes and polymer blends have been studied. Because organic heterojunctions are present throughout the entire active layer, nearly all absorbed photons result in charge photogeneration. The main challenge is to minimize carrier trapping and recombination so that devices have bipolar charge transport with minimal series resistance. The relative concentrations of the two materials can be adjusted so that both guests and hosts are above their percolation threshold. The morphology of the blend is a critical factor in the performance of bulk heterojunction devices.

4.3.2. Polymer–Dye Solar Cells

Buckminsterfullerene has attracted considerable attention since its discovery [66] and synthesis in bulk form [67]. One of the molecule's attractive electronic properties is that it is a strong electron acceptor, with a threefold degenerate LUMO capable of taking up to six electrons [68]. In 1992 Sariciftci et al. reported a strong photoinduced electron transfer reaction between C_{60} and the conjugated polymer MEH-PPV [64]. The PL intensity of MEH-PPV was quenched by almost three orders of magnitude in a 1:1 by weight blend of MEH-PPV and C_{60}. Light-induced electron spin resonance (LESR) spectra consisted of two clear signals with g values of 2.0000 and 1.9955. The former signal is due to positive polarons on MEH-PPV chains and the latter originates from C_{60}^- anions. Transient PL measurements show a decay within the time resolution of the experiment (60 ps); a later study showed that charge transfer occurs on a subpicosecond timescale [69]. The back-transfer rate, measured by photoinduced absorption spectroscopy, showed lifetimes on the order of milliseconds at 80 K [70]. Parallel work performed by Morita and co-workers showed evidence of photoinduced charge transfer in other polymer–fullerene[†] composites [71,72].

The combination of efficient, ultrafast charge transfer producing metastable photocarriers has resulted in fullerene–polymer composites becoming one of the most promising light-harvesting class of materials for organic solar cell technologies. The first such device was fabricated by spin-coating a layer of MEH-PPV onto an ITO-coated glass substrate, followed by vacuum evaporation of C_{60} to form a p–n heterojunction [73]. The parameters of the device, measured under $\approx 1\,mW/cm^2$ illumination at 514.5 nm, were relatively modest: $J_{sc} = 2.1\,\mu A/cm^2$ and $V_{oc} = 0.44\,V$, rising to 0.53 V under increasing illumination. The fill factor is about 20%, resulting in a power conversion efficiency of only 0.02%.

[†] We use the term "fullerene" to refer to the class of molecules, including C_{60}, C_{70}, and derivatives such as PCBM.

Drees et al. have shown that the performance of a bilayer polymer–C_{60} device can be dramatically enhanced by increasing the width of the heterojunction [74]. A bilayer was made by subliming 100 nm of C_{60} on top of a 90-nm thick spin-cast film of MEH-PPV. The bilayer film was then heated to the vicinity of the glass transition temperature of the polymer in order to created an interdiffused C_{60}–polymer layer at the organic–organic interface. Films were heated to 150°C or 250°C for 5 min inside a N_2 glovebox and then an aluminum electrode completed the device structure. The glass transition temperature of MEH-PPV used in this study was determined to be 230°C by differential scanning calorimetry. The PL intensity of MEH-PPV in a bilayer film was quenched by more than an order of magnitude following the annealing process.

The spectral photoresponse of MEH-PPV films and bilayer films (as deposited and annealed at 150°C or 250°C) was measured. Evaporation of C_{60} onto MEH-PPV enhanced the photoresponse by a factor of 20. The photocurrent spectrum contains peaks at 350 and 560 nm, characteristic of C_{60}, and has a depressed response between 400 and 550 nm, where MEH-PPV absorption is strongest. This response is characteristic of the internal filter effect; absorption by MEH-PPV occurs too far from the interface to contribute to the photocurrent, whereas absorption by C_{60} is strongest at the interface. Annealed devices show a much greater photocurrent than the untreated bilayer device and it can be clearly seen that absorption by MEH-PPV contributes to the photocurrent. The spectral response of the device annealed above the glass transition temperature of MEH-PPV has a photosensitivity of ~20 mA/W between 350 and 530 nm, roughly equivalent to an EQE of 5–7%.

Chen et al. [75] fabricated bilayer C_{60}–polymer devices using a polymer blend as one of the active layers. The most efficient polymer–fullerene devices use derivatives of PPV, which has an absorption edge at ~500 nm, depending upon the side-chain. Polythiophene has been investigated for use in solar cells as its optical gap is lower than that of PPV and regio-regular polythiophenes have excellent hole transport properties [76]. The absorption spectra of PPV and polythiophene are complementary to one another, resulting in improved coverage of the solar spectrum. Strong overlap of the emission spectrum of BEHP–PPV with polythiophene absorption results in efficient Förster energy transfer. The PL spectra of PPV–PT blends with equal weights showed complete energy transfer to polythiophene.

Bilayer devices were fabricated by evaporating a C_{60} film onto neat polymer films of BEHP–PPV or polythiophene or polymer blends. PEDOT–PSS-coated ITO was used as the anode and aluminum as the cathode. Figure 4.16 shows the short-circuit action spectra of devices made using three different polythiophenes. Neat polythiophene films outperformed those of BEHP–PPV with blend devices showing the best performance. The authors attribute the better performance of polythiophene to improved charge transfer at the PT–C_{60} interface than at the PPV–C_{60} interface. A peak quantum efficiency of nearly 40% was measured from a device based on a blend of BEHP–PPV and PTOPT. The open-circuit voltage of these devices is, however, quite low (0.2–0.4 V). Polymer blends may have significant potential for use in bulk heterojunction devices.

Bulk heterojunction devices made from a blend of MEH-PPV and C_{60} showed significantly better photoresponse than the first bilayer devices [77]. The device was prepared by dissolving 10 mg MEH-PPV with ~1 mg C_{60} in xylene and spin-casting onto ITO-coated glass. Under 2.8 mW/cm^2 illumination at 500 nm, the current density $J_{sc} \approx 15.3$ μA/cm^2, corresponding to an EQE of roughly 1.3%, and $V_{oc} \approx 0.8$ V were measured. It is notable that this device had a linear intensity response under biased illumination. The intensity dependence of the photocurrent and corresponding

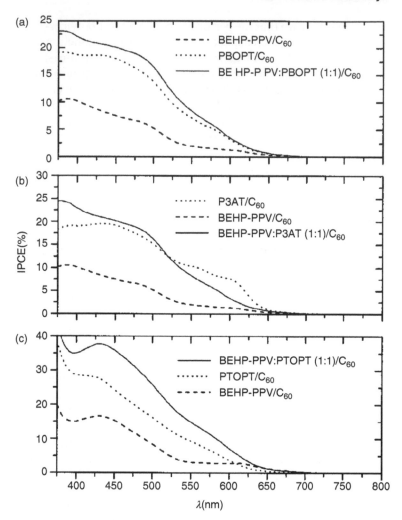

Figure 4.16. IPCE vs. wavelength of PT/C$_{60}$, BEHP–PPV/C$_{60}$, and (1:1) PT:BEHP–PPV devices. The polythiophene used is (a) PBOPT, (b) P3HT, and (c) PBOPT. (From Chen, L.; Roman, L. S.; Johansson, D. M.; Svensson, M.; Andersson, M. R.; Janssen, R. A. J.; Inganäs, O. *Adv. Mater.* **2000**, *12*(15), 1110–1114. Copyright Wiley-VCH, 2000. With permission.)

photosensitivity of the ITO/MEH-PPV–C$_{60}$/Ca photodiode were measured under 10 V reverse bias and illuminated at 500 nm. Lower intensity measurements were performed with a modulated light source to permit lock-in amplification of the photodiode signal. The device has a sensitivity of 0.2 A/W for illumination intensities ranging from 10^{-8} to 10^{-2} W/cm^2.

Bulk heterojunction devices made from C$_{60}$ and PPV improved upon the performance of bilayer devices, but still fell well short of needed levels. Efficient charge photogeneration is not the most serious problem as PL quenching is seen at low C$_{60}$ concentrations. The discrepancy between PL quenching and device EQE suggests that the problem was charge transport through the active layer. In particular, the concentration of both components of the blend must be high enough so that two percolation networks are formed in the active layer. This is problematic as C$_{60}$ has a tendency to crystallize during film formation and its solubility is relatively low in organic solvents used for spin-casting polymer films, for example, 2.8 mg/ml in toluene [78]. The solution to this problem lies in the field of fullerene chemistry. A wide variety of materials have been prepared through the modification of fullerenes [79], including functionalized fullerenes suitable for solution processing.

Yu et al. fabricated bulk heterojunction devices using the methanofullerene derivative 1-(3-methoxycarbonyl)-propyl-1-phenyl-(6,6)C_{61} or [6,6]-PCBM [80]. The synthesis of PCBM has been discussed separately in an earlier publication [81]. Films with up to 80 wt% PCBM could be fabricated, roughly one fullerene per polymer repeat unit. Figure 4.17 compares the intensity-dependent quantum and power conversion efficiencies for devices containing 50 and 80 wt% PCBM, 25 wt% C_{60}, and pure MEH-PPV. Illumination was performed at 430 nm. The polymer device shows a quantum efficiency of roughly 0.1–0.2% and the lower concentration devices (PCBM or C_{60}) have quantum efficiencies of about 15% at low intensities, falling to 5–8% at 20 mW/cm^2. The device containing 80 wt% [6,6]PCBM, in contrast, reaches an EQE of 45% at low intensities, falling to 29% at 20 mW/cm^2. The open-circuit voltage of the device was 0.82 V for Ca–ITO electrodes and 0.68 V for Al–ITO electrodes. Due to rise in the photovoltage with intensity, the power efficiency remained relatively constant; $\eta_e = 3.2\%$ at 10 μW/cm^2, falling to 2.9% at 20 mW/cm^2.

The Sariciftci group has recently reported the fabrication of fullerene–polymer bulk heterojunction devices with power efficiencies as high as 3.3% under simulated AM1.5 illumination. The performance of these devices illustrates the significance of

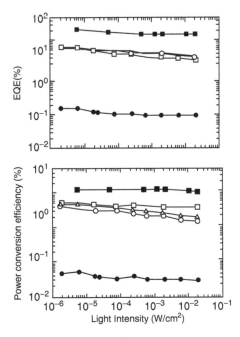

Figure 4.17. Intensity dependence of the (a) EQE and (b) power conversion efficiency of PCBM:MEH-PPV devices under monochromatic illumination at 430 nm. Solid squares: (1:4) MEH-PPV:[6,6]PCBM; open squares: (1:1) MEH-PPV:[6,6]PCBM; diamonds: (1:1) MEH-PPV:[6,6]PCBM; open circles: (1:1) MEH-PPV:[5,6]PCBM; triangles: (3:1) MEH-PPV:C_{60}; solid circles: MEH-PPV. All devices have calcium cathodes and ITO anodes. (Reprinted with permission from Yu, G.; Gao, J.; Hummelen, J. C.; Wudl, F.; Heeger, A. J. *Science* **1995**, *270*(5243), 1789–1791. Copyright AAAS, 1995. With permission.)

film morphology and interfaces in polymer photovoltaics. The active layer consisted of a blend of 80 wt% [6,6]-PCBM and 20 wt% poly[2-methyl,5-(3′,7′-dimethyloctyloxy)]-1,4-phenylene vinylene) (MDMO-PPV). Devices reported in both papers had the following construction: ITO/PEDOT–PSS/PCBM–MDMO-PPV/LiF/Al. Devices using a gold electrode or substituting silica for LiF were also fabricated. The first study compared devices cast from toluene and chlorobenzene solutions, finding that aggregation is suppressed in chlorobenzene-cast films and that devices using such films are more efficient [82]. The latter study concentrated on the effects of the LiF layer between the negative electrode and the polymer–PCBM film [83]. By optimizing the thickness of the LiF layer, devices with a power conversion efficiency of 3.3% were demonstrated.

Figure 4.18 shows AFM images of the surfaces of the blend film when cast from either toluene or chlorobenzene [82]. The toluene-cast film has features on the order of 0.5 μm wide and with surface roughness of up to 10 nm. The mechanical stiffness and adhesion properties were found to vary from the valleys and such features were not observed in neat films of MDMO-PPV cast from toluene. These features were therefore assigned to phase-segregated domains with different (presumably higher) fullerene concentrations. In contrast, the film cast from chlorobenzene is much smoother and no large-scale features can be discerned. The different film morphologies were attributed to the greater solubility of PCBM in chlorobenzene.

The film morphology is correlated with device performance. The transmission spectra are virtually identical, whereas the photocurrent action spectra show a quantum efficiency almost twice as high for the chlorobenzene-cast film [82]. *J–V*

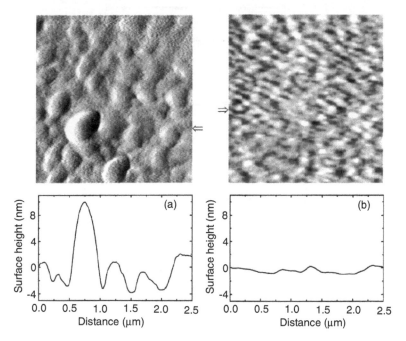

Figure 4.18. AFM images showing the surface morphology of MDMO-PPV:PCBM (1:4 by wt.) blend films with a thickness of approximately 100 nm and the corresponding cross-sections. (a) Film spin-coated from a toluene solution. (b) Film spin-coated from a chlorobenzene solution. (From Shaheen, S. E.; Brabec, C. J.; Sariciftci, N. S.; Padinger, F.; Fromherz, T.; Hummelen, J. C. *Appl. Phys. Lett.* **2001**, *78*(6), 841–843. Copyright American Institute of Physics, 2001. With permission.)

characteristics measured under white light illumination at an intensity of $80 \, \text{mW/cm}^2$ showed consistent improvements. The short-circuit current density increases from 2.33 to $5.25 \, \text{mA/cm}^2$ and the fill factor improves from 0.50 to 0.61. The corresponding power conversion efficiency, correcting for spectral mismatch, increases from 0.9% to 2.5%. The improved device performance was attributed to increased charge carrier mobility in chlorobenzene-cast films. Reduced aggregation of PCBM molecules reduces voids, which inhibit hopping, resulting in a continuous path for negative charge carriers. It was also suggested that polymer chains cast from chlorobenzene will assume an open conformation, resulting in significantly higher hole mobility.

Insertion of thin layers of LiF between the active layer and the cathode has been shown to improve electron injection in organic LEDs. Figure 4.19 shows the J–V characteristics of devices with LiF layers of 0, 3, 6, and $12 \, \text{Å}$ deposited onto the MDMO-PPV–PCBM layer [83]. For small thicknesses of LiF layers, a continuous LiF film is not expected to form on the photoactive layer. Deposition of only $3 \, \text{Å}$ of LiF improves both the fill factor and the current density; the white light power conversion efficiency was calculated to be 3.3%. Further deposition of LiF did not

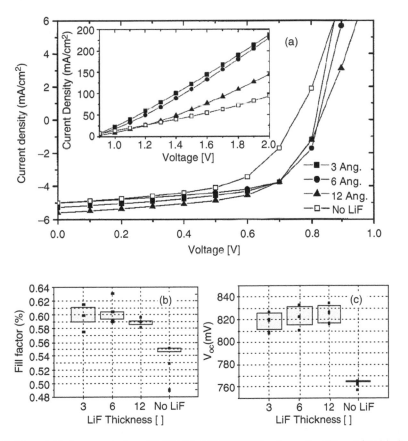

Figure 4.19. J–V characteristics of MDMO-PPV/PCBM solar cells with a LiF/Al electrode of varying LiF thickness (squares: $3 \, \text{Å}$, circles: $6 \, \text{Å}$, triangles: $12 \, \text{Å}$) compared to the performance of a MDMO-PPV/PCBM solar cell with a pristine Al electrode (open squares). The lower insets show box plots with the statistics of the (b) fill factor and (c) the V_{oc} from six separate solar cells. (From Brabec, C. J.; Shaheen, S. E.; Winder, C.; Sariciftci, N. S.; Denk, P. *Appl. Phys. Lett.* **2002**, *80*(7), 1288–1290. With permission.)

change the average value of the fill factor, but narrowed the distribution in device performance (see gray boxes in lower portion of Figure 4.19). A decrease in the fill factor was observed for a 12 Å-thick LiF film and at 20 Å, the high resistivity of the LiF degrades device performance.

A numerical analysis of device parameters was conducted using the following equation for a Shottky diode:

$$J(V) = J_0 \{\exp[q(V - JR_S)/nkT] - 1\} + (V - JR_S)/R_P + J_{sc} \tag{3}$$

where J_0 is the saturation current density, R_S and R_P are the serial and parallel resistivities, and n is the diode ideality factor. The shunt resistivity was approximately the same for all photodiodes (1.2 ± 0.1 kΩ), but the parallel resistance decreased from 10 Ω with no LiF to 4 Ω for LiF between 3 and 9 Å thick and increased to 5 Ω for LiF 12 and 15 Å thick. Brabec et al. proposed that the strong dipole moment of LiF causes a shift of the vacuum level at the LiF–Al interface, changing the effective work function of the electrode. Similar improvements in the performance of devices with a gold electrode were seen, resulting in a power conversion efficiency of 2.3%.

The highest-performance devices reported to date use PPV derivatives. Other polymers have been used, most notably polyfluorenes and polythiophenes. Svensson et al. [84] used an alternating polyfluorene copolymer (PFDTBT) and achieved peak EQE of ~40% and a power conversion efficiency of 2.2% under solar AM1.5 illumination. Although the power conversion efficiency is lower than the PPV-based devices discussed above, the open-circuit voltage is significantly higher (1.04 V). This opens the prospect for higher power conversion efficiencies from optimized devices. Brabec et al. fabricated devices from a novel polymer with a low bandgap to harvest near-IR light (PTPTB, see Figure 4.2). The fill factor of such devices was relatively low (37%), resulting in a power conversion efficiency of 1% under solar illumination.

Schilinsky et al. [85] have reported bulk heterojunction devices from blends of P3HT with [6,6]-PCBM that achieve comparable levels of performance. The device construction is similar to that reported above — a layer of PEDOT–PSS is spin-cast onto ITO-coated glass, followed by a blend of P3HT and PCBM (1:3 by weight for a 350 nm thick film, 1:2 by weight for a 70 nm thick film). The top metal electrode consisted of calcium capped by silver. The spectral response of the P3HT–PCBM device was measured and a peak EQE of 70% at ~530 nm was reported. The photocurrent of the P3HT–PCBM device is higher than that of PPV-based devices (8.7 mA/cm²) due to a better match of the polymer absorption and solar emission spectra, although the open-circuit voltage is somewhat lower (0.58 V). The fill factor is 55% and the power conversion efficiency is 2.8%, within striking distance of the best PPV–PCBM device.

We present progress on a technologically important issue: fabrication of large area devices. Although the performance of organic solar cells remains inferior to their inorganic counterparts, organic devices still have intrinsic advantages in device fabrication. Spin-coating is the standard technique for production of polymer optoelectronic devices and has been used to fabricate large area (6 cm × 6 cm) devices [86]. Such devices have been compared to similar small area devices (<1 cm²) and found to have comparable performance. There are limits to the maximum area of spin-coated devices, which has led several groups to investigate novel printing methods suitable for large area devices.

Padinger et al. [87] have reported the fabrication of large area solar cells using the "doctor blade" method in which a wire is used to spread solution onto a substrate, whereas Shaheen et al. have used screen printing to fabricate photovoltaic devices [88]. The polymer–full solution is loaded onto the substrate and a rubber "squeegee" is swept across the surface of the screen, contacting it to the surface and depositing the film. A blend of MDMO-PPV and 6,6-PCBM (1:4 by weight) was used for these OPVs. Films with a thickness of 40 nm and an rms surface roughness of 2.6 nm were deposited; these films are much thinner and more uniform than achieved by a doctor blade. The device parameters under 27 mW/cm^2 illumination at 488 nm are $J_{sc} = 3.16$ mA/cm^2 and $V_{oc} = 0.841$ V, with a fill factor of 44%. This gives a monochromatic EQE of 30% and power conversion efficiency of 4.3%. Inkjet printing, used with polymer LEDs and transistors, is another possible printing technique for producing devices with large active areas [89,90].

We conclude the discussion of molecule-sensitized polymer devices with a review of other molecular systems used to sensitize charge photogeneration in conjugated polymers. Several groups have investigated single wall nanotubes as a potential dopant in organic solar cells. The hope of such a system is to combine the high electronegativity of C_{60} with charge transport along the nanotube axis. Ago et al. [91] compared the spectral response of devices using a layer of nanotubes with that of devices with ITO as the hole-conducting electrode and found that the EQE of the nanotube device was roughly twice that without the nanotubes. Kymakis and Amartunga [92] tried blending nanotubes into P3OT (at 1% concentration) and found a significant enhancement of the photocurrent, though the short-circuit current density was only of the order of 0.1 mW/cm^2. Open-circuit photovoltages were increased from 0.35 to 0.75 V in a follow-up study [93], above the open-circuit voltage calculated from the electrode work function difference. The power conversion efficiency of this device (0.07%) is still quite low.

Molecular semiconductors other than fullerenes have also been used as dopants in polymer solar cells. Systems that have been investigated include viologen [94], phthalocyanines [95], and perylene derivatives [96–99]. Angadi et al. fabricated devices from blends of poly(1,4-bis(2-ethylhexyloxy)phenylene vinylene) (BEH-PPV) with naphthalene, N,N'-bis(octyl)-1,4,5,8-napthalenedicarboximide (NDI), and perylene, N,N'-bis(2,5-di-*tert*-butylphenyl)-3,4,9,10-perylenedicarboximide (PDI), derivatives [96]. Under 2.5 mW/cm^2 illumination at 514.5 nm, the short-circuit current density ($\sim$10 μA/cm^2) and the fill factor (25%) of such devices were low, resulting in a monochromatic power conversion efficiency of 0.8%, though open-circuit voltages of 1.2 V for NDI and 2.1 V for PDI were reported. Much better results were reported from a blend of P3HT with a similar perylene derivative N,N'-bis(1-ethylpropyl)-3,4,9,10-perylenetetracarboxylimide (EP-PTC) [96]. Quantum efficiencies of 7% were achieved from a 80% EP-PTC and 20% P3HT, and the PC spectral response matched the absorption spectrum of the blend. The power conversion efficiency is still only 0.4% at 540 nm.

The most promising approach taken by molecular solar cells used a bilayer structure with a conjugated polymer as the hole-transporting layer and either perylene benzimidazle (PBI) or magnesium phthalocyanine (MgPc) as the electron-transporting layer. The best bilayer device used PBI and a PPV derivative (M3EH-PPV), achieving a power conversion efficiency of 0.71% under 80 mW/cm^2 white light illumination. The short-circuit current density is 1.96 mA/cm^2, the open-circuit voltage is 0.63 V, and the fill factor is 46%.

OPVs based on polymer–fullerene blends with power conversion efficiencies of more than 5% were recently reported from a group led by Brabec at Siemens (recently acquired by Konarka). This is the highest value reported from a solid-state OPV, though has not been verified so far.

4.3.3. Polymer Blend and Multilayer Solar Cells

In principle, a polymer blend should possess ideal properties for use in organic solar cells: efficient charge photogeneration at interfaces between the two polymers and separate pathways for transport of electrons and holes through phase-segregated domains. A blend of similar polymers with different electron affinities will form a bulk heterojunction throughout a device. High electron affinity polymers have been developed by side-chain substitution with electron-withdrawing moieties such as cyano groups [100] or by polymerization of model compounds with high electron affinity such as pyridine [101]. Although polymer blends dramatically outperform single polymer devices, the performance of polymer blend devices substantially lags behind fullerene–polymer blends. We will first review early work on solar cells based on polymer blends and then describe recent efforts to improve performance by controlling the morphology of the blends.

The first efficient OPVs based on polymer blends were fabricated in parallel by the research groups at Cambridge (UK) [102] and Santa Barbara (USA) [103]. Both groups used a blend of MEH-PPV and a dialkoxy-substituted PPV with cyano side groups on the vinyl bond (CN–PPV). The electron-withdrawing character of the cyano side group gives CN–PPV an electron affinity 0.5 eV higher than MEH-PPV [104]. Addition of MEH-PPV reduced the PL quantum yield of CN–PPV from 32% for pure CN–PPV to less than 5% at equal weight fractions [102]. PL quenching in polymer blends is much weaker than in C_{60} doping, which led to the hypothesis that the doped film consists of phase-segregated domains of the two polymers. This hypothesis was confirmed by transmission electron microscopy (TEM) using $FeCl_3$ to selectively stain MEH-PPV.

Figure 4.20 compares the J–V characteristics of an MEH-PPV photodiode with the one based on a 1:14 (by weight) blend of CN–PPV and MEH-PPV [103]. The dark current characteristics of the devices are quite similar, whereas the reverse bias photocurrent is enhanced by two orders of magnitude. The open-circuit voltage V_{oc} decreases from 1.6 to 1.25 V upon blending. This is consistent with reports on other devices, where the maximum open-circuit voltage is limited by the difference between the ionization potential of the donor and the electron affinity of the acceptor. The photocurrent yield is roughly stable for doping concentrations between 10% and 80% by weight.

One of the most striking differences between polymer [58,61] and polymer blend devices is in the photocurrent action spectrum. Figure 4.21 shows the spectral response of photodiodes based on MEH-PPV, CN–PPV, and the polymer-blend photodiode along with the absorption spectrum of the polymer blend. The low-energy feature in the photocurrent action spectrum of MEH-PPV is characteristic of poor electron transport. When devices are illuminated through the ITO electrode, charge photogeneration takes place near the electrode for light significantly above the absorption edge. Electrons generated in this manner cannot traverse the device, become trapped, and do not contribute to the photocurrent. The penetration depth of light is greatest near the onset of absorption, resulting in photocurrent generation throughout the device. The spectral response of the polymer blend photodiode closely resembles the absorption spectrum of the polymer blend, with evidence of efficient charge photogeneration and transport throughout the bulk of the material. The photocurrent generation efficiency of the polymer blend device (up to 6%) was much improved as compared to the devices based on MEH-PPV (0.04%) or CN–PPV (10^{-3}%), though photocurrent–voltage characteristics show a low fill factor (~0.25) and an open-circuit voltage of 0.6 V.

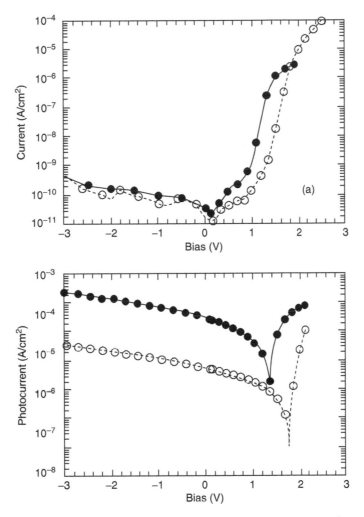

Figure 4.20. *J–V* characteristics of in the (a) dark and (b) under 20 mW/cm² illumination at 430 nm of OPVs based on MEH-PPV and a blend of MEH-PPV and CN–PPV. (From Yu, G.; Heeger, A. J. *J. Appl. Phys.* **1995**, *78*(7), 4510–4515. Copyright American Institute of Physics, 1995. With permission.)

Recent investigations have concentrated on improving the polymer morphology [105]. Polymer blends tend to phase separate into separate domains due to a low entropy of mixing [106]. Ideally, the polymer blend would consist of phase-segregated, interpenetrating networks. Absorption of light occurs within the exciton diffusion length of an interface between two phases, followed by creation of positive and negative charge carriers within separate domains. The morphology of polymer blends can be modified either during deposition by varying the polymer concentration, the solvent, or deposition conditions or by postdeposition treatments such as annealing at elevated temperatures.

Arias et al. [107–109] reported the performance of photovoltaic devices based on blends of poly(9,9′-dioctylfluorene-*co*-benzothiadiazole) (F8BT) and poly(9,9′-dioctylfluorene-*co*-bis-*N*,*N*′-(4-butylphenyl)-bis-*N*,*N*′-phenyl-1,4-phenylenediamine)

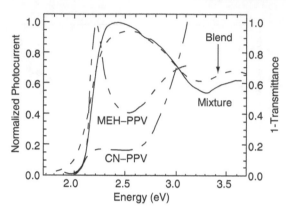

Figure 4.21. Spectral response of the short-circuit photocurrent of polymer OPVs based on MEH-PPV, CN–PPV, and an equal weight blend of MEH-PPV and CN–PPV. The absorption spectrum of the blend is shown for comparison. (From. Halls, J. J. M.; Walsh, C. A.; Greenham, N. C.; Marseglia, E. A.; Friend, R. H.; Moratti, S. C.; Holmes, A. B. *Nature* **1995**, *376*(6540), 498–500. Used with permission from Nature Publishing Group.)

(PFB). Polyfluorene and its copolymers have emerged as leading candidates for polymer LEDs and displays due to attractive optical and chemical properties such as chemical tunability via copolymerization, high fluorescence quantum yield in neat films, and trap-free charge transport [110]. Polyfluorene also exhibits superior thermal stability and photostability *vis-à-vis* PPV, suggesting its use in solar cells where stability has been problematic. PFB is a copolymer with alternating fluorene and triphenylamine subunits that has been used as a hole-transporting layer. Triphenylamines have excellent hole transport properties and have been used in photoconductors and LEDs. Time-of-flight mobility measurements of PFB performed by Redecker et al. have shown hole mobilities of up to 2×10^{-3} cm^2/V s [111]. F8BT, on the other hand, has a high electron affinity (3.53 eV) [108]. Because of the large difference between the highest occupied molecular orbital (HOMO; 0.8 eV) and LUMO (1.24 eV) levels of the two polymers, photoinduced charge transfer should be strong in a blend of PFB and F8BT.

Halls et al. investigated how the performance of a photocell can be affected by the morphology of a polymer blend [107]. Films were spin-cast from a xylene solution containing equal weights of PFB and F8BT. The substrate was mounted in a chuck for spin-coating and then heated with a halogen lamp up to 95°C prior to spin-coating. The temperature of the substrate affected the rate of solvent evaporation and, in turn, the film morphology. The substrate temperature was not uniform, resulting in a range of different film morphologies across the film. Fluorescence and topographical AFM images showed that a range of morphologies were distributed across the substrate. The morphology of films suggests the presence of two distinct phases. In the most coarse regions of the film, grains diameters were approximately 4.7 μm in size. This region of the film also had the lowest quantum efficiency of 0.25% under monochromatic illumination at 500 nm. In contrast, the region with the smallest grains (~0.86 μm diameter) had a quantum efficiency of 0.52%.

The quantum efficiency of given regions was correlated with morphology by looking into the relationship between the incident photon-to-electron conversion efficiency (IPCE) and light scattering. Figure 4.22 compares the EQE of pixels with

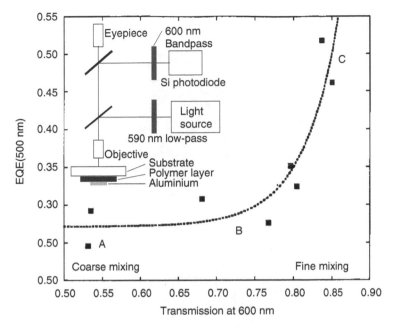

Figure 4.22. Comparison of the EQE and the transmittance of eight pixels distributed across an OPV substrate. The inset shows a schematic diagram of the transmission experiment. (From Halls, J. J. M.; Arias, A. C.; MacKenzie, J. D.; Wu, W.; Inbasekaran, M.; Woo, E. P.; Friend, R. H. *Adv. Mater.* **2000**, *12*(7), 498–502. Copyright Wiley-VCH, 2000. With permission.)

the transmittance of the same film at 600 nm, which is below the absorption edge of the film. Pixels with the highest EQE also have the highest transmission and, therefore, the lowest degree of scattering. The authors presumed that composite phase separation varies with the rate of solvent evaporation due to the local surface temperature. While this method of varying film morphology is rather uncontrolled, the results demonstrate how the morphology of a polymer blend affects device performance and points the way towards more systematic means of optimizing device performance.

In a follow-up study, Arias et al. [108] studied the film morphology and device performance of PFB–F8BT films cast from solutions of xylene and chloroform. Figure 4.23 shows topographical AFM images of spin-cast films of PFB–F8BT that have been cast from (a) xylene or (b) chloroform solutions (1:1 by weight; 14 mg/ml). Whereas the xylene-cast film shows clear phase segregation with domains on the order of microns, the chloroform-cast films show much less phase separation (<100 nm). Charge photogeneration is also much more efficient — the fluorescence yield drops from 16.7% for a film spun from xylene to 3.7% for the film spun from chloroform. The EQE has also more than doubled, to a maximum of ~4% at 3.2 eV.

In their most recent work on PFB–F8BT blends, surface treatment of substrates and a slow evaporation of the solvent were used to promote the formation of domains vertically oriented with respect to the substrate [109]. Films cast from isodurene solution onto a patterned substrate exhibited EQEs of ~10% at 2.6 eV (≈480 nm). These devices also used ITO anodes coated with a thin layer of PEDOT–PSS. We report a related work by the same group [112] in which PPV was combined with a polyfluorene derivative that has pendant perylene sidegroups. Such devices showed a peak EQE of ~7% at 500 nm.

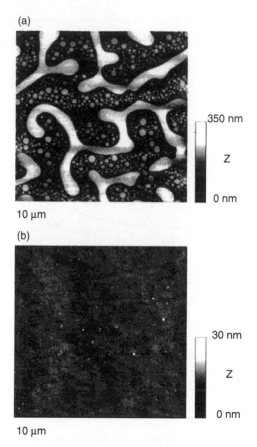

Figure 4.23. Topographic AFM images of PFB–F8BT films spun from (a) xylene or (b) chloroform solutions. (Reprinted with permission from Arias, A. C.; MacKenzie, J. D.; Stevenson, R.; Halls, J. J. M.; Inbasekaran, M.; Woo, E. P.; Richards, D.; Friend, R. H. *Macromolecules* **2001**, 34, 6005–6103. Copyright American Chemical Society, 2001. With permission.)

Kietzke et al. [113] have tried a novel approach; devices were fabricated using fused blends of polymer nanoparticles. Figure 4.24 illustrates the process used to create the nanoparticle devices. A polymer solution is added to an aqueous solution containing an appropriate surfactant and dispersed by ultrasonication to create an emulsion of small polymer-containing droplets in water. The solvent is then evaporated to create an aqueous dispersion of polymer nanoparticles in water. This dispersion is spin-coated onto a substrate and annealed at elevated temperatures to create a solid film. For the preparation of solar cells, the polymer solution consisted of a blend of F8BT and PFB, the same material system used by Arias et al. These devices produced peak quantum efficiencies of 1.7%, comparable to the best devices prepared by spin-coating F8BT–PFB from xylene, though below the quantum efficiency for layers deposited from chloroform.

Kietzke et al. recently fabricated solar cells from derivatives of MEH-PPV and CN–PPV [114]. Dispersions of M3EH–PPV and CN–ether–PPV were separately prepared and then mixed in equal ratios (by weight). The average diameter of nanoparticles was 54 nm for M3EH–PPV and 36 nm for CN–ether–PPV. Then, monolayers of the particles were prepared by spin-coating the mixed dispersion on

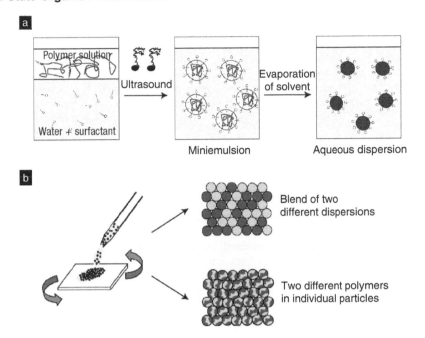

Figure 4.24. Illustration of the process to create nanoparticles. (Adapted from Kietzke, T.; Neher, D.; Landfester, K.; Montenegro, R.; Güntner R.; Scherf, U. *Nat. Mater.* **2003**, *2*, 408–412. With permission.)

a PEDOT–PSS-covered ITO substrate. Multilayer devices were further prepared from this mixture by repeating the spin-coating steps and annealing on a hot plate for 1 h at 200°C. The highest efficiency of 14% IPCE was obtained for the device consisting of only a monolayer of particles as the active layer (Figure 4.25) with performance decreasing steadily as the number of layers increased. The photocurrent action spectrum measured for the eight-layer device exhibited the typical signature of the internal filter effect. Only light penetrating through the multilayer leads to a considerable photocurrent.

These experimental results are consistent with particles randomly distributed in each layer without the existence of percolating path for charge carriers. Although results from multilayer devices are not encouraging, the high quantum yield of a spin-coated monolayer device suggests that an improved structure can be designed. For example, self-assembly of nanoparticles could lead to a device consisting of vertical domains of differing polymers. In this device, charge separation would take place laterally within a layer, but charge transport occurs vertically within a domain.

We conclude the section on polymer OPVs with a review of work on multilayer polymer OPVs. Despite progress reported in molecular solar cells with similar structures, relatively few groups have investigated polymer OPV systems. The greatest difficulty in fabricating these devices is to deposit the electron-transporting layer without redissolving the underlying hole-transporting layer (assuming solution-cast films). Tada et al. fabricated the first bilayer photocells, using poly(*p*-pyridyl vinylene) (PPyV) and poly(3-hexylthiophene) (P3HT) [115]. PPyV is an electron-conducting conjugated polymer with an electron affinity ~1 eV higher than P3HT. Furthermore, the two polymers have orthogonal solvents, i.e., PPyV is soluble in

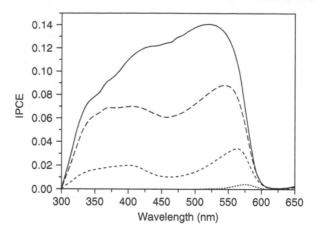

Figure 4.25. Spectral response (wavelength vs. IPCE) of nanoparticle devices for one (solid line), two (dashed line), three (dotted line), and eight (dashed-dotted line) layers of M3EH–PPV and CN–ether–PPV nanoparticles. (From Kietzke, T.; Neher, D.; Montenegro, R.; Landfester, K.; Montenegro, R.; Güntner R.; Scherf, U.; Hoerhold, H. H. *SPIE Proc.* **2004**, *5215*, 206–210. With permission.)

formic acid, but not in chloroform, whereas the reverse is true for P3HT. Finally, the optical gap of P3HT is relatively low (2.2 eV), resulting in a greater absorption of the total solar spectrum than higher-gap materials such as F8BT. Spin-coating a layer of P3HT onto PPyV quenched the PL intensity of PPyV by approximately two-thirds, indicative of exciton quenching at the P3HT–PPyV interface. Under illumination at 2.6 eV (intensity not given), the parameters of a single-layer P3HT device are $J_{sc} = 0.025\ \mu A/cm^2$ and $V_{oc} = 0.6\ V$ with ff $= 18\%$. The double layer showed superior performance in all three characteristics. Most notably, the short-circuit current improved by over two orders of magnitude. The bilayer device parameters are $J_{sc} = 6.0\ \mu A/cm^2$, $V_{oc} = 1.0\ V$ and ff $= 23\%$. The two-layer device had a higher open-circuit voltage, possibly due to Schottky barriers at the P3HT–electrode interfaces.

In a related report, Zhang et al. fabricated photovoltaic cells composed of MEH-PPV and a pyridine-containing polymer, poly(pyridopyrazine vinylene) (EHH–PPyPzV) [116]. Figure 4.26 shows the photocurrent action spectrum of single layer (unblended), single layer (blended), and bilayer structures (unblended and blended). Bilayer devices outperformed blend devices, with the peak EQE of ~7% between 500 and 550 nm reached for a bilayer device containing 19:1 and 1:19 blends of MEH-PPV and EHH–PPyPzV. The power conversion efficiency of the best device under AM1.5 conditions was only 0.03%. The fill factors of devices were inferior to the power conversion efficiency of fullerene-based devices, suggesting that the electron mobility of PPyPzV is less than that of C_{60}.

Granström et al. achieved a breakthrough by using lamination to prepare the organic heterojunction in a polymer OPV [117]. Devices were prepared by spin-casting films of MEH-CN-PPV (MEH-PPV with cyano-substitution on the vinyl bonds) and poly((3,4′-octyl)phenyl)thiophene) (POPT) onto separate substrates and laminating the two films to form an organic heterojunction. Following deposition of electrodes onto glass substrates (ITO or PEDOT on gold for the anode; calcium or aluminum for the cathode), a polymer film was spin-cast onto the substrate. The

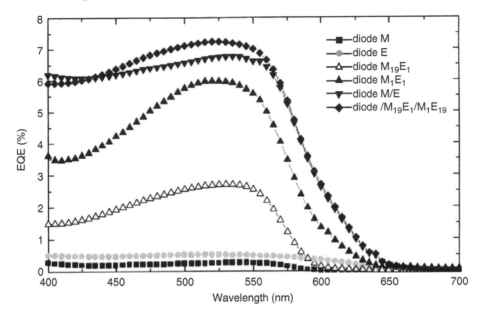

Figure 4.26. Spectral response single-layer OPVs using MEH-PPV (squares), EHH–PPyPz (circles), 19:1 (open up-triangles) and 1:1 (filled triangles) blends. The spectral response of bilayer OPVs with MEH-PPV/EHH–PPyPz (down-triangle) and (1:19)/(19:1) blends. (From Zhang, F.; Jonforsen, M.; Johansson, D. M.; Andersson, M. R.; Inganäs, O. *Synth. Met.* **2003**, *138*, 555–560. With permission.)

acceptor film was MEH-CN-PPV and the donor film was POPT doped by ~5% MEH-CN-PPV. The POPT film was heated to 200°C under vacuum and the device was laminated together by applying light pressure.

Figure 4.27 shows the EQE of the device, determined from the short-circuit current. This EQE is significantly higher than polymer blend devices, though still lower than that of fullerene–polymer blends. Devices were illuminated by monochromatic light through the anode. The open-circuit voltage was maximized by selecting electrodes with a large work function difference: PEDOT–PSS-coated gold as the anode and calcium as the cathode. The dependence of the open-circuit voltage and short-circuit current on intensity were measured under monochromatic illumination at 480 nm. J_{sc} depends linearly on intensity and V_{oc} rises rapidly with illumination intensity, saturating at ~2.2 V, the work function difference of the electrodes. The fill factor of laminated devices was roughly 30–35% and did not vary significantly with intensity.

Despite many efforts to improve the performance of materials and film morphology, the performance of polymer solar cells still lags behind bulk heterojunction devices. The scarcity of electron-conducting (n-type) polymers compounded with the difficulty of fabricating multilayer structures has led most investigators in other directions. There have been several investigations into the n-type ladder polymer poly(benzimidazo-benzophenanthroline ladder) (BBL). BBL is a robust polymer that is stable at high temperatures in air and has been shown to have promising n-type semiconducting properties [118]. The optical gap of BBL is also quite low, permitting a greater fraction of the solar spectrum to be harvested.

Jenekhe et al. prepared bilayer OPVs using PPV as the hole-conducting layer and BBL as the electron-conducting layer [119]. A two-layer device was prepared by

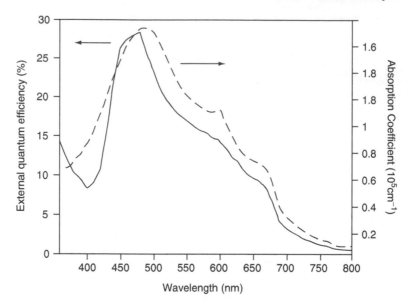

Figure 4.27. Optical absorption of a POPT/MEH-CN-PPV double layer and EQE vs. wavelength of a laminated OPV with polymer layers containing 19:1 and 1:19 blends of POPT and MEH-CN-PPV. (From Granström, M.; Petritsch, K.; Arias, A. C.; Lux, A.; Andersson M. R.; Friend, R. H. *Nature* **1998**, *395*, 257–260. Used with permission from Nature Publishing Group.)

first spin-casting a non-conjugated precursor polymer of PPV onto the substrate and thermally converting the film to a fully conjugated, non-soluble form by heating in vacuum at 250°C. A layer of BBL was then spin-coated from solution in GaCl₃–nitromethane and immersed in deionized water for 8 h to remove GaCl₃. Because BBL has a high electron affinity (5.9 eV), an air-stable aluminum electrode could be used as the cathode without a large energetic barrier to electron extraction. The parameters of devices measured under simulated solar AM1.5 illumination were $J_{sc} = 0.4$ mA/cm² and $V_{oc} = 0.7$ V at 10 mW/cm², increasing to $J_{sc} = 1.2$ mA/cm² and $V_{oc} = 1.2$ V at 100 mW/cm² with fill factors of 41–43%.

The spectral response of devices was measured as a function of the BBL layer thickness. The best performing photocell consisted of 50-nm thick layers of PPV and BBL, and had a maximum charge collection efficiency of 49% (equivalent to an EQE of 66%) between 400 and 500 nm. This is much better than the 29% peak reported for a laminated device [117], but still below the best fullerene–polymer devices [83]. The quantum efficiency increases with decreasing BBL thickness with a large jump in the efficiency of the thinnest device between 400 and 500 nm. This matches the absorption spectrum of PPV, although it is unclear why charge photogeneration within the PPV layer is so sensitive to the thickness of the BBL layer. Illumination is through the substrate and the PPV layer, so filtering by BBL cannot be related to this phenomenon. The power conversion efficiency of the best device was calculated to be 1.2% at 10 mW/cm² and 0.7% at 100 mW/cm². Work on BBL-based photocells indicates the promise of electronegative polymers with small optical gaps, particularly in tandem solar cells in which a second layer harvests light not absorbed by a higher-gap material.

4.4. HYBRID OPVs

Hybrid devices incorporating organic and inorganic materials are used due to the prospect of combining the advantages of both types of materials. One class of photovoltaic devices showing particular promise is based on composites of conjugated polymers and semiconducting nanocrystals, including quantum dots. Quantum dots are optical materials in which the crystal dimensions are reduced to the point where quantum confinement of excitations affects the electronic structure of the material. Quantum confinement leads to interesting effects such as a size-dependent optical gap and novel nonlinear optical properties. High quality, monodisperse nanocrystals can be prepared by chemical means [120], permitting the electronic structure to be tuned by varying the size of nanocrystals. Combining quantum dots with an organic material permits solution processing and eases device fabrication.

We also discuss the property of bulk heterojunction devices made from composite of conjugated polymers and TiO_2. Nanocrystalline TiO_2 has reasonably good electron transport properties and its conduction band lies below the LUMO level of typical organic semiconductors. Thus, TiO_2 is suitable to sensitize charge photogeneration. Bulk heterojunction devices have been prepared by spin-coating the blends of conjugated polymers and TiO_2 nanocrystals, conversion of a TiO_2 precursor embedded in a polymer film, and infiltration of a conjugated polymer into a porous TiO_2 film. Multilayer structures have also been tested in which a solid TiO_2 layer serves as the electron-transporting layer and charge photogeneration takes place at the organic–inorganic interface.

Dye-sensitized solar cells (DSSC; also known as Grätzel cells) have attracted considerable attention since the report of 10% power conversion efficiency in 1991 [121]. In these devices, TiO_2 nanoparticles are sensitized with a monolayer of a ruthenium-based dye. The nanoparticle film has a large surface area for dye absorption, resulting in efficient absorption of solar light. Charge transport occurs by hopping of holes through a mesoporous titania structure and electrons through a liquid electrolyte containing iodine. Such devices have great potential, but use of a liquid electrolyte has led to concern about device stability due to desorption of the dye, leakage of the electrolyte, or corrosion of the electrode. These concerns have led a number of groups to explore devices in which an organic semiconductor film replaces the electrolyte. We conclude this chapter with a review of progress towards a solid state DSSC.

4.4.1. Polymer–Quantum Dot Devices

Greenham et al. [122] reported on charge transfer in composites of the conjugated polymer MEH-PPV and CdS or CdSe nanocrystals in 1996. This work followed an early report of sensitization of poly(vinylcarbazole) by nanocrystals [123]. The electron affinities of CdS (3.8 eV) and CdSe (4.7 eV) are sufficiently high for these materials to sensitize charge photogeneration in polymer films. It should be noted that quantum size effects can modify these values by several tenths of an electron volt [124]. This can be seen in the decline of the PL efficiency of MEH-PPV with increasing nanocrystal concentration, as shown in Figure 4.28 [122]. Charge transfer is inhibited when the nanocrystals are coated with trioctylphosphineoxide (TOPO), which forms a 11 Å thick barrier between the core and the polymer, and is most effectively quenched by bare CdSe nanocrystals. These were prepared by washing the nanocrystals in methanol, repeated dissolution in pyridine, and precipitation by addition of hexane, and finally dissolution in chloroform. PL quenching is not

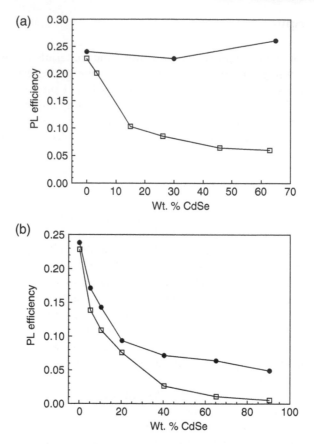

Figure 4.28. PL efficiency of MEH-PPV:CdSe blends vs. nanocrystal concentration. Both TOPO-coated (circles) and pyridine-treated (squares) nanocrystals are shown. (a) 4-nm-diameter CdS nanocrystals; (b) 5-nm diameter CdSe nanocrystals. (Reprinted with permission from Greenham, N. C.; Peng, X.; Alivisatos, A. P., *Phys. Rev. B* **1996**, *54*(24), 17628–17637. Copyright American Physical Society, 1996. With permission.)

complete, but is limited by phase segregation. TEM images of polymer–nanocrystal composites (at 65 wt.%) show polymer regions 70–120 nm in diameter, much larger than the diffusion radius of excitons (5–15 nm) [102,125]. A certain degree of aggregation is desirable due to the need to form interpenetrating networks above the percolation threshold for charge transport. TOPO-coated nanocrystals showed no aggregation at low concentrations, though TOPO inhibits charge transfer. Ideally, a surface treatment would both inhibit aggregation and sensitize the nanocrystals.

Photovoltaic devices were prepared from composites containing MEH-PPV and pyridine-treated, 5 nm diameter CdSe nanocrystals at concentrations of 0 to 90 wt.%. Devices were illuminated at 514 nm at an excitation intensity of 5 mW/cm². The EQE of a polymer-only device was 0.014% and increases steadily with nanocrystal content, reaching 0.8% at 40 wt.% CdSe. The EQE then sharply rises to 3.8% at 65 wt.% CdSe, 7% at 85 wt.% CdSe, and reaching a maximum of ∼12% at 90 wt.% CdSe. The sharp rise of the EQE above 40 wt.% CdSe is due to improved transport of electrons through the nanocrystal network. The PL efficiency of MEH-PPV is already reduced by a factor of 10 at 40 wt.% CdSe; additional doping has only a marginal effect on charge photogeneration. Increasing the nanocrystal fraction will, however, have a significant effect on charge transport. Few photogenerated electrons escape to the cathode at low doping and device efficiency is low. Nanocrystals create an interconnected network at high concentrations, permitting electron transport. The spectral response of the hybrid OPV followed the absorption spectrum of the composite, characteristic of bipolar charge transport through the film.

This early study showed significant promise, having achieved a quantum efficiency competitive with other types of organic photovoltaics. However, J_{sc} increases sublinearly with illumination intensity and the fill factor is only 0.26, resulting in an estimated power efficiency of 0.1% under solar AM1.5 conditions. These limitations were attributed to charge trapping of electrons due to "dead ends" in the nanocrystal network. Consistent results were obtained by Arici et al. who fabricated devices based on $CuInS_2$ nanoparticles; this material is notable in that the stoichiometry can be varied to produce n-type or p-type crystals. Bilayer devices were fabricated from a blend of $CuInS_2$, the fullerene derivative PCBM (Figure 4.1) and PEDOT–PSS. These bulk heterojunction devices achieved external quantum efficiencies of up to 20%. Although a short-circuit current density of 0.9 mA/cm^2 was obtained under 80 mW/cm^2 AM 1.5 illumination, both V_{oc} (0.2 V) and the fill factor (0.25) were quite low, suggesting that these devices are also limited by charge trapping.

The Alivisatos group has pursued composites of conjugated polymers and rod-shaped nanocrystals in order to improve charge transport in hybrid devices [126–129]. Figure 4.29 shows TEMs of CdSe nanocrystals (7 nm in diameter) which are 7, 30, and 60 nm long [127]. Nanorods tend to form directed chains, in which particles stack along the long axis. Alignment of the nanoparticle rods will provide a direct path for electrons to the cathode, which should improve both the current density and fill factor of devices. The initial report on nanorod–polymer solar cells used CdSe nanorods with dimensions of 4 × 7 nm and 8 × 13 nm [126]. P3HT was used as the host matrix because P3HT is a more efficient hole-transporting material than MEH-PPV. Devices achieved a monochromatic power efficiency of 2% at 514 nm, eight times better than previously reported.

Increasing the aspect ratio of CdSe nanorods led to a dramatic enhancement of device performance. Figure 4.30(a) compares the performance of hybrid devices

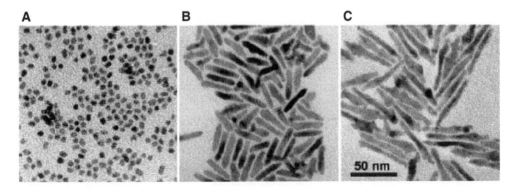

Figure 4.29. TEM images of CdSe nanorods with cylindrical cross-sections with dimensions (a) 7 × 7 nm, (b) 7 × 30 nm, and (c) 7 × 60 nm. (Reprinted with permission from Huynh, W. U.; Dittmer, J. J.; Alivisatos, A. P. Hybrid nanorod-polymer solar cells. *Science* **2002**, *295*, 2425–2427. Copyright AAAS, 2002.)

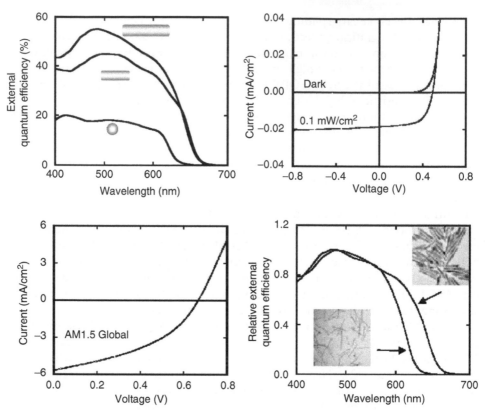

Figure 4.30. (a) Spectral response of OPVs using blends of P3HT and cylindrical nanorods with lengths 7, 30, and 60 nm. $J–V$ characteristics of an OPV with 7×60 nm nanorods under (b) 0.084 mW/cm^2 at 515 nm and (c) simulated AM1.5 illumination. Spectral response of OPVs with nanorods 60 nm long and having 3 and 7 nm diameters. (Reprinted with permission from Huynh, W. U.; Dittmer, J. J.; Alivisatos, A. P. Hybrid nanorod-polymer solar cells. *Science* **2002**, *295*, 2425–2427. Copyright AAAS, 2002. With permission.)

using composites of P3HT nanocrystals shown in Figure 4.29. These devices used an aluminum cathode and an ITO anode coated by PEDOT–PSS to improve hole extraction. The EQE rises with increasing aspect ratio with the best device using 7×60 nm nanorods. This device achieved a peak EQE of 55% at 485 nm under 0.1 mW/cm^2 illumination. The fill factor of this device is also notably high (0.6) under low-intensity illumination (0.1 mW/cm^2 at 515 nm), though is reduced to 0.4 under simulated solar illumination (Figure 4.41b and d). When tested under simulated AM1.5 illumination, the device parameters are $J_{sc} = 5.8$ mA/cm^2, $V_{oc} = 0.7$ V, and ff $= 0.4$, resulting in a power conversion efficiency of 1.7%. This is the best performance reported for a hybrid device, though it is half the performance of the best molecular [34] and fullerene–polymer [83] solar cells.

Subsequent to the report in 2002, Huynh et al. studied the effect of thermal treatment on organic–inorganic hybrid solar cells [128]. Spin-casting of treated nanorods requires the solution to contain a significant amount of pyridine; excess pyridine coating the nanorods will inhibit charge transport between the nanorods. The thermal treatment was particularly important for longer nanorods, which have a relatively large surface area and require higher concentrations of pyridine to be spin-cast. Huynh et al. showed that excess pyridine could be removed by heating the P3HT–nanorod film to 120°C. This resulted in improving the EQE of a P3HT/CdSe device from 6% to 15%. These devices are less efficient than devices discussed above [127], although PEDOT–PSS was not used.

4.4.2. Polymer-Sensitized TiO₂

A number of groups have explored using titania (TiO_2) in organic–inorganic hybrid devices. TiO_2 has been used to fabricate highly efficient DSSCs and this has naturally led to investigation of bulk heterojunction devices using a conjugated polymer and TiO_2. Salafsky et al. investigated charge photogeneration in films consisting of TiO_2 nanoparticles and PPV [130,131]. The size of the nanoparticles used in this study distinguishes these from the quantum dot devices discussed above. TiO_2 particles used are approximately 20 nm in diameter, well above the critical size for quantum confinement. Hence, it is the bulk properties of TiO_2 that are significant in these devices. The conduction (4.2 eV) and valence (7.4 eV) bands of TiO_2 lie well below the corresponding HOMO and LUMO of PPV, making this material suitable to sensitize charge photogeneration.

Films were prepared by spin-coating a PPV precursor polymer with TiO_2 nanocrystals (1:1 by weight); the film was heated to 220°C in vacuum for 10 h to convert the precursor to PPV. Photoexcitation dynamics were studied by TRMC, as shown in Figure 4.31.

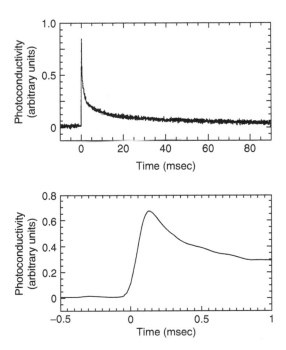

Figure 4.31. Time-resolved microwave photoconductivity of a 1:1 (by weight) PPV:TiO₂ nanocrystal composite after a flash of 2.5 eV (488 nm) light of energy 2 mJ and duration 10 nsec. The rise of the signal is shown in the lower figure and is limited by the detection electronics to about 100 nsec. (From Salafsky, J. S.; Lubberhuizen, W. H.; Schropp, R. E. I. *Chem. Phys. Lett.* **1998**, *290*(4–6), 297–303. With permission.)

The conductivity shows a fast rise (within the experimental resolution of 100 nsec) followed by a slow decay. Half of the initial TRMC signal decays within 600 nsec followed by a slow decay that can be fit to the sum of two exponentials with time constants of 4.3 and 80 sec. Back transfer of charge is much faster than in C_{60}^- polymer composites, resulting in poor device performance. The device parameters are $J_{sc} = 30 \,\mu\text{A/cm}^2$ and $V_{oc} = 0.65$ V under $100 \,\text{mW/cm}^2$ white light illumination. The authors concluded that device performance is limited by recombination.

One approach used to form a polymer/TiO$_2$ bulk heterojunction that does not use nanocrystals involves preparing a composite film containing a conjugated polymer and a precursor to TiO$_2$ [132]. Polymer–TiO$_2$ composites were prepared by spin-casting a film containing MDMO-PPV and titanium(IV) isopropoxide [Ti(OC$_3$H$_7$)$_4$]. The TiO$_2$ precursor was then converted in the dark via hydrolysis in air to form a TiO$_2$ phase in the polymer film. The resulting film is insoluble and resistant to scratching. Based on X-ray photoelectron spectroscopy (XPS) measurements of the polymer–titania composite and a similarly prepared pure TiO$_2$ film, the authors concluded that the TiO$_2$ conversion yield was at least 65%. Spectroscopic measurements of the MDMO-PPV/TiO$_2$ composite showed evidence for charge photogeneration. Incorporation of TiO$_2$ into a polymer film quenched the PL intensity by nine-tenths at 25% TiO$_2$ (by volume). The photoinduced absorption spectrum of the MDMO-PPV–TiO$_2$ thin film shows peaks at 0.42 and 1.32 eV, consistent with polaron photogeneration, and the LESR signal is enhanced.

Photovoltaic devices were prepared by sandwiching an MDMO-PPV/TiO$_2$ composite between PEDOT–PSS-coated ITO and LiF/Al electrodes. The photovoltaic properties of the device are shown in Figure 4.32. The peak EQE of 11% is reached between from 430 to 480 nm. The operating parameters of devices illuminated by a halogen lamp set at 0.7 sun intensity are $J_{sc} = 0.6 \,\text{mA/cm}^2$, $V_{oc} = 0.52$ V, and ff = 0.42. One interesting difference between this device and other bulk heterojunctions is that the peak EQE is reached at a loading of 20% by volume (roughly 40% by weight). This is much lower than peak EQE required for nanocrystals, suggesting that better dispersion of the sensitizer can be achieved via the precursor route than spin-coating a composite film. The photocurrent is, however, much lower than that obtained using other approaches.

Instead of incorporating TiO$_2$ into a polymer film, Coakley et al. have incorporated a conjugated polymer into a mesoporous titania film [133]. This approach ensures that an interconnected network exists for electron transport through TiO$_2$. Films were made by dip-coating substrates in a solution of titania sol–gel precursor and a structure-directing block copolymer. After an ordered structure is created, the polymer is removed as the films are calcined at 400–450°C. A conjugated polymer is spin-cast on top of the titania film (regioregular P3HT) and infiltrated by heating. XPS depth profiling was performed to prove that RR-P3HT penetrated the mesoporous titania. Figure 4.33 shows XPS measurements on titania films that have had RR-P3HT infiltrated by heating (a) at 200°C for 4 h or (b) at 100°C for 16 h [133]. Depth profiles of titania films showed only trace amounts of carbon. Depth profiles of infiltrated films show a carbon signal at all depths, proving that the polymer penetrates to the bottom of the film. The quantity of embedded polymer was estimated by measuring absorption spectra of films following washing off the surface layer. The optical absorbance of polymer–titania composites vs.

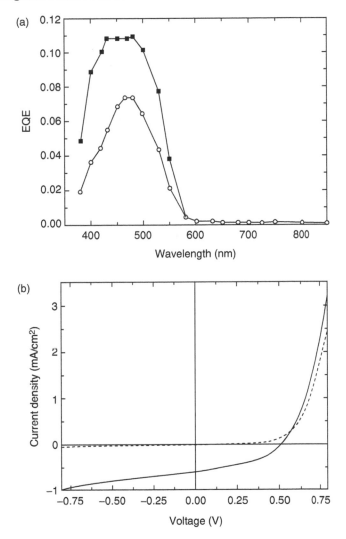

Figure 4.32. Photovoltaic properties of ITO/PEDOT:PSS/MDMO-PPV:TiO₂/LiF/Al device. (a) EQE vs. wavelength of 1:1 (circles) and 4:1 (squares) blends by volume. (b) *J–V* characteristics in the dark (dashed line) and under ∼70 mW/cm² illumination (solid line) of a 4:1 device. (From van Hal, P. A.; Wienk, M. M.; Kroon, J. M.; Verhees, W. J. H.; Slooff, L. H.; van Gennip, W. J. H.; Jonkheijum, P.; Janssen, R. A. J. *Adv. Mater.* **2003**, *15*(2), 118–121. Copyright Wiley-VCH, 2003. With permission.)

time for various annealing temperatures was also measured. At 200°C, the optical density of embedded polymer reached 0.5 after a few minutes of heating and then saturated. At lower temperatures, less polymer was observed to infiltrate the film and saturation of absorption took 4–8 h.

Following the successful demonstration of polymer infiltration, photovoltaic devices were prepared from titania–polymer composites [134]. Fluorine-doped tin oxide (SnO₂–F) was used instead of ITO as indium migrates through the film during calcination. The composite film was prepared by spin-coating a 40-nm thick P3HT layer onto a 100-nm thick TiO₂ layer and infiltrating the P3HT by heating at 200°C

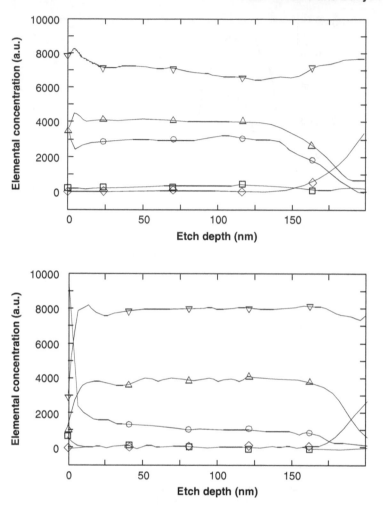

Figure 4.33. XPS depth profiles of P3HT-infiltrated mesoporous titania after (a) 4 h at 200°C and (b) 16 h at 100°C. The peaks used to detect elemental concentration are oxygen 1s (inverted triangles), titanium 2p (triangles), carbon 1s (circles) sulfur 2p (squares), and indium 3d (diamonds). (From Coakley, K. M.; Liu, Y.; McGehee, M. D.; Friendell, K. L.; Stucky, G. D. *Adv. Func. Mater.* **2003**, *13*(4), 301–306. With permission.)

for 1 min. A 30-nm thick P3HT overlayer remained on top of the TiO_2 film to prevent electrons from reaching the silver electrode. The operating parameters of a device under 33 mW/cm^2 monochromatic illumination at 514 nm are $J_{sc} = 1.4$ mA/cm^2, $V_{oc} = 0.72$ V, and ff $= 0.51$, yielding a power efficiency of 1.5%. The EQE corresponding to 1.4 mA/cm^2 photocurrent is 10%. The authors calculated a power efficiency of 0.45% under solar AM1.5 illumination from the spectral response of the device. The actual device performance is likely to be lower due to the intensity and wavelength dependence of the fill factor.

The last TiO_2–polymer device we discuss consists of a hole-transporting polymer spin-cast onto a TiO_2 film [135]. A complete, dense TiO_2 film was prepared by spin-coating a solution containing a $Ti(OCH_2CH_3)_4$, a TiO_2 precursor. The sample was heated under vacuum to 78°C for 45 min to allow further cross-linking and dehydration of the TiO_x matrix. The film was finally annealed at 450°C, resulting in the formation of a transparent, smooth film of anatase TiO_2. A PPV derivative was then spin-cast onto the TiO_2 film for light absorption and hole transport. A copolymer with alternating phenylene–vinylene and triphenylamine units was used: PA–PPV or poly(N-phenylamino-1,4-phenylene-1,2-ethenylene-1,4-(2,5-dioctoxy)-phenylene-1,2-ethenylene-1,4-phenylene). Figure 4.34 compares the spectral response of an $ITO/TiO_2/PA$–PPV/Au device with one using PEDOT–PSS-coated ITO and aluminum electrodes. The EQE is enhanced by a factor of 50 and closely follows the absorption spectrum of the polymer. The fill factor of the OPV devices was found to be dependent upon the layer thickness. Under $100\,mW/cm^2$ white light illumination, fill factors ranged from 0.24 for a 300 nm thick polymer film to 0.52 for an 80 nm thick polymer film. Based on this result, the authors estimate a diffusion length for excitons of 20 nm in PA–PPV. The short-circuit current density is ~$0.8\,mA/cm^2$, yielding a power of ~0.4% under white light illumination.

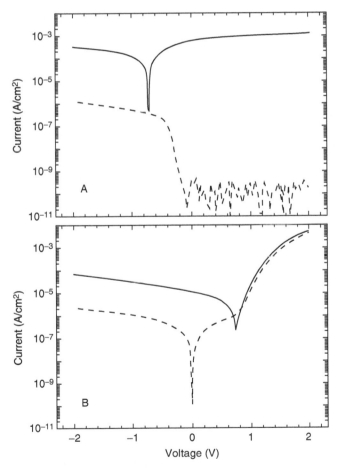

Figure 4.34. *J–V* characteristics of layered TiO_2/PA–PPV device in the dark (dashed line) and under ~$100\,mW/cm^2$ illumination (solid line) for (a) $ITO/TiO_x/PA$–PPV/Au and (b) ITO/PEDOT:PSS/PA–PPV/Al devices. (From Arango, A. C.; Johnson, L. R.; Bliznyuk, V. N.; Schlesinger, Z.; Carter, S. A.; Hörhold, H.-H. *Adv. Mater.* **2000**, *12*(22), 1689–1692. Copyright Wiley-VCH, 2000. With permission.)

4.4.3. Solid State Dye-Sensitized Solar Cells

We conclude this survey with a discussion of solid state dye-sensitized solar cells. DSSCs were first reported by O'Regan and Grätzel [121] in 1991 (see Hagfeldt and Grätzel [136] for a recent review). Two groups published reports of solid state DSSCs in 1997. Hagen et al. replaced the liquid electrolyte with a vacuum-deposited film of N,N'-diphenyl-N,N'-bis(3-methyl-phenyl)-[1,1'-biphenyl]-4,4'-diamine (TPD), a molecular semiconductor commonly used as a hole-transporting material in organic LEDs [137]. Murakoshi electropolymerized polypyrrole onto a dye-sensitized TiO_2 film [138,139]. In both cases, the quantum efficiency of resulting devices was disappointing (0.1–0.2%). The first breakthrough in solid state DSSCs came in 1998, when the Grätzel group reported a device that used a novel, amorphous hole-transporting material, 2,2',7,7'-tetrakis(N,N-di-p-methoxyphenyl-amine)-9,9'-spiro-bifluorene (spiro-MeOTAD) [140]. Excitation of ruthenium dye $Ru(III)L_2(SCN)_2$, where $L = 4,4'$-dicarboxy-2,2'-bipyridyl) results in ultrafast charge transfer of an electron to TiO_2; this process is followed by charge transfer to spiro-MeOTAD within 40 nsec (the experimental resolution).

Figure 4.35 shows the structure and spectral response of the device fabricated by Bach et al. [140] A compact film of TiO_2 was deposited onto an F-doped SnO_2 electrode, followed by screen printing a 4.2 μm thick film of nanocrystalline TiO_2. The nanoporous TiO_2 film was derivatized with $Ru(II)L_2(SCN)_2$ by absorption from acetonitrile, followed by spin-coating a layer of spiro-MeOTAD from chlorobenzene solution. The spectral response follows the absorption spectrum of the dye, reaching a maximum IPCE of 33%, more than two orders of magnitude better than previous solid state DSSCs and within a factor of two for a device using a liquid electrolyte. A key difference between these devices and the TPD-based device of Hagen is the addition of small quantities of $N(PhBr)_3SbCl_6$ and $Li[(CF_3SO_2)_2N]$ to spiro-MeOTAD (in 1:45:500 ratios). $N(PhBr)_3SbCl_6$ will oxidize spiro-MeOTAD, creating a light-doped organic salt. The second additive is a source of Li^+ ions, which are potential-determining for TiO_2 [141]. Murakoshi et al. observed a similar enhancement in device performance when doping polypyrrole [139].

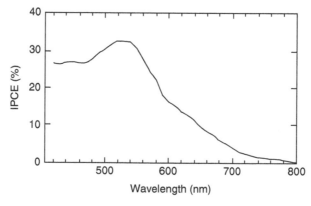

Figure 4.35. Spectral response (wavelength vs. IPCE) of a solid-state dye-sensitized solar cell. The device was fabricated from a 4.2-mm thick film of mesoporous TiO_2 sensitized with $Ru(II)L_2(SCN)_2$, spin-coated with a solution of 0.17 OMeTAD, 0.33 mM $N(PhBr)_3SbCl_6$, and 15 mM $Li[(CF_3SO_2)_2]N$. (From Bach, U.; Lupo, D.; Comte, P.; Moser, J. E.; Weissörtel, F.; Salbeck, J.; Spreitzer, H.; Grätzel, M. *Nature* **1998**, *395*(6702), 583–585. Used with permission from Nature Publishing Group.)

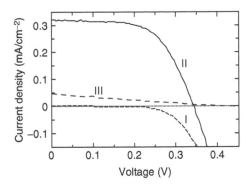

Figure 4.36. *J–V* characteristics of solid state DSSC in dark (I) and under white light illumination at 9.4 mW/cm² (II). The *J–V* characteristics of a cell without N(PhBr)$_3$SbCl$_6$ or Li[(CF$_3$SO$_2$)$_2$)N is also shown (III). (From Bach, U.; Lupo, D.; Comte, P.; Moser, J. E.; Weissörtel, F.; Salbeck, J.; Spreitzer, H.; Grätzel, M. *Nature* **1998**, *395*(6702), 583–585. Used with permission from Nature Publishing Group.)

Figure 4.36 shows the *J–V* characteristics of the device shown in Figure 4.35 [140]. Curve I shows the dark current density of a device and curves II and III show a device with and without the additives, respectively. The device that contains the hole conductor without additives performs poorly with $J_{sc} \cong 40\ \mu A/cm^2$ under white light illumination at 9.4 mW/cm². This device also has a resistive *J–V*-response, resulting in a fill factor of ~0.25. Addition of additives increases J_{sc} almost tenfold and greatly improves the fill factor. Consequently, the device efficiency increases from 0.04% to 0.74%. Under full sunlight (100 mW/cm², AM1.5), J_{sc} reached 3.2 mA/cm².

Nogueira et al. achieved even higher efficiencies by using a polymer electrolyte to fabricate solid state DSSCs [142]. The polymer electrolyte film was formed by repeatedly drop-casting onto dye-sensitized TiO$_2$ and drying. The electrolyte consisted of a poly(epichlorohydrin-*co*-ethylene oxide) elastomer that contained NaI and I$_2$ as mobile redox carriers. The top electrode was mechanically pressed against the polymer film. Transient absorption spectroscopy of bilayer films showed that the polymer electrolyte inhibited carrier recombination, indicating charge transfer at the TiO$_2$–polymer interface. A 150 W Xe lamp was used with an AM1.5 solar simulation filter and neutral density filters were used to adjust the intensity. Both the short-circuit current density and open-circuit voltage are improved than before, reaching a power efficiency of 1.6% under 1 sun and 2.6% under one-tenth sun. Photocurrent action spectra showed a peak IPCE of ~50% at 520 nm under illumination through the TiO$_2$ film.

Krüger, Grätzel, and co-workers recently reported solid state DSSCs with the highest power conversion efficiencies observed to date. A power conversion efficiency of 2.56% was achieved by controlling charge recombination across the heterojunction interface [143], later raised to 3.2% by using silver ions to improve dye absorption [144]. It is notable that these power efficiencies were certified by an independent authority using a AM1.5 solar simulator at the U.S. National Renewable Energy Laboratory.

The earlier report investigated the effect of adding 4-*tert*-butylpyridine (*t*BP) to the hole conductor matrix [143]. Devices were prepared by adding *t*BP to the spiro-MeOTAD solution or by exposing the dye-sensitized TiO$_2$ prior to deposition of the spiro-MeOTAD hole transport layer. The spiro-MeOTAD layer was also doped with a lithium salt as reported previously [140]. Incorporation of *t*BP into the spiro-MeOTAD layer increased the short-circuit current by 26% and doubled the open-circuit voltage. A series of devices were prepared with varying concentrations of tBP and lithium salt. The best device achieved $J_{sc} = 5\,mA/cm^2$ and $V_{oc} = 0.91\,V$. The device has a power conversion efficiency of 2.56% (corresponding to a fill factor of 56%) and was shown to be stable under 3 months when stored mostly in the dark. In contrast, exposure of the dye-sensitized TiO_2 layer to tBP yielded only marginal improvements in performance.

Improved device performance was correlated with carrier lifetimes in the hole conductor matrix. Transient absorption spectroscopy measurements were performed on bilayers of dye-sensitized TiO_2 and spiro-MeOTAD blends with no electrodes in order to measure the decay kinetics of the hole conductor matrix in the presence and absence of tBP [143]. The cation signal completely decayed within $50\,\mu s$ in the absence of tBP, but extended out to a millisecond in the presence of tBP. The decay kinetics were also sensitive to the lithium salt concentration. Reduced carrier trapping is a plausible explanation for the improved current density in tBP. The improvement of V_{oc} was not directly addressed, though could be related to the different effects of lithium and tBP on band edge shifts at the TiO_2 interface [145].

The performance of solid state DSSCs was further improved by adding silver ions to the solution used to stain nanocrystalline TiO_2. The best performance was previously obtained using TiO_2 films with a layer thickness of $\sim 2\,\mu m$. Using a thinner film reduces the series resistance and increases the charge collection efficiency due to reduced trapping. However, this also has the effect of reducing dye absorption. Krüger et al. addressed the problem of reduced absorption by performing dye adsorption in the presence of silver ions. TiO_2 films were dye-coated by soaking in a $5 \times 10^{-4}\,M$ solution of a ruthenium-containing dye; solutions were prepared with and without half the dye molar concentration of silver nitrate. XPS spectra of films showed peaks characteristic when silver nitrate was added. The TiO_2 film scatters too much light to perform a standard absorption measurement, so dye solution spectra were obtained after desorption from TiO_2 surfaces. Addition of silver ions substantially increases the absorption and changes the spectrum, suggesting the ligation of silver ions to the dye. Removing silver ions by adding an excess of iodide changes the spectrum back to its original lineshape, though with a 38% increase with respect to dye adsorbed in the absence of silver.

Figure 4.37 compares the J–V characteristics of devices using dye-sensitized TiO_2 layers prepared in the absence and presence of silver nitrate. The hole-conducting layer was a blend of spiro-MeOTAD, tBP, and lithium salt [144]. The incorporation of silver ions increases the short-circuit current by 31% and the open-circuit voltage by 13%, though the fill factor is slightly reduced. The overall power efficiency increases from 2.1% to 3.2% by addition of silver ions. The effect of silver ions on performance was studied by washing the TiO_2 film with iodide prior to depositing the hole-conducting layer. The same improvement in short-circuit current is seen, though both the open-circuit voltage and fill factor are reduced. Enhanced device performance was due to increased dye absorption, which led the authors to attribute the effect to the complexation of silver ions to the dye.

4.5. CONCLUDING REMARKS

Organic photovoltaics is a diverse and interesting field of research. We conclude with a comparison of the performance of the best OPVs with that of dye-sensitized and

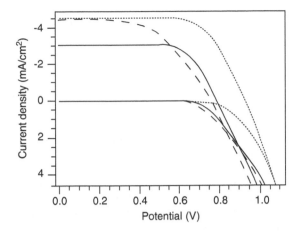

Figure 4.37. *J–V* characteristics in the dark and at under AM1.5 illumination at 100 mW/cm^2 for a standard device without silver (solid line), a device dyed in the presence of silver (dotted line), and a device dyed in the presence of silver and rinsed with iodide solution before depositing the hole conductor (dashed line). (From Krüger, J.; Plass, R.; Grätzel, M.; Matthieu, H.-J. *Appl. Phys. Lett.* **2002**, *81*(2), 367–369. Copyright American Institute of Physics, 2002. With permission.)

Table 4.2. Comparison of OPV performance with other technologies

Device	P_0	V_{oc}	I_{sc}	ff (%)	Efficiency (%)	Source
Single-crystal Si	100	0.696	42	83.6	24.4	Refs.
Amorphous Si	100	0.887	19.4	74.1	12.7	United Solar
DSSC	100	0.721	20.53	70.4	10.4	[136]
Molecular	150	0.58	18.8	52.0	3.6	[34]
Polymer	80	0.825	5.25	61.1	3.3	[83]
SSDSSC	100	0.931	4.6	71.0	3.2	[144]

inorganic solar cells, as shown in Table 4.2. While the performance of organic devices lags behind their inorganic counterparts, much progress has been made in the last several years. We have presented reports of a molecular OPV with 4.2% power conversion efficiency [39] from the Princeton group and more than 5% from an OPV based on a fullerene–polymer blend from a group at Siemens. The latter report has not yet been published and so should be taken as unconfirmed. Still, the fabrication of 10% efficient devices, long seen as an unachievable goal, now looks to be in the realm of the possible.

REFERENCES

1. Goetzberger, A.; Hebling, C., and Schock, H.-W. Photovoltaic materials, history, status and outlook. *Mater. Sci. Eng.* **2003**, *R40*, 1–46.
2. Zhao, J.; Wang, A.; Green, M. A., and Ferrazza, F. 19.8% Efficient "honeycomb" textured multicrystalline and 24.4% monocrystalline silicon solar cells. *Appl. Phys. Lett.* **1998**, *73*(14), 1991–1993.
3. Tang, C. W. Two-layer organic photovoltaic cell. *Appl. Phys. Lett.* **1986**, *48*(2), 183–185.
4. Wenham, S. R.; Green, M. A., and Watt, M. E. *Applied Photovoltaics*, Appendix B, Bridge Printery; Sydney, 1994.

5. Chamberlain, G. A. Organic solar cells — a review. *Sol. Cells* **1983**, *8*(1), 47–83.

6. Kampas, F. J. and Gouterman, M. *J. Phys. Chem.* **1977**, *81*(8): 690–695.

7. Ghosh, A. K.; Morel, D. L.; Feng, T.; Shaw, R. F., and Rowe, C. A. Photovoltaic and rectification properties of Al–Mg phthalocyanine-Ag Schottky-barrier cells. *J. Appl. Phys.* **1974**, *45*(1), 230s–236s.

8. Kearns, D. and Calvin, M. Photovoltaic effect and photoconductivity in laminated organic systems. *J. Chem. Phys.* **1958**, *29*(4), 950–951.

9. Fan, F. R. and Faulkner, L. R. Photo-voltaic effects of metal-free and zinc phthalocyanines. 2. Properties of illuminated thin-film cells. *J. Chem. Phys.* **1978**, *69*(7), 3341–3349.

10. Ghosh, A. K. and Feng, T. Merocyanine organic solar-cells. *J. Appl. Phys.* **1978**, *49*(12), 5982–5989.

11. Chamberlain, G. A.; Cooney, P. J., and Dennison, S. Photo-voltaic properties of merocyanine solid-state photocells. *Nature* **1981**, *289*(5793), 45–47.

12. Moriizumi, T. and Kudo, K. Merocyanine-dye photovoltaic cell on a plastic film. *Appl. Phys. Lett.* **1981**, *381*(2), 85–86.

13. Choong, V.; Park, Y.; Gao, Y.; Wehrmeister, T.; Müllen, K.; Hsieh, B. R., and Tang, C. W. Dramatic photoluminescence quenching of phenylene vinylene oligomer thin films upon submonolayer Ca deposition. *Appl. Phys. Lett.* **1996**, *69*(10), 1492–1494.

14. Meier, H. *Organic Semiconductors*, Verlag Chemie, Weinheim **1974**, p. 372, 429, 459.

15. Harima, Y.; Yamashita, K., and Suzuki, H. Spectral sensitization in an organic p–n junction photovoltaic cell, *Appl. Phys. Lett.* **1984**, *45*(10), 1144–1145.

16. Yamashita, K.; Matsumura, Y.; Harima, Y.; Miura, S., and Suzuki, H. Reaction of pyridinium phenacylides and related ylides with cyclopentadienone derivatives. *Chem. Lett.* **1984**, 4, 489–492.

17. Pope, M.; Braams, B. J., and Brenner, H. C. Diffusion of excitons in systems with non-planar geometry: theory. *Chem. Phys.* **2003**, *288*(2–3), 105–112.

18. Halls, J. J. M.; Pichler, K.; Friend, R. H.; Moratti S. C., and Holmes, A. B. Exciton diffusion and dissociation in a poly(p-phenylenevinylene)/C_{60} heterojunction photovoltaic cell. *Appl. Phys. Lett.* **1996**, *68*(22), 3120–3122.

19. Pettersson, L. A. A.; Roman, L. S., and Inganäs, O. Modeling photocurrent action spectra of photovoltaic devices based on organic thin films. *J. Appl. Phys.* **1999**, *86*(1), 487–496.

20. Stübinger, T. and Brütting, W. Exciton diffusion and optical interference in organic donor–acceptor photovoltaic cells. *J. Appl. Phys.* **2001**, *90*(7), 3632–3641.

21. Theander, M.; Yartsev, A.; Zigmantas, D. Sundström, V.; Mammo, W.; Andersson, M. R., and Inganäs, O. Photoluminescence quenching at a polythiophene/C_{60} heterojunction. *Phys. Rev. B*, **2000**, *61*(19), 12957–12963.

22. Hiramoto, M.; Fujiwara, H., and Yokoyama, M. p–i–n like behavior in three-layered organic solar cells having a co-deposited interlayer of pigments. *J. Appl. Phys.* **1992**, *72*(8), 3781–3787.

23. Rostalski, J. and Meissner, D. Monochromatic versus solar efficiencies of organic solar cells. *Sol. Energy Mater. Sol. Cells* **2000**, *61*(1), 87–95.

24. Tsuzuki, T.; Shirota, Y.; Rostalski, J., and Meissner, D. The effect of fullerene doping on photoelectric conversion using titanyl phthalocyanine and a perylene pigment. *Sol. Energy Mater. Sol. Cells*, **2000**, *61*(1), 1–8.

25. Gebeyehu, D.; Maennig, B.; Drechsel, J.; Leo, K., and Pfeiffer, M. Bulk-heterojunction photovoltaic devices based on donor–acceptor organic small molecule blends. *Sol. Energy Mater. Sol. Cells* **2003**, *79*, 81–92.

26. Geens, W.; Aernouts, T.; Poortmans, J., and Hadziioannou, G. Organic co-evaporated films of a PPV-pentamer and C_{60}: model systems for donor/acceptor polymer blends. *Thin Solid Films* **2002**, *403–404*, 438–443.

27. Wang, X. H.; Grell, M.; Lane, P. A., and Bradley, D. D. C. Determination of the linear optical constants of poly(9,9-dioctylfluorene). *Synth. Met.* **2001**, *119*(1–3), 535–536.

28. Peumans, P.; Uchida, S., and Forrest, S. R. Efficient bulk heterojunction photovoltaic cells using small molecular-weight organic thin films. *Nature* **2003**, *425*, 158–162.

29. Gregg, B. A.; Fox, M. A., and Bard, A. J. Photovoltaic effect in symmetrical cells of a liquid crystal porphyrin. *J. Phys. Chem.* **1990**, *94*(4), 1586–1598.

30. Gregg, B. A.; Fox, M. A., and Bard, A. J. J. 2,3,7,8,12,13,17,18-Octakis(beta-hydroxyethyl)porphyrin (octaethanolporphyrin) and its liquid-crystalline derivatives — synthesis and characterization. *J. Am. Chem. Soc.* **1989**, *111*(8), 3024–3029.

31. Schmidt-Mende, L.; Fechtenkötter, A.; Müllen, K.; Moons, E.; Friend, R. H., and MacKenzie, J. D. Self-organized discotic liquid crystals for high-efficiency organic photovoltaics. *Science* **2001**, *293*, 1119–1122.

32. van de Craats, A. M.; Warman, J. M.; Fechtenkötter, A.; Brand, J. D.; Harbison, M. A., and Müllen, K. Record charge carrier mobility in a room-temperature discotic liquid-crystalline derivative of hexabenzocoronene. *Adv. Mater.* **1999**, *11*(17), 1469–1472.

33. Peumans, P.; Bulovic, V., and Forrest, S. R. Efficient photon harvesting at high optical intensities in ultrathin organic double-heterostructure photovoltaic diodes. *Appl. Phys. Lett.* **2000**, *76*(19), 2650–2652.

34. Peumans, P. and Forrest, S. R. Very-high-efficiency double-heterostructure copper phthalocyanine/C_{60} photovoltaic cells. *Appl. Phys. Lett.* **2001**, *79*(1), 126–128.

35. Peumans, P.; Yakimov, A., and Forrest, S. R. Small molecular weight organic thin-film photodetectors and solar cells. *J. Appl. Phys.* **2003**, *93*(7), 3693–3723.

36. Hung, R. R. and Grabowski, J. J. A precise determination of the triplet energy of C_{60} by photoacoustic calorimetry. *J. Phys. Chem.* **1991**, *95*(16): 6073–6075.

37. Roman, L. S.; Mammo, W.; Pettersson, L. A. A.; Andersson, M. R., and Inganäs, O. High quantum efficiency polythiophene/C_{60} photodiodes. *Adv. Mater.* **1998**, *10*(10): 774–777.

38. Arias, A. C.; Gränstrom, M.; Petritsch, K., and Friend, R. H. Organic photodiodes using polymeric anodes. *Synth. Met.* **1999**, *102*(1–3), 953–954.

39. Xue, J.; Uchida, S.; Rand, B. P., and Forrest, S. R. 4.2% Efficient organic photovoltaic cells with low series resistances. *Appl. Phys. Lett.* **2004**, *16*(19), 3013–3015.

40. Wilkinson, C. I.; Lidzey, D. G.; Palilis, L. C.; Fletcher, R. B.; Martin, S. J.; Wang, X. H., and Bradley, D. D. C. Enhanced performance of pulse driven small area polyfluorene light emitting diodes. *Appl. Phys. Lett.* **2001**, *79*(2), 171–173.

41. Shockley, W. and Queisser, H. J. Detailed balance limit of efficiency of p–n junction solar cells. *J. Appl. Phys.* **1961**, *32*(3), 510.

42. Campbell, I. H.; Hagler, T. W.; Smith, D. L., and Ferraris, J. P. Direct measurement of conjugated polymer electronic excitation energies using metal/polymer/metal structures. *Phys. Rev. Lett.* **1996**, *76*(11), 1900–1903.

43. Nelson, J.; Kirkpatrick, J., and Ravirajan, P. Factors limiting the efficiency of molecular photovoltaic devices. *Phys. Rev. B* **2004**, *69*(3), 1–11.

44. Riedel, I.; Parisi, J.; Dyakonov, V.; Lutsen, L.; Vanderzande, D., and Hummelen, J. C. Effect of temperature and illumination on the electrical characteristics of polymer–fullerene bulk-heterojunction solar cells. *Adv. Func. Mater.* **2004**, *14*(1), 38–44.

45. Mihailetchi, V. D.; Blom, P. W. M.; Hummelen, J. C., and Rispens, M. T. Cathode dependence of the open-circuit voltage of polymer:fullerene bulk heterojunction solar cells. *J. Appl. Phys.* **2003**, *94*(10), 6849–6854.

46. Ramsdale, C. M.; Barker, J. A.; Arias, A. C. MacKenzie, J. D.; Friend, R. H., and Greenham, N. C. The origin of the open-circuit voltage in polyfluorene-based photovoltaic devices. *J. Appl. Phys.* **2002,** *92*(8), 4266–4270.

47. Gregg, B. A. and Hanna, M. C. Comparing organic to inorganic photovoltaic cells: theory, experiment, and simulation. *J. Appl. Phys.* **2003**, *93*(6), 3605–3614.

48. Takamoto, T.; Ikeda, E.; Kurita, H., and Ohmori, M. Over 30% efficient InGaP/GaAs tandem solar cells. *Appl. Phys. Lett.* **1997**, *70*(3), 381–383.

49. www.spectrolab.com (accessed February 2004).

50. Hiramoto, M.; Suezaki, M., and Yokoyama, M. Effect of thin gold interstitial-layer on the photovoltaic properties of tandem organic solar-cell. *Chem. Lett.* **1990**, (3), 327-330.

51. Yakimov, A. and Forrest, S. R. High photovoltage multiple-heterojunction organic solar cells incorporating interfacial metallic nanoclusters. *Appl. Phys. Lett.* **2002**, *80*(9), 1667–1669.

52. Tsukamoto, J.; Ohigashi, H.; Matsumura, K., and Takahashi, A. A Schottky-barrier type solar-cell using polyacetylene. *Jpn. J. Appl. Phys.* **1981**, *20*(2), L127–L129.

53. Kanicki, J. and Fedorko, P. Electrical and photovoltaic properties of *trans*-polyacetylene. *J. Phys. D: Appl. Phys.* **1984**, *17*, 805–817.

54. Glenis, S.; Horowitz, G.; Tourillon, G., and Garnier, F. Electrochemically grown polythiophene and poly(3-methylthiophene) organic photovoltaic cells. *Thin Solid Films* **1984**, *111*(2), 93–103.

55. Koezuka, H.; Hyodo, K., and MacDiarmid, A. G. Organic heterojunctions utilizing two conducting polymers: poly(acetylene)/poly(*N*-methylpyrrole) junctions. *J. Appl. Phys.* **1985**, *58*(3), 1279–1284.

56. Fang, Y.; Chen, S. A., and Chu, M. L. effect of side-chain length on rectification and photovoltaic characteristics of poly(3-alkylthiophene) Schottky barriers. *Synth. Met.* **1992**, *52*(3), 261–272.

57. Horowitz, G. and Garnier, F. Polythiophene–GaAs p–n-heterojunction solar-cells. *Sol. Energy Mater.* **1986**, *13*(1), 47–55.

58. Marks, R. N.; Halls, J. J. M.; Bradley, D. D. C.; Friend, R. H., and Holmes, A. B. *J. Phys.: Condens. Mat.* **1994**, *6*(7), 1379–1394.

59. Yu, G.; Pakbaz, K., and Heeger, A. J. Semiconducting polymer diodes: large size, low cost photodetectors with excellent visible-ultraviolet sensitivity. *Appl. Phys. Lett.* **1994**, *64*(25), 3422–3424.

60. Alvarado, S. F.; Seidler, P. F.; Lidzey, D. G., and Bradley, D. D. C. Direct determination of the exciton binding energy of conjugated polymers using a scanning tunneling microscope. *Phys. Rev. Lett.* **1998**, *81*(5): 1082–1085.

61. Harrison, M. G.; Grüner, J., and Spencer, G. C. W. Analysis of the photocurrent action spectra of MH-PPV polymer photodiodes. *Phys. Rev. B* **1997**, *55*(12), 7831–7849.

62. Blom, P. W. M.; de Jong, M. J. M., and Vleggaar, J. J. M. Electron and hole transport in poly(*p*-phenylene vinylene) devices. *Appl. Phys. Lett.* **1996**, *68*(23), 3308–3310.

63. Antoniadis, H.; Abkowitz, M. A., and Hsieh, B. R. Carrier deep-trapping mobility-lifetime products in poly(*p*-phenylenevinylene). *Appl. Phys. Lett.* **1994**, *65*(16), 2030–2032.

64. Sariciftci, N. S.; Smilowitz, L.; Heeger, A. J., and Wudl, F. Photoinduced electron transfer from a conducting polymer to buckminsterfullerene. *Science* **1992**, *258*(5087), 1474–1476.

65. Burn, P. L.; Grice, A. W.; Tajbakhsh, A.; Bradley, D. D. C., and Thomas, A. C. Insoluble poly[2-(2′-ethylhexyloxy)-5-methoxy-1,4-phenylenevinylene] for use in multilayer light-emitting diodes. *Adv. Mater.* **1997**, *9*(15), 1171.

66. Kroto, H. W.; Heath, J. R.; O'Brien, S. C.; Curl, R. E., and Smalley, R. E. C_{60} — buckminsterfullerne. *Nature* **1985**, *318*(6042), 162–163.

67. Krätschmer, W.; Lamb, L. D.; Fostiropocclos, K., and Huffman, D. R. Solid C-60 — a new form of carbon. *Nature* **1990**, *347*(6291), 354–358.

68. Ohsawa, Y. and Saji, T. Electrochemical detection of $C_{60}^{(6-)}$ at low-temperature. *J. Chem. Soc.-Chem. Commun.* **1992**, *10*, 781–782.

69. Kraabel, B.; Lee, C. H.; McBranch, D.; Moses, D.; Sariciftci, N. S., and Heeger, A. J. Ultrafast photoinduced electron-transfer in conducting polymer buckminsterfullerene composites. *Chem. Phys. Lett.* **1993**, *213*(3–4), 389–394.

70. Smilowitz, L.; Sariciftci, N. S.; Wu, R.; Gettinger, C.; Heeger, A. J.; Wudl, F. Photo-excitation spectroscopy of conducting-polymer–C(60) composites — photoinduced electron-transfer. *Phys. Rev. B* **1993**, *47*(20), 13835–13842.

71. Morita, S.; Zakhidov, A. A., and Yoshino, K. Doping effect of buckminsterfullerene in conducting polymer — change of absorption-spectrum and quenching of luminescence. *Solid State Commun.* **1992**, *82*(4), 249–252.

72. Morita, S.; Kiyomatsu, S.; Yin, X. H.; Zakhidov, A. A.; Noguchi, T.; Ohnishi, T., and Yoshino, K. Doping effect of buckminsterfullerene in poly(2,5-dialkoxy-*p*-phenylenevinylene). *J. Appl. Phys.* **1992**, *74*(4), 2860–2865.

73. Sariciftci, N. S.; Braun, D.; Zhang, C.; Srdanov, V. I.; Heeger, A. J.; Stucky, G., and Wudl, F. Semiconducting polymer-buckminsterfullerene heterojunctions: diodes, photodiodes, and photovoltaic cells. *Appl. Phys. Lett.* **1993**, *62*(6), 585–587.

74. Drees, M.; Premaratne, K.; Graupner, W.; Heflin, J. R.; Davis, R. M.; Marciu, D., and Miller, M. Creation of a gradient polymer–fullerene interface in photovoltaic devices by thermally controlled interdiffusion. *Appl. Phys. Lett.* **2002**, *81*(24), 4607–4609.

75. Chen, L.; Roman, L. S.; Johansson, D. M.; Svensson, M.; Andersson, M. R.; Janssen, R. A. J., and Inganäs, O. Excitation transfer in polymer photodiodes for enhanced quantum efficiency. *Adv. Mater.* **2000**, *12*(15), 1110–1114.

76. Sirringhaus, H.; Brown, P. J.; Friend, R. H.; Nielsen, M. M.; Bechgaard, K.; Langeveld-Voss, B. M. V.; Spiering, A. J. H.; Janssen, R. A. J.; Meijer, E. W.; Herwig, P., and de Leeuw, D. M. Two-dimensional charge transport in self-organized, high-mobility conjugated polymers. *Nature* **1999**, *401*(6754), 685–688.

77. Yu, G.; Pakbaz, K., and Heeger, A. J. Semiconducting polymer diodes: large size, low cost photodetectors with excellent visible-ultraviolet sensitivity. *Appl. Phys. Lett.* **1994**, *64*(25), 3422–3424.

78. Ruoff, R. S.; Tse, D. S.; Malhotra, R., and Lorents, D. C. Solubility of C_{60} in a variety of solvents. *J. Phys. Chem.* **1993**, *97*(13), 3379–3383.

79. Taylor, R. and Walton, D. R. M. The chemistry of fullerenes. *Nature* **1993**, *363*(6431), 685–693.

80. Yu, G.; Gao, J.; Hummelen, J. C.; Wudl, F., and Heeger, A. J. Polymer photovoltaic cells — enhanced efficiencies via a network of internal donor–acceptor heterojunctions. *Science* **1995**, *270*(5243), 1789–1791.

81. Hummelen, J. C.; Knight, B. W.; LePeq, F.; Wudl, F.; Yao, J., and Wilkins, C. L. Preparation and characterization of fulleroid and methanofullerene derivatives. *J. Org. Chem.* **1994**, *60*(3), 532–538.

82. Shaheen, S. E.; Brabec, C. J.; Sariciftci, N. S.; Padinger, F.; Fromherz, T., and Hummelen, J. C. 2.5% Efficient organic plastic solar cells. *Appl. Phys. Lett.* **2001**, *78*(6), 841–843.

83. Brabec, C. J.; Shaheen, S. E.; Winder, C.; Sariciftci, N. S., and Denk, P. Effect of LiF metal electrodes on the performance of plastic solar cells. *Appl. Phys. Lett.* **2002**, *80*(7), 1288–1290.

84. Svensson, M.; Zhang, F. L.; Veenstra, S. C.; Verhees, W. J. H.; Hummelen, J. C.; Kroon, J. M.; Inganas, O., and Andersson, M. R. High-performance polymer solar cells of an alternating polyfluorene copolymer and a fullerene derivative. *Adv. Mater.* **2003**, *15*(12), 988.

85. Schilinsky, P.; Waldauf, C., and Brabec, C. J. Recombination and loss analysis in polythiophene based bulk heterojunction photodetectors. *Appl. Phys. Lett.* **2002**, *81*(20), 3885–3887.

86. Brabec, C. J.; Padinger, F.; Hummelen, J. C.; Janssen, R. A. J., and Sariciftci, N. S. Realization of large area flexible fullerene–conjugated polymer photocells: a route to plastic solar cells. *Synth. Met.* **1999**, *102*(1–3): 861–864.

87. Padinger, F.; Brabec, C. J.; Fromherz, T.; Hummelen, J. C., and Sariciftci, N. S. Fabrication of large area photovoltaic devices containing various blends of polymer and fullerene derivatives by using the doctor blade technique. *Opto-Electr.* **2000**, *8*(4), 280–283.

88. Shaheen, S. E.; Radspinner, R.; Peyghambarian, N., and Jabbour, G. E. Fabrication of bulk heterojunction plastic solar cells by screen printing. *Appl. Phys. Lett.* **2001**, *79*(18), 2996–2998.

89. Bharathan, J. and Yang, Y. Polymer electroluminescent devices processed by inkjet printing: I. Polymer light-emitting logo. *Appl. Phys. Lett.* **1998**, *72*(21), 2660–2662.

90. Sirringhaus, H.; Kawase, T.; Friend, R. H.; Shimoda, T.; Inbasekaran, M.; Wu, W., and Woo, E. P. High-resolution inkjet printing of all-polymer transistor circuits. *Science* **2000**, *290*(5499), 2123–2126.

91. Ago, H.; Petritsch, K.; Shaffer, M. S. P.; Windle, A. H., and Friend, R. H. Composites of carbon nanotubes and conjugated polymers for photovoltaic devices. *Adv. Mater.* **1999**, *11*(15), 1281–1285.

92. Kymakis, E. and Amaratunga, G. A. J. Single-wall carbon nanotube'conjugated polymer photovoltaic devices. *Appl. Phys. Lett.* **2002**, *80*(1), 112–114.

93. Kymakis, E.; Alexandrou, I., and Amaratunga, G. A. J. High open-circuit voltage photovoltaic devices from carbon-nanotube polymer composites. *J. Appl. Phys.* **2003**, *93*(3), 1764–1768.

94. Park, J. Y.; Lee, S. B.; Park, Y. S.; Park, Y. W.; Lee, C. H.; Lee, J. I., and Shim, H. K. Doping effect of viologen on photoconductive device made of poly (*p*-phenylenevinylene). *Appl. Phys. Lett.* **1998**, *72*(22), 2871–2873.

95. Jenekhe, S. A. and Yi, S. Highly photoconductive nanocomposites of metallophthalocyanines and conjugated polymers. *Adv. Mater.* **2000**, *12*(17), 1274.

96. Angadi, M. A.; Gosztola, D., and Wasielewski, M. R. Characterization of photovoltaic cells using poly(phenylenevinylene) doped with perylenediimide electron acceptors. *J. Appl. Phys.* **1998**, *83*(11), 6187–6189.

97. Dittmer, J. J.; Marseglia, E. A., and Friend, R. H. Electron trapping in dye/polymer blend photovoltaic cells. *Adv. Mater.* **2000**, *12*(17), 1270–1274.

98. Sicot, L.; Geffroy, B.; Lorin, A.; Raimond, P.; Sentein, C., and Nunzi, J.-M. Photovoltaic properties of Schottky and p–n type solar cells based on polythiophene. *J. Appl. Phys.* **2001**, *90*(2), 1047–1054.

99. Breeze, A. J.; Saloman, A.; Ginley, D. S.; Gregg, B. A.; Tillmann, H., and Hörhold, H.-H. Polymer–perylene diimide heterojunction solar cells. *Appl. Phys. Lett.* **2002**, *81*(16), 3085–3087.

100. Greenham, N. C.; Moratti, S. C.; Bradley, D. D. C.; Friend, R. H., and Holmes, A. B. *Nature* **1993**, *365*(6447), 628–630.

101. Wang, Y. Z.; Gebler, D. D.; Lin, L. B.; Blatchford, J. W.; Jessen, S. W.; Wang, H. L., and Epstein, A. J. Alternating-current light-emitting devices based on conjugated polymers. *Appl. Phys. Lett.* **1996**, *68*(7), 894–896.

102. Halls, J. J. M.; Walsh, C. A.; Greenham, N. C.; Marseglia, E. A.; Friend, R. H.; Moratti, S. C., and Holmes, A. B. Efficient photodiodes from interpenetrating polymer networks. *Nature* **1995**, *376*(6540), 498–500.

103. Yu, G. and Heeger, A. J. Charge separation and photovoltaic conversion in polymer composites with internal donor/acceptor heterojunctions. *J. Appl. Phys.* **1995**, *78*(7), 4510–4515.

104. Moratti, S. C.; Cervini, R.; Holmes, A. B.; Baigent, D. R.; Friend, R. H.; Greenham, N. C.; Gruner, J., and Hamer, P. J. High electron-affinity polymers for LEDs. *Synth. Met.* **1995**, *71*(1–3), 2117–2120.

105. Moons, E. Conjugated polymer blends: linking film morphology to performance of light emitting diodes and photodiodes. *J. Phys.: Condens. Matter* **2002**, *14*(47), 12235–12260.

106. Bates, F. S. Polymer–polymer phase-behavior. *Science* **1991**, *251*(4996), 898–905.

107. Halls, J. J. M.; Arias, A. C.; MacKenzie, J. D.; Wu, W.; Inbasekaran, M.; Woo, E. P., and Friend, R. H. Photodiodes based on polyfluorene composites: influence of morphology. *Adv. Mater.* **2000**, *12*(7), 498–502.

108. Arias, A. C.; MacKenzie, J. D.; Stevenson, R.; Halls, J. J. M.; Inbasekaran, M.; Woo, E. P.; Richards, D., and Friend, R. H. Photovoltaic performance and morphology of polyfluorene blends: a combined microscopic and photovoltaic investigation. *Macromolecules* **2001**, 34, 6005–6103.

109. Arias, A. C.; Corcoran, N.; Banach, M.; Friend, R. H.; MacKenzie, J. D., and Huck, W. T. S. *Appl. Phys. Lett.* **2002**, *80*(10), 1695–1697.

110. Bernius, M. T.; Inbasekaran, M.; O'Brien, J., and Wu, W. Progress with light-emitting polymers. *Adv. Mater.* **2000**, *12*(23), 1737–1750.

111. Redecker, M.; Bradley, D. D. C.; Inbasekaran, M.; Wu, W. W., and Woo, W. P. High mobility hole transport fluorene-triarylamine copolymers. *Adv. Mater.* **1999**, *11*(3), 241.

112. Russell, D. M.; Arias, A. C.; Friend, R. H.; Silva, C.; Ego, C.; Grimsdale, A. C., and Müllen, K. Efficient light harvesting in a photovoltaic diode composed of a semiconductor conjugated copolymer blend. *Appl. Phys. Lett.* **2002**, *80*(12), 2204–2206.

113. Kietzke, T.; Neher, D.; Landfester, K.; Montenegro, R.; Güntner R., and Scherf, U. Novel approaches to polymer blends based on polymer nanoparticles. *Nat. Mater.* **2003**, *2*, 408–412.

114. Kietzke, T.; Neher, D.; Montenegro, R.; Landfester, K.; Montenegro, R.; Güntner R.; Scherf, U., and Hoerhold, H. H. Nanostructured solar cells based on semiconducting polymer nanospheres (SPNs) of M3EH–PPV and CN–ether–PPV. *SPIE Proc.* **2004**, *5215*, 206–210.

115. Tada, K.; Onoda, M.; Zakhidov, A. A., and Yoshino, K. Characteristics of poly(*p*-pyridyl vinylene)/poly(3-alkylthiophene) heterojunction photocell. *Jpn. J. Appl. Phys.* **1997**, *36*(3A), L306–L309.

116. Zhang, F.; Jonforsen, M.; Johansson, D. M.; Andersson, M. R., and Inganäs, O. Photodiodes and solar cells based on the n-type polymer poly(pyridopyrazine vinylene) as electron acceptor. *Synth. Met.* **2003**, *138*, 555–560.

117. Granström, M.; Petritsch, K.; Arias, A. C.; Lux, A.; Andersson M. R., and Friend, R. H. Laminated fabrication of polymeric photovoltaic diodes. *Nature* **1998**, *395*, 257–260.

118. Chen, X. L. and Jenekhe, S. A. Bipolar conducting polymers: blends of p-type polypyrrole and an n-type ladder polymer. *Macromolecules* **1997**, *30*(6): 1728–1733.

119. Jenekhe, S. A. and Yi, S. Efficient photovoltaic cells from semiconducting polymer heterojunctions. *Appl. Phys. Lett.* **2000**, *77*(17), 2635–2637.

120. Murray, C. B.; Norris, D. J., and Bawendi, M. G. Synthesis and characterization of nearly monodisperse CdE (E = S, Se, Te) semiconductor nanocrystallites. *J. Amer. Chem. Soc.* **1993**, *115*(19): 8706–8715.

121. O'Regan, B. and Grätzel, M. A. Low-cost, high-efficiency solar-cell based on dye-sensitized colloidal TiO$_2$ films. *Nature* **1991**, *353*(6346), 737–740.

122. Greenham, N. C.; Peng, X., and Alivisatos, A. P. Charge separation and transport in conjugated-polymer/semiconductor-nanocrystal composites studied by photoluminescence quenching and photoconductivity. *Phys. Rev. B* **1996**, *54*(24), 17628–17637.

123. Wang, Y. and Herron, N. Photoconductivity of CdS nanocluster-doped polymers. *Chem. Phys. Lett.* **1992**, *200*(1–2), 71–75.

124. Brus, L. E. Electron–electron and electron–hole interactions in small semiconductor crystallites — the size dependence of the lowest excited electronic state. *J. Chem. Phys.* **1984**, *80*(9), 4403–4409.

125. Yan, M.; Rothberg, L. J.; Papadimitrakopoulos, F.; Galvin, M. E., and Miller, T. M. Defect quenching of conjugated polymer luminescence. *Phys. Rev. Lett.* **1994**, *73*(5), 744–747.

126. Huynh, W. U.; Peng, X., and Alivisatos, A. P. CdSe nanocrystal rods/poly(3-hexylthiophene) composite photovoltaic devices. *Adv. Mater.* **1999**, *11*(11), 923–927.

127. Huynh, W. U.; Dittmer, J. J., and Alivisatos, A. P. Hybrid nanorod–polymer solar cells. *Science* **2002**, *295*, 2425–2427.

128. Huynh, W. U.; Dittmer, J. J.; Libby, W. C.; Whiting, G. L., and Alivisatos, A. P. Controlling the morphology of nanocrystal-polymer composites for solar cells. *Adv. Func. Mater.* **2003**, *13*(1), 73–79.

129. Huynh, W. U.; Dittmer, J. J.; Teclemariam, N.; Milliron, D. J.; Alivisatos, A. P., and Barnham, K. W. J. Charge transport in hybrid nanorod–polymer composite photovoltaic cells. *Phys. Rev. B* **2003**, *67*, 1–12.

130. Salafsky, J. S.; Lubberhuizen, W. H., and Schropp, R. E. I. Photoinduced charge separation and recombination in a conjugated polymer-semiconductor nanocrystal composite. *Chem. Phys. Lett.* **1998**, *290*(4–6), 297–303.

131. Salafsky, J. S. Exciton dissociation, charge transport, and recombination in ultrathin, conjugated polymer–TiO$_2$ nanocrystal intermixed composites. *Phys. Rev. B* **1999**, *59*(16), 10885–10894.

132. van Hal, P. A.; Wienk, M. M.; Kroon, J. M.; Verhees, W. J. H.; Slooff, L. H.; van Gennip, W. J. H.; Jonkheijum, P., and Janssen, R. A. J. Photoinduced electron transfer and photovoltaic response of a MDMO-PPV:TiO$_2$ bulk-heterojunction. *Adv. Mater.* **2003**, *15*(2), 118–121.

133. Coakley, K. M.; Liu, Y.; McGehee, M. D.; Friendell, K. L., and Stucky, G. D. Infiltrating semiconducting polymers into self-assembled mesoporous titania films for photovoltaic applications. *Adv. Func. Mater.* **2003**, *13*(4), 301–306.

134. Coakley, K. M. and McGehee, M. D. Photovoltaic cells made from conjugated polymers infiltrated into mesoporous titania. *Appl. Phys. Lett.* **2003**, *83*(16), 3380–3382.

135. Arango, A. C.; Johnson, L. R.; Bliznyuk, V. N.; Schlesinger, Z.; Carter, S. A., and Hörhold, H.-H. Efficient titanium oxide/conjugated polymer photovoltaics for solar energy conversion. *Adv. Mater.* **2000**, *12*(22), 1689–1692.

136. Hagfeldt, A. and Grätzel, M. Molecular photovoltaics. *Acc. Chem. Res.* **2000**, *33*(5), 269–277.

137. Hagen, J.; Schaffrath, W.; Otschik, P.; Fink, R.; Bacher, A.; Schmidt, H.-W., and Haarer, D. Novel hybrid solar cells consisting of inorganic nanoparticles and an organic hole transport material. *Synth. Met.* **1997**, *89*(3), 215–220.

138. Murakoshi, K.; Kogure, R.; Wada, Y., and Yanagida, S. Solid state dye-sensitized TiO$_2$ solar cell with polypyrrole as hole transport layer. *Chem. Lett.* **1997**, (5), 471–472.

139. Murakoshi, K.; Kogure, R.; Wada, Y., and Yanagida, S. Fabrication of solid-state dye-sensitized TiO$_2$ solar cells combined with polypyrrole. *Sol. Energy Mater. Sol. Cells* **1998**, *55*(1–2), 113–125.

140. Bach, U.; Lupo, D.; Comte, P.; Moser, J. E.; Weissörtel, F.; Salbeck, J.; Spreitzer, H., and Grätzel, M. Solid-state dye-sensitized mesoporous TiO$_2$ solar cells with high photon-to-electron conversion efficiencies. *Nature* **1998**, *395*(6702), 583–585.

141. Enright, B.; Redmond, G., and Fitzmaurice, D. Spectroscopic determination of flat-band potentials for polycrystalline TiO$_2$ electrodes in mixed-solvent systems. *J. Phys. Chem.* **1994**, *98*(24), 6195–6200.

142. Nogueira, A. F.; Durrant, J. R., and De Paoli, M. A. Dye-sensitized nanocrystalline solar cells employing a polymer electrolyte. *Adv. Mater.* **2001**, *13*(11), 826–830.

143. Krüger, J.; Plass, R.; Cevey, L.; Piccirelli, M.; Grätzel, M., and Bach, U. High efficiency solid-state photovoltaic device due to inhibition of interface charge recombination. *Appl. Phys. Lett.* **2001**, *79*(13), 2085–2087.

144. Krüger, J.; Plass, R.; Grätzel, M., and Matthieu, H.-J. Improvement of the photovoltaic performance of solid-state dye-sensitized device by silver complexation of the sensitizer *cis*-bis(4,4'-dicarboxy-2,2'bipyridine)-bis(isothiocyanato) ruthenium(II). *Appl. Phys. Lett.* **2002**, *81*(2), 367–369.

145. Haque, S. A.; Tachibana, Y.; Willis, R. L.; Moser, J. E.; Grätzel, M.; Klug, D. R., and Durrant, J. R. Parameters influencing charge recombination kinetics in dye-sensitized nanocrystalline titanium dioxide films. *J. Phys. Chem. B* **2000**, *104*(3), 538–547.

Section 2
Mechanisms and Modeling

5

Simulations of Optical Processes in Organic Photovoltaic Devices

Nils-Krister Persson and Olle Inganäs
Biomolecular and Organic Electronics, Department of Physics and Measurement Technology, Linköping University, Linköping, Sweden

Contents

Abstract By simulation of the optical processes in photovoltaic devices it is possible to enhance both the understanding of the physical processes occurring as a result of optical absorption and also to optimize the design of devices given this understanding. Calculations allow properties that are out of reach for direct measurements, such as absorption profiles, to be evaluated. Besides, in a relatively short time on a computer, thousands of different device geometries can be simulated. This is far beyond what can be performed experimentally, both from a time and a material householding perspective, and allows optimization of device design from an optical point of view. We present a modeling approach based on a number of assumptions such as homogeneous layers, sharp and planar interfaces, scattering free optics, coherence in the stack part of the device, and incoherence when adding energies across the substrate. The model has only layer thickness, layer dielectrical functions, and exciton diffusion length as input. Many different kinds of output are possible: absorption profile, reflectance, absorbance, absorption distribution, limits for quantum efficiencies, etc. The model is applied to devices with the active material being a pure polyfluorene–copolymer or blended with C_{60} as the acceptor.

Keywords optical modeling, polymer, organic, photovoltaic device, complex index of refraction, spectroscopic ellipsometry, polyfluorene, fullerene, blend

5.1. INTRODUCTION

The optical processes inside an organic photovoltaic device include electromagnetic wave transmission and reflection at all the interfaces found inside the device, and the decay of wave energy due to absorption inside the active photovoltaic layers, as well as other layers such as structural layers and electrodes. We can describe the processes in terms of incoupling and absorption of the incoming photon. As the layers found inside a polymer-based photovoltaic device (PPVD) are all thin, and the total thickness of the film part of a device may be of the order of one or two wavelengths of incoming light, wave phenomena are ubiquitous and must be properly accounted for when modeling the optical processes inside the PPVD. This can be done with a full dielectric function model of all the materials building the device layers together with geometrical description of the device. With this set of data, it is possible to calculate the electromagnetic mode structure of the device and the distribution of excited states in the active layers for monochromatic or polychromatic light impinging on the device. This allows calculations of the energy redistribution between layers and gives input to the determination of charge carrier distribution within the device. This, in turn, is a necessary input for models of electrical transport in the device under illumination, to build a full-fledged model of optical processes and electrical transport in PPVDs. The optical processes are the topic of this chapter.

    Simulations have sometimes been described as the third cornerstone of physics of equal merit with the traditional pillars of theory and experiment.[1] In simple terms it is the technique of performing experiments, not on the real system which ultimately is the interesting topic, but on its model.[2,3] The model is the result of mapping some aspects of real system — in our case a polymer-based photovoltaic device and its optical behavior — so that the system complexity is decreased to manageable levels but still leaves a model powerful enough to deliver nontrivial results. The mapping heavily depends on the understanding of the processes inside and how well the processes can be described in mathematical language, i.e., physical formulas and in program code, and a valuation and judgment of what is important and what is superfluous. Today's user-friendly software, where much of the complicated calculations are concealed to the user, makes a computerized simulation the standard choice.

The great benefit with simulation is avoidance of manipulating a set of real-world objects. This saves resources, both time and working efforts, as well as materials as often only very small quantities are available. During some tenths of hours on an ordinary PC, 20,000 different PPVD structures can be defined and simulated, irradiated with artificial solar light consisting of 500 wavelengths and then analyzed with respect to their optical response. This is simply not possible to achieve with experiments. When scanning this large number of structures, it is natural to pick out which in some sense is the best one, and by this entering into the realms of optimization. Through simulation, which exhausts all — within limits — parameter combinations, the classical obstacle in optimization is avoided, namely solving the inverse problem when one starts from a parameter-dependent objective function and goes backwards to the parameter space for the optimal combination.

Real experiments demand planning and thoughtful considerations. Simulation on the other hand gives an unsurpassed opportunity to test casual thoughts and examine innovative alternatives — with an answer within minutes. Finally, by simulation it is possible to reach otherwise inaccessible quantities, exemplified with the calculation of the square modulus of the optical electric field in every point inside the PPVD, or the distribution of excited states within the thin films inside a device.

The de-coupling of the model from the real system is also the source of the weakness of the procedure. Even if simulation has the ability to confirm the experimenter's prejudice, or predict the unforeseen, it must be tested on dummy examples and checked against experiments. To bring in physical justification makes simulation somewhat of an art and something more than brute programming.

In the following we will give an example of how a simulation of a PPVD, a "solar cell," can be done and go into details of assumptions and derivations besides delivering useful results. In general terms the simulation presented is simple and straightforward. It is *static* in that time-dependence is not relevant, and *deterministic* as no variables or processes are modeled as stochastic.

Based on the formalism developed, we are able to calculate the total electromagnetic (EM) field at any point inside the device, taking into account both the primary irradiated light and light reflected in different internal interfaces. In a thin film stack, which indeed is a part of the device, this gives rise to interference. As we simulate polychromatic solar-distributed irradiation, the model will allow us to calculate absorbed energies as absolute values. One of the great advantages of the model is that it offers the opportunity to optimize the structure so that maximum energy is absorbed. The free variables are the layer thicknesses, and by varying them the total reflectance, and the beneficial and nonbeneficial absorption is changed. From this information, an optimal set of thicknesses can be determined, given the materials and their optical properties. It is also possible to resolve the spatial energy dissipation and to construct EM field profiles. This is done both for monochromatic and polychromatic light. Simulation also admits calculation of reflectance and allows estimates of limits of quantum efficiencies.

After a short description of the physics involved in a PPVD, the matrix-based formalism underlying the simulation is presented in detail. Then results follow. These are first demonstrated for monochromatic irradiation and later for polychromatic solar light. Three types of structures are studied: (1) the cavity, (2) the double layer structure, and (3) the blend layer structure.

In the following, *geometry* will mean a certain choice of thicknesses, *morphology* refers to the structure of materials, preferentially the active layer, and whether it is in one phase or phase separated, *Topology* refers to the same thing but is restricted to the surface.

To simplify our model we avoid some possible complexities. Only normal incoupling of light is considered. Surface roughness is assumed to be so small that interfaces can be approximated by flat ideal surfaces. We assume linear optical behavior. Important aspects of real PPVDs are therefore excluded, i.e., no waveguiding will be discussed. Anisotropy of layers may be found in these materials, but due to uniaxiality and normal incidence the whole complexity of the anisotropy is not seen, as only the ordinary direction modulates the penetrating light. Both of these exclusions prevent us from analyzing the impact of patterning of layers, where diffractive optics can contribute to the absorption features of devices.

5.2. THE SEVEN PROCESSES OF POLYMER PHOTOVOLTAIC DEVICES

To go beyond the simple working principle for photovoltaic devices, "light in–current out," it is fruitful to leave the holistic system picture for the understanding of PPVDs, and divide and *analyze* the performance into seven processes. These are:

- Incoupling of the photon
- Photon absorption
- Exciton formation
- Exciton migration
- Exciton dissociation
- Charge transport
- Charge collection at the electrodes.

The processes will shortly be described below. The first two items in the list are the optical mechanisms of the device. By formulating the total performance into seven steps, it is possible to de-couple the optical and electrical behavior thereby more clearly finding possibilities for enhancing performance and identifying bottlenecks.

The device geometry in PPVDs is typically a multilayer stack of electrodes and absorber–photocurrent generator layers deposited on a common transparent substrate in a sandwich fashion. It may include special electrode buffer layers for adapting electric conditions at interfaces, layers of p-type and n-type organic conductors, and exciton blocking layers. The substrate can easily be 1000 times thicker than the total multilayer structure.

5.2.1. Incoupling of the Photon

Some interface bounds the device from the outside. Typically devices are manufactured such that the substrate, glass — assumed in the model below — or quartz or a polymeric material is the first material the photons encounter. The criterion is that the material should be as transparent as possible for light. The losses due to reflection at air–substrate interfaces should be minimized. Upon normal incidence from one layer to another, the higher the difference between optical refractive index, the higher the losses. At $n_{substrate} = n_{glass} = 1.5$ the reflectance will be $R = 0.04$ and for $n_{substrate} = 2$, $R = 0.11$. It is important to realize that the device reflectance is not the same as the reflectance of the air–substrate interface, but depends on all layers including substrate and stack. Manipulating the surfaces by suitable patterning such as downward pointing pyramids, globes, or one-dimensional gratings[4] can diminish losses. Antireflection properties can be controlled also by considering the layer ordering, layer thicknesses, and suitable choices of the dielectric functions of the layers. For this, simulation is an invaluable tool. Corrugation of interfaces[5] gives rise to diffraction, which certainly is part of the optics in the device but this theme will not be included in the present simulation. Any redirecting of photons from normal incidence to some degree of coplanarity with the film surface will help absorption as the path length increases, but is likewise excluded from the model.

5.2.2. Photon Absorption

As we define the active layer as the layer where beneficial absorption (for device function) takes place, it is important to focus as much energy as possible to this active layer. The distribution of the optical electrical field, E, is best described in terms of its square modulus, $|E|^2$ as $|E|^2$ is closely related to absorption. At a given point in the device, $|E|^2$ is dependent both on the local dielectric function at that point and on the global properties of device geometry, including the optical properties of the layers and interfaces.

From a materials perspective, it is important to have an absorption coefficient of the active material that matches the solar irradiation. During the years there has been a continuous striving for lowering the band gap of the polymers used.[6]

It should be kept in mind that PPVDs are in the realm of thin film optics. Therefore simple assumptions of Beer–Lambert absorption (given by the following equation) are inadequate.

$$S_t = S_{in}e^{-\alpha d} \tag{1}$$

where S is power per unit area, α the absorption coefficient, and d the distance calculated from the front interface, and the index "t" refers to transmission. S_{in} should refer to a point immediately inside the interface but is commonly replaced by the power per unit area immediately before the interface, thereby confusing reflectance and absorption. In the inorganic solar cell community, optical modeling based on Equation (1) is common.

5.2.1 and 5.2.2 together describe the optical part of the total PPVD mechanism. For processes 5.2.1 and 5.2.2 a joint efficiency, η_A, is formulated as the ratio between the number of absorbed photons in the device and the number of incoming photons to the air–substrate interface.

5.2.3. Exciton Formation

After optical absorption has occurred, the excitation of the organic solid is described in terms of the exciton. The exciton consists of a pair of Coulomb-attracted electrons and holes, is electrically neutral, and can be compared to the Frenkel exciton of solid state physics.

Only a fraction of the incoming photons turn into excitons. One can attribute the exciton formation efficiency η_{EC} with this step. Processes 5.2.2 and 5.2.3 are collectively called photogeneration.

5.2.4. Exciton Migration

In a diffusive three-dimensional migration the exciton moves through the material. A parameter describing this process is the exciton diffusion length, L_D, Typically it is of the order of 5–10 nm but is dependent on the structure of the materials and the dielectric environment.[7]

Excitons have a finite lifetime, and during their diffusion they end their existence via several beneficial as well as disadvantageous decay channels; in a radiative decay, a photon is re-emitted in luminescence, thereby constituting a loss mechanism for a PPVD. All vibronic and thermal decay routes are also loss of energy for PPVDs. The desired path is the transformation of the excitons into free electrons and holes, which is assumed to happen at certain *dissociation sites* where two materials with different electron affinities come into proximity of each other. Conditions for generating charge from excited states may also be found in inhomogeneities in materials, and at interfaces to electrodes, but are typically very inefficient compared to processes of photoinduced charge transfer at junctions between two dissimilar materials, and are therefore neglected here.

From the diffusion length L_D, we define a diffusion zone. This is the part of the device that has the ability to give charge carriers for a photocurrent. The diffusion zone can extend into one or several layers.

5.2.5. Exciton Dissociation

Solar cell dissociation of the exciton into free charge carriers (electron and hole) is the beneficial way of converting the energy emerging from the absorbed photon. At the dissociation site the electron and the hole are separated, free to move, or to move and then be trapped.

Processes counteracting the dissociation are the (geminate) recombination where separated electrons and holes merge back into an exciton, because the field is too weak for separation beyond electrostatic attraction, and nongeminate bimolecular recombination when an electron and a hole generated from different excitons recombines. A mobile hole may also recombine with a trapped electron.

5.2.6. Charge Transport

The free charges must be allowed to reach the electrodes where they constitute the photocurrent from the device.

The location of the dissociation site is important for the extracted photocurrent. Electrons and holes have different mobilities in the material. Trapping into localized states may occur. Irrespective of whether the trapping is permanent or temporary, the efficiency of charge transport is diminished. As the risk for trapping increases with the distance traveled, a thin layer is better than a thick layer, but optical absorption, which is proportional to thickness, is simultaneously reduced.

Recombination of free charge carriers into excitons, and between one trapped and one free carrier, is another loss mechanism.

5.2.7. Charge Collection

Even if an electron or a hole is present close to an electrode, whether they will pass into the outer circuit is not certain. The probability associated with all the barrier penetration mechanisms involved at the interfaces towards the metallic surfaces is a function of geometry, topology, and interface formation.

5.3. ROUTES TO OPTICAL MODELS OF PPVDs

In the literature on modeling of organic photovoltaic devices from Ghosh et al. and forward,[8–10] much of the focus has been on the processes 5.2.4 to 5.2.7. Often less attention is paid to the optics, with photon absorption commonly described as a simple source term in an exciton diffusion equation, or by assuming a Beer–Lambert (Equation (1)) decrease of absorption or production from the front interface. Realistic optical models must include reflection at internal interfaces and the subsequent interference in these thin film structures. Studies of this kind are in demand but not frequent.[7,11–14] Support from the inorganic solar cell literature is scarce as typically layers are thicker with these materials. Systems with both coherent and incoherent light addition has been discussed in a number of studies.[15–21]

Considerations of interference effects in PPVD was done by Halls[22] but best described in his doctoral dissertation.[23] The approach avoids cumbersome calculations because of intensity rather than addition of amplitude. Input parameters are the transmittances through the active layers upstream and downstream, reflectance at the aluminium interface, and thickness and absorption coefficient of the active layer.

One of the first thoroughly worked out studies was by Petterson.[24] In a matrix formalism, internal reflections and interference could be handled for calculating the electrical field at any given point inside the device. This in turn made it possible to calculate the absorbance as identified with the energy dissipation. In this calculation, input parameters were the thicknesses of the layers and the complex dielectric functions of the materials involved.

Peumans et al.[7] followed the same line in a large study mainly devoted to small molecular material (CuPc). Hoppe et al.[13] addressed blended active layers. Their case is the para-phenylene-vinylene derivative MDMO–PPV and [6,6]-phenyl-C_{61}-butyric acid methyl ester (PCBM) by 1:4 wt.%. The connection between model and experimental data is the photocurrent. This is also used by Rostalski and Meissner[14] for a double layer of small molecular materials with calculations based on a slightly different expression for the absorptance. All these four articles also aim for optimizing the performance by finding the right layer thickness combination and in the case of Peumans and Rostalski, and Meissner, experimenting with added layers.

A somewhat different approach to interference modeling was taken by Stübinger.[25] Still if the model was simple it was potent enough to give estimates of the optimal fullerene layer thickness by curve fitting.

It can be noted that the literature does not seem to include examples of two-dimensional models, allowing non-normal angle of incidence. Idealization prevails concerning layer homogeneity, thickness uniformity, and interface sharpness. More elaborate studies of electromagnetic transmission, for example using finite element methods, could be the next step.

5.4. THE MATRIX MODEL

5.4.1. General Assumptions

The primary purpose of the modeling is to find the optical electric field, E, at every point. E will be a complicated function of the dielectric functions and thicknesses of the materials.

Going by the conventional way of building PV devices in a stacked sandwich-like fashion, the structure lends itself to a one-dimensional model. The directions are indicated in Figure 5.1 where $+$ indicates down into the device, and $-$ upstream. The plane of incidence is defined by the plane normal, the incident, and the (specular) reflected ray. By "p", we indicate light with polarization parallel to the plane of incidence and by "s" (from German *senkrecht*) we mean vertical or perpendicular light. It is always possible to decompose light into these two directions. It is sufficient to follow only one ray hitting the surface at a certain point, which is the origin. A Cartesian coordinate system is introduced, $Oxyz$, with x and y in the plane and z into the device, perpendicular to the plane surface. x and y will sometimes be summarized with $\parallel$ and z with $\perp$.

Layers will be numbered starting with 0 for air, 1 for the substrate, and so on. The index of refraction is a complex quantity for which we will use the plus sign convention: $\tilde{n} = n + ik$.

Matrices are intimately related to linear processes. As linearity will be used as an assumption, in a way explained below, together with the inherent one-dimensionality and the separation into $+$ and $-$ direction a matrix formalism gives an effective

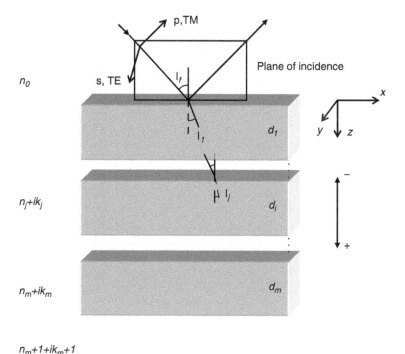

Figure 5.1. Modeling of the stack. Coordinate system is shown with plane of incidence and angle of incidence. j indicates a general layer and m is the last one in the stack. 0 and $m+1$ are semiinfinite.

description. The code is written in Matlab (The MathWorks, Inc), a mathematical software based on matrix manipulations.

We assume layers to be smooth and homogeneous, and interfaces to be parallel. We do not take scattering into account. But we make no assumptions of isotropy. In fact, some of the materials involved are found to be uniaxial anisotropic.

It is necessary for the optical treatment to divide the structure into two parts, a multilayer thin film *stack* and a comparatively much thicker glass *substrate*. Internally in the stack, the light is coherently added using a matrix formalism, as typical layer thicknesses and irradiated wavelengths are of the same order of magnitude. Whenever the millimeter-thick substrate is also taken into account, coherence is lost and irradiances rather than optical field amplitudes must be added.

5.4.2. Derivation — the Stack Model

We start by decomposing the optic electrical field into an upstream and a down-stream component

$$E = E^+ + E^- \quad \text{for s or p}$$

For a nonpatterned surface and isotropic or uniaxial anisotropic with optical axis normal to the plane, E, E^+, E^- does not depend on x and y. Therefore we write

$$E(z) = E^+(z) + E^-(z) \tag{2}$$

for each polarisation state, s and p. The indexation with s and p will be omitted in most cases in the following discussion, but is still relevant. For relating two nearby points, z_1 and z_2, we assume linearity

$$
\begin{aligned}
E^+(z_1) &= M_{11}E^+(z_2) + M_{12}E^-(z_2) \\
E(z_1) &= M_{21}E^+(z_2) + M_{22}E^-(z_2)
\end{aligned}
\tag{3}
$$

Introducing the generalized field vectors $(E^+, E^-)_x^{\mathrm{T}}$ and the scattering matrix, M, consisting of the numbers $M_{ij}, i,j = 1,2$, we comprehend these equations into a single vector–matrix relation

$$\begin{pmatrix} E^+ \\ E^- \end{pmatrix}_{z_1} = M \begin{pmatrix} E^+ \\ E^- \end{pmatrix}_{z_2}$$

The nearby points can be related in either of the two cases; if z_1 and z_2 are in the same layer without any interface in between, we write

$$\begin{pmatrix} E^+ \\ E^- \end{pmatrix}_{z_1} = L \begin{pmatrix} E^+ \\ E^- \end{pmatrix}_{z_2}$$

but if z_1 and z_2 are on opposite side of an interface, just inside the two layers, we write

$$\begin{pmatrix} E^+ \\ E^- \end{pmatrix}_{z_1} = I \begin{pmatrix} E^+ \\ E^- \end{pmatrix}_{z_2}$$

L is denoted as layer matrix and I as interface matrix. For layer j, we write L_j and for the interface between layer i and j, I_{ij}. Thus, for two points z_0 in layer 0 and z_{m+1} in layer $m+1$,

$$M = I_{01}L_1L_{12}L_2L_{23}L_3, \ldots, L_mI_{m,m+1} = \left(\prod_{v=1}^{m} I_{v-1,v}L_v \right) I_{m,m+1} \tag{4}$$

$(E^+, E^-)_0^{\mathrm{T}}$ is the same as $(E_{\mathrm{in}}, E_{\mathrm{refl}})^{\mathrm{T}}$ where 'in' indicates incoming and 'refl' the reflected field amplitude, and for $(E^+, E^-)_{m+1}^{\mathrm{T}}$ we put $(E_{\mathrm{trans}}, 0)^{\mathrm{T}}$.

The Fresnel complex transmission and reflection coefficients then relates the amplitudes as

$$t = \frac{E_{m+1}^+}{E_0^+} \tag{5}$$

$$r = \frac{E_0^-}{E_0^+} \tag{6}$$

Applying (3)

$$\begin{cases} E_0^+ = M_{11}E_{m+1}^+ + 0 \\ E_0^- = M_{21}E_{m+1}^+ + 0 \end{cases} \text{gives} \begin{cases} t = \frac{1}{M_{11}} \\ r = \frac{M_{21}}{M_{11}} \end{cases} \tag{7}$$

It is now possible to determine I_{ij} (Figure 5.2(a)). The relation

$$\begin{pmatrix} E^+ \\ E^- \end{pmatrix} = \begin{pmatrix} I_{11} & I_{12} \\ I_{21} & I_{22} \end{pmatrix} \begin{pmatrix} E^+ \\ E^- \end{pmatrix}_j$$

is true for all $(E^+, E^-)^{\mathrm{T}}$ in particular for the situation in Figure 5.2(b): $E_j^+ = t_{ij}E_i$, $E_j^- = 0$, and $E_i = r_{ij}E_i^+$ which gives I_{11} and I_{21} and for the situation in Figure 5.2(c): $E_j^+ = r_{ji}E_j^-$, $E_i^+ = 0$, and $E_i^- = t_{ji}E_j^-$. Fresnel formulae give $r_{ij} = -r_{ji}$ and

$$t_{ij} = \frac{1 - r_{ji}^2}{t_{ji}}$$

from which it is possible to derive the complete I_{ij}-matrix:

$$I_{ij} = \frac{1}{t_{ij}} \begin{pmatrix} 1 & r_{ij} \\ r_{ij} & 1 \end{pmatrix} \tag{8}$$

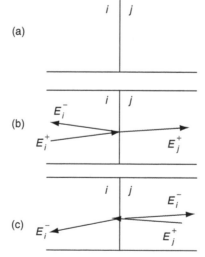

Figure 5.2. (a) Derivation of the I-matrix, i and j are two adjacent layers in the stack. (b) A ray from the left, which gives reflection and refraction. (c) A ray from the right, which is reflected and refracted.

Via Fresnel formulae for r_{ij} and t_{ij}, I_{ij} can be expressed in the indices of refraction and angle of incidence.

We now turn to L_j. For a passage through layer j (see Figure 5.3) of thickness d, index of refraction n, angle of refraction ϕ, and the phase difference β

$$\beta_j = \frac{2\pi d}{\lambda} n_j \cos \phi_1 = \xi_j d$$

where

$$\xi_j = \frac{2\pi}{\lambda} n_j \cos \phi_1.$$

Then $E_{d=0}^+ = e^{-i\beta} E_d^+$ and $E_{d=0}^- = e^{+i\beta} E_d^-$ which gives

$$L_j = \begin{pmatrix} e^{-i\beta_j} & 0 \\ 0 & e^{i\beta_j} \end{pmatrix} \tag{9}$$

Having both L_j and I_{ij} we are able to start to dwell upon the E-field in layer j.

The environment around layer j is partitioned into an upstream system indicated with $'$ and a downstream system indicated with $''$ (Figure 5.4(a)),

$$M = M_j' L_j M_j''$$

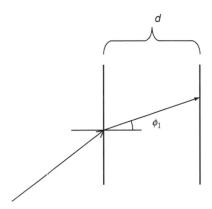

Figure 5.3. Deriving β. A ray passes through a medium of thickness d.

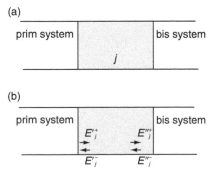

Figure 5.4. (a) Focusing on a certain layer j in the stack. The environment is indicated $'$ and $''$ respectively. (b) Output and input to layer j in the form of electrical field amplitudes. The arrows indicate the direction of propagation, not the field amplitudes.

where

$$M'_j = \left(\prod_{v=1}^{j-1} I_{v-1,v} L_v \right) I_{j-1,j}$$

$$M''_j = \left(\prod_{v=j+1}^{m} I_{v-1,v} L_v \right) I_{m,m+1}$$

The upstream and downstream systems have their own reflection and transmission coefficients for which bis quantities follow:

$$t''_j = \frac{1}{M''_{j,11}}$$

$$r''_j = \frac{M''_{j,21}}{M''_{j,11}} \tag{10}$$

derived in the same manner as (7).

It will turn out to be efficient to introduce certain transfer coefficients that couples between layer j and the incoming light:

$$t^+_j = \frac{E^+_j}{E^+_0} \tag{11}$$

$$t^-_j = \frac{E^-_j}{E^+_0} \tag{12}$$

Expressions can be derived with these transfer coefficients. For the upstream system

$$\begin{pmatrix} E^+_0 \\ E^-_0 \end{pmatrix} = \begin{pmatrix} M'_{j,11} & M'_{j,12} \\ M'_{j,21} & M'_{j,22} \end{pmatrix} \begin{pmatrix} E'^+_j \\ E'^-_j \end{pmatrix}$$

which gives

$$E^+_0 = M'_{j,11} E'^+_j + M'_{j,12} E'^-_j \tag{13}$$

The downstream system

$$\begin{pmatrix} E''^+_j \\ E''^-_j \end{pmatrix} = \begin{pmatrix} M''_{j,11} & M''_{j,12} \\ M''_{j,21} & M''_{j,22} \end{pmatrix} \begin{pmatrix} E'^+_{m+1} \\ 0 \end{pmatrix}$$

contains elements related such as

$$r''_j = \frac{M''_{j,21}}{M''_{j,11}} = \frac{E''^-_j}{E''^+_j}$$

Applying (9) gives

$$\begin{pmatrix} E'^+_j \\ E'^-_j \end{pmatrix} = \begin{pmatrix} e^{-i\beta_j} & 0 \\ 0 & e^{i\beta_j} \end{pmatrix} \begin{pmatrix} E''^+_j \\ E''^-_j \end{pmatrix}$$

i.e.,

$$r''_j = \frac{E''^-_j}{E''^+_j} = \frac{E'^-_j e^{-i\beta_j}}{E'^+_j e^{i\beta_j}} = -\frac{1}{r'_j} e^{-2i\beta_j}$$

Divide (13) by $E_j'^+$. Then

$$t_j^+ = \frac{1}{M_{j,11}' + M_{j,12}' r_j'' e^{i2\beta_j}} \tag{14}$$

and by dividing (13) with $E_j'^-$, it is easy to show that

$$t_j^- = t_j^+ r_j'' e^{i2\beta_j} \tag{15}$$

We are now ready for expressing the E-field inside layer j. From $E_j(z) = E_j^+(z) + E_j^-(z)$ we use (11), (12), (14), and (15), and finally expressing all coefficients in terms of matrix elements,

$$E_j(z) = \frac{M_{j,11}'' e^{i\xi_j(z-d_j)} + M_{j,21}'' e^{i\xi_j(d_j-z)}}{M_{j,11}' M_{j,11}'' e^{-i\xi_j d_j} + M_{j,12}' M_{j,11}'' e^{i\xi_j d_j}} E_0^+ \tag{16}$$

where z belongs to layer j.

Based on the Poynting theorem,[26-28] that can be interpreted as a statement of conservation of energy,

$$Q(z) = \langle -\nabla \cdot S \rangle \tag{17}$$

where Q is the (time average) energy flow dissipation per time unit at the point z, $\langle\ \rangle$ indicates time average and S is the Poynting vector,

$$S = E \times H$$

Because time averaging of a product of $\tilde{A} = A_0 e^{i(kx-wt)}$ and $B = B_0 e^{i(kx-wt)}$ can[28] be performed as $\frac{1}{2}\text{Re}(\tilde{A}\tilde{B}^*)$ and

$$-\nabla \cdot S = \left(H \frac{\partial B}{\partial t} + E \frac{\partial D}{\partial t} \right)$$

using Maxwell equations, one can show that

$$Q(z) = \frac{1}{2} c\varepsilon_0 \alpha n |E(z)|^2 \tag{18}$$

Q has the unit $W/(m^2\,nm)$ where it is indicated that z, the distance is measured in nm. c is speed of light, $3.00 \times 10^8\,m/s$, ε_0 is permittivity of vacuum, $8.85 \times 10^{-12}\,F/m$, n is the real index of refraction, α is the absorption coefficient, $\alpha = 4\pi k/\lambda$ with $\tilde{n} = n + ik$ and λ is the vacuum wavelength; and $E(z)$ is the total electrical optical field at the point z. All parameters are for the layer under consideration. The factor $1/2$ is due to averaging of the rapid frequency variation, ca. $10^{14}\,Hz$, in the optical field. Q is a discontinuous function of z, and Q at the points corresponding to the interfaces is not well defined.

When comparing our calculations with experiments, we will use the accumulated dissipation Q from an interval of z,

$$\int_{z\in \text{interval}} Q(z) \mathrm{d}z.$$

These Qs will be indexed as Q_{zone} and $Q_{\text{zonedouble}}$ as explained later.

For a given layer j in the stack, the absorptance A_j is related to Q as

$$A_j = \frac{1}{S_0} \int\limits_{z \in \text{layer } j} Q(z)\mathrm{d}z \tag{19}$$

where S_0 is the irradiance from air for a given wavelength.

5.4.3. Taking Into Account the Substrate

$E(z)$ is dependent on the incoming field to the stack. This is not the same as that coming to the glass substrate from air, but is moderated due to reflection at the air–glass interface and multiple reflection at glass–air interface (Figure 5.5). Light from the left is entering with a general angle of incidence of θ. Reflection occurs in the air–glass interface with some of the light reflected and some transmitted to the stack. Another division between reflected and transmitted light takes place at this interface. The reflected part goes back towards the glass–air interface, where some part of the reflected light re-enters into air and adds with the light reflected only once and some part of it turns back to the stack, and so on.

Using the Poynting vector, the irradiance in air B is

$$B = \frac{1}{2} c \varepsilon_0 |E_0|^2$$

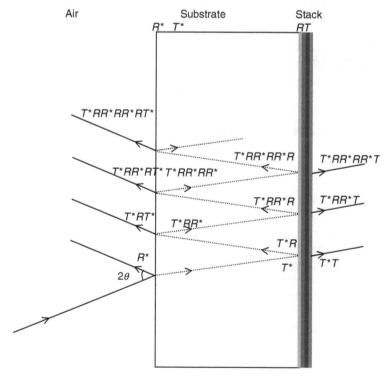

Figure 5.5. Optics in the structure. R represents reflectances and T transmittances. Star (*)-marked quantities refer to the glass substrate, unmarked to the stack. The derivation was done generally keeping a finite T although it is later assumed that T is zero due to thick enough aluminium. The angle if incidence is θ. Normal incidence is used throughout the study.

For an optical description of the diode, we cut the device into two optical elements — the *stack* consisting of the multilayer of thin films, indium tin oxide, ITO, poly(3,4-ethylene dioxythiophene)poly(4-styrenesulfonate), polymer blend, and the other layers, and the *substrate*, which in our case is glass. Typically the thicknesses for the stack layers are less than the irradiated wavelength ($\approx$0.5 µm), whereas the distance between the stack and the air–substrate interface is of the order of millimeters. This makes it necessary to add light coherently inside the stack and incoherently, using irradiances, when the substrate is involved.

Let the air–glass interface be characterized by a reflectance R^*. The substrate has a transmittance T^*. The reflectance and transmittance for the stack are R and T, respectively. All these quantities can be calculated from the complex indices of refraction of the materials and the different layer thicknesses through Fresnel equations. For normal incidence from a material i to j, the reflectance is

$$\text{reflectance} = \frac{(n_j - n_i)^2 + (k_j + k_i)^2}{(n_j + n_i)^2 + (k_j + k_i)^2} \tag{20}$$

The total transmittance for the whole structure (Figure 5.5) is calculated by adding energy (irradiance) quantities. Then

$$T_{\text{tot}} = T^* \left(\sum_{i=0}^{\infty} (RR^*)^i \right) T = T^* \frac{1}{1 - RR^*} T = \frac{T^*T}{1 - RR^*}$$

For the total reflectance,

$$R_{\text{tot}} = R^* + T^* \left(\sum_{i=0}^{\infty} (R^*R)^i \right) RT^* = R^* + T^* \frac{1}{1 - RR^*} RT^*$$

Explicitly including absence of absorption in the substrate i.e., $1 = R^* + T^*$ we reach

$$T_{\text{tot}} = \frac{(1 - R^*)T}{1 - RR^*} \tag{21}$$

and

$$R_{\text{tot}} = R^* + \frac{(1 - R^*)(1 - R^*)}{1 - RR^*} R = \frac{R^* + R - 2R^*R}{1 - RR^*} \tag{22}$$

The irradiance to the stack is

$$T \left(\sum_{i=0}^{\infty} (R * R)^i \right) B = T * \frac{1}{1 - RR*} B = \frac{1 - R*}{1 - RR*} B$$

Writing this as $\frac{1}{2} c\varepsilon_0 n_g |E_{o,\,g}|^2$ (where the subscript g indicates a quantity in glass), one reaches

$$|E_{o,g}|^2 = \frac{2(1 - R^*)B}{\varepsilon_0 c n_g (1 - RR^*)}$$

This is the square modulus amplitude that has to be used as input to the Q calculation. Compared to a stack in air the expression is modified with the factor

$$\frac{(1 - R^*)}{n_g (1 - RR^*)}.$$

Under the assumption of zero stack transmittance, fully transmitting glass and no scattering effects,

$$1 = R_{\text{tot}} + \sum_j A_j.$$

5.4.4. Solar Spectrum

A solar cell is to be exposed to polychromatic solar light rather than monochromatic. The simplest approximation is perhaps to treat the sun as a blackbody radiator. Modulating the peak position to 515 nm and the peak value ($1.5 \, \text{W/m}^2$) to resemble Air Mass 1.5 standard[29] (AM1.5) we introduce the quantity B proportional to the Planck expression

$$B \propto \frac{2hc^2}{\lambda^5} \frac{1}{e^{hc/\lambda kT} - 1} \tag{23}$$

with $T = 5630 \, \text{K}$ (equivalent solar blackbody temperature) and the Planck constant $h = 6.63 \times 10^{-34}$ Js. λ is the vacuum wavelength.

We will consider the wavelength interval 300 to 800 nm. The total amount of irradiated power is

$$S_0 = \int\limits_{\lambda=300\,\text{nm}}^{800} B(\lambda)\mathrm{d}\lambda = 632 \, \text{W/m}^2 \tag{24}$$

with our AM1.5 approximation. As a comparison, taking a more extended range,

$$S_0 = \int\limits_{\lambda=300\,\text{nm}}^{2500} B(\lambda)\mathrm{d}\lambda$$

this value would be $1131 \, \text{W/m}^2$, closer to the standard,[29] $952 \, \text{W/m}^2$.

5.4.5. Efficiencies

As absorptances are additive $A = \sum_{\text{all}j} A_j$ where we single out A_{a}, the absorptance in the active layer with the definition

$$\eta_{A_{\text{a}}} = A_{\text{a}} = \frac{\text{number of photons absorbed in active layer}}{\text{number of incoming photons to the structure}} \tag{25}$$

The external quantum efficiency η_{EQE} is experimentally determined, and is commonly defined as the number of electrons generated in the photovoltaic device (at short circuit) divided by the number of incoming photons to the structure. It is measured for monochromatic light. Of the seven processes mentioned in the introduction, some processes could be attributed with certain efficiency. We choose to single out only η_{A}, the ratio between the number of absorbed photons in the device and the number of incoming photons, but amalgamate the processes 5.2.3 to 5.2.7 in a geometry, topology, and interface formation dependent function. This function is identified with η_{IQE}, the internal quantum efficiency, a quantity cleared from reflection:

$$\eta_{\text{IQE}} = \frac{\text{number of electrons in an outer circuit}}{\text{number of photons absorbed}} \tag{26}$$

Then

$$\eta_{\text{EQE}} = \eta_{\text{IQE}} \eta_A \qquad (27)$$

or

$$\eta_{\text{IQE}} = \frac{\eta_{\text{EQE}}}{\eta_A} \qquad (28)$$

As the number of photons absorbed is a lower number than the number of incident photons $\eta_{\text{IQE}} \geq \eta_{\text{EQE}}$. Thus η_A is an upper limit for η_{EQE}. We identify η_A with A.

From $1 = R_{\text{tot}} + A + T_{\text{tot}}$ using additivity of A and assuming $T_{\text{tot}} = 0$ one gets $1 = R_{\text{tot}} + A_w + A_a$ from which η_{aap}, the active layer absorption part, is defined: $\eta_{\text{aap}} = A_a/(A_w + A_a) = A_a/A$. A_w is absorptance in parts of the device that do not generate photocurrent (electrodes, buffer layers). A measure of how efficient a certain layer is to transform a photon to charge carrier is $\eta_{\text{IQE}j}$, the layerwise internal quantum efficiency (in principle we only consider the active one(s)):

$$\eta_{\text{IQE},j} = \frac{\text{number of electrons in an outer circuit coming from } j}{\text{number of photons absorbed in } j} \qquad (29)$$

From the very definition of *active layer*, the numerator is the same as that of (26) when $j = a$. The denominator is A_a times the number of incoming photons to the structure giving

$$\eta_{\text{IQE},a} = \frac{\eta_A \eta_{\text{IQE}}}{A_a} = \frac{\eta_{\text{IQE}}}{\eta_{\text{aap}}} \qquad (30)$$

One sees that η_{aap} is an upper limit for η_{IQE}. Because A_a is less than $A = \eta_A$, $\eta_{\text{IQE}a} \geq \eta_{\text{IQE}}$.

Finally, the optical power efficiency, η_P, is a energy measure more relevant to device performance, and possible to define both for monochromatic and polychromatic illumination, but used here for the case of simulated solar light:

$$\eta_P = \frac{\text{absorbed optical power}}{\text{irradiated optical power}} \qquad (31)$$

5.5. SIMULATIONS AND RESULTS

The theory is now to be used for analyzing photodevices. Three types are to be encountered: (1) the cavity, (2) the double layer structure, and (3) the blend layer structure (Figure 5.6).

1. The cavity is the simplest structure consisting only of four layers: a substrate, a semitransparent electrode material, a polymer, and a mirror of aluminum. For this structure we just treat the *E*-field inside.
2. The double layer structure is a photodiode consisting of a polymer, in this case a copolymer containing polyfluorene. The class of polyfluorenes has evoked some interest as the active layer in recent time.[30–32] As the name indicates (*fluor*enes giving *fluor*escence), they were first noticed for their eminent luminescent properties[33,34] and have found an extensive use in light-emitting diodes. The copolymer is poly(2,7-(9,9-dioctyl-fluorene)-*alt*-5,5-(4′,7′-di-2-thienyl-2′,1′,3′-benzothiadiazole))[DiO-PFDTBT]. For a review, see Chapter 17 (Reference 35). For enhanced

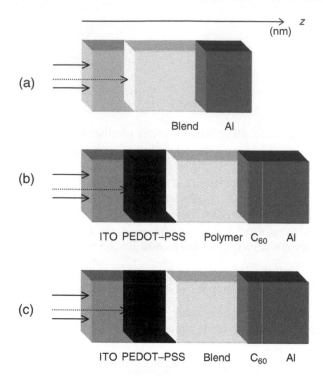

Figure 5.6. The structures assumed for the simulation. Interfaces are assumed to be sharp and layers to be homogeneous. Note that z is calculated from the point where the glass meets the thin film layers and have the unit nm. The glass substrate is many times thicker than any layer in the stack. (a) The cavity consist of only four layers; a substrate (not shown), a semitransparent mirror material, a polymer, and a mirror of aluminum. (b) The double layer structure (d_{ITO}, $d_{PEDOT-PSS}$, $d_{polymer}$, $d_{C_{60}}$) nm. (c) The blend layer structure (d_{ITO}, $d_{PEDOT-PSS}$, d_{blend}, $d_{C_{60}}$) nm.

performance, an acceptor-acting layer of C_{60} is added close to the active one. This has the double purpose of both transporting the electrons and, at the interface towards the polymer, providing sites for exciton dissociation. Optical simulation of the double layer structure is presented in Reference 12.

3. The blend layer structure has a blend of DiO-PFDTBT with the fullerene derivative PCBM in the weight proportions 1:4 as the central layer. Added to this is a pure fullerene layer. The blend layer structure and its optical simulation is found in Reference 36.

Chemical formulas of the substances are shown in Figure 5.7.

Double layer and blend layer photodevices have different extensions for their diffusion zones. In the double layer structure only a part close to the polymer–C_{60} interface constitutes the diffusion zone. Typically the exciton diffusion length is 10 nm[7,37,38] from which we created a diffusion zone of 10 nm into the polymer counted from the interface but also included 10 nm in the C_{60} layer.

For the blend layer structure, it is assumed that the whole of the blend layer contributes to the photocurrent. This layer is where the bulk heterojunction is

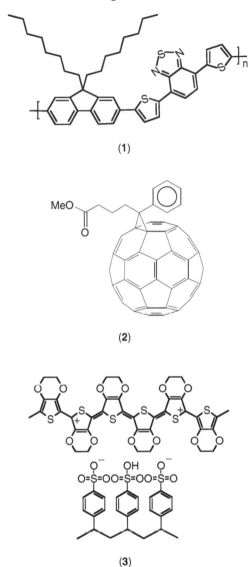

(1)

(2)

(3)

Figure 5.7. Chemical formulas for the constituents in the structures: (**1**) is poly(2,7-(9,9-dioctyl-fluorene)-*alt*-5,5-(4′,7′-di-2-thienyl-2′,1′,3′-benzothiadiazole)) [DiO-PFDTBT], an alternating polyfluorene copolymer, (**2**) is the fullerene derivative [6,6]-phenyl-C_{61}-butyric acid methyl ester (PCBM), (**3**) is poly(3,4-ethylenedioxythiphene)–poly(4-styrenesulfonate) [PEDOT–PSS].

formed, and we focus our attention on this. But we also include the possibility of a layer of C_{60} on top of the blend in some of these calculations.

The criterion for optimization is then to maximize the energy absorption in the diffusion zone.

The input to the calculation is

Dielectric functions for the layers
Layer thicknesses
Diffusion length.

where the dielectric functions for the DiO-PFDTBT is extracted by spectroscopic ellipsometry[12] and is found in Figure 5.8. The same method was used for deriving the $\tilde{\varepsilon}$ for the blend[36] (Figure 5.9). Other material data were taken from previous measurements[12,39] or literature.[40,41] The properties of ITO can vary depending on the manufacturing protocol. We used values previously measured in our group which also agreed with literature findings.[41]

We start by discussing the monochromatic case and then turn to polychromatic irradiation.

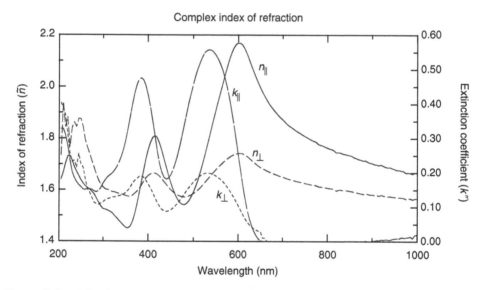

Figure 5.8. $\tilde{n}$ for (**1**) as extracted from spectroscopic ellipsometry. Note that the material is anisotropic. (From N.-K. Persson, M. Schubert, and O. Inganäs. *Solar Energy Materials and Solar Cells*, **83**(2–3), 169–186, 2004. With permission.)

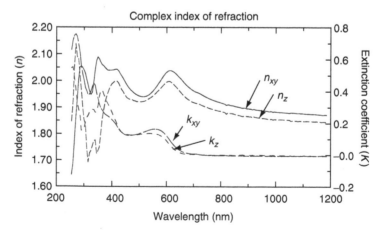

Figure 5.9. $n(\lambda)$ and $k(\lambda)$ for the blend of 4:1 weight ratio PCBM and the copolymer DiO-PFDTBT from ellipsometric investigation in the wavelength interval 240 to 1200 nm. Note that the material is anisotropic. *xy* is parallel to the surface plane (ordinary direction) and *z* is normal to the sample plane (extraordinary).

5.5.1. Simulation of the Optical Electric Field Inside the Device

It is appropriate to start with discussion of monochromatic irradiation. The first case to study is the optic electric field distribution inside a simple device, in this case a cavity (Figure 5.10). Normal incidence is assumed.

$|E(z)|^2$ is continuous at the layer interfaces. The high wavelengths 700 and 800 nm create a standing wave behavior as the polymer is not absorbing. The others show different behavior. What is common is the marked undulation. Reflection and interference are indeed marked features. We note that it is possible to choose the illumination wavelength in order to focus the E-field to certain positions inside the film.

5.5.2. Q-Profile for Different Wavelengths

Calculating the distribution of the E-field using the formalism is only the first step. The time-averaged energy flow dissipation per time unit at the point z is a more interesting quantity as it describes the absorption, and acts like a source term for the exciton production. For the case of structure c, the spatially resolved absorption profile, $Q(z)$, for a certain structure (d_{ITO}, $d_{PEDOT-PSS}$, d_{blend}, $d_{C_{60}}$) is shown in Figure 5.11 for three different wavelengths. Discontinuity at the layer interfaces is seen caused by the sharp change of α and n when going between the materials. The profiles are very different depending on the wavelength. For $\lambda = 300$ nm, the behavior in the blend layer ($z = 140$ to 350 nm) is almost Beer–Lambert-like with only a small superimposed undulation. This is due to the high absorption coefficient. For $\lambda = 550$ nm, the situation is very different.

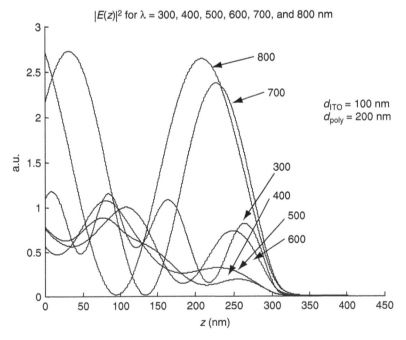

Figure 5.10. The cavity. The distribution of square modulus of electric field for different wavelengths. The geometry is $d_{ITO} = 100$ nm, $d_{polymer} = 200$ nm and aluminum.

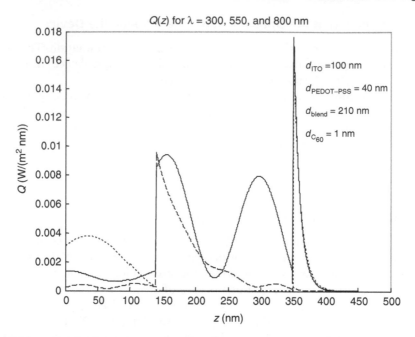

Figure 5.11. $Q(z)$ for the geometry $d_{ITO} = 100$ nm, $d_{PEDOT\text{-}PSS} = 40$ nm, $d_{blend} = 210$ nm, and $d_{C_{60}} = 1$ nm. Irradiated wavelengths are 300 nm, (---) 550 nm (——), and 800 nm (······). As follows from the optimization part, this set corresponds to the optimal geometry for structure c.

Now reflectance in the aluminum and interference with a standing wave behavior is established. Two major peaks inside the blend are seen at $z = 160$ and 300 nm. A small amount of photons are absorbed in the middle of the blend film. $\lambda = 800$ nm is not absorbed at all in the blend. In all three cases, the field is quickly (within 55 nm) diminished in aluminum.

5.5.3. Q-Profile for Different Thicknesses, Monochromatic Illumination

In Figure 5.12, the calculations in 5.5.2 are repeated with only a change of d_{blend} from 210 to 100 nm. The profiles are different from each other but also different compared with profiles shown in Figure 5.11. The blend layer is now so thin that the Beer–Lambert behavior for $\lambda = 300$ nm is lost and $\lambda = 550$ nm now gives a peak in the middle of the blend layer.

5.5.4. Polychromatic Q-Profile

From monochromatic irradiation, the next step is irradiation with polychromatic light. In Figure 5.13, $Q(z)$ is summed over all wavelengths in the interval from 300 to 800 nm. An undulating curve with two hills is seen. This may be compared with the profile for a different blend thickness (Figure 5.14), with the major part of absorption in the middle of the blend layer. The Q distribution is indeed dependent on the geometry. In both cases, the density of excited states is not constant throughout the blend layer.

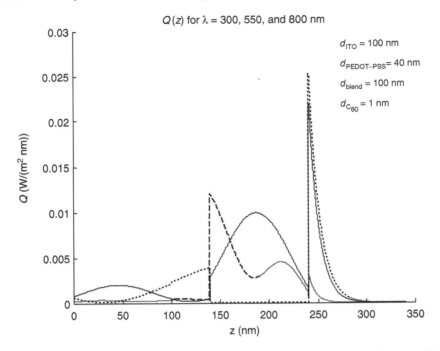

Figure 5.12. $Q(z)$ for the geometry $d_{ITO} = 100\,\text{nm}$, $d_{PEDOT\text{-}PSS} = 40\,\text{nm}$, $d_{blend} = 100\,\text{nm}$ and $d_{C_{60}} = 1\,\text{nm}$. Irradiated wavelengths are 300 nm, (- - -) 550 nm (——) and 800 nm (· · · · · ·) for structure 3.

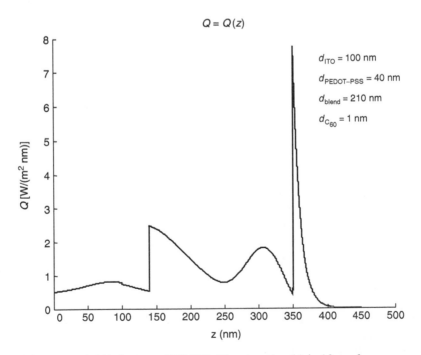

Figure 5.13. Sum of $Q(z)$ for $\lambda = [300,800]$. The structure (c) is $(d_{ITO}, d_{PEDOT\text{-}PSS}, d_{blend}, d_{C_{60}}) = (100, 40, 210, 1)$ nm.

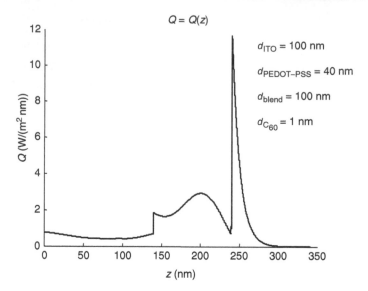

Figure 5.14. Sum of $Q(z)$ for $\lambda = [300,800]$. Structure c. The structure is (d_{ITO}, $d_{PEDOT-PSS}$, d_{blend}, $d_{C_{60}}$) = (100, 40, 100, 1) nm.

5.5.5. Device Optimization

The criterion for optimization is to maximize the energy absorption in the diffusion zone. We can then formulate functional relationships such as (suppressing n-dependence)

$$Q_{zone} = Q_{zone}(d_{polymer \ or \ blend}, d_{C_{60}}, d_{PEDOT-PSS}, d_{ITO}) \qquad (32)$$

with a diffusion zone in the polymer or blend only and $Q_{zonedouble}$ with an extended zone into C_{60}.

As the diffusion zone in the double layer and the blend layer structures are different, we discuss them separately.

5.5.5.1. Optimizing the Double Layer Structure

As Q_{zone} is a function of many variables, only projections are possible to illustrate, as done in Figure 5.15. In this figure, Q_{zone} as a function of $d_{polymer}$ and $d_{C_{60}}$ for a certain choice of PEDOT–PSS and ITO thicknesses is shown. The ranges of thicknesses are broad, giving us the necessary overview in order to try to find the global maximum. The pictures also illustrate the modest complexity of Q_{zone}.

Details of the optimization of structure 2 can be found elsewhere.[12] In short, an area of high Q_{zone} values can be identified. With respect to the C_{60}-layer, this occurs for values around 50 nm. With respect to the PEDOT–PSS-layer, values around 100 nm are optimal, which agrees with the value used experimentally.[42] The best ITO-thicknesses are those in the interval 70 to 120 nm.

We note that our assumption of a definite 10-nm dissociation zone gives the result that the polymer layer should be as thin as possible. In fact, only counting absorption in the polymer as beneficial and not including parts of PEDOT–PSS in the dissociation zone shows that Q_{zone} decreases if $d_{polymer}$ is less than 10 nm (Figure 5.15). Thus a suitable polymer thickness corresponds to the dissociation zone width. The very value of 10 nm is just an approximate

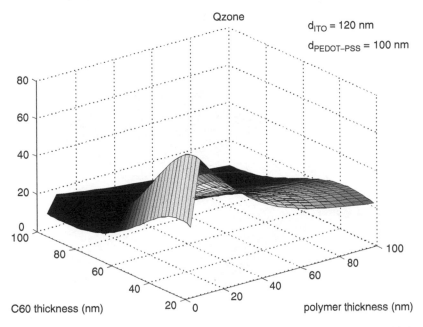

Figure 5.15. Q_{zone} as a function of $d_{polymer}$ and $d_{C_{60}}$ for structure b. The thickness of PEDOT–PSS is 100 nm and that of ITO 120 nm. Arbitrary values on the ordinate. (From N.-K. Persson, M. Schubert, and O. Inganäs. *Solar Energy Materials and Solar Cells*, **83**(2–3): 169–186 (2004). With permission.)

estimate taken from the literature and should not be given too much emphasis. Other values could have been applied or a probability distribution with a tail further down in the polymer been implemented. However, our simulations point towards the conclusion that extra polymeric material in excess of filling the dissociation zone do not tune the optical electric field into something beneficial for the performance.

To summarize; one example of a thickness combination of the materials studied that gives a high Q_{zone} for structure b is ITO (120 nm), PEDOT–PSS (100 nm), polymer equal to the exciton diffusion length, and C_{60} (50 nm). Of course, this set of optimal values comes from optimization restricted to the optical processes only.

5.5.5.2. Optimizing the Blend Layer Structure

If we now turn to the blend layer structure we take, as was said above, the diffusion zone, Z, to be the whole of the blend layer and 10 nm into C_{60}. The absorbed power per unit area, the integrated Q, which we here call $Q_{zonedouble} = \int_{x \in Z} Q(z)dz$ is again calculated for each structure. Optimization is done by varying the thickness of the layers in the range $d_{ITO} \in \{70,110,150\}$, $d_{PEDOT-PSS} \in \{40,60,\cdots,120\}$, $d_{blend} \in \{5,20,\cdots,290\}$, $d_{C_{60}} \in \{70,110,150\}$. In Figure 5.16, $Q_{zonedouble}$ is presented as a function of d_{blend} and $d_{C_{60}}$ for two fixed values of d_{ITO} and $d_{PEDOT-PSS}$. Note that energies here are given in terms of absolute values (W/m^2). The overall trend is that $Q_{zonedouble}$ is increasing for increasing d_{blend} and $d_{C_{60}}$. But some small features are seen on the surface. A faint local maximum of approximately 240 W/m^2 exists around

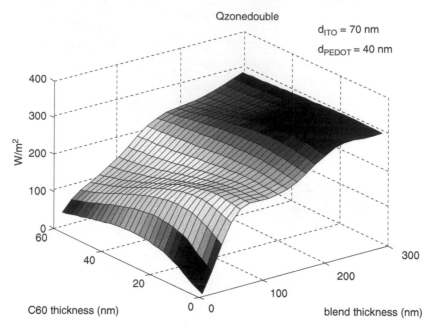

Figure 5.16. (**Color figure follows page 348**) Absorbed power per unit area, i.e., $\int Q(z)dz$ as a function of $d_{blend} \in [5,290]$ nm and $d_{C_{60}} \in [0,58]$ nm for a zone comprising all of the blend layer, and extending $\min(d_{C_{60}}, 10)$ nm into the C_{60} layer. d_{ITO} and $d_{PEDOT-PSS}$ are fixed 70 and 40 nm, respectively.

$d_{blend} = 60$ nm and $d_{C_{60}} = 10$ nm. And an even less distinct maximum somewhat above 300 W/m^2 is found around $(d_{blend}\ d_{C_{60}}) = (250, 10)$ nm. For more pronounced blend thicknesses, $Q_{zonedouble}$ is more or less insensitive to changes in $d_{C_{60}}$. This means that the contribution from absorption in C_{60} is negligible compared to that in the blend because of the large differences in thicknesses. But this does not mean that the *E*-field is weak in C_{60} for the same size of $d_{C_{60}}$ and d_{blend}.

In principle, it is immaterial if the diffusion zone also comprises C_{60}. The curve for this case is very similar to the curve for large d_{blend}, as seen in Figure 5.16.

For high d_{blend} and $d_{C_{60}}$ values, $Q_{zonedouble}$ seems to saturate. This is confirmed by extending (not shown) the d_{blend} interval studied. $Q_{zonedouble}$ is around 370 W/m^2 for $d_{blend} = 1000$ nm.

A variation in the ITO thickness has a low impact on $Q_{zonedouble}$. Surfaces are almost overlapping for $d_{ITO} = 70$, 110, and 150 nm. Also for variation in PEDOT–PSS thickness, the impact on $Q_{zonedouble}$ is relatively low. Again surfaces as shown in Figure 5.16 for $d_{PEDOT-PSS} \in \{40,60,80,100,120\}$ nm have very similar forms. Yet, compared to the ITO case, there are larger gaps in between the surfaces. The conclusion is that $Q_{zonedouble}$ is more sensitive to the PEDOT–PSS layer, which is closer to the blend.

Thus, some conclusions can be drawn about the optimal geometry for structure c. As the dependence of $Q_{zonedouble}$ on ITO thickness is low, it is possible to take $d_{ITO} = 100$ nm because this is in accordance with the value that manufacturers state as the nominal value for the glass–ITO substrates. The PEDOT–PSS layer should be thin; therefore $d_{PEDOT-PSS} = 40$ nm is chosen. The blend layer should be as thick as possible as far as the optical optimization states. $d_{blend} = 210$ nm is a balance between this and the finite charge mobility. As the C_{60} layer is unimportant, $d_{C_{60}} = 1$ nm is used for blend layer of this thickness.

5.5.6. Sensitivity Analysis

One of the important evaluation tools of a model is an analysis studying the sensitivity to variations in parameters.[43] This is important for understanding the numerics but could also generate useful results. For structure b, Figure 5.17 shows the curvature in Q_{zone} and also the sensitivity for variation in thickness of the C_{60}-layer. This is valuable information for the device maker. For the structure as in the legend, the maximum value is given by a C_{60} thickness of around 30 nm. A 10 nm deviation from this C_{60} thickness renders 9% change in the ordinate.

5.5.7. Quantum Efficiency

Our model also allows us to reach estimates of quantum efficiencies. It follows from Equation (28) that η_A is an upper limit for the external quantum efficiency, η_{EQE}, which could be experimentally determined. The efficiency of the active layer to absorb the irradiated photons is η_{Aa}. η_{aap} is an upper limit for the internal quantum efficiency. All measures are shown in Figure 5.18 for a particular choice of thickness that most closely resembles the form of the experimental curve for structure c. $\eta_A \geq \eta_{Aa}$. For the optimal thickness combination with $d_{blend} = 210$ nm, η_A is much higher, above 90% for many wavelengths whereas EQE values can be around 50%. Obviously there is a potential for enhancing device performance if more of the absorbed photons can be utilized.

5.5.7.1. Optical Power Efficiency

As we are working with quantities measured in real units, we can also calculate how much of the total incoming light in the interval 300 to 800 nm is absorbed, a distant

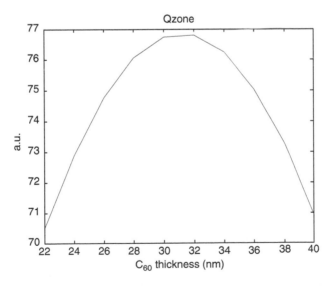

Figure 5.17. The sensitivity in Q_{zone} for variation in C_{60}-thickness for structure 2. The layer thicknesses are air/glass/ITO(90 nm)/PEDOT–PSS(90 nm)/polymer(10 nm)/C_{60}/Al. (From N.-K. Persson, M. Schubert, and O. Inganäs. *Solar Energy Materials and Solar Cells,* **83**(2–3): 169–186 (2004). With permission.)

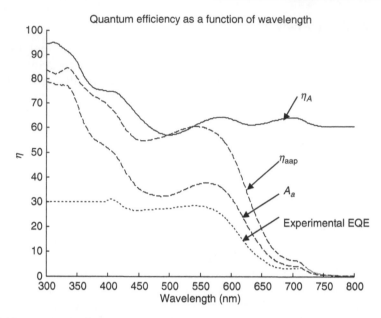

Figure 5.18. η_A (unbroken line), η_{Aa} (dashed line), η_{aap} (dashed-dotted line), and η_{EQE} (dotted) as functions of wavelength. η_A is an upper limit for η_{EQE}. η_{EQE} is experimentally determined.[35] It was not measured below 400 nm. We were aiming at (d_{ITO}, $d_{PEDOT-PSS}$, d_{blend}, $d_{C_{60}}$) = (100, 90, 160, 1) nm for the experimental device. However, we find that simulations best agree in shape for the set (100, 90, 100, 1) nm, which are the shown simulation curves.

upper limit of the power conversion coefficient. The sum of the absorbed energy in the blend layer for $\lambda = [300,800]$ nm is 303 W/m^2. Comparing with Equation (24), one gets $\eta_P = 0.48$.

5.5.8. Energy Redistribution

From a purely energetic perspective, the solar cell is a device capable of redistributing energy spatially and directionally. Radiation energy goes to radiation energy by reflection, with a change of energy flow. It is also converted to heat and other non-useful forms and finally, however beyond the present treatment, to electrical energy via the process of transport of charge. A mapping of all this is found in Figure 5.19 for structure 3. All the way up to where the polymer absorption ends, around 620 nm, most of the incoming energy is absorbed in the blend layer. Absorption in the fullerene component in the blend extends the tail to 720 nm. Reflection is moderate in this region, except for a peak at 450 nm, where it reaches 30%, coinciding with a dip in the absorption of the blend. The general view is that the polymer and its $k(\lambda)$ dependence manifest itself in the energy redistribution behavior of the whole component. $k(\lambda)$ for the blend layer was seen in Figure 5.9. Reflection losses are growing when the absorption of the blend is ceasing. Disadvantageous absorption in ITO and PEDOT–PSS is low for lower wavelengths, but increases in the band gap (longer wavelengths) of the blend. Noteworthy is the great part of absorption taken by the ITO above 650 nm. It can be explained to be due to Beer–Lambert absorption in the material that is foremost to the irradiation, but might also be due to an

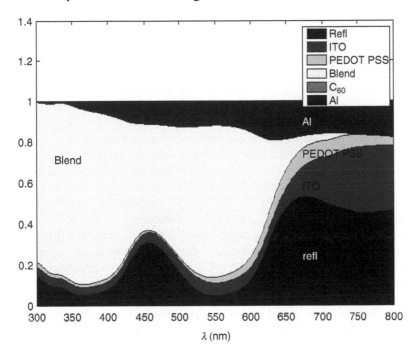

Figure 5.19. (Color figure follows page 348) Redistribution of the incoming irradiation on reflection and layerwise absorption for structure 3. The structure is $(d_{ITO}, d_{PEDOT\ PSS}, d_{blend}, d_{C_{60}}) = (100, 40, 210, 1)$ nm. As the C_{60} is very thin, it is not seen in the diagram. The curve is free from differences due to different number of irradiated photons.

overestimated α_{ITO}. Aluminium is increasing its share with increasing wavelength. It is as high as 18% for 620 nm.

5.6. SUMMARY

Optical modeling is important in two aspects — it gives insight into properties that are out of reach for direct measurements such as the $|E|^2$ profile, and provides guidance for the device manufacturer. Besides, in a couple of hours on an ordinary Macintosh or PC, the computer runs through thousands of different device geometries. This is far beyond what can be performed experimentally.

By computer-based simulation, it has been possible to examine the optical processes of the PPVD. The modeling has been based on a number of assumptions such as homogeneous layers, sharp and planar interfaces, scattering free optics, coherence in the stack part of the device, and incoherence when adding energies across the substrate. The model, which has layer thickness, layer dielectrical functions, and exciton diffusion length as input, allows us to optimize the performance by examining a large number of geometries, i.e., a set of thicknesses. The criterion has been to reach as high energy absorption in the exciton diffusion zone layer as possible.

In this chapter, we have reported on modeling of two kinds of photovoltaic devices having a layered geometry, both with a polyfluorene–copolymer as the active material and C_{60} as the acceptor. Structure b consisted of a pure polymer layer and a pure C_{60} layer. Structure c had a blended polymer–C_{60} derivative layer in the weight ratio 1:4 with an added pure C_{60} layer.

It is assumed that due to the proximity of polymer and fullerene, the whole of the blend layer, and 10 nm in the C_{60} contribute to the photocurrent. It is also assumed that every point in the blend and in the relevant part of C_{60} contribute equally well, according to the square modulus field distribution and α and n. For structure b, a diffusion zone is defined, expanding 10 nm into the polymer and 10 nm into the C_{60} counted from the interface between these materials.

Thin film theory in a matrix formalism enables the extraction of the impact of reflection and interference on the optical electric field. $|E|^2$ as a function of depth into the device is presented. The model lets us predict, for example, an optimal C_{60}-thickness when light is assumed to be both polychromatic and distributed following solar irradiation. The value reached is 50 nm for structure 2 and 0 nm for structure 3. It is also possible to see the sensitivity for the variation from this value. The curves, such as that in Figure 5.17, are relatively flat making the diode tolerant towards mistakes in the process step of adding the C_{60} layer. In the same manner, the other layers can be analyzed.

For structure c, having only optical constraints, i.e., maximizing absorption in the active layer, the result is that the thicker the blend layer the more the absorption. This is not enough for finding an upper limit of d_{blend}. Therefore, electrical modeling is also necessary for the overall maximum performance.

A good illustration of the potential of simulation is the energy redistribution diagram that summarizes all the layers' share of absorbed energy. For wavelengths up to where the polymer ends absorbing, the major part is indeed taken by the blend layer. The total structure related reflectance is calculated. For some wavelength intervals it is very low, 5% for $\lambda = 550$ nm. It is in many parts of the spectrum below 10%, but reaches above 50% in the high-wavelength range where the copolymer has ceased to absorb.

REFERENCES

1. K. Hartt. Mathematical modeling. In *Encyclopedia of Applied Physics*. VCH Publishers, 1994, pp. 417–420.
2. B. E. Gillett. *Introduction to Operations Research*. McGraw-Hill, 1976.
3. D. T. Philips, A. Ravindran, and J. J. Solberg. *Operations Research: Principles and Practice*. John Wiley & Sons, 1976.
4. C. Heine and R. Morf. Submicrometer gratings for solar energy applications. *Applied Optics* **34**(14):2476–2482 (1995).
5. L. S. Roman, O. Inganäs, T. Granlund , et al. Trapping light in polymer photodiodes with soft embossed gratings. *Advanced Materials* **12**(3):189–195 (2000).
6. J. Nelson. Organic photovoltaic films. *Current Opinion in Solid State & Materials Science* **6**(1):87–95 (2002).
7. P. Peumans, A. Yakimov, and S. R. Forrest. Small molecular weight organic thin-film photodetectors and solar cells. *Journal of Applied Physics* **93**(7):3693–3723 (2003).
8. A. K. Ghosh, D. L. Morel, T. Feng, R. F. Shaw, and C. A. Rowe Jr. Photovoltaic and rectification properties of Al/Mg phthalocyanine/Ag Schottky-barrier cells. *Journal of Applied Physics* **45**(1):230–236 (1974).
9. M. G. Harrison, J. Grüner, and G. C. W. Spencer. Analysis of the photocurrent action spectra of MEH-PPV polymer photodiodes. *Physical Review B (Condensed Matter)* **55**(12):7831–7849 (1997).

10. A. K. Ghosh and T. Feng. Cyanine organic solar cells. *Journal of Applied Physics* **49**(12):5982–5989 (1978).

11. L. A. A. Pettersson, L. S. Roman, and O. Inganäs. Quantum efficiency of exciton-to-charge generation in organic photovoltaic devices. *Journal of Applied Physics* **89**(10):5564–5569 (2001).

12. N.-K. Persson, M. Schubert, and O. Inganäs. Optical modeling of a layered photovoltaic device with a polyfluorene derivative/fullerene as the active layer. *Solar Energy Materials and Solar Cells* **83**(2–3):169–186 (2004).

13. H. Hoppe, N. Arnold, N. S. Sariciftci, and D. Meissner. Modeling the optical absorption within conjugated polymer/fullerene-based bulk-heterojunction organic solar cells. *Solar Energy Materials and Solar Cells* **80**(1):105–113 (2003).

14. J. Rostalski and D. Meissner. Photocurrent spectroscopy for the investigation of charge carrier generation and transport mechanisms in organic p/n junction solar cells. *Solar Energy Materials and Solar Cells* **63**:37–47 (2000).

15. H. A. MacLeod. *Thin Film Optical Filters*. Institute of Physics Publishing, 2001.

16. L. Vriens and W. Rippens. Optical constants of absorbing thin solid films on a substrate. *Applied Optics* **22**(24):4105–4110 (1983).

17. B. Harbecke. Coherent and incoherent reflection and transmission of multilayer structures. *Applied Physics B* **39**(3):165–170 (1986).

18. E. Elizalde and F. Rueda. On the determination of the optical constants $n(\lambda)$ and $\alpha(\lambda)$ of thin supported films. *Thin Solid Films* **122**(1):45–57 (1984).

19. R. F. Potter. Basic parameters for measuring optical properties. In E. D. Palik (ed.), *Handbook of Optical Constants of Solids I*. Academic Press, 1985, pp. 11–34.

20. E. D. Palik. Errata. In E. D. Palik (ed.), *Handbook of Optical Constants of Solids II*. Academic Press, 1991, pp. 8–11.

21. A. H. M. Holtslag and P. M. L. O. Scholte. Optical measurement of the refractive index, layer thickness, and volume changes of thin films. *Applied Optics* **28**(23):5095–5104 (1989).

22. J. J. M. Halls, K. Pichler, R. H. Friend, S. C. Moratti, and A. B. Holmes. Exciton diffusion and dissociation in a poly(p-phenylenevinylene)/C_{60} heterojunction photovoltaic cell. *Applied Physics Letters* **68**(22):3120–3122 (1996).

23. J. J. M. Halls. Photoconductive Properties of Conjugated Polymers. PhD dissertation, St John's College, Cambridge, UK, 1997.

24. L. A. A. Pettersson, L. S. Roman, and O. Inganäs. Modeling photocurrent action spectra of photovoltaic devices based on organic thin films. *Journal of Applied Physics* **86**(1):487–496 (1999).

25. T. Stubinger and W. Brutting. Exciton diffusion and optical interference in organic donor–acceptor photovoltaic cells. *Journal of Applied Physics* **90**(7):3632–3641 (2001).

26. J. D. Jackson. Electrodynamics, Classical. In *Encyclopedia of Applied Physics*. VCH Publishers.1994, pp. 283–285.

27. W. Greiner. *Classical Electrodynamics*. Springer, 1996.

28. L. D. Landau, E. M.Lifshitz, and L. P. Pitaevskii. *Electrodynamics of Continous Media*. Pergamon Press, 1984.

29. Key Center of Photovoltaic Engineering UNSW, Australia. UNSW Air Mass 1.5 Global Spectrum. www.pv.unsw.edu.au/am1.5.html

30. A. C. Arias, J. D. MacKenzie, R. Stevenson, J. J. M. Halls, M. Inbasekaran, E. P. Woo, D. Richards, and R. H. Friend. Photovoltaic performance and morphology of polyfluorene blends: a combined microscopic and photovoltaic investigation. *Macromolecules* **34**(17):6005–6013 (2001).

31. M. Svensson, F. Zhang, S. Veenstra, W. Verhes, J. C. Hummelen, J. Kroon, O. Inganäs, and M. R. Andersson. High performance polymer solar cells of an alternating polyfluorene copolymer and a fullerene derivative. *Advanced Materials* **15**(12):988–991 (2003).

32. R. Pacios, D. D. C. Bradley, J. Nelson, and C. J. Brabec. Efficient polyfluorene based solar cells. *Synthetic Metals* **137**(1–3):1469–1470 (2003).

33. M. Leclerc. Polyfluorenes: twenty years of progress. *Journal of Polymer Science: Part A: Polymer Science* **39**(17):2867–2873 (2001).
34. A. Meisel, T. Miteva, H. G. Nothofer, W. Knoll, D. Sainova, D. Neher, G. Nelles, A. Yasuda, F. C. Grozema, T. J. Savenije, B. R. Wegewijs, L. D. A. Siebbeles, J. M. Warman, and U. Scherf. 2001. Anisotropy of the optical and electrical properties of highly-oriented polyfluorenes. In *Polytronic 2001, Proceedings*. IEEE New York, pp. 284–290.
35. O. Inganäs, F. Zhang, X. Wang, A. Gadisa, N.-K. Persson, M. Svensson, E. Perzon, W. Mammo, and M. R. Andersson. Alternating fluorene copolymer-fullerene blend solar cells. In S.-S. Sun and N. S. Sariciftci (eds), *Organic Photovoltaics: Mechanisms, Materials and Devices*, CRC Press, Boca Raton, FL, USA, 2005, pp. 389–404.
36. N.-K. Persson, H. Arwin, and O. Inganäs. Optical optimization polyfluorene–fullerene blend photodiodes. Submitted for publication.
37. A. Haugeneder, M. Neges, C. Kallinger, W. Spirkl, U. Lemmer, J. Feldmann, U. Scherf, E. Harth, A. Gugel, and K. Mullen. Exciton diffusion and dissociation in conjugated polymer fullerene blends and heterostructures. *Physical Review B (Condensed Matter)* **59**(23):15346–15351 (1999).
38. M. Theander, A. Yartsev, D. Zigmantas, V. Sundström, W. Mammo, M. R. Andersson, and O. Inganäs. Photoluminescence quenching at a polythiophene/C60 heterojunction. *Physical Review B (Condensed Matter)* **61**(19):12957–12963 (2000).
39. L. A. A. Pettersson, F. Carlsson, O. Inganäs, and H. Arwin. Spectroscopic ellipsometry studies of the optical properties of doped poly(3,4-ethylenedioxythiophene); an isotropic metal. *Thin Solid Films* **313–314**:356–361 (1998).
40. D. Y. Smith. The optical properties of metallic aluminium. In *Handbook of Optical Constants of Solids*. Academic Press, 1998.
41. J. Bartella, J. Schroeder, and K. Witting. Characterization of ITO- and TiO_xN_y films by spectroscopic ellipsometry, spectraphotometry and XPS. *Applied Surface Science* **179**(1–4):182–191 (2001).
42. L. Roman, W. Mammo, L. A. A. Pettersson, M. R. Andersson, and O. Inganäs. High quantum efficiency polythiophene/C_{60} photodiodes. *Advanced Materials* **10**(10):774–777 (1998).
43. R. Beare. *Mathematics in Action: Modeling in the Real World Using Mathematics.* Chartwell-Bratt, 1997.

6

Coulomb Forces in Excitonic Solar Cells

Brian A. Gregg
National Renewable Energy Laboratory, Cole Boulevard, Golden,
CO, USA

Contents

Abstract Organic-based solar cells convert light to electricity by a mechanism involving exciton formation, transport, and dissociation. The same factors that cause exciton formation upon light absorption, as opposed to the formation of free electron–hole pairs, also control the doping process and carrier transport in organic semiconductors (OSCs). A unified approach has been developed to describe the excitonic processes, doping, and transport in these "excitonic" semiconductors. A simple equation is proposed that semiquantitatively distinguishes between excitonic semiconductors (XSCs) and conventional semiconductors (CSCs). The most essential qualitative difference between them is that Coulomb forces can often be neglected in CSCs, whereas they often dominate the behavior of XSCs. The Coulomb force between photogenerated electrons and holes in XSCs causes exciton formation. Mobile excitons dissociate into free electrons and holes primarily at heterojunctions, thus producing a large interfacial chemical potential energy gradient that drives the photovoltaic (PV) effect even in the absence of, or in opposition to, a bulk electric field. This force is usually insignificant in CSCs. Electrostatic considerations also control the doping process in XSCs: most added charge carriers are not free but rather are electrostatically bound to their conjugate dopant counterions. A superlinear increase in conductivity with doping density is thus expected to be a universal attribute of XSCs. An analogy is drawn between purposely doped XSCs and adventitiously doped XSCs such as π-conjugated polymers: in both cases the number of free carriers is a small, and field-dependent, fraction of the total carrier density. The Poole–Frenkel (PF) mechanism accounts naturally for the interactions between carriers bound in a Coulomb well and an applied electric field. Together with a field-dependent mobility, this mechanism is expected to semiquantitatively describe the conductivity in doped XSCs.

Keywords organic solar cells, excitons, coulomb forces, charge generation, doping, transport

6.1. THE ESSENCE OF EXCITONIC SOLAR CELLS

The use of organic semiconductors in photovoltaic cells has become a promising area of recent research. Presently available solar cells, based on silicon or other inorganic semiconductors (ISCs) [1], can be highly efficient and stable but they are still quite expensive. On the other hand, OSCs are usually inexpensive and can be processed by low-temperature, high-throughput techniques that are not available to ISCs [2–5]. The challenge is to make organic solar cells far more efficient than they currently are without losing their main potential advantage over ISCs, i.e., their low-cost. This chapter covers some of the fundamental aspects of OSCs and solar cells made from them while contrasting them to the much better understood ISCs.

6.1.1. Differences Between Conventional and Excitonic Semiconductors

There are fundamental differences between CSCs, which are usually, but not always, ISCs, and XSCs, which are usually OSCs. The most obvious distinction between CSCs and XSCs is that free charge carriers (electrons and holes) are created directly upon light absorption in CSCs, while electrostatically bound charge carriers (excitons) are formed in XSCs. This leads to quite different mechanisms for photoconversion between CSCs and XSCs as described in recent reviews [6,7]. One of the two primary reasons for this difference in mechanism is the low dielectric constant, ε, of XSCs compared to CSCs. The dielectric constant determines the magnitude of the electrostatic attraction between electrons and holes and also between these charge carriers and any fixed ionic charges in the lattice such as dopant ions. As described later, the photogeneration of free charge carriers in CSCs, or excitons in XSCs, depends on the square of ε.

The other primary difference is the small Bohr radius of carriers, r_B, in XSCs compared to CSCs. For a hydrogen atom in its ground state, the Bohr radius, $r_0 = 0.53\,\text{Å}$, is the average distance between the electron wavefunction and the positively charged nucleus. In a semiconductor with hydrogen-like wavefunctions (such as silicon), the Bohr radius of the lowest electronic state is [8]

$$r_B = r_0 \varepsilon (m_e / m_{eff}) \tag{1}$$

where m_e is the mass of a free electron in vacuum and m_{eff} is the effective mass of the electron in the semiconductor (usually less than m_e in ISCs but greater than m_e in OSCs). The effective mass decreases as the carrier becomes more delocalized and its transport becomes more wave-like. Thus increasing ε and decreasing m_{eff} lead to a greater average distance between the charges.

A charge carrier becomes "free" from its Coulomb attraction to an opposite charge if the energy of attraction is less than $k_B T$, the average thermal energy of the carrier. This occurs when

$$E = (q^2 / 4\pi\varepsilon\varepsilon_0)(1/r_c) = k_B T \tag{2}$$

where q is the electronic charge, ε_0 (8.85×10^{-14} C/V cm) is the permittivity of free space, and r_c is the critical distance between the two charges. Rearranging Equation (2) gives

$$r_c = (q^2 / 4\pi\varepsilon\varepsilon_0 k_B T) \tag{3}$$

6.1.2. Characteristics of Excitonic Semiconductors

Excitonic behavior is observed if $r_c > r_B$. In this regime, excitons are formed upon light absorption rather than free electron–hole pairs, and dopants and other charged impurities tend to remain unionized because of the strong Coulomb attraction between the charge carrier and its conjugate dopant or impurity ion. Excitonic behavior is also observed when r_B is greater than the particle radius, but these "quantum confined" structures will not be considered here.

We define a quantity, γ, that approximately distinguishes between conventional and excitonic semiconductors:

$$\gamma = \frac{r_c}{r_B} \approx \left(\frac{q^2}{4\pi\varepsilon_0 k_B r_0 m_e} \right) \left(\frac{m_{eff}}{\varepsilon^2 T} \right) \tag{4}$$

$\gamma > 1$ indicates an excitonic semiconductor and $\gamma < 1$ indicates a conventional semiconductor.

The inverse temperature-dependence shows that if CSCs are cooled to sufficiently low temperatures, they will become XSCs. Equation (4) is only a rough approximation for several reasons. First, OSCs conduct mainly through π-orbitals rather than σ-orbitals, thus they are usually low-dimensional materials: conductivity along one axis is often far greater than along other axes. Therefore, the parameters r_c, r_B, ε, and m_{eff} do not have the spherical spatial symmetry implied by the derivation of Equation (4). To accurately describe XSCs, and many CSCs, Equation (4) would have to be a tensor equation rather than a simple algebraic equation. Furthermore, the effective carrier mass, m_{eff}, is not well-defined in XSCs in which carrier transport typically occurs via a hopping mechanism rather than via delocalized band transport [2]. Finally, ε is a bulk quantity and is valid only over distances of many lattice spacings; however, for the tightly bound wavefunctions that may occur in some XSCs, the effective ε can approach its molecular value of 1.0 [8]. Despite these rather drastic approximations, Equation (4) is useful to show which parameters are responsible for the distinct differences in behavior between the two types of semiconductors. Although the uncertainty in γ means it cannot distinguish between semiconductors in which γ is near 1, such situations are rare. In general, $\gamma \gg 1$ for XSCs. A schematic

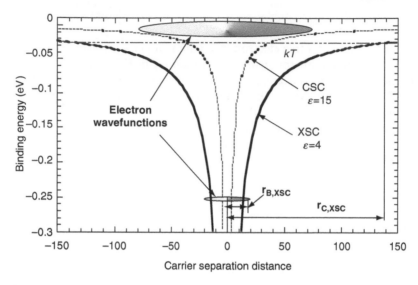

Figure 6.1. A schematic representation of the fundamental differences between CSCs and XSCs. The calculations assume Coulomb's law with the positive charge at 0 Å. The relevant distinction is between the size of the wavefunction (r_B) and the width of the Coulomb potential well at kT (r_c). When the wavefunction "fits" deep inside the potential well, that is when $\gamma = r_c/r_B > 1$, excitonic behavior is observed. This general scheme applies to all electrostatic attractions in the semiconductor, not just to photogenerated carriers.

representation of the Coulomb attractions in XSCs and CSCs illustrating these concepts is shown in Figure 6.1.

The conventional description of solar cells [1] relies on several approximations that are appropriate for Si cells ($\gamma < 1$) but are not valid for XSCs ($\gamma > 1$) [6,7]. This results in a number of important distinctions between CSCs and XSCs. We begin with the most fundamental difference: the charge carrier photogeneration mechanism.

6.2. CHARGE CARRIER PHOTOGENERATION IN CSCs AND XSCs

Because $\gamma < 1$ in CSCs, electrons and holes are photogenerated instantaneously upon light absorption and they are free from electrostatic attraction to each other. Therefore, electrons and holes are photogenerated throughout the bulk with a Beer's law (exponentially decaying) spatial distribution. In XSCs, however, it is electrically neutral excitons that are generated in a Beer's law distribution. They must then diffuse to a type 2 heterojunction (Figure 6.2) in order to dissociate into electron–hole pairs. This generates electrons in one chemical phase and holes on the opposite side of the interface, in a separate chemical phase. Thus, the spatial distribution of photogenerated carriers is very different in CSCs and XSCs.

It is important to understand the difference between the electrochemical *potential* energy of charge carriers and the exciton *energy*. To define an electrochemical potential energy, it requires an electrical charge that can be added or removed from a solid (or molecule, etc). Electrically neutral excited species such as excitons cannot have potentials, only energies, until they dissociate into electrically charged species.

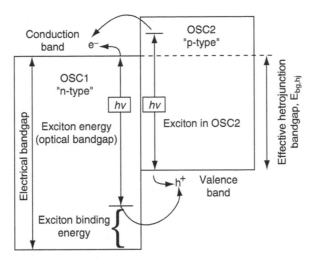

Figure 6.2. Potential energy level diagram for a typical XSC solar cell. Excitons created by light absorption in organic semiconductors 1 and 2 do not possess enough energy to dissociate in the bulk (except at trap sites). But the (type 2) band offset at the interface between OSC1 and OSC2 provides an exothermic pathway for dissociation of excitons in both phases, producing electrons in OSC1 and holes in OSC2. The band offset must be greater than the exciton binding energy for dissociation to occur. The three relevant bandgaps are shown: the electrical bandgap, E_{bg}, the optical bandgap, E_{opt}, and the effective bandgap of the heterojunction, $E_{bg,hj}$.

When an exciton dissociates into an electron and hole, one of these carriers must remain in the original phase in an available potential energy level, and this defines the potential energy of the injected carrier. For example, when an exciton in OSC1 (Figure 6.2) injects a hole into OSC2, the hole's initial potential is an exciton energy below the conduction band of OSC1; and when an exciton in OSC2 injects an electron into OSC1, the electron's initial potential is an exciton energy above the valence band of OSC2. In both cases, the carriers then quickly relax to the potential energy of the band edge.

6.2.1. Forces and Fluxes in XSCs

The difference in the spatial distribution of photogenerated charge carriers results in a powerful driving force for carrier separation in XSCs that does not exist in CSCs. To explain this, we briefly review the generalized forces that drive a flux of electrons through a solar cell. The electrochemical potential energy, E, is the sum of the electrical and chemical potential energies, $E = U + \mu$. The spatial gradient of a potential energy is a force, thus, ∇E is the fundamental force that drives the charge carrier fluxes through solar cells. In equilibrium solar cells (and other solid-state devices) E is called the Fermi level, E_f, and $\nabla E_f = 0$ by the definition of equilibrium. Away from equilibrium, we must also distinguish between the Fermi levels of electrons and holes. The spatial gradient of these "quasi" Fermi levels, ∇E_{Fn} and ∇E_{Fp}, is the force that drives the fluxes of electrons and holes, respectively, and each is made up of the two quasi-thermodynamic forces, ∇U and $\nabla \mu$. The general kinetic expression for the one-dimensional current density of electrons, $J_n(x)$, through a device is:

$$J_n(x) = n(x)\mu_n\{\nabla U(x) + \nabla\mu(x)\} = n(x)\mu_n\nabla E_{Fn}(x) \qquad (5)$$

where $n(x)$ is the electron density and μ_n is the electron mobility, not to be confused with the chemical potential energy. While $n(x)$ and μ_n influence the magnitude of the electron flux, $\nabla U(x) + \nabla\mu(x)$ controls its direction. We employ the symbols ∇U_{hv} and $\nabla\mu_{hv}$ to denote the two fundamental forces in a solar cell under illumination.

6.2.2. The Chemical Potential Energy Gradient in XSCs

Equation (5) shows that the electrical potential energy gradient, ∇U, and the chemical potential energy gradient, $\nabla\mu$, are *equivalent* forces. This equivalence is sometimes overlooked because of the predominant importance of ∇U in CSCs which results primarily from two factors that are specific to CSCs: (1) the photogeneration of free carriers throughout the bulk and (2) the high carrier mobilities that allow them to quickly "equilibrate" their spatial distributions regardless of their point of origin. Both of these factors minimize the influence of $\nabla\mu_{hv}$. However, in XSCs, almost all carriers are photogenerated in a narrow region near the heterointerface via exciton dissociation (Figure 6.2), leading to a photoinduced carrier concentration gradient (proportional to $\nabla\mu_{hv}$) that is much larger and qualitatively distinct from that in conventional photovoltaic (PV) cells (Figure 6.3). This effect, and the spatial separation of the two carrier types across the interface upon photogeneration, combine to produce a powerful photovoltaic driving force. This force is further enhanced by the generally low equilibrium charge carrier density, often making $\nabla\mu_{hv}$ the *dominant* driving force in XSCs. For example, ∇U can be ~ 0 in the bulk and a highly efficient solar cell can be made based almost wholly on $\nabla\mu_{hv}$. This is how dye-sensitized solar cells function [9,10]. These are, to date, the most efficient organic PV cells ($\sim 10\%$), which may suggest that internal (photoinduced) electric fields are a major efficiency-limiting factor in solid-state organic PV cells. In cells that

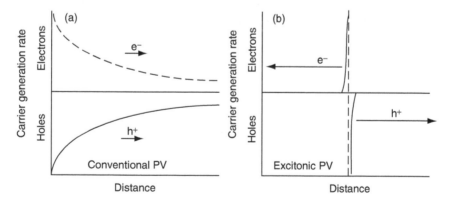

Figure 6.3. A schematic illustrating the fundamental difference in charge carrier generation mechanisms in conventional (a) and in excitonic (b) solar cells. (a) The photoinduced chemical potential energy gradient, $\nabla\mu_{hv}$ (represented by arrows), drives both carrier types in the *same* direction in CSCs (although it has a greater influence on minority carriers because of their lower concentration). (b) In XSCs, carrier generation is simultaneous to, and identical with, carrier separation across the interface in XSC cells; $\nabla\mu_{hv}$ therefore drives electrons and holes in *opposite* directions.

do not contain a high concentration of mobile electrolyte, both ∇U and $\nabla \mu$ must be taken into account [6,11].

One conclusion from this analysis is that when $\gamma > 1$, $\nabla \mu_{hv}$ plays an important, often dominant, role in the photovoltaic effect. This is also true for organic PV cells with highly convoluted heterointerfaces such as dye-sensitized solar cells and bulk heterojunction cells, although it is more difficult to visualize the three-dimensional cases. On the other hand, when $\gamma < 1$ it is sometimes possible to ignore the effect of $\nabla \mu_{hv}$ as is commonly done when describing mobile charge carriers in CSCs.

6.2.3. Open Circuit Photovoltage

An important corollary of the preceding discussion is that the open circuit photovoltage, V_{oc}, will be determined by different forces in CSCs and XSCs. Because CSCs are driven almost entirely by the electrical potential energy gradient, V_{oc} is limited to the equilibrium electrical potential energy difference across the cell, ϕ_{bi} or "band bending" [7]. However, the photoinduced chemical potential energy gradient, $\nabla \mu_{hv}$, often plays the dominant role in XSCs. Thus, V_{oc} is routinely greater than ϕ_{bi} and can even produce a V_{oc} of opposite polarity to ϕ_{bi} in XSCs [6]. The electrostatic factors contained in γ play a fundamental role in the charge carrier photogeneration process of all semiconductors. Not surprisingly, γ plays an equally fundamental role in the generation of dark charge carriers via doping.

6.3. DOPING OSCs

The utility of ISCs in solar cells as well as in integrated circuits derives almost entirely from the ability to precisely dope them, that is, to add atoms that produce free electrons (n-type doping) or free holes (p-type doping) in precisely defined spatial arrangements. This important area is just beginning to be explored in OSCs. We first describe doping in a well-characterized, highly crystalline OSC model system and then suggest that the same factors must also control adventitiously doped materials such as π-conjugated polymers. Because of their weak intermolecular forces, doping *molecular* OSCs is quite difficult compared to doping ISCs. In ISCs, the strong covalent, or covalent–ionic, interatomic bonds, along with the relatively high ε and large r_B ($\gamma < 1$) contribute to the ease of doping [12]. Bending or breaking the high-energy interatomic bonds at crystal defects and grain boundaries, or incorporating impurities of a valence different from the host, often produce electronic states near enough to the bandedges to generate free carriers. On the other hand, molecular OSCs are van der Waals solids: bending or breaking these low-energy intermolecular bonds, or adding different molecules into the lattice, only inefficiently produces free carriers. Despite concentrations of chemical impurities and structural defects that are often much larger than in ISCs, most molecular OSCs are nearly intrinsic (i.e., undoped in the sense of having very few free carriers) [12]. The so-called molecularly "doped" polymers (e.g., triphenylamine "doped" into polycarbonate) are similar to *molecular* OSCs in this regard, despite the unfortunate use of the term "doping," which in this case has nothing to do with producing free charge carriers.

6.3.1. Adventitiously Doped π-Conjugated Polymers

Conducting (or π-conjugated) polymers, such as poly(phenylene vinylene) or PPV, are quite different from molecular OSCs because their conducting backbone is made up of a semi-infinite chain of high-energy covalent π-bonds. Entropy, and the morphological constraints of thin film formation, inevitably lead to a large density of electroactive defects caused by twists and bends in the polymer backbone that generate electronic states in the bandgap. Another source of defect carriers comes from the difficulties in purifying high molecular weight polymers. When discussing doping in XSCs, therefore, we must always distinguish between molecular semiconductors (and molecularly doped polymers), which are expected to have a low free charge density, and π-conjugated polymers in which the free charge density is often quite high (10^{15}–10^{17} cm^{-3}) [13]. As a very rough analogy, π-conjugated polymers are somewhat comparable to amorphous silicon with its many broken and distorted bonds, while molecular semiconductors are more comparable to polycrystalline silicon.

6.3.2. Purposely Doped Perylene Diimide Films

To return to the simpler case, the weak and unintentional doping observed in most molecular OSCs results primarily from structural defects that tend to preferentially trap electrons (resulting in p-type "doping") or holes (n-type "doping"). Most OSC films have a mixture of trap sites, but one type of "doping" usually predominates. We wished to study the behavior of purposely doped, semiconducting molecular OSCs. In some ways, the *semiconducting* state of organic materials is more difficult to study than either the intrinsic or metallic states [14]. One reason is that the dopants must be *bound* in the organic lattice in order to achieve stable semiconducting properties. If the dopants are mobile, the electrical properties of the material will change with time, voltage, illumination intensity, etc., and p–n junctions will disappear rapidly as the electrons and holes, as well as their oppositely charged dopant ions, diffuse together and recombine [15,16]. Because of the weak lattice (intermolecular) forces, it is difficult to immobilize dopants in OSCs. Most "dopants" employed so far consist of small molecules such as O_2 and Br_2 that diffuse rapidly through organic films [2,17–19]. Larger molecular dopants such as tetracyanoquinodimethane (TCNQ) will also diffuse through the lattice, but much more slowly [16,20–23].

   The Dresden group of K. Leo recently initiated a series of studies of doping XSCs via cosublimation of an XSC with a dopant molecule such as perfluorinated TCNQ (p-type) [16,22,23] and Pyronin B (n-type) [24]. The long-term spatial stability of such doping may be open to question, but this approach has provided a wealth of information. Employing conductivity, Seebeck, and field-effect measurements as a function of temperature, this group has provided a detailed model of doping in several amorphous and polycrystalline XSCs. In their studies, as in others (see below), the conductivity is observed to increase superlinearly with dopant density as the activation energy decreases. They interpret this behavior as supporting their model of shallow (i.e., mostly ionized) dopants. They do not consider the Coulomb force between carriers and dopant ions in their model.

   Our approach to study doping in OSCs employs n-doped films of a (liquid) crystalline organic semiconductor, PPEEB (Figure 6.4) [12,15]. PPEEB thin films are well-characterized, highly crystalline (3–10 μm crystallites), and have very low defect

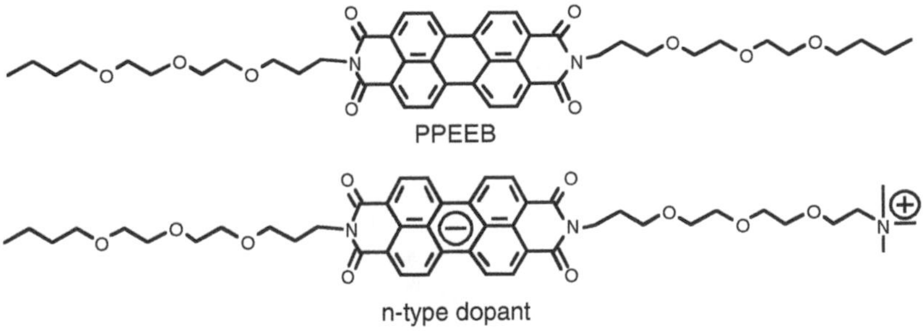

PPEEB

n-type dopant

Figure 6.4. The chemical structure of the host semiconductor, PPEEB, a liquid crystal perylene diimide (that is solid at room temperature), and its associated n-type dopant, a reduced derivative of PPEEB with a covalently attached counterion. There are no mobile ions in these doped films.

densities [25]. To obtain spatially and temporally stable doping in PPEEB films, we synthesized a zwitterionic dopant molecule that is a reduced (for n-type doping) derivative of the host molecule, which contains a covalently bound positive counter-charge (Figure 6.4). This molecule is a *substitutional* dopant, that is, it occupies a crystal lattice site identical to the host PPEEB molecule. In contrast, the usual dopants in OSCs are *interstitial*; they reside in normally unoccupied lattice sites and therefore perturb the crystal symmetry. Interstitial dopants are usually far more mobile than substitutional dopants. It is expected therefore, and is consistent with all of our observations that the n-type dopant in PPEEB remains tightly bound in the lattice of its host. Only in such a case can stable electrical junctions can be achieved by doping. The characteristics observed in this model system are expected to also pertain qualitatively to other purposely or adventitiously doped OSCs as described later. In most of these cases though, the observable phenomena will be further complicated by the presence of mobile ions and by energetic and structural disorder.

6.3.3. Superlinear Increase in Conductivity with Doping Density

We investigated a large number of PPEEB films with dopant concentrations ranging from 34 ppm to 1% ($\sim 10^{16} - 10^{19}\,cm^{-3}$). Somewhat surprisingly, at least compared to ISCs at room temperature, the conductivity of these doped films increased *super-linearly* with increasing dopant concentration (Figure 6.5) [12–15]. Upon further study of the literature, it became apparent that this superlinear increase has been observed in all quantitative studies of doping in monomeric [16,22,23,26] and polymeric OSCs [3,27–30] of which we are aware. Something fundamental must underlie this behavior that is independent of crystalline order and OSC type since it appears in doped highly crystalline PPEEB as well as in doped amorphous conducting polymers and doped quasi-amorphous molecular semiconductors. We present here a simple model based on our results with the doped PPEEB films that explains the apparent universality of the superlinear increase in conductivity with doping density in OSCs. We assume an isotropic dielectric constant, $\varepsilon = 4$. Our discussion from here on considers only n-doped OSCs, but it is equally applicable to p-doped OSCs.

It has been known since the work of Pearson and Bardeen in 1949 [31] that the activation energy for free carrier production, E_a, in doped ISCs decreases with increasing dopant concentration. This occurs because of the many-body electrical

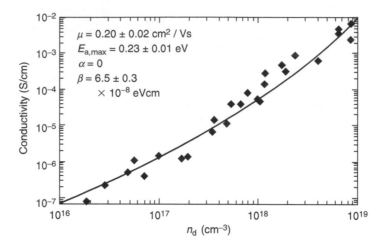

Figure 6.5. Experimental conductivity of PPEEB films (at a field $F = 0.9$ V/µm) versus dopant concentration; data fit to Equation (9). Data are from Reference 12.

interactions between neutral donors, ionized donors, free electrons, and ionized acceptors in crystalline (n-type) doped ISCs. This effect was analyzed further by many groups [32–34]. To discuss this effect (and others) in OSCs, we define four quantities: the added dopant density, n_d; the number of free electrons, n_f; the maximum activation energy that occurs for the case of dilute, non-interacting dopants, $E_{a,max}$; and the concentration-dependent activation energy, E_a.

The experimentally observed decrease in E_a with increasing doping density leads to a superlinear increase in n_f and therefore conductivity ($\sigma = q\mu_e n_f$). This is observed only at relatively low temperatures in ISCs [32–34] because $E_{a,max}$ is on the order of 10 meV. However, the low dielectric constant and small Bohr radii of OSCs (Figure 6.1, $\gamma > 1$) generally result in $E_{a,max}$ of hundreds of meV. Therefore, E_a will almost always control the free carrier density in OSCs, whether purposely or adventitiously doped. Decreases in E_a for conductivity in OSCs with increasing n_d have been observed by us and by others [3,15,16,22,23,26–30], but unambiguous studies of n_f versus n_d, ideally by Hall effect measurements, are not yet available. In the earlier work on ISCs, values of n_f were known directly from Hall effect measurements; however, we know only n_d and must calculate n_f.

There are three conceptually distinct mechanisms that can lead to a decrease in E_a with increasing doping density [32]:

(1) The attraction of free electrons to ionized dopant cations lowers the potential energy of electrons (the conduction band).

(2) The electrical screening by free electrons reduces the potential well depth around the cation.

(3) The polarizability of the film is increased by the number of electrons in bound but delocalized states near dopant cations, resulting in an increase in the effective dielectric constant.

The first two mechanisms are governed by n_f, the third by $n_d - n_f$ ($\approx n_d$ in our experiments). The third mechanism is depicted schematically in Figure 6.6.

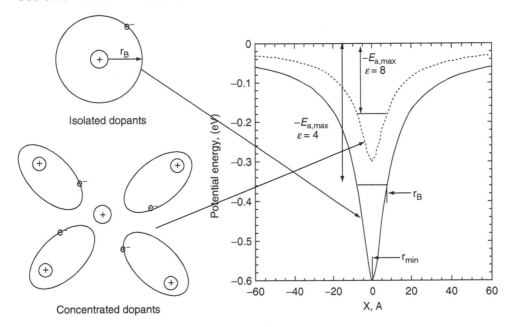

Figure 6.6. A schematic describing mechanism (3) showing the Coulomb potential around a dopant ion in the LC perylene diimide films. The potential does not go to $-\infty$ at 0 Å because the cation is offset from the electron-conducting pathway (Figure 6.4). For isolated dopants, the dipole formed by the bound electron orbiting the dopant cation is approximately isotropically oriented. At high concentrations, the ionization of one dopant is energetically stabilized by the ability of surrounding neutral dopants to polarize around the resulting cation. Thus, the polarizability of the unionized dopants increases the effective dielectric constant of the film. This explains why the number of free carriers, n_f, and the film conductivity (Figure 6.5) increase superlinearly with the dopant concentration.

Following Debye and Conwell [32], we assume that

$$E_a = E_{a,\,max} - \alpha n_f^{1/3} - \beta n_d^{1/3} \tag{6}$$

where α and β describe the dependence of E_a on the average distances between free electrons, $n_f^{1/3}$ (mechanisms 1 and 2), and neutral dopants, $(n_d - n_f)^{1/3} \approx n_d^{1/3}$ (mechanism 3), respectively. Ignoring the miniscule number of intrinsic carriers [12,15] and assuming Boltzmann statistics, the free electron density is

$$n_f = n_d e^{-E_a/kT} \tag{7}$$

and the conductivity is

$$\sigma = q\mu_n n_f = q\mu_n n_d e^{-E_a/kT} \tag{8}$$

where μ_n is the field-dependent mobility of the electron (discussed in Section 4.5). Combining Equations (6) and (8) leads to

$$\sigma = q\mu_n n_d \exp\{(-E_{a,\,max} + \alpha n_f^{1/3} + \beta n_d^{1/3})/kT\} \tag{9}$$

This model is expected to be valid for low to moderate doping levels where $n_d^{-1/3} \gg r_B$. Equation (9) is nonlinear in n_f but can be approximated self-consistently by the use of Equation (8). Over the range of our experiments $\sigma \sim n_d^{1.84}$, so we can replace n_f in Equations (8) and (9) by $n_d^{1.84}$ for fitting. A four-parameter fit of Equation (9) to the data yielded good results but α was negative and very small, suggesting that mechanism (3) dominates over (1) and (2). An equally good three-parameter fit was obtained with $\alpha = 0$, and this is shown in Figure 6.5.

The fact that the superlinear increase of σ with n_d (Figure 6.5) seems to depend only on the concentration of neutral dopants, rather than on ionized dopants, may be explained by the large $E_{a,max} = 0.23\,eV$. Over the range of our data, n_d is 100 to 10,000 times greater than n_f, therefore the two mechanisms involving n_f are insignificant. These two mechanisms may become important at higher dopant concentration and temperature. The reasonable values of the fitting parameters provide confidence in the analysis. From Equation (6) and Coulomb's law, β should be approximately $q^2/4\pi\varepsilon_d\varepsilon_0$ where ε_d is the effective dielectric constant caused by the presence of the neutral dopants (Figure 6.6). The fit value of $\beta = 6.5 \times 10^{-8}\,eV\,cm$ leads to $\varepsilon_d = 7.2$, substantially larger than the undoped value of $\varepsilon \approx 4$. This also suggests why mechanism (3) is unimportant in crystalline Ge where $\varepsilon = 16$ [32]. The obtained free carrier mobility value of $0.2\,cm^2/V\,s$ is reasonable for a highly ordered XSC at $F = 0.9\,V/\mu m$ and is similar to the values obtained from our ongoing Hall effect and transport measurements. Finally, the value of $E_{a,max} = 0.23\,eV$ at $F = 0.9\,V/\mu m$ is close to that expected from an extrapolation of our calculation at zero field strength, $E_{a,max} = 0.36\,eV$ [12]. To a first approximation, $E_{a,max}$ at zero field may be of similar magnitude to the singlet exciton binding energy because they both arise from the same forces, although their geometrical constraints are somewhat different.

6.3.4.   No Shallow Dopants in XSCs

The superlinear increase in conductivity observed with increasing doping concentration in the LC perylene diimide films is expected to be, and apparently is, a universal attribute of excitonic semiconductors. This follows directly from the characteristics that define XSCs [6,7]: the strong, long-range, electrostatic attraction between opposite charges, and the relatively localized wavefunctions of the charge carriers (leading to $\gamma > 1$). The binding energy between charges in PPEEB films decreases with increasing dopant concentration because of the increasing polarizability of the film (Figure 6.6). One consequence of these results is that it may be impossible to produce shallow (i.e., mostly ionized) dopants in XSCs, *independent of the chemistry or potentials of the dopant and host*. This follows because the binding energy of the charge carrier is determined by the electrostatic effects contained in γ, not by the difference in potential between the dopant and its respective band edge. In our previous example (Section 3.2), the dopant redox potential was exactly at the band edge (Figure 6.4) but nevertheless, the charge carrier was localized near the dopant molecule because of the Coulomb force. If the dopant redox potential lies *outside* the bandgap of the host, the charge carrier will still relax to the band edge and then be localized near neighboring molecules because of the Coulomb force. Dopants can become shallow in typical XSCs only at high dopant density where the film's effective dielectric constant increases and the film's behavior approaches that of a CSC or a molecular metal rather than a semiconductor.

6.4. CARRIER TRANSPORT IN XSCs

The electrostatic effects expressed in γ have a major influence on the charge generation mechanism in XSCs, both via light absorption and via doping. It is anticipated that γ should also play an important role in the carrier transport characteristics of XSCs by controlling the number of *free* charge carriers, n_f. Our experimental studies in this area are still in progress and thus we restrict our discussion to general considerations. The electrical current in an XSC can be limited by many factors [2]: carrier transport through the bulk [35–38], potential barriers at crystallite grain boundaries [39], injection from the electrodes [40], space charge limitations [4,13,41–43], etc. Models of carrier transport in XSCs can be highly complex and no single model can yet describe all cases. Consistent with what was described earlier, we suggest that one important concept has been overlooked in existing transport models: *the number of free bulk carriers is expected to be an increasing function of applied field in many XSCs*, whether purposely or adventitiously doped because $\gamma >$ 1. We are not referring to injected carriers, n_{inj}, whose number is obviously field-dependent, but to n_f. In some cases, this can have a major impact on the interpretation of experimental results.

6.4.1. The $n\mu$ Product

When measuring the conductivity, n_f and μ_n occur together in the relevant equation (e.g., Equation (8)) in most instances, and they are often simply referred to as the "$n\mu$ product." It can be difficult to distinguish experimentally between changes in n_f and μ_n. A concept called "effective mobility" is sometimes introduced that, in essence, assumes n_f is fixed and that therefore any changes in conductivity with temperature, applied voltage, etc. can be ascribed to the effective mobility. This unfortunate assumption can result in serious errors in interpreting the experimental results. When characterizing a newly synthesized organic material for example, decreases in the equilibrium n_f compared to a known material may be misinterpreted as a decrease in μ_n; while the former would be beneficial for photovoltaic applications, the latter would be detrimental. Misinterpreting a beneficial change as a detrimental one is far worse than admitting that only the $n\mu$ product can be measured. Ideally, an unambiguous distinction between n_f and μ_n should be made, but this is not always possible with existing techniques. Arguably the best technique would be measurements of the Hall effect [2]. However, Hall effect measurements have not been reliable in OSCs so far because of sample geometry problems, low carrier density, and anisotropic conductivity effects. Mobility measurements in a field effect transistor (FET) geometry [35,44,45], and measurements of the time-of-flight (TOF) transition time [46] are some of the best available methods to estimate μ_n in OSCs. However, they are accurate primarily in materials where n_f is very small. The interpretation becomes more ambiguous as n_f increases.

6.4.2. Adventitiously Doped XSCs

In some nominally undoped materials such as π-conjugated polymers, n_f may be far larger than expected [13]. There are several possible sources for free charge carriers at equilibrium: the number of intrinsic carriers that are created by thermal excitation across the bandgap, n_i, the number of purposely added carriers from dopants, n_d, and the number of carriers produced from adventitious doping, n_a, caused by impurities, crystal imperfections, grain boundaries, surface states, bent or twisted bonds in a conducting polymer backbone, etc. The concentration of intrinsic carriers is usually so small that it can be neglected, except when studying single crystals or highly purified polycrystalline OSCs. In most cases, n_d, n_a, or both, dominate the production of free carriers and thus control the electrical properties. Away from equilibrium, carriers can also be produced by injection from the electrodes, n_{inj}, and by light absorption.

Although charge generation from crystal imperfections, impurity molecules, etc. is less efficient in molecular OSCs than in ISCs, as described earlier, it is still significant. Such defect carrier generation is expected to be especially prominent in π-conjugated polymers. It is estimated, for example, that $n_f \approx 10^{15}-10^{17}\,cm^{-3}$ in poly(phenylene vinylene) (PPV)-type polymers [13]. There are no purposely added dopants, $n_d = 0$, in these materials and $n_i \approx 10^3\,cm^{-3}$. The electrical properties of these nominally undoped conducting polymers are therefore entirely controlled by n_a, at least at low fields where n_{inj} is insignificant. Not surprisingly, their J–F curves [3,4,42,43] are very similar to those obtained from purposely doped PPEEB films, and we suggest below that they can be interpreted in a similar way.

By generalizing from the results of the doped PPEEB films, we reach an important conclusion about carrier density in most XSCs: *each equilibrium charge carrier* produced in the bulk, whether free or bound, whether originating from n_i, n_d, or n_a, *must be counterbalanced by an opposite charge* if the XSC is to maintain overall electrical neutrality. This opposite charge may be fixed in the lattice (such as a dopant ion), have a very low mobility (such as a trapped countercharge), or be another mobile charge carrier. Regardless of the origin of the carriers, the electrostatic attraction between the opposite charges must be taken into account because $\gamma > 1$. This leads immediately to the conclusion that *most equilibrium charge carriers are not free in XSCs, but rather will be bound ("self-trapped") in the Coulomb well of oppositely charged ions.* In the previous example of a PPV-type polymer, this means that $n_a \gg n_f$. This is almost exactly the case we treated in Section 3 when considering the purposely doped perylene diimides in which $n_d \gg n_f$. There appears to be no qualitative difference between the electrostatic interactions of charges in adventitiously doped XSCs and in purposely doped XSCs. In other words, Equation (7) should be approximately valid for all XSCs, although n_d should be replaced by $n_i + n_d + n_{a1} + n_{a2} + \cdots$ for the general case, and there may be a distribution of E_as.

6.4.3. The Poole–Frenkel Mechanism

The quantity of interest in XSCs for transport studies is the number of *free* carriers, n_f, since it determines the observable current density. As discussed, $n_d + n_a \gg n_f$ because $\gamma > 1$. The Poole–Frenkel mechanism proposed in 1938 [2,37,40] treats the effect of an applied electric field, F, on carriers trapped in Coulomb potential wells in an isotropic solid (Figure 6.7). We expect this mechanism (slightly modified below) to semiquantitatively describe the J–F characteristics of many purposely and adventitiously doped XSCs, even if it cannot account for the spatial and energetic complexities of some specific materials.

The dark current in the PF mechanism is expected to increase with $\exp(F^{1/2}/kT)$ as the field decreases the height of the electrostatic potential well surrounding the bound carrier pairs (Figure 6.7 and Equation (10)):

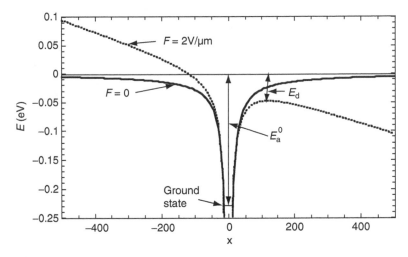

Figure 6.7. The activation energy for thermal emission of a charge carrier out of a Coulomb potential well is expected to decrease with the square root of the applied field, $F^{1/2}$ (Poole–Frenkel mechanism, Equation (10)). Therefore, the free carrier density, n_f, should be proportional to $\exp(F^{1/2}/kT)$.

$$J = q\mu_n Fn_d \exp\left(\left(-E_a^0 + (q^3/\pi\varepsilon_0\varepsilon)^{1/2}F^{1/2}\right)/kT\right) \qquad (10)$$

where ε is the high-frequency dielectric constant. Equation (10) can be derived easily by considering the factors shown in Figure 6.7; it is written for the simplest case where n_d dominates J and where there is only a single E_a^0. The increase in current with $\exp(F^{1/2}/kT)$ occurs because n_f is proportional to $\exp(F^{1/2}/kT)$. This point is commonly misunderstood in the recent literature where it is asserted that the PF mechanism predicts an $\exp(F^{1/2}/kT)$ dependence of the *mobility*. However, Frenkel explicitly stated that, "the increase in electrical conductivity in intense fields is due to the number of free electrons, and not to their mobility" [37]. The failure to distinguish between the field dependence of n_f and μ_n is a serious problem in many models. The change in current density with applied field predicted by Equation (10) is shown in Figure 6.8. Such J–F curves are observed in many XSCs, including some inorganic insulators [4,13,37,40,47].

The PF mechanism assumes that J is controlled by the number of carriers trapped in Coulomb potential wells. It is thus not expected to apply to molecularly doped polymers or pure crystalline XSCs in which n_f is very low, because the large density of Coulomb-trapped charges assumed by PF should not be present [46,47]. However, in the doped perylene diimide films described earlier, where one dopant electron occurs for every cationic dopant molecule, the PF mechanism is the obvious choice for describing (at least approximately) the bulk transport-limited conductivity. As described earlier, it may also apply to other XSCs in which $n_d + n_a \gg n_f$, because there appears to be no fundamental electrostatic distinction between these systems and the doped perylene diimides.

A mechanism very similar to the PF mechanism is commonly employed to describe contact-limited conductivity. The "Schottky effect" is described by an equation almost identical to Equation (10); however, the second term in the exponential is just half that shown in Equation (10), and E_a^0 is now interpreted as the energy barrier between the electrode Fermi level and the conduction band of the film [40].

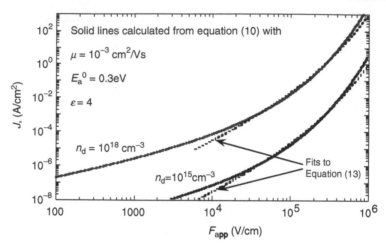

Figure 6.8. Calculated current–field curves on a log–log plot using Equation (10) (solid lines, PF mechanism) with the parameters shown in the figure. Once the slope reached 2.0, where Equation (13) (modified SCLC model) becomes applicable, Equation (13) was fit to the "data" generated by Equation (10). In this range these two models are almost indistinguishable experimentally; however the mobility values obtained from them are very different: For $n_d = 10^{15}\,\text{cm}^{-3}$, the calculated zero field mobility from Equation (13) assuming $d = 1\,\mu\text{m}$ was $3 \times 10^{-8}\,\text{cm}^2/\text{V s}$; while for $n_d = 10^{18}\,\text{cm}^{-3}$, it was $3 \times 10^{-5}\,\text{cm}^2/\text{V s}$; in both cases the actual mobility input into Equation (10) was $10^{-3}\,\text{cm}^2/\text{V s}$.

6.4.4. Space Charge Limited Currents

Another well-known mechanism describes the J–F curves when the electrodes are capable of injecting into the film a carrier density that completely overwhelms the film's free carrier density, $n_{\text{inj}} \gg n_f$. Such space charge limited currents (SCLCs) may be observed at high F so long as only one carrier type is injected into the film [2,4,41]. The theory of SCLC was developed originally to describe the current flowing in a vacuum between a cathode and an anode (Child's law) [2] in which, of course, $n_f = 0$. In XSCs, it was employed first for studies of high-quality single crystals such as anthracene and tetracene [2] and later in studies of molecularly doped polymers such as the hole conductors employed in electrophotography and electroactive polymers without a conjugated backbone [46,48]. In all these cases, n_f is negligibly small at high F. An ohmic regime is assumed at low voltages where n_f dominates the conductivity. At higher F, SCLCs are expected:

$$J_{\text{SCL}} = 9\mu_n \varepsilon \varepsilon_0 F^2 / 8d \tag{11}$$

where d is the film thickness. This is the simplest, trap-free formulation of J_{SCL}. One advantage of SCLC in the characterization of XSCs is that n_f is not included in Equation (11). This occurs because J_{SCL} is controlled entirely by n_{inj}, which is the maximum charge density theoretically capable of existing in the film at that field and can be calculated from the film capacitance [2]. Therefore, this is one of the few techniques that can provide a measure of μ_n without knowing n_f. On the other hand, if SCLC is assumed where it is not actually operating, the calculated μ_n will *under-estimate* the actual μ_n because the SCLC model assumes that the maximum possible number of carriers is contributing to the current (Figure 6.8). The opposite problem may occur in FET mobility measurements where it is assumed that the only carriers present are those induced by the applied field. If other carriers are present, μ_n will be *overestimated*.

Experimentally, the current in many XSCs is observed to increase approximately linearly (ohmically) at low F, then superlinearly with increasing slope through F^2 (where SCLC might apply) and then on to ever higher powers of F, similar to what is shown in Figure 6.8 for the PF mechanism. To describe the experimental results, a field and temperature dependence of the mobility is almost universally postulated when interpreting SCLC, TOF, FET, etc. measurements in disordered XSCs. A mobility of the empirical form

$$\mu = \mu_0 \exp(aF^{1/2}/kT) \tag{12}$$

is often assumed (although it may not be physically realistic, see below). With the addition of this field- and temperature-dependent mobility, the approximate equation for (modified) SCL currents becomes

$$J_{SCL} = 9\mu_0\varepsilon\varepsilon_0 F^2/8d \exp(aF^{1/2}/kT) \tag{13}$$

A more correct approach, which treats the nonlinear coupling between μ and J, can only be solved numerically [49].

Considering the usually limited range of experimental data and the presence of experimental noise, it now becomes very difficult to distinguish PF behavior from the modified SCLC model. For example, when "data" are generated by Equation (10) (PF model) and then fit to Equation (13) (SCLC model), the SCLC model can fit these data well (Figure 6.8). However, Equation (13) gives a much lower mobility value than what was used to generate the data (see caption to Figure 6.8) because it assumes that $n_{inj} \gg n_f$ while this is not true in our example. The thickness dependence of the two models is quite similar and can probably not distinguish between them; the same argument applies to temperature dependence. In general, the two models can fit the same data over the range where J is superlinear in F. However, the PF and SCLC models are not compatible because they disagree about the relevant carrier density; the PF mechanism assumes it is n_f, while the SCLC mechanism assumes it is n_{inj}.

SCLCs are easily observed in single-crystal OSCs and in molecularly doped polymers and similar materials [2,46,48] because n_f is so low that it is easily overwhelmed by n_{inj}. However, as n_f increases due to crystal defects, added dopants, etc., it becomes even more difficult to achieve SCLC. The injected charge density under SCLC conditions is [2]

$$n_{inj} = F\varepsilon\varepsilon_0/qd \tag{14}$$

If we take typical values of $F = 10^3$–10^6 V/cm, $\varepsilon = 4$, and $d = 1\ \mu m$, n_{inj} takes on values between $n_{inj} = 2 \times 10^{13}$ and $2 \times 10^{16}\ cm^{-3}$. In order to observe pure SCLC, n_f must be at least an order of magnitude below these values at the applied field strength. In other words, observation of SCLC normally requires quite pristine XSCs. SCLCs can be easily obscured by n_f in disordered or doped materials, especially at low fields, and may not be observable all in materials such as conducting polymers. Unfortunately, a PF-like current is easily mistaken for SCLC. At present, we can only measure the $n\mu$ product in these materials, unless, for example, n_f can be estimated via a known doping density, n_d, or if μ_n is known from independent experiments.

6.4.5. Field-Dependent Carrier Mobilities

The mobility of charge carriers in many XSCs increases with increasing electric field. This is independent of the field-assisted increase in n_f (e.g., the PF mechanism). To clearly understand transport in XSCs, therefore, we must distinguish between these two independent mechanisms which both result in a super-linear increase of J with F and which may both occur in the same material. The field-dependent mobilities are most clearly observed in disordered materials such as molecularly doped polymers [46,50]. Models of charge transport in disordered XSCs justify such a field-dependence of the carrier mobility by assuming an approximate Gaussian distribution of hopping site energies [46,47,50] or by assuming a random distribution of fixed dipoles in the film, or both [36,38,44]. These mechanisms have also been applied to the description of conducting polymers [36,42,43]. They are not expected to apply, however, to undoped crystalline XSCs [2] because these materials do not possess the requisite energetic and morphological disorder.

Despite their highly crystalline nature, the "random dipole" mechanism may also apply to the mobility of carriers in the doped PPEEB films. Dipoles are formed by the electrons electrostatically bound at an average distance of r_B (Equation (1)) to their dopant counterions. They are free to orient in almost any direction (Figure 6.6) and therefore will minimize any existing *static* electric fields. In this sense, they are not the random permanent dipoles treated by the existing models. When discussing transport, however, it is important to consider the *rate* at which the dipoles reorient with respect to the charge hopping rate. Since the dipoles reorient by what is, in essence, a charge hopping process, we expect these two rates to be similar. Therefore, the dipoles will begin to reorient on the timescale of charge hopping but they will not have time to reach an equilibrium configuration. The dipole field will lag the charge hopping process. Thus the charge carrier will experience a not-quite-random distribution of dipoles and there will be an energetic correlation between the site energies visited by the mobile charge [38,47]. Although this mechanism has not been previously considered to our knowledge, it would appear to result in the same field-dependent mobility as described in the various existing models [36,38,44,46,47,50]. Thus it seems likely that in most cases where there is a substantial dipole density in the XSC, the carrier mobility will be field-dependent. The internal dipole density should be $\approx n_d + n_a$. Therefore, nominally undoped π-conjugated polymers are expected to have a large dipole density.

The empirical expression for the field-dependent mobility, Equation (12), does not seem to be physically realistic because it predicts that the mobility will increase exponentially with decreasing temperature. This is not compatible with the assumed hopping mechanism of carrier transport. The use of Equation (12) may be based on the failure to distinguish between n_f and μ_n. A number of more sophisticated models for μ have been proposed [2,38,47], but they still overlook the crucial field- and temperature-dependence of n_f. Most models treat n_f as a constant (which we argue is incorrect) and attempt to describe the changes in conductivity primarily as a function of the mobility. We believe therefore that a viable description of transport in disordered or doped XSCs is not yet available. However, in order to provide a reasonable, but only semiquantitative, description of transport in doped (purposely or not) XSCs, we employ a more general equation for the mobility [38,47] in place of Equation (12):

$$\mu = \mu_0(T) \exp\left(\alpha(T)F^{1/2}\right) \tag{15}$$

where the T-dependencies of μ_0 and α are unknown and probably material-dependent.

This equation then, together with Equation (10), leads to a modified form of the Poole–Frenkel equation that should account semiquantitatively for the J–F characteristics expected of doped XSCs including the field-dependencies of both n_f and μ_n:

$$J = qFn_d\mu_0(T)\exp\left(\alpha(T)F^{1/2}\right)\exp\left((-E_a^0 + (q^3/\pi\varepsilon_0\varepsilon)^{1/2}F^{1/2})/kT\right) \quad (16)$$

This equation does not account for the field-mediated coupling between μ and n_f, which would result in a series of coupled nonlinear equations solvable only by numerical simulation.

6.5. SUMMARY

Conventional semiconductor theory is only appropriate for materials in which the Coulomb forces between a charge carrier and its countercharge can be neglected. In this case, free electron–hole pairs are generated upon light absorption and mostly ionized dopants are produced upon doping. As described here, this corresponds to $\gamma < 1$. However, excitonic semiconductors ($\gamma > 1$), in which the influence of electrostatic attractions is dominant, are fundamentally different. In these materials, excitons are generated upon light absorption and electrostatically bound carriers are created upon doping. The interfacial generation of charge carriers via exciton dissociation at the heterojunction produces a powerful chemical potential energy gradient in XSCs that drives electrons away from holes. This force is usually negligible in CSCs. Because of this additional force, the photovoltage in XSC solar cells often exceeds, or even counteracts, the built-in equilibrium potential difference, ϕ_{bi}; while qV_{oc} is usually limited to ϕ_{bi} in CSCs.

The generation of dark charge carriers via n-type doping was studied in polycrystalline films of a liquid crystal perylene diimide. This is the first example of substitutional doping in XSCs, instead of the more common interstitial doping, and probably the first example where doping is compatible with stable junction formation. It became apparent that the same Coulomb forces that cause exciton formation in XSCs also control the doping process. Most dopant electrons are electrostatically bound to their conjugate dopant counterions; only a tiny fraction, about 1 in 10^4, are free in weakly doped PPEEB films of $n_d = 10^{16}\,\text{cm}^{-3}$. However, the fraction of free carriers increases with increasing dopant concentration because the highly polarizable electrons bound to dopant cations cause an increase in the effective dielectric constant of the film. When the doping level reaches $n_d = 10^{19}\,\text{cm}^{-3}$, the fraction of free electrons increases to about 1 in 100. Both of these values are for an applied electric field of $F = 0.9\,\text{V}/\mu\text{m}$. It is expected that the free carrier density, n_f, in doped XSCs should increase with increasing field approximately as $\exp(F^{1/2}/kT)$ as predicted by the Poole–Frenkel mechanism.

In many organic materials that are not purposely doped, there is a substantial density of extrinsic charge carriers created, for example, by morphological disorder or, in the case of conducting polymers, by defects in, or rotations around, the conducting polymer backbone. At equilibrium, these charge carriers will necessarily be counterbalanced by opposite charges. The Coulomb attraction between the opposite charges, mediated by the effective dielectric constant and the Bohr radius of carriers, will determine n_f. There appears to be no qualitative difference between the electrostatic interactions that govern the intentional doping process in highly puri-fied, polycrystalline perylene diimides and the adventitious doping in most XSCs caused by morphological and structural defects. Therefore, we expect the majority of charge carriers to remain bound to their countercharges by the Coulomb attraction also in adventitiously doped XSCs. It follows that n_f should *always* be a function of the applied potential, often increasing approximately as $\exp(F^{1/2}/kT)$. This fact appears to have been overlooked in earlier treatments of XSCs. The lower the adventitious doping density and the lower the carrier mobility, or the higher the activation energy for free carrier production, the more extensive the J–F region will be in which there is approximately ohmic (linear) behavior. However, true ohmic behavior is not expected in XSCs because $\gamma > 1$.

The increase in carrier mobility, μ_n, with electric field is expected to be a general feature in all materials with a substantial dipole density derived either from "fixed" dipoles, as in molecularly doped polymers, or from slowly orienting dipoles, as in purposely doped or adventitiously doped XSCs. The current–potential curves are thus expected to contain the field-dependencies of both n_f and μ_n.

ACKNOWLEDGMENT

I am grateful for many helpful discussions with Howard Branz, Pauls Stradins, and Sean Shaheen and thank the U.S. DOE, Office of Science, Division of Basic Energy Sciences, and Chemical Sciences Division for supporting this research.

REFERENCES

1. A. L. Fahrenbruch and R. H. Bube, *Fundamentals of Solar Cells. Photovoltaic Solar Energy Conversion*, Academic Press, New York, 1983.
2. M. Pope and C. E. Swenberg, *Electronic Processes in Organic Crystals and Polymers*, 2nd ed., Oxford University Press, New York, 1999.
3. A. J. Heeger, *J. Phys. Chem. B*, 105, 8475–8491 (2001).
4. P. W. M. Blom and M. C. J. M. Vissenberg, *Mater. Sci. Eng.*, 27, 53–94 (2000).
5. P. Peumans, A. Yakimov, and S. R. Forrest, *J. Appl. Phys.*, 93, 3693–3723 (2003).
6. B. A. Gregg, *J. Phys. Chem. B*, 107, 4688–4698 (2003).
7. B. A. Gregg and M. C. Hanna, *J. Appl. Phys.*, 93, 3605–3614 (2003).
8. R. A. Smith, *Semiconductors*, Cambridge University Press, Cambridge, 1978.
9. A. Hagfeldt and M. Grätzel, *Acc. Chem. Res.*, 33, 269–277 (2000).
10. B. A. Gregg, The essential interface: Studies in dye-sensitized solar cells in: *Molecular and Supramolecular Photochemistry*, Vol. 10, K. S. Schanze and V. Ramamurthy (eds.), Marcel Dekker, New York, 2002, pp. 51–87.
11. A. Zaban, A. Meier, and B. A. Gregg, *J. Phys. Chem. B*, 101, 7985–7990 (1997).
12. B. A. Gregg, S.-G. Chen, and H. M. Branz, *Appl. Phys. Lett.*, 84, 1707–1709, (2004).
13. S. C. Jain, W. Geens, A. Mehra, V. Kumar, T. Aernouts, J. Poortmans, R. Mertens, and M. Willander, *J. Appl. Phys.*, 89, 3804–3810 (2001).
14. T. J. Marks, *Science*, 227, 881–889 (1985).
15. B. A. Gregg and R. A. Cormier, *J. Am. Chem. Soc.*, 123, 7959–7960 (2001).
16. M. Pfeiffer, A. Beyer, B. Plönnigs, A. Nollau, T. Fritz, K. Leo, D. Schlettwein, S. Hiller, and D. Wörhle, *Sol. Eng. Mater. Sol. Cells*, 63, 83–99 (2000).
17. D. Wöhrle and D. Meissner, *Adv. Mater.*, 3, 129–138 (1991).
18. K.-Y. Law, *Chem. Rev.*, 93, 449–486 (1993).
19. J. Simon and J.-J. Andre, *Molecular Semiconductors*, Springer-Verlag, Berlin, 1985.

20. D. R. Kearns, G. Tollin, and M. Calvin, *J. Chem. Phys.*, 32, 1020–1025 (1960).
21. P. Leempoel, F.-R. F. Fan, and A. J. Bard, *J. Phys. Chem.*, 87, 2948–2955 (1983).
22. M. Pfeiffer, A. Beyer, T. Fritz, and K. Leo, *Appl. Phys. Lett.*, 73, 3202–3204 (1998). Erratum: *Appl. Phys. Lett.*, 74, 2093 (1999).
23. B. Maennig, M. Pfeiffer, A. Nollau, X. Zhou, K. Leo, and P. Simon, *Phys. Rev. B*, 64, 195208-1–195208-9 (2001).
24. A. G. Werner, F. Li, K. Harada, M. Pfeiffer, T. Fritz, and K. Leo, *Appl. Phys. Lett.*, 82, 4495–4497 (2003).
25. S.-G. Liu, G. Sui, R. A. Cormier, R. M. Leblanc, and B. A. Gregg, *J. Phys. Chem. B*, 106, 1307–1315 (2002).
26. Y. Shen, K. Diest, M. H. Wong, B. R. Hsieh, D. H. Dunlap, and G. G. Malliaras, *Phys. Rev. B*, 68, 081204 (2003).
27. D. M. deLeeuw, *Synth. Met.*, 57, 3597–3602 (1993).
28. J. A. Reedijk, H. C. F. Martens, H. B. Brom, and M. A. J. Michels, *Phys. Rev. Lett.*, 83, 3904–3907 (1999).
29. F. Zuo, M. Angelopoulos, A. G. MacDiaramid, and A. J. Epstein, *Phys. Rev. B*, 39, 3570–3578 (1989).
30. C. P. Jarrett, R. H. Friend, A. R. Brown, and D. M. de Leeuw, *J. Appl. Phys.*, 77, 6289–6294 (1995).
31. G. L. Pearson and J. Bardeen, *Phys. Rev.*, 75, 865–883 (1949).
32. P. P. Debye and E. M. Conwell, *Phys. Rev.*, 93, 693–706 (1954).
33. G. F. Neumark, *Phys. Rev. B*, 5, 408–417 (1972).
34. G. W. Castellan and F. Seitz, in: *Semiconducting Materials*, H. K. Henisch (ed.), Butterworth, London, 1951.
35. L. Torsi, A. Dodabalapur, L. J. Rothberg, A. W. P. Fung, and H. E. Katz, *Science*, 272, 1462–1464 (1996).
36. V. I. Arkhipov, P. Heremans, E. V. Emilianova, G. J. Adriaenssens, and H. Bässler, *Appl. Phys. Lett.*, 82, 3245–3247 (2003).
37. J. Frenkel, *Phys. Rev.*, 54, 647–648 (1938).
38. S. V. Novikov, D. H. Dunlap, V. M. Kenkre, P. E. Parris, and A. V. Vannikov, *Phys. Rev. Lett.*, 81, 4472–4475 (1998).
39. S. Verlaak, V. I. Arkhipov, and P. Heremans, *Appl. Phys. Lett.*, 82, 745–747 (2003).
40. J. G. Simmons, *Phys. Rev.*, 155, 657–660 (1967).
41. W. Helfrich, in: *Physics and Chemistry of the Organic Solid State*, Vol. 3, D. Fox, M. M. Labes, and A. Weissberger (eds.), Interscience, New York, 1967, pp. 1–58.
42. G. G. Malliaras, J. R. Salem, P. J. Brock, and C. Scott, *Phys. Rev. B*, 58, R13411–R13414 (1998).
43. L. Bozano, S. A. Carter, J. C. Scott, G. G. Malliaras, and P. J. Brock, *Appl. Phys. Lett.*, 74, 1132–1134 (1999).
44. H. E. Katz and Z. Bau, *J. Phys. Chem. B*, 104, 671–678 (2000).
45. F. Garnier, *Chem. Phys.*, 227, 253–262 (1998).
46. M. Van der Auweraer, F. C. De Schryver, P. M. Borsenberger, and H. Bässler, *Adv. Mater.*, 6, 199–213 (1994).
47. S. V. Rakhmanova and E. M. Conwell, *Appl. Phys. Lett.*, 76, 3822–3824 (2000).
48. J. A. Freire, M. G. E. da Luz, D. Ma, and I. A. Hümmelgen, *Appl. Phys. Lett.*, 77, 693–695 (2000).
49. P. W. M. Blom, M. J. M. de Jong, and M. G. van Munster, *Phys. Rev. B*, 55, R656–R659 (1997).
50. P. M. Borsenberger and J. J. Fitzgerald, *J. Phys. Chem.*, 97, 4815–4819 (1993).

7

Electronic Structure of Organic Photovoltaic Materials: Modeling of Exciton-Dissociation and Charge-Recombination Processes

Jérôme Cornil[a,b], Vincent Lemaur[a], Michelle C. Steel,[a] Hélène Dupin[a], Annick Burquel[a], David Beljonne[a,b], and Jean-Luc Brédas[a,b]

[a] *Laboratory for Chemistry of Novel Materials, Center for Research in Molecular Electronics and Photonics, University of Mons-Hainaut, Mons, Belgium*

[b] *School of Chemistry and Biochemistry, Georgia Institute of Technology, Atlanta, USA*

Contents

Abstract In this chapter, we describe at the quantum-chemical level, the main parameters governing the exciton-dissociation and charge-recombination processes that determine the efficiency of organic photovoltaic materials. We take as example a blend made of phthalocyanine as the electron donor and perylene bisimide as the acceptor. On the basis of the theoretical results, various strategies are discussed in order to increase the number of generated charge carriers in organic blends.

Keywords quantum-chemical calculations, electron-transfer theory, photoinduced electron-transfer, exciton dissociation, charge recombination, organic conjugated materials, organic photovoltaics, Marcus theory, Weller equation, reorganization energy, electronic coupling, driving force.

7.1. INTRODUCTION

Since the early days of organic electronics, quantum chemistry has proven to be a very valuable tool to understand the electronic and optical properties of the conjugated materials used in the devices and to define new strategies towards enhanced efficiencies. State-of-the-art calculations are no longer limited to a single conjugated molecule and have addressed over the recent years important interchain processes, such as solid-state luminescence properties [1,2], charge hopping [1,3], resonant energy transfer [4,5], or photoinduced charge transfer [6]. Thus, the mechanisms governing the operation of organic-based devices, and in particular solar cells, can be adequately described at the quantum-chemical level. This is what we illustrate in this contribution, where we focus on two key processes in organic solar cells, namely the dissociation of photogenerated electron–hole pairs (which has to be maximized) and the geminate recombination of charge carriers (which has to be prevented) at organic–organic interfaces. In order to provide the proper context for our contribution, we first recall in a simple way the working principle of organic solar cells.

An organic solar cell is fabricated by sandwiching an organic layer between two electrodes of different nature, typically indium tin oxide (ITO, which is metallic and transparent) and aluminum. This device is intended to convert the light emitted by the sun into electrical charges, which can then be collected by an external circuit [7]. To do so, the photons initially penetrate into the organic layer through the transparent side of the device and are absorbed by the organic material. The key aspect is then to have the photogenerated electron–hole pairs (excitons) dissociate into charge carriers. It appears that the efficiency of the charge-generation process is extremely low if a single material is incorporated into the organic layer. This is explained by the fact that the binding energy of the exciton, i.e., the strength of the Coulomb attraction between the electron and hole, is significant in conjugated materials (it is generally considered to be around 0.3–0.4 eV in conjugated polymers and even larger in small molecules [8]), thus making excitons very stable species. As a result, the organic devices built nowadays rely either on multilayer structures or blends made from an electron-donor component and an electron-acceptor component. The two components can be polymers (for instance, phenylenevinylene chains (PPV) with different substituents [9,10]), two different molecules [11,12], or a mixture of polymer and molecule (for instance, a conjugated polymer as the donor and C_{60} as the acceptor [13,14]).

In such organic blends, the photons are absorbed by the donor or the acceptor, or by both. If the donor absorbs most of the incident light, electron–hole pairs are generated on the donor moieties following (in a simple one-electron picture) the promotion of a π-electron from their highest occupied molecular orbital (HOMO) level to their lowest unoccupied molecular orbital (LUMO) level. In order to dissociate into charge carriers, the excitons have then to migrate towards a donor–acceptor interface. This points to the predominant role of the morphology of the blends in defining the efficiency of the exciton-dissociation process. If each of the two components are deposited in separate layers, only the excitons that are generated close to the interface or that can reach it during their lifetime are likely to be converted into electrical charges; since the exciton diffusion range is generally limited in organic thin films (on the order of 10 nm [15]), such a double-layer structure is far from optimal. A largely distributed contact area between the two partners is thus required for efficient devices; this can be promoted by mixing homogeneously the two components or by vertical segregation of the two phases

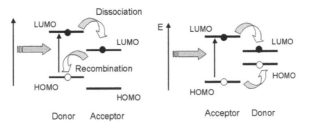

Figure 7.1. Illustration of the exciton-dissociation and charge-recombination processes (left) and of an energy transfer (right) in a donor–acceptor pair.

[16]. The key role played by the morphology of the blends is further attested by the fact that the efficiency of solar cells can be improved by a postproduction treatment of the devices [17].

Once an exciton reaches the donor–acceptor interface, the electron promoted in the LUMO level of the excited donor can be transferred to the lower-lying LUMO level of the acceptor (see Figure 7.1). If the hole then remains on the HOMO level of the donor, light is converted by this photoinduced *electron-transfer* process into electrical charges (i.e., one negative polaron on the acceptor and one positive polaron on the donor after the fast nuclear relaxation upon charge formation has occurred). Note that the same final charge-separated state can be reached when the acceptor is initially photoexcited, following a photoinduced *hole-transfer* from the HOMO level of the acceptor to the HOMO level of the donor. Finally, the charges have to escape from their mutual Coulomb attraction before migration towards the electrodes can occur, typically via a hopping mechanism and possibly with the help of the built-in potential created by connecting the two different metallic electrodes.

7.2. THE FAILURE OF THE STATIC VIEW

This widely accepted description of the operation of organic solar cells suggests that an offset of the frontier electronic levels of the donor and acceptor units is just what it takes to induce exciton dissociation. However, the situation is not that simple: first, exciton dissociation can take place only if the energy gained by the electron [hole] when transferred from the LUMO of the donor to the LUMO of the acceptor [from the HOMO of the acceptor to the HOMO of the donor] compensates for the binding energy of the intrachain exciton; second, it is essential to make sure that the charge-separated state is the lowest in energy in the blend. The latter is not always the case, as demonstrated in a recent work in collaboration with the Cambridge group [18]. In that study, we characterized the electronic properties of two blends made of two different PPV-substituted derivatives, namely a DMOS-PPV/CN-PPV blend and a MEH-PPV/CN-PPV blend (see chemical structures in Figure 7.2). According to standard semiempirical intermediate neglect of differential overlap (INDO) [19] calculations, there is a significant offset between the frontier electronic levels of DMOS-PPV or MEH-PPV (acting as donor) and those of CN-PPV (acting as acceptor); we thus expect a priori efficient exciton dissociation in both blends. This is, however, contradicted by the experiments, which show very efficient charge generation in the MEH-PPV/CN-PPV blend (evidenced by a quenching of the photoluminescence and an increase in the photocurrent) but indicate energy transfer

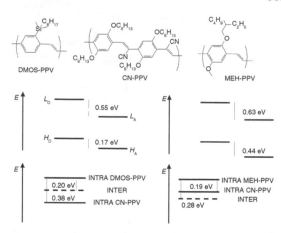

Figure 7.2. Description of the frontier electronic levels and ordering of the lowest intramolecular (INTRA) versus intermolecular (INTER) excited states in DMOS-PPV/CN-PPV and MEH-PPV/CN-PPV blends, as calculated at the INDO/SCI level.

in the DMOS-PPV/CN-PPV blend, with a luminescence signal characteristic of the CN-PPV chains.

These experimental observations cannot be explained at the one-electron level but can be rationalized on the basis of a model where the key idea is to look at the ordering of three relevant excited states: (i) the lowest intramolecular excited state of the donor (essentially described by a HOMO–LUMO excitation); (ii) the lowest intramolecular excited state of the acceptor (originating similarly from an electronic transition between its HOMO and LUMO levels); and (iii) the charge-transfer (CT) excited state that can be seen as the result of an electron excitation from the HOMO of the donor to the LUMO of the acceptor. Figure 7.2 provides the estimated ordering of these states in the two blends. In going from DMOS-PPV to CN-PPV, we calculate the lowest intramolecular excited state to be redshifted by 0.38 eV due to the asymmetric stabilization of the HOMO and LUMO levels in CN-PPV. The stabilization of the LUMO level by 0.55 eV induced by the cyano groups significantly lowers the energy of the lowest charge-transfer excited state; however, this stabilization is partly offset by the energy required to transform the intrachain exciton into an interchain exciton (considered here to be equal to 0.35 eV). Therefore, we estimate the lowest charge-transfer excited state in the PPV/CN-PPV blend to be located 0.20 eV below the lowest excited state of PPV, and hence 0.18 eV *above* the lowest intrachain excitation of CN-PPV. Thus, upon excitation of DMOS-PPV, we expect an energy transfer towards the CN-PPV chains to take place in the DMOS-PPV/CN-PPV blend, which is supported by the experimental data.

In the case of the MEH-PPV/CN-PPV pair, analysis of the one-electron structure reveals that the lowest optical transition of CN-PPV is redshifted by only 0.19 eV with respect to that of MEH-PPV while the LUMO level of CN-PPV is 0.63 eV lower than that of MEH-PPV. Since the stabilization of the lowest charge-transfer excited state has to take into account the energy required to separate the electron and the hole, the lowest charge-transfer excited state in the MEH-PPV/CN-PPV blend is estimated to be ca. 0.28 eV (0.63–0.35) *below* the lowest intrachain transition of MEH-PPV, that is some 0.10 eV *below* the lowest excited state of CN-PPV. Thus, in this blend, the occurrence of a charge-transfer process is predicted at the polymer–polymer interface, in full agreement with the experimental observations. Interestingly, these results demonstrate that a seemingly minor change in the substitution pattern of conjugated chains can decide whether charge transfer or energy transfer takes place between the two chains.

Though this model is instructive, it suffers from strong limitations (the excitations are described in a one-electron picture and the exciton binding energy is introduced somewhat arbitrarily). As a result, it cannot be readily applied to any donor–acceptor pair. Moreover, it is critical to go beyond the static view based on the ordering of various excited states and to address the dynamics of the processes of importance in organic solar cells. This is motivated by the fact that the exciton-dissociation process competes with a charge-recombination mechanism in which the charge-separated state decays back into the ground state of the blend (i.e., the electron in the LUMO level of the acceptor is transferred to the HOMO level of the donor, see Figure 7.1); this mechanism should be limited as much as possible to ensure an efficient generation of charge carriers in the device.

In this context, it is the goal of this chapter to illustrate that quantum-chemical calculations can provide estimates of all the parameters governing the rates of exciton dissociation and charge recombination in donor–acceptor blends. Our approach is described by considering a blend made of a free-base phthalocyanine (Pc) as the donor and perylene bisimide (PTCDI) as the acceptor (see chemical structures in Figure 7.3). This choice is motivated by the fact that closely related blends, typically incorporating a metal phthalocyanine, have been the subject of recent experimental studies [20,21]; moreover, phthalocyanine molecules are low-energy absorbers, with the lowest absorption band peaking around 1.8 eV [22], thus closely matching the region where the solar emission is the most intense. We have chosen not to consider here the widely used C_{60} derivatives since they would make the analysis much more complex; this is related to the fact that the C_{60} molecule possesses a large number of low-lying unoccupied orbitals which all create possible pathways for exciton dissociation [23].

The remainder of this chapter is structured as follows. We introduce our theoretical approach in Section 3 and present our preliminary results for the Pc/PTCDI blend in Section 4. On the basis of these results, we give in Section 5 an overview of various possible strategies to increase the number of generated free carriers in organic blends.

7.3. THEORETICAL APPROACH

It is not the goal of this section to describe standard quantum-chemical techniques that are well documented in textbooks and in the cited references but rather to

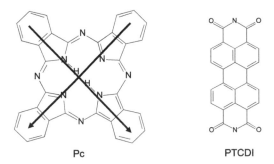

Figure 7.3.   Chemical structures of phthalocyanine (Pc, left) and perylene bisimide (PTCDI, right); the arrows represent the orientation of the lowest two unoccupied orbitals of Pc.

illustrate the way these have been exploited here to calculate the parameters relevant in the description of exciton dissociation and charge recombination in solar cells. Since both these processes correspond to an electron-transfer reaction, we have estimated their rates within the framework of Marcus theory and extensions thereof [24]. This formalism has been developed in the weak coupling limit, which in the case of dissociation means that the excitation is initially localized on the donor and ultimately yields a positive [negative] polaron localized on the donor [acceptor], as expected in the organic blends. We will not consider here situations where the excitations and charges are delocalized over several molecules, as might occur in highly crystalline phases. In the semiclassical limit of Marcus theory, the charge-transfer rate is expressed as a typical Arrhenius law and can be written as [25]:

$$k = A \exp^{(-\Delta G^{\#}/kT)} = \left(\frac{4\pi^2}{h}\right) V_{RP}^2 \left(\frac{1}{\sqrt{4\pi\lambda kT}}\right) \exp^{\left(-(\Delta G^{\circ}+\lambda)^2/4\lambda kT\right)} \tag{1}$$

where ΔG° represents the free enthalpy of the reaction, $\Delta G^{\#}$ the free enthalpy needed to reach the transition state, V_{RP} the electronic coupling between the initial and final states, and λ the reorganization energy (see Figure 7.4a). The latter parameter encompasses two contributions: (i) the internal part λ_i, which describes the changes in the geometry of the donor and acceptor moieties upon charge transfer; this relates to the fact that the geometry of a conjugated molecule is significantly modified upon excitation or charging due to the strong electron–phonon coupling, characteristic of these systems [26]; and (ii) the external part λ_s linked to the change in the electronic polarization of the surrounding medium (the reaction is usually so fast that it does not allow for the nuclear reorganization of the surrounding molecules). The square

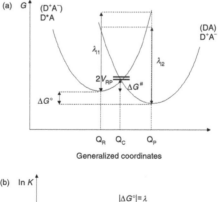

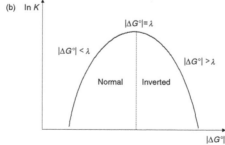

Figure 7.4. (a) Illustration of the various parameters entering into Marcus theory to estimate the transfer rate for exciton dissociation ($D^*A \rightarrow D^+A^-$) or charge recombination ($D^+A^- \rightarrow DA$); (b) evolution of the transfer rate as a function of [$\Delta G^{\circ} + \lambda$], which shows a peak profile.

dependence in $(\Delta G° + \lambda)$ in the exponential implies that the transfer rate displays a peak profile as a function of this term and reaches a maximum when $\Delta G°$ (taken in the example here as a negative number) is equal in absolute value to λ (a positive number). The rate is smaller when $|\Delta G°| < \lambda$ (normal region) and when $|\Delta G°| > \lambda$ (inverted region) (see Figure 7.4b). The semiclassical formalism is based on the assumption that the system has to reach the transition state for the transfer to occur; it neglects the tunneling effects that can assist the transfer, especially at low temperatures. These can be treated quantum mechanically by introducing into the expression for the transfer rates the thermally averaged density of vibrational modes in the initial and final states and their overlap. This has been accounted for in our approach by adopting Marcus–Levich–Jortner's formalism in which the electron-transfer rate is written as

$$k = \left(\frac{4\pi^2}{h}\right) V_{RP}^2 \left(\frac{1}{\sqrt{4\pi\lambda kT}}\right) \sum_{v'} \exp^{-S} \frac{S^{v'}}{v'!} \exp^{(-(\Delta G°+\lambda_s+v'h\langle\omega_v\rangle)^2/4\lambda_s kT)} \tag{2}$$

In our calculations, a single effective high-frequency mode of energy $h\omega$ (set here equal to 0.20 eV), which promotes the charge transfer, is treated quantum-mechanically. In Equation (2), the Huang–Rhys factor S is directly related to the internal reorganization energy $(S = \lambda_i/h\omega)$ and the summation runs over the vibrational levels.

All the parameters $(\Delta G°, \lambda_i, \lambda_s, \text{ and } V_{RP})$ involved in Equation (2) can, in principle, be estimated from quantum-chemical calculations:

- $\Delta G°$ has been estimated from Weller's equation [27] as the energy difference between the constituents in the initial and final states, accounting for the Coulomb attraction between the two polarons in the charge-separated state. Thus, for exciton dissociation and when neglecting the entropy contributions, $\Delta G°$ is written as:

$$\Delta G_{dis}° = E(D^+) + E(A^-) - E(D^*) - E(A) + E_{coul} \tag{3}$$

with

$$E_{coul} = \sum_D \sum_A \frac{q_D q_A}{4\pi\varepsilon_0\varepsilon_s r_{DA}} \tag{4}$$

where $E(D^*)$, $E(D^+)$, $E(A)$, and $E(A^-)$ represent the total energy of the donor in the equilibrium geometry of the lowest excited state and of the cationic state, and that of the acceptor in the equilibrium geometry of the ground state and of the anionic state, respectively. q_D and q_A correspond to the atomic charges on the donor and the acceptor, respectively (as calculated at the AM1-CI/COSMO level following a Mulliken population analysis), which are separated by a distance r_{DA}; ε_s is the static dielectric constant of the medium.

In order to compute the first four terms, we have first optimized the geometry of the molecules in a given state with the help of the Austin Model 1 (AM1) method [28] coupled with a full configuration interaction (FCI) scheme within an active space built from a few frontier electronic levels, as implemented in the AMPAC package [29]; note that the geometry of all systems discussed in this chapter have been optimized with this same technique. The size of the active space has been systematically chosen to ensure the convergence of the results. The influence of the dielectric properties of the medium has also been taken into account by means of the COSMO software [30] implemented in AMPAC. The Coulomb attraction term has been estimated by summing the paired interactions between the atomic charges of the donor and the acceptor, as obtained from a Mulliken population analysis performed on the AM1-CI/COSMO results. The free enthalpy for the charge recombination is estimated from a similar expression, involving this time the charge-separated state and the ground state. In both cases, we have varied the static dielectric constant ε_s in the range 2.5–5, which is typical for organic thin films [31].

- λ_i corresponds to the difference between the energy of the reactants [products] in the geometry characteristic of the products [reactants] and in their equilibrium geometry. These two ways of estimating λ_i provide the same value only if the two parabolas representing the reactants and the products have the same curvature. Since this is often not the case, λ_i is actually estimated as the average of λ_{i1} and λ_{i2} defined for exciton dissociation as:

$$\lambda_{i1} = E(D^*_{Q_P}) + E(A_{Q_P}) - E(D^*_{Q_R}) - E(A_{Q_R}) \tag{5}$$

$$\lambda_{i2} = E(D^+_{Q_R}) + E(A^-_{Q_R}) - E(D^+_{Q_P}) - E(A^-_{Q_P}) \tag{6}$$

where $E(D^*)$, $E(D^+)$, $E(A)$, and $E(A^-)$ represent the total energy of the donor in the lowest excited state and in the cationic state, and that of the acceptor in the ground state and in the anionic state, respectively; Q_R and Q_P refer to the equilibrium geometry of the reactants and products, respectively. We have evaluated all these terms at the AM1-CI level.

- λ_s has been estimated by the classical dielectric continuum model developed by Marcus [32]. We thus make the assumption here, which is reasonable for amorphous blends, that the electron transfer occurs in an isotropic dielectric environment. The reorganization term is given by

$$\lambda_s = \frac{e^2}{8\pi\varepsilon_0} \left(\frac{1}{\varepsilon_{op}} - \frac{1}{\varepsilon_s} \right) \left(\frac{1}{R_D} + \frac{1}{R_A} + 2 \sum_i^{acceptor} \sum_j^{donor} \frac{q_i q_j}{r_{ij}} \right) \tag{7}$$

where e is the transferred electrical charge, ε_s the static dielectric constant of the medium, and ε_{op} the optical dielectric constant set in all cases equal to a typical value of 2.25 [33]; R_D ($= 4.06 \, \text{Å}$) and R_A ($= 3.45 \, \text{Å}$) are the effective radii of the phthalocyanine and perylene molecules estimated from their total volume when assimilated to a sphere.

- The electronic coupling V_{RP} appearing in Equation (2) is to be evaluated in a diabatic description where the initial and final states do not interact. This is not the situation considered in our correlated quantum-chemical calculations, which explicitly take into account the interaction between the two states, and hence provide an adiabatic description of the system. However, V_{RP} can be estimated from quantities given directly by the CI calculations performed on the donor–acceptor pair in the framework of the generalized Mulliken–Hush (GMH) formalism, which refers to a vertical process between the two states. V_{RP} then is written as

$$V_{RP} = \frac{\mu_{RP} \Delta E_{RP}}{\sqrt{(\Delta \mu_{RP})^2 + 4(\mu_{RP})^2}} \tag{8}$$

where ΔE_{RP}, $\Delta \mu_{RP}$, and μ_{RP} correspond to the energy difference, the dipole moment difference, and the transition dipole moment between the initial and final states. On one hand, we have computed these parameters for exciton dissociation using the nuclear positions corresponding to the initial state (and without including surrounding medium effects) by means of the INDO Hamiltonian coupled to a single configuration interaction (SCI) scheme. Since the calculated energy of the charge-transfer state is not often reliable due to the neglect of the intermolecular polarization effects, we have applied a static electric field along the charge-transfer direction to match ΔE_{RP} to the ΔG°_{dis} value provided by Weller's equation. This is clearly an approximation for a vertical process for which we should actually estimate the energy of the CT state in the geometry of the D*A (or DA*) state, considering the *optical* constant of the medium (thus without the slow nuclear reorganization of the surrounding medium included in the *static* dielectric constant); however, this simple choice is relevant for charge recombination and is justified for exciton dissociation by the fact that the V_{RP} amplitude is only weakly sensitive to changes in the actual energy of the CT state. On the other hand, since the charge recombination starts from the fully equilibrated CT state, it makes sense to calculate $\Delta \mu_{RP}$ and μ_{RP} in the D*A geometry to first approximation and with the same electric field in order to position the CT state at an energy describing reasonably well its mixing with the lowest intramolecular excited states; we then set ΔE_{RP} equal to ΔG°_{rec} to evaluate the electronic coupling using Equation (8).

7.4. DYNAMICAL ASPECTS

We have considered initially a dimer made of a PTCDI molecule superimposed on a Pc molecule with a typical intermolecular distance of 4 Å. In Figure 7.5, the evolution of ΔG° is plotted for exciton dissociation (ΔG°_{dis}, assuming that the donor is initially excited) and charge recombination (ΔG°_{rec}) as a function of the inverse dielectric constant of the medium in the range between 2 and 5. Exciton dissociation ΔG°_{dis} is

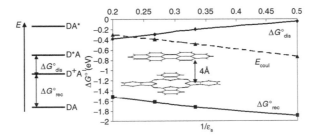

Figure 7.5. Evolution of ΔG° for exciton dissociation (ΔG°_{dis}) and charge recombination (ΔG°_{rec}) and of the Coulomb attraction between the polarons of opposite sign (E_{coul}) as a function of the inverse static dielectric constant ε_s in a Pc/PTCDI dimer (with the two molecules superimposed and separated by 4 Å), as calculated at the AM1-CI/COSMO level.

always negative, indicating that exciton dissociation occurs irrespective of the polarity of the medium. Its value becomes more negative (going from -0.20 eV at $\varepsilon_s = 3$ to -0.38 eV at $\varepsilon_s = 5$) when the dielectric constant is increased even though stabilizing Coulomb term is reduced by screening effects (from -0.48 eV at $\varepsilon_s = 3$ to -0.30 eV at $\varepsilon_s = 5$). Similar changes in the energy of the charge-transfer state upon variation of the medium polarity have been observed experimentally for electron-transfer reactions in solution [34].

A detailed analysis of our data reveals that exciton dissociation would not occur or would be only weakly thermodynamically favored if the Coulomb attraction between the two polarons of opposite sign was neglected. Thus, it has to be realized that the two generated charges cannot easily escape their mutual attraction to yield free carriers (the two charges must be separated by 11 nm at room temperature to compensate for the Coulomb attraction for $\varepsilon_s = 5$). It is clear that this Coulomb attraction is very detrimental for the operation of the solar cells; in fact, it would be expected from the previous considerations that most of the geminate polaron pairs (that can also be referred to as exciplexes) decay either by non-radiative recombination or by light (exciplex) emission [21,35,36]. This raises the important question of the nature of the mechanism allowing for the separation of the bound polaron pairs into free carriers. The mechanism of separation has been tentatively attributed to: (i) the role of interfacial dipoles between the donor and the acceptor units in their ground state, which would facilitate the dissociation of the polaron pairs [37]; and (ii) the role of disorder intrinsic to organic thin films; in this view, the energy that the hole or the electron gained by transfer to a segment of lower energy would help to compensate for the Coulomb attraction [38]. The separation could also be facilitated if it occurs prior to the thermalization of the polaron pairs. Conversely, ΔG°_{rec} becomes less negative when the polarity of the medium increases, as a result of the progressive stabilization of the charge-separated state. Interestingly, the calculated values for ΔG°_{rec} at $\varepsilon_s = 3$ (-1.72 eV) and $\varepsilon_s = 5$ (-1.52 eV) agree with the experimental value inferred from the energy of the exciplex emission in Pc/PTCDI-like blends (1.63 eV) [21]; we note that this good agreement gives confidence in the reliability of the theoretical approach we have followed. The sum of ΔG°_{rec} and ΔG°_{dis} has an almost constant value (on the order of 1.9 eV, consistent with the energy of the lowest excited state of Pc) owing to the fact that the energies of the ground state and the intramolecular excited states are only slightly affected by the medium polarity.

The internal reorganization energy λ_i is found to be slightly larger for the exciton dissociation (0.33 eV) than for the charge recombination (0.26 eV). The external reorganization energy λ_s has a similar order of magnitude and increases with the polarity of the medium (from 0.24 eV at $\varepsilon_s = 3$ to 0.52 eV at $\varepsilon_s = 5$), as expected. Since the sum of the two λ contributions is larger than $|\Delta G^\circ|$ for exciton dissociation, this process operates in the normal region of Marcus; in contrast, $|\Delta G^\circ|$ is larger than the total reorganization energy for charge recombination which thus occurs in the inverted region. The electronic coupling V_{RP} evolves from 447 to 445 cm^{-1} for exciton dissociation and from 1600 to 642 cm^{-1} for charge recombination when going from $\varepsilon_s = 3$ to $\varepsilon_s = 5$. That V_{RP} is almost constant for exciton dissociation results from a compensation of the ΔE_{RP} and μ_{RP} terms in Equation (8), which does not prevail for charge recombination. Note that these V_{RP} values actually encompass two different pathways; this occurs because the lowest absorption band of phthalocyanine is made of two closely separated excited states that are mostly described by an electronic transition from the HOMO level to the LUMO or LUMO+1 level, respectively (the lowest two unoccupied orbitals have very similar shapes and are oriented perpendicularly along the two branches of the conjugated core, see Figure 7.3) [22]. We have thus defined V_{RP} as the square root of the sum of the individual contributions ($V_{RP1}^2 + V_{RP2}^2$) of the two excited states, for which we have considered the same $\Delta G°$ and λ values.

All the parameters we have discussed can be entered into Equation (2) to estimate the corresponding transfer rates. In doing so, we obtain values of 2.23×10^{13} and 1.20×10^{13} s^{-1} for the exciton-dissociation rate for $\varepsilon_s = 3$ and 5, respectively. The transfer rate would increase if $|\Delta G°|$ and λ_s were to converge towards a similar value; this does not occur here since the absolute value of both parameters increases with medium polarity. The reduction in transfer rate with ε_s is thus explained by the fact that the increase in λ_s is slightly larger than that in $|\Delta G°|$. In contrast, charge recombination occurs faster when the dielectric constant is increased (from 4.98×10^{11} to 4.79×10^{12} s^{-1} on going from $\varepsilon_s = 3$ to $\varepsilon_s = 5$). This is due to the opposite evolutions of $|\Delta G°|$ and λ_s that reduce the gap between their absolute values. It is also of interest to analyze the evolution of the ratio k_{dis}/k_{rec} as a function of the dielectric constant; strikingly, the results indicate that this ratio goes from 45 at $\varepsilon_s = 3$ to 2.5 at $\varepsilon_s = 5$. Thus, *an increase in the polarity of the medium leads to a situation where charge recombination increasingly competes with exciton dissociation.* Note that, by extension, this result suggests that the high dielectric constant typical of the inorganic semiconductors present in hybrid devices might have a detrimental impact on the efficiency of such devices. These results thus demonstrate that the dielectric properties of the medium can play a key role in defining the efficiency of free carrier generation in organic solar cells. They also contrast with the common understanding that exciton dissociation is much faster than charge recombination in donor–acceptor pairs; the latter case was found for instance in recent measurements on oligo(phenylenevinylene)–perylene copolymers that yield a k_{dis}/k_{rec} ratio on the order of a few hundreds [39]. The main difference is that we are dealing here with a Pc donor presenting a lowest excited state at much lower energy than in oligo(phenylenevinylene)s, which results in a much lower $|\Delta G°_{rec}|$ (see the left side of Figure 7.5). This is confirmed by the fact that the charge-recombination rate in the Pc/PTCDI blend for $\varepsilon_s = 5$ is decreased by a factor of 40 or 3000 when $|\Delta G°_{rec}|$ is increased artificially by 0.5 or 1 eV, respectively (which makes it go deeper into the inverted Marcus regime).

If we assume now that the acceptor is initially excited, we calculate the following parameters for exciton dissociation and $\varepsilon_s = 3$ [$\varepsilon_s = 5$]: $\lambda_i = 0.11$ eV; $\Delta G° = -1.11$ [-1.31] eV, corresponding to the lowest excited state of PTCDI at about 0.9 eV above that of Pc, in agreement with experimental values (~ 1.8 eV for Pc [22] versus ~ 2.5 eV for PTCDI [38]); $V_{RP} = 585$ [559] cm^{-1}; and $k_A = 4.94 \times 10^{11}$ [1.91×10^{12}] s^{-1}. This translates into a ratio k_D/k_A (where k_D corresponds to the rate when the donor is initially excited and k_A to that when acceptor is initially excited) of 45 and 6.3 for $\varepsilon_s = 3$ and 5, respectively. The reduction in the exciton-dissociation rate when exciting the acceptor is associated with the increase in energy gap between $|\Delta G°|$ and λ, the dissociation actually taking place in the inverted region. The nature of the excited species thus also dictates the dynamics of charge generation in the device.

Next, we have analyzed the way the transfer rates vary as a function of the relative positions of the donor and acceptor units. First we have computed the change in the exciton-dissociation rate (assuming that the donor is initially excited) when modulating the intermolecular separation in the range between 4 and 8 Å (see Figure 7.6). Analysis of the evolution of the various parameters entering into the rate expression shows that: (i) $|\Delta G°|$ slightly decreases due to the attenuation of

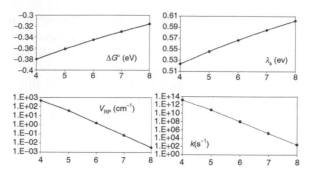

Figure 7.6. Evolution of the free enthalpy $\Delta G°$, external reorganization energy λ_s, electronic coupling V_{RP}, and transfer rate k for exciton dissociation in a Pc/PTCDI dimer as a function of the intermolecular separation in angstroms (Å) between the donor and acceptor units; ε_s is set equal to 5.

the Coulomb attraction between the electron and the hole; (ii) λ_s slightly increases when the two net charges are separated; and (iii) V_{RP} decreases exponentially as a result of the reduction in the degree of overlap between the wavefunctions of the electronic levels of the two molecules. The distance dependence of the electron-transfer rate is clearly governed by the latter parameter and is only slightly affected by the evolution of $\Delta G°$ and λ_s. These data fit the typical expression for the distance dependence of the electron-transfer rate via a through-space, i.e., superexchange mechanism (whereby the charge tunnels from the donor to the acceptor) [6,25,40]:

$$k \approx k_0 \exp(-\beta d) \qquad (9)$$

where β is the decay factor estimated here to be $6.2 \, \text{Å}^{-1}$. Similar considerations prevail for the charge-recombination process, which becomes extremely slow at large intermolecular distances.

We have also investigated the way in which the dissociation rate is affected when rotating the perylene molecule on top of the phthalocyanine molecule, with the intermolecular distance fixed at 4 Å. We observe a weak sensitivity of the electronic coupling to the rotational disorder irrespective of the value of the dielectric constant; V_{RP} fluctuates on an average by a factor of 2 around the mean value (see Figure 7.7). Similar trends also apply to the charge-recombination process. This behavior is attributed to the two-dimensional character of the Pc molecule, which promotes two low-lying unoccupied orbitals with perpendicu-

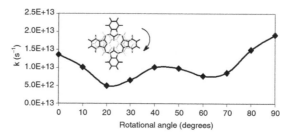

Figure 7.7. Evolution of the exciton-dissociation rate in a Pc/PTCDI dimer as a function of the rotational angle between the two molecules (the intermolecular distance is set equal to 4 Å and ε_s to 5).

lar orientations. Since V_{RP} is primarily driven by the overlap between the LUMO levels of the donor and acceptor units, its nearly constant value is explained by the fact that one orbital of Pc mostly contributes to V_{RP} when this orbital is aligned with the molecular axis of PTCDI, while the other comes into play when the PTCDI molecule is rotated by 90°; both levels participate for intermediate rotational angles. This insensitivity to rotational disorder would not occur if the donor and acceptor were rigid rods; the dimensionality of the chemical structures used in the organic blends is thus another parameter of importance for the solar cells.

Finally, we have characterized the impact of translating one molecule laterally in the dimer. As expected, both the dissociation and recombination rates drop rapidly as the degree of spatial overlap between the two molecules is reduced.

7.5. STRATEGIES FOR EFFICIENT CHARGE GENERATION

Despite the remarkable achievements in the field of organic photovoltaics, it has to be recognized that up to now the donor and acceptor moieties used in solar cells have often been chosen mostly by empiricism rather than on the basis of well-defined guidelines. Thus, the results of quantum-chemical calculations can prove useful and suggest the synthetic chemists some strategies for a rational design of donor–acceptor pairs. It is from this perspective that further theoretical calculations are now presented, suggesting routes that could be explored in order to maximize exciton dissociation and limit charge recombination in the devices.

7.5.1. Specifically Designed Supramolecular Architectures

Many recent experimental works have exploited covalently linked donor–acceptor pairs (i.e., dyads or tryads) to ensure a good dispersion of the two components in the organic phase [41,42]. The concept of a double-cable copolymers based on polymer chains with a large number of acceptor units as substituents has also been introduced [43]. We demonstrate here that the nature of the spacer connecting the donor and acceptor moieties as well as the nature of the anchoring points can be modulated to promote higher electronic couplings for exciton dissociation. To do so, we consider supramolecular architectures made of a perylene molecule connected to a three-ring phenylenevinylene oligomer (Figure 7.8), systems very reminiscent of those recently synthesized by the Eindhoven group [39]. When the two units are connected by the two ends in an extended configuration (see left side of Figure 7.8), the electronic coupling for exciton dissociation is very small regardless of the nature of the spacer we considered (oxygen, methylene, dimethylene, ester, or amide). Since the amplitude of this coupling is mostly governed by the overlap between the LUMO levels of the two units, the results can be understood by the fact that there is no electronic density on the nitrogen atom in the LUMO level of PTCDI. This is confirmed by the huge increase in V_{RP} calculated when anchoring the spacer on one carbon atom of PTCDI that carries a significant electronic density in the LUMO level. The coupling can be further enhanced by a judicious choice of the spacer; this can lead to an enhancement of the coupling by up to about a factor of 40 with respect to the extended configuration. The exciton-dissociation rate can also be amplified with respect to the

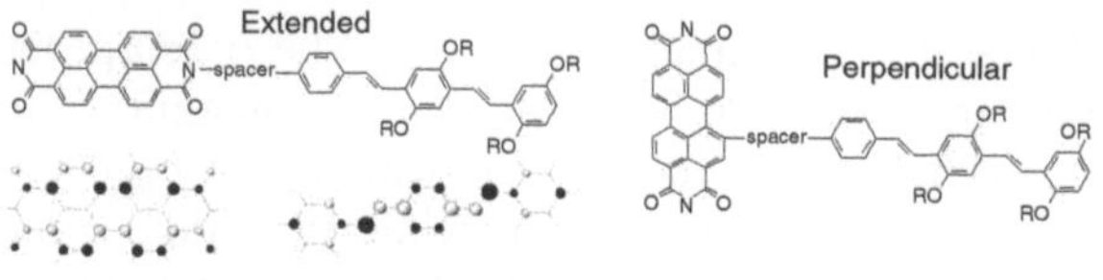

$D^*A \rightarrow D^+A^-$	V_{RP} (cm^{-1})
no spacer	29.5
—O—	24.4
$-CH_2-$	25.2
$-CH_2-CH_2-$	5.7
—C(=O)—O—	2.3
—C(=O)—NH—	58.9

$D^*A \rightarrow D^+A^-$	V_{RP} (cm^{-1})
—O—	225
$-CH_2-$	929
$-CH_2-CH_2-$	16
—C(=O)—O—	56
—C(=O)—NH—	393

Figure 7.8. Electronic couplings calculated for exciton dissociation (with the donor initially excited) in a supramolecular architecture made of a perylene and a methoxy-substituted three-ring oligophenylenevinylene connected, by various spacers, to different anchoring points, leading to extended and perpendicular conformations. The V_{RP} values are obtained from the GMH formalism with all the parameters calculated at the INDO/SCI level without any static electric field applied. The shape of the LUMO orbitals of perylene and of a three-ring unsubstituted OPV, whose overlap governs the amplitude of the electronic coupling, is also depicted; the size and color of the balls represent the amplitude and sign of the linear combination of atomic orbitals (LCAO) coefficients.

extended configuration by building an architecture where the two molecules are superimposed [39].

7.5.2. Donor–Bridge–Acceptor Architectures

The use of a short spacer does not guarantee that the generated polaron pairs can easily escape from their mutual Coulomb attraction. Thus, it is desirable to separate the donor and acceptor moieties while conserving significant transfer rates. It has long been known that electrons can transfer over large distances in supramolecular architectures where the donor is linked to the acceptor by means of a long saturated bridge [44]. In such systems, the charge transfer operates via a superexchange mechanism; its rate decreases exponentially with distance, with the decay factor β typically around 1 instead of the value of 6.2 found earlier [40]. This β-value is reduced by lowering the energy separation between the LUMO of the donor [HOMO of the acceptor] and the LUMO [HOMO] of the bridge for a photoinduced electron [hole] transfer process [45,46]; this can be achieved by replacing the saturated bridge by a conjugated backbone [47]. The transfer rate at large distances is further increased by creating a resonance between the frontier electronic levels of the bridge and the LUMO of the donor or HOMO of the acceptor. In this situation, the charge transfer can operate via two sequential hopping processes, first from the donor [acceptor] to the bridge then from the bridge to the acceptor [donor] for photoinduced electron [hole] transfer (see Figure 7.9); such a sequential hopping mechanism for instance plays a role in the conducting properties of DNA [48]. Another possible mechanism is that the LUMO of the donor [HOMO of the acceptor] gets delocalized over the bridge and that the transfer takes place by a superexchange mechanism from such a "super" donor [acceptor].

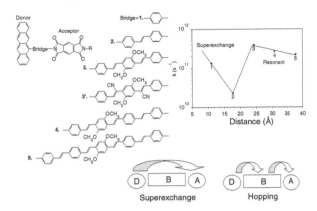

Figure 7.9. Illustration of the superexchange versus hopping mechanisms in the D–B–A architectures synthesized by Wasielewski and co-workers [49], made of tetracene (donor), pyromellitimide (acceptor), and p-phenylenevinylene oligomers of increasing size (bridge). The evolution of the measured transfer rate for exciton dissociation (with the donor initially excited) as a function of bridge size, as extracted from [49], is also shown.

These concepts are nicely illustrated by recent measurements performed on D–B–A systems built from tetracene (D), pyromellitimide (A), and oligophenylene-vinylenes (B) (see the chemical structures in Figure 7.9) [49]. Upon selective photo-excitation of the donor, the experimental data show that the electron-transfer rate first drops abruptly when the size of the bridge is increased before reaching much higher values for longer bridges. This has been rationalized by quantum-chemical calculations indicating that the transfer first operates via a superexchange mechanism before entering into a resonant regime [50]. The synthesis of D–B–A architectures including conjugated bridges is thus an interesting route to explore in order to generate loosely bound polaron pairs in organic solar cells. This could also be promoted in D–A_1– A_2–A_n systems, with A_1, A_2, and A_n representing acceptors of increasing strength (i.e., with a progressively stabilized LUMO level).

7.5.3. Symmetry Effects

Symmetry has rarely been exploited to design efficient molecular architectures despite clear demonstrations of its impact on charge-transfer rates [40]. This can be illustrated by considering two series of D–B–A systems with dimethoxynaphtalene as the donor, norbornene bridges of increasing size, and dicyanovinyl or dimethylmale-ate as the acceptor, respectively (Figure 7.10) [51,52]. We have calculated the electronic coupling for exciton dissociation and charge recombination for the two series within the GMH formalism. Using the nuclear configuration of the initial state (D^*A and D^+A^-, respectively) optimized at the AM1-CI level, all the parameters entering into the GMH expression were evaluated from INDO/SCI calculations performed with an electric field locating the CT state at a constant value of 3.15 eV (i.e., in the typical range suggested by the experimental measurements). This procedure is justi-fied by the fact that: (i) the energy of the CT states becomes exceedingly high for the longer bridges in the absence of polarization effects; and (ii) the V_{RP} value for these systems is rather insensitive to small changes in the energy of the CT state.

Electronic coupling: V_{RP} (cm^{-1})		
1(n)	D*A→D$^+$A$^-$	D$^+$A$^-$→DA
1(4)	710	707
1(6)	154	130
1(8)	48	36
1(10)	26	9.5
1(12)	8.8	1

2(n)	D*A→D$^+$A$^-$	D$^+$A$^-$→DA
2(3)	269	813
2(5)	25.3	488
2(7)	8.8	145
2(9)	1.7	48
2(11)	0.25	13

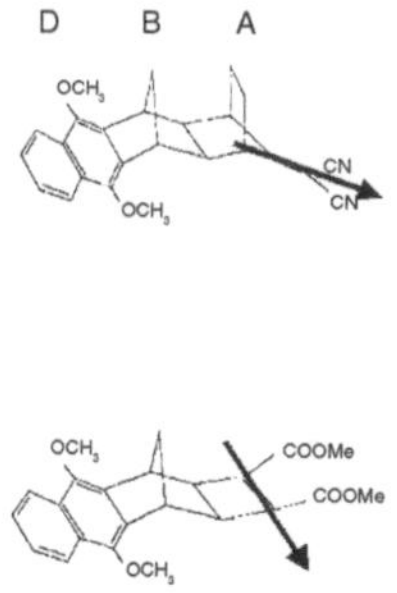

Figure 7.10. Chemical structures of the D–B–A systems synthesized by Paddon-Row and co-workers [50,51], made of dimethoxynaphtalene as the donor, norbornene bridges of increasing size, and dicyanovinyl or dimethylmaleate as the acceptor for series 1 and 2, respectively; n represents the number of methylene units connecting the donor to the acceptor along the bridge. The calculated electronic couplings for exciton dissociation and charge recombination in these systems are also given.

The results show that the exciton-dissociation rate is much larger in series 1 than in series 2, while opposite trends prevail for charge recombination (see Figure 7.10). These results can be primarily understood by symmetry arguments (see Figure 7.11). In fact, the LUMO of the donor and that of dicyanovinyl (acceptor 1) are found to be symmetric with respect to the plane of symmetry of these molecules while the LUMO of dimethylmaleate (acceptor 2) and the HOMO of the donor are antisymmetric. Since V_{RP} for exciton dissociation is driven by the overlap between the LUMO levels of the donor and acceptor units, the rate for acceptor 2 is expected to be small due to the involvement of two levels of different symmetry, leading to a cancellation of the global overlap; in contrast, the interaction between two levels of same parity yields a net global overlap, and hence a large electronic coupling, in agreement with the calculations for acceptor 1. On the other hand, the rate of charge recombination depends on the overlap between the LUMO level of the acceptor and the HOMO level of the donor. The same symmetry arguments rationalize why V_{RP} is much larger for acceptor 2 than for acceptor 1.

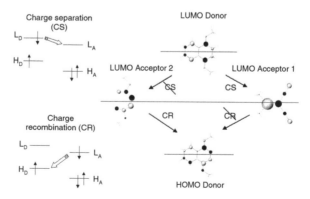

Figure 7.11. Illustration of the symmetry of the electronic levels involved in the exciton-dissociation and charge-recombination processes in the D–B–A systems described in Figure 7.10.

7.5.4. Low-Bandgap Polymers

Another way to increase the number of photogenerated free charges is simply to improve the match between the solar emission spectrum and the absorption of the materials. This is far from getting optimized in many cases (e.g., in the widely used PPV/C_{60} blends) and has motivated the replacement of C_{60} by C_{70} derivatives which absorb at longer wavelengths, an approach that has led to an increase in power conversion efficiency up to 3% [53]. Likewise, an attractive strategy would be to design organic blends involving an easily processible low-bandgap conjugated polymer as donor and a traditional acceptor, such as perylene or C_{60} [54]. The potential drawback here is that the frontier electronic levels of the low-bandgap material become located in the gap of the acceptor, and that both the hole and the electron therefore transfer to the donor when the acceptor is initially excited (see Figure 7.1); such an energy-transfer process leads to the confinement of the excitons in one component of the blend is of course unwanted for the efficient operation of solar cells.

For this reason, the location of the CT state in donor–acceptor pairs made of conjugated polymers with increasingly small bandgaps and PTCDI was calculated by means of Weller's equation. We have considered poly-*p*-phenylenevinylene (PPV), polythienylenevinylene (PTV), polythiophene (PT), polycyclopenta[2,1-b;3, 4-b']dithiophen-4-one (PT2O), and the derivative where the carbonyl group is replaced by dicyanovinyl (PT2CN) (see Figure 7.12). The lowest optical transition of these chains, assumed to be coplanar, is estimated at the INDO/SCI level to be 2.7, 1.8, 1.6, 0.9, and 0.8 eV, respectively; this compares very well with the measured absorption onsets at 2.4 eV for PPV [55], 1.7 eV for PTV [56], 1.8 eV for PT [57], 1.1 eV for PT2O [58], and less than 1.0 eV for PT2CN [59]. Analysis of the position of the frontier electronic levels of the chains (obtained at the INDO level by extrapolating the data calculated for oligomers of increasing size) shows a progressive stabilization of the LUMO level when the bandgap is reduced; since the LUMO energy of the low-bandgap polymers becomes similar to that of PTCDI, the results

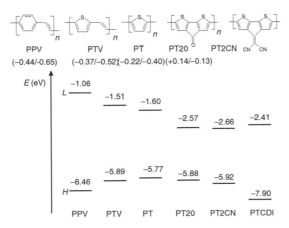

Figure 7.12. Description of the relative positions of the frontier electronic levels of conjugated polymers with increasingly small bandgaps (PPV → PTV → PT → PT2O → PT2CN) with respect to the perylene molecule, as calculated at the INDO level. The chemical structures of the various polymers are also shown and, between parentheses, the $\Delta G°$ values (in eV) calculated from Weller's equation for exciton dissociation between donor oligomers and PTCDI at $\varepsilon_s = 2.8$ and 4.8, respectively.

suggest that the free enthalpy of reaction will be gradually diminished when going from PPV to PT2CN. This is corroborated by the ΔG° values calculated for PTCDI located on top of the center of five-ring PPV, PTV, PT, and ten-ring PT2O, PT2CN oligomers, with the intermolecular distance fixed at 4 Å. They vary from -0.67 eV for PPV to $+0.11$ eV for PT2CN for $\varepsilon_s = 2.8$, thus indicating that the exciton-dissociation process becomes thermodynamically less favored when the bandgap is reduced. The situation can be improved by increasing the polarity of the medium, in which case ΔG° is always negative, varying from -0.89 to -0.16 eV for $\varepsilon_s = 4.8$. Thus, the incorporation of low-bandgap polymers in solar cells requires a proper selection of the two components. The driving force for exciton dissociation could be further increased by designing new low-bandgap polymers with frontier electronic levels higher in energy or by using acceptors with a larger electron affinity than PTCDI.

7.5.5. Triplet Excitons

The exploitation of triplet excitons in solar cells is another route that has remained mostly unexplored. However, two characteristics of triplet excitons should be of great interest for solar cells: (i) they have a long lifetime and can thus migrate over large distances to reach an interface, as evidenced by recent experimental studies [60]; and (ii) they dissociate into bound triplet polaron pairs, provided that the CT state is located below the lowest triplet excited state of the pumped material, thus limiting a priori the efficiency of the recombination channel into the singlet ground state. Dissociation of triplet excitons has been observed, for instance, in blends of C_{60} and platinum-polyynes [61].

7.6. CONCLUSIONS

In summary, we have described in this chapter our theoretical approach to describe at the full quantum-chemical level exciton-dissociation and charge-recombination processes in organic solar cells. The application of our methodology has allowed us to pinpoint a large number of parameters that control the rate of charge generation in solar cells, in particular the medium polarity, the dimensionality of the molecules, the donor–acceptor separation with or without intervening bridges, and symmetry and spin effects. We emphasize that any comparison between the theoretical predictions and the corresponding experimental data should focus on trends rather than on absolute values of transfer rates due to theoretical uncertainties in the estimates of the various parameters. There is clearly room for improvement of our approach in the future, for instance, by describing more appropriately solid-state polarization effects or the role of vibronic couplings.

ACKNOWLEDGMENTS

The work in Mons is partly supported by the Belgian Federal Government "Interuniversity Attraction Pole in Supramolecular Chemistry and Catalysis, PAI 5/3" and "Technological Attraction Pole SOLTEX," the Région Wallonne (Programme PIMENT—SOLPLAST), the European Commission in the framework of "Objectif 1: Materia Nova" and Project DISCEL (Growth project GRD1-2000-25211), and the Belgian National Fund for Scientific Research (FNRS/FRFC). The work at

Georgia Tech is partly supported by the National Science Foundation through the STC Program under Award Number DMR-0120967 and through grant CHE-0342321, the Office of Naval Research, the Department of Energy (NREL), and the IBM Shared University Research Program. J. Cornil and D. Beljonne are FNRS Research Associates; V. Lemaur acknowledges a grant from "Fonds pour la Formation à la Recherche dans l'Industrie et dans l'Agriculture (FRIA)."

REFERENCES

1. J. Cornil, D. Beljonne, J. P. Calbert, and J. L. Brédas, Interchain interactions in organic π-conjugated materials: impact on electronic structure, optical response, and charge transport, *Adv. Mater.* **13**, 1053–1067 (2001).
2. S. Tretiak and S. Mukamel, Density matrix analysis and simulation of electronic excitations in conjugated and aggregated molecules, *Chem. Rev.* **102**, 3171–3212 (2002).
3. Y. A. Berlin, G. R. Hutchinson, P. Rempala, M. A. Ratner, and J. Michl, Charge hopping in molecular wires as a sequence of electron-transfer reactions, *J. Phys. Chem. A* **107**, 3970–3980 (2003).
4. G. D. Scholes, X. J. Jordanides, and G. R. Fleming, *J. Phys. Chem. B* **105**, 1640–1651 (2001).
5. D. Beljonne, G. Pourtois, C. Silva, E. Hennebicq, L. M. Herz, R. H. Friend, G. D. Scholes, S. Setayesh, K. Müllen, and J. L. Brédas, Interchain vs. intrachain energy transfer in acceptor-capped conjugated polymers, *Proc. Natl. Acad. Sci. USA* **99**, 10982–10987 (2002).
6. M. D. Newton, Quantum chemical probes of electron-transfer kinetics—the nature of donor–acceptor interactions, *Chem. Rev.* **91**, 767–792 (1991).
7. C. J. Brabec, N. S. Sariciftci, and J. C. Hummelen, Plastic solar cells, *Adv. Funct. Mater.* **11**, 15–26 (2001).
8. I. G. Hill, A. Kahn, Z. G. Soos, and R. A. Pascal, Charge-separation energy in films of pi-conjugated organic molecules, *Chem. Phys. Lett.* **327**, 181–188 (2000).
9. J. J. M. Halls, C. A. Walsh, N. C. Greenham, E. A. Marseglia, R. H. Friend, S. C. Moratti, and A. B. Holmes, Efficient photodiodes from interpenetrating networks, *Nature* **376**, 498–500 (1995).
10. G. Yu, J. Gao, J. C. Hummelen, F. Wudl, and A. J. Heeger, Polymer photovoltaic cells enhanced efficiencies via a network of internal donor–acceptor heterojunctions, *Science* **270**, 1789–1791 (1995).
11. C. W. Tang, Two-layer organic photovoltaic cell, *Appl. Phys. Lett.* **48**, 183–185 (1986).
12. D. Wöhrle and D. Meissner, Organic solar cells, *Adv. Mater.* **3**, 129–138 (1991).
13. N. S. Sariciftci, L. Smilowitz, A. J. Heeger, and F. Wudl, Photoinduced electron-transfer from a conducting polymer to buckminsterfullerene, *Science* **258**, 1474–1476 (1992).
14. M. Svensson, F. L. Zhang, S. C. Veenstra, W. J. H. Verhees, J. C. Hummelen, J. M. Kroon, O. Inganas, and M. R. Andersson, High-performance polymer solar cells of an alternating polyfluorene copolymer and a fullerene derivative, *Adv. Mater.* **15**, 988–991 (2003).
15. J. J. M. Halls, K. Pichler, R. H. Friend, S. C. Moratti, and A. B. Holmes, Exciton diffusion and dissociation in a poly(*p*-phenylenevinylene)/C-60 heterojunction photovoltaic cell, *Appl. Phys. Lett.* **68**, 3120–3122 (1996).
16. N. Corcoran, A. C. Arias, J. S. Kim, J. D. MacKenzie, and R. H. Friend, Increased efficiency in vertically segregated thin-films conjugated polymer blends for light-emitting diodes, *Appl. Phys. Lett.* **82**, 299–301 (2003).
17. F. Padinger, R. S. Rittberger, and N. S. Sariciftci, Effects of postproduction treatment on plastic solar cells, *Adv. Funct. Mater.* **13**, 85–88 (2003).

18. J. J. M. Halls, J. Cornil, D. A. dos Santos, R. Silbey, D. H. Hwang, A. B. Holmes, J. L. Brédas, and R. H. Friend, Charge- and energy-transfer processes at polymer/polymer interfaces: a joint experimental and theoretical study, *Phys. Rev. B* **60**, 5721–5727 (1998).

19. J. Ridley and M. C. Zerner, An intermediate neglect of differential overlap technique for spectroscopy: pyrroles and the azines, *Theor. Chim. Acta* **32**, 111–134 (1973).

20. D. M. Adams, J. Kerimo, E. J. C. Olson, A. Zaban, B. A. Gregg, and P. F. Barbara, Spatially-resolving nanoscopic structure and excitonic-charge-transfer quenching in molecular semiconductor heterojunctions, *J. Am. Chem. Soc.* **119**, 10608–10619 (1997).

21. R. Aroca, T. Del Cano, and J. A. de Saja, Exciplex formation and energy transfer in mixed films of phthalocyanine and perylene tetracarboxylic diimide derivatives, *Chem. Mater.* **15**, 38–45 (2003) and references therein.

22. L. Edwards and M. Gouterman, Porphyrins. XV. Vapor absorption spectra and stability: phthalocyanines, *J. Mol. Spectrosc.* **33**, 292–310 (1970).

23. M. Braga, S. Larsson, A. Rosen, and A. Volosov, Electronic-transitions in C-60—on the origin of the strong interstellar absorption at 217 nm, *Astron. Astrophys.* **245**, 232–238 (1991).

24. R. A. Marcus, Electron transfer reactions in chemistry. Theory and experiment, *Rev. Mod. Phys.* **65**, 599–610 (1993).

25. P. F. Barbara, T. J. Meyer, and M. A. Ratner, Contemporary issues in electron transfer research, *J. Phys. Chem.* **100**, 13148–13168 (1996).

26. J. L. Brédas and G. B. Street, Polarons, bipolarons, and solitons in conducting polymers, *Acc. Chem. Res.* **18**, 309–315 (1985).

27. D. Rhem and A. Weller, Kinetics of fluorescence quenching by electron and H-atom transfer, *Isr. J. Chem.* **8**, 259–271, 1970.

28. M. J. S. Dewar, E. G. Zoebisch, E. F. Healy, and J. J. P. Stewart, AM1: a new general purpose quantum mechanical molecular model, *J. Am. Chem. Soc.* **107**, 3902–3909 (1985).

29. AMPAC 6.55, created by Semichem Inc., 7128 Summit, Shawnee, KS 66216, USA, 1997.

30. A. Klamt and G. Schürmann, COSMO—a new approach to dielectric screening in solvents with explicit expressions for the screening energy and its gradient, *J. Chem. Soc., Perkin Trans.* **2**, 799–805 (1993).

31. D. Y. Zang, F. F. So, and S. R. Forrest, Giant anisotropies in the dielectric properties of quasi-epitaxial crystalline organic semiconductor thin films, *Appl. Phys. Lett.* **59**, 823–825 (1991).

32. R. A. Marcus, On the theory of electron-transfer reactions. VI. Unified treatment of homogeneous and electrode reactions, *J. Chem. Phys.* **43**, 679–701 (1965).

33. D. R. Lide and H. P. R. Frederikse, Eds., *Handbook of Chemistry and Physics*, 76th ed., CRC Press, Boca Raton, FL, 1995.

34. P. A. van Hal, R. A. J. Janssen, G. Lanzani, G. Cerullo, M. Zavelani-Rossi, and S. De Silvestri, Two-step mechanism for the photoinduced intramolecular electron transfer in oligo(*p*-phenylene vinylene)–fullerene dyads, *Phys. Rev. B* **64**, 075206 1–7 (2001).

35. S. A. Jenekhe and J. A. Osaheni, Excimers and exciplexes of conjugated polymers, *Science* **265**, 765–768 (1994).

36. A. C. Morteani, A. S. Dhoot, J. S. Kim, C. Silva, N. C. Greenham, C. Murphy, E. Moons, S. Cina, J. H. Burroughes, and R. H. Friend, Barrier-free electron-hole capture in polymer blend heterojunction light-emitting diodes, *Adv. Mater.* **15**, 1708–1712 (2003).

37. V. I. Arkhipov, P. Heremans, and H. Bässler, Why is exciton dissociation so efficient at the interface between a conjugated polymer and an electron acceptor? *Appl. Phys. Lett.* **82**, 4605–4607 (2003).

38. T. Offermans, S. C. J. Meskers, and R. A. J. Janssen, Charge recombination in a poly(*para*-phenylene vinylene)–fullerene derivative composite film studied by transient, nonresonant, hole-burning spectroscopy, *J. Chem. Phys.* **119**, 10924–10929 (2003).

39. E. Neuteboom, S. C. J. Meskers, P. A. van Hal, J. K. J. van Duren, E. W. Meijer, R. A. J. Janssen, H. Dupin, G. Pourtois, J. Cornil, R. Lazzaroni, J. L. Brédas, and D. Beljonne, Alternating oligo(p-phenylene vinylene)–perylene bisimide copolymers: synthesis, photophysics, and photovoltaic properties of a new class of donor–acceptor materials, *J. Am. Chem. Soc.* **125**, 8625–8638 (2003).

40. J. W. Verhoeven, From close contact to long-range intramolecular electron transfer, *Adv. Chem. Phys.* **106**, 603–644 (1999).

41. P. A. van Hal, E. H. A. Beckers, S. C. J. Meskers, R. A. J. Janssen, B. Jousselme, P. Blanchard, and J. Roncali, Orientational effect on the photophysical properties of quaterthiophene-C-60 dyads, *Chem. Eur. J.* **8**, 5415–5429 (2002).

42. M. A. Loi, P. Denk, H. Hoppe, H. Neugebauer, C. Winder, D. Meissner, C. Brabec, N. S. Sariciftci, A. Gouloumis, P. Vazquez, and T. Torres, Long-lived photoinduced charge separation for solar cell applications in phthalocyanine–fulleropyrrolidine dyad thin films, *J. Mater. Chem.* **13**, 700–704 (2003).

43. S. Luzzati, M. Scharber, M. Catellani, N. O. Lupsac, F. Giacalone, J. L. Segura, N. Martin, H. Neugebauer, and N. S. Sariciftci, Tuning of the photoinduced charge transfer process in donor–acceptor double-cable copolymers, *Synth. Met.* **139**, 731–733 (2003).

44. M. N. Paddon-Row, Investigating long-range electron transfer processes with rigid, covalently linked donor–{norbornylogous bridge}–acceptor systems, *Acc. Chem. Res.* **27**, 18–25 (1994).

45. F. C. Grozema, Y. A. Berlin, and L. D. A. Siebbeles, Mechanism of charge migration through DNA: molecular wire behaviour, single-step tunnelling or hopping? *J. Am. Chem. Soc.* **122**, 10903–10909 (2000).

46. D. Beljonne, G. Pourtois, M. A. Ratner, and J. L. Brédas, Pathways for photoinduced charge separation in DNA hairpins, *J. Am. Chem. Soc.* **125**, 14510–14517 (2003).

47. D. J. Wold, R. Haag, M. A. Rampi, and C. D. Frisbie, Distance dependence of electron tunnelling through self-assembled monolayers measured by conducting probe atomic force microscopy: unsaturated versus saturated molecular junctions, *J. Phys. Chem. B* **106**, 2813–2816 (2002).

48. C. Dekker and M. A. Ratner, Electronic properties of DNA, *Physics World*, August, 29–33 (2001).

49. W. B. Davis, W. A. Svec, M. A. Ratner, and M. R. Wasielewski, Molecular-wire behaviour in p-phenylene vinylene oligomers, *Nature* **396**, 60–63 (1998).

50. G. Pourtois, D. Beljonne, J. Cornil, M. A. Ratner, and J. L. Brédas, Photoinduced electron-transfer processes along molecular wires based on phenylenevinylene oligomers: a quantum-chemical insight, *J. Am. Chem. Soc.* **124**, 4436–4447 (2002).

51. H. Oevering, J. W. Verhoeven, M. N. Paddon-Row, and J. M. Warman, Charge-transfer absorption and emission resulting from long-range through-bond interaction — exploring the relation between electronic coupling and electron-transfer in bridged donor–acceptor systems, *Tetrahedron* **45**, 4751–4766 (1989).

52. A. M. Oliver, M. N. Paddon-Row, J. Kroon, and J. W. Verhoeven, Orbital symmetry effects on intramolecular charge recombination, *Chem. Phys. Lett.* **191**, 371–377 (1992).

53. M. M. Wienk, J. M. Kroon, W. J. H. Verhees, J. Knol, J. C. Hummelen, P. A. van Hal, and R. A. J. Janssen, Efficient methano[70]fullerene/MDMO–PPV bulk heterojunction photovoltaic cells, *Angew. Chem. Int. Ed.* **42**, 3371–3375 (2003).

54. L. Goris, M. A. Loi, A. Cravino, H. Neugebauer, N. S. Sariciftci, I. Polec, L. Lutsen, E. Andries, J. Manca, L. De Schepper, and D. Vanderzande, Poly(5,6-dithiooctylisothianaphtene), a new low band gap polymer: spectroscopy and solar cell construction, *Synth. Met.* **138**, 249–253 (2003).

55. E. K. Miller, C. Y. Yang, and A. J. Heeger, Polarized ultraviolet absorption by a highly oriented dialkyl derivative of poly(paraphenylene vinylene), *Phys. Rev. B* **62**, 6889–6891 (2000).

56. J. J. Apperloo, C. Martineau, P. A. van Hal, J. Roncali, and R. A. J. Janssen, Intra- and intermolecular photoinduced energy and electron transfer between oligothienyleneviny-lenes and N-methylfulleropyrrolidine, *J. Phys. Chem. A* **106**, 21–31 (2002).

57. T. A. Chen and R. D. Rieke, Polyalkylthiophenes with the smallest bandgap and the highest intrinsic conductivity, *Synth. Met.* **60**, 175–177 (1993).

58. T. L. Lambert and J. P. Ferraris, Narrow-band gap polymers — polycyclopenta[2,1-B-3,4-B']dithiophen-4-one, *J. Chem. Soc. Chem. Commun.* **11**, 752–754 (1991).

59. J. P. Ferraris and T. L. Lambert, Narrow bandgap polymers— poly-4-dicyanomethylene-4H-cyclopenta[2,1-B-3,4-B'] dithiophene (PCDM), *J. Chem. Soc. Chem. Commun.* **18**, 1268–1270 (1991).

60. J. E. Kroeze, T. J. Savenije, and J. M. Warman, Efficient charge separation in a smooth-TiO$_2$ palladium-porphyrin bilayer via long-distance triplet-state diffusion, *Adv. Mater.* **14**, 1760–1763 (2002).

61. A. Kohler, H. F. Wittmann, R. H. Friend, M. S. Khan, and J. Lewis, Enhanced photocurrent response in photocells made with platinum-poly-yne/C60 blends by photo-induced electron transfer, *Synth. Met.* **77**, 147–150 (1996).

8

Optimization of Organic Solar Cells in Both Space and Energy–Time Domains

Sam-Shajing Sun and Carl E. Bonner
Center for Materials Research and Chemistry Department, Norfolk State University, Norfolk, VA, USA

Contents

Abstract The optimization of organic solar cells in both space and energy–time domains have been preliminarily investigated, either experimentally or theoretically, in order to achieve high efficiency photoelectric energy conversion. Specifically, in the spatial domain, a "tertiary" block copolymer supramolecular nanostructure has been designed using a –DBAB- type of block copolymer, where D is a conjugated donor block, A is a conjugated acceptor block, and B is a nonconjugated and flexible bridge unit. Several –DBAB- type block copolymers have already been designed, synthesized, characterized, and preliminarily examined for target photovoltaic functions. In comparison to a simple donor–acceptor (D–A) blend film, a corresponding –DBAB- block copolymer film exhibited much better photoluminescence (PL) quenching and photoconductivity. These are mainly attributed to the improvement in the spatial domain for charge carrier generation and transportation. With respect to the energy levels and electron transfer dynamics of these materials, the photoinduced charge separation appears to be most efficient when the donor–acceptor frontier orbital energy offset approaches the sum of the charge separation reorganization energy and the exciton binding energy. Other donor–acceptor frontier orbital energy offsets have also been identified where the charge recombination becomes most severe, and where the ratio of the charge separation rate constant over the charge recombination rate constant becomes largest. Implications and ways of achieving these optimized energy levels are also briefly discussed.

Keywords   organic and polymer photovoltaic materials and devices, plastic solar cells, conjugated block copolymers, block copolymer self-assembly, donor–bridge–acceptor, supramolecular nanostructures, spatial optimizations, frontier orbital levels, Gibbs free energies, optimal energy levels and energy offsets, electron transfer dynamics, Marcus theory.

8.1.   INTRODUCTION

Sunlight is a clean and renewable energy source conveniently available on planet Earth and in the outer space nearby the sun or other shining stars. Photovoltaic materials and devices can convert light (or photon) into electricity (or mobile charges such as electrons) [1]. In addition to solar energy conversion, photovoltaic materials and devices also can be used in photodetector applications such as in photoelectric signal transducers in optical communication or optical imaging systems. The key difference in these different applications is that, in photodetector applications, the optical excitation energy gap (optical gap) of the photovoltaic materials must match the energy of the optical signal (e.g., $1.5\,\mu m$ or $0.8\,eV$ IR light signal in optical communications). In the case of solar cells, the optical excitation gap of the material should match the solar spectrum with maximum photon flux between 1.3 and $2.0\,eV$ on the surface of the Earth (air mass 1.5), or 1.8 and $3.0\,eV$ in outer space (air mass 0) [1–4]. Though certain inorganic semiconductor-based photovoltaic materials and devices can convert about 30% of solar energy into electric power [1], in order to effectively and economically utilize sunlight for general energy needs, particularly in remote areas where large spaces are available, low cost and large area organic or polymer-based solar panels or sheets are more attractive [1–14]. Though power conversion efficiencies for purely organic and "plastic" photovoltaic devices are still less than 5% [2–14], in comparison to inorganic materials, semiconducting and conducting conjugated polymers exhibit some inherent advantages such as: (1) lightweight, (2) flexible shape, (3) ultrafast (up to femtoseconds) optoelectronic response, (4) nearly continuous tunability of materials energy levels and bandgaps via molecular design and synthesis, (5) versatile materials processing and device fabrication schemes, and (6) low cost on large-scale industrial manufacturing [15]. Additionally, as research in organic and polymeric photovoltaic materials are rapidly growing, key bottleneck factors, such as the "photon losses," the "exciton losses," and the "carrier losses" that hinder organic and polymeric photovoltaic performance become clear [14], high-efficiency organic photovoltaic systems appear to be feasible, as all these "losses" can be minimized by systematic optimization in space, energy and time domains. In this chapter, some fundamental mechanisms and current problems of organic photovoltaic materials and devices are briefly presented first, then spatial domain optimizations using a "tertiary" nanostructured –DBAB- type block copolymer, and energy–time domain optimizations identifying optimal donor–acceptor energy levels and offsets are described.

8.2. FUNDAMENTALS AND CURRENT PROBLEMS OF ORGANIC PHOTOVOLTAICS

To develop high-efficiency organic or polymeric photovoltaic materials and devices, a brief review and comparison of the inorganic solar cells (such as first inorganic "Fritts Cell" [16]) versus the organic solar cells (such as first organic "Tang Cell" [6]) is necessary. The first inorganic solar cell was described by Charles Fritts in 1885 [16]. As illustrated in Figure 8.1a, the "Fritts" cell was composed of a semiconducting selenium thin layer sandwiched between two different metal electrodes, one very thin and semitransparent gold layer acting as a large work function electrode (LWFE) to collect photogenerated positive charges (holes), and the other copper layer acting as a small work function electrode (SWFE) to collect photogenerated negative charges (electrons). In this cell, when an energy-matched photon strikes selenium, a loosely bound electron–hole pair was generated, the electron and hole can be separated easily by room temperature thermal energy kT (less than $0.05\,eV$), where the free electron would be in a conduction band (CB), and the free hole is left in the valence band (VB), as shown in Figure 8.2a. The free electrons and holes (also called charged "carriers" or simply "carriers") can then diffuse to the respective and opposite electrodes driven by a field created by the two different work function metal electrodes.

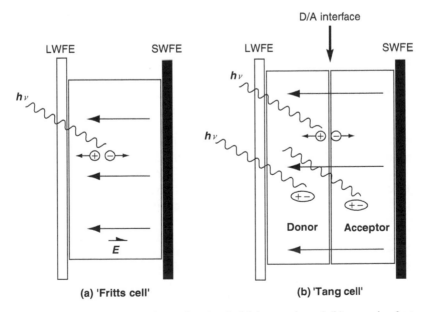

Figure 8.1.   A schematic comparison of a classic (a) inorganic and (b) organic photovoltaic cells.

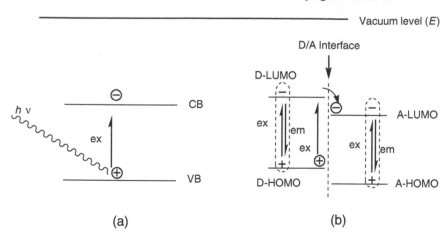

Figure 8.2. Schematic optoelectronic transfer processes in (a) inorganic and (b) organic photovoltaic cells. D-HOMO refers to the donor highest occupied molecular orbital, and D-LUMO refers to the donor lowest unoccupied molecular orbital, A-HOMO refers to the acceptor highest occupied molecular orbital, and A-LUMO refers to the acceptor lowest unoccupied molecular orbital. Ex means photoexcitation, and Em means photoemission.

In contrast, in the first organic solar cell ("Tang Cell") as shown in Figure 8.1b, when an energy-matched photon strikes an organic unit (mainly low bandgaped π electron unit), it only generates a strongly bound and polarizable neutral electron–hole pair called "exciton." The energy required to separate the electron from the hole in an exciton, also called exciton binding energy E_B (typically in a range of 0.05–1.5 eV) is much higher than room temperature energy kT [17–19]. Such an exciton can diffuse (e.g., via energy transfer) randomly within a distance defined by its lifetime of typically picoseconds to nanoseconds. The average exciton diffusion length (AEDL) for organic conjugated materials is typically 5–70 nm [17–19]. The schematic frontier orbital energy levels are shown in more detail in Figure 8.3 and Figure 8.4. Figure 8.3 shows the schematic diagram of the band structure of an organic donor–acceptor binary light harvesting system. Figure 8.4 shows the same system from the perspective of the Gibbs free energy. As shown in Figure 8.1b, Figure 8.2b, and Figure 8.3, if two different organic materials with different frontier electronic orbitals are present and in direct contact to each other, one material, the "donor" has a smaller ionization potential (IP), and the other material, the "acceptor," has a larger electron affinity (EA) (Figure 8.2b and Figure 8.3), when an exciton (in either donor or acceptor) diffuses to a donor–acceptor interface, the frontier orbital level offset between the donor and the acceptor would induce electron transfer across the interface. If the exciton is at donor side, the electron at the donor lowest unoccupied molecular orbital (LUMO) will quickly transfer into the acceptor LUMO (transfer #3 in Figure 8.3 and Figure 8.4). If the exciton is at acceptor side, the hole at acceptor highest occupied molecular orbital (HOMO) will jump quickly into the donor HOMO (corresponding to an electron back transfer #7 in Figure 8.3 and Figure 8.4), thus an exciton now becomes a free electron (at acceptor LUMO) and a free hole (at donor HOMO), resulting in an electron–hole charge separation. Now the free electrons and holes (charged carriers) can diffuse to their respective electrodes, hopefully in two separate donor and acceptor phases, so the chance of electron–hole recombination would be minimal.

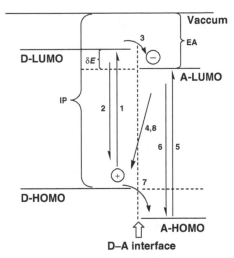

Figure 8.3. Scheme of molecular frontier orbitals and photoinduced charge separation and recombination processes in an organic donor–acceptor binary light harvesting system.

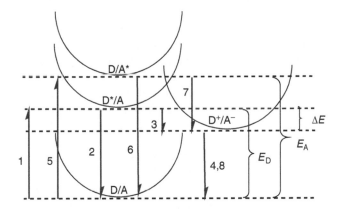

Figure 8.4. Scheme of standard Gibbs free-energy potential wells of photoinduced charge separation and recombination processes in an organic donor–acceptor light harvesting system.

Thus, a donor–acceptor binary system appears very critical for organic photovoltaic function [6].

For an organic solar cell, the overall power conversion efficiencies are determined by at least following five steps:

1. Photon absorption and exciton generation.
2. Exciton diffusion to donor–acceptor interface.
3. Exciton split or charged carrier generation at donor–acceptor interface.
4. Carrier diffusion to respective electrodes.
5. Carrier collection by the electrodes.

For all currently reported organic or polymeric photovoltaic materials and devices, none of the above mentioned five steps have been optimized. Therefore, it is not surprising that the power conversion efficiencies of all currently reported organic or polymeric solar cells are relatively low in comparison to their inorganic counterparts.

8.2.1. Photon Absorption and Exciton Generation

In this first step of organic photovoltaic conversion, a basic requirement is that the optical excitation energy gap (optical gap) of the materials should be equal or close to the incident photon energy. In most amorphous organic materials, it is difficult to form electronic band structures due to the lack of both long-range and short-range molecular order. The energy gap defaults to the difference between the frontier orbitals, i.e., the HOMO and the LUMO. In organic conjugated system, HOMO is typically an occupied π bonding orbital and LUMO is typically an unoccupied π^* antibonding orbital. Since an organic LUMO–HOMO excitation basically generates a tightly bound exciton instead of a free electron and a hole, the "optical energy gap" is therefore used instead of the conventional "electronic energy gap" that typically refers to the energy gap between the free holes at VB and the free electrons at CB in inorganic semiconducting materials (Figure 8.2a). In organics, the relationship of "optical gap (E_{go})" versus "electronic gap (E_{ge})" may be expressed as $E_{ge} = E_{go} + E_B$, where E_B is called exciton binding energy that represents a minimum energy needed to separate the electron from the hole in an exciton into a radical ion pair [17]. E_{go} values are usually estimated directly from optical absorption band edge and absolute E_{ge} values may be estimated by electrochemical redox analysis. Absolute HOMO–LUMO levels may also be estimated from a "half" electrochemical analysis in combination with the optical absorption spectroscopy. For a widely used conjugated semiconducting polymer poly-p-phenylenevinylenes (PPV), the exciton binding energy has been reported to be in the range of 0.05–1.1 eV [17]. If VB is defined as containing "free" holes, and CB is defined as containing "free" electrons, then for a donor–acceptor binary organic system, the self-organized or well aligned acceptor LUMO bands may then be called conduction band, and self-organized donor HOMO bands may then be called valence band. Unfortunately, these "bands" have not yet been materialized so far.

For solar cell applications, solar radiation spans a wide range of wavelengths, with largest photon flux between 600 and 1000 nm (1.3–2.0 eV, on surface of the earth or 1.5 air mass) or 400 and 700 nm (1.8–3.0 eV, in space or air mass 0) [1–4]. For terrestrial applications, it is desirable that the bandgaps of a solar cell span a range from 1.3 to 2.0 eV. This may be achieved by incorporating a series of different bandgap donor–acceptor or organic dyes that absorb light in that radiation range. However, while the solar photon loss can be minimized in this manner, due to energy transfer processes where all high-energy excitons will eventually become lowest energy excitons [19], the open circuit voltage (V_{oc}) of the cell will also be reduced accordingly, as experimental studies have revealed a close correlation between the V_{oc} and the gap between the lowest acceptor LUMO and highest donor HOMO levels [20]. In reality, several widely used conjugated semiconducting polymers used in organic solar cell studies have optical gaps higher than 2.0 eV [15]. Widely used alkyloxy derivatized poly-p-phenylenevinylenes (PPV) has a typical optical gap of about 2.3 to 2.6 eV, well above the maximum solar photon flux range. This is why the photon absorption (or exciton generation) for PPV-based solar cells are far from getting optimized at AM1.5. This "photon loss" problem is in fact very common in almost all currently reported organic photovoltaic materials and devices. However, one advantage of organic materials is the flexibility of its energy levels. They can be fine-tuned via molecular design and synthesis. Therefore, ample opportunity exists for improvement. A number of recent studies on the developments of low bandgap conjugated polymers are such examples [21–23].

8.2.2. Exciton Diffusion

Once an organic exciton is photogenerated, it typically diffuses (e.g., via intrachain or interchain energy transfer or "hopping," including Förster energy transfer for a singlet exciton) to a remote site. At the same time, the exciton can decay either radiatively or nonradiatively to its ground state with typical lifetimes from picoseconds to nanoseconds [18,19]. Alternatively, in condensed phases, some excitons may be trapped in defect or impurity sites. Both exciton decay and trapping would contribute to the "exciton loss." The average distance an organic exciton can travel within its lifetime is called AEDL. In noncrystalline and amorphous materials, the AEDL depends heavily on the spatial property (morphology) of the materials. For most conjugated organic materials, the AEDL is typically in the range of 5–70 nm [1–3,18,19]. The AEDL for PPV is around 10 nm [18]. Since the desired first step of photovoltaic process is that each photogenerated exciton will be able to reach the donor–acceptor interface where charge separation can occur, one way to minimize the "exciton loss" would be to make a defect-free and donor–acceptor phase separated and ordered material. One example would be a donor–acceptor phase separated tertiary nanostructure such that an exciton generated at any site of the material can reach a donor–acceptor interface in all directions within the AEDL [14]. This can also be called a "bulk heterojunction" structure [7,8]. One limitation of the first organic bilayer solar cell "Tang Cell" was that, if the donor or acceptor layer is thicker than the AEDL, excitons do not reach the interfacial region to separate before decay. This "exciton loss" is a serious problem. On the other hand, if the photovoltaic active layer is too thin or much shorter than the penetration depth of the light in the material, then "photon loss" due to poor light absorption would result. This is also why "bulk hetero-junction" type solar cells are attractive, as they not only minimize the exciton loss by increasing the donor–acceptor interface, but they can also offer enough thickness for effective photon harvesting.

8.2.3. Exciton Separation and Charge Carrier Generation

Once an exciton arrives at a donor–acceptor interface, the potential field at the interface due to the donor–acceptor frontier orbital energy level offsets, i.e., δE as shown in Figure 8.3 can then separate the exciton into a free electron at acceptor LUMO and a free hole at donor HOMO, provided this field or energy offset is close to its optimal value or range as discussed in Section 4 of this chapter. This photoinduced charge separation process is also called "photo-doping," as it is a photoinduced (in contrast to chemical or thermal induced) redox reaction between the donor and the acceptor. For a derivatized PPV donor and fullerene acceptor binary system, it has been experimentally observed that the photoinduced charge separation process at the PPV–fullerene interface was orders of magnitude faster than either the PPV exciton decay or the charge recombination [7,8]. This means that optoelectronic quantum efficiency at this interface is near unity and a high-efficiency organic photovoltaic system is feasible.

8.2.4. Carrier Diffusion to the Electrodes

Once the carriers, either free electrons or holes, are generated, holes need to diffuse towards the LWFE, and electrons need to diffuse towards the SWFE. The driving forces for the carrier diffusion may include the field created by the work function difference between the two electrodes, and a "chemical potential" driving force [24]. "Chemical potential" driving force may be interpreted as a density potential driving force, i.e., particles tend to diffuse from a higher density domain to a lower density domain. In an organic donor–acceptor binary photovoltaic cell, the high-density electrons at the acceptor LUMO nearby the donor–acceptor interface tend to diffuse to lower electron density region within the acceptor phase, and high-density holes at the donor HOMO nearby the donor–acceptor interface tend to diffuse to the lower holes density region within the donor phase. In the "Tang Cell" as shown in Figure 8.1b, once an exciton was separated into a free electron at acceptor side and a free hole at donor side of the D–A interface, the electron will be "pushed" away from the interface toward the negative electrode by both the "chemical potential" and by the field formed from the two electrode work functions. The holes will be "pushed" toward the positive electrode by the same forces but in the opposite direction. With this chemical potential driving force, even if the two electrodes are the same, asymmetric photovoltage could still be achieved (i.e., the donor HOMO would yield the positive and acceptor LUMO would yield the negative electrodes) [24]. Mid-gap state species, either impurities and defects, or intentionally doped redox species, may also facilitate the carrier diffusion and conductivities by providing "hopping" sites for the electrons or holes. However, right after electron–hole is separated at the interface, they can also recombine due to both potential drop of A-LUMO/D-HOMO and the Coulomb force between the free electron and hole. Fortunately, the charge recombination rates in most cases are much slower than the charge separation rates (charge recombination rates are typically in micro- to milliseconds as compared to femto- and picoseconds charge separation rate) [7,8,36], so there is an opportunity for the carriers to reach the electrodes before they recombine. Yet, in most currently reported organic solar cells, the diffusion of electrons and holes to their respective electrodes are not really fast due to poor morphology. If donor and acceptor phases are perfectly "bicontinuous" between the two electrodes, and that all LUMO and HOMO orbitals are nicely aligned and overlapped to each other in both donor and acceptor phases, like in a molecularly self-assembled thin films or crystals, then the carriers would be able to diffuse smoothly in "bands" towards their respective electrodes. Currently, carrier thermal "hopping" and "tunneling" are believed to be the dominant diffusion and conductivity mechanism for most reported organic photovoltaic systems; therefore, the "carrier loss" is believed to be another key factor for the low efficiency of organic photovoltaic materials and devices.

8.2.5. Carrier Collection at the Electrodes

It has been proposed [9] that when the acceptor LUMO level matches the Fermi level of the SWFE, and the donor HOMO matches the Fermi level of the LWFE, an ideal "Ohmic" contact would be established for efficient carrier collection at the electrodes. So far, there are no organic photovoltaic cells that have achieved this desired "Ohmic" alignment due to the availability and limitations of materials and electrodes involved. There were a number of studies, however, focusing on the open circuit voltage (V_{oc}) dependence on LUMO–HOMO level changes, electrode Fermi levels, and chemical potential gradients of the materials [20,24]. The carrier collection mechanisms at electrodes are relatively less studied and are not well understood. It is believed that the carrier collection loss at the electrodes is also a critical contributing factor for the low efficiency of existing organic solar cells.

8.3. OPTIMIZATION IN THE SPATIAL DOMAIN VIA A –DBAB- TYPE BLOCK COPOLYMER

8.3.1. Block Copolymers and Self-Assembled Supramolecular Nanostructures

Block copolymer solid melts are well known to exhibit behavior similar to conventional amphiphilic systems such as lipid–water mixtures, soap, and surfactant solutions [25,26]. The covalent bond connection between distinct or different blocks imposes severe constraints on possible equilibrium states; this results in unique supramolecular nanodomain structures such as lamellae (LAM), hexagonally (HEX) packed cylinders or columns, spheres packed on a body-centered cubic lattice (BCC), hexagonally perforated layers (HPL), and at least two bicontinuous phases: the ordered bicontinuous double diamond phase (OBDD) and the gyroid phase [25,26]. The morphology of block copolymers is affected by chemical composition, block size, temperature, processing, and other factors. For a triblock copolymer, a variety of even more complex and unique morphologies can be formed and are shown in Figure 8.5. Clearly, the block copolymer approach to photovoltaic function offers some intrinsic advantages over the bilayer or composite–blend systems [13,14,27–35]. An MEH-PPV/polystyrene (with partial C_{60} derivatization on polystyrene block) donor–acceptor diblock copolymer system has been synthesized, and phase separation between the two blocks was indeed observed [13]. However, the

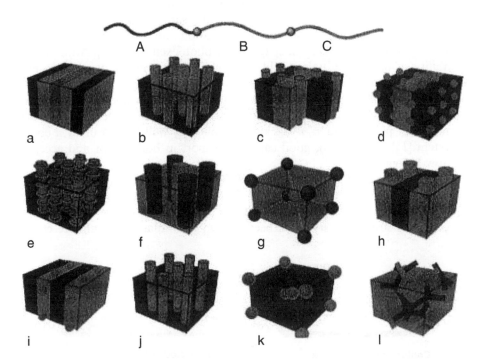

Figure 8.5. **(Color figure follows page 348)**. Representative self-assembled supramolecular nanostructures from a triblock copolymer (From N. Hadjichristidis, S. Pispas, and G. Floudas, eds., *Block Copolymers: Synthetic Strategies, Physical Properties, and Applications*, John Wiley & Sons, New York, 2003. With permission.)

polystyrene–C_{60} acceptor block is not a conjugated chain system and the poor electron mobility or "carrier loss" problem in polystyrene phase still remains as an issue. On the other hand, when a conjugated donor block was linked directly to a conjugated acceptor block to form a direct p–n type conjugated diblock copolymer, while energy transfer from higher gap block to lower gap block were observed, no charge separated states (which is critical for photovoltaic functions) were detected [27].

8.3.2. Design and Development of a –DBAB- Type Block Copolymer for a "Tertiary" Supramolecular Nanostructure

To address the several loss problems of organic photovoltaics discussed earlier, particularly the "exciton loss" and the "carrier loss" problems, optimization in spatial domain of the donor and acceptor materials has been investigated. Using this rationale, a photovoltaic device based on a –DBAB- type of block copolymer and its potential "tertiary" supramolecular nanostructure was designed (Figure 8.6–Figure 8.11) [14], where D is a π-electron conjugated donor block, A is a conjugated acceptor block, and B is a nonconjugated and flexible bridge unit. In this structure, the HOMO level of the bridge unit is lower than the acceptor HOMO, and the bridge's LUMO level is higher than the donor LUMO (Figure 8.7). With this configuration, a wide bandgap energy barrier is formed between the donor and acceptor conjugated blocks on the polymer chain (Figure 8.7). This potential energy barrier separates the energy levels of the donor and acceptor blocks, retarding the electron–hole recombination encountered in the case of directly linked p–n type diblock conjugated copolymer system [27]. At the same time, intramolecular or intermolecular electron transfer or charge separation can still proceed effectively through bridge σ-bonds or through space under photoexcitation [36]. Additionally, the flexibility of the bridge unit would also enable the rigid donor and acceptor conjugated blocks more easily to self-assemble, phase separate, and become less susceptible to distortion of the conjugation. Since both donor and acceptor blocks are π-electron conjugated chains, if they are self-assembled in planes perpendicular to the molecular plane like a π–π stacking morphology well-known in all π conjugated system [15] (Figure 8.8), good carrier transport in both donor and acceptor phases now become feasible.

While the –DBAB- block copolymer backbone structure may be called "primary structure" (Figure 8.6), the conjugated chain π orbital closely stacked and ordered morphology may therefore be called "secondary structure" (Figure 8.8). This "secondary structure" style has been known to dramatically enhance carrier

"Primary structure"

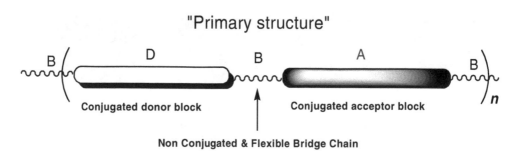

Figure 8.6. Scheme of a –DBAB- type of block copolymer "primary structure."

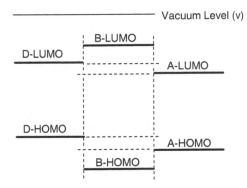

Intrachain energy level schematic diagram
of –DBAB-type block copolymer

Figure 8.7. Scheme of a –DBAB- type of block copolymer relative energy levels.

"Secondary structure"

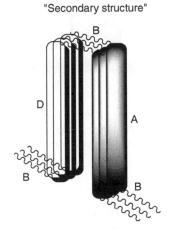

Figure 8.8. Scheme of a –DBAB- type of block copolymer "secondary structure."

"Tertiary structure"

'HEX' Columnar morphology

Figure 8.9. Scheme of a –DBAB- type of block copolymer "tertiary structure."

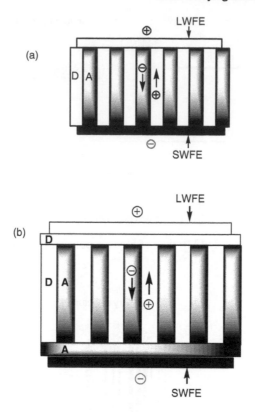

Figure 8.10. Scheme of a −DBAB- type of block copolymer solar cell in the form of (a) columnar structure directly sandwiched between two electrode layers and (b) with terminal asymmetric active material layers.

mobility due to improved π orbital overlap as demonstrated in ordered discotic type liquid crystalline phases [10,37], or in derivatized and self-assembled regio-regular polythiophenes [38], or template aligned poly-*p*-phenylenevinylenes [39]. Most importantly, this "secondary structure" as shown in Figure 8.8 is in fact favorable for the exciton diffusion in horizontal direction and charge transport in vertical direction as has been experimentally observed [39]. Finally, through the adjustment of block size, block derivatization, and thin film processing protocols, a "tertiary structure" (Figure 8.9) where a "bicontinuous," such as columnar (or "HEX") type of morphology of the donor and acceptor blocks is vertically aligned on top of the substrate and sandwiched between two electrodes (Figure 8.10a) can be obtained. Even better, a thin donor layer may be inserted between ITO and active "HEX" layer, and a thin acceptor layer is inserted between metal electrode and active layer (Figure 8.10b). A diblock copolymer where a "honey comb" type columnar structure was formed with either top or bottom of the "honey comb" completely covered by one block has already been observed [40]. The terminal donor and acceptor layers would enable a desired asymmetry and favorable chemical potential gradient for asymmetric (selective) carrier diffusion and collection at respective electrodes [6,14,24]. Since the diameter of each donor or acceptor block column can be conveniently controlled via synthesis and processing to be within the organic AEDL of 5–70 nm, so that every photoinduced exciton will be in convenient

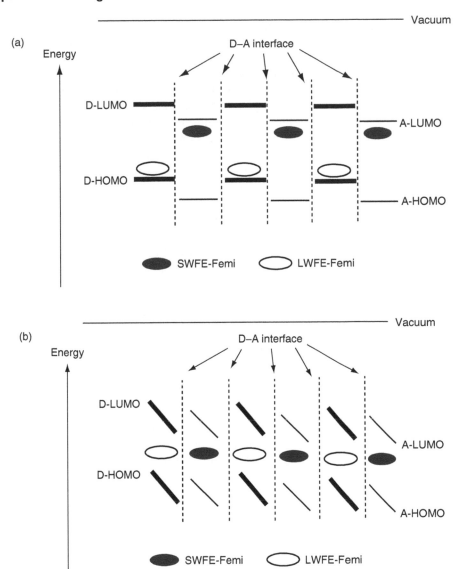

Figure 8.11. Schematic energy diagram of a "HEX" tertiary block copolymer solar cell in (a) open circuit situation and (b) short circuit situation.

reach of a donor–acceptor interface along the direction perpendicular to the columnar. At the same time, photogenerated charge carriers can diffuse more smoothly to their respective electrodes via a truly "bicontinuous" block copolymer columnar morphology. The energy domain diagram of such a spatial tertiary structure can be represented in Figure 8.11. Figure 8.11a shows the open circuit situation, where each nanometer-sized donor phase (or column) is in contact with an acceptor phase, as depicted in Figure 8.9. SWFE-Fermi refers to the Fermi level of SWFE, and LWFE-Femi refers to the Fermi level of LWFE. Figure 8.11b shows the short circuit situation, where "band bending" occurs within each "HEX" column. This "band bending" also drives the charge diffusion towards their respective electrodes.

While the increased donor and acceptor interface area and phase morphology will dramatically minimize the exciton and carrier losses, it may also increase the carrier recombination at the same interfaces. However, by proper energy level manipulation via molecular engineering, the charge recombination rate can be reduced in comparison to the charge separation as discussed in Section 4 of this chapter. In many of the reported organic photovoltaic systems, the charge recombination typically occurs on the microseconds or slower timescale, which is in contrast to the ultrafast pico- or femtoseconds charge separation rate at the same interface [8,36]. Therefore, the charge carrier recombination does not appear to be of a major concern for solar cell applications where the radiation is continuous. This block copolymer photovoltaic device may be to a certain degree mimic a dye-sensitized solar cell (DSSC) yet with whole donor–acceptor interface covered by photosensitizing dyes, and that both donor and acceptor phases are solids and "bicontinuous." Additionally, with appropriate adjustment of donor and acceptor block sizes and their substituents, energy levels, or with attachment of better photon energy-matched sensitizing dyes on the polymer backbone, it is expected that the photon loss, the exciton loss, and the carrier loss (including charge recombination) issues can all be addressed and optimized simultaneously in one such block copolymer photovoltaic device. In order to examine the feasibility of this block copolymer solar cell design [14], a novel –DBAB- type of block copolymer (Figure 8.12) has been synthesized and characterized, and some optoelectronic studies are already in progress [28–35].

8.3.3. Materials and Equipment, Experimental

All starting chemicals, materials, reagents, and solvents were purchased from commercial sources and used directly except noted otherwise. Proton and carbon nuclear magnetic resonance (NMR) data were obtained from a Bruker Avance 300 MHz spectrometer. Elemental analyses were done at Atlantic Microlab. HR-MASS and MALDI data were obtained from mass spectrometry facility at Emory University. Perkin–Elmer DSC-6/TGA-6 systems were used to characterize the thermal property of the materials. Gel-permeation chromatography (GPC) analysis was done using a Viscotek T60A/LR40 Triple-Detector GPC system with mobile phase of THF at ambient temperature (universal calibration based on polystyrene standards is used). Ultraviolet and visible spectroscopy (UV–VIS) spectra were collected from a Varian Gary-5 spectrophotometer. Luminescence spectra were obtained from a SPEX Fluoromax-3 spectrofluorometer. Electrochemical analysis was done on a BAS Epsilon-100 unit. Film thicknesses were measured on a Dektak-6M profilometer. Thin film metal electrodes were deposited in high vacuum using a BOC-360 metal vapor deposition system. For dynamic spectroscopic studies, an Ar ion pumped and mode locked Ti–Sapphire laser system was used to create optical pulses at 800 nm and 120 fsec at 76 MHz. The emission from the solutions or films were spectrally filtered with a monochromator and directed on the photocathode of a streak camera with 2 psec resolution.

Figure 8.12 shows the chemical structures of the RO-PPV donor block (D), the SF-PPV-I acceptor block (A), the bridge units (B), and the synthetic coupling scheme of the target –DBAB- block copolymer. Specifically, the donor block RO-PPV is an alkyloxy derivatized poly-p-phenylenevinylene, and the acceptor block SF-PPV-I is an alkyl-sulfone derivatized poly-p-phenylenevinylene. Two bridge units were investigated; the first one is a long dialdehyde terminated bridge unit 1 containing

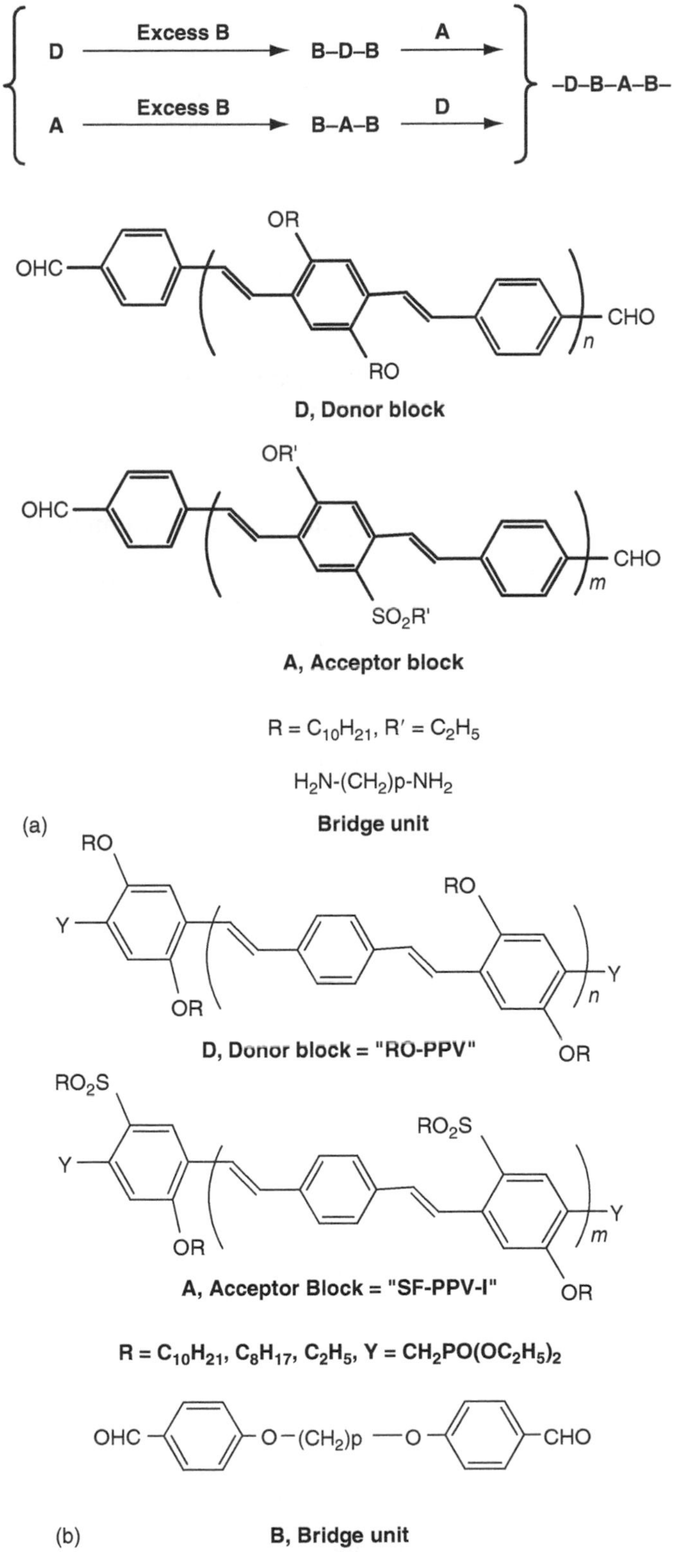

Figure 8.12. −DBAB- type conjugated block copolymer system already studied with (a) a diamine bridge unit and (b) a dialdehyde bridge unit.

10 methylene units, and the second is a short diamine terminated bridge unit 2 containing two methylene units. When bridge unit 1 was used, both donor and acceptor blocks were synthesized with terminal phosphate groups. When diamine terminated bridge unit 2 was used, both donor and acceptor blocks were synthesized with terminal aldehyde groups. The alkyl derivatives (R) investigated includes branched 2-ethyl-hexyl group (C_8H_{17}), the ethyl (C_2H_5), and linear decacyl ($C_{10}H_{21}$) groups. The RO-PPV/SF-PPV-based block copolymer syntheses and chemical characterizations have been reported separately [28,29,31] or are going to be published [34]. Though GPC shows a molecular size corresponding to one DBAB unit, however, due to the fact that THF-insoluble higher molecular weight fractions were filtered off during GPC measurement, –DBAB- (instead of DBAB) is used for single as well as possible multiple DBAB repeating units. In this chapter, only some critical comparisons of the –DBAB- (with a two-carbon diamine bridge unit 2) with a D–A blend system is presented.

8.3.4.   Results and Discussion on Spatial Domain Optimization

As elaborated earlier, the first critical step in organic photovoltaics is a photoinduced electron transfer from the donor to the acceptor (photo-doping) as shown in Figure 8.1–Figure 8.4, and this process can be characterized by a number of techniques, including photoluminescence quenching for radiative exciton decay, light-induced conductivity (photocurrent) measurements, light-induced electron spin resonance (LIESR) spectroscopy, etc. [1–7]. Figure 8.13 shows the solution absorption spectra of the RO-PPV donor block, the SF-PPV-I acceptor block, and the –DBAB- block copolymer. Since no obvious new bands were observed in the –DBAB- absorption spectrum in comparison to D and A, there was no evidence of ground state charge transfer or "chemical doping" in the synthesized –DBAB-. Figure 8.14 shows the solution PL emission spectra of the donor block, the acceptor block, and

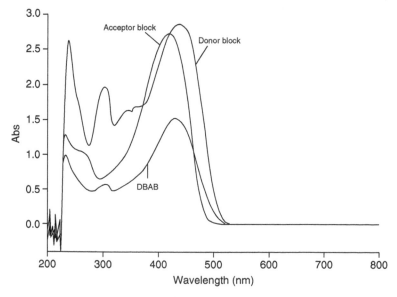

Figure 8.13.   UV–VIS absorptions of RO-PPV (donor), SF-PPV-I (acceptor), and –DBAB- in dichloromethane dilute solution.

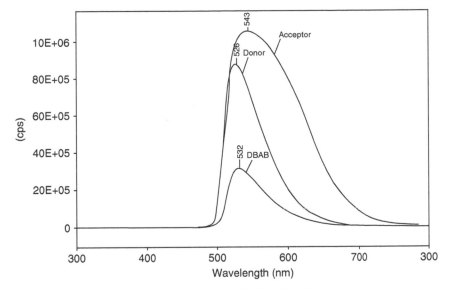

Emission spectra of donor, acceptor, and DBAB (solution)

Figure 8.14. PL emissions of RO-PPV (D), SF-PPV-I (A), and –DBAB– in dichloro-methane dilute solution. The PL intensity (*y*-axis) is arbitrary.

–DBAB– block in arbitrary units (because the PL emission from –DBAB– was too weak to be seen if on a same scale with D or A). From a molecular density calibrated analysis, it was found that the PL of –DBAB– was quenched by over 80% relative to pristine donor or acceptor block in dilute solution [29]. This PL quenching of –DBAB– was also confirmed by a much faster PL emission decay (687 psec) of –DBAB– versus the pristine donor or acceptor decay (1600 psec), as shown in Figure 8.15. Since the solutions were very dilute, the probability of intermolecular photoinduced charge separation or defect trapping is very small as the polymer chains do not interact with each other. Therefore, more than 80% PL quenching can be attributed mainly to intrachain charge separation through the two-carbon bridge unit. This intrachain electron transfer through a bridged energy barrier has been in fact widely observed before [36,41,42]. These results demonstrate that a short two-carbon bridge would be sufficient to separate the electronic structures of the RO-PPV donor and the SF-PPV-I acceptor block, yet it still allows effective electron transfer. Figure 8.16 shows the thin-film absorption spectra of the RO-PPV donor block, the SF-PPV-I acceptor block, and –DBAB–. Again, no ground state electron transfer was observed. Figure 8.17 shows the thin-film PL emission spectra of the donor block, the acceptor block, and the final –DBAB-block copolymer in arbitrary units. The PL emission of D–A blend films was similar to –DBAB– though the amount of emission quenching varies from sample to sample, i.e., very sensitive to processing conditions. Again from a molecular density calibrated PL emission analysis [29], it was found that the PL intensity of the blend films were typically quenched in the range of 10–70% relative to pristine donor or acceptor blocks, while the PL emission of the final –DBAB– films were typically quenched at 90–99%. This strong PL quenching in –DBAB– film was also confirmed by a much faster PL emission decay of the –DBAB– films as compared to the D–A blend or pristine donor or acceptor films as shown in Figure 8.18. It is expected that this PL quenching

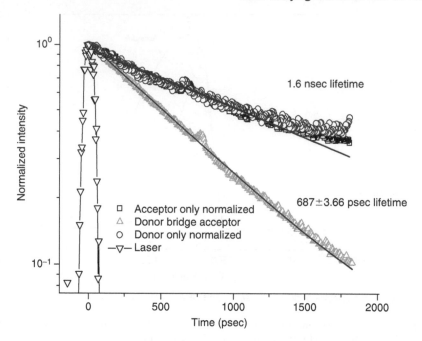

Figure 8.15. PL emission dynamics of RO-PPV (donor), SF-PPV-I (acceptor), and –DBAB- in dichloromethane dilute solution.

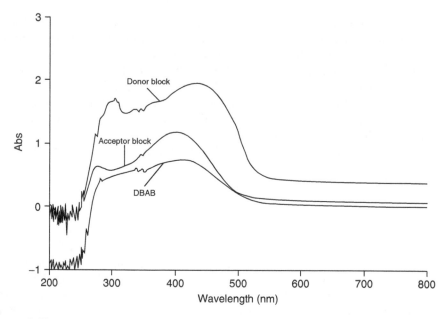

Figure 8.16. UV–VIS of RO-PPV (donor), SF-PPV-I (acceptor), and –DBAB- films on glass substrates.

enhancement of –DBAB- film was mainly due to the photoinduced interchain electron transfer from a donor block to a nearby acceptor block via close spatial contact. Clearly, such an interchain electron transfer enhancement is mainly due to the increase of the intermolecular donor–acceptor interface and the improvement of

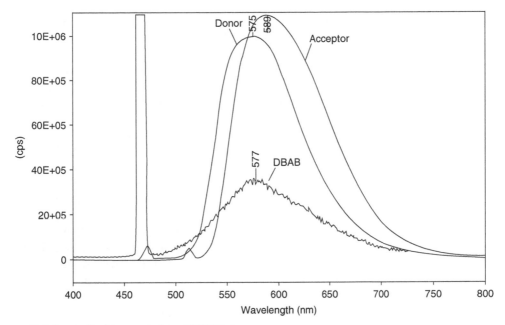

Emission spectra of donor, acceptor and DBAB (film)

Figure 8.17. PL emissions of RO-PPV (D), SF-PPV (A), and −DBAB- in thin films on glass substrates. The PL intensity (*y*-axis) is arbitrary for better view. (Note: the spikes at 470 and 510 nm are due to reflected excitation beam).

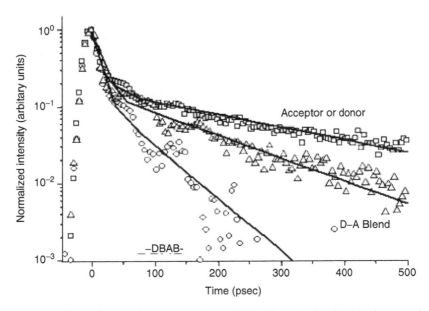

Figure 8.18. PL emission dynamics of RO-PPV (donor), SF-PPV-I (acceptor), and −DBAB- in films on glass substrates.

the morphology of the –DBAB- block copolymer thin film. Atomic force microscopy (AFM) and scanning tunneling microscopy (STM) studies did not reveal any regular pattern in a D–A blend film, yet some interesting regular pattern can be seen in –DBAB- block copolymer films [30,32,34]. Though details or mechanism of such pattern and its formation are unclear and are still under investigation, however, it is known that block copolymer morphology can be affected or controlled by many factors, such as chemical composition, block size, film substrates, processing conditions, etc. [25,26]. Finally, a few optoelectronic thin-film devices have also been fabricated and studied from these polymers. The dark current–voltage (I–V) curves (Figure 8.19) of –DBAB- and D–A blend films were compared under identical fabrication and measurement conditions (i.e., same thickness, same density, etc). Figure 8.19 shows that the biased current (as well as calculated carrier mobility) of –DBAB- was at least two orders of magnitude better than the simple D–A blend [35]. Figure 8.20 shows the photoconductivity (zero bias) comparisons of –DBAB-, D–A simple blend, the commercially available MEH-PPV/C_{60}, and at dark current; all film devices were fabricated and measured under identical conditions with same molecular densities. As Figure 8.20 shows, the photocurrent density of –DBAB- was almost doubled than that of the D–A simple blend at absorption peak of around 400 nm. The dark current is also shown at the bottom. Both the open circuit voltage (V_{oc}) and short circuit current (I_{sc}) of these devices were very small, and it was possibly due to several causes, such as unoptimized device fabrication, e.g., no charge collection or injection layers were used, and probable PPV photo-oxidative degradation in the air. PPV is well-known for photo-oxidative degradation [9]. Additionally, much larger photocurrents were initially observed when the film was irradiated in a freshly fabricated device. The values reported in the plot are the lower steady-state photocurrents. It is interesting to note that the voltage-biased dark current of –DBAB- was at least two orders of magnitude better than the D–A blend [35], while the photocurrent of –DBAB- was only twice that of D–A. This may be explained as follows:

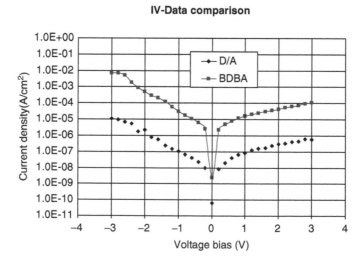

Figure 8.19. Voltage biased electric current density from thin films of –DBAB- block copolymer and D–A blend. Both films have the same material density, thickness, and electrode area.

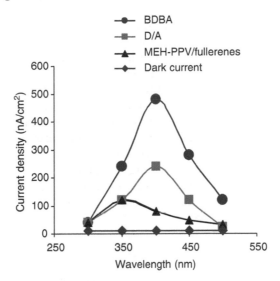

Figure 8.20. Photocurrent comparisons of several organic thin-film photovoltaic cells.

in biased current, sufficient and same amount of carriers were injected from the electrodes for both –DBAB- films and D–A blend films. Therefore, the orders of magnitude current density differences can only be attributed to the main difference of the two films, i.e., the much better carrier transport pathways in the –DBAB- film than in the D–A blend film. However, in photocurrent measurements, even if the –DBAB- film has a much better carrier transport pathway than the D–A blend film, the photogenerated carriers may be limited in both –DBAB- and D–A films due to either limited interface sizes, or improper energy levels, etc.; therefore, the photocurrent difference was not as large as in biased situation. Optimizations of structures, energy levels, morphological controls, device fabrication, and measurements of these materials are ongoing and will be discussed in the near future.

8.4. OPTIMIZATION IN THE ENERGY–TIME DOMAIN

8.4.1. Background

To address the optimal energy levels in a paired donor–acceptor organic light harvesting system, first, both the donor and acceptor optical excitation energy gaps should match the intended photon energy. In solar light harvesting applications, maximum photon flux is between 1.3 and 2.0 eV on surface of the Earth (air mass 1.5) and 1.8 and 3.0 eV in outer space (air mass 0) [1–4]. For optical telecommunications and signal processing, an optical bandgap of 0.8 eV (for 1.55 μm IR signal) is needed. Energy gaps in both donor and acceptor should be fine-tuned via molecular engineering to match the photon energy, as both can absorb photon and incur charge separation at donor–acceptor interface as shown in Figure 8.3 and Figure 8.4. A critical remaining question is the magnitude of energy offset between the donor and the acceptor that is assumed to drive the charge separation. A current widely cited view is that the frontier orbital energy offset between the donor and the acceptor should be at least over the exciton binding energy E_B (i.e., the minimum energy needed to overcome the electric Coulomb forces and separate the tightly bound and neutral exciton into a separate or "free" electron and hole) [1–4,17]. Indeed when the LUMO energy offset is too small, charge separation appears to become less efficient [43]. However, even in many positive energy offset situations (such as in electron transfer from donor to acceptor via a higher energy level bridge unit as in many DBA systems), electron transfer or charge separation still occurs effectively [36,41,42,44,45]. On the other hand, if the energy offset is too large, Marcus "inverted" region may slow down charge separation [41,42,44,45], and thermal ground state charge separation without photoexcitation may also occur, and these are not desirable for light harvesting functions. Large energy offset also reduces open circuit voltage [20]. Therefore, an analysis of optimal donor–acceptor energy offsets is necessary in order to achieve efficient charge separation, particularly in consideration of exciton decay, charge separation, and recombination processes in both donor and acceptor [46–48].

8.4.2. Formulation

In an "ideal" organic donor–acceptor binary solar cell, both donor and acceptor should harvest photon and contribute to photovoltaic functions. The processes may be simplified as following (also illustrated in Figure 8.3 and Figure 8.4):

1. Photoexcitation at donor (D/A + $h\nu_1$ → D*/A, D* designates a donor exciton, $h\nu_1$ is the absorbed photon energy and can be estimated from absorption or excitation spectra).
2. Donor exciton decay to its ground state (D*/A→D/A + $h\nu_2$) corresponding to a standard Gibbs free energy change of E_D, decay rate constant of k_{dD}, and a reorganization energy of λ_{dD} [41,42]. $h\nu_2$ is the emitted photon energy and can be estimated from emission spectra.
3. Charge separation or electron transfer from donor LUMO to acceptor LUMO (D*/A→D$^+$A$^-$) corresponding to a standard free energy change of ΔE, electron transfer rate constant of k_{sD}, and a reorganization energy of λ_{sD}.
4. Charge recombination or electron back transfer from acceptor LUMO to donor HOMO (D$^+$A$^-$→D/A) corresponding to a standard free energy change of $E_D - \Delta E$ (see Figure 8.4), transfer rate constant of k_r, and a reorganization energy change of λ_r.
5. Photoexcitation at acceptor (D/A + $h\nu_3$ → D/A*, A* designates an acceptor exciton).
6. Acceptor exciton decay to its ground state (D/A*→DA + $h\nu_4$) corresponding to a free energy change of E_A, decay rate constant of k_{dA}, and a reorganization energy of λ_{dA}.
7. Charge separation or electron transfer from donor HOMO to acceptor HOMO (D/A*→D$^+$A$^-$) corresponding to a free energy change of $E_{sA} = E_A - E_D + \Delta E$ (see Figure 8.4), transfer rate constant of k_{sA}, and a reorganization energy of λ_{sA}.
8. Charge recombination, same as in process 4.

For organic solar cells, the charge separated state is the desired starting point. However, charge separation (steps 3 and 7 in Figure 8.3) is also competing with exciton decay (steps 2 and 6). The ratio of charge separation rate constant versus exciton decay rate constant can therefore be defined as exciton quenching parameter (EQP, mathematically represented as Y_{eq}) as:

$$Y_{eqD} = k_{sD}/k_{dD} \tag{1}$$

$$Y_{eqA} = k_{sA}/k_{dA} \tag{2}$$

for donors and acceptors, respectively. The parameter Y_{eq} reflects the efficiency of exciton $\rightarrow$ charge conversion. It was experimentally observed that the charge separation could be orders of magnitude faster than the exciton decay in a MEH-PPV/fullerene donor–acceptor binary pair [7]. Second, charge separation (steps 3 and 7) is also competing with charge recombination (steps 4 and 8). The ratio of charge separation rate constant over charge recombination rate constant may therefore be defined as recombination quenching parameter (RQP, mathematically represented as Y_{rq})

$$Y_{rqD} = k_{sD}/k_r \tag{3}$$

$$Y_{rqA} = k_{sA}/k_r \tag{4}$$

for donors and acceptors, respectively. For any light harvesting applications, such as solar cell applications, it is desirable that both Y_{eq} and Y_{rq} parameters are large.

From classical Marcus electron transfer theory [41,42], the electron transfer rate constants may be simplified as

$$k_{dD} = A_{dD} \exp\left[-(E_D + \lambda_{dD})^2/4\,\lambda_{dD}\,kT\right] \tag{5}$$

$$k_{sD} = A_{sD} \exp\left[-(\Delta E + \lambda_{sD})^2/4\,\lambda_{sD}\,kT\right] \tag{6}$$

$$k_r = A_r \exp\left[-(E_D - \Delta E + \lambda_r)^2/4\,\lambda_r\,kT\right] \tag{7}$$

$$k_{dA} = A_{dA} \exp\left[-(E_A + \lambda_{dA})^2/4\,\lambda_{dA}\,kT\right] \tag{8}$$

$$k_{sA} = A_{sA} \exp\left[-(E_A - E_D + \Delta E + \lambda_{sA})^2/4\,\lambda_{sA}\,kT\right] \tag{9}$$

$$A_y = (2\pi H_y^2/h)(\pi/\lambda_y\,kT)^{1/2} \tag{10}$$

where $y = $ dD, sD, r, dA, and sA. H_y is an electronic coupling term between two electron transfer sites and can be estimated from molecular energy and dipole parameters using Mulliken–Hush model [41–45], λ_y is the reorganization energy containing contributions from molecular motions, vibrations, solvent effects, etc., and can be estimated from molecular vibrational spectroscopy or from excitation–emission spectroscopy under certain conditions [41–45]. T is temperature and k is the Boltzmann constant. The standard free energy of E_D and E_A can be estimated from spectroscopic, electrochemical, and thermodynamic analyses [41–45]. When ground and photoexcited states free energy potential wells have similar shapes, the following apply [36]:

$$\lambda_{dD} = (h\nu_1 - h\nu_2)/2 \tag{11}$$

$$E_D = -(h\nu_1 + h\nu_2)/2 \tag{12}$$

$$\lambda_{dA} = (h\nu_3 - h\nu_4)/2 \tag{13}$$

$$E_A = -(h\nu_3 + h\nu_4)/2 \tag{14}$$

The charge separation free energy ΔE can be approximated using Weller's equation [44,45,49] and may be simplified as:

$$\Delta E = \delta E + E_B \tag{15}$$

where the driving force δE is the frontier orbital (LUMO–LUMO) energy offset between the donor and the acceptor (negative values), and E_B (positive values) includes all counter-driving force terms, mainly electric Coulomb forces that need to be overcome in order to separate the exciton into a stable radical ion pair. If the device external applied electric fields are negligibly smaller than the exciton binding energy or the frontier orbital energy offset, and since the exciton binding energy is generally defined as the energy needed to separate an intramolecular exciton (D* or A*) into an intermolecular electron–hole radical ion pair (D^+A^-), E_B therefore can also be approximated as the exciton binding energy [17].

The donor exciton quenching parameter (EQP) can thus be expressed as:

$$Y_{eqD} = k_{sD}/k_{dD} = (H_{sD}/H_{dD})^2(\lambda_{dD}/\lambda_{sD})^{1/2} \exp(Z_{eqD}) \tag{16}$$

where

$$Z_{eqD} = -(\delta E + E_B + \lambda_{sD})^2/4\lambda_{sD}kT + (E_D + \lambda_{dD})^2/4\,\lambda_{dD}\,kT \tag{17}$$

The donor RQP can be expressed as:

$$Y_{rqD} = k_{sD}/k_r = (H_{sD}/H_r)^2(\lambda_r/\lambda_{sD})^{1/2} \exp(Z_{rqD}) \tag{18}$$

where

$$Z_{rqD} = -(\delta E + E_B + \lambda_{sD})^2/4\lambda_{sD}kT + (E_D - \delta E - E_B + \lambda_r)^2/4\,\lambda_r\,kT \tag{19}$$

8.4.3. Results and Discussion

If the frontier orbital energy offset δE is set as variable, for convenience demonstration, using temperature $T = 300\,K$, $k = 0.000086\,eV/K$, and calculated (from Equations (11) to (14)) and estimated RO-PPV and SF-PPV-I data of $E_D = -2.6\,eV$, $E_A = -2.7\,eV$, $E_B = -0.4\,eV$, $\lambda_{sD} = \lambda_{sA} = 0.2\,eV$, $\lambda_r = 0.5\,eV$, $\lambda_{dD} = 0.4\,eV$, $\lambda_{dA} = 0.6\,eV$ [29–34,46–48], and arbitrary values $H_x = 1$ ($x = sD$, dD, sA, dA, and r), a plot of normalized Y_{eqD}, k_r, and Y_{rqD} versus δE are shown in Figure 8.21.

As Figure 8.21 shows, when frontier LUMO orbital offset δE varies, k_r, Y_{eqD}, and Y_{rqD} all exhibit their own maximum values. Using $\partial Y_{eqD}/\partial \delta E = 0$, this gives

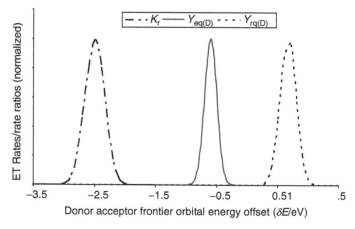

Figure 8.21. Donor RO-PPV exciton quenching parameter ($Y_{eq(D)} = k_{s(D)}/k_{d(D)}$, middle solid curve), charge recombination rate constant (K_r, left long dashed curve), and charge recombination quenching parameter ($Y_{rq(D)} = k_{s(D)}/k_{r(D)}$, right short dashed curve) versus LUMO offset of RO-PPV/SF-PPV-I pair.

$$\delta E_{eqD} = -\lambda_{sD} - E_B = -0.6 \, \text{eV} \tag{20}$$

which corresponds to maximum donor exciton-charge conversion.

Using $\partial Y_{rqD}/\partial \delta E = 0$, this gives

$$\delta E_{rqD} = [2 + E_D/\lambda_r]/(1/\lambda_r - 1/\lambda_{sD}) - E_B = 0.67 \, \text{eV} \tag{21}$$

which corresponds to a maximum k_{sD}/k_r value.

Using $\partial k_r/\partial \delta E = 0$ from Equation (7), this gives

$$\delta E_r = E_D + \lambda_r - E_B = -2.5 \, \text{eV} \tag{22}$$

As can be seen in Figure 8.21, the fastest photoinduced charge separation occurs at $-0.6 \, \text{eV}$ when the RO-PPV/SF-PPV-I LUMO offset (driving force) equals the sum of the charge separation reorganization energy (0.2 eV) and the exciton binding energy (0.4 eV, counter forces). Also, the fastest charge recombination occurs at a LUMO offset of $-2.5 \, \text{eV}$, far away from optimum charge separation offset ($-0.6 \, \text{eV}$) as well as the actual RO-PPV/SF-PPV-I offset ($-0.9 \, \text{eV}$, [32]). Therefore, charge recombination in RO-PPV/SF-PPV-I pair does not seem to be of a major concern as long as the LUMO offset is nearby the δE_{eqD}. One interesting observation was that, during the charge separation, E_B represents counter-Coulomb forces, while in charge recombination, E_B represents driving Coulomb forces. Figure 8.21 shows that the recombination quenching parameter Y_{rqD} (k_{sD}/k_r) does not reach its maximum until 0.67 eV, i.e., at a positive energy offset. At this positive energy offset, the photoinduced charge separation might be too slow to be attractive for efficient photovoltaic function, therefore, the δE_{rqD} value does not appear to be important in this particular case. It is desirable that the δE_{rqD} is coincident with or close to δE_{eqD}, and that δE_{rD} is far away from δE_{eqD}.

Similarly, for acceptor, the exciton quenching parameter (Y_{eqA}) can be expressed as:

$$Y_{eqA} = k_{sA}/k_{dA} = (H_{sA}/H_{dA})^2 (\lambda_{dA}/\lambda_{sA})^{1/2} \exp(Z_{eqA}) \tag{23}$$

where

$$Z_{eqA} = -(E_A - E_D + \delta E + E_B + \lambda_{sA})^2/4\lambda_{sA}kT + (E_A + \lambda_{dA})^2/4\lambda_{dA}kT \tag{24}$$

Using $\partial Y_{eqA}/\partial \delta E = 0$, this gives

$$\delta E_{eqA} = E_D - E_A - \lambda_{sA} - E_B = -0.5 \, \text{eV} \tag{25}$$

corresponding to the most effective acceptor photoinduced charge separation. As a matter of fact, since the donor–acceptor HOMO offset is $E_A - (E_D - \delta E)$ (see Figure 8.3), from Equation (25), this means most effective photoinduced charge separation at acceptor occurs where the HOMO offset equals the sum of the exciton binding energy and the acceptor charge separation reorganization energy.

For a donor–acceptor pair where both can harvest light, the exciton quenching parameter for the pair can be expressed as:

$$Y_{eq(D+A)} = Y_{eqD} Y_{eqA} \tag{26}$$

using $\partial Y_{eq(D+A)}/\partial \delta E = 0$, this gives

$$\delta E_{eq(D+A)} = [(E_D - E_A)/\lambda_{sA} - 2]/(1/\lambda_{sD} + 1/\lambda_{sA}) - E_B = -0.55 \, \text{eV} \tag{27}$$

The exciton quenching parameters $Y_{eq}(A)$, $Y_{eq}(D)$, and $Y_{eq}(D+A)$ versus the LUMO offset are plotted in Figure 8.22. As shown in Figure 8.22, $Y_{eq}(D+A)$ represents an

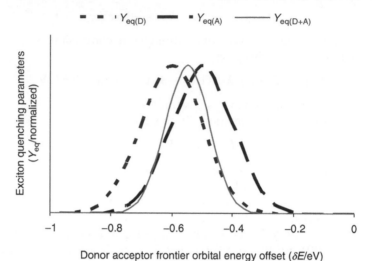

Figure 8.22. Exciton quenching parameters of the donor RO-PPV (left long dashed curve), acceptor SF-PPV-I (right short dashed curve), and their product $Y_{eq(D)}Y_{eq(A)}$ (middle solid curve) versus the frontier LUMO orbital energy offset.

overlap area where both donor and acceptor would harvest light efficiently, and the optimum offset is around -0.55 eV. The actual RO-PPV/SF-PPV-I LUMO offset of -0.9 eV appears a little larger than this optimum. Further improvement of photo-induced charge separation can be achieved via either reducing the LUMO level of RO-PPV, or increasing the LUMO level of SF-PPV-I via molecular engineering.

Likewise, the charge recombination quenching parameter for acceptor can be expressed as:

$$Y_{rqA} = k_{sA}/k_r = (H_{sA}/H_r)^2 (\lambda_r/\lambda_{sA})^{1/2} \exp(Z_{rqA}) \tag{28}$$

where

$$Z_{rqA} = -(E_A - E_D + \delta E + E_B + \lambda_{sA})^2/4\lambda_{sA}kT + (E_D - \delta E - E_B + \lambda_r)^2/4\lambda_r kT \tag{29}$$

Using $\partial Y_{rq(A)}/\partial \delta E = 0$, this gives

$$\delta E_{rq(A)} = [2 + (E_A - E_D)/\lambda_{sA} + E_D/\lambda_r]/(1/\lambda_r - 1/\lambda_{sA}) - E_B = 0.83 \, eV \tag{30}$$

This corresponds to a large k_s/k_r ratio for light harvesting at SF-PPV-I acceptor.

For both donor and acceptor, the recombination quenching parameter can be expressed as:

$$Y_{rq(D+A)} = k_{sA}k_{sD}/k_r k_r \tag{31}$$

Using $\partial Y_{rq(D+A)}/\partial \delta E = 0$, this gives

$$\delta E_{rq(D+A)} = [4 + (E_A - E_D)/\lambda_{sA} + 2E_D/\lambda_r]/(2/\lambda_r - 1/\lambda_{sD} - 1/\lambda_{sA}) - E_B = 0.75 \, eV \tag{32}$$

Again, the hybrid δE value is between the donor and acceptor δE of highest Y_{rq}, and the positive value indicates the charge separation might be very slow at this offset. The plot of $Y_{rq(D)}$, $Y_{rq(A)}$, and $Y_{rq(D+A)}$ versus δE are shown in Figure 8.23.

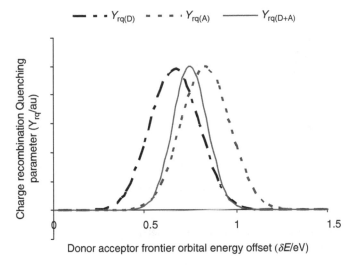

Figure 8.23. Charge recombination quenching parameters of the donor RO-PPV (left long dashed curve), acceptor SF-PPV-I (right short dashed curve), and their product $Y_{rq(D+A)} = Y_{rq(D)} Y_{rq(A)}$ (middle solid curve) versus the frontier LUMO orbital energy offset.

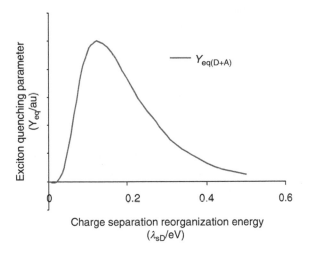

Figure 8.24. Exciton quenching parameter $Y_{cs(D+A)}$ of RO-PPV/SF-PPV-I pair versus donor RO-PPV charge separation (λ_{sD}) reorganization energy.

When energy offset δE is fixed (e.g., at its optimum value of $-0.55\,\text{eV}$), and donor RO-PPV charge separation reorganization energy (λ_{sD}) is varied, as Figure 8.24 shows, the exciton quenching parameter $Y_{eq(D+A)}$ also experiences a maximum value. Using $\partial Y_{eq(D+A)}/\partial \lambda_{sD} = 0$, this gives

$$\lambda_{sD}(\text{where } Y_{eq(D+A)} = \text{maximum}) = [(\delta E + E_B)^2 + k^2 T^2]^{1/2} - kT = 0.13\,\text{eV} \quad (33)$$

This result implies the more closer the charge separation reorganization energy toward 0.13 eV, the larger the $Y_{eq(D+A)}$ would be.

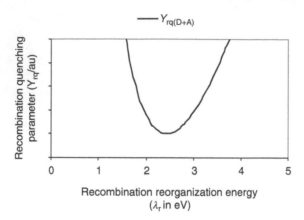

Figure 8.25. Recombination quenching parameter $Y_{rq(D+A)}$ of RO-PPV/SF-PPV-I pair versus charge recombination reorganization energy of (λ_r).

When both energy offset and charge separation reorganization energy are fixed (with $\lambda_{sD} = 0.13\,\text{eV}$), and charge recombination reorganization energy (λ_r) is varied, as shown in Figure 8.25, the cell recombination quenching parameter $Y_{rq(D+A)}$ would experience a minimum value. Using $\partial Y_{rq\,(D+A)}/\partial \lambda_r = 0$, this gives

$$\lambda_r(\text{where } Y_{cs\,(D+A)} = \text{minimum}) = [(E_D - \delta E - E_B)^2 + k^2 T^2]^{1/2} - kT$$
$$= 2.42\,\text{eV} \tag{34}$$

This result implies that the more far away charge recombination reorganization energy from 2.42 eV, the larger the $Y_{rq(D+A)}$ would be.

Like in any modeling or simulation studies, the numbers used here may not be accurate or important, rather, it is the trend that is the most important and meaningful. In order to further examine this model and its predictions, a series and systematic experiments need to be designed and performed. For Y_{eq} trend tests, when a donor (or acceptor) is fixed, as δE only changes k_s and not k_d, it can be regarded as a special case of "Marcus inversion" case, and the trend has already been verified by experiments [41,42,50]. For Y_{rq} trend tests, since δE will change both k_s and k_r, therefore, a series donor–acceptor pairs where a donor (or an acceptor) is fixed first, and then a series acceptors (or a series of donors) with different δE in relation to the fixed donor (or acceptor) need to be experimentally evaluated for their charge separation and recombination rate constants. The type of experiments described in Ref. [50] is a good example, though no charge recombination rates and related reorganization energies were given. In these experiments, it is also important that the molecular structures of the changing acceptors (or donors) are similar, so that the Coulomb force terms (or exciton binding energy) and reorganization energies are similar, and that the electronic withdrawing (or donating) strength (or δE) would be the only or major variable. In this way, δE versus the Y_{eq}, the Y_{rq} and the k_r can all be evaluated at the same time. The overall power conversion efficiency of the solar cell is expected to follow Y_{eq} more closely when both Y_{rq} and k_r are far away from Y_{eq}, and the cell efficiency can be evaluated at the same time, provided the charge transport and collection at electrodes are also similar. Unfortunately, these type of experiments have not yet been systematically performed (or are not able to be performed) at the moment due to lack of suitable materials. Finally, additional parameters and competing processes (such as other electron and energy transfer processes) may also need to be taken into account in order to have a more accurate simulation. Systematic and expanded studies, including effects from excitation energy gap changes, experimental case studies, etc., are underway and will be reported in the near future.

8.5. CONCLUSIONS AND FUTURE PERSPECTIVES

The current low photoelectric power conversion efficiencies of organic photovoltaic materials and devices can be attributed mainly to the "photon loss," the "exciton loss," and the "carrier loss" due to improper donor–acceptor energy levels or offsets and poor morphologies (spatial geometries) for the charge carrier generation, transportation, and collection at electrodes. However, high-efficiency organic photovoltaic materials and devices can be achieved via optimization in both space and energy–time domains.

For optimization in the spatial domain, the key is a donor–acceptor phase separated and "bicontinuous" morphology, where the dimension of each phase should be within the AEDL (e.g., 5–70 nm). For this reason, a –DBAB- type of block copolymer system and its potential self-assembled "tertiary" supramolecular nanostructure has been designed and preliminarily examined. In this system, along the carrier transport direction, it is "bicontinuous" between the two electrodes. Yet, in the plane perpendicular to the carrier transport direction, it is donor–acceptor phase separated morphology on the nanoscale, and each phase diameter is within the exciton drift distance. The much improved PL quenching (from less than 70% to 99%), biased conductivity (two orders of improvement), and photoconductivity (twofold increase) of the synthesized –DBAB- over the simple D–A blend system is attributed mainly to morphology (spatial) improvement. The target "tertiary" nanostructured photovoltaic device is expected to improve the photovoltaic power conversion efficiency significantly in comparison to the existing organic photovoltaic devices due to the reduction of the "exciton loss" and the "carrier loss" via three-dimensional spatial optimizations (via block copolymer supramolecular structural and morphological control).

On energy–time domain optimizations, first of all, the optical excitation energy gaps in both donor and acceptor should match the intended photon energy, and that optimal donor–acceptor energy offsets which correspond to most efficient photo-induced charge separation should be identified and materialized. Specifically, in electron transfer dynamic regime and based on Marcus theory, this study has found that, there exists an optimal donor–acceptor energy offset where exciton-charge conversion is most efficient (or EQP reaches its maximum), and a second optimal energy offset where charge recombination is relatively slow compared to charge separation (or RQP become largest). If the maximum RQP is too far away from maximum EQP, then this optimum RQP is insignificant as the charge separation at this maximum RQP might be too slow. The molecules should be designed and developed such that the maximum RQP is close to or coincides with maximum EQP. There also exists a third energy offset where charge recombination becomes fastest. The molecules should be designed and developed such that this maximum charge recombination is far away from maximum EQP. For a donor–acceptor binary photovoltaic system, there exists a fourth optimal donor–acceptor energy offset, where the EQP product of both donor and acceptor become largest, so that both donor and acceptor can effectively contribute to photoinduced charge separation. This final optimal donor–acceptor energy offset is related to the exciton binding energy, the optical excitation energy gaps, and the reorganization energies of the charge separation of both donors and acceptors. While there exist a desired donor (or acceptor) charge separation reorganization energy where Y_{eq} has maximum value, there also exists an undesired charge recombination reorganization energy value where Y_{rq} becomes minimum. Both the desired energy offset, the desired charge separation reorganization energy, and the undesired charge recombination reorganization energy values are critically important for molecular structure and energy level fine-tuning in developing high efficiency organic light harvesting systems, including organic photovoltaic cells, photodetectors, or any artificial photo-charge synthesizers and converters.

ACKNOWLEDGMENTS

The authors are very grateful for research and educational grant supports from a number of funding agencies including NASA (Awards NAG3-2289, NCC3-1035), Air Force Office of Scientific Research (Awards F-49620-01-1-0485, F-49620-02-1-0062), National Science Foundation (Award HRD-0317722), Department of Education (Title III award), and Dozoretz foundation. Professor Sun is also very grateful to Professors M. Wasielewski, J. Bredas, and in particular, to Professor R. Marcus for stimulating discussions on "Marcus theory."

REFERENCES

1. M. D. Archer and R. Hill, eds., *Clean Electricity From Photovoltaics*, Imperial College Press, London, UK, 2001.
2. S. Sun and N. S. Sariciftci, eds., *Organic Photovoltaics: Mechanisms, Materials and Devices*, CRC Press, Boca Raton, FL, 2005.
3. Z. Kafafi and P. Lane, eds., *Organic Photovoltaics IV*, SPIE, Bellingham, 2004.
4. C. Brabec, V. Dyakonov, J. Parisi, and N. Sariciftci, *Organic Photovoltaics: Concepts and Realization*, Springer, Berlin, 2003.
5. A. Hagfeldt and M. Graetzel., *Molecular Photovoltaics, Acc. Chem. Res.*, **33**, 269 (2000).
6. C. Tang, Two-layer organic photovoltaic cell, *Appl. Phys. Lett.*, **48**, 183–185 (1986).
7. N. S. Sariciftci, L. Smilowitz, A. J. Heeger, and F. Wudl, Photoinduced electron transfer from a conducting polymer to buckminsterfullerene, *Science*, **258**, 1474 (1992).
8. B. Kraabel, J. Hummelen, D. Vacar, D. Moses, N. Sariciftci, A. Heeger, and F. Wudl, Subpicosecond photoinduced electron transfer from counjugated polymers to functionalized fullerenes, *J. Chem. Phys.*, **104**, 4267–4273 (1996).
9. G. Yu, J. Gao, J. Hummelen, F. Wudl, and A. Heeger, Polymer photovoltaic cells: enhanced efficiencies via a network of internal donor–acceptor heterojunctions, *Science*, **270**, 1789–1791 (1995).
10. L. Schmidt-Mende, A. Fechtenkötter, K. Müllen, E. Moons, R. H. Friend, and J. D. MacKenzie, Self-organized discotic liquid crystals for high-efficiency organic photovoltaics, *Science*, **293**, 1119 (2001).
11. M. Granström, K. Petritsch, A. Arias, A. Lux, M. Andersson, and R. Friend, Laminated fabrication of polymeric photovoltaic diodes, *Nature*, **395**, 257–260 (1998).
12. L. S. Roman, M. Anderson, T. Yohannes, and O. Inganas, Photodiode performance and nanostructure of polythiophene/C_{60} blends, *Adv. Mater.*, **9**, 1164 (1997).

13. B. Boer, U. Stalmach, P. Hutten, C. Melzer, V. Krasnikov, and G. Hadziioannou, Supramolecular self-assembly and opto-electronic properties of semiconducting block copolymers, *Polymer*, **42**, 9097 (2001).

14. S. Sun, Design of a block copolymer solar cell, *Sol. Energy Mater. Sol. Cells*, **79**, 257–264 (2003).

15. T. A. Skotheim, R. L. Elsenbaumer, and J. R. Reynolds, eds., *Handbook of Conducting Polymers*, 2nd ed., Marcel Dekker, New York, 1998.

16. J. Perlin, *From Space to Earth — The Story of Solar Electricity*, AATEC Publications, Ann Arbor, MI, 1999.

17. M. Knupfer, Exciton binding energies in organic semiconductors, *Appl. Phys. A*, **77**, 623–626 (2003).

18. T. Stübinger and W. Brütting, Exciton diffusion and optical interference in organic donor–acceptor photovoltaic cells, *J. Appl. Phys.*, **90**, 3632 (2001).

19. H. Amerongen, L. Valkunas, and R. Grondelle, eds., *Photosynthetic Excitons*, World Scientific, Singapore, 2000.

20. C. Brabec, et al., Origin of the open circuit voltage of plastic solar cells, *Adv. Funct. Mater.*, **11**, 374–380 (2001).

21. C. Brabec, C. Winder, N. Sariciftci, J. Hummelen, A. Dhanabalan, P. Hal, and R. Janssen, A low-bandgap semiconducting polymer for photovoltaic devices and infra-red emitting diodes, *Adv. Funct. Mater.*, **12**, 709–712 (2002).

22. N. Sariciftci, et al., Convenient synthesis and polymerization of 5,6-disubstituted dithiophthalides toward soluble poly(isothianaphthene): an initial spectroscopic characterization of the resulting low-bandgap polymers, *J. Poly. Sci., A.*, **41**, 1034–1045 (2003).

23. V. Seshadri and G. Sotzing, Progress in optically transparent conducting polymers, in: *Organic Photovoltaics: Mechanisms, Materials and Devices*, S. Sun and N. S. Sariciftci, eds., CRC Press, Boca Raton, FL, 2005.

24. B. Gregg, Excitonic solar cells, *J. Phys. Chem. B.*, **107**, 4688–4698 (2003).

25. N. Hadjichristidis, S. Pispas, and G. Floudas, eds., *Block Copolymers: Synthetic Strategies, Physical Properties, and Applications*, John Wiley & Sons, New York, 2003.

26. M Lazzari and M. Lopez-Quintela, Block copolymers as a tool for nanomaterials fabrication, *Adv. Mater.*, **15**, 1584–1594, (2003).

27. X. L. Chen and S. A. Jenekhe, Block conjugated copolymers: toward quantum-well nanostructures for exploring spatial confinement effects on electronic, optoelectronic, and optical phenomena, *Macromolecules*, **29**, 6189 (1996).

28. S. Sun, Z. Fan, Y. Wang, C. Taft, J. Haliburton, and S. Maaref; Design and synthesis of novel block copolymers for efficient optoelectronic applications, in *Organic Photovoltaics II*, *SPIE Proc.*, **4465**, 121 (2002).

29. Z. Fan, Synthesis and Characterization of a Novel Block Copolymer System Containing RO-PPV And SF-PPV-I Conjugated Blocks, M.S. thesis, Norfolk State University, Norfolk, Virginia, July 2002.

30. S. Sun, Z. Fan, Y. Wang, J. Haliburton, C. Taft, K. Seo, and C. Bonner, Conjugated block copolymers for opto-electronic functions, *Syn. Met.*, **137**, 883–884 (2003).

31. S. Sun, Z. Fan, Y. Wang, C. Taft, J. Haliburton, and S. Maaref, Synthesis and characterization of a novel –DBAB-block copolymer system for potential light harvesting applications, in: *Organic Photovoltaics III*, *SPIE*, **4801**, 114–128 (2003).

32. S. Sun, Design and development of conjugated block copolymers for use in photovoltaic devices, in: *Organic Photovoltaics IV*, *SPIE*, **5215**, 195–205 (2004).

33. S. Sun and J. Haliburton, Spectroscopic properties of a novel –D-B-A-B-type block copolymer and its component blocks, to be submitted.

34. S. Sun, Z. Fan, Y. Wang, J. Haliburton, S. Vick, M. Wang, S. Maaref, K. Winston, A. Ledbetter, and C. E. Bonner, Synthesis and characterization of a –donor–bridge–acceptor–bridge type block copolymer containing poly-*p*-phenylenevinylene conjugated donor and acceptor blocks, to be submitted.

35. S. Sun, Z. Fan, Y. Wang, K. Winston, and C. E. Bonner, Morphological effects to carrier mobility in a RO-PPV/SF-PPV donor–acceptor binary thin film opto-electronic device, *Mater. Sci. Eng. B.*, **116**, 279–282 (2005).

36. D. Gosztola, B. Wang, and M. R. Wasielewski, Factoring through-space and through-bond contributions to rates of photoinduced electron transfer in donor–spacer–acceptor molecules, *J. Photochem. Photobiol., (A)*, **102**, 71–80 (1996).

37. D. Adam, P. Schuhmacher, J. Simmerer, L. Haussling, K. Siemensmeyer, K. Etzbach, H. Ringsdorf, and D. Haarer, Fast photoconduction in the highly ordered columnar phase of a discotic liquid crystal, *Nature*, **371**, 141 (1994).

38. Z. Bao, A. Dodabalapur, and A. J. Lovinger, Soluble and processable regio-regular poly(3-hexylthiophene) for thin film field-effect transistor applications with high mobility, *Appl. Phys. Lett.*, **69**, 4108 (1996).

39. T. Nguyen, J. Wu, V. Doan, B. Schwartz, and S. H. Tolbert, Control of energy transfer in oriented conjugated polymer-mesoporous silica composites, *Science*, **288**, 652–656 (2000).

40. Private communication from Professor Marc Hillmyer.

41. V. Balzani, ed., *Electron Transfer in Chemistry*, Wiley-VCH, New York, 2000.

42. R. Marcus, et al., Charge transfer on the nanoscale: current status, *J. Phys. Chem., B*, **107**, 6668–6697 (2003).

43. S. Sensfuss, et al., Characterization of potential donor acceptor pairs for polymer solar cells by ESR, optical and electrochemical investigations, in *Organic Photovoltaics IV*, Z. Kafafi and P. Lane, eds., *SPIE Proc.*, **5215**, 129–140 (2004).

44. E. Peeters, P. Hal, J. Knol, C. Brabec, N. Sariciftci, J. Hummelen, and R. Janssen, Synthesis, photophysical properties, and photovoltaic devices of oligo(*p*-phenylenevinylene)–fullerene dyads, *J. Phy. Chem. B*, **104**, 10174–10190 (2000).

45. J. Bredas, et al., Alternating oligo(*p*-phenylene vinylene)–perylene bisimide copolymers, *J. Am. Chem. Soc.*, **125**, 8625–8638 (2003).

46. S. Sun, Optimum energy levels for organic solar cells, project briefings to AFOSR and NASA, and *Mater. Sci. Eng., B*, **116**, 251–256 (2005).

47. S. Sun, Optimal energy offsets for organic donor/acceptor binary solar cells, *Sol. Energy Mater. Sol. Cells*, **85**, 261–267 (2005), and published online on June 20, 2004.

48. S. Sun, Z. Fan, Y. Wang and J. Haliburton, Organic solar cell optimizations, *J. Mater. Sci.*, **40**, 1429–1443 (2005).

49. A. Weller, Photoinduced electron transfer in solution, *Z. Phys. Chem.*, **133**, 93–98 (1982).

50. J. Miller, L. Calcaterra, and G. Closs, Intramolecular long-distance electron transfer in radical anions, *J. Am. Chem. Soc.*, **106**, 3047–3049 (1984).

Section 3
Materials and Devices

9

Bulk Heterojunction Solar Cells

Harald Hoppe and Niyazi Serdar Sariciftci
Linz Institute for Organic Solar Cells (LIOS), Physical Chemistry, Johannes Kepler University of Linz, Linz, Austria

Contents

Abstract   Today, AM1.5 white light solar power conversion efficiencies reaching 4% have been reported with organic photovoltaic devices. The rapidly increasing power conversion efficiencies of these thin-film organic devices and the perspective of cheap roll-to-roll manufacturing drive their development in a dynamic way. In this chapter, an introductory overview is presented on the bulk heterojunction concept with a special emphasis on the open circuit potential of such devices.

Keywords   bulk heterojunction solar cells, ultrafast photoinduced charge separation, derivatized poly(*p*-phenylene vinylene) (PPV) and fullerenes, open circuit voltage, short circuit current, double cable molecular systems, quantum efficiency.

9.1.   INTRODUCTION

Within the past three decades there has been a tremendous effort to develop organic photovoltaics [1–12], starting with the application of small organic molecules (pigments) [2,3,9,10], and later with semiconducting polymers (see Refs. [13–16]), remarkable improvements have been reported within the last 4 years [4,5,8–10,17]. Organic semiconductors show poor charge carrier mobility (usually $<10^{-2}$ to 10^{-3} cm^2/V s) [15], which has a large limiting effect on the short circuit photocurrent.

However, strong optical absorption coefficients (usually $\geq 10^5$ cm^{-1}) allow for using <100 nm thin devices, which somehow circumvents the problem of low mobilities. The small diffusion length of primary photoexcitations (excitons) in these amorphous and disordered organic semiconductors is also an important limiting factor [9,19–27]. As a consequence of relatively large exciton binding energies (>100 meV) exceeding those of most of the inorganic semiconductors (<<20 meV), such excitons created in the bulk material of organic semiconductors do not dissociate into free charge carriers quantitatively with thermal excitations [28,29]. These excitons dissociate in strong electric fields as observed at metal–semiconductor interfaces or at donor–acceptor interfaces, where the electronic potential has nearly discontinuous jumps.

The first generation of organic photovoltaic solar cells was based on single organic layers sandwiched between two metal electrodes of different work functions [2,3]. The rectifying behavior of single-layer devices was attributed to the asymmetry in the electron and hole injection into the molecular π^* and π-orbitals [30] and to the formation of a Schottky-barrier (e.g., Refs. [3,31–33]) between the p-type (hole conducting) organic layer and the metal with the lower work function. The first power conversion efficiencies reported were up to 0.7% [34,35]. In this case, the organic layer was sandwiched between a metal–metal oxide and a metal electrode, thus enhancing the Schottky-barrier effect (metal–insulator–semiconductor (MIS) devices [36]). Tang reported in 1986 about 1% power conversion efficiency for a stacked bilayer of two organic materials (a phthalocyanine derivative as p-type semiconductor and a perylene derivative as n-type semiconductor) sandwiched between a transparent conducting oxide anode and a metal cathode [37]. This result was for many years the outstanding benchmark and was surmounted only at the turn of the millennium [38,39]. Hiramoto and coworkers did pioneering work, introducing the concept of an organic tandem cell structure by stacking two bilayer devices [40] and by developing a trilayer configuration with an intermixed layer between the two layers of different materials [41,42].

9.2. PHOTOINDUCED ELECTRON TRANSFER FROM CONJUGATED POLYMERS ONTO FULLERENES

Conjugated polymer single-layer devices exhibited rather low efficiencies [33,43–45]. Independently, the Santa Barbara group and the Osaka group reported studies on the photophysics of mixtures of conjugated polymers with C_{60} [46–51]. The observations clearly evidenced an ultrafast, reversible, metastable photoinduced electron transfer from conjugated polymers onto Buckminsterfullerenes in solid films. A schematic description of this phenomenon is displayed in Figure 9.1. Once the photoexcited electron is transferred to an acceptor unit, the resulting cation radical (positive polaron) species on the conjugated polymer backbone is known to be highly delocalized and stable as shown in electrochemical and chemical oxidative doping studies. Especially the long lifetime of the charge transferred state (around milliseconds at 80 K) and thus, the high quantum efficiency of this process (~100%) in conjugated polymer–fullerene composites compared to pristine conjugated polymer films initiated the development of bulk heterojunction solar cells.

Figure 9.2 shows the luminescence spectrum of MEH-PPV/C_{60} composites compared to MEH-PPV alone. The strong photoluminescence of MEH-PPV is quenched by a factor of 10^3. The luminescence decay time is reduced

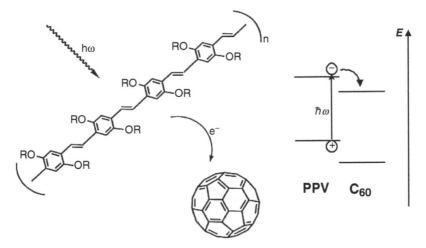

Figure 9.1. Illustration of the photoinduced charge transfer (left) with a sketch of the energy level scheme (right). After excitation in the PPV polymer the electron is transferred to the C_{60} due to its higher electron affinity.

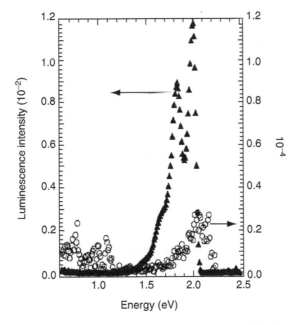

Figure 9.2. Quenching of photoluminescence in MEH-PPV: C_{60} blends: pristine MEH-PPV (filled triangles) and 1:1 (by weight) blend with C_{60} (open circles).

from $\tau_0 = 550$ psec to $\tau_{rad} \ll 1$ psec, indicating the existence of a rapid quenching process; e.g., subpicosecond electron transfer [47].

The strong quenching of the luminescence of another conjugated polymer P3OT reported by Morita et al. [49,50] is also consistent with efficient photoinduced electron transfer. Photoinduced electron transfer from the singlet excited state is a prominent photoluminescence quenching mechanism; however, other nonradiative relaxation channels have to be ruled out to assign the luminescence quenching to an electron transfer.

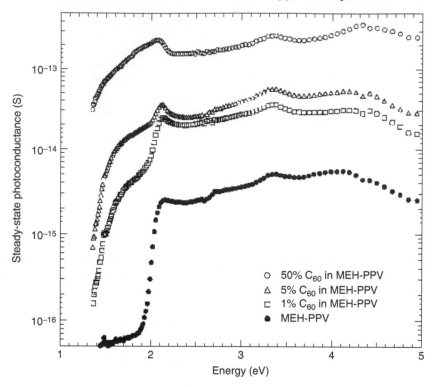

Figure 9.3. Sensitization of photoconductivity as a function of fullerene content. The photoconductivity of the conjugated matrix can be increased by several orders of magnitude.

Figure 9.3 shows the photoconductivity as a function of optical energies by increasing the fullerene content in the conjugated polymer matrix. It is clearly observable that the addition of even few percent fullerenes into the polymer matrix increases the overall photoconductivity by nearly one order of magnitude [48].

9.3. THE BULK HETEROJUNCTION CONCEPT

Photoinduced electron transfer from optically excited conjugated polymers to the C_{60} molecule [46–50] resulted in development of bilayer heterojunction [19,51,52] and bulk heterojunction [53,54] devices. Since the electron is transferred from a p-type hole conducting polymer onto the n-type electron conducting C_{60} molecule, the notation of donor (D) and acceptor (A) with respect to the electron transfer was introduced. The bulk heterojunction concept, also valid for the co-evaporated molecular structures [41,42], was introduced by blending two organic semiconductors having donor (D) and acceptor (A) properties in solution [55–57]. Spin-cast films from such binary solutions then resulted in solid-state mixtures of both components. Another approach was the lamination of polymer–polymer layers [58].

The small exciton diffusion length of organic semiconductors limits the photo-activity within a small volume with dimensions of around 10–20 nm, which are typical diffusion lengths of excitons in conjugated polymers [19–26]. Although the quantum efficiency for photoinduced charge separation is near unity for donor–

acceptor pairs, the photoelectric conversion efficiency is limited by small regions around these interfaces between donors and acceptors.

This turned out to be a main limitation for photocurrent generation in bilayer devices, and consequently an interpenetrating phase-separated D–A network composite, i.e., "bulk heterojunction," where the donor phase intimately intermixed with the acceptor phase would appear to be the ideal organic photovoltaic material. Since any point in the solid-state composite film is within a few nanometers of a neighboring D–A interface, such a composite is a "bulk D–A heterojunction" material. If the network in a device is bicontinuous, the collection efficiency can be quite high. A bulk heterojunction device is displayed schematically in Figure 9.4. Lowering the content of one phase in the other phase in bulk heterojunctions below percolation would hinder the continuous path formation to the respective electrodes and stop the photocurrent flow. Therefore, it is necessary that both phases are percolated in the volume. For spherical objects embedded in a matrix this percolation limit is around 17 vol.%. Assuming the formation of semispherical nanostructures for both donor and acceptor phases (see below), this would imply that a minimum of roughly 20% of a phase is needed for getting a short circuit current in bulk heterojunctions. Indeed, such a percolation behavior has been observed in polymer–fullerene bulk heterojunction solar cells (Figure 9.5).

Another important issue is the symmetry breaking in the device to enable the directed transport of charges. As the ideal bulk heterojunction has a homogenous blend everywhere in the volume, there will be also an equal concentration of positive and negative charges throughout the bulk. To give the transport direction, metals with different work functions have to be used on either side of the junction. This will create an internal built-in field (see section on "Metal–Insulator–Metal (MIM) picture") and tilt the energy levels of the insulator in between. Thus, the holes will be transported towards the higher work function electrode and the electrons will be transported towards the lower work function electrode. A selective contact of the respective electrodes can also create a diffusion gradient by depleting exclusively one sign of charge near them.

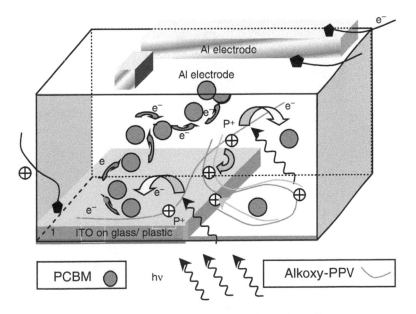

Figure 9.4. Schematic description of bulk heterojunction solar cells.

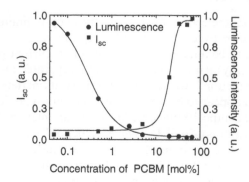

Figure 9.5. The photoluminescence (right axis) as well as the short circuit current (left axis) of a bulk heterojunction as a function of the acceptor content in donor-conjugated polymer matrix. In accordance with percolation theory the short circuit current sets in after a filling of around 20%.

This selective transport (check-valve type) through a "membrane" also introduces a symmetry breaking by the respective electrodes. Such selective contacts, enabling one sign of charge carriers to be transported and the other sign blocked, might be important for devices where the charge carrier diffusion plays a major role.

The bulk heterojunction devices used today consist, in general, of a PEDOT–PSS-covered ITO substrate coated with a single, photoactive bulk heterojunction layer (e.g., polymer–fullerene mixtures) closed by a low work function cathode such as Al (see Figure 9.6). Today conversion efficiencies up to 4% have been achieved [10, 59–72].

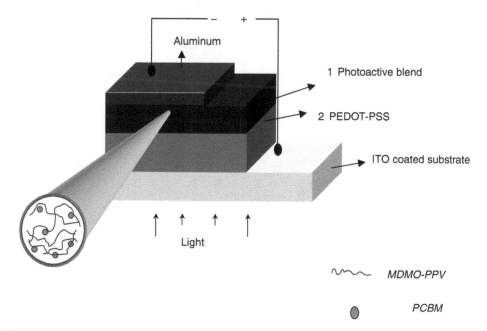

Figure 9.6. The schematic description of the bulk heterojunction solar cell devices. A single photoactive layer is sandwiched between two electrodes. PEDOT–PSS is an organic conductor used for manipulating ITO surface and work function.

9.4. METAL–INSULATOR–METAL (MIM) PICTURE

To understand the rectifying behavior of an intrinsic (undoped) semiconductor device in the dark, the MIM model is useful [36]. In Figure 9.7 a thin-film semiconductor, sandwiched between two metal electrodes with different work functions, is depicted for several scenarios. The metals are represented by their Fermi levels, while for the semiconductor the valence and conduction bands, corresponding to the molecular lowest unoccupied molecular orbital (LUMO) and the highest occupied molecular orbital (HOMO) levels, are shown. In Figure 9.7a there is no voltage applied, i.e., short circuit conditions. Hence there is no net current flowing in the dark, and the built-in electric field, resulting from the difference in the work functions. Under illumination, separated charge carriers can drift in this electric field to the respective contacts: the electrons move to the lower and the holes to the higher work function metal. The device then works as a solar cell. In Figure 9.7b the situation is shown for open circuit conditions, also known as "flat band condition." The applied voltage is called open circuit voltage V_{OC}, which corresponds in this case to the difference in the work functions of metals and balances the built-in field with no net driving force for the charge carriers; the current in this circuit is zero. In Figure 9.7c the situation is shown for an applied reverse bias and only a very small injected dark current j_0 can flow. Under illumination, the generated charge carriers drift under strong electric fields to the respective electrodes and the diode is working as a photodetector. If a forward-bias larger than the open circuit voltage is applied (Figure 9.7d), the contacts can efficiently inject charges into the semiconductor. If these can recombine radiatively, the device works as a light-emitting diode (LED). The asymmetric diode behavior results basically from the different injection of the two metals into the HOMO and LUMO levels, respectively, which depends exponentially on the energy barrier between them [30].

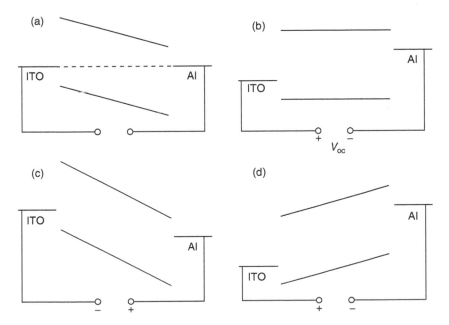

Figure 9.7. Metal–insulator–metal (MIM) picture of organic diode device function: (a) short circuit condition, (b) flat band or open circuit condition, (c) reversed-bias, and (d) forward-bias larger than V_{OC}.

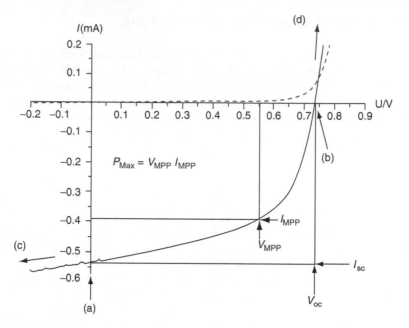

Figure 9.8. Current–voltage (*I–V*) curves of a solar cell (dark: dashed, illuminated: full line). The characteristic intersections with the ordinate and the abscissa are the open circuit voltage (V_{OC}) and the short circuit current (I_{SC}). The largest power output (P_{max}) is determined by the point where the product of voltage and current is maximized. Division of P_{max} by the product of I_{SC} and V_{OC} yields the filling factor FF.

In Figure 9.8 the current–voltage characteristics are shown for a solar cell in the dark and under illumination conditions. In the dark, there is almost no current flowing until the contacts start to inject heavily at forward-bias for voltages larger than the open circuit voltage. Under illumination, the current flows in the opposite direction to the injected currents. At (a), the maximum generated photocurrent flows under short circuit conditions; at (b), the photogenerated current is balanced to zero (flat band condition). Between (a) and (b), in the fourth quadrant, the device generates power, i.e., current × voltage. At some point, denoted as maximum power point (MPP), the product between current and voltage and hence the power output is largest. To determine the efficiency of a solar cell, this power needs to be compared with the incident light intensity. Generally the fill factor is calculated as $FF = V_{MPP}I_{MPP}/(V_{OC}I_{SC})$ to denote the part of the product of V_{OC} and I_{SC} that can be used for external loads. With this, the power conversion efficiency can be written as

$$\eta_{POWER} = \frac{P_{OUT}}{P_{IN}} = \frac{I_{MPP}V_{MPP}}{P_{IN}} = \frac{FF \cdot I_{SC}V_{OC}}{P_{IN}} \tag{1}$$

The organic solar cells using a single photoactive layer (mono component) sandwiched between two metal electrodes of different work functions have been schematically explained by the MIM model (for undoped insulators as active layer). Upon illumination or in the case of (impurity) doping, however, Schottky barriers with a depletion region W can be formed close to the contacts, resulting in a band bending. This corresponds to an electric field in which excitons can be dissociated. Therefore illumination from two different sides may result in different spectral photocurrent spectra, reflecting the electrode location where the photocurrent is generated

[3,32,73]. Since the exciton diffusion length for most organic solar cell materials is below 20 nm, again only those excitons generated in a small region within <20 nm from the active contacts contribute to the photocurrent.

9.5. BILAYER HETEROJUNCTION DEVICES

The bilayer consists of an organic donor layer and a subsequent acceptor layer, and is sandwiched between two electrodes matching the donor HOMO and the acceptor LUMO, respectively, for efficient extraction of the corresponding charge carriers. The bilayer device structure is schematically depicted in Figure 9.9, neglecting all kinds of possible band bending due to energy level alignments. Classical p/n-junctions require doped semiconductors with free charge carriers to form the electric field in the depleted region but the bilayer heterojunction formed between undoped donor and acceptor materials is due to the differences in the ionization potential and electron affinity of the adjacent materials. In one polarity of the device the hole injection to the HOMO of the donor layer from a high work function electrode and the electron injection into the LUMO level of the acceptor from the low work function electrode are both easily possible and the device carries large injected currents (forward-bias). In the opposite polarity of the device, the electron injection to the donor LUMO level from a high work function electrode as well as the hole injection to the acceptor HOMO level from a low work function electrode are energetically heavily unfavored and strong barriers hinder the charge injection in both materials. The device is in reverse bias. Such rectification is conceptually intrinsic and does not need a charge carrier exchange at the interface like in classical p–n junctions. In many discussions the organic donor–acceptor bilayer devices are misleadingly denoted as organic p–n junctions. Only in the case of doping (eg. chemically) true p- and n-type organic semiconductors may be achieved resulting in the classical p–n junction formation.

Photogenerated excitons in bilayer devices can only be quantitatively dissociated in a thin layer at the heterojunction and thus the device is exciton diffusion limited. The excitons created in either phases have to travel within their lifetimes to reach the D–A interface and dissociate. This is also confirmed by photoluminescence quenching [9,23] and photocurrent modeling [9,24,26], which indicate that only

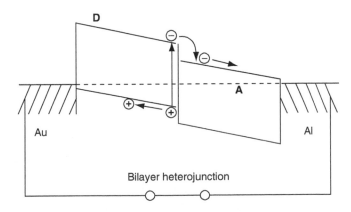

Figure 9.9.   Schematic of a bilayer heterojunction device.

photogenerated excitons in proximity to the D–A interface within less than the exciton diffusion length can be dissociated. Thus only a part of the absorbed photons contributes to the charge generation.

There are experimental indications [74–76] supported by theoretical considerations [77] for the formation of an interfacial dipole between the donor and acceptor phases, independent of illumination. This can stabilize the charge-separated state by a repulsive interaction between the interface and the free charges [77] and result in the large lifetimes observed.

9.6. BULK HETEROJUNCTION DEVICES

The bulk heterojunction device is similar to the bilayer device with respect to the donor—acceptor concept, but it exhibits a vastly increased interfacial area dispersed throughout the bulk. While in the bilayer heterojunction the donor and acceptor phase contact the anode and cathode selectively, the bulk heterojunction requires percolated pathways for both phases throughout the volume, i.e., a bicontinuous and interpenetrating network. Therefore, the bulk heterojunction devices are much more sensitive to the complicated nanoscale morphology in the blend, which is demonstrated using a cross-section scanning electron microscope picture in Figure 9.10. From this figure it is observable that the nanostructure is dominated by large nanoclusters embedded into a matrix (skin). These nanoclusters have been shown to consist of fullerenes [78–80].

To achieve high quantum efficiency, excitons have to dissociate at a donor–acceptor interface and created charges have to reach the respective electrodes. This limits the size of nanoparticles within the blend to approximately 10–20 nm (typical exciton diffusion lengths) for optimizing the charge generation efficiency.

Generally, bulk heterojunctions may be achieved by co-deposition of donor and acceptor pigments [41,67,68,81] or solution-casting of either polymer–polymer [55–57], polymer–molecule [20,53,54,82,83], or molecule–molecule [84,85] blends.

The use of toluene as solvent in the MDMO-PPV:PCBM system led to a coarser phase separation (~200–500 nm grain size) than for chlorobenzene (~50 nm grain size). This was shown with atomic force microscopy (AFM) [59] and transmission electron microscopy (TEM) measurements on this system [78]. Scanning electron microscopy (SEM) measurements show that the polymer phase exhibited a domain size of about 20 nm (Figure 9.10, taken from Ref. [80]). Two AFM images of MDMO-PPV:PCBM spin-cast films are shown in Figure 9.11 for comparison. The large-scale phase separation in the case of toluene is clearly visible. In contrast, using chlorobenzene as solvent having a large fullerene content of up to 80% by weight did not show a coarse phase separation and resulted in improved power conversion efficiencies by almost a factor of 3 [59]. When these results are considered together, they suggest that an optimum domain size of the phase separation between donor and acceptor is needed to balance exciton dissociation and percolated charge transport. This leads immediately to the idea for self-organized nanostructures as a clever way of organizing the bulk structure (tertiary structure) by manipulating the molecular structure (primary structure). This bio-mimicking strategy might be important for supramolecular chemistry groups working in this field of research. An optimum nanostructure is difficult to draw but for vertical transport of charge carriers a vertical self-organization of the donor–acceptor phases between the electrodes might lead to enhanced transport properties (see Figure 9.12). In this nanostructure it is important

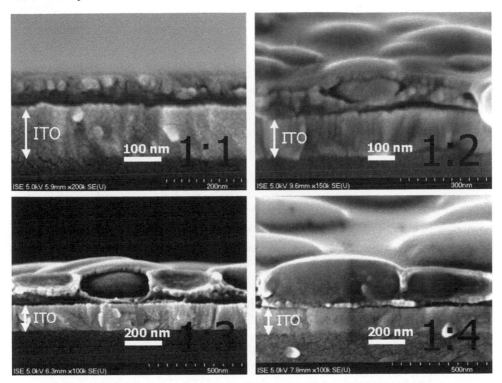

Figure 9.10. Side view of the cross section of MDMO-PPV:PCBM blended films cast from toluene with varying mixing ratios (by weight). For the ratios 1:4, 1:3, and 1:2, the nanoclusters in the form of discs are surrounded by another phase, i.e., skin that contains smaller spheres of about 30 nm in size. The magnifications used are $100,000\times$ (1:4 and 1:3), $150,000\times$ (1:2), and $200,000\times$ (1:1). (Reproduced from H. Hoppe, M. Niggemann, C. Winder, J. Kraut, R. Hiesgen, A. Hinsch, D. Meissner, and N.S. Sariciftci, *Adv. Funct. Mater.*, 14, 1005 (2004)).

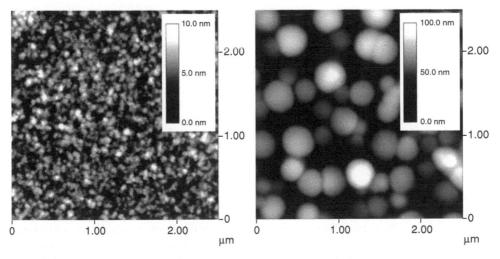

Figure 9.11. AFM topography scans of MDMO-PPV:PCBM 1:4 (by weight) blended films, spin-cast from chlorobenzene (left) and from toluene solution (right). The toluene-cast film shows a tenfold increased roughness as compared to the chlorobenzene-cast film. Features of a few hundred nanometers in size are visible in the right, while features at the left are around <50 nm in size. The scan size is 2.5 μm in both cases.

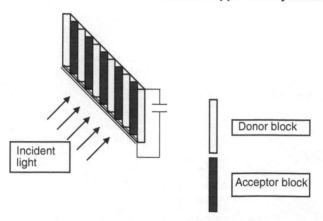

Figure 9.12. The schematic cartoon of a "molecular highway" structure using donor–acceptor blocks, which connect the two electrodes.

to consider the orientation of the molecules with their absorption axis (in linear molecules the molecule axis) parallel to the polarization of the incoming light. Perpendicularly standing linear molecules will not absorb the incoming light but deliver the maximum mobility. This contradiction might be solved by tilting the molecules in the columns or orienting them within these nanochannels. Another important point is the recombination of the respective charge carriers at the opposite electrode, e.g., a negative charge arrives at the positive electrode (ITO). To prevent this loss mechanism, charge blocking layers can be introduced between the molecular highway structure and the respective electrode, preventing the opposite sign charge carriers from reaching the electrodes. In a way, in today's bulk heterojunction solar cells, the use of PEDOT–PSS on the ITO electrode as well as using the LiF underlayer at the aluminum electrode may both facilitate such a blocking layer function.

Several chapters in this book are denoted to orientation of block co-polymers and this approach might have a significant impact in future for a priori determining the nanomorphology via the primary structure.

9.7.   THE OPEN CIRCUIT POTENTIAL, V_{OC}

The potential energy stored within one pair of separated positive and negative charges is equivalent to the difference in their respective quasi-Fermi levels, or often denoted as difference in their electrochemical potentials [27]. When the quasi-Fermi level splitting persists during charge transport through the interfaces at the contacts, the open circuit voltage (V_{OC}) will approach this Fermi level splitting. While for ideal (ohmic) contacts no loss is expected, energy level offsets or band bending at nonideal contacts will influence the V_{OC}.

To reach the electrodes, the charge carriers need a net driving force, which, in general, results from two "forces": internal electric fields and concentration gradients of the respective charge carrier species. The first leads to a field-induced drift and the other to a diffusion current. Without a detailed analysis, one can generally assume that thin-film devices (<100 nm) are mostly field drift dominated whereas thick devices, having effective screening of the electrical fields inside the bulk, are more dominated by the diffusion of charge carriers in concentration gradients created by the selective contacts.

The quasi-Fermi level of photoinduced holes on the conjugated polymer donor phase in the bulk heterojunction will be denoted to the positive polaronic level just above the HOMO of the polymer. On the other side, the quasi-Fermi level of the photoinduced electrons on the fullerene phase will be denoted to the LUMO level of the fullerene. Thus, their splitting (energetic distance from HOMO of the donor to the LUMO of the acceptor) shall conceptually determine the maximum open circuit potential of the organic heterojunction devices.

To investigate this hypothesis, several studies have been reported varying the HOMO level of the donor polymer by using similar compound with different oxidation potentials [86–88] as well as by varying the LUMO level of the acceptor fullerene by using similar fullerene derivatives with different first reduction potentials [89]. The results in Figure 9.13 and Figure 9.14 clearly show a rather linear dependence

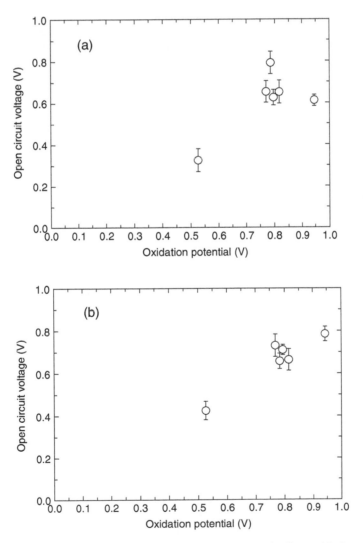

Figure 9.13. The dependence of the open circuit potential on the first oxidation potential of the donor polymer (reproduced from A. Gadisa, M. Svensson, M.R. Andersson, O. Inganas, *Appl. Phys. Lett.* 84, 1609 (2004)).

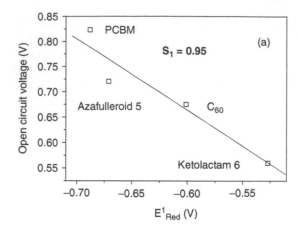

Figure 9.14. The dependence of the open circuit potential on the first reduction potential of the fullerene acceptor. Different names in this figure denote derivatives of fullerenes with different first reduction potentials without major change in the size or shape of the fullerene sphere.

of the V_{OC} or the first oxidation (reduction) potential of the donor (acceptor) materials.

The MIM model would predict the maximum V_{OC} to be determined by the difference in the work functions of the two asymmetrical electrodes [30]. Experimental data, however, showed strong deviations where the V_{OC} exceeded largely the expected difference between the electrode work functions [89]. Fermi level pinning between the fullerene and the gold electrode has been accounted for this, but cannot be generalized for all metal cathodes [90]. Thus, the individual energy level alignments between organic–metal interfaces are critical [91–98]. Interfacial dipoles formed at these organic semiconductor–electrode interface change the effective metal work function and thus affect V_{OC} [91,94,95,99,100]. The ITO anode can be modified by plasma etching [101,102] or by coating with a higher work function organic hole transport layer [103–105] to achieve a better matching between the energy levels of the anode and the HOMO of the hole-conducting material. But even monolayers of polar molecules were found to modify the work function of ITO by up to 0.9 eV [106]. A commonly applied modification of the cathode is the deposition of a very thin LiF layer between the metal electrode and the organic semiconductor. This was found to improve charge injection in LEDs [107,108] and also resulted in some cases in a higher V_{OC} for organic solar cells [109,110]. Another influence may arise from aluminum electrodes, which can form a thin oxide layer at the interface to the organic materials [111–113], with possible changes in the effective work function.

In addition, a dependence of the V_{OC} on the fullerene content in the bulk heterojunction blend was observed [80,114–117] and proposed to originate from the partial coverage of the cathode by fullerenes [115]. Other experiments show that changing the electrode work function on either side also influences the V_{OC} [90,118]. Furthermore, dependencies on temperature [119–121] and light intensity [9,55,120–123] are also observed for the V_{OC}s.

In conclusion, the open circuit potential is a sensitive function of the energy levels of the materials as well as of the interfaces and contacts. To maximize the V_{OC}, molecular engineering is necessary to manipulate the HOMO level of the donor phase as well as the LUMO level of the acceptor phase. Further improvement can be expected by introducing interfacial layers, more selective electrodes, dipoles at interfaces, etc.

9.8. DOUBLE CABLE POLYMERS

The covalent linking of electron accepting and conducting moieties to a electron-donating and hole-transporting conjugated polymer backbone appears to be an interesting idea for the preparation of bipolar conducting "double cable" polymers (see Figure 9.15) [124–128]. In this way, the secondary–tertiary structures of the macromolecule might be possibly dictated by its primary structure. The effective donor–acceptor interfacial area will be maximized, and phase separation and clustering phenomena should be prevented. In addition, the interaction between the donor-conjugated backbone and the acceptor moieties may be tuned by varying the chemical structure (nature and length) of their connecting bridge. Basically, the realization of effective double cable polymers will bring the donor–acceptor heterojunction at a molecular level. Their primary structure should prevent the occurrence of phase separation since the material is basically one macromolecule with two different chains (cables) for different signs of charges, thus forcing the formation of continuous, interconnected pathways for the transport of both holes and electrons at the same time. Further functionalities (such as amphiphilic properties) may also be added to this "double cable" primary structure by chemical synthesis, resulting in interesting self-organized secondary and tertiary structures like in biological systems. First devices were realized by attaching fulleropyrrolidine groups onto alkyl substituted polythiophene backbones [124–128]. Since the solubility of such complicated structures is very limited, the practical handling for device fabrication is cumbersome. To overcome this problem, attaching long alkyl chain substitutions is necessary [124–128]. As such, these materials are a challenge for synthetic organic chemists since the

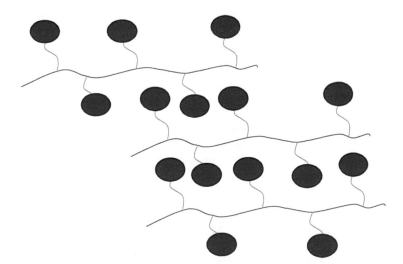

Figure 9.15. Schematic description of a "double cable" polymer with chemically grafted acceptor moieties (here fullerenes are displayed as spheres, negative cable) via a spacer to the conjugated backbone (donor, positive cable).

different specializations of organic synthesis should work together in an interdisciplinary manner (e.g., thiophene chemists collaborating with fullerene chemists, etc).

9.9. OUTLOOK

There is a remarkable development of organic solar cells worldwide today. With this speed and increasing effort of research and development, one can expect the increase of power conversion efficiencies within the next years up to >10% from the value of ≈4% today. To achieve this jump, new materials with improved absorption in the red and infrared regions of the solar emission spectrum (low bandgap materials) as well as a controlled nanoscale morphology for optimizing the charge generation and transport is needed. The design and synthesis of these new materials will require an interdisciplinary effort of synthetic and physical chemists and device physicists.

The manufacturing advantages and the low-cost materials with upscaling possibilities used in the organic solar cells will guarantee their successful marketing in future. Another important issue is the integrability of processing of such polymeric solar cells. By integration of plastic electronic devices in general and plastic solar cells in particular into chip cards, textiles, packaging materials, consumer goods, etc. will create new applications and markets that are not accessible by today's silicon technology.

REFERENCES

1. A. Goetzberger, C. Hebling, and H.-W. Schock, *Materials Science and Engineering R* 40, 1–46 (2003).
2. G. A. Chamberlain, *Solar Cells* 8, 47–83 (1983).
3. D. Wöhrle and D. Meissner, *Advanced Materials* 3, 129–138 (1991).
4. C. J. Brabec, N. S. Sariciftci, and J. C. Hummelen, *Advanced Functional Materials* 11, 15–26 (2001).
5. J. J. M. Halls and R. H. Friend, in *Clean Electricity from Photovoltaics*, edited by M. D. Archer and R. Hill (Imperial College Press, London, 2001).
6. J. Nelson, *Current Opinion in Solid State and Materials Science* 6, 87–95 (2002).
7. J.-M. Nunzi, *C. R. Physique* 3, 523–542 (2002).
8. *Organic Photovoltaics: Concepts and Realization*, Vol. 60, edited by C. J. Brabec, V. Dyakonov, J. Parisi, and N. S. Sariciftci (Springer, Berlin, 2003).
9. P. Peumans, A. Yakimov, and S. R. Forrest, *Journal of Applied Physics* 93, 3693–3723 (2003).
10. B. Maennig, J. Drechsel, D. Gebeyehu, P. Simon, F. Kozlowski, A. Werner, F. Li, S. Grundmann, S. Sonntag, M. Koch, K. Leo, M. Pfeiffer, H. Hoppe, D. Meissner, N. S. Sariciftci, I. Riedel, V. Dyakonov, and J. Parisi, *Applied Physics A* 79, 1–14 (2004).
11. H. Spanggaard and F. C. Krebs, *Solar Energy Materials & Solar Cells* 83, 125–146 (2004).
12. H. Hoppe and N. S. Sariciftci, *Journal of Materials Research* 19, 1924–1945 (2004).
13. *Handbook of Conducting Polymers*, Vols. 1 and 2, edited by T. A. Skotheim (Marcel Dekker, New York, 1986).
14. *Handbook of Organic Conductive Molecules and Polymers*, Vols. 1–4, edited by H. S. Nalwa (John Wiley & Sons Ltd., Chichester, 1997).
15. *Handbook of Conducting Polymers*, edited by T. A. Skotheim, R. L. Elsenbaumer, and J. R. Reynolds (Marcel Dekker, New York, 1998).

16. *Semiconducting Polymers*, edited by G. Hadziioannou and P. F. v. Hutten (Wiley-VCH, Weinheim, 2000).

17. C. Winder and N. S. Sariciftci, *Journal of Materials Chemistry* 14, 1077–1086 (2004).

18. C. D. Dimitrakopoulos and D. J. Mascaro, *IBM Journal of Research & Development* 45, 11–14 (2001).

19. J. J. M. Halls, K. Pichler, R. H. Friend, S. C. Moratti, and A. B. Holmes, *Applied Physics Letters* 68, 3120–3122 (1996).

20. J. J. M. Halls and R. H. Friend, *Synthetic Metals* 85, 1307–1308 (1997).

21. H. R. Kerp, H. Donker, R. B. M. Koehorst, T. J. Schaafsma, and E. E. v. Faassen, *Chemical Physics Letters* 298, 302–308 (1998).

22. T. J. Savanije, J. M. Warman, and A. Goossens, *Chemical Physics Letters* 287, 148–153 (1998).

23. A. Haugeneder, M. Neges, C. Kallinger, W. Spirkl, U. Lemmer, J. Feldmann, U. Scherf, E. Harth, A. Gügel, and K. Müllen, *Physical Review B* 59, 15346–15351 (1999).

24. L. A. A. Pettersson, L. S. Roman, and O. Inganäs, *Journal of Applied Physics* 86, 487–496 (1999).

25. M. Stoessel, G. Wittmann, J. Staudigel, F. Steuber, J. Blässing, W. Roth, H. Klausmann, W. Rogler, J. Simmerer, A. Winnacker, M. Inbasekaran, and E. P. Woo, *Journal of Applied Physics* 87, 4467–4475 (2000).

26. T. Stübinger and W. Brütting, *Journal of Applied Physics* 90, 3632–3641 (2001).

27. J. E. Kroeze, T. J. Savanije, M. J. W. Vermeulen, and J. M. Warman, *Journal of Physical Chemistry B* 107, 7696–7705 (2003).

28. *Primary Photoexcitations in Conjugated Polymers: Molecular Exciton versus Semiconductor Band Model*, edited by N. S. Sariciftci (World Scientific, Singapore, 1997).

29. B. A. Gregg and M. C. Hanna, *Journal of Applied Physics* 93, 3605–3614 (2003).

30. I. D. Parker, *Journal of Applied Physics* 75, 1656–1666 (1994).

31. A. K. Gosh, D. L. Morel, T. Feng, R. F. Shaw, and C. A. Rowe Jr., *Journal of Applied Physics* 45, 230–236 (1974).

32. D. Meissner, S. Siebentritt, and S. Günster, in *International Symposium on Optical Materials Technology for Energy Efficiency and Solar Energy Conversion XI: Photovoltaics, Photochemistry and Photoelectrochemistry*, Vol. 1729, edited by A. H.-L. Goff, C. G. Granquist, and C. M. Lampert (Toulouse, 1992).

33. S. Karg, W. Riess, V. Dyakonov, and M. Schwoerer, *Synthetic Metals* 54, 427–433 (1993).

34. D. L. Morel, A. K. Gosh, T. Feng, E. L. Stogryn, P. E. Purwin, R. F. Shaw, and C. Fishman, *Applied Physics Letters* 32, 495–497 (1978).

35. A. K. Gosh and T. Feng, *Journal of Applied Physics* 49, 5982–5989 (1978).

36. S. M. Sze, *Physics of semiconductor devices* (John Wiley & Sons, New York, 1981).

37. C. W. Tang, *Applied Physics Letters* 48, 183–185 (1986).

38. J. Rostalski and D. Meissner, *Solar Energy Materials & Solar Cells* 61, 87–95 (2000).

39. P. Peumanns, V. Bulovic, and S. R. Forrest, *Applied Physics Letters* 76, 2650–2652 (2000).

40. M. Hiramoto, M. Suezaki, and M. Yokoyama, *Chemistry Letters*, 327–330 (1990).

41. M. Hiramoto, H. Fujiwara, and M. Yokoyama, *Applied Physics Letters* 58, 1062–1064 (1991).

42. M. Hiramoto, H. Fujiwara, and M. Yokoyama, *Journal of Applied Physics* 72, 3781–3787 (1992).

43. R. N. Marks, J. J. M. Halls, D. D. C. Bradley, R. H. Friend, and A. B. Holmes, *Journal of Physics: Condensed Matter* 6, 1379–1394 (1994).

44. G. Yu, C. Zhang, and A. J. Heeger, *Applied Physics Letters* 64, 1540–1542 (1994).

45. H. Antoniadis, B. R. Hsieh, M. A. Abkowitz, S. A. Jenekhe, and M. Stolka, *Synthetic Metals* 62, 265–271 (1994).

46. N. S. Sariciftci, L. Smilowitz, A. J. Heeger, and F. Wudl, *Science* 258, 1474–1476 (1992).
47. L. Smilowitz, N. S. Sariciftci, R. Wu, C. Gettinger, A. J. Heeger, and F. Wudl, *Physical Review B* 47, 13835–13842 (1993).
48. C. H. Lee, G. Yu, D. Moses, K. Pakbaz, C. Zhang, N. S. Sariciftci, A. J. Heeger, and F. Wudl, *Physical Review B* 48, 15425–15433 (1993).
49. S. Morita, A. A. Zakhidov, and K. Yoshino, *Solid State Communications* 82, 249–252 (1992).
50. S. Morita, S. Kiyomatsu, X. H. Yin, A. A. Zakhidov, T. Noguchi, T. Ohnishi, and K. Yoshino, *Journal of Applied Physics* 74, 2860–2865 (1993).
51. N. S. Sariciftci, D. Braun, C. Zhang, V. I. Srdanov, A. J. Heeger, G. Stucky, and F. Wudl, *Applied Physics Letters* 62, 585–587 (1993).
52. L. S. Roman, W. Mammo, L. A. A. Petterson, M. R. Andersson, and O. Inganäs, *Advanced Materials* 10, 774–777 (1998).
53. G. Yu, J. Gao, J. C. Hummelen, F. Wudl, and A. J. Heeger, *Science* 270, 1789–1791 (1995).
54. C. Y. Yang and A. J. Heeger, *Synthetic Metals* 83, 85–88 (1996).
55. G. Yu and A. J. Heeger, *Journal of Applied Physics* 78, 4510–4515 (1995).
56. J. J. M. Halls, C. A. Walsh, N. C. Greenham, E. A. Marseglia, R. H. Friend, S. C. Moratti, and A. B. Holmes, *Nature* 376, 498–500 (1995).
57. K. Tada, K. Hosada, M. Hirohata, R. Hidayat, T. Kawai, M. Onoda, M. Teraguchi, T. Masuda, A. A. Zakhidov, and K. Yoshino, *Synthetic Metals* 85, 1305–1306 (1997).
58. M. Granström, K. Petritsch, A. C. Arias, A. Lux, M. R. Andersson, and R. H. Friend, *Nature* 395, 257–260 (1998).
59. S. E. Shaheen, C. J. Brabec, N. S. Sariciftci, F. Padinger, T. Fromherz, and J. C. Hummelen, *Applied Physics Letters* 78, 841–843 (2001).
60. J. M. Kroon, M. M. Wienk, W. J. H. Verhees, and J. C. Hummelen, *Thin Solid Films* 403–404, 223–228 (2002).
61. T. Munters, T. Martens, L. Goris, V. Vrindts, J. Manca, L. Lutsen, W. D. Ceunick, D. Vanderzande, L. D. Schepper, J. Gelan, N. S. Sariciftci, and C. J. Brabec, *Thin Solid Films* 403–404, 247–251 (2002).
62. T. Aernouts, W. Geens, J. Portmans, P. Heremans, S. Borghs, and R. Mertens, *Thin Solid Films* 403–404, 297–301 (2002).
63. P. Schilinsky, C. Waldauf, and C. J. Brabec, *Applied Physics Letters* 81, 3885–3887 (2002).
64. F. Padinger, R. S. Rittberger, and N. S. Sariciftci, *Advanced Functional Materials* 13, 1–4 (2003).
65. M. Svensson, F. Zhang, S. C. Veenstra, W. J. H. Verhees, J. C. Hummelen, J. M. Kroon, O. Inganäs, and M. R. Andersson, *Advanced Materials* 15, 988–991 (2003).
66. M. M. Wienk, J. M. Kroon, W. J. H. Verhees, J. Knol, J. C. Hummelen, P. A. v. Hall, and R. A. J. Janssen, *Angewandte Chemie International Edition* 42, 3371–3375 (2003).
67. W. Geens, T. Aernouts, J. Poortmans, and G. Hadziioannou, *Thin Solid Films* 403–404, 438–443 (2002).
68. P. Peumans, S. Uchida, and S. R. Forrest, *Nature* 425, 158–162 (2003).
69. D. Gebeyehu, M. Pfeiffer, B. Maennig, J. Drechsel, A. Werner, and K. Leo, *Thin Solid Films* 451–452, 29–32 (2004).
70. J. Krüger, R. Plass, L. Cevey, M. Piccirelli, M. Grätzel, and U. Bach, *Applied Physics Letters* 79, 2085–2087 (2001).
71. J. Krüger, R. Plass, M. Grätzel, and H.-J. Matthieu, *Applied Physics Letters* 81, 367–369 (2002).
72. W. U. Huynh, J. J. Dittmer, and A. P. Alivisato, *Science* 295, 2425–2427 (2002).

73. C. W. Tang and A. C. Albrecht, *The Journal of Chemical Physics* 62, 2139–2149 (1975).

74. M. Murgia, F. Biscarini, M. Cavallini, C. Taliani, and G. Ruani, *Synthetic Metals* 121, 1533–1534 (2001).

75. G. Ruani, C. Fontanini, M. Murgia, and C. Taliani, *Journal of Chemical Physics* 116, 1713–1719 (2002).

76. T. Toccoli, A. Boschetti, C. Corradi, L. Guerini, M. Mazzola, and S. Iannotta, *Synthetic Metals* 138, 3–7 (2003).

77. V. I. Arkhipov, P. Heremans, and H. Bässler, *Applied Physics Letters* 82, 4605–4607 (2003).

78. T. Martens, J. D'Haen, T. Munters, Z. Beelen, L. Goris, J. Manca, M. D'Olieslaeger, D. Vanderzande, L. D. Schepper, and R. Andriessen, *Synthetic Metals* 138, 243–247 (2003).

79. X. Yang, J. K. J. v. Duren, R. A. J. Janssen, M. A. J. Michels, and J. Loos, *Macromolecules* 37, 2151–2158 (2004).

80. H. Hoppe, M. Niggemann, C. Winder, J. Kraut, R. Hiesgen, A. Hinsch, D. Meissner, and N. S. Sariciftci, *Advanced Functional Materials* 14, 1005–1011 (2004).

81. D. Gebeyehu, B. Maennig, J. Drechsel, K. Leo, and M. Pfeiffer, *Solar Energy Materials & Solar Cells* 79, 81–92 (2003).

82. J. J. Dittmer, R. Lazzaroni, P. Leclere, P. Moretti, M. Granström, K. Petritsch, E. A. Marseglia, R. H. Friend, J. L. Bredas, H. Rost, and A. B. Holmes, *Solar Energy Materials & Solar Cells* 61, 53–61 (2000).

83. J. J. Dittmer, E. A. Marseglia, and R. H. Friend, *Advanced Materials* 12, 1270–1274 (2000).

84. K. Petritsch, J. J. Dittmer, E. A. Marseglia, R. H. Friend, A. Lux, G. G. Rozenberg, S. C. Moratti, and A. B. Holmes, *Solar Energy Materials & Solar Cells* 61, 63–72 (2000).

85. L. Schmidt-Mende, A. Fechtenkötter, K. Müllen, E. Moons, R. H. Friend, and J. D. MacKenzie, *Science* 293, 1119–1122 (2001).

86. H. Kim, S.-H. Jin, H. Suh, and K. Lee, in *Proceedings of the SPIE: The International Society for Optical Engineering* (San Diego, CA, 2003).

87. A. Gadisa, M. Svensson, M. R. Andersson, and O. Inganäs, *Applied Physics Letters* 84, 1609–1611 (2004).

88. H. Hoppe, D. A. M. Egbe, D. Mühlbacher, and N. S. Sariciftci, *Journal of Materials Chemistry* 14, in print (2004). (Published on the Internet: 27th September 2004.)

89. C. J. Brabec, A. Cravino, D. Meissner, N. S. Sariciftci, T. Fromherz, M. T. Rispens, L. Sanchez, and J. C. Hummelen, *Advanced Functional Materials* 11, 374–380 (2001).

90. V. D. Mihailetchi, P. W. M. Blom, J. C. Hummelen, and M. T. Rispens, *Journal of Applied Physics* 94, 6849–6854 (2003).

91. I. H. Campbell, S. Rubin, T. A. Zawodzinski, J. D. Kress, R. L. Martin, D. L. Smith, N. N. Barashkov, and J. P. Ferraris, *Physical Review B* 54, 14321–14324 (1996).

92. C. M. Heller, I. H. Campbell, D. L. Smith, N. N. Barashkov, and J. P. Ferraris, *Journal of Applied Physics* 81, 3227–3231 (1996).

93. Y. Hirose, A. Kahn, V. Aristov, P. Soukiassian, V. Bulovic, and S. R. Forrest, *Physical Review B* 54, 13748–13758 (1996).

94. H. Ishii, K. Sugiyama, E. Ito, and K. Seki, *Advanced Materials* 11, 605–625 (1999).

95. L. Yan and Y. Gao, *Thin Solid Films* 417, 101–106 (2002).

96. N. Koch, A. Kahn, J. Ghijsen, J.-J. Pireaux, J. Schwartz, R. L. Johnson, and A. Elschner, *Applied Physics Letters* 82, 70–72 (2003).

97. D. Cahen and A. Kahn, *Advanced Materials* 15, 271–277 (2003).

98. S. C. Veenstra and H. T. Jonkman, *Journal of Polymer Science: Part B: Polymer Physics* 41, 2549–2560 (2003).

99. S. C. Veenstra, A. Heeres, G. Hadziioannou, G. A. Sawatzky, and H. T. Jonkman, *Applied Physics A: Materials Science & Processing* 75, 661–666 (2002).

100. C. Melzer, V. V. Krasnikov, and G. Hadziioannou, *Applied Physics Letters* 82, 3101–3103 (2003).

101. C. C. Wu, C. I. Wu, J. C. Sturm, and A. Kahn, *Applied Physics Letters* 70, 1348–1350 (1997).

102. K. Sugiyama, H. Ishii, Y. Ouchi, and K. Seki, *Journal of Applied Physics* 87, 295–298 (2000).

103. J. C. Scott, S. A. Carter, S. Karg, and M. Angelopoulos, *Synthetic Metals* 85, 1197–1200 (1997).

104. Y. Cao, G. Yu, C. Zhang, R. Menon, and A. J. Heeger, *Synthetic Metals* 87, 171–174 (1997).

105. T. M. Brown, J. S. Kim, R. H. Friend, F. Cacialli, R. Daik, and W. J. Feast, *Applied Physics Letters* 75, 1679–1681 (1999).

106. C. Ganzorig and M. Fujihira, in *Organic Optoelectronic Materials, Processing and Devices*, edited by S. C. Moss, (Mater. Res. Soc. Symp. Proc. **708**, Warrendale, PA, 2002), P. 83.

107. L. S. Hung, C. W. Tang, and M. G. Mason, *Applied Physics Letters* 70, 152–154 (1997).

108. G. E. Jabbour, Y. Kawabe, S. E. Shaheen, J. F. Wang, M. M. Morrell, B. Kippelen, and N. Peyghambarian, *Applied Physics Letters* 71, 1762–1764 (1997).

109. C. J. Brabec, S. E. Shaheen, C. Winder, N. S. Sariciftci, and P. Denk, *Applied Physics Letters* 80, 1288–1290 (2002).

110. F. L. Zhang, M. Johansson, M. R. Anderson, J. C. Hummelen, and O. Inganäs, *Synthetic Metals* 137, 1401–1402 (2003).

111. N. Koch, A. Pogantsch, E. J. W. List, G. Leising, R. I. R. Blyth, M. G. Ramsey, and F. P. Netzer, *Applied Physics Letters* 74, 2909–2911 (1999).

112. J. K. J. v. Duren, J. Loos, F. Morrissey, C. M. Leewis, K. P. H. Kivits, L. J. v. IJzendoorn, M. T. Rispens, J. C. Hummelen, and R. A. J. Janssen, *Advanced Functional Materials* 12, 665–669 (2002).

113. C. W. T. Bulle-Lieuwma, W. J. H. v. Gennip, J. K. J. v. Duren, P. Jonkheijm, R. A. J. Janssen, and J. W. Niemantsverdriet, *Applied Surface Science* 203–204, 547–550 (2003).

114. J. Gao, F. Hide, and H. Wang, *Synthetic Metals* 84, 979–980 (1997).

115. J. Liu, Y. Shi, and Y. Yang, *Advanced Functional Materials* 11, 420–424 (2001).

116. M. C. Scharber, N. A. Schulz, N. S. Sariciftci, and C. J. Brabec, *Physical Review B* 67, 085202 (2003).

117. J. K. J. v. Duren, X. Yang, J. Loos, C. W. T. Bulle-Lieuwma, A. B. Sieval, J. C. Hummelen, and R. A. J. Janssen, *Advanced Functional Materials* 14, 425–434 (2004).

118. H. Frohne, S. E. Shaheen, C. J. Brabec, D. C. Müller, N. S. Sariciftci, and K. Meerholz, *ChemPhysChem* 9, 795–799 (2002).

119. E. A. Katz, D. Faiman, S. M. Tuladhar, J. M. Kroon, M. M. Wienk, T. Fromherz, F. Padinger, C. J. Brabec, and N. S. Sariciftci, *Journal of Applied Physics* 90, 5343–5350 (2001).

120. V. Dyakonov, *Physica E* 14, 53–60 (2002).

121. I. Riedel, J. Parisi, V. Dyakonov, L. Lutsen, D. Vanderzande, and J. C. Hummelen, *Advanced Functional Materials* 14, 38–44 (2004).

122. C. M. Ramsdale, J. A. Barker, A. C. Arias, J. D. MacKenzie, R. H. Friend, and N. C. Greenham, *Journal of Applied Physics* 92, 4266–4270 (2002).

123. P. Schilinsky, C. Waldauf, J. Hauch, and C. J. Brabec, *Journal of Applied Physics* 95, 2816–2819 (2004).

124. A. Cravino, G. Zerza, M. Maggini, S. Bucella, M. Svensson, M. R. Andersson, H. Neugebauer, and N. S. Sariciftci, *Chemical Communications*, 2487–2488 (2000).

125. A. Cravino and N. S. Sariciftci, *Journal of Materials Chemistry*, 1931–1943 (2002).
126. A. Cravino, G. Zerza, M. Maggini, S. Bucella, M. Svensson, M. R. Andersson, H. Neugebauer, C. J. Brabec, and N. S. Sariciftci, *Chemical Monthly* 134, 519–527 (2003).
127. A. Cravino and N. S. Sariciftci, *Nature Materials* 2, 360–361 (2003).
128. G. Zerza, A. Cravino, H. Neugebauer, N. S. Sariciftci, R. Gómez, J. L. Segura, N. Martín, M. Svensson, and M. R. Andersson, *Journal of Physical Chemistry A* 105, 4172–4176 (2001).

10

Organic Solar Cells Incorporating a p–i–n Junction and a p–n Homojunction

Masahiro Hiramoto

Material and Life Science, Graduate School of Engineering, Osaka University, Yamadaoka, Suita, Osaka, Japan

Contents

Abstract The concept of an organic p–i–n junction in which i-layer is a co-deposited layer of two different organic semiconductors was proposed and demonstrated. Co-deposited layers have a vast number of heteromolecular (donor–acceptor) contacts acting as efficient photocarrier generation sites. In order to extract the photogenerated carriers from the bulk of the co-deposited films, the nanostructure of the films should be controlled. A crystalline–amorphous nanocomposite with spatially separated pathways for electrons and holes revealed a high efficiency for photocurrent generation. Although perylene pigment/metal-free phthalocyanine in the present study is not the best possible combination, a power conversion efficiency of 1% was obtained. On the other hand, alteration of conduction type from n-type to p-type in a single organic semiconductor of perylene pigment by doping technique was demonstrated. p–n homojunction was successfully fabricated. The concepts and results summarized here offer possible solutions for the realization of practical organic solid-state solar cells that possess a conversion efficiency of over 5%.

Key words: p–i–n junction, three-layered cells, co-deposited i-layer, donor–acceptor heteromolecular contacts, crystalline–amorphous nanocomposite, spatially separated pathways, pn-control, p–n homojunction, doping

10.1. INTRODUCTION

The two-layer organic photovoltaic cell reported by Tang in 1986 [1] had an impact on the field of organic solar cells, together with his work in 1987 on organic electroluminescence [2]. In 1988, the author switched his research field to organic semiconductors and soon encountered the two-layer cell. My first question was how such a large photocurrent density (of the order of mA cm^{-2}) could be obtained, because most previous examples of organic solar cells had shown very small photocurrents, typically less than several μA cm^{-2} [3,4]. I suspected some unknown effect was hidden at the interface of the two different organic semiconductors and decided to try to solve the question posed above. After a while, I recognized that direct contacts between different organic molecules [phthalocyanine/perylene] could act as a site for the generation of photocarriers, and in 1991 I proposed the concept of an organic p–i–n junction in which the i-layer is a mixture of two different organic semiconductors (Section 2) [5,6]. Later, it was revealed that control of the nanostructure of the mixed layer is crucial for efficient photocarrier generation (Section 3) [7].

On the other hand, the author was also convinced that conduction-type control is crucially important, i.e., pn-control of organic semiconductors, especially for solar cell applications. In the case of amorphous silicon, the establishment of a technique for pn-control by hydrogen termination of the dangling bonds offered a route to a practical conversion efficiency of over 5% by making a p–i–n junction [8]. Therefore, I decided to find a way of achieving conduction-type control in organic semiconductors by a purification and doping technique. I demonstrated a way of making an organic pn-homojunction cell in 1995 (Section 4) [9,10].

We need to overcome two major hurdles in order to realize practical organic solid-state solar cells that possess a conversion efficiency of over 5%, namely, how to generate sufficient photocurrent, and how to create sufficient built-in potential. The concepts and results summarized here offer possible solutions.

10.2. p–i–n JUNCTION CELLS

10.2.1. Motivation

I reasoned that an inherently low capability for photocarrier generation is the most serious problem for organic solar cells, although there are other problems such as the origin of the photovoltage, the high resistance of organic semiconductors, and so on. A two-layer organic photovoltaic cell [1] composed of phthalocyanine pigment (CuPc, Figure 10.1) and perylene pigment (Im-PTC, Figure 10.1) generates an exceptionally large photocurrent of several mA cm^{-2} under solar illumination. Since single-layered Schottky junction-type cells fabricated using phthalocyanine or perylene pigments exhibited very low photocurrents, there should be a clue for efficient photocarrier generation at the interface between phthalocyanine and perylene pigments.

10.2.2. Direct Heteromolecular Contact as a Photocarrier Generation Site

As a first step, we tried to observe the formation process of an organic heterojunction by means of the gradual insertion of phthalocyanine at the Schottky junction between perylene pigment and a metal electrode [6]. We used a combination of perylene pigment, N-methyl-3,4,9,10-perylenetetracarboxyl-diimide (Me-PTC, Figure 10.1) and metal-free phthalocyanine (H$_2$Pc, Figure 10.1). Me-PTC/H$_2$Pc cells were fabricated with various thicknesses (x) of H$_2$Pc (Figure 10.2). The pigments were purified three times by the train sublimation technique [11]. The cells were fabricated by successive vacuum depositions of Me-PTC and H$_2$Pc under 10^{-3} Pa pressure. Finally, a semitransparent Au electrode (20 nm) was deposited on the organic films.

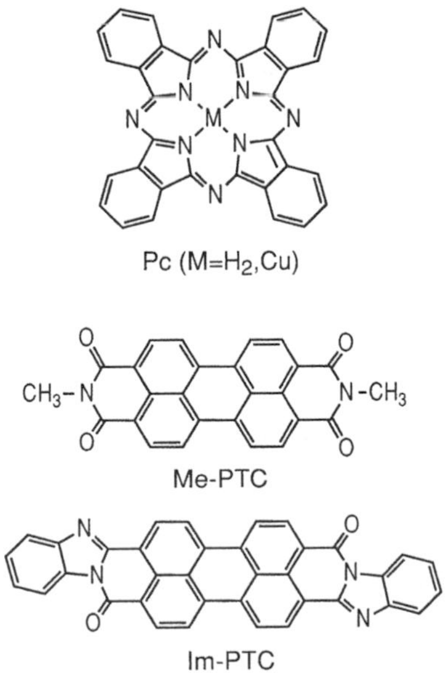

Pc (M=H$_2$,Cu)

Me-PTC

Im-PTC

Figure 10.1. Chemical structures of the phthalocyanine and perylene pigments used in our discussion.

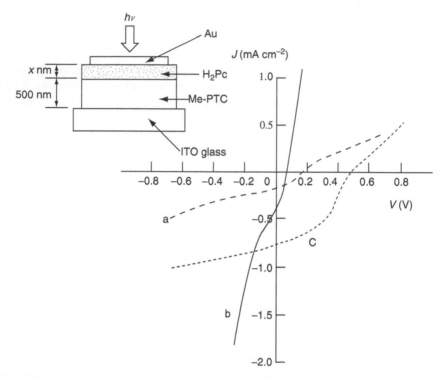

Figure 10.2. Photocurrent–voltage (*J–V*) curves for H_2Pc/Me-PTC cells with various H_2Pc thicknesses (*x*). (a) $x = 0$ nm, (b) $x = 1.5$ nm, and (c) $x = 50$ nm. The light intensity transmitted through the Au electrode was the same ($50\,mW\,cm^{-2}$) for all cells. (From M. Hiramoto, H. Fujiwara, and M. Yokoyama, *J. Appl. Phys.* **72**, 3781–3787 (1992). With permission.)

In Figure 10.2, typical current–voltage (*J–V*) characteristics under white light irradiation through the semitransparent Au electrode are shown. When $x = 0$ nm (no H_2Pc layer), a Schottky junction is formed between the Me-PTC layer and the Au electrode. The Au electrode exhibited positive photovoltage with respect to the indium tin oxide (ITO) electrode (curve a), indicating that the Me-PTC film behaves as an n-type semiconductor. Interestingly, the insertion of a monolayer-like H_2Pc layer ($x = 1.5$ nm) between the Au and the Me-PTC brings about a complete change in the *J–V* characteristics (curve b). The photocurrent increased dramatically and became much more dependent on the applied voltage. A thicker H_2Pc layer ($x = 50$ nm) shifts the photocurrent curve (curve c) to the right, increasing the short-circuit photocurrent (J_{sc}) and the open-circuit photovoltage (V_{oc}).

The action spectra of J_{sc} for cells of this type are shown in Figure 10.3. In the spectral region for wavelengths longer than 550 nm, where there is strong absorption of H_2Pc (see Figure 10.8(a)), the value of J_{sc} first increased and then decreased, owing to the increase in the masking effect of the thicker H_2Pc film. Even in the region between 400 and 550 nm, where there is little absorption of H_2Pc, clear photocurrent enhancement can be seen, depending on the thickness of the deposited H_2Pc film. The dramatic increase of J_{sc} at 480 nm reveals that a small amount of H_2Pc on the Me-PTC film apparently enhances the carrier generation efficiency of the Me-PTC (Figure 10.4(a)). The value of V_{oc}, which represents the potential between the Au electrode and the Me-PTC layer, was always smaller than that of the cell without H_2Pc

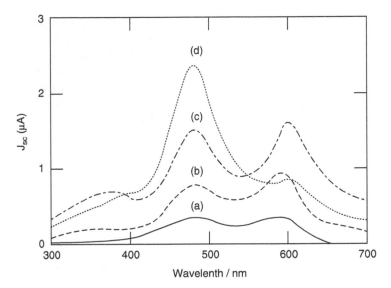

Figure 10.3. Action spectra of short-circuit photocurrent (J_{sc}) for the cells shown in Figure 10.2 for thickness of (a) $x = 0$ nm, (b) $x = 1.5$ nm, (c) $x = 10$ nm, and (d) $x = 50$ nm. Monochromatic light was irradiated through the Au electrode. (From M. Hiramoto, H. Fujiwara, and M. Yokoyama, *J. Appl. Phys.* **72**, 3781–3787 (1992). With permission.)

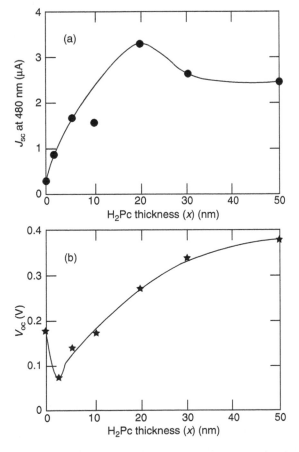

Figure 10.4. Dependence of short-circuit photocurrent (J_{sc}) (a) and open-circuit photovoltage (V_{oc}) (b) on H_2Pc film thickness (x). ((b) is from M. Hiramoto, H. Fujiwara, and M. Yokoyama, *J. Appl. Phys.* **72**, 3781–3787 (1992). With permission.)

($x = 0$ nm) when the layer thickness was less than 10 nm (Figure 10.4(b)). Therefore, it is difficult to explain the increase in the photocurrent observed in Figures 10.2, 10.3, and 10.4(a) by an increase in the internal electric field at the interface.

In practice, an H_2Pc thickness of about one monolayer ($x = 1.5$ nm) can be regarded as interfacial doping with H_2Pc molecules (Figure 10.5(a)). Thus, the increase in photocurrent is related to the increase in the sites where the H_2Pc and the Me-PTC molecules make direct contact. Moreover, in a similar experiment where the Me-PTC was deposited inversely on the H_2Pc layer (Figure 10.5(b)), the carrier generation efficiency of the H_2Pc film also increased drastically with only a monolayer-like deposition of Me-PTC. Thus, it can be concluded that direct intermolecular contact between Me-PTC and H_2Pc plays an important role in the process of charge carrier photogeneration.

10.2.3. Three-Layered Cells

The fact that direct molecular contact between Me-PTC and H_2Pc acts as a site for photocarrier generation suggests that a cell containing a layer of mixed pigments would show higher conversion efficiency. Based on this consideration, I decided to fabricate some three-layered organic cells [5,6]. The configurations of the (a) three- and (b) two-layered organic solar cells are depicted schematically in Figure 10.6. Two types of pigment combinations of H_2Pc/Me-PTC and CuPc/Im-PTC were examined. In these cells, the PTC film was fixed at a constant thickness of 40 nm in order to make a strict comparison of the photovoltaic properties of these cell configurations. The two-layered cell forms a p–n heterojunction at the interface of the two pigment layers. Charge carrier photogeneration only occurred near the Pc–PTC interface. It was confirmed by the masking effect that the photoresponse was observed mainly in the absorption regions of the PTC and Pc for irradiation from the Pc and PTC sides, respectively [1,12]. The three-layered cells contain a mixed interlayer consisting of Pc and PTC pigments

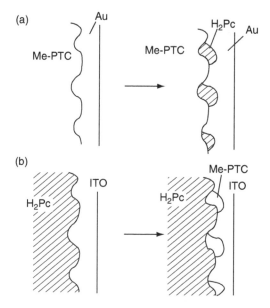

Figure 10.5. Schematic illustrations of interfacial doping. (a) H_2Pc molecules are doped at the Me-PTC/Au interface. (b) Me-PTC molecules are doped at the H_2Pc/ITO interface.

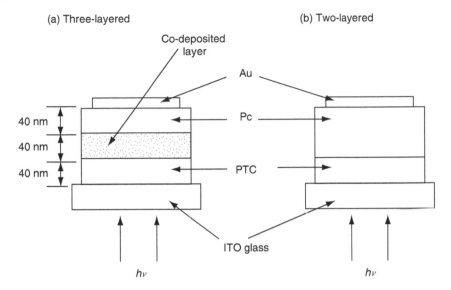

Figure 10.6. Structures of (a) three-layered and (b) two-layered organic solar cells. (From M. Hiramoto, H. Fujiwara, and M. Yokoyama, *J. Appl. Phys.* **72**, 3781–3787 (1992) and M. Hiramoto, H. Fujiwara, and M. Yokoyama, *Appl. Phys. Lett.* **58**, 1062–1064 (1991). With permission.)

sandwiched between the two pigment layers. Preparation of the mixed interlayer was accomplished by the co-deposition of PTC and Pc from two separate controlled sources, while each deposition rate was monitored by an oscillating quartz thickness meter.

Typical current–voltage (J–V) curves are shown in Figure 10.7 for two combinations of pigment (a) H$_2$Pc/Me-PTC and (b) CuPc/Im-PTC. The two-layered cells containing the p–n heterojunctions demonstrated good performances as organic

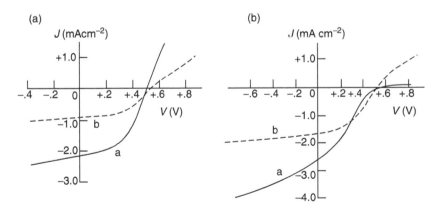

Figure 10.7. Typical current–voltage (J–V) characteristics for a combination of H$_2$Pc/Me-PTC (a) and CuPc/Im-PTC (b). Three-layer cells, ITO/PTC (40 nm)/co-deposited layer (40 nm)/Pc (40 nm)/Au (curves a), and two-layer cells, ITO/PTC (40 nm)/Pc (80 nm)/Au (curves b), are compared. The pigment ratios of PTC and Pc in the co-deposited layers are 1:2 (a) and 1:1 (b). White light (100 mW cm^{-2}) was irradiated through the ITO substrate. (From M. Hiramoto, H. Fujiwara, and M. Yokoyama, *J. Appl. Phys.* **72**, 3781–3787 (1992) and M. Hiramoto, H. Fujiwara, and M. Yokoyama, *Appl. Phys. Lett.* **58**, 1062–1064 (1991). With permission.)

Table 10.1. Performance of the cells shown in Figure 10.7

Pigment combination	Cell type	J_{sc} (mA cm^{-2})	V_{oc} (V)	ff	Efficiency (%)
Me-PTC/H$_2$Pc	Two-layered	0.94	0.54	0.48	0.29
Me-PTC/H$_2$Pc	Three-layered	2.14	0.51	0.48	0.63
Im-PTC/CuPc	Two-layered	1.61	0.53	0.42	0.43
Im-PTC/CuPc	Three-layered	2.56	0.57	0.25	0.44

J_{sc}, V_{oc}, ff, and efficiency denote the short-circuit photocurrent, the open-circuit photovoltage, the fill factor, and the power conversion efficiency, respectively, for the light energy absorbed by the organic layers. White light (100 mW cm^{-2}; metal halide lamp (Toshiba Co.)) was irradiated through the ITO substrate.

solar cells (curves b), while both combinations of three-layered cells with the co-deposited interlayer produced approximately twice the amount of photocurrent (curve a) under the same conditions (100 mW cm^{-2}, white light through the ITO substrate). The performance of the cells is summarized in Table 10.1. The photo-voltage remained almost unchanged for both the two- and three-layered configur-ations. The smaller fill factor in the three-layered cell with the Im-PTC/CuPc combination was attributed to the poor film quality of the co-deposited layer. The short-circuit photocurrent for all cells showed a linear dependence on the light intensity, up to 100 mW cm^{-2}.

Figure 10.8 shows the spectral dependence of the internal quantum efficiency for two- and three-layered cells with the Me-PTC–H$_2$Pc combination, along with the absorption spectra for each pigment film. The value of the internal quantum effi-ciency is defined as the ratio of the number of carriers collected under the short-circuit condition to the number of photons absorbed by the organic layers. The quantum efficiency of the three-layered cell (curve a) was about twice as large as that of the two-layered cell (curve b) over the entire spectrum. Similar photocurrent-doubling was obtained for the Im-PTC–CuPc combination. Since the masking effect due to the PTC film is the same for both two- and three-layered cells, it can be concluded that photocurrent enhancement of this type is due to an increase in the charge carrier generation efficiency by the insertion of the co-deposited interlayer, in which a large number of direct molecular contacts between the Pc and the PTC are formed.

Enhancement of the photocurrent enhancement due to introducing a co-depos-ited layer further illustrated that the molecular contact between PTC and Pc is an active site for charge separation. The values of the internal quantum efficiency tend to exhibit maxima near a pigment ratio of 1:1, irrespective of which pigment is excited (Figure 10.9). This result implies that a 1:1 molecular pair is an active site for charge separation. A 1:1 molecular pair is corresponds to the donor–acceptor pair in Figure 10.13(b).

10.2.4. p–i–n Energy Structure

In this section, the photovoltaic properties of the three-layered organic cell are discussed on the basis of the conventional energy band model for the Me-PTC–H$_2$Pc system. Figure 10.10(a) shows energy diagrams of the H$_2$Pc and Me-PTC films before contact. The valence band (VB) and the conduction band (CB) levels were estimated from the ionization potentials of these films when measured by atmospheric

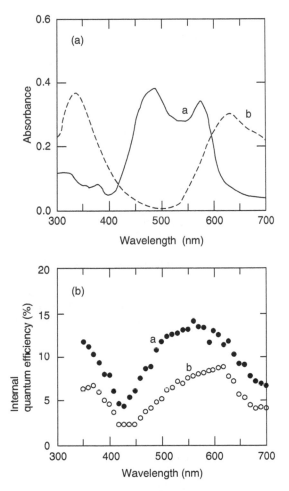

Figure 10.8. (a) Absorption spectra for films of Me-PTC (40 nm) shown by curve a (solid line) and H_2Pc (40 nm) shown by curve b (dashed line). (b) The spectral dependence of the internal quantum efficiency of J_{sc} for the three-layered cells (curve a: solid circles) and the two-layered cells (curve b: open circles) for the combination of Me-PTC/H_2Pc system. The cells are the same as in Figure 10.7(a). Light was irradiated through the ITO electrode. (From M. Hiramoto, H. Fujiwara, and M. Yokoyama, *J. Appl. Phys.* **72**, 3781–3787 (1992) and M. Hiramoto, H. Fujiwara, and M. Yokoyama, *Appl. Phys. Lett.* **58**, 1062–1064 (1991). With permission.)

photoelectron emission analysis [13] and from the optical bandgaps. The Fermi levels (E_F) were evaluated from Mott–Schottky plots based on capacitance measurements in the Schottky junctions using Au ($\phi_m = 5.05$ eV) metal for the Me-PTC and Al (4.28 eV) metal for the H_2Pc. Both the H_2Pc and the Me-PTC behave as p- and n-type semiconductors, as suggested by the position of the E_Fs. The energy structure of the two-layered cell can be drawn as in Figure 10.10(b-(i)), which represents an organic p–n heterojunction between p-type H_2Pc and n-type Me-PTC films.

In order to draw the energy structure of a three-layered cell, measurements of the dependence of J_{sc} and V_{oc} on the thickness of the co-deposited layer were taken. The results are shown in Figure 10.11. The pigment mixing ratio in the layer was maintained at 1:1. An important result shown in Figure 10.11 is that V_{oc} remained

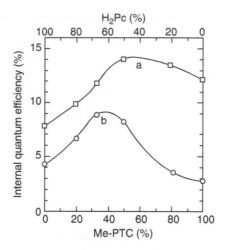

Figure 10.9. Dependence of internal quantum efficiency on the pigment ratio in the co-deposited layer of a three-layer cell. The curves (a) 630 nm (H_2Pc excitation) and (b) 480 nm (Me-PTC excitation) are for light irradiated through the ITO electrode. The cell structure is the same as that shown in Figure 10.6(a). (From M. Hiramoto, H. Fujiwara, and M. Yokoyama, *J. Appl. Phys.* **72**, 3781–3787 (1992) and M. Hiramoto, H. Fujiwara, and M. Yokoyama, *Appl. Phys. Lett.* **58**, 1062–1064 (1991). With permission.)

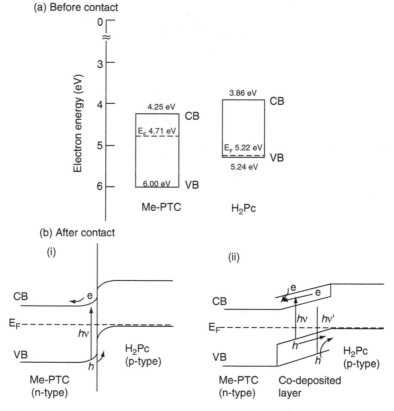

Figure 10.10. (a) Energy band diagram of H_2Pc and Me-PTC films before contact. CB, VB, and E_F denote the conduction band, the valence band, and the Fermi level, respectively. (b) The energy structure of (i) two-layered and (ii) three-layered organic solar cells after contact. (From M. Hiramoto, H. Fujiwara, and M. Yokoyama, *J. Appl. Phys.* **72**, 3781–3787 (1992). With permission.)

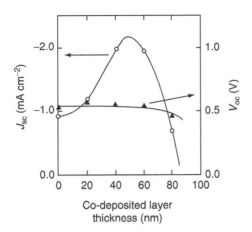

Figure 10.11. The dependence of J_{sc} and V_{oc} of a three-layered cell on the thickness of the co-deposited layer. The pigment ratio in the co-deposited layer (Me-PTC:H$_2$Pc) was kept constant (1:1). The cell structure is the same as shown in Figure 10.6(a) except for the interlayer thickness. White light (100 mW cm^{-2}) was irradiated through the ITO electrode. A co-deposited layer thickness of zero denotes the two-layered cell configuration. (From M. Hiramoto, H. Fujiwara, and M. Yokoyama, *J. Appl. Phys.* **72**, 3781–3787 (1992). With permission.)

constant, even in the case where there was no co-deposited layer, i.e., a two-layered cell. Moreover, V_{oc} for the three-layered cell exhibited a remarkable decrease when the front Me-PTC layer was removed. These results indicate that the built-in potential of the three-layered cell is mainly determined by the difference in the Fermi levels of both sides of the p-type H$_2$Pc and n-type Me-PTC films.

On the other hand, J_{sc} was highly dependent on the thickness of the interlayer, exhibiting a maximum around 50 nm. For interlayers thinner than 50 nm, the photocurrent appears to be decreased by lower light absorption within the interlayer and fewer active sites provided by the Me-PTC–H$_2$Pc molecular contacts. For thicker interlayers, it is thought that the photocurrent was decreased due to lower charge carrier generation efficiency caused by the lowering of the electric field within the interlayer. If this is the case, the latter observation indicates that most of the built-in potential created by the difference in Fermi levels is distributed across the interlayer. The electric field across the interlayer drives efficient charge carrier generation and charge transport. This speculation can be rationalized by introducing the concept that positive and negative charges consisting of donors and acceptors in n-type Me-PTC and p-type H$_2$Pc, respectively, are compensated by each other, and the resulting co-deposited interlayer behaves like an intrinsic semiconductor. Thus, the energy diagram of the three-layered cell can be depicted as shown in Figure 10.10(b-(ii)). The proposed energy structure of the three-layered cell can be regarded as a p–i–n junction, which is well known in amorphous silicon solar cells.

10.2.5. Application of Inorganic Semiconductors to the n-Type Layer

In the three-layered cells, the n-type and p-type pigment layers on both sides effectively provide a built-in potential to the photoactive interlayer, and also play an important role as carrier transport layers. From this standpoint, the application of inorganic semiconductors is attractive, especially in the n-type layer, since few n-type

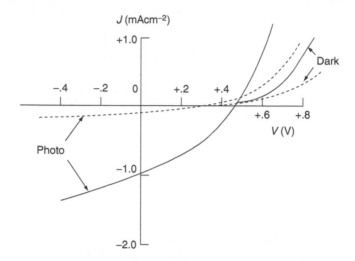

Figure 10.12. *J–V* curves for a three-layered cell, ITO/ZnO(40 nm)/co-deposited layer of Me-PTC and H₂Pc(40 nm)/Au (solid curve), and a two-layered cell, ITO/ZnO(40 nm)/H₂Pc(80 nm)/Au (broken curve). The pigment ratio of Me-PTC and H₂Pc in the co-deposited layer is 1:2. White light (100 mW cm⁻²) was irradiated through the ITO electrode. (From M. Hiramoto, H. Fujiwara, and M. Yokoyama, *J. Appl. Phys.* **72**, 3781–3787 (1992). With permission.)

organic pigments have been found. In addition, by using a wide bandgap semiconductor film, the masking effect due to the PTC, which is used as the n-layer, can be removed.

The n-type Me-PTC layer in the cells shown in Figure 10.6 was replaced with n-type ZnO or CdS films. In Figure 10.12, the *J–V* curves for the three-layered and the two-layered configurations are shown. Obviously, the introduction of the co-deposited interlayer of H₂Pc and Me-PTC caused an enhancement of the photocurrent by a factor of about 20 compared to the ZnO/H₂Pc heterojunction cell. It is clear that the co-deposited layer acts as an efficient carrier generation layer, even if an inorganic semiconductor is used as the n-layer. The use of CdS, which has more negative energy in the conduction band and the Fermi level than ZnO [14], resulted in an increase in V_{oc} from 0.47 to 0.61 V. This result strongly proposed a p–i–n energy structure whose built-in potential is determined by the difference in Fermi energy between the p- and n-layers. These are typical examples of the free choice of materials for each layer in the present three-layered cell. Recent report on C₆₀–phthlocyanine system [15] further supports the validity of this consideration.

10.2.6. Sensitization Mechanism of Photocarrier Generation at Heteromolecular Contacts

The co-deposited film of H₂Pc and Me-PTC exhibited no specific absorption that could be attributed to charge transfer (CT) interaction. The absorption spectrum of the ultra-thin H₂Pc–Me-PTC layered film completely agreed with the sum of the spectra of the respective H₂Pc and Me-PTC pigment films with the same thickness (2.5 nm). It is difficult, therefore, to consider any interaction under non-illumination condition, i.e., in the ground-state condition.

As a possible mechanism for carrier generation of heteromolecular contacts, we referred to the two possibilities that were discussed in 1992 [6]. The first of these is charge photogeneration via an exciplex of two pigments (Me-PTC$^-$ $\cdots$ H$_2$Pc$^+$)* since such an exciplex appears to act as an efficient precursor for the generation of an ion-pair that can dissociate into free carriers, depending on the strength of electric field [16]. The other is efficient charge carrier generation via a donor–acceptor pair, which had been reported in a tetraphenylamine (TPD)– Im-PTC system [17]. Pure organic semiconductors can produce little photocurrent due to the difficulty of dissociation of the Frenkel exciton, which is bound strongly by a Coulomb interaction (Figure 10.13(a)). Actually, organic semiconductor films fabricated and measured without exposure to air, and which therefore hardly contain any oxygen and water molecules, showed little photocurrent [18]. When phthalocyanine and perylene pigments are mixed, the energetic relationship between the HOMO and the LUMO levels is suitable for charge separation (Figure 10.13(b)). When an acceptor molecule, i.e., Me-PTC or Im-PTC, is photoexcited, electron transfer occurs between the HOMO of a donor molecule and the HOMO of an acceptor molecule, which is vacant due to the excitation. On the other hand, when a donor molecule, i.e., H$_2$Pc or CuPc, is photoexcited, electron transfer occurs between the LUMO of the donor molecule and the LUMO of the acceptor molecule. A CT exciton is formed irrespective of the kind of molecules that are excited, i.e., a donor molecule is charged positively and an acceptor molecule is charged negatively. Since the CT exciton is much easier to dissociate into a free electron and a hole compared to the Frenkel exciton, photocurrent generation is enhanced by mixing donor and acceptor molecules. At the present time, the donor–acceptor mechanism is widely accepted after a report of a polymer donor–acceptor heterojunction between poly(phenylene vinylene) and C$_{60}$ [19]. Recently, the phthalocyanine–C$_{60}$ heterojunction was revealed to be a very efficient donor–acceptor system [20].

Sensitization effects are universally observed for carrier generation at heteromolecular contacts. We observed similar sensitization in co-deposited films of

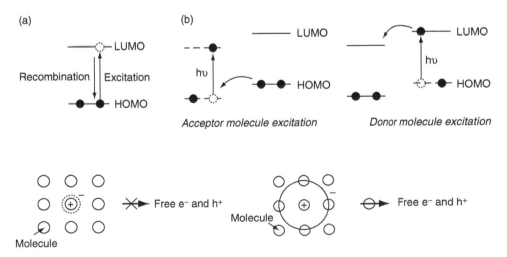

Figure 10.13. (a) Frenkel exciton photogenerated in pure organic semiconductor. It disappears easily due to recombination. (b) Charge transfer (CT) exciton photogenerated in mixed organic semiconductors acting as an electron donor and an electron acceptor. It easily dissociates into free carriers.

various combinations of acceptor molecules [Im-PTC, Me-PTC, C_{60}] and donor molecules [H_2Pc, CuPc, TiOPc, TPD, merocyanine, quinacridone, perylene] [21,22]. When two kinds of semiconductors are co-deposited, a photocurrent of several milliamperes per square centimeter under solar light illumination can be obtained rather easily. However, to obtain a larger photocurrent density, which is sufficient for practical solid-state solar cells, the nanostructure of the co-deposited films should be designed and controlled, as discussed in the next section.

10.3.   CONTROL OF THE NANOSTRUCTURE OF CO-DEPOSITED FILMS

10.3.1.   Motivation

It is quite probable that when two pigments are intermingled in the co-deposited layer, the efficient recombination of photogenerated electrons and holes may occur. Actually, transient microwave photoconductivity measurements had already suggested the occurrence of the recombination of photogenerated electrons and holes in co-deposited films of Me-PTC:H_2Pc [23]. Thus, I suspected that the molecular-level structure of the co-deposited film is decisive for the generation of photocurrent in the co-deposited films. In spite of such considerations, we had not been able to be clarify the nanostructure of co-deposited films by scanning electron microscopy (SEM) observations in the early stages of research [5,6].

We have recently succeeded in characterizing the nanostructure of co-deposited films by using atomic force microscopy (AFM) and we can also now control the nanostructure via the substrate temperature during co-deposition [7]. The recombination of photogenerated electrons and holes in the co-deposited film has been considerably suppressed. A power conversion efficiency of 1% was observed in a three-layered p–i–n cell, incorporating a co-deposited interlayer fabricated under low-temperature conditions.

10.3.2.   Photovoltaic Properties vs. Substrate Temperature

First, we investigated the effects of substrate temperature during co-deposition on the photovoltaic properties of co-deposited films between Me-PTC and H_2Pc by using single-layered cells. Figure 10.14 shows the photocurrent–voltage ($J-V$) characteristics of cells fabricated on substrates whose temperature was controlled at $-167°C$ (curve A), $-80°C$ (curve B), or $25°C$ (curve C). Interestingly, the photocurrent density was increased considerably by cooling the substrate during co-deposition. The co-deposited film fabricated at $-167°C$ was able to generate 15 times more photocurrent than the film that was fabricated at room temperature (Figure 10.18, curve A). This obviously shows the superior potential for photocarrier generation in co-deposited films fabricated on cooled substrates.

For this low-temperature-fabricated film, irrespective of whether the irradiation was applied from the side of the Au electrode or the ITO electrode, the shapes of the action spectra coincide well. This results is in marked contrast to those of the individual Me-PTC or H_2Pc films, which usually show different action spectra depending on the irradiation side due to the presence of a dead layer in the film, causing a masking effect. Therefore, we concluded that the present low-temperature-fabricated co-deposited films have no dead layer, i.e., the entire bulk of the film behaves as an active layer for photocarrier generation.

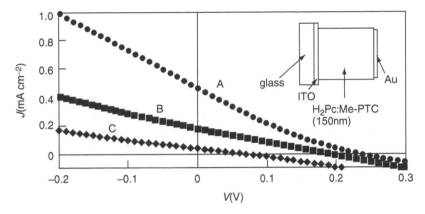

Figure 10.14. Photocurrent–voltage (J–V) curves for ITO/H$_2$Pc:Me-PTC/Au cells. The pigment ratio of the co-deposited layer [H$_2$Pc:Me-PTC] is 1:1. The pigments were co-deposited on the substrates at $-167°C$ (curve A), at $-80°C$ (curve B), and at $25°C$ (curve C). The ITO electrode was irradiated with simulated solar light (AM1.5, 100 mW cm^{-2}). Voltage is for Au electrode with respect to ITO electrode. (From M. Hiramoto, K. Suemori, and M. Yokoyama, *Jpn. J. Appl. Phys.* **41**, 2763 (2002). With permission.)

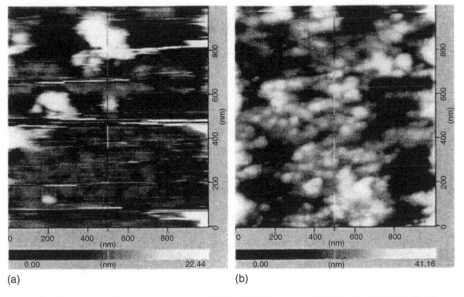

(a) (b)

Figure 10.15. (Color figure follows page 348). AFM images of H$_2$Pc:Me-PTC (1:1) films co-deposited on ITO glass substrates at $25°C$ (a) and $-167°C$ (b). The thickness of the films was 500 nm. (From M. Hiramoto, K. Suemori, and M. Yokoyama, *Jpn. J. Appl. Phys.* **41**, 2763 (2002). With permission.)

10.3.3. Nanostructure vs. Substrate Temperature

We then investigated the nanostructure of the co-deposited films to determine the cause of the photocurrent enhancement. Figure 10.15 shows AFM images of the surfaces of H$_2$Pc:Me-PTC films co-deposited on substrates at $25°C$ (a) and $-167°C$ (b). The former film has a noticeably flat surface. In contrast, the latter film contains many nanoparticles of approximately 20 nm in diameter. X-ray diffraction (XRD) patterns for H_2Pc:Me-PTC films co-deposited on substrates at 25°C (a) and −167°C (b) are shown in Figure 10.16. Since the former film showed no x-ray diffraction peak, we concluded that the co-deposited film fabricated at room temperature is amorphous. In contrast, the latter film showed a peak at 27.5. Individual films of Me-PTC and H_2Pc fabricated at 25°C were confirmed to have the polycrystalline structure by AFM observation and showed sharp diffraction peaks at 27.5 (Figure 10.16(c)) and 7 (Figure 10.16(d)), respectively. Thus, we concluded that the nanoparticles observed in the co-deposited film fabricated at −167°C (Figure 10.15(b)) are Me-PTC, while H_2Pc exists in the amorphous state in this film. This conclusion coincides with the fact that individual films of Me-PTC and H_2Pc fabricated at −167°C were confirmed as polycrystalline and amorphous films, respectively.

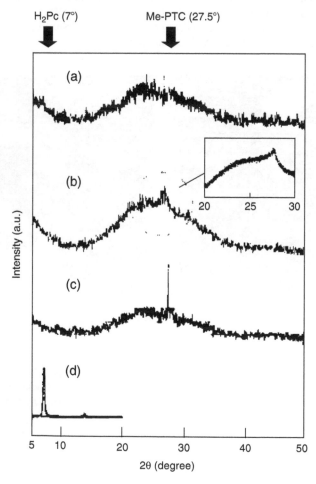

Figure 10.16. X-ray diffraction patterns for H_2Pc:Me-PTC (1:1) films co-deposited on the substrates at 25°C (a) and −167°C (b). The patterns for individual films of Me-PTC (c) and H_2Pc (d) fabricated at 25°C are also shown. The inset is a narrow scan from 20° to 30° for pattern (b). (From M. Hiramoto, K. Suemori, and M. Yokoyama, *Jpn. J. Appl. Phys.* **41**, 2763 (2002). With permission.)

10.3.4. Photocurrent Generation in Co-Deposited Films

Based on the above observations, in Figure 10.17, we depict some schematic illustrations of the nanostructure of the co-deposited films. In the case of room-temperature co-deposition, the Me-PTC and H_2Pc are mixed at a molecular level and consequently form an amorphous film [(a) molecular mixture]. On the other hand, in the case of low-temperature co-deposition, the Me-PTC and H_2Pc aggregate themselves and form a structure in which Me-PTC nanocrystals are surrounded by amorphous H_2Pc [(b) crystalline–amorphous nanocomposite]. This is a clear example of the successful control of the nanostructure of co-deposited organic films.

Since direct contact between the Me-PTC molecules and the H_2Pc molecules give rise to efficient photocarrier generation sites, co-deposited films have the potential for photocurrent generation (Section 2). Further photocurrent enhancement in low-temperature fabricated films can be attributed to the formation of individual routes for both electrons and holes, which enables carrier transport from the generation sites to the electrodes (Figure 10.17(b)). Thus, the entire bulk of the co-deposited films can participate in the photocurrent generation. It is difficult for electrons and holes to get close to each other, due to the p–n junction barriers formed at the interfaces where the Me-PTC nanocrystals are covered by amorphous H_2Pc. This may assist effective hole and electron transport through different channels in the H_2Pc and Me-PTC pigments, respectively. In contrast, in the case of a molecular-level mixture formed at room temperature (Figure 10.17(a)), photogenerated electrons and holes easily encounter each other and disappear due to recombination in the absence of suitable transport routes.

The crystalline–amorphous nanocomposite was universally revealed as capable of generating an enhanced photocurrent. We then investigated some other combinations of organic semiconductors, i.e., Im-PTC:CuPc and C_{60}:H_2Pc [24]. Figure 10.18 summarizes the dependence of J_{sc} on the substrate temperature during co-deposition for Me-PTC:H_2Pc (curve A), Im-PTC:CuPc (curve B), and C_{60}:H_2Pc (curve C). The

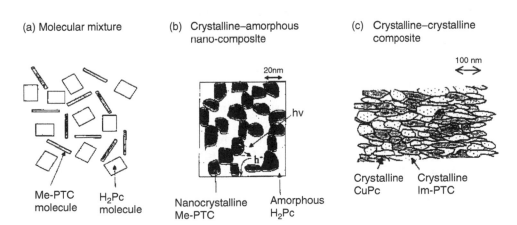

Figure 10.17. Schematic illustrations of the nanostructures of the co-deposited films. (a) Molecular mixture, (b) crystalline–amorphous nanocomposite and (c) crystalline–crystalline composite. Illustrations (b) and (c) are of the Me-PTC:H_2Pc crystalline–amorphous nanocomposite formed at −167°C and for the Im-PTC:CuPc crystalline–crystalline composite formed at +120°C. ((a) and (b) are from M. Hiramoto, K. Suemori, and M. Yokoyama, *Jpn. J. Appl. Phys.* **41**, 2763 (2002). With permission.)

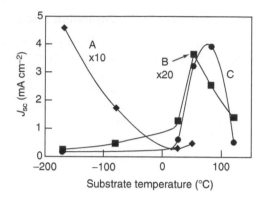

Figure 10.18. Dependence of J_{sc} on substrate temperature during co-deposition. Curves A, B, and C are for ITO/Me-PTC:H$_2$Pc/Au, ITO/Im-PTC:CuPc/Ag, and ITO/C$_{60}$:H$_2$Pc/Ag cells, respectively. The ratio of the pigments was 1:1. Simulated solar light (AM1.5, 100 mW cm^{-2}) was irradiated with the ITO electrode.

optimum temperature is dependent on the particular combination of organic semiconductors, i.e., the maximum values of J_{sc} were observed at $-170°$C, $+50°$C, and $+80°$C for Me-PTC:H$_2$Pc, Im-PTC:CuPc, and C$_{60}$:H$_2$Pc, respectively. For Im-PTC:CuPc and C$_{60}$:H$_2$Pc, at the peak temperature for J_{sc}, crystalline–amorphous nanocomposites (Figure 10.17(b)) were again confirmed as formed by cross-sectional SEM images of co-deposited films and XRD measurements. Moreover, a molecular mixture (Figure 10.17(a)) and crystalline–crystalline composites (Figure 10.17(c)) were confirmed as forming on the low- and high-temperature sides, respectively. In the case of the crystalline–crystalline composite, the increase in the number of distinct grain boundaries formed between the nanocrystals seems to seriously obstruct carrier transport. Although the optimum temperature is dependent on the particular semiconductor combination, a crystalline–amorphous nanocomposite was always formed when the largest photocurrent was observed. Therefore, we concluded that the crystalline–amorphous nanocomposite is the most suitable to extract photogenerated electrons and holes separately to the respective electrodes.

10.3.5. Three-Layered Cells Incorporating Crystalline–Amorphous Nanocomposite Films

We attempted to incorporate a crystalline–amorphous nanocomposite of Me-PTC:H$_2$Pc co-deposited film in the three-layered cell (Section 2). Figure 10.19 shows the photocurrent–voltage $(J–V)$ curves for two kinds of three-layered cells incorporating co-deposited films fabricated at $-167°$C (curve A) and at room temperature (curve B). The thicknesses of the co-deposited films (80 nm) and the all other films were the same for both cells. The effect of cooling the substrate during the co-deposition clearly appeared again in the three-layered cells, namely, J_{sc} increased by approximately fivefold and reached 2.42 mA cm^{-2}.

Figure 10.20 shows the dependence of J_{sc} on the co-deposited layer thickness in the three-layered cells. The advantage of low-temperature-fabricated films of crystalline–amorphous nanocomposites is conspicuous, especially for the thicker region. Although J_{sc} decreased abruptly above 40 nm for room-temperature-fabricated films of the molecular mixture, it reached a maximum at 80 nm and hardly decreased up to

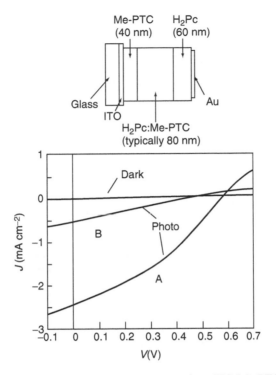

Figure 10.19. Photocurrent–voltage (J–V) curves for ITO/Me-PTC(40 nm)/H$_2$Pc:Me-PTC(80 nm)/H$_2$Pc(60 nm)/Au three-layered cells. The H$_2$Pc:Me-PTC ratio is 1:1. Individual Me-PTC and H$_2$Pc films were deposited at room temperature. Co-deposited films were fabricated on the substrates at $-167°C$ (curve A) and at room temperature (curve B). The ITO electrode was irradiated with simulated solar light (AM1.5, 100 mW cm^{-2}). The voltage is for the Au electrode with respect to the ITO electrode. (From M. Hiramoto, K. Suemori, and M. Yokoyama, *Jpn. J. Appl. Phys.* **41**, 2763 (2002). With permission.)

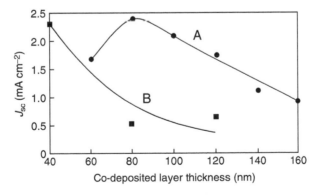

Figure 10.20. Dependence of J_{sc} on the co-deposited layer thickness in three-layered cells. The substrate temperatures during the co-deposition were $-167°C$ (curve A) and room temperature (curve B). The ITO electrode was irradiated with simulated solar light (AM1.5, 100 mW cm^{-2}). (From M. Hiramoto, K. Suemori, and M. Yokoyama, *Jpn. J. Appl. Phys.* **41**, 2763 (2002). With permission.)

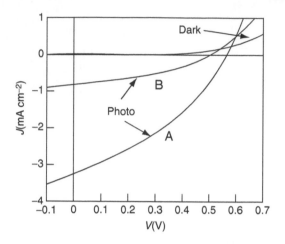

Figure 10.21. Photocurrent–voltage (*J–V*) curves for an ITO/Me-PTC(40 nm)/H$_2$Pc:Me-PTC(100 nm)/H$_2$Pc(60 nm)/Au three-layered cell. The pigment ratio of the co-deposited layer fabricated at −167°C is 6:4 (H$_2$Pc:Me-PTC). Individual Me-PTC and H$_2$Pc films were deposited at room temperature. The voltage is for Au electrode with respect to the ITO electrode. The ITO electrode (curve A) and the Au electrode (curve B) were irradiated with simulated solar light (AM1.5, 100 mW cm^{-2}). The light intensity transmitted through the Au electrodes is 17 mW cm^{-2}. (From M. Hiramoto, K. Suemori, and M. Yokoyama, *Jpn. J. Appl. Phys.* **41**, 2763 (2002). With permission.)

160 nm for low-temperature-fabricated films. It should be noted that most of the incident solar light can be absorbed by a 100-nm-thick Me-PTC:H$_2$Pc film. Thus, we concluded that in the latter films, high photocarrier generation efficiency is compatible with sufficient thickness for solar light absorption. This is a very important feature for making good use of the entire solar light.

The best performance so far was observed for a three-layered cell incorporating a 100-nm-thick crystalline–amorphous nanocomposite layer with a pigment ratio of 6:4 (H$_2$Pc:Me-PTC). Figure 10.21 shows the photocurrent–voltage (*J–V*) curves for films under irradiation with simulated solar light (AM1.5, 100 mW cm^{-2}). Cell performance is summarized in Table 10.2. For ITO-side irradiation (curve A), a J_{sc} of 3.26 mA cm^{-2} and a power conversion efficiency of 0.73% were obtained. For Au-side irradiation (curve B), a power conversion efficiency of 1.0% was obtained for the light transmitted through the Au electrode. Figure 10.22 shows the spectral dependence of the internal quantum efficiency of J_{sc}. For ITO-side irradiation (curve A), the values of the quantum efficiency were around 20% throughout the visible region. For Au-side irradiation (curve B), a maximum quantum efficiency of 55% was observed for monochromatic light at 420 nm. These results strongly suggest that the present

Table 10.2. Performance of the cell shown in Figure 10.21

Irradiation	J_{sc} (mA cm^{-2})	V_{oc} (V)	ff	Efficiency (%)
ITO-side	3.26	0.56	0.35	0.73
Au side	0.82	0.51	0.40	1.0

Simulated solar light was irradiated (AM1.5, 100 mW cm^{-2}). The efficiency is calculated for the light intensity transmitted through the ITO and Au electrodes.

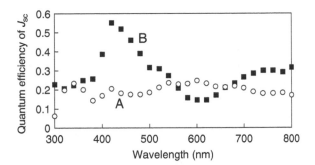

Figure 10.22. Spectral dependence of internal quantum efficiency of J_{sc} for the cell in Figure 10.23. The ITO electrode (curve A) and the Au electrode (curve B) were irradiated with monochromatic light. (From M. Hiramoto, K. Suemori, and M. Yokoyama, *Jpn. J. Appl. Phys.* **41**, 2763 (2002). With permission.)

crystalline–amorphous nanocomposite films are promising for the fabrication of efficient photovoltaic cells. Performance of the present p–i–n cells did not change during the measurements. As the next step, test of the long-term stability ought to be performed from the standpoint of practical application.

10.3.6. Nanostructure Design

The essence of the problem for solar energy conversion utilizing heteromolecular contacts of organic semiconductors is the simultaneous satisfaction of two severely conflicting requirements, i.e., that light should be absorbed only by an extremely thin active layer near the heteromolecular contacts, and that the whole of the incident light should be absorbed and utilized. The mixing of two organic semiconductors is a possible solution. However, it should be pointed out that once two kinds of molecular semiconductors are mixed, spatially separated routes are indispensable for extracting the photogenerated electrons and holes from the bulk films to the respective metal electrodes. Recently, influence of the annealing treatment for the co-deposited films was reported [25]. However, for now, we should depend on routes formed accidentally by percolation. If one intended to design such a structure, structural control technique down to extremely small dimensions (about 5 to 10 nm), which is not based on the percolation process, would be necessary [26]. This situation is essentially the same as a dye sensitization system in which only one monolayer of dye molecules is directly adsorbed on a titanium dioxide surface in order to generate photocurrent [27]. The reason why the solid-state system showed far lower conversion efficiency than the solution system ought to be rigorously examined.

10.4. p–n HOMOJUNCTION CELLS

10.4.1. Motivation

The photovoltaic properties of organic pigment films depend on their Fermi level, which seems to be strongly influenced by various unidentified impurities. Most of organic semiconductors show p-type character presumably due to the doping by oxygen molecules from the atmosphere. A few kinds of organic semiconductors, such as Me-PTC and Im-PTC, show n-type character. However, no information appeared initially about the reason why they showed n-type behavior. Fortunately, Whitlock et al. [28] reported a new purification technique for organic pigments called "reactive train sublimation," in which conventional train sublimation under a flow of inert gas [11] was modified to be performed under a flow of reactive gas. Taking the enhanced purification efficiency of this new technique into consideration, we expected that a detailed study regarding the effect of this reactive sublimation on the photovoltaic properties of Me-PTC films would offer some insight into the origin of the conduction type of Me-PTC [9,10].

As a next step, we tried to control the Fermi level by intentional doping with known molecules into the highly purified pigments [9,10]. Taking into consideration the unpractical conversion efficiency of amorphous silicon solar cells before the accomplishment of their pn-control by doping [8], it is also crucially important to apply organic semiconductors to solar cells. I reasoned that if the original conduction type of n-type perylene pigments could be changed to p-type, then the possibility of pn-control of organic semiconductors would have been clearly demonstrated. Thus, we tried doping with an electron acceptor, which strengthens the p-type character, and we adopted halogens, which have strong oxidizing characteristics, as suitable dopants. These were expected to act as acceptors in Me-PTC films, i.e., they may draw an electron from an Me-PTC molecule into the ground state and, in turn, thermally liberate a hole into the valence band of the pigment film since halogens are well known to form a conductive complex with perylene [29], which forms the skeleton of Me-PTC.

10.4.2. Efficient Purification by Reactive Sublimation

Me-PTC was purified using train sublimation [11,28], which was performed in a Pyrex glass tube inserted through an electric furnace with a gentle temperature gradient of about $4°/cm$. After evacuation to 0.1 Pa, an inert (N_2) or a reactive carrier gas (CH_3NH_2) was introduced at a pressure of 133 Pa. A methanolic solution of methylamine (40%) was used as a reactive gas source. Me-PTC powder was placed in position and heated to 500°C. The sublimed Me-PTC and the impurities appeared at separate parts of the tube. Figure 10.23 shows the photocurrent–voltage ($J–V$) curves for ITO/In(10 nm)/Me-PTC(300 nm)/Au sandwich cells, without sublimation (curve A) and after either a single sublimation (curve B) or after four sublimations (curve C) under CH_3NH_2. Indium was inserted to make an ohmic contact between ITO and Me-PTC. In this cell, a photoactive Schottky junction is formed at the interface between the n-type Me-PTC film and the Au electrode, which has a large work function. Interestingly, the open-circuit photovoltage (V_{oc}) significantly increased upon repetition of reactive sublimation. On the other hand, little increase in V_{oc} was observed with inert sublimation using N_2 as a carrier gas. Since the magnitude of V_{oc} is primarily determined by the built-in potential, which is equivalent to the energy difference between the work function of Au and the Fermi level of Me-PTC film, then the increase in V_{oc} suggests a shift in the Fermi level of the Me-PTC upon reactive sublimation.

In order to clarify this result, the energy position of the Fermi level of the Me-PTC film was evaluated by determining the difference in contact potential between an Me-PTC film and an Au reference plate by the Kelvin–Zisman vibrating capacitor method (Figure 10.24) [30]. The null point in the alternating current (AC) induced by the vibration (50 Hz) was detected precisely by a lock-in amplifier. Since the Me-PTC

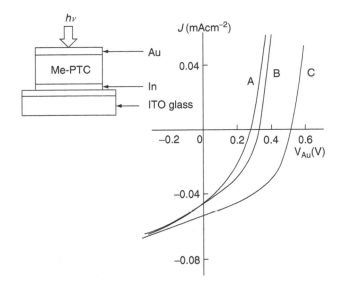

Figure 10.23. Photocurrent–voltage (*J–V*) curves for ITO/In(10 nm)/Me-PTC(300 nm)/Au(20 nm) sandwich cells using Me-PTC without sublimation (curve A), with a single sublimation (curve B), and with four sublimations (curve C) under CH_3NH_2 (133 Pa). White light (100 mW cm^{-2}) was irradiated onto the Au electrode. The direction of the observed photovoltage (V_{Au}) was (+)Au/Me-PTC/In/ITO(−). (From M. Hiramoto, K. Ihara, and M. Yokoyama, *Jpn. J. Appl. Phys.* **34**, 3803–3807 (1995). With permission.)

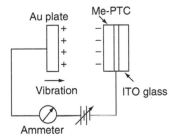

Figure 10.24. Schematic diagram of the Kelvin–Zisman vibrating capacitor method. (From M. Hiramoto, K. Ihara, and M. Yokoyama, *Jpn. J. Appl. Phys.* **34**, 3803–3807 (1995) and M. Hiramoto, K. Ihara, H. Fukusumi, and M. Yokoyama, *J. Appl. Phys.* **78**, 7153–7157 (1995). With permission.)

film has n-type character, the observed Fermi level was easily shifted in the positive direction due to downward band bending at the film surface caused by the adsorption of oxygen in air. However, this effect was suppressed under vacuum conditions (10^{-1} Pa) and similar relative shifts induced by the reactive sublimation were observed both in vacuum and in air when the measurements were performed immediately after breaking the vacuum. Curve A in Figure 10.25 shows the change ($\Delta\Phi$) in the contact potential difference between the Au reference plate and the Me-PTC film. $\Delta\Phi$ shifted in the negative direction, depending on the number of repetitions of the reactive sublimation. This result was clear evidence for the negative shift of the Fermi level induced by the reactive sublimation. Moreover, the increase in V_{oc} from 0.31 to 0.49 V corresponded well to the change in $\Delta\Phi$ (Figure 10.25, curve B). Obviously, the increase in V_{oc} can be attributed to the increased barrier height

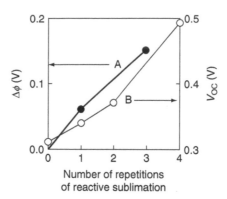

Figure 10.25. Dependence of the change in the contact potential difference ($\Delta\Phi$) between a Au reference plate and the Me-PTC films measured by the Kelvin vibrating capacitor method (curve A) and of the open-circuit photovoltage (V_{oc}) for ITO/In/Me-PTC/Au cells (curve B) on the number of repetitions of reactive sublimation. Values of $\Delta\Phi$ are plotted by taking that of unpurified Me-PTC to be zero. (From M. Hiramoto, K. Ihara, and M. Yokoyama, *Jpn. J. Appl. Phys.* **34**, 3803–3807 (1995) and M. Hiramoto, K. Ihara, H. Fukusumi, and M. Yokoyama, *J. Appl. Phys.* **78**, 7153–7157 (1995). With permission.)

at the Au/Me-PTC junction caused by the negative shift of the Fermi level in the Me-PTC film.

There is a possibility that the remaining methylamine acted as a donor and caused the negative shift of the Fermi level after the reactive sublimation. The results presented here, however, were observed for samples that were subjected to identical severe conditions during the vacuum deposition process under a pressure of 10^{-3} Pa for fabrication of the cell. Thus, we concluded that the above observation resulted from the removal of some impurities during reactive sublimation.

The shift in the Fermi level of the Me-PTC might arise as follows. The purity of the organic pigments is obviously inferior to that of inorganic semiconductors such as silicon to which strict purification techniques have been applied. A relatively large amount of impurities are present, including molecules which act as donors and acceptors (which unfortunately are not identified), and the position of the Fermi level in Me-PTC films is determined by the balance of their concentrations. Consequently, it is located near the center of the bandgap (Figure 10.26). The reason why the Me-PTC film behaves as a "weak" n-type semiconductor is that the overall donor concentration is slightly higher than the overall acceptor concentration. Thus, the negative shift in the Fermi level observed using the present technique is assumed to result from the more efficient removal of impurities, which act as electron acceptors compared to impurities acting as electron donors during sublimation under a reducing atmosphere.

10.4.3. pn-Control of a Single Organic Semiconductor by Doping

The shift in the Fermi level resulting from the purification of the pigment does imply the existence of unknown but specific impurities acting as donors or acceptors in organic pigments. Therefore, we concluded that the Fermi level of an organic pigment film could also be controlled by fundamentally the same method as in inorganic semiconductors, i.e., intentional doping with a known molecule.

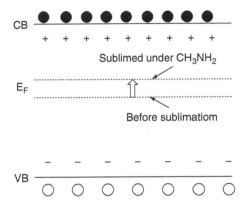

Figure 10.26. Energy diagram of Me-PTC film. CB, VB, and E_F denote the conduction band, the valence band, and the Fermi level, respectively. The filled and open circles are electrons and holes, respectively. + and − represent ionized donors and acceptors, respectively. (From M. Hiramoto, K. Ihara, and M. Yokoyama, *Jpn. J. Appl. Phys.* **34**, 3803–3807 (1995). With permission.)

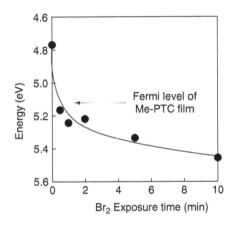

Figure 10.27. Dependence of the energy position of the Fermi level of the Me-PTC film on the Br$_2$ exposure. The energy is plotted against the vacuum level. Bromine water (3%) was used as the Br$_2$ gas source. (From M. Hiramoto, K. Ihara, and M. Yokoyama, *Jpn. J. Appl. Phys.* **34**, 3803–3807 (1995) and M. Hiramoto, K. Ihara, H. Fukusumi, and M. Yokoyama, *J. Appl. Phys.* **78**, 7153–7157 (1995). With permission.)

Figure 10.27 shows the effect of exposure to Br$_2$ on the energy of the Fermi level of an Me-PTC film, which was measured by the Kelvin–Zisman vibrating capacitor method. Me-PTC purified four times by train sublimation under a reactive atmosphere of methylamine gas at a pressure of 2.7×10^3 Pa was used. The exposure of the Me-PTC films to Br$_2$ gas was performed in a glass vessel using bromine water (3%) as the gas source. Energy is plotted against the vacuum level, taking the work function of the Au reference plate as 4.9 eV. Upon exposure for only 5 min, the Fermi level of the pigment film shifted significantly in the positive direction and nearly attained the energy position of the valence band of Me-PTC (5.4 eV), as measured by atmospheric electron photoemission analysis [13]. This effect was maintained under a vacuum produced by a rotary pump. Since a thin indium metal film deposited between the ITO electrode and the Me-PTC film was corroded within 10 sec by

Br$_2$ exposure, Br$_2$ seems to diffuse throughout the bulk film. Though a positive shift in the Fermi level was also observed when a pure Br$_2$ source was used, the film was damaged with a detectable change in the shape of the absorption spectrum due to CT. Since no change in absorption was observed under the present mild condition using bromine water, we concluded that a very small number of Br$_2$ molecules (which are hard to detect as an absorption change) doped the film without damaging it and caused a dramatic shift in the Fermi level. In the case of iodine gas exposure, however, little shift was observed in the Fermi level.

The positive shift of Fermi level reaching the nearby valence band strongly suggests the conduction-type alteration of Me-PTC film, which is originally n-type, to p-type. In order to confirm this, we evaluated the photovoltaic properties of sandwich cells incorporating an Au Schottky junction.

Figure 10.28 shows the spectral dependence of short-circuit photocurrent for sandwich cells incorporating Me-PTC films (a) without and (b) with treatment by Br$_2$ gas. In order to make the ohmic contacts, In or Pt film (2 nm thick) was inserted

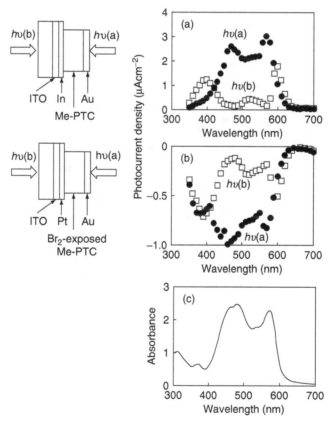

Figure 10.28. Spectral dependence of the short-circuit photocurrent for sandwich cells incorporating Me-PTC films (300 nm thick) (a) without and (b) with Br$_2$ treatment. The cell structures are also shown. $h\nu$(a) and $h\nu$(b) indicate light irradiation on the Au electrode and on the ITO electrode, respectively. The incident light intensity was around 1 mW cm^{-2}. Photocurrent flow from ITO to Au is regarded as positive. (c) Absorption spectrum of Me-PTC film with a thickness of 300 nm. (From M. Hiramoto, K. Ihara, and M. Yokoyama, *Jpn. J. Appl. Phys.* **34**, 3803–3807 (1995) and M. Hiramoto, K. Ihara, H. Fukusumi, and M. Yokoyama, *J. Appl. Phys.* **78**, 7153–7157 (1995). With permission.)

between the ITO and the Me-PTC for the respective cases. The absorption spectrum of the Me-PTC film (300 nm thick) is also shown in Figure 10.28(c). In the case of Me-PTC without Br_2 exposure (Figure 10.28(a)), the profile of the action spectrum closely resembled the absorption spectrum when light was irradiated on the Au electrode ($h\nu$(a)), whereas photocurrent peaks were observed at the absorption edges when light was irradiated on the ITO electrode ($h\nu$(b)). The latter is typical of a masking effect, indicating that the photoactive region is located near the Au/Me-PTC interface. Taking the observed direction of the photocurrent flow in the Me-PTC film from the ITO to the Au and the positive photovoltage on the Au electrode into account, it is clear that the Me-PTC behaves as an n-type semiconductor and that a Schottky junction was formed at the Au/Me-PTC interface. Very interestingly, in the case of Me-PTC exposed to Br_2 gas (Figure 10.28(b)), the sign of the observed photocurrent was negative, i.e., the photocurrent in the cell flowed from Au to ITO, which is the opposite to that observed for n-type material. In addition, the observed profiles of the action spectra were essentially the same as n-type Me-PTC, i.e., they closely resembled the absorption spectrum produced by the irradiation on Au ($h\nu$(a)) and showed a masking effect by irradiation on the ITO ($h\nu$(b)). In summary, although photoactive interfaces are located near the Au electrode irrespective of Br_2 exposure, the direction of the photocurrent flow arising from the Schottky junctions with the Au was reversed upon Br_2 exposure. The observed photovoltage was also reversed by Br_2 doping (Figure 10.31). The photovoltage at which the Au electrode was positive with respect to the In/ITO electrode was observed for the n-type Me-PTC, while the photovoltage at which the Au electrode was negative with respect to the Pt/ITO electrode was observed for the p-type Me-PTC.

In order to explain the present results, one should consider that n-type Me-PTC film changed to p-type upon Br_2 exposure. That is, as shown in Figure 10.29, the

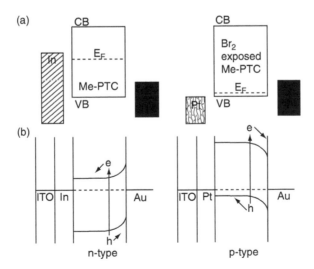

Figure 10.29. (a) Energy diagram of Me-PTC, Br_2-exposed Me-PTC, In, Pt, and Au films before contact. CB, VB, and E_F denote the conduction band, the valence band, and the Fermi level, respectively. (b) The energy structure of sandwich cells, ITO/In/Me-PTC/Au (left) and ITO/Pt/Br_2-exposed Me-PTC/Au (right), after contact. (From M. Hiramoto, K. Ihara, and M. Yokoyama, *Jpn. J. Appl. Phys.* **34**, 3803–3807 (1995) and M. Hiramoto, K. Ihara, H. Fukusumi, and M. Yokoyama, *J. Appl. Phys.* **78**, 7153–7157 (1995). With permission.)

direction of band bending formed after contact with Au becomes reversed by Br_2 exposure, due to the positive shift in the Fermi level to near the valence band. As a result, the photogenerated carriers near the photoactive Au/Me-PTC interface flow in the opposite direction to each other. Thus, a controlled change in the conduction type from n- to p-type for perylene pigment film was confirmed by photovoltaic measurements.

The present results show that the photovoltaic properties seem to be reasonably well explained on the basis of the band model and the shift in the Fermi level for the Me-PTC/Au system. However, the general applicability of the band model (which has been developed for inorganic semiconductors) to other organic systems should be attentively investigated. Especially for organic–organic junction such as the Im-PTC/CuPc, we have a experimental sign that the observed photovoltage is not only determined by the difference of Fermi level for respective pigment films but also related to spatial nanostructure of the interface [31,32].

10.4.4. p–n Homojunction in Perylene Pigment Film

Since both conduction types of Me-PTC films were available, an organic homojunction between p- and n-type films was constructed. The cell shown in Figure 10.30 was fabricated. The lower layer is p-type Me-PTC that was treated by Br_2 for 7 min. The upper layer is n-type Me-PTC that was deposited subsequently. In order to avoid

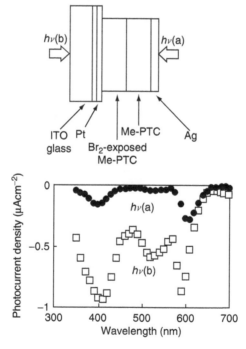

Figure 10.30. Spectral dependence of short-circuit photocurrent for an ITO/Pt(2 nm)/Br_2-exposed Me-PTC(500 nm)/Me-PTC(100 nm)/Ag(20 nm) cell. The cell structure is also shown. $h\nu$(a) and $h\nu$(b) indicate light irradiation on the Ag electrode and on the ITO electrode, respectively. The incident light intensity was around 1 mW cm^{-2}. A photocurrent flow from ITO to Ag is regarded as positive. (From M. Hiramoto, K. Ihara, and M. Yokoyama, *Jpn. J. Appl. Phys.* **34**, 3803–3807 (1995). With permission.)

the efflux of bromine from the film and to keep the Me-PTC underlayer p-type, the deposition of an n-type overlayer was performed in a relatively low vacuum of 0.13 Pa by introducing N_2.

This cell showed photocurrent flow (cell internal) from Ag to ITO/Pt and photovoltage of the polarity that the Ag electrode was negative with respect to the ITO/Pt electrode (Figure 10.31). Observed photovoltage of about 0.4 V is roughly the sum of the positive photovoltage of 0.2 V for an ITO/In/n-type Me-PTC/Au cell and the negative photovoltage of 0.17 V of an ITO/Pt/p-type Me-PTC/Au cell. Interestingly, the observed profiles of the action spectra showed a clear masking effect for light irradiation on both the Ag ($h\nu$(a)) and ITO ($h\nu$(b)) electrodes, as shown in Figure 10.30. The photocurrent in both cases showed peaks at the absorption edges of the Me-PTC film (see Figure 10.28(c)) where the light can penetrate into the bulk of the film. This means that the photoactive region is not located at the interfaces with

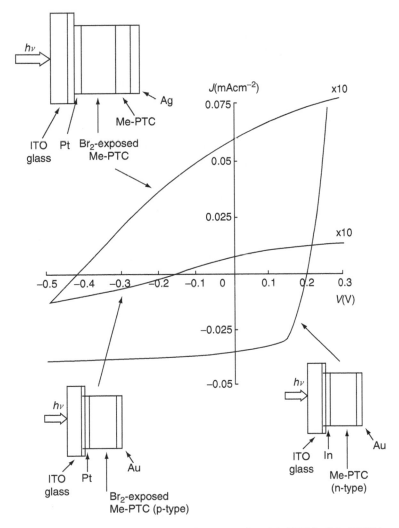

Figure 10.31. Photocurrent–voltage (J–V) curves for (a) ITO/In/Me-PTC/Au and (b) ITO/Pt/Br$_2$-exposed Me-PTC/Au, and (c) ITO/Pt(2 nm)/Br$_2$-exposed Me-PTC(500 nm)/Me-PTC(100 nm)/Ag(20 nm) cells.

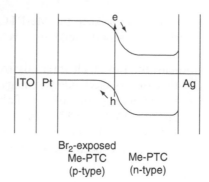

Figure 10.32. Possible energy structure of ITO/Pt/Br$_2$-exposed Me-PTC/Me-PTC/Ag cell. (From M. Hiramoto, K. Ihara, and M. Yokoyama, *Jpn. J. Appl. Phys.* **34**, 3803–3807 (1995). With permission.)

the Ag and the ITO electrodes, but is deep in the bulk of the Me-PTC film. Therefore, this result strongly suggests the existence of a p–n homojunction (Figure 10.32) formed between p- and n-type Me-PTC films. The directions of the photocurrent and the photovoltage can be explained reasonably well by this energy structure.

10.4.5. p–n Control Technique

Me-PTC is the first example of a single photoconductive organic pigment showing both n- and p-type characters. Br$_2$-doping also seems to be applicable to other perylene derivatives. The results in this section indicate that pn-control fundamentally based on the same technique used for inorganic semiconductors is also valid for organic semiconductors. For complete pn-control, a shift in the Fermi level in the negative direction due to doping by donors should be demonstrated. Alkali metals are promising candidates for donors. From the standpoint of long-term stability, the replacement of mobile and volatile small molecules such as Br$_2$ is desirable. Larger organic molecules such as F$_4$-TCNQ [33] are candidates for stable dopants. Even at the present stage, however, an organic pn homojunction composed of p- and n-type Me-PTC films can be fabricated. It should be pointed out that the formation of sufficient built-in potential is indispensable for solar cell applications. Therefore, the establishment of a complete pn-control technique for organic semiconductors is one of the main keys to realizing practical solid-state organic solar cells.

10.5. CONCLUSION

The concept of an organic p–i–n junction in which i-layer is a co-deposited layer of two different organic semiconductors was proposed and demonstrated. Co-deposited layers have a vast number of heteromolecular (donor–acceptor) contacts acting as efficient photocarrier generation sites. In order to extract the photogenerated carriers from the bulk of the co-deposited films, the nanostructure of the films should be controlled. A crystalline–amorphous nanocomposite with spatially separated pathways for electrons and holes revealed a high efficiency for photocurrent generation. Although Me-PTC/H$_2$Pc is not the best possible combination, a power conversion efficiency of 1% was obtained.

Alteration of conduction type from n-type to p-type in a single organic semiconductor of perylene pigment by doping technique was demonstrated. p–n Homojunction was successfully fabricated.

ACKNOWLEDGMENT

The author is grateful to the deceased S. Ooki of Dainichi Seika Kogyo Inc. for kindly donating the perylene pigments.

REFERENCES

1. C. W. Tang, Two-layer organic photovoltaic cell, *Appl. Phys. Lett.* **48**, 183–185 (1986).
2. C. W. Tang and S. A. VanSlyke, Organic electroluminescent diodes, *Appl. Phys. Lett.* **51**, 913–915 (1987).
3. G. A. Chamberlain, *Solar Cells* **8**, 47 (1983), and references therein.
4. D. Wohrle and D. Meissner, Organic solar cells, *Adv. Mater.* **3**, 129 (1991), and references therein.
5. M. Hiramoto, H. Fujiwara, and M. Yokoyama, Three-layered organic solar cell with a photoactive interlayer of codeposited pigments, *Appl. Phys. Lett.* **58**, 1062–1064 (1991).
6. M. Hiramoto, H. Fujiwara, and M. Yokoyama, p–i–n like behavior in three-layered organic solar cells having a co-deposited interlayer of pigments, *J. Appl. Phys.* **72**, 3781–3787 (1992).
7. M. Hiramoto, K. Suemori, and M. Yokoyama, Photovoltaic properties of ultramicrostructure-controlled organic co-deposited films, *Jpn. J. Appl. Phys.* **41**, 2763 (2002).
8. W. E. Spear and P. E. Lecomber, *Solid State Commun.* **17**, 1193 (1975).
9. M. Hiramoto, K. Ihara, and M. Yokoyama, Fermi level shift in photoconductive organic pigment films measured by Kelvin vibrating capacitor method, *Jpn. J. Appl. Phys.* **34**, 3803–3807 (1995).
10. M. Hiramoto, K. Ihara, H. Fukusumi, and M. Yokoyama, Conduction type control from n to p type for organic pigment films purified by reactive sublimation, *J. Appl. Phys.* **78**, 7153–7157 (1995).
11. H. J. Wagner, R. O. Loutfy, and C. Hsiao, Purification and characterization of phthalocyanines, *J. Mater. Sci.* **17**, 2781 (1982).
12. M. Hiramoto, Y. Kishigami, and M. Yokoyama, Doping effect on the two-layer organic solar cell, *Chem. Lett.* **1990**, 119–122 (1990).
13. H. Kirihata and M. Uda, Externally quenched air counter for low-energy electron emission measurements, *Rev. Sci. Instrum.* **52**, 68–70 (1981).
14. R. A. Vanden Berghe and W. P. Gomes, *Ber Bunsenges. Phys. Chem.* **76**, 481 (1972).
15. D. Gebeyehu, B. Maennig, J. Drechsel, K. Leo, and M. Pfeiffer, Bulk-heterojunction photovoltaic devices based on donor–acceptor organic small molecule blends, *Sol. Energy Mater. Sol. Cells*, **79**, 81–92 (2003).
16. M. Yokoyama, S. Shimokihara, A. Matsubara, and H. Mikawa, Extrinsic carrier photogeneration in poly-*N*-vinylcarbazole. III. CT fluorescence quenching by an electric field, *J. Chem. Phys.* **76**, 724 (1982).
17. Z. D. Popovic, A.-M. Hor, and R. O. Loutfy, A study of carrier generation mechanism in benzimidazole perylene/tetraphenyldiamine thin film structures, *Chem. Phys.* **127**, 451–457 (1988).
18. M. Hiramoto, K. Fujino, M. Yoshida, and M. Yokoyama, Influence of oxygen and water on photocurrent multiplication in organic semiconductor films, *Jpn. J. Appl. Phys.* **42**, 672–675 (2003).

19. G. Yu, J. Gao, J. C. Hummelen, F. Wudl, and A. J. Heeger, Polymer photovoltaic cells: enhanced efficiencies via a network of internal donor–acceptor heterojunctions, *Science* **270**, 1789–91 (1995).

20. P. Peumans and S. R. Forrest, Very-high-efficiency double-heterostructure copper phthalocyanine/C_{60} photovoltaic cells, *Appl. Phys. Lett.* **79**, 126–128 (2001).

21. Y. Oishi, M. Hiramoto, and M. Yokoyama, Photovoltaic properties of C_{60}–Phthalocyanine Co-deposited Thin Film, Extended Abstracts of the 48th Spring Meeting 2001, Japan Society of Applied Physics and Related Societies, 29a-ZG-7, Meiji University, Tokyo, March 28–31 (2001).

22. G. Matsunobu, Y. Oishi, M. Yokoyama, and M. Hiramoto, High-speed multiplication-type photodetecting device using organic co-deposited films, *Appl. Phys. Lett.* **81**, 1321–1322 (2002).

23. M. Hiramoto, Y. Sakaue, and M. Yokoyama, Carrier generation in organic pigment films and microwave photoconductivity, *Nihon Kagaku Kaishi* **1992**, 1180–1185 (1992) [in Japanese].

24. K. Suemori, T. Miyata, M. Hiramoto, and M. Yokoyama, Enhanced photovoltaic performance in fullerene: phthalocyanine codeposited films deposited on heated substrate, *Jpn. J. Appl. Phys.*, **43**, L1014–L1016 (2004).

25. P. Peumans, S. Uchida, and S. R. Forrest, Efficient bulk heterojunction photovoltaic cells using small-molecular-weight organic thin films, *Nature*, **425**, 158–162 (2003).

26. T. Yamaga and M. Hiramoto, unpublished results.

27. B. O'Regan and M. Grätzel, A low-cost, high-efficiency solar cell based on dye-sensitized colloidal TiO_2 films, *Nature* **353**, 737–740 (1991).

28. J. B. Whitlock, P. Panayotatos, G. D. Sharma, M. D. Cox, R. R. Sauers, and G. R. Bird, *Opt. Eng.*, **32**, 1921 (1993).

29. H. Akamatsu, H. Inokuchi, and Y. Matsunaga, Electrical conductivity of the perylene–bromine complex, *Nature* **173**, 168–169 (1954).

30. S. Saito, T. Soumura, and T. Maeda, *J. Vac. Sci. Technol.*, **A2**, 1389 (1984).

31. K. Aoi, M. Hiramoto, and M. Yokoyama, Mechanism of Built-In Field Formation at Perylene/Phthalocyanine Heterojunction, Extended Abstracts of the 56th Autumn Meeting 1995, Japan Society of Applied Physics and Related Societies, 29a-S-5, Kanazawa Institute of Technology, Kanazawa University, Kanazawa, August 26–29 (1995).

32. M. Hiramoto, K. Nakayama, T. Katsume, and M. Yokoyama, Field-activated structural traps at organic pigment/metal interfaces causing photocurrent multiplication phenomena, *Appl. Phys. Lett.* **73**, 2627–2629 (1998).

33. J. Blochwitz, M. Pfeiffer, and T. Frits, Low voltage organic light emitting diodes featuring doped phthalocyanine as hole transport material, *Appl. Phys. Lett.* **73**, 729–731 (1998).

11

Liquid-Crystal Approaches to Organic Photovoltaics

Bernard Kippelen,[a] Seunghyup Yoo,[a] Joshua A. Haddock,[a] Benoit Domercq,[a] Stephen Barlow,[b] Britt Minch,[c] Wei Xia,[c] Seth R. Marder,[b] and Neal R. Armstrong[c]

[a] *School of Electrical and Computer Engineering, Georgia Institute of Technology, Atlanta, GA, USA*
[b] *School of Chemistry and Biochemistry, Georgia Institute of Technology, Atlanta, GA, USA*
[c] *Department of Chemistry, University of Arizona, Tucson, AZ, USA*

Contents

Abstract This chapter provides a review of the semiconducting properties of organic molecules that self-assemble into ordered liquid-crystalline mesophases, and discusses their performance in organic solar cells. The chapter is organized as follows: in the first section, the need for high-mobility materials in efficient solar cells is discussed; in the second section, various techniques used to determine the charge mobility in thin films are presented and compared; the third section reviews the basic properties of liquid-crystalline materials; then, the current state-of-the-art developments in semiconducting liquid crystals are discussed. Finally, the properties of solar cells that are based on these self-ordering materials are presented.

Keywords calamitic, Child's law, cholesteric, clearing temperature, columnar discotic, cross-linking, discostic, disorder formalism, dispersive transport, energetic disorder, field-effect transistor, grain boundaries, liquid crystal, lyotropic, mesophase, mobility, mobility pre-factor, nematic, nematic discotic, order parameter, phase transition, phase, transition temperature, positional disorder, pulsed radiolysis time-resolved microwave, conductivity, smectic, space–charge limited current, time-of-flight, thermotropic

11.1. INTRODUCTION

Energy needs per person in the world are steadily rising as is the consumption of the fossil fuels that supply us with most of our energy. The world supplies of fossil fuels such as coal, oil, and natural gas are, however, limited and their use to produce energy causes severe environmental problems associated with the emission of carbon dioxide and other harmful greenhouse gases that contribute to global warming. Hence, there is an urgent need to develop new energy sources. Solar energy is potentially an inexpensive continuous source of energy, provided that costs of energy harvesting, storage, and transmission can be lowered to be competitive with other energy sources, the costs of which are generally rising. Research and development in photovoltaics (PV) has grown rapidly during the second half of the 20th century and was enabled by advances made in semiconducting materials. Solar conversion efficiencies well in excess of 10% can be achieved in several materials including silicon, gallium arsenide, indium phosphide, cadmium telluride, copper indium diselenide, and copper(I) sulfide. Today's PV technology is dominated by Si, a material with low optical absorption, requiring thick high-purity layers with crystalline perfection and high-temperature processing. These limitations keep the cost of crystalline Si technologies high and prevent use on lightweight, flexible substrates. Significant efforts have recently been targeted towards producing cost-effective thin-film (1 μm) amorphous Si, $CuIn_{1-x}Ga_xSe_2$, and CdTe technologies [1]. These technologies have reached efficiencies >10% (see Figure 11.1), and are entering pilot production.

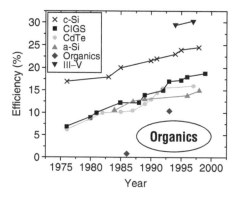

Figure 11.1. Evolution of the PV cell efficiencies for different material technologies (adapted from A. Goetzberger and C. Hebling, *Sol. Energy Mater. Sol. Cells* **62**, 1 (2000). With permission).

However, the manufacturing, use, and disposal of these materials raise health, safety, and environmental issues.

In recent years, several approaches to organic photovoltaic (OPV) cells have emerged. They currently exhibit power conversion efficiencies of up to a few percent and can be divided into the following categories: (i) small molecule multilayer devices [2–4]; (ii) polymeric mixtures [5–8]; (iii) hybrid organic–inorganic semiconductor nanostructure mixtures [9]; and (iv) dye-sensitized nanoporous oxide (Grätzel) cells [10–12]. In each type of OPV the efficiency is limited by the absorption coefficient as a function of wavelength, the probability of exciton dissociation into mobile charge carriers, the transport of those charges to the anode and cathode, and the collection of the carriers at the electrode–organic interfaces. In addition, the cell must have a built-in voltage and an electrical characteristic with a large fill factor. OPVs based on technologies (i) to (iii) exhibit high exciton dissociation efficiencies but all suffer from low charge mobility. Grätzel cells (iv) have efficiencies approaching 10% but these efficiencies have only been achieved with liquid electrolytes containing a redox couple (I^-/I_3^-), which limits their manufacturability and durability.

It is anticipated that OPV technologies will have the following attributes:

- The component organic materials can be processed from the vapor phase or from solution at low temperatures.
- The materials can potentially be processed into large-area devices at low cost.
- The OPV cells will be intrinsically light in weight.
- The materials and devices are likely to have increased functionality due to their inherent structural flexibility.
- OPVs will be amenable to patterning and processing using soft lithography, printing, and embossing.

Technologies combining these properties are anticipated to pave the way to low-cost, lightweight, large-area, flexible, and conformable solar panels. The ultimate goal is to develop materials that can be processed into cells by a roll-to-roll process as illustrated by Figure 11.2, or by using ink-jet printing or screen printing. Such cells would potentially find applications in on-grid power generation, and in numerous off-grid applications to power the ever-increasing number of portable electronic digital devices and sensors.

Except for dye-sensitized Grätzel cells, most organic solar cells fabricated to date are based on p–n organic–organic heterojunctions formed between hole-transporting and electron-transporting molecules or polymers. These junctions can be formed in bilayer devices or by blending. The organic materials are usually

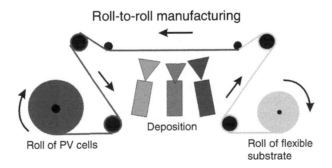

Figure 11.2. Schematics of the roll-to-roll manufacturing potential of organic solar cells.

sandwiched between a transparent electrode such as indium tin oxide (ITO) and a metal electrode with a low work function. Differences in work functions between the electrodes, and between electron affinities and ionization potentials of the organic materials provide a built-in field that assists exciton dissociation and transport of holes and electrons to their respective collecting electrodes. Light absorption in organic materials leads to the creation of excitons that possess a strong binding energy, which is typically on the order of 0.5 eV. Some of them can migrate to the junction by a diffusion process, and efficient charge separations occur when energy band (frontier orbital) offsets between hole and electron transport materials are large enough (>0.5 eV) to dissociate the exciton into an electron–hole pair. These separated charges are then transported by drift under the influence of the built-in field to the electrodes and collected.

In this chapter, we discuss an alternative approach to excitonic OPV cells that is based on materials that self-assemble into mesophases. The motivation to use self-assembling materials is driven by the need for new semiconductors which can be processed from solution and which have higher charge mobility than amorphous materials; the increased order, and, therefore, potentially increased electronic coupling between nearest neighbors possible in self-assembled systems can potentially lead to such high mobilities. Attractive examples of such self-assembly are found in liquid crystals, materials that have been extensively studied for their optical properties and are widely used in the display industry. Studies pioneered by Haarer in the 1990s have shown that liquid crystals can indeed have high charge mobilities. An analysis of the requirements for high-efficiency OPV cells shows that high-mobility materials could provide high efficiency by reducing the series resistance of solar cells. Furthermore, the unique properties of liquid-crystal mesophases can lead to nanostructured films that can also improve the PV performance by increasing the contact area between hole- and electron-transporting materials at which exciton dissociation takes place.

The chapter is organized as follows. First, the equivalent circuit derived from the p–n diode model for solar cells is briefly reviewed and used to illustrate the importance of reducing series resistance (R_s) in OPV cells. Next, the transport properties of organic semiconductors are discussed and different charge-mobility measurement techniques are presented. In Section 4, the basic properties of liquid crystals necessary to understand the semiconducting properties of these materials are discussed and the transport properties of various liquid-crystal materials are reviewed. Finally, the performance of OPV cells fabricated from semiconducting liquid crystals is discussed.

11.2. MODELING OF SOLAR CELLS

The challenge is to increase the conversion efficiency of solid-state OPV cells to above 10% by developing organic semiconductors with environmental stability that exhibit optimized light harvesting and dissociation properties and form durable ohmic contacts with metallic and transparent electrodes on flexible substrates, which are good barriers to moisture and oxygen. To understand the parameters that affect the performance of a solar cell, let us first consider the power conversion efficiency η which is defined as:

$$\eta = FF \frac{J_{sc} V_{oc}}{P_S} \tag{1}$$

where J_{sc} is the short circuit current density (mA/cm^2), V_{oc} is the open circuit voltage (V), P_S is the optical irradiance of the incident light from the sun (1 sun $= 100$ mW/cm^2 for AM 1.5G). FF is the fill factor and is defined as the ratio of the maximum power delivered by the solar cell to the product $J_{sc}V_{oc}$ and can be evaluated from the current–voltage (J–V) characteristics of a solar cell. The magnitude of the short circuit current is proportional to the portion of the solar spectrum (considering an illumination of AM 1.5G) absorbed by the materials forming the solar cell, and is dependent on the dissociation efficiency of the excitons created upon optical excitation into electrons and holes, which are collected by the electrodes. Employing light-harvesting materials with a low-energy bandgap can increase the photocurrent by extending the light harvesting to the infrared, but it can also lead to a decrease in photovoltage available in the solar cell. Therefore, the absorption spectrum of the organic semiconductors should be optimized in a manner that maximizes the product of the photogenerated current and the open-circuit voltage. Figure 11.3 shows an example of typical J–V characteristics measured in a multilayer organic PV cell in the dark and under illumination. The response is typical of a p–n junction and is, therefore, very similar to that of inorganic PV cells. Due to this similarity, it is generally assumed that we can apply the equivalent circuit model shown in Figure 11.4 to organic PV cells [13–15]. From this equivalent circuit, the following J–V characteristic is easily derived:

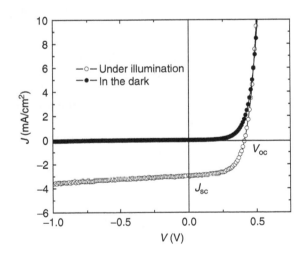

Figure 11.3. Example of J–V characteristics measured in multilayered organic PV cells with structure ITO/CuPC/C$_{60}$/BCP/Al measured in the dark and under illumination of white light (42 mW/cm^2; 350–800 nm) from a xenon arc lamp.

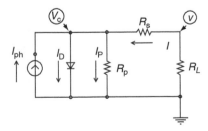

Figure 11.4. Schematics of the equivalent circuit used to model PV cells.

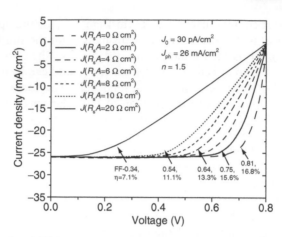

Figure 11.5.   Calculated fill factors and device efficiencies as a function of series resistance.

$$J = \frac{1}{1 + R_s/R_p} \left[J_0 \left\{ \exp\left(\frac{V - JR_sA}{nkT/e} \right) - 1 \right\} - \left(J_{Ph} - \frac{V}{R_pA} \right) \right] \qquad (2)$$

where R_s and R_p are the series and shunt resistance, respectively, n is the diode ideality factor (typically $1 < n < 2$), e is the elementary charge, kT is the thermal energy, J_0 is the diode saturation current density, J_{Ph} is the photogenerated current density, and A is the area of the cell. The series resistance in solar cells (R_s) is attributed to the finite conductivity of the semiconducting material, the contact resistance between the semiconductors and the adjacent electrodes, and the resistance associated with electrodes and interconnections. The shunt resistance (R_p) is introduced to take into account the loss of carriers via leakage paths that may be present; these may include pinholes in the film and the recombination and trapping of the carriers during their transit through the cell.

Analysis of Equation (2) in various regimes of photocurrent [13] shows that the series resistance strongly affects the maximum attainable fill factor, especially where the photocurrent, J_{Ph}, is high. Large R_s can further decrease the efficiency by reducing the short circuit current as well as the fill factor. To illustrate this behavior, J–V characteristics calculated from Equation (2) are shown in Figure 11.5 for various values of R_s. The reduction in fill factor caused by high series resistance is clearly illustrated.

11.3.   TRANSPORT IN ORGANIC SEMICONDUCTORS

The performance of an OPV cell largely depends on the bulk electronic and optical properties of the conjugated organic materials from which it is composed, and on the properties of the homogeneous and heterogeneous interfaces formed between two organic materials, and between organic materials and metal or oxide electrode materials. Charge mobility, in particular, plays a major role in several fundamental processes that govern the operation of a solar cell. High charge mobility will, for instance, influence the dissociation efficiency of excitons into electron–hole pairs by decreasing geminate recombination. Likewise, device parameters, such as series resistance or the ability to form nearly ohmic contacts between organics and metals, will rely on the optimization of charge mobility.

The modeling and the characterization of the transport properties in organic materials is a challenging task because they depend strongly on the nature of the molecular building block, the coupling to phonons, and the degree of ordering that can range from essentially amorphous to crystalline [16]. Furthermore, the electrical properties of organic materials will be influenced, if not dominated, by the presence of impurities and defects. Charge mobility in organic solids is generally orders of magnitude lower than that in inorganic semiconductors and is generally much lower in amorphous materials, such as conjugated polymers or molecular glasses in which charge carriers are localized onto single chains or on single molecules, than in high-quality single crystals, in which the relative positions of molecular units is well controlled over macroscopic distances. In highly ordered crystals, transport can be highly anisotropic and can sometimes be described by band-type models, while in amorphous materials transport operates by a thermally activated hopping mechanism. From a technological standpoint, high-mobility organic materials for photovoltaic applications need to be processed into large area films, and efficient charge collection and transport have to take place in a direction perpendicular to the substrate. In view of these requirements, the use of self-assembling materials such as columnar discotic liquid crystals is an approach that can provide the ordering required to increase the transfer integral between adjacent molecules relative to that in an amorphous material, while retaining the ability to be processed from solution (see Figure 11.6). Provided the desired orientation of the liquid crystalline molecules can be achieved by controlling and tailoring the interaction with the substrate, photopolymerization techniques could be used to "lock-in" this orientation. In this section, we will discuss the transport properties of amorphous organic semiconductors and review the various experimental techniques used to measure mobility.

11.3.1. Disorder Formalism for Transport in Amorphous Materials

Many recent transport studies in doped polymers and molecular glasses have been analyzed using the disorder formalism developed by Bässler, Borsenberger, and co-workers [17–20]. In the disorder formalism, it is assumed that charge transport occurs by hopping through a manifold of localized states with superimposed energetic and positional disorder. The distributions of hopping site energies and distances are Gaussian and characterized by their widths, σ and Σ, respectively. In this framework, the mobility is given by

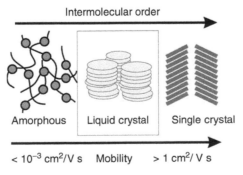

Figure 11.6. Schematics of the correlation between intermolecular order and charge mobility (adapted from F. Closs et al. [21]).

$$\mu = \mu_0 \exp\left[-\left(\frac{2\hat{\sigma}}{3}\right)^2\right] \exp[C(\hat{\sigma}^2 - \Sigma^2)E^{1/2}] \qquad (3)$$

where μ_0 is the disorder-free mobility (cm^2/V s), E is the applied electric field (in V/cm), T is the temperature, C is an empirical constant with a value of 2.9×10^{-4} (cm/V)$^{1/2}$, Σ is the width of the positional disorder distribution, $\hat{\sigma} = \sigma/k_B T$ where σ is the width of the energetical disorder distribution, and k_B is the Boltzmann constant. The units for charge mobility are cm^2/V s. According to the disorder formalism, the charge mobility increases with temperature and in most cases also with applied field in the high-field limit. Although good agreement with this model was found in various doped polymers and in molecular glasses, deviations from this model can be anticipated in liquid-crystalline materials in which charge transport is not three-dimensional but can be confined to one-dimensional columns.

11.3.2. Mobility Measurement Techniques

Charge mobility is a material property that is measured by a range of techniques operating over different length scales and that require different sample geometries. Sometimes, it becomes difficult, therefore, to directly compare the results obtained from these various techniques. The most common techniques are: (i) the time-of-flight (TOF) technique; (ii) the space-charge limited current (SCLC) technique; (iii) the field-effect transistor technique; and (iv) the pulse radiolysis time-resolved microwave conductivity (PR-TRMC) technique.

In the TOF geometry, the material is sandwiched between a transparent electrode (typically ITO) and a metal electrode. The sample is biased, with an appropriate field direction for transporting the carriers of interest, and mounted inside a temperature-controlled unit. As shown in Figure 11.7, a short laser pulse is transmitted through the ITO electrode and is strongly absorbed by the material. The excited states that are formed dissociate into electron–hole pairs. In hole (electron) transport materials, the injected holes (electrons) move towards the metal electrode under the influence of the applied electric field by drift. During their transit, the work produced by the displacement of the charges under the action of the electrical force in the sample leads to a transient electrical power in the circuit that manifests itself as a transient current. When the injected charges are collected, the transient current vanishes. By measuring the duration of the photocurrent, $\tau = L/v$, in a sample of thickness L, the drift velocity $v = \mu E$ can be calculated, from which mobility is derived. Because of dispersion in the arrival times of the carriers, the transient current is not a step

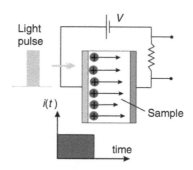

Figure 11.7. Schematics of the time-of-flight experiment.

function as shown in the ideal situation in Figure 11.7, but is typically followed by a tail. Typical transient currents measured in 20 μm-thick samples of 4,4′-bis(phenyl-*m*-tolylamino) biphenyl (TPD) doped into polystyrene (1:1 wt.%) are shown in Figure 11.8. For these experiments a low-noise high-voltage power supply was used for applying the bias voltage. The photocurrent was generated by irradiation with 6 nsec pulses from a N_2 laser (337 nm) and was amplified using a low-noise preamplifier and monitored with a digital oscilloscope. In order to keep $RC_p << t_t$ (transit time), resistance values were $R = 10^2–10^4$ Ω. C_p represents the total capacitance of the electrical circuit. Sample capacitance values were on the order of 10 pF. Neutral density filters were used when necessary to avoid excess charge accumulation in the samples that can create nonuniformity in the electrical field across the sample. As shown in Figure 11.8, the field dependence and the temperature dependence of the mobility are consistent with the disorder formalism. This suggests that the performance of organic solar cells based on typical amorphous organics should increase with temperature. Since the effective absorption length of the short pulse that produces the

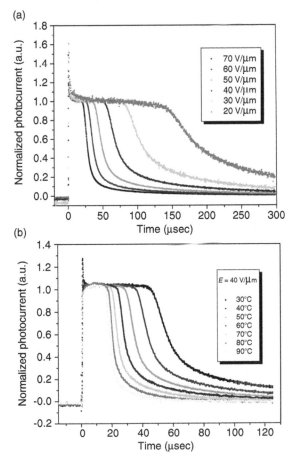

Figure 11.8. (**Color figure follows page 348**). (a) Field dependence of the transient photocurrents measured in a guest–host sample of 4,4′-bis(phenyl-*m*-tolylamino)biphenyl (TPD) doped into polystyrene (1:1 wt.%). (b) Temperature dependence of the transient photocurrents measured in the same sample.

photocarriers must remain small compared to the thickness of the sample, TOF experiments are generally carried out in thick films that are produced by blading thick solutions of the material, or by melting solid material on a hot plate. Furthermore, since the electric field in the sample is considered uniform and, therefore, the total amount of charge in the sample should remain small and injection of carriers from the metal contacts should be minimized.

For many applications in organic electronics (including photovoltaics) the organic films are thin (<1 μm). It is difficult to measure mobility by TOF experiments in these types of systems. Furthermore, charge injection of collection from metal contacts is desired for many applications. Therefore, another technique that is employed is the so-called SCLC technique in which the mobility is derived from the current–voltage characteristics of thin organic films between injecting electrodes. According to SCLC theory [22], the current–voltage characteristics of the sample should be ohmic at low values of electric field. When the injected charge density becomes comparable to the charge density on the electrodes, the field between the electrodes is no longer constant and the current becomes space–charge limited. If the contact is not injection limited and can provide enough charges, a trap-free semiconductor will carry a current described by Child's law [22]:

$$J = \frac{9}{8}\varepsilon_0\varepsilon_r\mu\frac{V^2}{L^3} \tag{4}$$

where J is the current density (A/m^2), V is the bias voltage between the electrodes (V), ε_0 is the free-space permittivity (8.85×10^{-12} F/m), ε_r is the dielectric constant of the material, L is the film thickness (m), and μ is the mobility (m^2/V s). This expression is derived for materials in which the mobility is independent of the electric field. As discussed above, in amorphous organic materials, the charge mobility is often well-described by the disorder formalism (see Equation (3)) and has a functional dependence on the electric field of the simplified form:

$$\mu = \mu_0 \exp(\gamma\sqrt{E}) \tag{5}$$

where μ_0 is the charge mobility at zero electric field and γ is a constant. In this case, the expression of the SCLC can be approximated by [23]:

$$J \cong \frac{9}{8}\varepsilon_r\varepsilon_0\mu_0 \exp\left(0.891\gamma\sqrt{\frac{V}{L}}\right)\frac{V^2}{L^3} \tag{6}$$

Values of μ_0 and γ can be estimated by fitting the experimental J–V curves to Equation (6). The knowledge of these parameters determines the mobility according to Equation (5).

Another technique used to measure mobility in organic semiconductors is based on the current–voltage analysis of metal oxide semiconductor field-effect transistor (MOSFET) devices in which the semiconductor is organic. MOSFET structures have become the building blocks of today's semiconductor industry. Hence, their organic-based counterparts, referred to as organic field-effect transistors (OFETs), are the subject of numerous studies [24] and have become a convenient means of screening the electronic properties of organic semiconductors. The geometries used to fabricate OFETs are shown in Figure 11.9. In these structures the current flowing between the source and drain electrodes is modulated by applying a voltage to the gate electrode. Two different fabrication processes are generally employed: (i) the top-contact approach, in which the source and drain electrodes

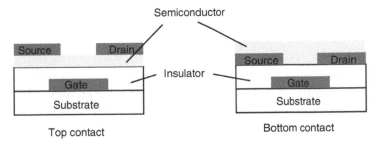

Figure 11.9. Schematics of a field-effect transistor geometry used for the measurement of charge mobility.

are deposited on the organic semiconductor, and (ii) the bottom-contact approach, where the deposition of the organic semiconductor is the last step in the process. The evaluation of carrier mobility in these geometries relies on the approximation that the current–voltage characteristics of these OFETs can be described by the relationships established for traditional MOSFET structures. One distinguishes the current–voltage response at low drain voltage V_D, where the response is linear, and the high-voltage response, in which the drain current I_D is saturated. In the linear regime, the current–voltage response is given by:

$$I_D = \frac{W C_{ox} \mu}{L} \left(V_G - V_T - \frac{V_D}{2} \right) V_D \qquad (7)$$

where W is the channel width, L is the distance between source and drain electrodes (channel length), C_{ox} is the capacitance per unit area of the insulator, V_T is the threshold voltage, V_G is the gate voltage, and μ is the "effective" field-effect mobility, which can be calculated in this regime from the transconductance defined by:

$$g_m = \left. \frac{\partial I_D}{\partial V_G} \right|_{V_D = \text{const.}} = \frac{W C_{ox}}{L} \mu V_D \qquad (8)$$

Likewise, for large drain voltages, the saturated drain current $I_{D\,sat}$ is given by the so-called "square-law":

$$I_{D\,sat} = \frac{W C_{ox} \mu}{2L} (V_G - V_T)^2 \qquad (9)$$

In this regime, mobility can be extracted from the slope of the plot of the square root of the drain current versus gate voltage. Mobility values obtained from such indirect measurements are approximate since the analysis is based on numerous approximations: (i) any field dependence of the charge mobility has not been taken into account in the derivation of Equations (7) and (9); (ii) the mobility measured might reflect that of a thin layer near the oxide surface that might not reflect the morphology of the entire thin film; (iii) the mobility in MOSFET structures is known to be an "effective" mobility because it is not only limited by the lattice scattering and impurity scattering also present in the bulk, but is also lowered by additional surface scattering mechanisms; and (iv) the contact resistance at the source (drain)–organic semiconductor interface might be high and modify the voltage drop across the channel.

Finally, charge mobility in liquid crystals can also be determined using PR-TRMC. In these experiments, high-intensity electron pulses (nsec) produced by an accelerator irradiate the organic semiconductor and lead to the transient formation of electron–hole pairs through ionization. Mobile carriers give rise to a change in conductivity in the sample at microwave frequencies given by

$$\Delta\sigma(t) = e \sum N_i(t)\mu_i \tag{10}$$

where e is the elementary charge, $N_i(t)$ is the time-dependent concentration of a given charged species i, and μ_i is its mobility. From this change in conductivity, the sum of hole and electron mobilities can be deduced after making several assumptions about the number of free carriers produced and yields of geminate recombination. Furthermore, one should note that the mobility values obtained using this technique should be considered intrinsic and reflect trap-free transport in organized domains within the material or on isolated polymer chains in dilute solutions [25]. Mobility values obtained from such indirect measurements are indicative of the maximum value that can be obtained in perfectly oriented materials without defects or grain boundaries.

Due to the differences in sample geometries and assumptions required to derive mobility values, significant discrepancies are anticipated between mobility values measured using the various techniques described above in liquid-crystal-based films in which control of order is crucial. For instance, PR-TRMC electron and hole mobilities measurements on isolated polymer chain of poly(phenylene vinylene) yielded mobilities of 0.5 and 0.2 cm^2/V s, respectively, which are several orders of magnitude higher than those measured in bulk films of the same material [26].

11.4. SEMICONDUCTING LIQUID CRYSTALS

Liquid crystals have been known to the scientists since the end of the 19th century; Austrian botanist Friedrich Reinitzer (1857–1927) first noted in 1888 that cholesteryl benzoate appeared to have two distinct melting points. The term "liquid crystal" was introduced by German physicist Otto Lehmann, who determined that some molecules do not melt directly, but instead first pass through a phase in which they have the ability to flow like a liquid, while retaining some of the molecular order and associated optical properties of a solid crystal. It was only in the late 1960s that liquid crystals started to be used in display applications due to their electro-optic properties. Today, these materials are widely used in all kinds of display applications from laptop computers to in-car navigation screens, mobile phones, personal TVs, and liquid-crystal display (LCD) projectors. The liquid crystals used in these applications are required to be good insulators. It was not until the 1990s that liquid crystals started to be recognized as potential photoconductors.

In this section, we will first review the fundamentals of liquid crystals to familiarize the reader with the different building blocks and the phases that they form. Then, we will review the properties of photoconducting liquid crystals [27,28] that were developed in recent years with an emphasis on discotic liquid crystals [29].

11.4.1. Fundamentals of Liquid Crystals

Liquid crystals, also referred to as mesophases, are a phase of matter that is between a liquid and a crystalline solid [30–36]. In a solid crystal, molecules exhibit positional and orientational order that in most of the cases leads to anisotropic optical and electrical properties, while liquids are usually isotropic. In the liquid crystalline phase, long-range orientational order can lead to physical properties that are anisotropic and resemble those of crystalline solids. The degree of long-range orientational order is measured by the order parameter S defined as

$$S = \frac{1}{2} \langle 3 \cos^2 \theta - 1 \rangle \tag{11}$$

where θ is the angle formed between individual molecules and their direction of orientation. The brackets denote a statistical average. For a perfect crystal $S = 1$, for an isotropic liquid $S = 0$, and for most liquid crystals the order parameter is in the range $0.4 < S < 0.6$.

Liquid crystalline materials are classified by considering the mechanism for mesophase formation, molecular structure, or the liquid-crystal phases formed and their respective symmetry. The two mechanisms for the formation of mesophases are dependent on either the concentration in solution or the temperature. Mesophases that result from the concentration of a material in a suitable solvent are referred to as lyotropic, while those formed as a function of temperature are called thermotropic. The latter are the focus of most research and development for optical and electronic applications. Thermotropic mesophases only exist over well-defined temperature ranges. Above a given temperature, commonly referred to as the clearing temperature, the material will become an isotropic fluid and all order is lost. At low temperature, the material forms a crystalline solid. Between these temperatures the material can form different liquid crystalline phases with increasing symmetry as the temperature is decreasing. Within thermotropic materials, liquid-crystal-forming molecules, also called mesogens, are classified as calamitic or discotic according to whether they are rod-shaped or disc-shaped, respectively. For calamitic materials the three most common mesophases formed are the nematic, cholesteric, and smectic. The nematic mesophase (labeled N) is by far the most common; these are the materials used for making twisted nematic and super-twisted nematic LCDs. Figure 11.10 shows an example of a molecule that forms a nematic phase in the temperature range $24°C < T < 35°C$. For temperatures above 35°C, the material is an isotropic liquid (labeled I) and at temperatures below 24°C, it forms a polycrystalline solid (labeled K). The cholesteric (or chiral nematic) phase is similar to the nematic phase but with the addition of chirality due to the director following a helical path through the bulk. This helix is defined by a characteristic pitch, the distance over which the director undergoes a 2π rotation. When this pitch is of the same order of magnitude as optical wavelengths, the material exhibits selective reflection properties. The smectic phase is the most ordered of the calamitic phases and is characterized by a layered structure. There is a high degree of variation within the smectic phase depending upon how the molecules arrange themselves within a single layer and with respect to adjacent layers. Schematic representations of calamitic nematic and smectic phases are shown in Figure 11(a) and (b).

Discotic materials generally fall into one of the two distinct phases, the discotic nematic phase or the columnar phase (see Fig. 11(c) and (d)). The discotic nematic phase (Figure 11(c)) is similar to the calamitic nematic phase where the disc-shaped molecules all tend to lie within the same plane but their distribution in one plane relative to another is random. In the columnar phase (Figure 11(d)), molecules exhibit a higher degree of order, arranging themselves into columns which can then be arranged within a two-dimensional lattice. It is this columnar phase which is of particular interest for OPVs as charge mobilities along the direction of the columns

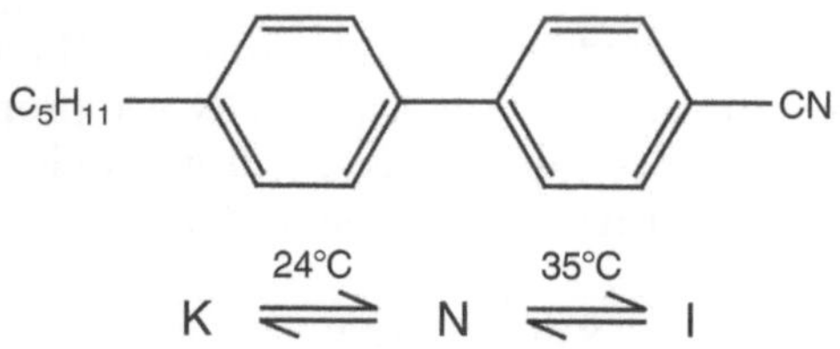

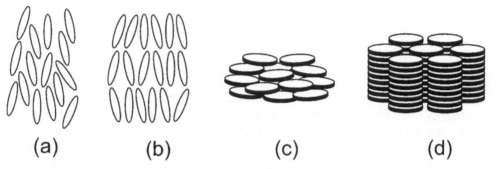

Figure 11.10. Chemical structure of pentylcyanobiphenyl (5CB) that forms a nematic phase in the temperature range 24°C < T < 35°C. For temperatures above 35°C, the material is isotropic and for temperature below 24°C is forms a solid.

Figure 11.11. Schematic representations of the order in different phases of liquid crystals: (a) nematic, (b) smectic, (c) discotic nematic, and (d) discotic columnar.

of these materials can potentially be high due to large electronic coupling between adjacent molecules.

11.4.2. Transport in Liquid Crystals

Early investigations of the transport properties of calamitic nematic liquid crystals provided evidence for the migration of photogenerated ions and ionic impurities under the influence of an applied field. This mass transport led to low mobility values. Ionic transport was considered a nuisance and efforts were geared towards minimizing this conduction to make liquid crystals compatible with active-matrix display technologies. Electronic charge transport in liquid crystals was first reported in Haarer et al.'s work on discotic liquic crystals in 1994 [37] and was followed by a report of electronic mobility in calamitic smectic liquid crystals by Funahashi and Hanna [38]. Since that seminal work, various new photoconducting liquid crystals have been developed both for electron and hole transport [27]. In the following, we discuss the hole, electron, and ambipolar transport properties of selected examples of calamitic and discotic liquid crystals.

11.4.2.1. Calamitic Materials

In Hanna's pioneering work, electronic conduction was found in the calamitic liquid-crystal phases of 2-(4'-heptyloxyphenyl)-6-dodecylthiobenzothiazole (7O-PBT-S12), the chemical structure of which is shown in Figure 11.12. This molecule forms a smectic A phase in the temperature range 90°C < T < 100°C and exhibits a hole mobility of 5 × 10^{-3} cm^2/V s in that temperature range. These mobility values were measured by TOF experiments and were found to be independent of the applied field. For temperatures

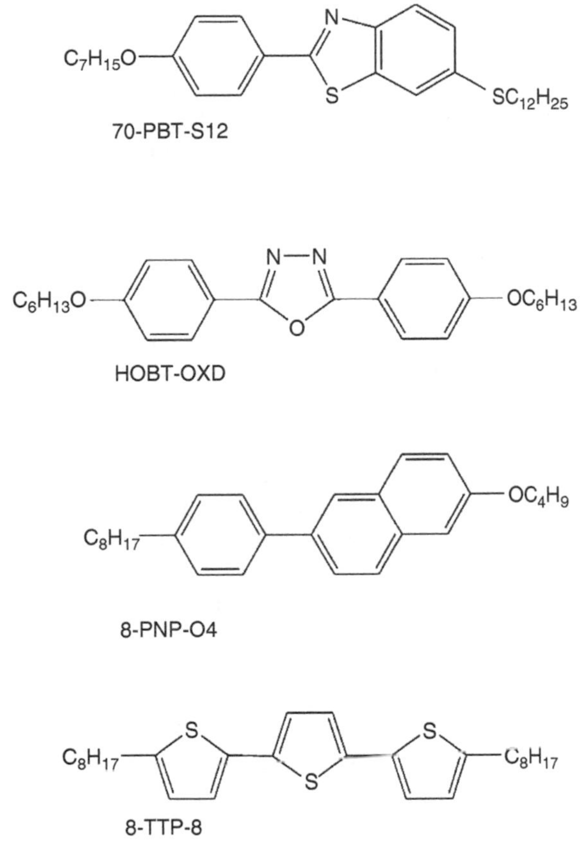

70-PBT-S12

HOBT-OXD

8-PNP-O4

8-TTP-8

Figure 11.12. Examples of calamitic liquid crystals with electronic charge mobility discussed in the text.

below 90°C, where the material is in a polycrystalline phase, charge transport could not be observed and was attributed to trapping at the grain boundaries. Above the clearing point, in the isotropic phase, charge mobilities dropped to 10^{-5} cm^2/V s.

An electron mobility of 8×10^{-4} cm^2/V s for temperatures of $41°C < T < 70°C$ was obtained using TOF experiments in the calamitic liquid-crystalline oxadiazole, hexyloxyphenyl-hexyloxybiphenyl-oxadiazole (HOBP-OXD) [39], also shown in Figure 11.12. With a homogeneous alignment layer, upon cooling, the material underwent a phase transition from a smectic A phase to a highly ordered smectic S_X phase at 70°C and the mobility jumped from a value of 10^{-4} cm^2/V s to 8×10^{-4} cm^2/V s. The mobility was nearly temperature independent within the smectic S_X phase and dropped when the temperature reached 41°C, which is the smectic to crystalline phase-transition temperature. At room temperature, where the material is in a polycrystalline phase, transport was dominated by traps at grain boundaries as discussed above, and transit times could not be determined.

Ambipolar carrier transport properties were investigated by TOF experiments in the different liquid-crystalline phases of 2-(4-octylphenyl)-6-n-butoxynaphthalene [40]. Carrier mobilities were found to increase stepwise when phase transitions took place as the temperature was decreased. The smectic E phase in the range of 55–125°C exhibited nondispersive ambipolar carrier transport with an electron and hole carrier mobility of 1.0×10^{-2} cm^2/Vs. In the smectic A phase between 125°C and 129°C, mobility decreased to values of 4×10^{-4} cm^2/Vs. Ambipolar transport properties were also found in dialkylterthiophene derivatives (8-TTP-8), the structures of which are shown in Figure 11.12 [41]. The liquid crystal exhibited a smectic G phase between 63.9°C and 72°C, a smectic F phase between 72°C and 87.8°C, and a smectic C phase between 87.8°C and 91.3°C. Electron and hole mobilities measured by TOF were found to increase from 5×10^{-4} to 2×10^{-3} cm^2/Vs when the molecular alignment was changing from smectic C to smectic F, and to 1×10^{-2} cm^2/Vs when changed from smectic F to smectic G. Ambipolar transport with charge mobilities for holes and electrons of 2×10^{-3} cm^2/Vs were also measured at temperatures between 65°C and 95°C by TOF in lamello-columnar mesophases formed by [1]benzothieno[3,2-*b*][1]benzothiophene-2,7-dicarboxylate (BTBT) [42].

In the examples of liquid crystals described above, the highly ordered liquid-crystal phases in which high charge mobility is observed are obtained at elevated temperatures. For device applications such as solar cells, high mobility is required at room temperature and over a wide operating temperature range. One approach to this problem is to develop mesogens that incorporate cross-linkable groups. This way, materials can be processed at elevated temperature to reach an optimum self-assembled structure that can be "frozen-in" by cross-linking the material. Calamitic photopolymerizable liquid crystals based on a penta-1,4-dien-3-yl derivative with polymerizable diene end-groups have shown a room temperature hole mobility of 1.8×10^{-5} cm^2/Vs measured by TOF. Radiation from an argon ion laser at 300 nm was found to photopolymerize and cross-link the diene end-groups of the mesogens. However, mobility could not be measured in cross-linked samples [43].

11.4.2.2.   Discotic Materials

In Haarer's seminal work, the transport properties of 2,3,6,7,10,11-hexahexylthio-triphenylene (HTT6) (see Figure 11.13) molecules were investigated [37]. These disc-like molecules form discotic columnar phases. When the material is cooled down from the isotropic phase (93°C), the liquid crystal forms a columnar hexagonal phase (see Figure 11.11(d)) in which hole mobilities in the range 10^{-3}–10^{-2} cm^2/Vs were measured using TOF experiments. When the material is cooled further (70°C), the molecules in adjacent columns become coupled and form a helical superstructure referred to as an H phase. In this highly ordered phase, hole mobility values of 0.1 cm^2/Vs were found at temperatures between 40°C and 70°C. Below 40°C, the transition from the H phase to the polycrystalline K phase takes place and the transport of charges becomes very dispersive. Transport in discotic liquid crystals was investigated by van de Craats and Warman in a series of derivatives that differ in the nature of their central core, in the type of side chains, and in the element that couples the peripheral chain to the core [44]. Mobilities were measured using PR-TRMC in numerous compounds with triphenylene, porphyrin, coronene mono-imide, azocarboxyldiimido-perylene, phthalocyanine, and hexabenzocoronene (HBC) cores. In phthalocyanine and triphenylene compounds, the nature and the length of the side chains was found to influence the phase-transition temperatures between crystalline, hexagonal columnar, and isotropic phases, but had little influence on the magnitude of the charge mobility. Compounds in which the central core was coupled to the alkyl chains via an oxygen atom generally exhibited lower

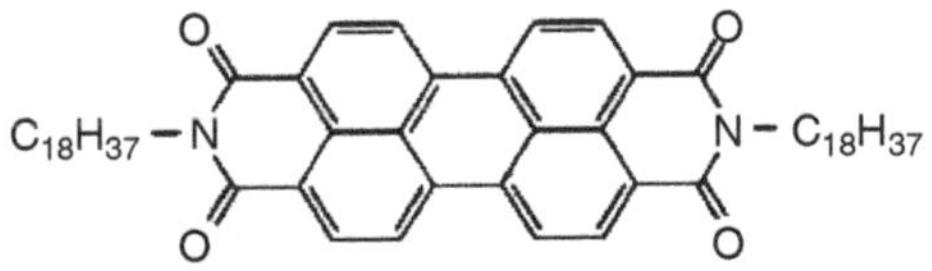

Figure 11.13. Examples of molecules that can form discotic liquid crystalline phases. (a) HTT6 with triphenylene core; (b) mixtures of PTP9 and HAT11; (c) perylene derivative.

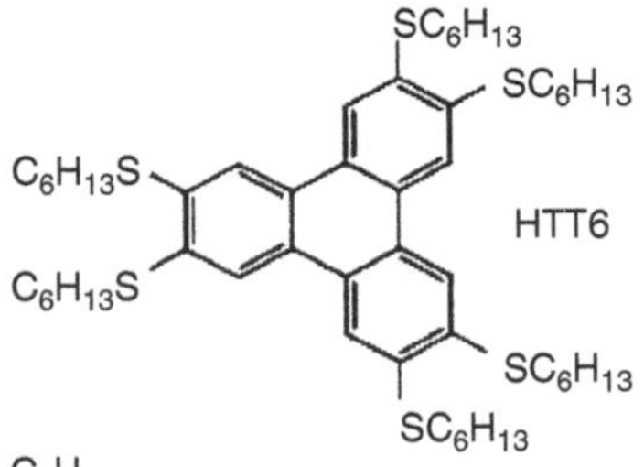

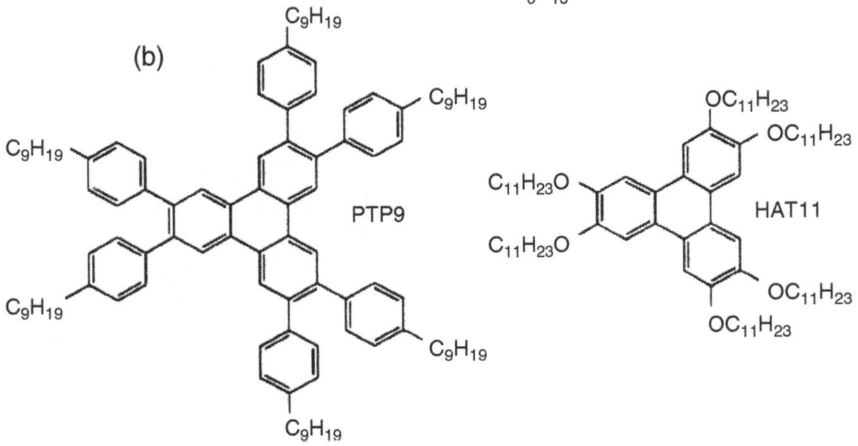

mobility than those where the akyl group was attached directly or via a sulfur or paraphenylene group. This study, conducted on a large number of compounds, showed that the maximum value of the mobility had a tendency to increase with core size. The following empirical formula could be established:

$$\Sigma\mu_i = 3\exp\left[-83/n\right] \text{ cm}^2/\text{V s} \tag{12}$$

where n is the number of second-row elements (C, O, and N) contained in the central core, and $\Sigma\mu_i$ is the sum of intracolumnar electron and hole mobilities. In the hexagonal columnar phase, mobility values between 0.4 and 0.002 cm^2/V s were deduced from the PR-TRMC experiments, with the largest values found for the compounds having a HBC core. These studies suggest that the extent of cofacial contact between the disc-like mesogens controls the transport of charge along the column axis. Brédas and co-workers [16,45] have argued that the proximity and precise location of adjacent π-conjugated molecules can exert significant control over the charge mobility in these materials. According to this hypothesis, it would appear that discotic mesophase materials which maximize the π–π overlap between adjacent mesogens, achievable by perfect cofacial columnar aggregation, will provide the highest charge mobilities along the aggregate column axis. We have recently shown for phthalocyanine materials possessing side chains, which can be polymerized, that conductivities and photoconductivities significantly increase after polymerization, which is also known to lead to an enhancement in the π–π cofacial interactions of these materials [46,47].

Recent studies have shown that mixing discotic mesogens with complementary large-core aromatic compounds can lead to materials with increased order, and consequently higher carrier mobility, compared to single-component systems [48–50]. A 209-fold increase in hole mobility was found in a mesophase formed from mixtures (1:1) of 2,3,6,7,10,11-hexakis-(undecyloxy)-triphenylene (HAT11) with 2,3,6,7,10,11-hexakis-(4-n-nonylphenyl)-triphenylene (PTP9), which is not a mesogen, compared to that in the pure HAT11 mesophase (structures are shown in Figure 11.13). A discotic mesophase is formed at temperatures between 60°C and 170°C. Below 60°C, the mixture forms a glassy phase in which a hole mobility of 1.6 $\times$ 10^{-2} cm^2/V s was measured by TOF experiments at 40°C. Temperature-dependent studies of the mobility in these mixtures showed that the mobility is almost independent of temperature in the range 30–170°C where the material is found in the glassy phase or in a mesophase [51].

Electron transport was measured in N-alkyl-substituted perylene diimides that form highly ordered liquid-crystalline phases, which are both smectic and columnar discotic in nature [52]. Using the PR-TRMC technique, mobility values of 0.1 and 0.2 cm^2/V s were measured in the liquid crystalline and the crystalline phases, respectively, of the perylene diimide compound shown in Figure 11.13(c). Due to the high structural ordering in these materials, the mobility values measured in the liquid crystalline phase and in the crystalline phase differ only by a factor of two. This small difference can be attributed to the technique that is used to measure mobility, providing trap-free values associated with well-organized domains within the material. Note the difference compared with TOF experiments in which large differences in mobility are generally observed between liquid-crystalline and crystalline phases, and in which mobilities are often higher in the liquid-crystalline phases. Furthermore, these materials form liquid crystalline phases only at high temperature (>180°C).

Another building block widely used for electron transport is the oxadiazole molecule. We have recently developed a range of discoid species with benzene or triazine cores and three (trialkoxyaryl)oxadiazole arms [53]. Molecule 1,3,5-tris{5-[3,4,5-tris(octyloxy)phenyl]-1,3,4-oxadiazol-2-yl}benzene (DLCOR), shown in Figure 11.14(a), exhibits a columnar discotic liquid-crystalline (DLC) mesophase between 38°C and ca. 210°C. The optical texture, observed above 38°C, is consistent with a discotic columnar mesophase. The similarities between the optical textures observed at temperature above and below the phase change (Figure 11.14(b)) indicate that the columnar structure of the mesophase is retained in the low-temperature phase (presumably a solid). The electron mobility of a sample cooled to room temperature (ca. 20°C) from above the isotropic melting point was measured in air by the conventional TOF method. Electron mobility was measured at room temperature in air and was found to vary from 10^{-3} to 10^{-4} cm^2/V s, decreasing with increasing electric field as shown in Figure 11.15.

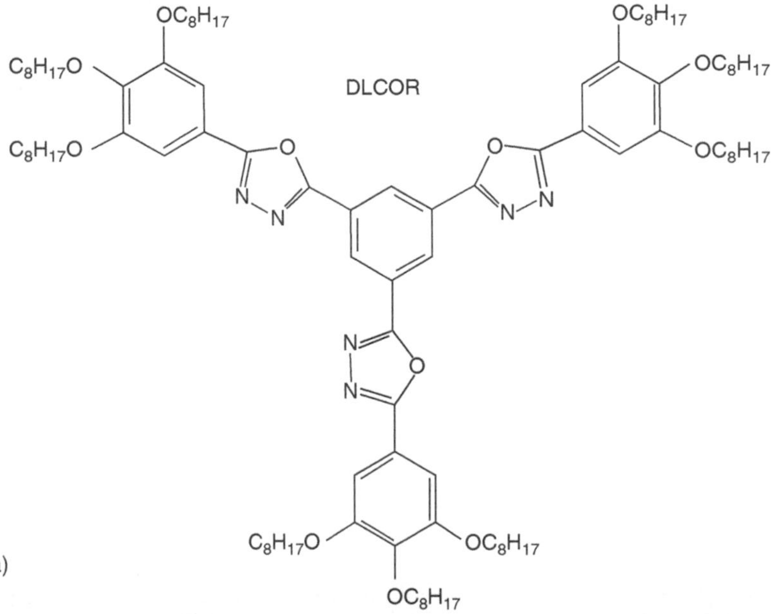

(a)

(b)

Figure 11.14. (**Color figure follows page 348**). (a) Structure of the oxadiazole derivative DLCOR which forms a discotic columnar mesophase and (b) photograph showing the optical texture of an DLCOR sample taken at 35°C during the warming of the sample.

11.4.3. Supramolecular Architectures

Charge transport has also been investigated in supramolecular materials that form columnar mesophases such as phthalocyaninato rare-earth metal sandwich complexes [54], for instance. Larger-scale nanostructures can be formed by self-assembly through a variety of weak interactions such as hydrogen bonding, π–π interactions, dipolar interactions, or donor–acceptor complex interactions. For instance, Percec et al. [55] have developed supramolecular materials that self-organize into homeotropically aligned columnar liquid-crystal phases. The building blocks are dendrons containing donor

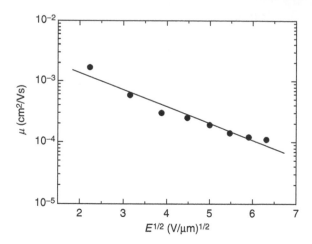

Figure 11.15. Field dependence of the mobility measured by TOF experiments in LC oxadiazole samples at room temperature.

or acceptor groups in the center of the column and fluorinated groups at the periphery. Charge mobilities in these materials were determined by the TOF method. Hole mobilities in the liquid-crystal state were ranging from 10^{-4} to 10^{-3} cm^2/V s, and a mobility of 10^{-3} cm^2/V s was found for electrons in dendrons containing nitrofluorenone.

Self-assembly was also used by Wasielewski et al. [56] to produce ordered nanoparticles that are formed from large molecules in which four perylene molecules with an acceptor character are attached to a central zinc porphyrin electron donor. Such supramolecular structures exhibit the characteristics of artificial photosynthetic systems and can potentially lead to efficient light-harvesting materials.

11.5. OVERVIEW OF LIQUID-CRYSTAL-BASED PHOTOVOLTAIC CELLS

Large-core HBC molecules that form DLC phases were combined with perylene dyes to form photovoltaic cells by Friend et al. [57,58]. Mixtures of the donor-like HBC molecules and the acceptor-like perylene dye (see Figure 11.16) were spin-coated and produced thin films with vertically segregated domains with a large interfacial surface area. When sandwiched between ITO and Al electrodes, blends of 40:60 HBC–perylene formed photodiodes with external quantum efficiency of 34% at 490 nm. The action spectra of the external quantum efficiency showed that efficient exciton dissociation takes place from either of the components with an efficiency for light absorbed by the perylene dye somewhat higher than that by HBC. However, the photovoltaic performance of these devices under standard solar illumination conditions was limited by their small absorption in the red region of the spectrum, with HBC absorbing mainly in the blue region and the perylene dye in the green-yellow region. The largest limitation of this system was the saturation of the photocurrent at light intensities higher than 1 mW/cm^2.

To get larger photocurrents from discotic liquid-crystal solar cells, we recently developed a series of substituted Cu–phthalocyanine (DL-CuPc) molecules that form columnar discotic mesophases. CuPc has been often used in small-molecule OPV cells due to its excellent absorption in the visible spectrum,

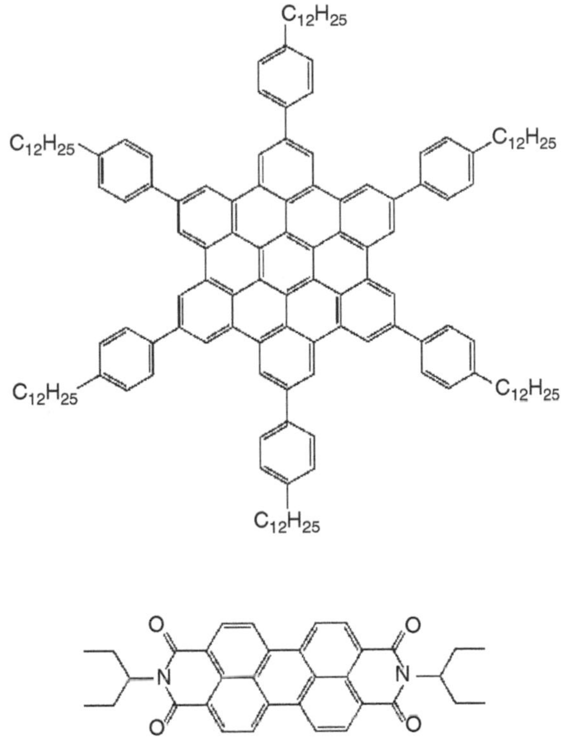

Figure 11.16. Structure of HBC and the soluble perylene dye used in the solar cells of Ref. [57].

relatively good hole-transporting properties, and its thermal stability. The molecule,   2,3,9,10,16,17,23,24-octakis(2-benzyloxyethylsulfanyl)phthalocyanatocopper(II) (DL-CuPc), shown in Figure 11.17, has CuPc in its core with symmetrical substitution of benzyloxyethyl groups linked by sulfur atoms. While the core provides most of the optoelectronic properties, the peripheral group enables the mesogenic properties and solubilities. These molecules exhibit a DLC phase-transition temperature of 134°C (DLC to solid) and are soluble in most common organic solvents. Moreover, this molecule has a high extinction coefficient that is similar to its parent molecule, CuPc. Thermal annealing of these films allows the molecules to self-organize into liquid-crystalline mesophases. To test the photovoltaic properties of these materials, we incorporated them into bilayer geometry with the C_{60} as an electron-transport layer. Devices with structure ITO/PEDOT:PSS (30 nm)/DL-CuPc (20 nm)/C_{60} (30 nm)/bathocuproine (BCP) (10 nm)/Al were fabricated. First, thin films of DL-CuPc layer were deposited on top of the PEDOT:PSS-coated ITO glass by spin-coating from a solution in chloroform. Then, some of the devices were annealed at 180°C, while other were not. Subsequently, C_{60}, BCP, and Al were deposited by vacuum sublimation without breaking the vacuum. The typical active area (A) of these devices was 0.1 cm^2. The device in which the layer of DL-CuPc was annealed exhibited a short-circuit current density (J_{sc}) of 0.42 mA/cm^2 and a fill factor (FF) of 0.39 under white light illumination of 46 mW/cm^2 from a 150 W Xe lamp with AM1.5D correction filters. Under the same conditions, a device with a nonannealed hole-transporting layer yielded short-circuit currents of $J_{sc} = 0.12$ mA/cm^2 and FF $= 0.32$. Increase in fill factor was attributed to a

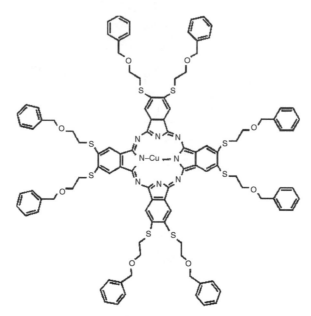

Figure 11.17. Structure of DL-CuPc molecule that forms discotic columnar mesophases.

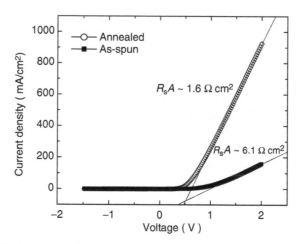

Figure 11.18. Estimation of series resistance from electrical characteristics under illumination. Solid lines are linear fits for $V \gg V_{oc}$. The inverse slope of the linear fit provide $R_s A$.

decrease of series resistance. The values of the products of series resistance and diode area ($R_s A$) were estimated from the inverse slopes of the J–V characteristics measured for $V \gg V_{oc}$ to be 6.1 Ω cm^2 (without annealing) and 1.6 Ω cm^2 (with annealing), respectively (see Figure 11.18). Reduction in the series resistance of devices with annealed films is attributed to an increase in charge mobility associated with a high degree of ordering of the DL-CuPc molecules following the annealing process, assuming that the contact resistance remains unchanged upon annealing. Atomic force microscopy (AFM) images (see Figure 11.19) show that thermal annealing process, consisting of heating at 180°C for 2 h followed by slow cooling

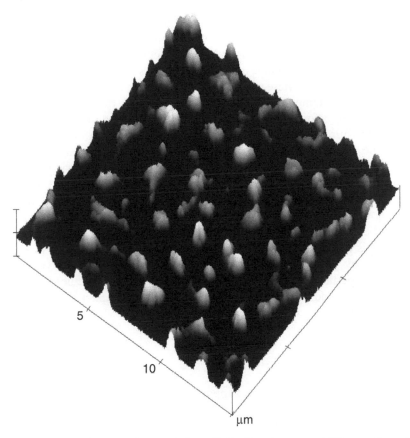

Figure 11.19. AFM image of annealed DL-CuPc film (spun at 4000 rpm, from 8 mg/ml solution) on PEDOT:PSS/ITO. 5 μm/div in *X*, *Y* axes, and 150 nm/div in *Z* axis.

to room temperature, results in 'rough' surfaces with micron- to sub-micron size features. This texturing can explain the increase in short-circuit current density that was observed in cells fabricated from devices with the annealed layer, as the textured surface can contribute to improvement of charge separation due to increased interfacial area.

11.6. CONCLUSION

In the quest for organic solar cells with high efficiency, the development of new materials with optimized optical, electrical, and structural properties plays a central role. The analysis of equivalent circuit models for solar cells teaches us that low series resistance is important to achieve the large fill factors that are required for high conversion efficiency. To minimize series resistance, good ohmic contacts are required between electrodes and high-mobility organic semiconductors. Based on the current level of understanding of charge mobility in these materials, high charge mobility can be achieved in materials that are highly ordered and in which the electronic coupling between adjacent molecular units is optimized. Order with various levels of symmetry can be achieved in liquid crystals and electronic transport has been demonstrated and studied in these materials during the last decade.

Of particular interest are discotic columnar mesophases that can self-organize from solution into columnar stacks in which one-dimensional hole, electron, or ambipolar transport can take place. Measurements of upper values for charge mobility in these materials have shown that values approaching 1 cm^2/Vs can, in principle, be achieved. These mobility values are three to four orders of magnitude higher than those obtained in amorphous films. Despite these advances, numerous challenges remain before these materials can be used at their full potential in solar cells. One hurdle is to control and tailor the orientation of the molecule with respect to the substrate, another is to "freeze-in" the optimized molecular order obtained by self-assembly at a particular temperature by using cross-linking such that this order can be preserved at room temperature and over a wide operational temperature range of the solar cell. During this process, high charge mobility should be preserved. Recent studies show that the orientation of liquid crystal can be controlled and locked into place as exemplified by the optical compensation films developed by Fuji. Surface energies of electrodes and substrates can be controlled by using self-assembled monolayers with varying hydrophobic or hydrophilic properties. Future advances in self-organizing molecules with optimized optical, structural, and electrical properties are likely to impact the field of organic photovoltaics.

ACKNOWLEDGMENTS

This material is based upon work supported in part by the STC Program of the National Science Foundation under Agreement Number DMR-0120967. We also gratefully acknowledge the Office of Naval Research, the National Renewable Energy Laboratory, NASA (through the University of Alabama at Huntsville), and the National Science Foundation for other financial support.

REFERENCES

1. A. Goetzberger and C. Hebling, Photovoltaic materials, past, present, future, *Sol. Energy Mater. Sol. Cells* **62**, 1 (2000).
2. C. W. Tang, Two-layer organic photovoltaic cell, *Appl. Phys. Lett.* **48**, 183–5 (1986).
3. P. Peumans, V. Bulovic, and S. R. Forrest, Efficient photon harvesting at high optical intensities in ultrathin organic double-heterostructure photovoltaic diodes, *Appl. Phys. Lett.* **76**, 2650 (2000).
4. P. Peumans and S. R. Forrest, Very high efficiency double heterostructure copper phthalocyanine/C_{60} photovoltaic cells, *Appl. Phys. Lett.* **79**, 126 (2001).
5. G. Yu, J. Gao, J. C. Hummelen, F. Wudl, and A. J. Heeger, Polymer photovoltaic cells: enhanced efficiencies via a network of internal donor–acceptor heterojunctions, *Science* **270**, 1789 (1995).
6. M. Granström, K. Petrisch, A. C. Arias, A. Lux, M. R. Andersson, and R. H. Friend, Laminated fabrication of polymeric photovoltaic diodes, *Nature* **395**, 257 (1998).
7. S. E. Shaheen, C. J. Brabec, S. Sariciftci, F. Padinger, T. Fromherz, and J. C. Hummelen, 2.5% Efficient organic plastic solar cells, *Appl. Phys. Lett.* **78**, 841 (2001).
8. F. Padinger, R. S. Rittberger, and N. S. Sariciftici, Effects of postproduction treatment on plastic solar cells, *Adv. Funct. Mater.* **13**, 85 (2003).
9. W. U. Huynh, J. J. Dittmer, and A. P. Alivisatos, Hybrid nanorod-polymer solar cells, *Science* **295**, 2425 (2002).

10. M. K. Nazeeruddin, A. Kay, I. Rodicio, R. Humphry-Baker, E. Müller, P. Liska, N. Vlachopoulos, and M. Grätzel, Conversion of light to electricity by cis-X_2bis(2,2'-bipyridyl-4-4'-dicarboxylate)ruthenium(II) charge transfer sensitizers ($X = Cl^-$, Br^-, I^-, CN^- and SCN^-) on nanocrystalline TiO_2 electrodes, *J. Am. Chem. Soc.* **115**, 6382 (1993).

11. A. Hagfeldt and M. Grätzel, Molecular photovoltaics, *Acc. Chem. Res.* **33**, 269 (2000).

12. U. Bach, D. Lupo, P. Comte, J. E. Moser, F. Weissörtel, J. Salback, H. Speitzer, and M. Grätzel, Solid-state dye sensitized mesoporous TiO_2 solar cells with high photon-to-electron conversion efficiencies, *Nature* **395**, 583 (1998).

13. M. A. Green, *Solar Cells*, Prentice-Hall, Upper Saddle River, NJ, 1982.

14. S. J. Fonach, *Solar Cell Device Physics*, Academic Press, New York, 1981.

15. H. S. Rauschenbach, *Solar Cell Array Design Handbook: the Principles and Technology of Photovoltaic Energy Conversion*, Van Nostrand Reinhold Co., New York, 1980.

16. J. L. Brédas, J. P. Calbert, D. A. da Silva Filho, and J. Cornil, Organic semiconductors: a theoretical characterization of the basic parameters governing charge transport, *Proc. Natl. Acad. Sci.* **99**, 5804 (2002).

17. H. Bässler, *Phys. Stat. Sol. (b)* **175**, 15 (1993).

18. P. M. Borsenberger, E. H. Magin, M. van der Auweraer, F. C. de Schryver, *Phys. Stat. Sol. (a)* **9**, 140, (1993).

19. H. Bässler, *Mol. Cryst. Liq. Cryst.* **11**, 252 (1994).

20. P. M. Borsenberger and D. S. Weiss, *Organic Photoreceptors for Imaging Systems*, Marcel Dekker, New York, 1993.

21. F. Closs, K. Siemensmeyer, T. H. Frey, D. Funhoff, Liquid crystalline photoconductors, *Liquid Crystals* **14**, 629 (1993).

22. M. A. Lampert and P. Mark, *Current Injection in Solids*, Academic Press, New York, 1970.

23. P. N. Mergatroyl, Theory of space-charge limited current enhanced by Frenkel effect, *J. Phys. D: Appl. Phys.* **3**, 151 (1970).

24. C. D. Dimitrakopoulos and P. R. L. Malenfant, Organic thin film transistors for large area electronics, *Adv. Mater.* **14**, 99 (2002).

25. R. J. O. Hoofman, M. P. de Haas, L. D. A. Siebbeles, and J. W. Warman, Highly mobile electrons and holes on isolated chains of the semiconducting polymer poly(phenylene vinylene), *Nature* **392**, 54 (1998).

26. P. W. M. Blom, M. J. M. de Jong, and J. J. M. Vleggaar, Electron and hole transport in poly(p-phenylene) devices, *Appl. Phys. Lett.* **68**, 3308 (1996).

27. R. J. Bushby and O. R. Lozman, Photoconducting liquid crystals, *Curr. Opin. Sol. Stat. Mater. Sci.* **6**, 569 (2002).

28. M. O'Neill and S. M. Kelly, Liquid crystals for charge transport, luminescence, and photonics, *Adv. Mater.* **15** 1135 (2003).

29. R. J. Bushby and O. R. Lozman, Discotic liquid crystals 25 years on, *Curr. Opin. Sol. Stat. Mater. Sci.* **7**, 343 (2002).

30. S. Chandrasekhar, *Liquid Crystals*, 2nd ed., Cambridge University Press, Cambridge, 1992.

31. P. Yeh and C. Gu, *Optics of Liquid Crystal Displays*, John Wiley & Sons, New York, 1999.

32. P. G. De Gennes and J. Prost, *The Physics of Liquid Crystals*, 2nd ed., Oxford University Press, Oxford, 2001.

33. L. M. Blinov, *Electro-Optical and Magneto-Optical Properties of Liquid Crystals*, John Wiley & Sons, Chicester, 1983.

34. I. C. Khoo and S. T. Wu, *Optics and Nonlinear Optics of Liquid Crystals*, World Scientific, Singapore, 1993.

35. P. J. Collings and J. S. Patel (eds.), *Handbook of Liquid Crystal Research*, Oxford University Press, Oxford, 1997.

36. B. Bahadur, (ed.), *Liquid Crystals — Applications and Uses*, World Scientific, Singapore, 1990.

37. D. Adam, P. Schuhmacher, J. Simmerer, L. Häussling, K. Siemensmeyer, K. H. Etzbach, H. Ringsdorf, and D. Haarer, Fast photoconduction in the highly ordered columnar phase of a discotic liquid-crystal, *Nature* **371**, 141 (1994).

38. M. Funahashi and J. -I. Hanna, Fast hole transport in a new calamitic liquid crystal of 2-(4′-Heptyloxyphenyl)-6-dodecylthiobenzothiazole, *Phys. Rev. Lett.* **78**, 2184 (1997).

39. H. Tokuhisa, M. Era, and T. Tsutsui, Novel liquid cristalline oxadiazole with high electron mobility, *Adv. Mater.* **10**, 404 (1998).

40. M. Funahashi and J. Hanna, Anomalous high carrier mobility in smectic E phase of a 2-phenylnaphthalene derivative, *Appl. Phys. Lett.* **73**, 3733 (1998).

41. M. Funahashi and J. Hanna, High ambipolar carrier mobility in self-organizing terthiophene derivative, *Appl. Phys. Lett.* **76**, 2574 (2000).

42. S. Mery, D. Haristoy, J. F. Nicoud, D. Guillon, S. Diele, H. Monobe, and Y. Shimizu, Bipolar carrier transport in a lamello-columnar mesophases of a sanidic liquid crystal, *J. Mater. Chem.* **12**, 37 (2002).

43. P. Vlachos, S. M. Kelley, B. Mansoor, and M. O'Neill, Electron-transporting and photopolymerisable liquid crystals, *Chem. Commun.* 874 (2002).

44. A. M. van de Craats and J. M. Warman, The core-size effect on the mobility of charge in discotic liquid crystalline materials, *Adv. Mater.* **13**, 130 (2001).

45. J. Cornil, V. Lemaur, J. P. Calbert, and J. L. Brédas, Charge transport in discotic liquid crystals: a molecular scale description, *Adv. Mater.* **14**, 726 (2002).

46. C. Donley, W. Xia, B. Minch, R. A. P. Zangmeister, A. S. Drager, K. Nebesny, D. F. O'Brien, and N.R. Armstrong, Thin films of polymerized rodlike phthalocyanine aggregates, *Langmuir* **19**, 6512 (2003).

47. C. Donley, R. A. P. Zangmeister, W. Xia, B. Minch, A. S. Drager, S. K. Cherian, L. LaRussa, B. Kippelen, B. Domercq, D. L. Mathine, D. F. O'Brien, and N. R. Armstrong, Anisotropies in the electrical properties of rod-like aggregates of liquid crystalline phthalocyanines: dc conductivities and field-effect mobilities, *J. Mater. Res.* **19**, 2087 (2004).

48. N. Boden, R. J. Bushby, G. Cooke, O. R. Lozman, and Z. Lu, CPI: A recipe for improving applicable properties of discotic liquid crystals, *J. Am. Chem. Soc.* **123**, 7915 (2001).

49. E. O. Arikainen, N. Boden, R. J. Bushby, O. R. Lozman, J. G. Vinter, and A. Wood, Complimentary polytopic interactions, *Angew. Chem. Int. Ed.* **39**, 2333 (2000).

50. T. Kreouzis, K. Scott, K. J. Donovan, N. Boden, R. J. Bushby, O. R. Lozman, and Q. Liu, Enhanced electronic transport properties in complementary binary discotic liquid crystal systems, *Chem. Phys.* **262**, 489 (2000).

51. T. Kreouzis, K. J. Donovan, N. Boden, R. J. Bushby, O. R. Lozman, and Q. Liu, Temperature-dependent hole mobility in discotic liquid crystals, *J. Chem. Phys.* **114**, 1797 (2001).

52. C. W. Struijk, A. B. Sieval, J. E. J. Dakhorst, M. van Dijk, P. Kimkes, R. B. M. Koehorst, H. Donker, T. J. Schaafsma, S. J. Picken, A. M. van de Craats, J. M. Warman, H. Zuilhof, and E. J. R. Sudhölter, Liquid crystalline perylene diimides: architecture and charge carrier mobilities, *J. Am. Chem. Soc.* **122**, 11057 (2000).

53. Y.-D. Zhang, K. G. Jespersen, M. Kempe, J. A. Kornfield, S. Barlow, B. Kippelen, and S. R. Marder, Columnar discotic liquid-crystalline oxadiazoles as electron-transport materials, *Langmuir* **19**, 6534 (2003).

54. K. Ban, K. Nishizawa, K. Ohta, A. M. van de Craats, J. M. Warman, I. Yamamoto, and H. Shirai, Discotic liquid crystals of transition metal complexes 29: mesomorphism and charge transport properties of alkylthio-substituted phthalocyanine rare-earth metal sandwich complexes, *J. Mater. Chem.* **11**, 321 (2001).

55. V. Percec, M. Glodde, T. K. Bera, Y. Miura, I. Shiyanovskaya, K. D. Singer, V. S. K. Balagurusamy, P. A. Heiney, I. Schnell, A. Rapp, H.-W. Spiess, S. D. Hudson, and H. Duan, Self-organization of supramolecular helical dendrimers into complex electronic materials, *Nature* **417**, 384 (2002).

56. T. van der Boom, R. T. Hayes, Y. Zhao, P. J. Bushard, E. A. Weiss, and M. R. Wasielewski, Charge transport in photofunctional nanoparticles self-assembled from zinc 5,10,15,20-tetrakis(perylenediimide)porphyrin building blocks, *J. Am. Chem. Soc.* **124**, 9582 (2002).

57. L. Schmidt-Mende, A. Fechtenkotter, K. Mullen, E. Moons, R. H. Friend, and J. D. MacKenzie, Self-organized discotic liquid crystals for high-efficiency organic photovoltaics, *Science* **293**, 1119 (2001).

58. L. Schmidt-Mende, A. Fechtenkotter, K. Mullen, R. H. Friend, and J. D. MacKenzie, Efficient organic photovoltaics from soluble discotic liquid crystalline materials, *Physica E* **14**, 263 (2002).

12

Photovoltaic Cells Based on Nanoporous Titania Films Filled with Conjugated Polymers

Kevin M. Coakley and Michael D. McGehee
Department of Materials Science and Engineering, Stanford University, Stanford, CA, USA

Contents

Abstract   While bulk heterojunction photovoltaic (PV) cells made from conjugated polymers and titania have not yet matched the efficiencies observed in other types of organic photovoltaic cells, they do possess several advantages over other organic photovoltaic cells and have the potential to achieve high efficiency if a few key loss mechanisms can be overcome. In this chapter, we describe the work that has been done to fabricate polymer–titania cells and to characterize their performance. In addition, we highlight some of the physical mechanisms responsible for the generation of open circuit voltage (V_{OC}) and the production of photocurrent in these cells. We conclude with an outlook for the future of polymer–titania photovoltaic cells and describe the improvements that will be required to achieve high efficiency.

Keywords   titania, conjugated polymer, photovoltaic cell.

12.1. INTRODUCTION

In a bulk heterojunction photovoltaic cell, two semiconductors with offset energy band edges are patterned around each other at the nanometer length scale

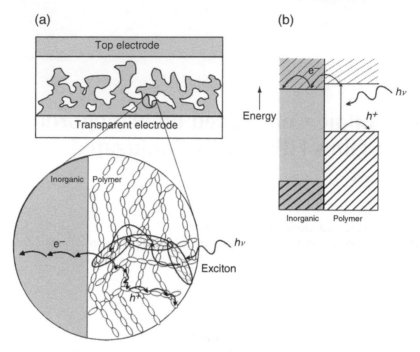

Figure 12.1. Schematic illustration of (a) device geometry and (b) energy band alignment in a polymer and inorganic bulk heterojunction PV cell. Similar cells could be made with two organic components also.

(Figure 12.1). One of the semiconductors is usually a conjugated polymer or some other organic semiconductor. When the organic semiconductor absorbs a photon, a bound exciton is formed. For the cell to function efficiently, the exciton must diffuse to the interface between the semiconductors and be split by electron transfer before it decays. In most organic semiconductors, the distance over which an exciton can diffuse is between 4 and 20 nm.[1–3] After electron transfer occurs, the electron must travel through one semiconductor to an electrode and the hole must travel through the other semiconductor to the other electrode. There are many choices of materials that can be used to accept electrons from typical organic semiconductors, including other organic semiconductors,[4,5] C_{60} derivatives,[6–8] CdSe nanoparticles,[9,10] and titania (TiO_2).[11–14] In this chapter, we will focus on cells made with conjugated polymers and titania. Furthermore, we will focus on cells in which the polymer is designed to absorb light and transport charge, as opposed to cells where a dye is used to sensitize titania and an organic semiconductor is only used to carry holes to an electrode[15–19] or cells where the polymers are only used as a sensitizer.[20] Advantages of titania for bulk heterojunction PV cells are that titania nanoporous films and nanocrystals can be easily fabricated, titania accepts electrons from virtually all organic semiconductors since the bottom of its conduction band lies lower than the lowest unoccupied molecular orbital (LUMO) of almost all organic semiconductors,[1,21,22] titania is nontoxic, many molecules can be attached to the surface of titania, and titania has been studied extensively by those who make dye-sensitized solar cells.[15,23]

There are two general approaches to making bulk heterojunctions with polymers and titania. One is to cast solutions that contain the polymer and titania nanocrystals or a titania precursor,[11,24] and the other is to first make a nanoporous titania film and then fill the pores with the polymer.[11,14,25,26] The first approach is very tempting, because it would be possible to make PV cells at an extremely low cost if one could simply cast the materials and deposit an electrode. Unfortunately, this approach has not resulted in efficient photovoltaic cells because it is difficult to achieve the ideal device architecture by simply blending the two components together. If the two components phase separate, even only slightly, then some excitons in the polymer might not reach the titania before recombining. Furthermore, if some titania nanocrystals are not connected to an electrode through other titania nanocrystals, then electrons can be trapped. For these reasons, we think the second approach is more attractive. In the next two sections of this chapter, we will discuss methods for making nanoporous titania films and filling these films with polymers. We will then address the performance of the PV cells that have been made and conclude with ideas on how to make the cells better.

12.2. NANOPOROUS TITANIA FILMS

Before we look at the methods that have been used to make nanoporous films for PV applications, it is worth thinking about the structure these films should have and the technological constraints on how they can be made. The most obvious design rule is that nanoporous titania films should have pores whose radius is smaller than an exciton diffusion length in the polymer, unless the polymer only coats the walls, in which case the pores could be wider. In addition, the titania films should have a high porosity so that a large amount of polymer can be infiltrated into them. The thickness should be in the range of 100–300 nm, depending on the amount of polymer infiltrated, so that there can be sufficient polymer in the film to absorb most of the incident light. The films should not be thicker than necessary for absorption since extra thickness in a PV cell leads to recombination and increased series resistance. The pores in titania should run straight to the top surface of the films so that holes in the polymer have the straightest possible pathway to the electrode. Furthermore, the films should be made at low cost on plastic substrates, which implies that the deposition and thermal treatments should be done at modest temperatures. In addition to these relatively straightforward design rules, we will explain later in this chapter that it is also important for the pores to have a size and shape that enables the polymer chains to pack in a way that promotes exciton diffusion and charge transport.

No one has made the ideal nanoporous titania film for polymer bulk heterojunction PV cells, but several research groups have taken important steps towards making it. Carter and coworkers[11] made the first polymer–titania PV cell by doctor blading a paste of titania nanocrystals and then sintering the particles together at 500°C. This method had been used extensively for making dye-sensitized solar cells and has now been used by others as well to make polymer cells.[12,25] A scanning electron microscopy (SEM) of a film is shown in Figure 12.2a. Typically the pores have a diameter of approximately 20 nm, but it should be kept in mind that the pores have an irregular shape and that the pore size varies widely. Films are typically 4–6 μm thick. Making films with uniform thickness less than 1 μm with doctor blading is difficult.

Huisman and coworkers[26] have made nanoporous titania films with spray pyrolysis. In this technique, droplets of titanium isopropoxide are formed with an ultrasonic nebulizer and then carried through a furnace with argon carrier gas. Anatase nanocrystals form as the droplets travel through the furnace. As these

(a)   (b)

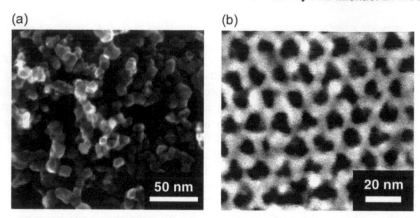

Figure 12.2. (a) Nanocrystalline titania, made by sintering together anatase nanocrystals. The average pore size is approximately 20 nm. (b) Mesoporous titania, made by evaporation induced self-assembly with a structure-directing block copolymer. The average pore diameter is 8 nm. The pores shown in (b) do not go straight down to the bottom of the film.

crystals exit the furnace, they are sprayed onto a substrate, which can be kept in air at room temperature. Since the nanocrystals are still hot when they hit the substrate, they tend to connect to each other. An advantage of this technique over the previous technique is that the substrate, which might some day be made of plastic, does not need to be heated. Another advantage is that the film thickness can easily be varied, even in the submicrometer range.

We have made mesoporous titania films with relatively well-ordered arrays of pores using evaporation induced self-assembly.[13] In this technique, titania ethoxide is mixed with an amphiphilic structure-directing block copolymer.[27,28] When a substrate is pulled from the solution, a thin film forms. As the solvent evaporates, the block copolymer and titania precursor self-assemble into an ordered mesostructure, which has the titania precursor and hydrophilic blocks separated from the hydrophobic blocks. After the films dry, they are heated to temperatures in the range of 350–450°C in order to condense the titania and remove the polymer. Figure 12.2b shows an example of a film with 8-nm diameter pores. The thickness of the films can be easily varied in the range of 50–300 nm by adding solvent to the precursor solution or adjusting the rate at which the substrates are pulled from the solution. Attempts to increase the film thickness beyond 300 nm have yielded films with cracks. Fortunately, 300-nm thick films can take in enough polymer to absorb most of the light that is incident on a PV cell. The films we have made are not perfectly ordered, but nearly have a body centered cubic lattice of ellipsoidal pores. It would be advantageous to have a hexagonal array of cylindrical pores oriented perpendicular to the substrate so that the pores would be as straight as possible. Unfortunately, no one has yet developed a method for making such a structure. When mesostructures with hexagonal arrays of pores have been synthesized, the pores were always found to lie in the plane of the substrate.

12.3. FILLING NANOPORES WITH CONJUGATED POLYMERS

Carter and coworkers[11] made a polymer–titania PV cells by simply spin-casting a solution of the polymer over a film of sintered titania nanocrystals. They inferred that the polymer at least partially penetrated the titania since PV cells made with nanoporous titania were more efficient than ones made with solid titania, but did not report on any measurements that could reveal the extent of polymer penetration. Over the last several years, multiple research groups have tried to infiltrate various polymers into films of sintered titania nanocrystals by spin-casting polymer solutions over the films or by dipping the films in polymer solutions and have found that only certain polymers go into the pores.[29] Recently, Huisman and coworkers[26] introduced a simple method for determining whether or not a polymer reaches the bottom of a film. They spin-cast poly(3-octyl) thiophene (P3OT) over several types of nanoporous films and then heated the films in air at 60°C, which lowers the polymer's resistance, possibly by doping it. When they measured the current–voltage characteristics of diodes made with these films, they observed linear $I-V$ curves if the polymer penetrated all the way to the bottom electrode and nonlinear diode-like curves if the polymer did not reach the bottom. Using this test, they determined that spin-casting P3OT was sufficient to penetrate up to 1 μm of 50-nm nanocrystals deposited by their spray pyrolysis method. They also made films by sintering together 9-nm-diameter and 50-nm-diameter titania nanocrystals. They found that P3OT could penetrate at least 2 μm into the film with 50-nm nanocrystals, but that it could not penetrate 300 nm into the film with 9-nm nanocrystals. This confirms that pore size is an important variable for the infiltration process. For the films made by spray pyrolysis, they measured the absorbance of the film and concluded that only approximately 10% of the pore volume was filled with polymer. Given that the polymer reaches the bottom of the film, but only fills up a small fraction of the volume, we speculate that the polymer probably coats the pore walls.

Our research group has developed other methods for filling nanopores with conjugated polymers and determining the degree of filling. One method is to electropolymerize a monomer, such as 3-methyl thiophene within the pores. We have shown that a 120-nm thick mesoporous film of titania with 8-nm diameter pores can be filled with poly(3-methyl thiophene) in approximately 2 min. After electropolymerization, the films are immersed in an ammonia solution for 2 h to dedope the polymer.[30] A possible problem with cells made by this route is that the polymer touches both electrodes, providing a pathway for dark current that runs in the opposite direction of the desired photocurrent.[11] The other method we have developed for filling pores is melt infiltration, which involves spin-casting a polymer film over a nanoporous film and then heating the film to soften or melt the polymer.[13] Using x-ray photoelectron spectroscopy (XPS) depth profiling (Figure 12.3), we have proven that regioregular poly(3-hexyl thiophene) (P3HT) penetrates all the way to the bottom of 200-nm thick mesoporous films of titania with 8-nm diameter pores when heated at temperatures in the range of 100–200°C. At 200°C, the infiltration only takes several minutes. If shorter heating times are used, it is possible to infiltrate the polymer only part of the way into a film, which is useful for gaining insight on how the cells work.[14] We have also shown that when lower infiltration temperatures are used, less polymer is incorporated into the films. We think the most likely explanation for this observation is that the polymer coats the titania walls and that the coating is thicker at higher temperatures due to increased twisting in the P3HT chains. Blue shifts in the absorption and photoluminescence (PL) spectra (Figure 12.4) with increasing infiltration temperature provide evidence for the twisted conformation of the chains. Later in this chapter, we will discuss the effect that the twisted chain conformation has on exciton diffusion and charge transport.

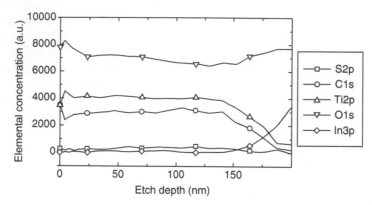

Figure 12.3. X-ray photoelectron spectroscopy (XPS) depth profile of a 180-nm titania film that had been infiltrated with P3HT by heating to 200°C. The carbon profile is flat and reaches the bottom of the titania film, indicating full infiltration of the polymer. In XPS depth profiling, the elemental concentration is measured as the film is sputtered away by an argon ion beam. (Adapted from Coakley, K. M., Liu, Y., McGehee, M. D., Frindell, K. M., and Stucky, G. D., *Adv. Funct. Mater.* **13,** 301 (2003) with the permission of John Wiley & Sons.)

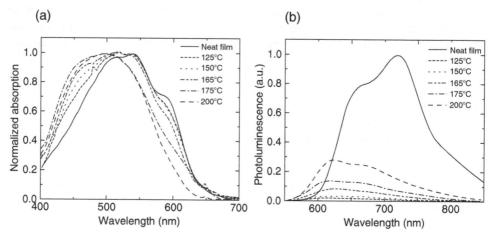

Figure 12.4. Absorption (a) and photoluminescen spectra of P3HT in a neat film and following infiltration into mesoporous titania at various temperatures. All spectra were recorded at room temperature following infiltration and a rinsing step to remove excess polymer. In both (a) and (b), a blue shift in the spectrum is observed when the polymer is infiltrated into the pores, which is an indication that the polymer is coiled and poorly packed. In (b), the PL spectra are normalized by the optical density of the polymer in the film, so that the degree of PL quenching in the polymer can be inferred from the plot. (Adapted from Coakley, K. M., Liu, Y., McGehee, M. D., Frindell, K. M., and Stucky, G. D., *Adv. Funct. Mater.* **13,** 301 (2003) with the permission of John Wiley & Sons.)

The science of infiltrating conjugated polymers into nanopores is still in its infancy. Each research group working on this subject has used different polymers, titania films, and infiltration techniques, so it is difficult to compare the work they have done. The results obtained so far suggest that the polymer's affinity to a titania surface, glass transition temperature, melting temperatures, and molecular weight are important variables, along with the titania pore size and shape. It is still not clear whether it is best to diffuse chains from solution, infiltrate them from a melt (or softened solid), or polymerize them directly in the pores. It is clear, however, that it is possible to incorporate high molecular weight polythiophene into pores as small as 8 nm in diameter with melt infiltration and electropolymerization. In the future, it will be important to determine how the various infiltration techniques affect the polymer chain packing, and how this in turn affects absorption, exciton diffusion, electron transfer, and charge transport.

12.4. PERFORMANCE OF PHOTOVOLTAIC CELLS AND CHARACTERIZATION OF POLYMER–TITANIA FILMS

12.4.1. Photovoltaic Cells with Non-Interpenetrating Semiconductors

In this section, we will review the basic models used to understand single and bilayer polymer PV cells. When an undoped conjugated polymer is sandwiched between two metals with different work functions, the Fermi levels of the metals align and there is a built-in electric field across the entire thickness of the film (Figure 12.5a). The short circuit current density (J_{SC}) is generally quite low since most of the excitons formed in the film decay instead of getting split. In the absence of selective contacts which remove one carrier type more effectively than the opposite type,[31] the V_{OC} can be as large as the difference between the work functions of the metals divided by the charge of an electron, unless the Fermi level of one of the electrodes gets pinned at the highest occupied molecular orbital (HOMO) or LUMO the polymer. At room temperature, V_{OC} is typically slightly less than the electrode work function difference because of dark current.[32]

Photovoltaic cells with two layers of undoped polymers having offset energy bands can have much higher quantum efficiencies for charge collection than single layer cells, since excitons formed near the interface are split by electron transfer. Furthermore, they can also have a larger V_{OC} because electrons selectively diffuse to one electrode and holes selectively diffuse to the other electrode, even if there is no electric field to provide drift current.[31,33] To understand why V_{OC} is larger than the difference of the work functions of the metals divided by the charge of an electron, which is the flat band voltage (V_{FB}), it is helpful to think about why there is a current at V_{FB} even though there is no electric field to cause the charge carriers to drift. At V_{FB} there is a gradient of electron density in the electron accepting polymer and a gradient of hole density in the hole-accepting polymer since electrons and holes are constantly generated at the charge splitting polymer–polymer interface and removed by the electrodes. This gradient causes the charge carriers to diffuse away from the interface to their respective electrodes (Figure 12.5b). To eliminate the current in the cell, the voltage must be larger so that there is a drift current to cancel the diffusion current. Greenham and coworkers[33] have measured the enhancement in V_{OC} that arises from the diffusion current for one combination of polymers and found it to be 1.0 V, which shows that this effect is very important and must be taken into account when modeling heterojunction PV cells.

Bilayer cells made with a titania layer and an undoped conjugated polymer are different from polymer bilayers devices because the titania is usually doped due to oxygen vacancies or impurities. The open circuit voltage appears to be primarily determined by the work function of the top metal electrode and the quasi-Fermi level of the titania, rather than the work function of the transparent conducting oxide

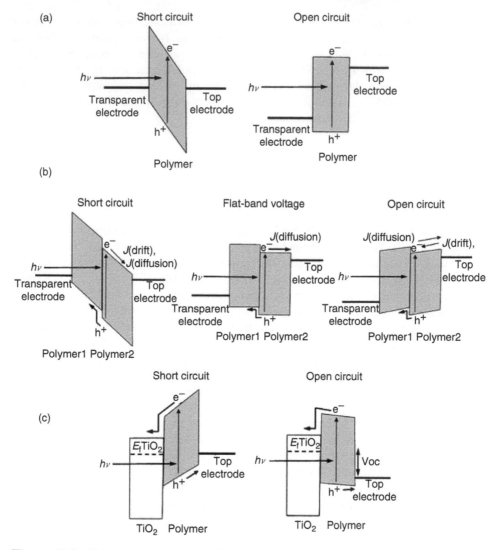

Figure 12.5. Schematic energy band diagrams highlighting the roles of drift and diffusion photocurrents in organic PV cells. (a) A single layer organic PV cell at short circuit and open circuit. (b) An organic bilayer PV cell at short circuit, at the flat-band voltage, and at open circuit. The drift and diffusion currents for electrons are labeled. Similar currents for holes exist also. (c) A titania–polymer bilayer PV cell at short circuit and open circuit. The role of the dark current is not included in these models for the open circuit voltage.

electrode (Figure 12.5c).[3,12] To the best of our knowledge, there is not yet, however, a model that completely describes the dependence of the open circuit voltage on the doping level in the titania, the recombinative dark current, and the drift and diffusion photocurrents. It is clear that excitons are split at the polymer–titania interface, that electrons travel through the titania to the transparent conducting oxide, and that holes travel to the top metal. Most researchers use a high work function metal for the top electrode, such as gold or silver.[3,14,25,26] Silver electrodes probably have a higher work function than one would expect for this metal since they have a tendency to oxidize at the polymer interface.

12.4.2. Interpenetrating Polymer–Titania Nanostructures

As described earlier, Carter and coworkers[11,12] made polymer–titania PV cells by sintering together titania nanocrystals and then spin-casting MEH–PPV on top. They used a gold top electrode to extract holes from the polymer. They also made similar devices with solid titania instead of nanoporous titania. They found that the external quantum efficiency (EQE) at the peak absorption wavelength of the polymer was 2% and 6% for the devices with solid and nanoporous titania, respectively. They attributed the enhancement that arises with the nanoporous titania to the increased interfacial area between the two semiconductors for exciton splitting. The exciton diffusion length in a pure film of MEH-PPV is 20 nm.[1] Thus we would expect that only excitons formed within about 20 nm of the solid titania would be absorbed. With a thick layer of titania filled with polymer, we would expect a significantly larger number of excitons to be split and the efficiency to be much higher. The fact that the efficiency only increased by a factor of three, indicates that either there was very little polymer in the pores, that excitons in the pores were not split, or that charge carriers were not able to escape the region of interpenetrating titania and polymer, due to back electron transfer. Gebeyehu, Sariciftci, and coworkers have made photovoltaic cells with very similar titania, but with polythiophene derivatives instead of MEH-PPV. In some cases they attached a ruthenium dye to the titania before spin-casting the polymer. They observed an energy conversion efficiency of 0.16% under simulated solar irradiation, which is also much lower than one would expect to measure if every exciton were split and every charge carrier collected.

To try and improve the efficiency of polymer–titania bulk heterojunction PV cells and create a system that would be easier to characterize, we made polymer–titania PV cells by melt infiltrating regioregular P3HT into mesoporous titania films, which have reasonably well-ordered arrays of 8-nm diameter pores. As described earlier, XPS was performed to confirm that the polymer penetrates deep into the film. Inganäs and coworkers[2] have shown that the exciton diffusion length of one polythiophene derivative is 5 nm. With this in mind, one might expect that almost every exciton formed in a pore with a radius of 4 nm would diffuse to the interface and be split by electron transfer. To see if excitons are split by electron transfer, we measured the quenching of photoluminescence. Somewhat surprisingly, we found that the photoluminescence inside the pores was not fully quenched. If the polymer was infiltrated at 100°C, 95% of the photoluminescence was quenched. If it was infiltrated at 200°C, only 68% was quenched. We do not yet know if exciton splitting was prevented by insufficient exciton diffusion or slow electron transfer at the interface. When the polymer is infiltrated at higher temperatures, the coating on the walls is thicker and the chains are more twisted. Both of these factors could keep excitons from reaching the interface. The thicker coating results in some excitons getting formed farther away from the titania. The chain twisting could slow down exciton diffusion. An important point from this photoluminescence quenching study is that the exciton diffusion length inside the pores is probably different than it is in a pure film of polymer since the chain packing is different. Despite the fact that not all excitons are split in the polymer-filled titania films, a 68% exciton splitting is still much better than could be accomplished with a bilayer of solid titania and a polymer film thick enough to absorb most of the light incident on a cell.

If a silver electrode is deposited directly onto a polymer-filled titania film, it typically contacts the titania and shorts the device. For this reason, we typically cover the titania film with 30 nm of extra P3HT when making a PV cell (Figure 12.6a).

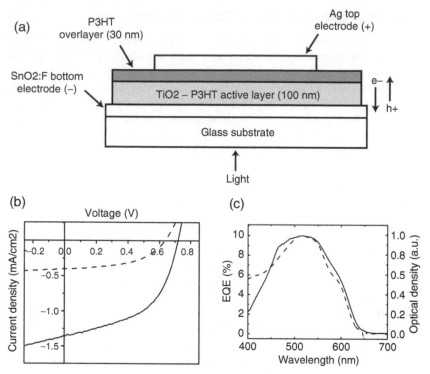

Figure 12.6. Device geometry and photovoltaic performance for a PV cell made using P3HT and mesoporous titania: (a) device geometry; (b) $I-V$ curves under $33\,\mathrm{mW/cm^2}$ 514-nm illumination for PV cells made using mesoporous titania (solid line) and nonporous titania (dashed line); (c) external quantum efficiency (solid line) and optical density (dashed line) for a PV cell made from mesoporous TiO_2 and P3HT. (Reprinted with permission of the American Institute of Physics from Coakley, K. M. and McGehee, M. D., *Appl. Phys. Lett.* **83,** 3380 (2003).).

Figure 12.6b shows the $I-V$ curve for such a PV cell under $33\,\mathrm{mW/cm^2}$ 514-nm monochromatic illumination. Data for a control sample with a solid titania film is also shown. The device with the mesoporous titania is approximately three times more efficient. The EQE corresponding to $1.4\,\mathrm{mA/cm^2}$ photocurrent is approximately 10%. Multiplying the photocurrent by the fill factor (0.51) and open circuit voltage (0.72) of the PV cell yields a power efficiency of 1.5% for the device with mesoporous titania. By integrating the spectral response (Figure 12.6c) of the PV cell over the solar spectrum, we estimate that the power efficiency of the cell would be $0.45 \pm 0.05\%$ under AM1.5 conditions.

In order to determine the relative contribution of excitons formed in the polymer overlayer and in the infiltrated polymer region to the photocurrent, we measured the EQE under 514-nm illumination from a series of PV cells with varying amounts of infiltrated polymer. In these PV cells, the optical density of infiltrated polymer was controlled by infiltrating P3HT into 200-nm titania films for periods of time ranging between 0 and 10 min at 170°C. Although PV cells made from these relatively thick titania films produce smaller photocurrents than PV cells made from thin films, a wider range of infiltrated polymer optical densities can be achieved using thick films. In order to achieve a constant total device thickness following infiltration of the polymer, the remaining polymer overlayer was rinsed off using toluene and replaced

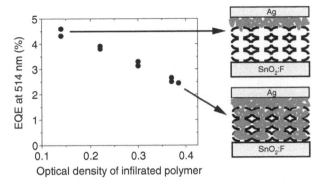

Figure 12.7. External quantum efficiency for illumination at 514 nm of PV cells made from mesoporous titania and P3HT with varying degrees of polymer infiltration. The infiltration depth of the polymer was controlled by heating for 0–10 min at 170°C. For each PV cell the total device thickness was 250 nm. (Adapted from Coakley, K. M. and McGehee, M. D., *Appl. Phys. Lett.* **83,** 3380 (2003) with the permission of the American Institute of Physics.)

by a spin-cast 50-nm P3HT layer. Figure 12.7 shows that there is a monotonic decrease in EQE as the optical density of infiltrated polymer is increased. This is an indication that the charge generation in the cell occurs predominantly near the top of the titania film, and that polymer chains located in the bottom of the titania layer effectively act as a filter that prevents light from reaching the top interface.[34] This strongly suggests that the photocurrent in these cells is limited by the transport of holes to the top electrode of the device, and that holes generated more than 10–20 nm below the top of the titania film undergo back recombination with electrons in the titania before escaping the polymer–titania region.

The key issue for improving these PV cells, and many other kinds of bulk heterojunction cells as well, is getting carriers out of the device before back electron transfer occurs. We do not know the rate at which back electron transfer occurs from mesoporous titania to P3HT, but it is estimated that it takes place in approximately one microsecond based on measurements performed on similar polymers, and on either titania nanocrystals or C_{60} derivatives.[21,22,35] Regioregular P3HT is known to have a hole mobility as high as 0.1 cm^2/V s in field effect transistors (FETs).[36] The high mobility arises because the chains form fairly well-ordered lamellar structures over the gate dielectric. The straightness of the chains promotes charge transfer along the chains and the good π-stacking between chains promotes interchain charge transport.[37] If the mobility were 0.1 cm^2/V s inside the pores, holes would be able to escape the pores before back electron transfer occurred. Unfortunately, absorption and PL spectra indicate that the chains do not π-stack on each other inside the pores,[13] and preliminary space–charge-limited current measurements of the mobility of RR P3HT in mesoporous silica, which has a structure very similar to the mesoporous titania, indicate that the mobility is on the order of 10^{-9} cm^2/V s. This once again shows that the properties of a polymer inside a nanopore are very different than in a pure film.

12.5. FUTURE OUTLOOK

The primary factor limiting the internal quantum efficiency of polymer–titania bulk heterojunction PV cells is poor hole transport in the polymer, which causes holes to recombine with electrons in the titania before escaping the device. This problem can be avoided either by putting a thin barrier layer on the titania to prevent back electron transfer or improving the hole mobility in polymer. Given that the hole mobility of regioregular P3HT in nanopores is extremely low and that this polymer is capable of having a mobility as high as $0.1\,cm^2/V\,s$, it is likely that the easiest pathway to substantially raising the efficiency of the cells will be adjusting the pore size, shape, and surface properties so that the polymer chains can pack the way they do in FETs. It should be possible to increase the hole mobility by up to eight orders of magnitude and enable efficient extraction of charge from polymer–titania films.

A second factor limiting the internal efficiency of the cells is incomplete exciton splitting. Most, but not all, of the excitons in polymer–titania cells with 8-nm-diameter pores are split. Since it might be necessary to increase the pore size in order to give polymer chains room to move around and pack in an ordered way, it could be important to find ways to increase the exciton diffusion length inside the pores. It might also be necessary to use polymers or molecules that attach to the titania in a certain way so that the overlap of wavefunctions will be sufficient for fast electron transfer.

The EQE of the cells is also limited by incomplete absorption of the solar spectrum. So far, most research has been done with polymers that have band gaps of 2.0 eV or larger. Consequently, a significant part of the solar spectrum is not absorbed. If a band gap of approximately 1.4 eV were used, it might be possible to substantially raise the photocurrent and still obtain a high open circuit voltage.

To achieve the maximum possible energy conversion efficiency, it is necessary to generate a large V_{OC}. Approximately 1.0 eV of energy is lost when an electron transfers from the LUMO of P3HT to the bottom of the conduction band of titania. In order to achieve higher V_{OC}, it will be necessary to reduce this energy loss by adjusting the energy levels. This could involve tuning the organic energy levels or switching to other inorganic semiconductors. If other inorganic semiconductors are used, much of what has been learned by studying polymer–titania PV cells will still apply.

REFERENCES

1. Savenije, T. J., Warman, J. M., and Goossens, A., Visible light sensitisation of titanium dioxide using a phenylene vinylene polymer, *Chem. Phys. Lett.* **287**, 148 (1998).
2. Theander, M., Yartsev, A., Zigmantas, D., Sundström, V., Mammo, W., Anderson, M. R., and Inganäs, O., Photoluminescence quenching at a polythiophene/C_{60} heterojunction, *Phys. Rev. B* **61**, 12957 (2000).
3. Arango, A. C., Johnson, L. R., Bliznyuk, V. N., Schlesinger, Z., Carter, S., and Hörhold, H. H., Efficient titanium oxide/conjugated polymer photovoltaics for solar energy conversion, *Adv. Mater.* **12**, 1689 (2000).
4. Halls, J. J. M., Walsh, C. A., Greenham, N. C., Marseglia, E. A., Friend, R. H., Moratti, S. C., and Holmes, A. B., Efficient photodiodes from interpenetrating polymer networks, *Nature* **376**, 498 (1995).
5. Granström, M., Petritsch, K., Arias, A. C., Lux, A., Andersson, M. R., and Friend, R. H., Laminated fabrication of polymeric photovoltaic diodes, *Nature* **395**, 257 (1998).
6. Yu, G., Gao, J., Hummelen, J. C., Wudl, F., and Heeger, A. J., Polymer photovoltaic cells: enhanced efficiencies via a network of internal donor–acceptor heterojunctions, *Science* **270**, 1789 (1995).

7. Shaheen, S. E., Brabec, C. J., Sariciftci, N. S., Padinger, F., Fromherz, T., and Hummelen, J. C., 2.5% Efficient organic plastic solar cells, *Appl. Phys. Lett.* **78,** 841 (2001).

8. Padinger, F., Rittberger, R. S., and Sariciftci, N. S., Effect of postproduction treatment on plastic solar cells, *Adv. Funct. Mater.* **13,** 85 (2003).

9. Greenham, N. C., Peng, X., and Alivisatos, A. P., Charge separation and transport in conjugated-polymer/semiconductor-nanocrystal composites studied by photoluminescence quenching and photoconductivity, *Phys. Rev. B* **54,** 17628 (1996).

10. Huynh, W. U., Dittmer, J. J., and Alivisatos, A. P., Hybrid nanorod-polymer solar cells, *Science* **295,** 2425 (2002).

11. Arango, A. C., Carter, S. A., and Brock, P. J., Charge transfer in photovoltaics consisting of polymer and TiO_2 nanoparticles, *Appl. Phys. Lett.* **74,** 1698 (1999).

12. Breeze, A. J., Schlesinger, Z., Carter, S. A., and Brock, P. J., Charge transport in TiO_2/MEH-PPV polymer photovoltaics, *Phys. Rev. B* **64,** 1252051 (2001).

13. Coakley, K. M., Liu, Y., McGehee, M. D., Frindell, K. M., and Stucky, G. D., Infiltrating semiconducting polymers into self-assembled mesoporous titania films for photovoltaic applications, *Adv. Funct. Mater.* **13,** 301 (2003).

14. Coakley, K. M. and McGehee, M. D., Photovoltaic cells made from conjugated polymers infiltrated into mesoporous titania, *Appl. Phys. Lett.* **83,** 3380 (2003).

15. Bach, U., Lupo, D., Comte, P., Moser, J. E., Weissortel, F., Salbeck, J., Spreitzer, H., and Gratzel, M., Solid-state dye-sensitized mesoporous TiO_2 solar cells with high photon-to-electron conversion efficiencies, *Nature* **395,** 583 (1998).

16. Kruger, J., Plass, R., Cevey, L., Piccirelli, M., and Gratzel, M., High efficiency solid-state photovoltaic device due to inhibition of interface charge recombination, *Appl. Phys. Lett.* **79,** 2085 (2001).

17. Wang, P., Zakeeruddin, S. M., Moser, J. E., Nazeeruddin, M. K., Sekiguchi, T., and Gratzel, M., A stable quasi-solid-state dye-sensitized solar cell with an amphiphilic ruthenium sensitizer and polymer gel electrolyte, *Nat. Mater.* **2,** 402 (2003).

18. Murakoshi, K., Kogure, R., Wada, Y., and Yanagida, S., Solid state dye-sensitized TiO_2 solar cell with polypyrrole as hole transport layer, *Chem. Lett.* **5,** 471 (1997).

19. Smestad, G. P., Spiekermann, S., Kowalik, J., Grant, C. D., Schwartzberg, A. M., Zhang, J., Tolbert, L. M., and Moons, E., A technique to compare polythiophene solid-state dye sensitized TiO_2 solar cells to liquid junction devices, *Sol. Energy Mater. Sol. Cells* **76,** 85 (2003).

20. Kim, Y.-G., Walker, J., Samuelson, L. A., and Kumar, J., Efficient light harvesting polymers for nanocrystalline TiO_2 photovoltaic cells, *Nano Lett.* **3,** 523 (2003).

21. van Hal, P. A., Christiaans, M. P. T., Wienk, M. M., Kroon, J. M., and Janssen, R. A. J., Photoinduced electron transfer from conjugated polymers to TiO_2, *J. Phys. Chem. B* **103,** 4352 (1999).

22. Anderson, N. A., Hao, E., Ai, X., Hastings, G., and Lian, T., Ultrafast and long-lived photoinduced charge separation in MEH-PPV/nanoporous semiconductor thin film composites, *Chem. Phys. Lett.* **347,** 304 (2001).

23. O'Regan, B. and Gratzel, M., A low-cost, high-efficiency solar cell based on dye-sensitized colloidal TiO_2 films, *Nature* **353,** 737 (1991).

24. van Hal, P. A., Wienk, M. M., Kroon, J. M., Verhees, W. J. H., Slooft, L. H., van Gennip, W. J. H., Jonkheijm, P., and Janssen, R. A. J., Photoinduced electron transfer and photovoltaic response of a MDMO-PPV: TiO_2 bulk-heterojunction, *Adv. Mater.* **15,** 118 (2003).

25. Gebeyehu, D., Brabec, C. J., Sariciftci, N. S., Vanmaekelbergh, D., Kiebooms, R., Vanderzande, F., and Schindler, H., Hybrid solar cells based on dye-sensitized nanoporous TiO_2 electrodes and conjugated polymers as hole transport materials, *Synth. Met.* **125,** 279 (2002).

26. Huisman, C. L., Goossens, A., and Schoonman, J., Aerosol synthesis of anatase titanium dioxide nanoparticles for hybrid solar cells, *Chem. Mater.* **15,** 4617 (2003).

27. Alberius-Henning, P., Frindell, K. L., Hayward, R. C., Kramer, E. J., Stucky, G. D., and Chmelka, B. F., General predictive syntheses of cubic, hexagonal, and lamellar silica and titania mesostructured thin films, *Chem. Mater.* **14,** 3284 (2002).

28. Crepaldi, E. L., Soler-Illia, G. J. D. A., Grosso, D., Cagnol, F., Ribot, R., and Sanchez, C., Controlled formation of highly organized mesoporous titania thin films: from mesostructured hybrids to mesoporous nanoanatase TiO_2, *J. Am. Chem. Soc.* **125,** 9770 (2003).

29. Luzzati, S., Basso, M., Catellani, M., Brabec, C. J., Gebeyehu, D., and Sariciftci, N. S., Photo-induced electron transfer from a dithieno thiophene-basedpolymer to TiO_2, *Thin Solid Films* **403,** 52 (2002).

30. Liu, Y., Coakley, K. M., and McGehee, M. D., Electropolymerization of conjugated polymers in mesoporous titania for photovoltaic applications, *Proc. SPIE Meeting* (in press).

31. Gregg, B. A., Excitonic solar cells, *J. Phys. Chem. B* **107,** 4688 (2003).

32. Malliaras, G., Salem, J. R., Brock, P., and Scott, J., Photovoltaic measurement of the built-in potential in organic light emitting diodes and photodiodes, *J. Appl. Phys.* **84,** 1583 (1998).

33. Ramsdale, C. M., Barker, J. A., Arias, A. C., Mackenzie, J. D., Friend, R. H., and Greenham, N. C., The origin of the open-circuit voltage in polyfluorene-based photovoltaic devices, *J. Appl. Phys.* **92,** 4266 (2002).

34. Harrison, M. G., Gruner, J., and Spencer, G. C. W., Analysis of the photocurrent action spectrum of MEH-PPV polymer photodiodes, *Phys. Rev. B* **55,** 7831 (1996).

35. Nogueira, A. F., Montanari, I., Nelson, J., Durrant, J. R., Winder, C., Sariciftci, N. S., and Brabec, C. J., Charge recombination in conjugated polymer/fullerene blended films studied by transient absorption spectroscopy, *J. Phys. Chem. B* **107,** 1567 (2003).

36. Sirringhaus, H., Tessler, N., and Friend, R., Integrated optoelectronic devices based on conjugated polymers, *Science* **280,** 1741 (1998).

37. Sirringhaus, H., Brown, P., Friend, R., Nielsen, M., Bechgaard, K., Langeveld-Voss, B., Spiering, A., Janssen, R., Meijer, E., Herwig, P., and de Leeuw, D., Two-dimensional charge transport in self-organized, high-mobility conjugated polymers, *Nature* **401,** 685 (1999).

13

Solar Cells Based on Cyanine and Polymethine Dyes

He Tian and Fanshun Meng
Institute of Fine Chemicals, East China University of Science & Technology, Shanghai, China

Contents

Abstract As an alternative to heavy metal-based bipyridyl complexes, organic dyes have been studied as sensitizer in dye-sensitized nanocrystalline TiO_2 solar cells (DSSCs), among which cyanine and polymethine dye is a very important class. Remarkable progress has been made for cyanine and polymethine dyes sensitized solar cells in recent years. The formation of H- and J-aggregation of cyanine and polymethine dyes on TiO_2 surface significantly widens the absorption spectra and photoaction spectra of DSSCs in the visible spectra. The structure of cyanine and polymethine dyes affects the adsorption amount on TiO_2 electrode and therefore affects the performance of the solar cells. The dyes with highest occupied molecular orbital (HOMO) level make it possible to use 4-*tert*-butylpyridine to improve the open circuit voltage of DSSCs. The highest overall conversion efficiency of DSSCs based on cyanine and polymethine dyes has reached 7.7%. The potential use of cyanine dyes in thin-film heterojunction all-organic photovoltaic devices has also been investigated.

Keywords solar cell, dye, cyanine, polymethine

13.1. INTRODUCTION

Commercial solar cells are based mainly on inorganic semiconductors such as silicon and gallium arsenide. While these inorganic solar cells have relatively high light conversion efficiencies, they are quite expensive in the market place. As a result, considerable research has been devoted to reduce the cost of inorganic solar cells. Organic materials have been studied as a low-cost alternative since the 1970s. Nevertheless, the power conversion efficiencies of organic solar cells hardly reached 1%. In 1991, Grätzel and coworkers [1] developed dye-sensitized nanocrystalline TiO_2 solar cells based on Ru complex. The dye is directly attached onto the TiO_2 surface through anchor groups such as carboxylic, sulfonic, and phosphoric acids. The nanostructured TiO_2 electrode affords an enormous internal surface area, expanding the optical path length of the surface-anchored dye and the liquid-junction area, thus enhancing the conversion efficiencies greatly. The overall conversion efficiency was as high as over 10% and incident-photon-to-current conversion efficiency (IPCE) almost reached unity [2].

As an alternative to expensive heavy metal-based bipyridyl complexes, organic dyes have also been studied as sensitizer in DSSCs due to their large absorption coefficients, easier preparation, and low cost. Among organic dye sensitizers, cyanine and polymethine dyes are very important ones. They have been used as photographic sensitizers for more than 100 years and are still used because they can sensitize silver halide efficiently upon absorption of light. The high extinction coefficients can make them absorb enough light using thin film, which will benefit the electron transfer and injection into the nanocrystalline electrode. The absorption spectra of cyanine and polymethine dyes can be tuned easily in the whole spectrum by tailoring their structures. Therefore, it is of great interest to study their sensitization of nanocrystalline TiO_2 solar cells. Recently, the performance of cyanine and polymethine dye-sensitized solar cells has been improved rapidly and the highest overall power conversion efficiency has reached 7.7% [3]. In addition, the potential use of cyanine dyes in thin-film heterojunction all-organic photovoltaic devices has also been investigated [4].

13.2. CYANINE AND POLYMETHINE DYE SENSITIZATION

13.2.1. Cyanine Dyes

Cyanine dyes have very large absorption extinction coefficients ($\sim 10^{-5} \, M^{-1} \, cm^{-1}$). However, the absorption peaks of monomers in solution are very sharp and exhibit typical vibronic shoulder absorption. The fluorescent quantum yield of cyanines is usually low and the Stokes shift is relatively small. It seems that cyanine dyes are not suitable for use as solar cell sensitizers, since solar cell materials need to have strong absorption in the whole spectrum in order to harvest the light. Indeed when cyanine dyes are adsorbed on the nanocrystalline TiO_2 film, they can form J- and H-aggregate (about J- and H-aggregate, see Ref. [5]), which broadens the absorption spectra of the electrode in the visible region. So cyanine dyes with carboxyl groups in the alkyl chain (Scheme 1) [6,7] and in the aromatic ring (Scheme 2) [8,9] were synthesized to study their performance in DSSCs.

The structure of cyanine dyes has a great effect on the optical and electrochemical properties of the dye-adsorbed nanocrystalline TiO_2 solar cells. When attached to TiO_2 film, cyanines 1a and 1b form two absorption peaks in the spectra, which are blue- and redshifted compared to the spectra in dilute solution [6]. The two peaks,

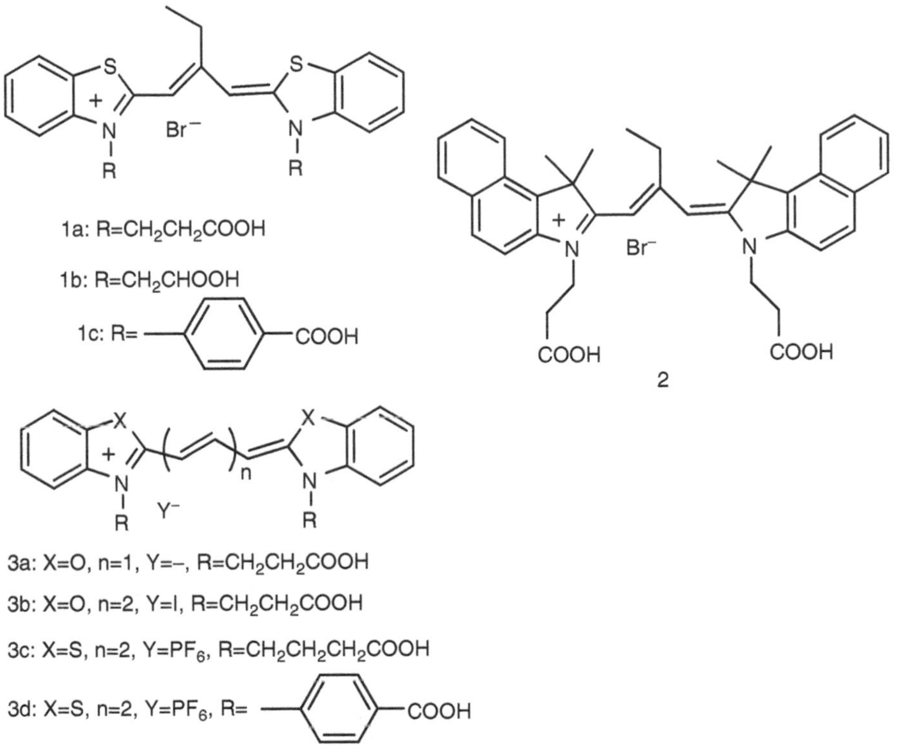

1a: R=CH$_2$CH$_2$COOH

1b: R=CH$_2$CHOOH

1c: R= —⟨benzene⟩—COOH

2

3a: X=O, n=1, Y=–, R=CH$_2$CH$_2$COOH

3b: X=O, n=2, Y=I, R=CH$_2$CH$_2$COOH

3c: X=S, n=2, Y=PF$_6$, R=CH$_2$CH$_2$CH$_2$COOH

3d: X=S, n=2, Y=PF$_6$, R= —⟨benzene⟩—COOH

Scheme 1.

4a: n=1   4b: n=2

5

6

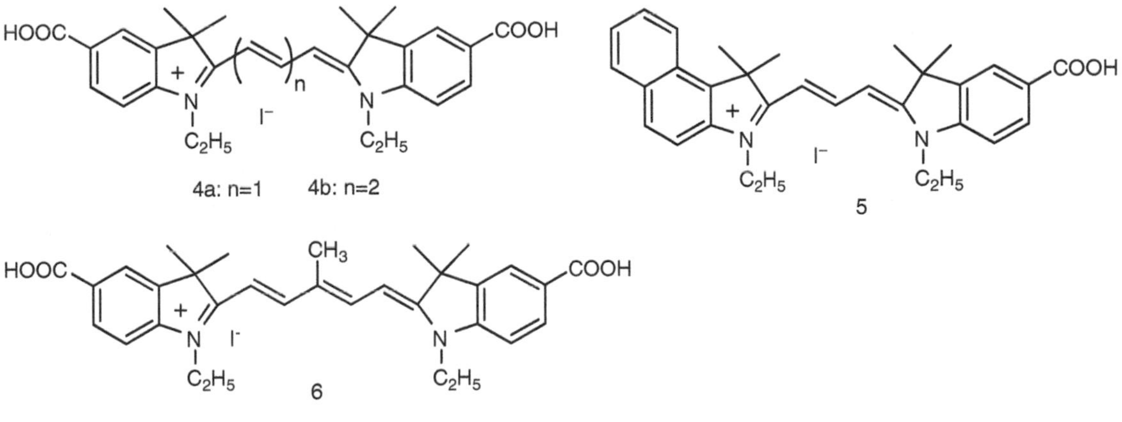

Scheme 2.

corresponding to H-aggregate and monomer absorption, imply the presence of heterogeneous mixture of monomer and aggregate on the surface of the TiO$_2$ film. On the other hand, the bulky methyl benzoic acid of cyanine dyes 1c group prevents the formation of aggregate and the absorption spectrum exhibits only one absorption peak, which represents a redshift compared to its solution. While the spectra of penta-methine cyanine 3c and 3d show predominantly blueshifted peaks and smaller monomer peaks, the featureless aggregate peaks are heterogeneous in nature and represent different kinds of H-aggregates. No J-aggregates were observed with the cyanine dyes bearing carboxyl groups in the alkyl chains. On the contrary, the

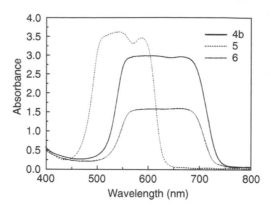

Figure 13.1. The absorption spectra of 4b, 5, and 6 on TiO$_2$ electrode. (From F. S. Meng, Y. J. Ren, E. Q. Gao, S. M. Cai, K. C. Chen, H. Tian, *Proc. SPIE, Vol. 4465 Organic Photovoltaic II*, Z. Kafafi, ed., SPIE, San Diego, USA, 2001, pp. 143–148).

cyanine dyes with carboxyl group in the aromatic rings can form both H- and small J-aggregate except cyanine 4a (Figure 13.1) [8]. In this case, the absorption spectra of the electrode are even broader, which is beneficial for collecting the light. Whether the cyanine dyes are attached to TiO$_2$ surface in any way (monomer, H- and J-aggregates), the photoaction spectra of the cyanine dyes in DSSCs match their absorption spectra well. This means that the sensitization of the TiO$_2$ electrode from all forms of the dye, aggregates or not, can be achieved with equal yields for charge separation and electron injection. The IPCE of cyanine-dye-sensitized nanocrystalline TiO$_2$ solar cells increased with the decreasing distance between the skeleton of the dye and the TiO$_2$ surface [10]. The maximum IPCE for cyanine 3c was measured to be 70% [7] and 73% for cyanine 5 [9]. The maximum overall conversion efficiency of 3.9% was achieved for cyanine 5 sensitized nanocrystalline solar cell under 27 mW/cm^2 white light irradiation [9].

   Although the absorption spectra of cyanine-adsorbed TiO$_2$ electrode are expanded, each of these cyanine dyes absorbs only a fraction of the solar spectrum, which is still far from the requirements for solar cell materials. So a combination of several dyes is necessary to absorb light over the entire visible spectral region. The possibility of sensitization of the electrode with several cyanine dyes was studied [7]. The photoaction spectra can nearly span the whole spectra when using three different dyes and the short circuit current can be improved greatly. For example, the short circuit currents of 3a and 1b co-sensitized solar cells can be improved to 8.3 mA/cm^2 from 2.23 mA/cm^2 for 3a and to 5.9 mA/cm^2 for 1b. Therefore, it is possible to improve the overall conversion efficiency of DSSC through the design of cyanine to form different aggregate and the use of several dyes to co-sensitize the electrode.

   Squarylium cyanine dyes with different attaching groups were synthesized and their performances in DSSCs were studied (Scheme 3) [11]. The dye 7c displays the best values among the three squarylium dyes. The photoelectric conversion efficiency of 7c is 2.17%, which represents an increase of roughly 100% to 200% as compared to that of 7b (1.07%) and 7a (0.84%), respectively. This implies that the adsorption difference of the squarylium dyes on TiO$_2$ electrode have dramatic influence on photovoltaic behaviors. The sensitization performance is improved with the increase in adsorption ability.

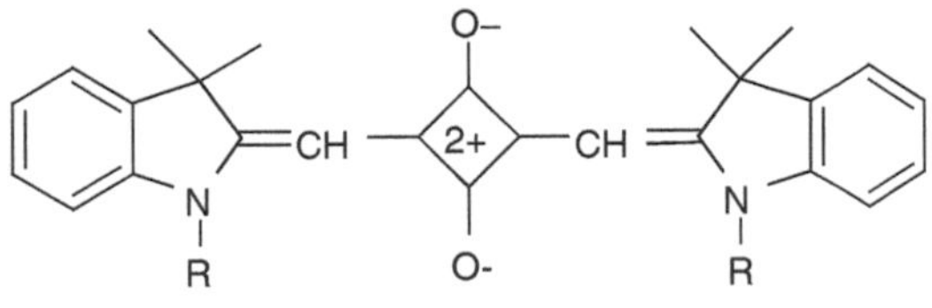

7a: R=CH$_3$

7b: R=CH$_2$CH$_2$OH

7c: R=(CH$_2$)$_3$SO$_3^-$

Scheme 3.

13.2.2. Hemicyanine Dyes

A general method for the synthesis of hemicyanine dyes involves the condensation of a quaternary salt having a reactive methyl group with a substituted aldehyde in solvent in the presence of catalytic amount base (Scheme 4). The hemicyanine dye

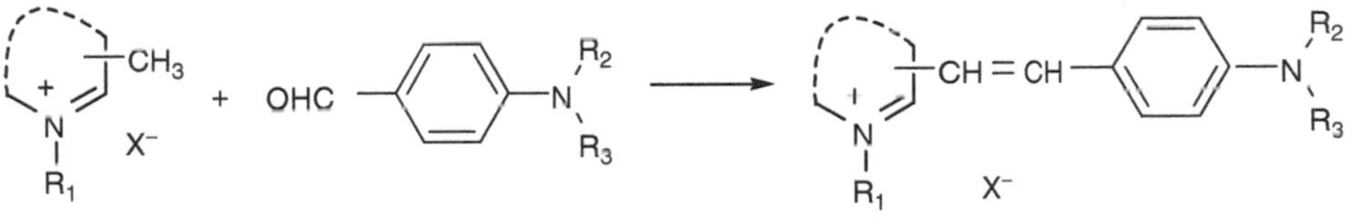

Scheme 4.

is usually composed of a strong donor and an acceptor linked with a π-conjugation bridge. This molecular structure results in a large dipole moment which facilitates efficient charge separation, producing high quantum yield for photon-to-current conversion (The chemical structures and redox properties at some hemicyanine are shown in Scheme 5 and Table 13.1).

The absorption spectra of hemicyanine in solution are broader than that of cyanine dyes, but hemicyanine has no vibronic shoulder absorption. The absorption peaks for hemicyanines 8a, 8b, 9a, 9b, 10a, and 10b attached to TiO$_2$ films are blueshifted compared to their corresponding absorption in solution [12–14]. Even though hemicyanine tends to form H-aggregate either in the presence or absence of external forces, it does not form H-aggregate. These blueshifts could be attributed to the interaction between adsorbed dyes and the TiO$_2$ surface. For better adsorption on the nanocrystalline TiO$_2$ film, hemicyanine dyes with multi adsorbing groups were synthesized [15–18]. In comparison to 8, 9, 10a, and 10b, hemicyanines 10c, 10d, 11, 12, and 14 have broader absorption in the blue and red region of the spectra due to H- and J-aggregates (Figure 13.2). The absorption spectra are structureless and have high absorption in a broad wavelength range, which is beneficial for the light collection. One thing particularly noticeable is that the absorption spectra of hemicyanine with two attaching groups on the sensitized electrode are broader than that of the corresponding hemicyanine with only one attaching group.

The maximum IPCE for 8a, 8b, 9a, and 9b are 38.5%, 17.4%, 33.8%, and 5.7%, respectively [12]. The large differences of these values are due to the differences in

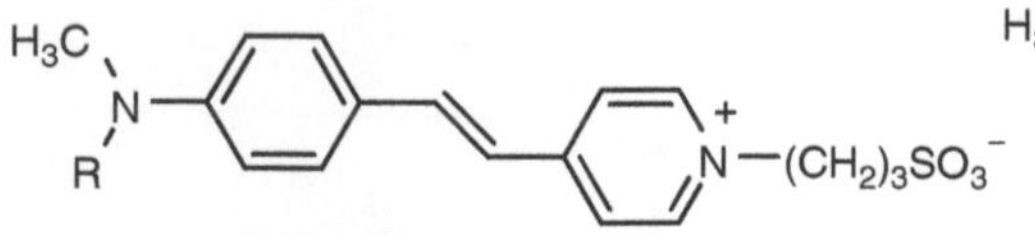

8a: R=CH$_3$

8b: R=C$_{16}$H$_{33}$

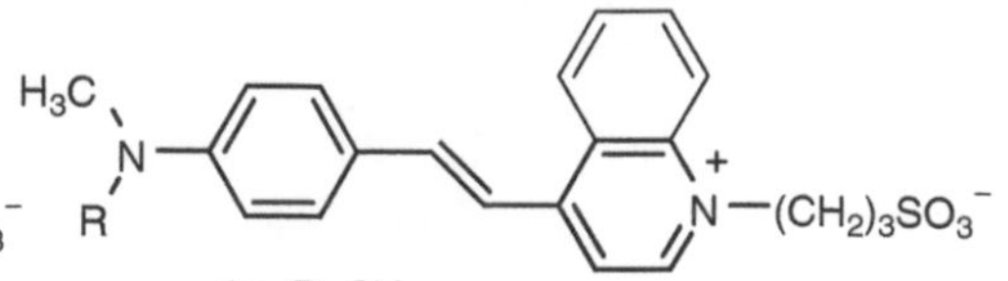

9a: R=CH$_3$

9b: R=C$_{16}$H$_{33}$

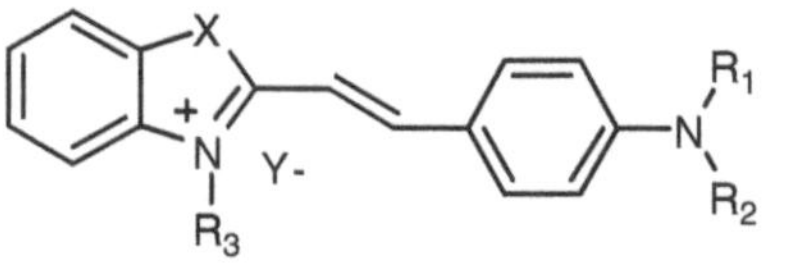

10a: X=S, Y=-, R$_1$=CH$_3$, R$_2$=CH$_3$, R$_3$=(CH$_2$)$_3$SO$_3^-$

10b: X=C(CH$_3$)$_2$, Y=-,R$_1$=CH$_3$, R$_2$=CH$_3$, R$_3$=(CH$_2$)$_3$SO$_3^-$

10c: X=C(CH$_3$)$_2$, Y=I$^-$, R$_1$=C$_2$H$_5$, R$_2$=C$_2$H$_4$COOH, R$_3$=CH$_3$

10d: X=C(CH$_3$)$_2$, Y=I$^-$, R$_1$=C$_2$H$_4$COOH, R$_2$=C$_2$H$_4$COOH, R$_3$=CH$_3$

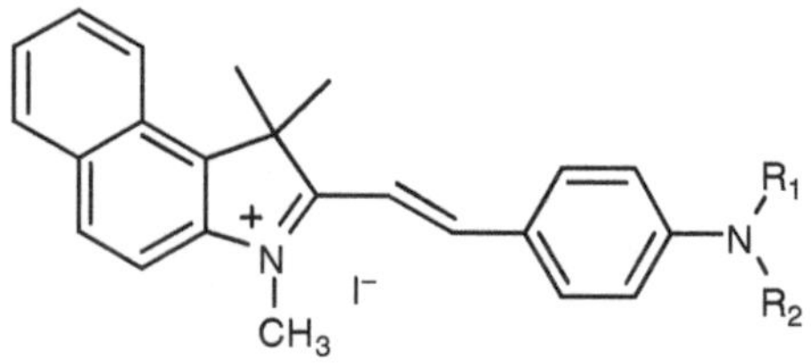

11a: R$_1$=C$_2$H$_5$, R$_2$=C$_2$H$_4$COOH

11b: R$_1$=C$_2$H$_4$COOH, R$_2$=C$_2$H$_4$COOH

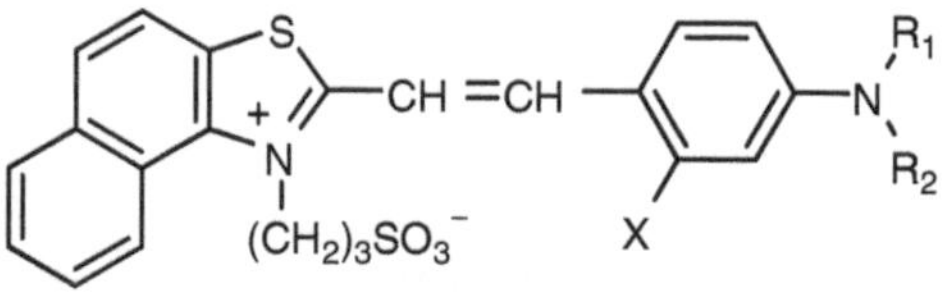

12a: R$_1$=CH$_3$, R$_2$=CH$_3$, X=H

12b: R$_1$=C$_2$H$_5$, R$_2$=C$_2$H$_5$, X=H

12c: R$_1$=C$_2$H$_5$, R$_2$=C$_2$H$_5$, X=OH

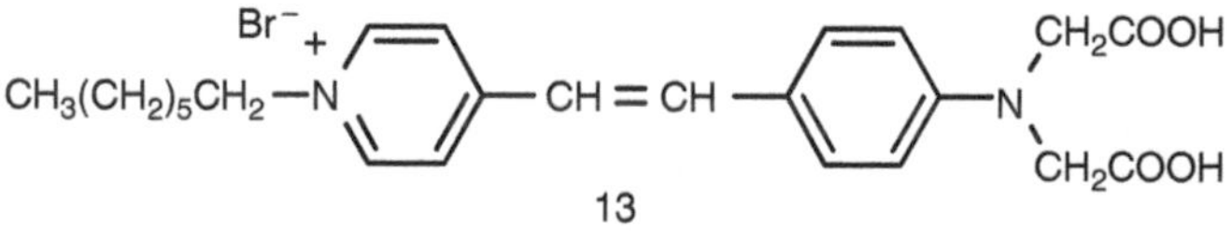

13

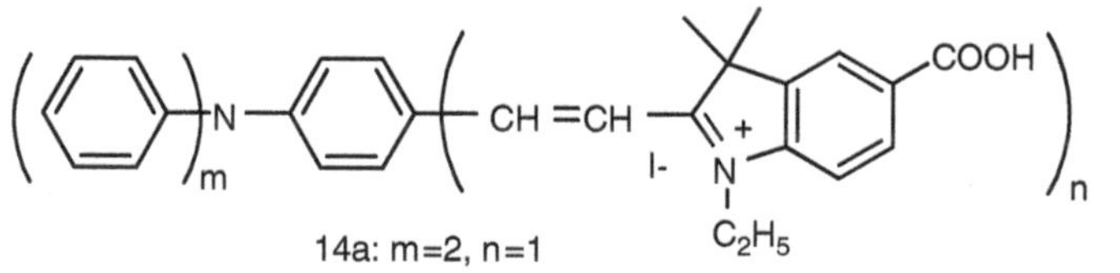

14a: m=2, n=1

14b: m=1, n=2

14c: m=0, n=3

Scheme 5.

Table 13.1. The HOMO and LUMO levels of some hemicyanine dyes

	Compounds			
	10d	11b	12b	12c
HOMO (eV)	−3.9	−3.7	−3.34	−3.17
LUMO (eV)	−5.8	−5.7	−5.54	−5.52

Note: The data of 10d and 11b are from Q. H. Yao, F. S. Meng, F. Y. Li, H. Tian, and C. H. Huang, *J. Mater. Chem.* **13**, 1048–1053 (2003) and 12b and 12c are from Q. H. Yao, L. Shan, F. Y. Li, D. D. Yin, and C. H. Huang, *New J. Chem,* **27**, 1277–1283 (2003).

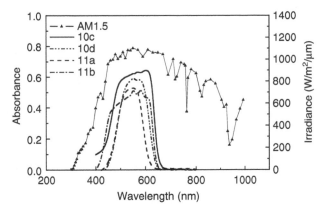

Figure 13.2. The absorption spectra of 10c, 10d, 11a, and 11b on TiO_2 electrode. (From Q. H. Yao, F. S. Meng, F. Y. Li, H. Tian, C. H. Huang, *J. Mater. Chem.* **13**, 1048–1053 (2003)) and the solar spectrum under AM1.5 (from http://www.mse.arizona.edu/~birnie/solar/).

their adsorption in electrode. A long alkyl chain is generally necessary for LB film formation, but for solar cell it is a waste of surface and space because of its large size and small positive effect on photocurrent generation. In addition, the quinoline ring is slightly larger than pyridine ring. This difference in ring size can affect the adsorption of the dye and consequently, the photoelectrochemical performance. Correspondingly, the short circuit currents for 8a and 9a are much higher than that of 8b and 8c. Hemicyanine 8a has the highest short circuit current and overall conversion efficiency (about 2%) in the four dyes due to its higher IPCE values in a broad range. The steric effect on the amount of dyes adsorbed on the electrode and the conversion yield are also observed in other hemicyanine dyes.

Dyes with attached sulfonic acid are adsorbed on the TiO_2 electrode surface through electrostatic interaction between sulfonic group and the surface Ti^{4+} ion. A surface without Ti^{4+} ion may not be beneficial to adsorption, which is one of the key factors limiting the adsorbed amount and conversion efficiency. Hydrochloric acid has been applied to treat the TiO_2 film to improve the conversion efficiency for hemicyanine dyes with sulfonic groups [13,14]. When treated with hydrochloric acid at pH 2 for 2 h, the amount of adsorbed dye molecules increased dramatically. The amount of dye adsorbed increased by 83% and 350% for 10a and 10b, respectively. Both hemicyanine 10a and 10b show outstanding charge transfer properties with nearly 100% IPCE at their maximum absorption wavelength. The short circuit current was improved 99% and 329% for 10a and 10b, respectively. The overall conversion efficiencies were also improved as a result of short circuit current increase: 10a increased by 66% from 3.1% to 5.1% and 10b increased by 260% from 1.3% to 4.8%. After treatment with hydrochloric acid the size of nanocrystalline particles became smaller, increasing the overall surface area of the nanocrystalline electrode. More importantly, a great deal of hydrogen ions was adsorbed on the surface of TiO_2 film. These newly formed surface sites (H^+) are also active for dye adsorption besides the Ti^{4+} adsorbing site. As a consequence, dye molecules can be adsorbed in a more compact way, resulting in a remarkable increase in the amount of adsorbed dyes. In addition, the flat band potential of TiO_2 electrode was shifted more than 200 mV towards the positive direction after treatment with hydrochloric acid. This favors the electron injection from the dye and reduces charge recombination between dye cations and injected electrons.

The attaching group is another factor that affects the amount of adsorbed dyes and photoelectrochemical properties of the DSSCs. The hemicyanine dyes with two carboxyl groups (10d, 11b) absorb two or three times more than the dyes with only one carboxyl group (10c, 11a). The spectra of 10d and 11b on the electrode are broader than that of 10c and 11a. As a result, the DSSCs based on 10d and 11b have broader IPCE spectra with 10d having the higher IPCE values ($>60\%$) in the region from 460 to 640 nm. The short circuit currents of 11b reached a surprisingly high value of 21.4 mA/cm^2 under the irradiation of 90.0 mW/cm^2 white light. The overall conversion efficiencies of DSSCs based on 10c, 10d, 11a, and 11b were 4.0%, 4.6%, 4.4%, and 4.9%, respectively [17]. Compared to the adsorption amount of dyes with short circuit current, carboxyl groups have two different effects on the adsorption and photoelectric conversion properties of the dye. On the positive side, the adsorbed amount of dye molecules on the TiO$_2$ film is greatly increased, which is favorable to the overall light to electricity conversion. On the negative side, the second carboxyl group impedes the charge separation of the molecules because it is attached to the electron donor side and hence reduces the polarity of the molecules to some extent. The hemicyanine dye (12c) bearing sulfonic acid and hydroxyl as adsorbing groups has a remarkably high overall conversion efficiency of 6.3% [18]. Although the introduction of hydroxyl group reduces the amount of the dye adsorbed on TiO$_2$ electrode, it can greatly enhance the electron injection efficiency overcoming the decrease in the amount of adsorption. Hemicyanines 14b and 14c have two and three carboxyl groups in the molecules. However, their large steric hindrance impedes their adsorption on the electrode. The overall conversion efficiency of 14a (2.12%) is much higher than those of 14b (0.85%) and 14c (0.51%) [16].

13.2.3. Merocyanine Dyes

Merocyanine dyes (Scheme 6) [19,20] 15(a–h) have a narrow absorption band in ethanol with the monomer peak located at around 500 nm. When merocyanine dyes are adsorbed on the TiO$_2$ electrode, the absorption spectrum was broadened and the absorption threshold redshifted. The redshift and broadening of the absorption is suggested to be the interaction between neighboring dyes, which forms small J-like aggregate. But the chemical anchoring of the dye on electrode restricted the formation of large and highly oriented J-aggregate.

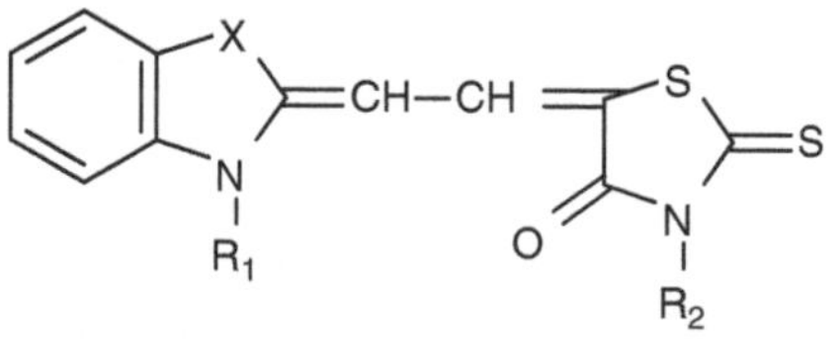

15a: X=CMe$_2$, R$_1$=C$_2$H$_5$, R$_2$=CH$_2$COOH
15b: X=S, R$_1$=C$_2$H$_5$, R$_2$=CH$_2$COOH
15c: X=S, R$_1$=C$_5$H$_{11}$, R$_2$=CH$_2$COOH
15d: X=S, R$_1$=C$_{10}$H$_{21}$, R$_2$=CH$_2$COOH
15e: X=S, R$_1$=C$_{18}$H$_{37}$, R$_2$=CH$_2$COOH
15f: X=S, R$_1$=C$_2$H$_5$, R$_2$=CH$_3$
15g: X=S, R$_1$=CH$_2$COOH, R$_2$=C$_2$H$_5$
15h: X=CMe$_2$ R$_1$=CH$_2$COOH, R$_2$=CH$_2$COOH

Scheme 6.

All merocyanine-containing carboxyl groups adsorb strongly on the TiO_2 surface and show high efficiencies. On the other hand, the merocyanine dye 14f containing no carboxyl group adsorbs scarcely on TiO_2 and the cell efficiency is negligible, suggesting that the presence of absorbing anchoring groups on the semi-conductor surface is essential for efficient dye sensitization. The overall conversion efficiency increases with the increase in side chain length except 14e. It is suggested that the intrinsic sensitization efficiency, i.e., quantum efficiency, increases with the increase in side chain length. This could be due to the fact that the long alkyl chains and aggregate restrict the rotation around the methane chain, thus preventing the isomerization from taking place. This could also be due to repulsion between long alkyl chain and TiO_2 surface, favoring the formation of predominant isomer. Mer-ocyanine 14 g with a carboxyl group at another position shows almost the same efficiency as 14b. Meanwhile, the photocurrent of merocyanine 14 h with two carb-oxyl groups is very low due to restricted configuration and aggregation by two anchors. Under optimized conditions, the maximum solar energy efficiency for DSSC based on 14e can reach 4.2% with open circuit voltage of 0.62 V, short circuit current of 9.7 mA/cm², and fill factor of 0.69.

13.2.4. Other Polymethine Dyes

Arakawa and coworkers [21] designed and synthesized a novel polyene dye that contains a *N,N*-dimethylaniline (DMA) moiety as the donor part and a methine unit connecting the cyano and carboxyl groups as the acceptor part (Scheme 7). The introduction of a DMA moiety and expansion of a methine unit both contributed to a redshift of the absorption spectra. Such redshift in the absorption spectra is desirable for harvesting absorption of the solar spectrum. The reduction potential of 16, 17, and 18 in DMF are −1.10, −0.90, and −1.0 V vs. normal hydrogen electrode (NHE), respectively (Table 13.2). These relatively large energy gaps between the lowest un-occupied molecular orbital (LUMO) of the dyes and the conduction band edge level of TiO_2 electrode enable addition of 4-*tert*-butylpyridine (TBP) in the electrolyte to improve the photovoltage and consequently total efficiency. IPCE values higher than 70% were observed in the range from 460 to 600 nm with a maximum value of 82% at 468 nm for DSSC based on 16. High efficiencies of more than 5% were attained

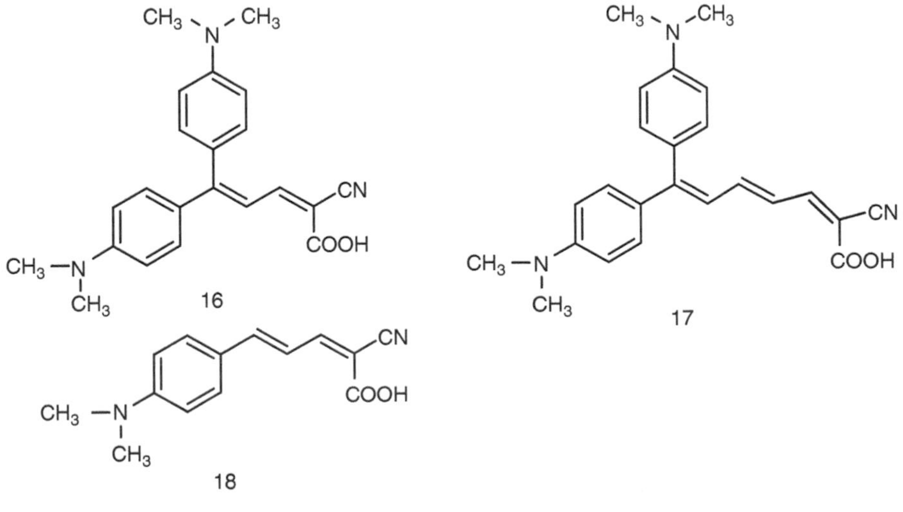

Scheme 7.

Table 13.2. The redox properties of some polymethine dyes (vs. NHE)

Compounds	16	17	19	21a	21c	21d	24	25	27	28
E_{ox} (V)	—	—	1.21	1.35	1.28	1.15	1.18	1.15	—	—
E_{red} (V)	−1.1	−0.9	−1.23	−0.85	−0.82	−0.83	−0.35	−0.57	−0.91	−0.87

Note: The data of 16 and 17 are from K. Hara, M. Kurashge, S. Ito, A. Shinpo, S. Suga, K. Sayama, and H. Arakawa, *Chem. Commun.* 252–253 (2003), 19, 21c, 21d, 24, and 25 are from K. Hara, T. Sato, R. Katoh, A. Furube, Y. Ohga, A. Shinpo, S. Suga, K. Sayama, H. Sugihara, and H. Arakawa, *J. Phys. Chem. B* **107**, 597–606 (2003), and 23, 27, and 28 are from K. Hara, M. Kurashige, Y. Dan-oh, C. Kasada, A. Shinpo, S. Suga, K. Sayama, and H. Arakawa, *New J. Chem*, **27**, 783–785 (2003).

based on dyes 16–18. A maximum η value of 6.8% was achieved under AM 1.5 irradiation based on 17 ($J_{sc} = 12.9$ mA/cm^2, $V_{oc} = 0.71$ V, and ff = 0.74).

Novel coumarin dyes with polymethine chain have been designed and synthesized to absorb in the visible range from 400 to 750 nm (Scheme 8) [22–24]. Compared to the conventional coumarin dye 19, the absorption spectra of dyes 20–23 are

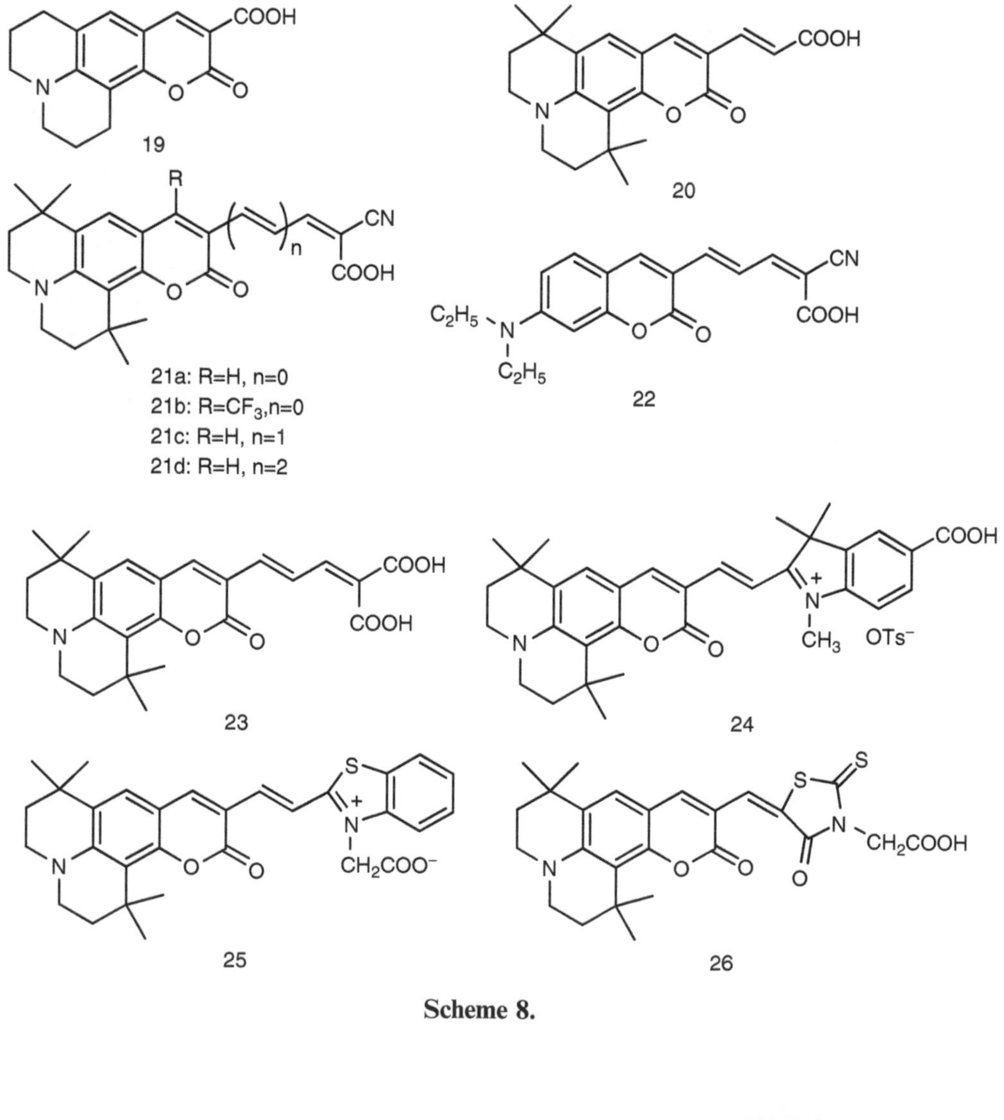

Scheme 8.

considerably redshifted by connecting the cyano acetic acid moiety to the coumarin framework via a methane chain. The introduction of heterocyclic moieties in π-conjugation length with coumarin framework also results in a redshift in the absorption spectra (dyes 24–26) due to their strong electron-withdrawing abilities. The introduction of electron-withdrawing groups (cyano, trifluoromethyl, or heterocyclic) shifted the reduction potential of the dye in the positive direction, while lengthening the methine unit shifted the oxidation potential of the dye negatively. The redshifts associated with the expansion of the methine unit can be attributed to negative shifts in the oxidation potentials of the dyes rather than to positive shifts in the reduction potentials. The absorption spectra of these coumarin dyes on TiO_2 electrode are much broader than the spectra in solution due to an interaction between the dyes and TiO_2.

The IPCE spectra of solar cell based on novel coumarin dyes with polymethine chain are remarkably redshifted compared to the spectra of 19. The onset of the IPCE spectrum of a DSSC based on 21c is 750 nm and high IPCE performance (>70%) was observed in the range from 460 to 600 nm, with a maximum value of 80% at 470 nm. However, the DSSC cells based on 21b and 26 showed lower IPCE with maxima of 42% and 38%, respectively. The short circuit current values for the solar cell depend markedly on the dyes; these values are reflected in their IPCE performances. The short circuit current based on 21c and 21d can reach 15 mA/cm^2. The overall conversion efficiency for a DSSC based on 20c can reach 5.2%. The DSSCs based on dyes 22 and 23 show relatively low open circuit voltage compared to 21c, which results in lower efficiency. A maximum efficiency of 6.0% was achieved by optimization under AM 1.5 with a DSSC based on 21c ($J_{sc} = 14.0$ mA/cm^2, $V_{oc} = 0.60$ V, and ff $= 0.71$) [24].

The effect of the counter cation of the carboxyl group in the coumarin dye system on the photovoltaic performance was investigated by using a proton (H^+), a piperidinium ion, and a triethyl hydroammonium ion ($NHEt_3^+$) as the counter cation for the carboxyl group in dye 21c [23]. The dark currents for DSSCs based on dyes with piperidinium and $NHEt_3^+$ cations are shifted negatively compared to that for a DSSC based on a proton-type dye, which indicates the dark current is suppressed in DSSCs for dyes with piperidinium and $NHEt_3^+$ cations. Consequently the open circuit voltages for these solar cells were increased relative to the values for a DSSC based on a proton-type dye. The ammonium cations exist around the TiO_2 surface after dye adsorption and the resulting suppression of dark current, owing to a blocking effect, leads to an improved open circuit voltage. Although the open circuit voltage can be improved when ammonium cations are employed, the short circuit currents for DSSCs based on dyes with piperidinium and triethyl hydroammonium cations are lower than the values for a DSSC based on a proton-type dye. The dissociation of protons from the carboxyl group shifted the conduction band of TiO_2 positively, thus lowering the open circuit voltage. A positive shift of the conduction band of TiO_2 does not occur in the ammonium cation type dyes and the lack of positive shift leads to a smaller driving force for electron injection from the dye and consequently decreased short circuit current.

In order to further improve the photovoltaic performance of DSSCs, coumarin with thiophene moieties have been synthesized (Scheme 9) [3]. The introduction of the thiophene moieties produced no remarkable changes in the absorption spectra of the dyes in the solution. The maximum absorption wavelengths of 21c, 27, and 28 are 507, 507, and 510 nm, respectively. In contrast, the absorption spectra of 27 and 28 on the TiO_2 surface are remarkably broadened relative to that of 21c. The thiophene

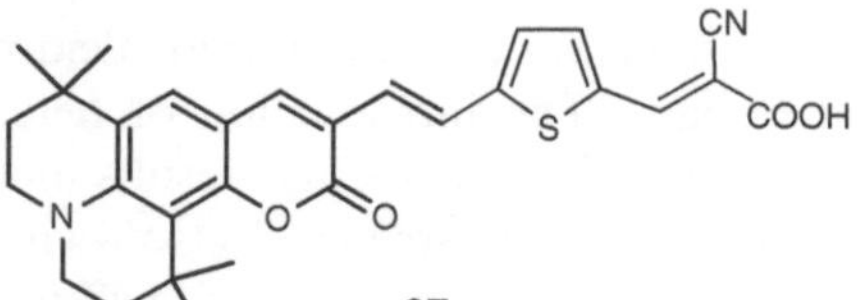

Scheme 9.

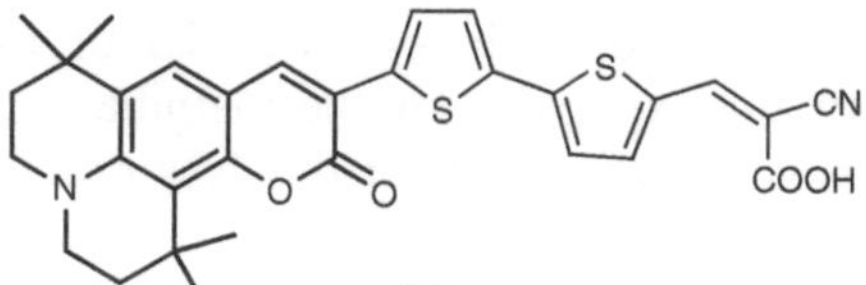

moieties seem to contribute to the broadening of the absorption spectrum of dye on the TiO_2 surface. Correspondingly, the onset of IPCE spectra (850 nm) for DSSCs based on 27 and 28 is redshifted relative to the onset for the DSSC based on 21c (750 nm). The IPCE spectra clearly indicate that introducing thiophene moieties into the methine unit of 21c broadens the IPCE spectrum, leading to an improved photocurrent under white light irradiation. Aside from the broadening of the absorption spectra of the dyes on an electrode, the strong electron donating ability of thiophene moieties makes the reduction potentials of the dyes 27 (-0.91 V) and 28 (-0.87 V) more negative than that of 21c (-0.82 V), which allows the use of TBP to improve the open circuit voltage. An overall conversion efficiency of 7.2% was achieved using 27 under AM1.5 radiation ($J_{sc} = 14.7$ mA/cm^2, $V_{oc} = 0.67$ V, and ff $= 0.73$). The efficiency of the DSSC based on 28 under AM1.5 radiation was as high as 7.7% ($J_{sc} = 14.3$ mA/cm^2, $V_{oc} = 0.73$ V, and ff $= 0.74$), the largest value to date among organic dyes [3].

13.3.   THIN FILM HETEROJUNCTION PHOTOVOLTAIC DEVICES

The potential usage of cyanine dyes in thin film heterojunction photovoltaic devices has been investigated [4]. Heterojunction multilayer devices were made from cyanine Cy-5, polymer M2, and fullerene C_{60} (Scheme 10). The cyanine and polymer were spin coated from solution and C_{60} was evaporated under high vacuum. When the cyanine formed films on ITO glass or polymer, cyanine monomer absorption around

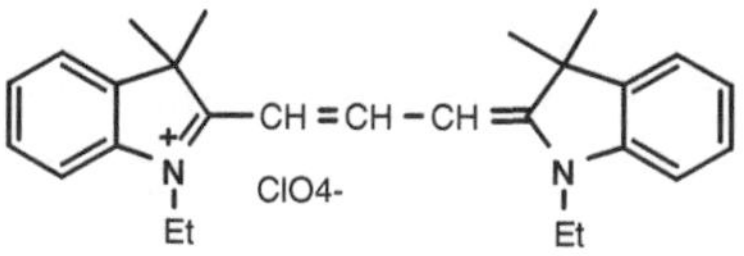

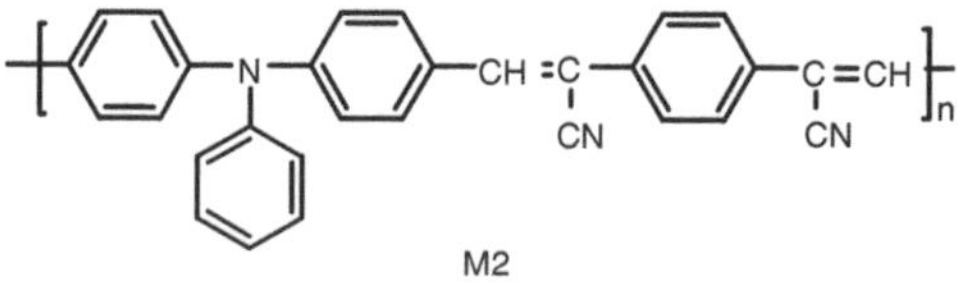

Scheme 10.

530 nm is broadened due to aggregation. A bathochromic J-aggregate peak appears at 580 nm, while the H-aggregate shows up on the hypsochromic side.

The photocurrent spectrum for the device (a), ITO/CY-5/C_{60}/Al, has an IPCE of more than 2% in the range of 450–600 nm where CY-5 absorbs. In device (b), ITO/M2/C_{60}/Al, the cyanine donor is replaced by polymer M2 and IPCE values as high as 6% are obtained in a spectral domain where M2 absorbs. In device (c), ITO/M2/CY-5/Al, an important photocurrent in the M2 absorption domain is indeed observed. However, no photocurrent matching the CY-5 absorption spectrum is observed. This phenomenon can be attributed to the presence of J-aggregate, which makes the HOMO level of cyanine dye shift towards higher energies. The difference between the HOMO level of polymer and that of cyanine is not large enough to dissociate the excitons formed in the cyanine layer. While in device (d), ITO/M2/CY-5/C60/Al, the photocurrent contribution from both M2 and CY-5 occurred; this resulted in a fairly broad photocurrent spectrum. Excitons generated in M2 will dissociate at the M2/CY–5 interface, while excitons generated in CY-5 will dissociate at the CY-5/C_{60} interface. In this device architecture, CY-5 acts both as an acceptor and a donor at the same time and the maximum IPCE is over 10%. Another advantage of using cyanine as acceptor is that the device has high open circuit voltage. The open circuit voltage of device (c) is as high as 1.28 and that of device (d) is 0.83 V, and both are higher than devices (a) and (b).

13.4. FUTURE PROSPECTS

The design and synthesis of new cyanine and polymethine dyes resulted in a rapid improvement in the overall light to power conversion efficiencies of DSSCs and the highest efficiency has reached 7.7%. The development of highly efficient dye-sensitized solar cells using pure organic compounds appears to be very promising. For practical applications, however, improvement in the long-term stability of the solar cells is necessary. In the meantime, improvements are needed for the open circuit voltage and the short circuit current.

Since the DSSCs based on hemicyanine dyes usually have high short circuit current but low open circuit voltage, the treatment of the electrode to improve the open circuit voltage is of great importance for efficiency improvement. The introduction of thiophene moieties into the hemicyanine makes the HOMO level of the materials higher, which helps to improve the open circuit voltage. Another approach was to shift the conduction band of TiO_2 by using some additives adsorbed on the surface of the TiO_2. For example, Arakawa and coworkers [21] added TBP into the electrolyte, and Frank and coworkers [25,26] conducted a study on an I^-/I_3^- electrolyte with NH_3 and pyridine derivatives as additives in acetonitrile. Relatively large energy gaps between the LUMO level and the conduction band enable addition of TBP or pyridine derivatives in the electrolyte improving the photovoltage and consequent total efficiency. TBP adsorbed on the TiO_2 surface shifts the conduction band level of TiO_2 negatively and prevents recombination between injected electrons and I_3^- ions on the TiO_2 surface, resulting in an improved photovoltage [2,21,27]. However, employing TBP in the system which has a small energy gap between the LUMO level and the conduction band led to substantial decrease in the photocurrent due to a decreasing electron injection yield. Recently, Kusama and Arakawa [28] have investigated the influence of pyrimidine additives in electrolytic solution on dye-sensitized solar cell performance in detail. The results suggest that the electron donor ability of pyrimidine additives influenced the interaction between TiO$_2$ electrode and the electrolyte. Comparing the two approaches mentioned above, it seems that the design and the introduction of some moieties into cyanine dyes, which would change the LUMO level of the materials, will remain much more space to improve the open circuit voltage effectively.

The introduction of multiple anchoring groups into the molecule can provide better adsorption, broaden the absorption spectra of the electrode, and enhance short circuit current. However, the anchoring groups, same or different, should be positioned as close as possible to shorten the distance between the skeleton of molecules and the electrode. These anchoring groups should be introduced into the acceptor part instead of the donor part to facilitate the charge separation and electron injection. In general, expansion of the methine moiety of cyanine dyes, which results in a redshift in the adsorption spectrum of the dye, would simultaneously cause several problems in terms of a complicated synthesis procedure and chemical instability of the dye. In order to improve the light absorption efficiency further, co-adsorption of different dyes on the surface of TiO$_2$ is another approach. By using the blue and red squarilium cyanines co-adsorption on the surface of TiO$_2$ in the presence of a co-adsorbent, the efficiency was improved to 4.53% in the Research Laboratory of Fuji Photo Film Co. Ltd., Japan. The IPCE curve was shown as the summation of the curves of two used dyes [29]. Therefore, co-sensitization and co-adsorption with different dyes covering whole visible spectral region might be effective and economic way to increase the efficiency of the solar cells, in which some aggregates may be suppressed to form due to the fact that the co-adsorbent prevents the formation of aggregation in content. Final confirmation should be proved by the stability measurements under continuous illumination containing ultraviolet light.

The cyanine dyes are promising in heterojunction photovoltaic devices fabricated from the solution. The efficiencies of the devices can be improved considerably when cyanine acts both as donor and acceptor at the same time. Polymers with higher HOMO level should be used as the donor in order to take advantage of the hole injection from the cyanine to polymer. Some new copolymers containing anchoring groups and cyanine dye units should be explored to improve electron transporting ability in the bulk and absorption region, and used for fabrication of solid-state dye-sensitized solar cells. The replacement of liquid electrolyte with solid hole conductors is interesting as it might offer practical advantages over a liquid-junction cell. CuI, a kind of p-type semiconductor, is the most popular solid hole conductor used for solid-state DSSC. But, solar energy conversion efficiency is still much lower than a liquid-junction cell. One of the reasons is the serious interfacial recombination in solid-state DSSC due to the lack of substantial built-in electric field at the interface of TiO$_2$ and CuI. In addition, solid-state DSSC also met with the problem of instability under continuous illumination of sunlight. It was believed that the deterioration of rectification at the interface of the heterojunction, which was induced by ultraviolet light, should be responsible for this instability.

It was found in liquid-junction cells that an interfacial blocking layer of insulator could suppress the interfacial recombination between TiO$_2$ and excited dye molecules or I$^-$/I$_3^-$ couple, which led to an improvement on cell performance [30–32]. Employing this kind of ultrathin blocking layer to solid-state solar cells would suppress the strong photooxidation capability of TiO$_2$ when exposed to UV light, which would improve the stability of the solid-state solar cell against longtime illumination. Fujishima and coworkers [33] have demonstrated a simple method to deposit MgO ultrathin layer on TiO$_2$ porous film, and have shown that such a thin layer improves both long-term stability and solar energy conversion efficiency to a remarkable extent.

ACKNOWLEDGMENTS

The authors acknowledge the support from NSFC/China, the Scientific Committee of Shanghai and Education Committee of Shanghai. H.T. thanks Dr. Lisheng Xu for very helpful assistance in manuscript preparation.

REFERENCES

1. B. O'Regan and M. Grätzel, A low-cost, high-efficiency solar cell based on dye-sensitized colloidal TiO_2 films, *Nature* **353**, 737–740 (1991).
2. M. K. Nazeeruddin, A. Kay, I. Rodicio, R. Humphry-Baker, E. Muller, P. Liska, N. Vlachopoulos, and M. Grätzel, Conversion of light to electricity by *cis*-X_2bis(2,2′-bipyridyl-4,4′-dicarboxylate)ruthenium(II) charge-transfer sensitizers ($X = Cl^-$, Br^-, I^-, CN^-, and SCN^-) on nanocrystalline TiO_2 electrode, *J. Am. Chem. Soc.* **115**, 6382–6390 (1993).
3. K. Hara, M. Kurashige, Y. Dan-oh, C. Kasada, A. Shinpo, S. Suga, K. Sayama, and H. Arakawa, Design of new coumarin dyes having thiophene moieties for highly efficient organic dye-sensitized solar cells, *New J. Chem*, **27**, 783 785 (2003).
4. F. S. Meng, K. C. Chen, H. Tian, L. Zuppiroli, and F. Nüesch, Cyanine dye acting both as donor and acceptor in heterojunction photovoltaic devices, *Appl. Phys. Lett.* **82**, 3788–3790 (2003).
5. A. Mishra, R. K. Behera, P. K. Behera, B. K. Mishra, and G. B. Behera, Cyanines during the 1990s: a review, *Chem. Rev.* **100**, 1973–2011 (2000).
6. A. Ehret, L. Stuhl, and M. T. Spitler, Variation of carboxylate-functionalized cyanine dyes to produce efficient spectral sensitization of nanocrystalline solar cells, *Electrochim. Acta* **45**, 4553–4557 (2000).
7. A. Ehret, L. Stuhl, and M. T. Spitler, Spectral sensitization of TiO_2 nanocrystalline electrode with aggregated canine dyes, *J. Phys. Chem.* **105**, 9960–9965 (2001).
8. F. S. Meng, Y. J. Ren, E. Q. Gao, S. M. Cai, K. C. Chen, and H. Tian, High efficient cyanine dyes used for nanocrystalline TiO_2 electrode, in *Proc. SPIE,* Vol. 4465 "Organic Photovoltaic II," Z. Kafafi, ed., SPIE, San Diego, USA, 2001, pp. 143–148.
9. Y. J. Ren, F. S. Meng, H. Tian, and S. M. Cai, Highly efficient photosensitization of mesoporous TiO_2 electrode with a cyanine dye, *Chinese Chem. Lett.* **13**, 379–380 (2002).
10. K. Sayama, K. Hara, Y. Ohga, A. Shinpou, S. Suga, and H. Arakawa, Significant effects of the distance between the cyanine dye skeleton and the semiconductor surface on the photoelectrochemical properties of the dye-sensitized porous semiconductor electrodes, *New J. Chem*, **25**, 200–202 (2001).
11. W. Zhao, Y. J. Hou, X. S. Wang, B. W. Zhang, Y. Cao, R. Yang, W. B. Wang, and X. R. Xiao, Study on squarylium cyanine dyes for photoelectric conversion, *Sol. Energy Mater. Sol. Cells* **58**, 173–183 (1999).
12. Z. S. Wang, F. Y. Li, C. H. Huang, L. Wang, M. Wei, L. P. Jin, and N. Q. Li, Photoelectric conversion properties of nanocrystalline TiO_2 electrode sensitized with hemicyanine derivatives, *J. Phys. Chem. B* **104**, 9676–9682 (2000).
13. Z. S. Wang, F. Y. Li, and C. H. Huang, Highly efficient sensitization of nanocrystalline TiO_2 films with styryl benzothiazolium propylsulfonate, *Chem. Commun.* 2063–2064 (2000).
14. Z. S. Wang, F. Y. Li, and C. H. Huang, Photocurrent enhancement of hemicyanine dyes containing RSO_3^- group through treating TiO_2 films with hydrochloric acid, *J. Phys. Chem. B* **105**, 9210–9217 (2001).

15. E. Stathatos, P. Lianos, Synthesis of a hemicyanine dye bearing two carboxylic groups and its use as a photosensitizer in dye-sensitized photoelectrochemical cells, *Chem. Mater.* **13**, 3888–3892 (2001).

16. F. S. Meng, Q. H. Yao, J. G. Shen, F. L. Li, C. H. Huang, K. C. Chen, and H. Tian, Novel cyanine dyes with multi-carboxyl groups and their sensitization on nanocrystalline TiO_2 electrode, *Synth. Met.* **137**, 1543–1544 (2003).

17. Q. H. Yao, F. S. Meng, F. Y. Li, H. Tian, and C. H. Huang, Photoelectric conversion properties of four novel carboxylated hemicyanine dyes on TiO_2 electrode, *J. Mater. Chem.* **13**, 1048–1053 (2003).

18. Q. H. Yao, L. Shan, F. Y. Li, D. D. Yin, and C. H. Huang, An expanded conjugation photosensitizer with two different adsorbing groups for solar cells, *New J. Chem*, **27**, 1277–1283 (2003).

19. K. Sayama, K. Hara, N. Mori, M. Satsuki, S. Suga, S. Tsukagoshi, Y. Abe, H. Sugihara, and H. Arakawa, Photosensitization of a porous TiO_2 electrode with merocyanine dyes containing a carboxyl group and a long alkyl chain, *Chem. Commun.* 1173–1174 (2000).

20. K. Sayama, S. Tsukagoshi, K. Hara, Y. Ohga, A. Shinpou, Y. Abe, S. Suga, and H. Arakawa, Photoelectrochemical properties of J aggregates of benzothiazole merocyanine dyes on a nanostructured TiO_2 film, *J. Phys. Chem. B* **106**, 1363–1371 (2002).

21. K. Hara, M. Kurashge, S. Ito, A. Shinpo, S. Suga, K. Sayama, and H. Arakawa, Novel polyene dyes for highly efficient dye-sensitized solar cells, *Chem. Commun.* 252–253 (2003).

22. K. Hara, K. Sayama, Y. Ohga, A. Shinpo, S. Suga, and H. Arakawa, A coumarin-derivative dye sensitized nanocrystalline TiO_2 solar cell having a high solar-energy conversion efficiency up to 5.6%, *Chem. Commun.* 569–570 (2001).

23. K. Hara, T. Sato, R. Katoh, A. Furube, Y. Ohga, A. Shinpo, S. Suga, K. Sayama, H. Sugihara, and H. Arakawa, Molecular design of coumarin dyes for efficient dye-sensitized solar cells, *J. Phys. Chem. B* **107**, 597–606 (2003).

24. K. Hara, Y. Tachibana, Y. Ohga, A. Shinpo, S. Suga, K. Sayama, H. Sugihara, and H. Arakawa, Dye-sensitized nanocrystalline TiO_2 solar cells based on novel coumarin dyes, *Sol. Energy Mater. Sol. Cells* **77**, 89–103 (2003).

25. S. Y. Huang, G. Schlichthorl, A. J. Nozik, M. Grätzel, and A. J. Frank, Charge recombination in dye-sensitized nanocrystalline TiO_2 solar cells, *J. Phys. Chem. B.* **101**, 2576–2582 (1997).

26. G. Schlichthorl, S. Y. Huang, J. Sprague, and A. J. Frank, Band edge movement and recombination kinetics in dye-sensitized nanocrystalline TiO_2 solar cells: a study by intensity modulated photovoltage spectroscopy, *J. Phys. Chem. B* **101**, 8141–8155 (1997).

27. A. Hagfeldt and M. Grätzel, Light-induced redox reactions in nanocrystalline systems, *Chem. Rev.* **95**, 49–68 (1995).

28. H. Kusama and H. Arakawa, Influence of pyrimidine additives in electrolytic solution on dye-sensitized solar cell performance, *J. Photochem. Photobiol. A: Chem.* **160**, 171–179 (2003).

29. M. Okazaki, Sensitizing dyes for Grätzel type photovoltaic cells. *The First Asia Symposium on Functional Dyes and Advanced Materials*, pp. 27–36 (IL-02), Osaka, Japan, October 6–8, 2003.

30. A. Kay, M. Grätzel, Dye-sensitized core-shell nanocrystals: improved efficiency of mesoporous tin oxide electrodes coated with a thin layer of an insulating oxide, *Chem. Mater.* **14**, 2930–2935 (2002).

31. E. Palomares, J. N. Clifford, S. A. Haque, T. Lutz, J. R. Durrant, Control of charge recombination dynamics in dye sensitized solar cells by the use of conformal deposited metal oxide blocking layers, *J. Am. Chem. Soc.* **125**, 475–482 (2003).

32. E. Palomares, J. N. Clifford, S. A. Haque, T. Lutz, and J. R. Durrant, Slow charge recombination in dye-sensitised solar cells (DSSC) using Al_2O_3 coated nanoporous TiO_2 films, *Chem. Commun.* 1464–1465 (2002).

33. T. Tguchi, X.-T. Zhang, I. Sutanto, K.-I. Tokuhiro, T. N. Rao, H. Watanabe, T. Nakamori, M. Uragami, A. Fujishima, Improving the performance of solid-state dye-sensitized solar cell using MgO-coated TiO_2 nanoporous film. *Chem. Commun.* 2480–2481 (2003).

14

Semiconductor Quantum Dot Based Nanocomposite Solar Cells

Marvin H. Wu, Akira Ueda, and Richard Mu
Nanoscale Materials and Sensors Group, Department of Physics,
Fisk University, Nashville, TN, USA

Contents

Abstract Solar cells incorporating semiconductor quantum dots and organic polymers may provide a lightweight, flexible, and cheaply produced alternative to conventional bulk semiconductor solar cells. We present here an overview of the advantageous properties of quantum dots and a review of the development of quantum dot–polymer composite and quantum dot sensitized photoelectrochemical cells. In these cell designs, quantum dots are employed as light absorbers and components of charge transport networks. Power conversion efficiencies in these devices are currently very low (less than 2% for quantum dots–polymer composite solar cells), with limitations arising mainly from charge transport inefficiencies. The primary sources of these losses and prospects for improvement in both types of devices are discussed. This chapter, along with the cited references, is intended to provide readers with a current picture of this rapidly expanding field of research and assist them in initiating quantum dot based solar cell research.

Keywords   quantum dots, nanowires, nanocomposite solar cells, photoelectro-chemical cells

14.1.   INTRODUCTION

Semiconductor quantum dots form a relatively new class of materials with optical and electronic properties that differ greatly from those of bulk semiconductors and organic dyes currently used in photonic and photovoltaic devices. Intensive research efforts since the initial description of the effects of quantum confinement on semi-conductors by Brus [1] and Efros et al. [2] have led to significant advances in fabrication and characterization of quantum dots. These techniques are now suffi-ciently mature to enable the employment of quantum dots in devices including infrared photodetectors [3], lasers [4], and biomedical imagers [5]. Quantum dots are also of interest to the photovoltaic research community, with several new designs of quantum dot based solar cells proposed in the previous decade. Quantum dots match or surpass bulk semiconductors in the optical properties necessary for photovoltaic power conversion, while removing constraints placed on device designs due to the high-vacuum environments, high processing temperatures, and lattice-matched substrates required to manufacture bulk semiconductor solar cells. Quan-tum dots are thus able to provide semiconductor-like optical and electronic perform-ance to solar cells previously restricted to use of organic dyes. This introduces the possibility of inorganic quantum dot–organic polymer composite solar cells with low production costs and reasonable efficiencies.

The unique properties of quantum dots arise from quantum confinement. When the dimensions of a semiconductor crystal are reduced to the order of the extent of an exciton wavefunction (typically of the order of a few nanometers) measurable changes of its properties occur due to the boundary conditions imposed by the particle surface. Quantum dots with diameters significantly smaller than the bulk exciton Bohr radius are considered to be in the "strong confinement" regime and exhibit large changes in optical properties. The physical principles at the root of these changes can be seen in the simplified example of a particle in a one-dimensional potential well (particle-in-a-box). Realistic calculations of quantum dot properties require inclusion of semiconductor-appropriate band structure, electron–hole correl-ation, and surface effects, making accurate estimation of optical properties difficult [6], but the particle-in-a-box model illustrates two significant differences between quantum dots and bulk semiconductors. First, energy levels of a particle-in-a-box system are discrete, unlike the continuum of energy levels of free particles. An analogous change is observed in quantum confined semiconductors: energy levels in the conduction and valence bands become discrete. While this leads to some concentration of the oscillator strength at well-defined energy bands, inhomogeneous broadening due to size dispersion and phonon-assisted transitions leads to significant absorption for all photons with energy greater than the lowest excited state. At higher photon energies, the absorption spectra of quantum dots with finite potential barriers resemble bulk semiconductors [7]. The large absorption cross sections of quantum dots at photon energies greater than the bandgap lead to lower weight and material requirements for photovoltaic devices. Enhanced overlap of confined electron and hole wavefunctions may also help to increase absorption cross sections in strongly confined quantum dots [8], further improving device performance.

The second difference between bulk and quantum-confined semiconductors can be observed when the dimension of the potential well is reduced. The energy of the lowest state in the box increases as the box size decreases. A similar change is observed for quantum dots: the bandgap energy increases as the size of the dot is reduced. Solutions of quantum dots with controlled sizes provide striking visual confirmation of this fact, with luminescence from solutions of CdSe quantum dots changing in color from blue to red as the size of the dot is varied from 1.5 to 5 nm [7]. The bandgap energy of quantum dots in a photovoltaic device can therefore be tuned by varying the size of the dots, instead of varying the type of semiconductor in the cell. The optimum bandgap energy (1.4 eV) of a photovoltaic device under AM1.5 conditions calculated by Shockley and Qucisser [9] can be reached by nanometer-sized In As [23], InP [24], Si [25] or Pb [26] quantum dots. Bandgap ranges covered by quantum dots commonly used in photovoltaic devices are shown in Figure 14.1.

Optimized tandem (or multiple active region) solar cells can also be produced, with the precise bandgap energy for each layer easily obtained by use of quantum dots of different sizes. This greatly reduces the complexity of cell production, which requires two different processing temperatures and elemental environments for current tandem cells.

A third difference due to quantum confinement that may significantly impact the performance of quantum dots in photovoltaic devices is the slowed cooling of carriers, due to the "phonon bottleneck" [10]. Bulk semiconductors in higher excited states typically relax to the lowest excited state via cascaded optical phonon emission. However, in strongly confined quantum dots, the separation between the discrete energy levels is often much larger than the optical phonon energy. Since there are no real intermediate states available for phonon transitions, relaxation from higher excited states in quantum dots can only occur via multiphonon processes involving virtual intermediate states. Experimental results concerning this phenomenon are mixed, however, with slower relaxation time observed in CdSe quantum dots modified to prevent energy transfer from the electron to holes, which have much more closely spaced energy levels [11]. Measurements of InAs [12] and PbSe [13] quantum

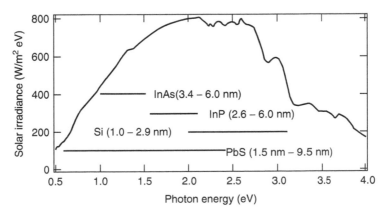

Figure 14.1. Spectral range of the bandgap of InAs [23], InP [24], Si [25], and PbS [26] quantum dots and the solar photon energy distribution.

dots have shown no evidence of phonon bottleneck effects. Circumvention of the bottleneck has been attributed to Auger (electron to hole) and defect-mediated processes [14,15]. The existence of the bottleneck may depend critically on the electron and hole level spacings as well as the presence of defects in the dot core, dot surface, or the surrounding matrix. If dots with phonon bottlenecks can be reliably engineered, long-lived, higher lying excited state populations in quantum dots open the possibility of separation and extraction of "hot" carriers from the dots. This will result in increased photovoltage and a reduction in the excess photon energy transferred to the cell as heat, which has been shown to contribute to cell degradation. Nozik [16] has also proposed an increase in efficiency due to impact ionization processes, in which multiple lower energy carriers can be generated from absorption of a single higher energy photon.

Quantum dots also provide the positive attributes of semiconductors, but with unprecedented processing flexibility. Unlike conventional bulk crystalline semiconductors, or even thin films of crystalline or amorphous semiconductors, quantum dots can be easily be dispersed in liquid or solid solutions. Conventional considerations for semiconductor device design, such as lattice matching or stability under high-temperature processing, are unnecessary for quantum dot based devices. Chemically synthesized quantum dots can be simply mixed with the desired solvent or polymer and spin-coated, dropped, cast, or screen printed [17] onto a chosen substrate. Two novel approaches to solar cell design, the quantum dot sensitized photoelectrochemical cell [18] and the quantum dot–organic polymer composite cell [19], take full advantage of the processability of quantum dots by incorporating them in systems with liquid electrolytes or conducting organic polymers. Quantum dots may be superior to organic dyes traditionally used as photosensitizers in these cell designs, as all photons greater than the bandgap energy are absorbed. Quantum dots have also shown greater resistance to photobleaching and thermal degradation than commonly used organic dyes [20]. These cell designs may dramatically reduce manufacturing costs while maintaining power conversion efficiencies large enough to economically justify widespread adoption of solar technology.

The development of quantum dot based solar cells depends heavily on the availability of quantum dots with desired properties. Since the optical properties of quantum dots depend critically on size, the size distribution of the dots must be controllable. Dots with low densities of defects, which trap carriers and reduce conversion efficiency, must also be available. Capping layers of organic ligands or semiconductor nanoshells used to passivate surface defects that hinder transport of photogenerated charge carriers are also undesirable. Fabrication techniques that require high temperature or lattice-mismatch effects, such as Stranski–Krastonow growth methods, are not compatible with quantum dot–organic composite device structures. Wet chemical synthetic techniques do show great promise, however. Quantum dots of many elemental compositions with extremely narrow size distributions can now be produced in solution. Capping layers, which must be used to prevent aggregation in solution, may be removed by low-temperature annealing or may be designed to facilitate charge transfer to the appropriate network [21]. Pulsed laser ablation, another technique used to produce quantum dots, results in quantum dots without capping layers at the cost of vacuum-based processing and wider dot size distributions. Scaling production to bulk quantities of quantum dots presents challenges for both techniques, as high-quality quantum dots have only recently become commercially available [22].

14.2. QUANTUM DOT–ORGANIC POLYMER COMPOSITE SOLAR CELLS

Novel solar cells consisting of semiconductor quantum dots embedded in organic polymer hosts offer a promising alternative to traditional crystalline semiconductor cells. These organic–inorganic hybrid nanocomposite cells have been designed to take full advantage of the properties of the constituent materials. Organic polymers can be easily synthesized and formed into thin films by methods far cheaper than typical semiconductor thin-film deposition techniques. Polymers can be deposited on flexible substrates by spin-coating, casting, or other methods which do not require vacuum environments, in contrast to crystalline or polycrystalline semiconductors. Organic polymers can also be structurally altered by the powerful arsenal of synthetic chemical techniques to aid in device optimization. Quantum dot–polymer composite systems have significant advantages compared to other photoabsorbers or charge transfer agents, such as fullerenes [27,28] and organic dyes, which are commonly mixed with polymers to form photovoltaic devices. Semiconductor quantum dots, as detailed in previous sections, possess many attributes that are advantageous to development of high-efficiency solar cells. Quantum dots have tunable bandgaps and thus can be tailored to match the solar radiation spectrum in various environments (AM1.5, AM0, etc.). They also possess exceedingly large surface-to-volume ratios, which enhance charge transfer rates. Finally, quantum dots have large absorption cross sections, so incoming light can be efficiently absorbed and converted into electron–hole pairs. Combining semiconductor quantum dots and polymers may pave the way for large-scale production of low-cost, lightweight photovoltaic cells with adequate efficiencies. The credit for both the concept and the realization of quantum dot–organic polymer solar cells must be given to Alivisatos and co-workers, who have done the bulk of the research in this subject area.

Initial studies of photovoltaics based on these principles have illustrated the potential of these techniques and have also provided a view of the challenges which need to be addressed before inorganic–organic hybrid nanocomposites can effectively compete with traditional inorganic semiconductor photovoltaic technology [19,21,29–32]. The efficiencies of hybrid cells produced by research laboratories have thus far been limited to a few percent under AM1.5 conditions [29]. These values, while impressive for such limited research, fall far short of what is currently achievable with bulk inorganic semiconductors [33]. Limitations have been imposed primarily by inefficiencies in the charge transfer processes necessary to transport electron–hole pairs generated in the quantum dot to the electrical contacts. Losses, which will be discussed later, occur at nanocrystal surfaces and along both electron and hole conduction paths [21].

Quantum dot–polymer composite cells described in reports published thus far share common design principles: nanoparticles are blended with conducting polymers and coated onto a conductive, transparent substrate. A metallic contact is evaporated on the top surface. A typical cell design is illustrated in Figure 14.2 [29]. Incident photons are absorbed in the active region, resulting in generation of charge carriers. Since conjugated organic polymers typically possess hole mobilities significantly larger than their electron mobilities, hole transport occurs primarily along the polymer network. Electron transport occurs by hopping events along a percolation pathway of quantum dots.

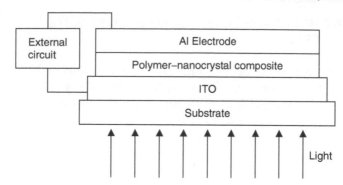

Figure 14.2. Diagram of typical quantum dot–polymer composite photovoltaic cell.

14.2.1. Published Results

Research into the photovoltaic properties of hybrid nanocomposite materials is still in its early stages. Relatively few studies of working cell prototypes have thus far been published. Selected results are given in Table 14.1.

The efficiencies listed in the table are far below those routinely achieved by laboratory prototypes of conventional semiconductor photovoltaic cells. Conventional pn-junction crystalline silicon solar cells have efficiencies of well over 20% under AM1.5, one sun conditions [33]. While quantum dot–polymer composite cells probably need not match the efficiencies of conventional semiconductor cells due to their extremely low processing costs, efficiencies must be significantly increased before widespread application of these cells becomes economically feasible. Efficiencies listed in Table 14.1 are also below the ~3% conversion efficiencies possible with polymer-based cells employing fullerenes or fullerene derivatives as electron acceptors [49]. This is likely due to limitations in charge transport due to poor dot surface conditions and phase separation in quantum dot–polymer composites, as discussed below. However, researchers have only recently begun to obtain a physical picture of charge transport in these quantum dot–polymer composite structures, and relatively little effort has gone into optimization of cell materials and parameters. Major sources of inefficiency in light absorption, carrier generation, and carrier transport remain in the cell structures listed in Table 14.1.

Table 14.1. Selected data from published reports of the performance of quantum dot/rod–polymer composite solar cells

Quantum Dot Material	Polymer Concentration	EQE[a]	Conversion Efficiency (%)	Ref.
CdSe nanosphere (5 nm)	MEH-PPV	90 wt.% 12%[b]	0.2	[32]
CdSe nanorods (3–8 nm diameter, 7–60 nm length)	P3HT	90 wt.% 59%[c]	1.7	[29]
CuInS$_2$ nanosphere (unreported)	PEDOT:PSS 0.06 M	20%[d]	<0.1	[31]

[a] EQE, external quantum efficiency.
[b] 514 nm, 0.5 mW/cm^2.
[c] 450 nm, 0.1 mW/cm^2.
[d] 450 nm, 0.2 mW/cm^2.

14.2.2. Power Conversion in Quantum Dot–Polymer Composite Solar Cells

The overall efficiency of a photovoltaic device is affected by many parameters, with potential losses occurring at each step of the conversion process. This process can be broadly divided into three steps. First, the active layer of the cell must absorb light. Photogenerated charge carriers must next be separated and then transported without undergoing recombination to the proper electrode. Effective cell designs, including optimum choice of electrode, polymer, and quantum dot materials, must be implemented to maximize the extracted power. Analysis of each of the steps for hybrid organic–inorganic nanocomposite cells provides a clear view of both the advantages of this type of photovoltaics and the challenges that must be overcome before these cells can effectively compete with conventional inorganic semiconductor solar cells.

14.2.2.1. Light Absorption and Generation of Charge Carriers

The first step of the process, absorption of light and generation of charge carriers, is expected to be quite efficient for these systems. The increased density of states of semiconductor quantum dots due to quantum confinement leads to exceptionally strong absorption cross sections [34]. Films largely composed of direct bandgap semiconductor quantum dots are expected to absorb nearly all of the incoming light with thicknesses of well under 1 μm for high quantum dot loading values [21]. Thicknesses of bulk semiconductor films needed to absorb most of the incident photons range from 10 μm (direct bandgap) to over 100 μm for silicon, an indirect bandgap semiconductor [33]. This could result in significant reductions in the weight of future photovoltaic cells, which may be important for space power and miniature power applications. Material costs are also significantly reduced, making these cells economically more feasible.

Quantum efficiencies of charge carrier generation approach unity in these systems. Photoluminescence quantum yields of well-passivated quantum dots, such as ZnS-coated CdSe quantum dots, are exceptionally high [35]. This reflects the efficiency of both carrier generation and subsequent radiative recombination, which competes with carrier extraction from the quantum dots. The high photoluminescence efficiency of quantum dots is maintained under high-excitation light intensities, unlike dyes, which suffer from significant photoluminescence degradation over time [36]. Quantum dots thus appear to be excellent candidates for the role of photoabsorber and charge carrier generator in photovoltaic devices, provided carriers can be extracted from the quantum dots before radiative recombination occurs.

Quantum dot and polymer properties may not be fully optimized for this role, however. CdSe quantum dots or rods, which can be chemically synthesized with narrow size distributions and a wide variety of surface passivating ligands, have been employed in most studies of quantum dot–polymer composite solar cells [21,29–32]. These quantum structures have optical bandgaps greater than the ~1.7 eV bandgap of bulk CdSe. This bandgap is well above the optimum value of ~1.4 eV for single active layer photovoltaics under one sun AM1.5 illumination, resulting in a significant source of overall device inefficiency [11]. Predicted ideal solar cell efficiency for a bandgap value of 2.0 eV, a value consistent with CdSe quantum dots with diameters of a few nanometers, is ~22%. The ideal efficiency for a 1.4 eV semiconductor is close to 30%. However, organic synthetic routes have been found for a wide variety of group IV (e.g., Si, Ge) [37,38], group III–V (e.g., InP, InAs) [39,40], and ternary (e.g., $CuInS_2$) [31] semiconductors possessing bandgaps closer to the ideal value. Other quantum dot growth methods, such as pulsed laser deposition [41] and Stranski–Krastonow [42] growth methods have also been used to form quantum dots with bandgaps in the near infrared region of the solar spectrum. These growth methods require high vacuum and add significantly to the cost of a quantum dot–polymer system. Adoption of quantum dots with bandgaps in the near infrared region is also limited by the availability of narrow bandgap conducting polymers [47], as band mismatch considerations will limit charge transfer from the quantum dot to the polymer network. Use of quantum dots in the infrared region will also result in reduced photovoltages. The beneficial optical properties of quantum dots may be best leveraged through tandem or spectrally dispersed cell designs. This approach takes advantage of both the low cost of polymer processing techniques and the excellent control of quantum dot size. In these approaches, quantum dot sizes may be tuned to varying portions of the solar spectrum, increasing the number of absorbed photons. Low-temperature polymer processing techniques decrease the amount of interdiffusion between cell components, a major concern for traditional bulk semiconductor tandem cells [33].

The role of the polymer in photoabsorption should not be neglected. Conjugated polymers with good charge mobilities often have broad absorption bands in the visible region. Since a dense network of polymer chains is necessary for efficient charge transfer between the quantum dots and the polymer network, polymers can often absorb a significant fraction of the incident photons. In a recent study of a CdSe quantum rod–polymer composite solar cell, the polymer was estimated to account for more than half of the total number of photons absorbed by the device [21]. This is an important source of losses, as recombination in polymers has not suppressed to the same degree as recombination in the quantum dots.

Absorption of light by nanocomposite solar cells is also strongly influenced by the cell morphology. Care must be taken during the blending of the quantum dots and the polymers, and during the spin-coating process to avoid significant phase separation of the components. This limits absorption by producing micron-size index of refraction domains in the active layer of the film, promoting scattering and reducing absorption [31]. It is important to note that none of the steps taken to reduce reflection in bulk semiconductor solar cells, such as surface texturing, antireflection coatings, or reduction of the surface area of the metal coating, have been taken for these nanocomposite cells. Reflection of incident light can be a significant source of loss, even with the relatively low dielectric constants of the conducting polymer medium.

14.2.2.2. Separation and Extraction of Carriers from Quantum Dots

Carriers generated by photon absorption must be efficiently separated and removed from the quantum dots to ensure proper device operation. In most devices developed thus far, polymers play the role of hole transporters, while electron transportation occurs by dot-to-dot hopping [19,21,29–32]. Separation and extraction of the charge carriers from the quantum dot must be accomplished before radiative or nonradiative recombination occurs. Carrier dynamics occur on extremely rapid timescales in semiconductor quantum dots, with measured relaxation times in the subpicosecond range for both electrons and holes in CdSe quantum dots [43]. Charge transfer may occur in both directions at the quantum dot–polymer interface, as electrons generated by photoabsorption in the polymer host can be transferred to the quantum dot, and holes generated in the quantum dot can be transferred to the polymer. Exciton transfer can also occur to the lower bandgap material in the system, followed by transfer of the appropriate charge to the higher bandgap material [32]. These charge transfer processes require close contact between the polymer network and the quantum dots, since the exciton diffusion length of conjugated polymers is on the order of nanometers [44]. Regions of the active layer where phase separation occurs will have significant light absorption, but will not contribute to power conversion. The morphology of polymer films can be controlled through careful exploration of optimum quantum dot concentration, solvent choice, and film deposition conditions [49]. The large surface area to volume ratio of the quantum dots or rods is beneficial to these charge transport mechanisms, all of which rely on direct quantum dot or rod–polymer surface contact.

Charge transfer between components of the quantum dot or rod–polymer composite materials is complicated by the presence or organic ligands on the surface of the nanoparticles. The presence of these ligands is necessary during the synthetic process to maintain size control and prevent aggregation. Ligands must also be present while the quantum dots or rods are blended with the conducting polymer and spun onto a substrate to ensure that the nanoparticles are evenly distributed throughout the film. These ligands may not be compatible with the polymer, however, resulting in phase separation for high ligand concentrations. Efforts must be made to find an intermediate ligand concentration that avoids phase separation through nanoparticle or polymer aggregation [21].

Evidence for fast carrier separation and extraction in quantum dot or rod–polymer composites has been found by photoluminescence quenching and photoconductivity studies [45]. Efficient separation of the charge carriers prevents radiative recombination and thus quenches photoluminescence of both the polymer and nanoparticle components of the system. Photoluminescence quantum yields measured as a function of nanoparticle weight percentage show that exciton disassociation occurs more efficiently with pyridine than the more commonly used surface passivation agent, trioctylphosphine oxide (TOPO). 90 wt.% CdSe quantum dots–polymer blends with pyridine surface coverage exhibited photoluminescence efficiencies reduced by over an order of magnitude compared to low wt.% quantum dot–polymer blends [32]. This reduction is much greater than that observed in colloidal CdSe quantum dot films without the presence of a conducting polymer, illustrating the critical role of the polymer as a hole conduction pathway [46]. Although the magnitude of the photoluminescence degradation is partially due to efficient polymer-to-quantum dot exciton transfer, this illustrates that exciton disassociation occurs at rates much faster than radiative recombination. The small remaining amount of photoluminescence from the blended samples is ascribed to the photoluminescence from the polymer indicating that phase separation between the components of the blend has taken place [21].

Photoconductivity studies of cell structures containing nanoparticle–polymer blends provide evidence that pyridine surface coverage, while significantly quenching photoluminescence, still hinders charge transfer [21]. Low-temperature (120°C) annealing of high surface-to-volume ratio nanoparticles (i.e., those which require a large amount of pyridine to passivate all surface traps) removes pyridine from the surface of the nanoparticle and from the interior of the active region of the cell [21]. This has been demonstrated to increase photoconductivities of thin ($\sim$ 200 nm) quantum rod–polymer samples by more than a factor of two [21]. This is ascribed to closer nanoparticle surface–polymer and nanoparticle–nanoparticle contact, as well as reducing nonradiative recombination in areas of higher local pyridine concentration. Annealing polymers above the glass transition temperature has also been shown to increase the hole conductivity and photovoltaic peformance [50]. Increasing annealing times or temperatures for thicker films may not be desirable due to diffusion of the metallic contact material into the active region of the cell [21]. Research into alternative surface passivation agents based on monomers of conductive polymers may yield nanoparticles with increased solubility and superior charge transfer properties [47]. Other approaches currently under investigation are direct production of quantum dots in a polymer matrix and direct linkage of polymers to quantum dots via chemical bonds [48].

14.2.2.3.   Carrier Transport

Inefficient transport of photogenerated charge carriers is the largest source of inefficiency in polymer quantum dot composite photovoltaic devices. This difficulty arises from the relatively poor transport properties along both carrier pathways. Hole transport primarily occurs through the polymers used in prototype devices thus far include poly(2-methoxy,5-(2′-ehtyl)-hexyloxy-p-phenylenevinylene) (MEH-PPV) [19] or poly(3-hexylthiophene) (P3HT) [29]. Carrier mobilities in these polymers (typically ∼0.1 cm^2/V s) are far below those found for bulk semiconductors, such as Si (10–1000 cm^2/V s, doping level dependent), used in conventional solar cell designs. Electron transport occurs through the quantum dots, with carriers hopping from one dot to the next. Losses are introduced by the inherent inefficiencies of this process, with "dead end" conduction pathways prevalent even in high weight percentage quantum dot films [29]. Recent attempts have been made to improve the electron transport of the quantum structures by replacing quantum dots with quantum rods, which offer improved charge transport properties while retaining the optical characteristics of quantum dots. CdSe quantum rods with diameters ranging from 1 to 5 nm exhibit significant optical bandgap shifts due to quantum confinement at lengths up to 50 nm [51]. Carrier transport along the long axis of the rods, while more inefficient than transport in bulk semiconductors due to scattering at the outer diameter of the rod, is much improved over dot-to-dot hopping [29]. Carriers still must hop from one rod to another, but the length of the rods is typically more than ten times the diameter, resulting in fewer hopping events necessary to reach the contacts if the rods are effectively aligned. This has improved electron transport efficiency to such a degree in photovoltaics cells with thin active regions that hole transport along the polymer network is believed to be the limiting factor for quantum rod–polymer photovoltaics [29]. This improvement comes at the price of solubility of the nanoparticles, which makes solution phase production of films without phase separation a significant barrier to overcome.

  Significantly lower efficiencies have been observed in quantum rod–polymer composite cells under full one sun illumination than at lower light intensities [30]. This has been mainly ascribed to two sources: the increase in photoconductivity of the active region as light intensity increases and increased bimolecular recombination processes. The former is caused by an increase in the conductivity of pathways between the metal and transparent oxide electrodes, caused by an increase in the number of free charges under increased illumination. This is similar to the increased charge mobility as doping levels are increased in bulk semiconductors. The latter is caused by recombination between charge carriers arising from absorption of different photons. As the charge density in the active region increases, the probability of recombination at various interfaces will also increase. This effect has lowered the efficiency of a quantum rod–polymer device by up to 40% [29]. It has been suggested that incorporation of electron- and hole-blocking layers with appropriate energy levels at the back and front electrodes, respectively, will improve performance by minimizing nonproductive photoconductive losses.

14.2.4. Conclusions

Semiconductor nanoparticle–polymer solar cells are expected to possess many advantages over conventional bulk semiconductor cells. Low cost, low-temperature processing resulting in solar cells deposited on flexible substrates can potentially make solar cells economically competitive and open new areas of application for solar power conversion. Initial research efforts concerning these composite cells have produced impressive progress. Researchers have demonstrated of power conversion under standard AM1.5 one sun illumination of nearly 2%, despite the use of materials with carrier mobilities orders of magnitude lower than those of bulk semiconductors. Many optimizations of cell constituent materials and cell designs have not yet been investigated, leaving a host of research opportunities. In particular, the development of a nanoparticle–polymer combination that can effectively convert near infrared solar radiation will lead to significant increases in the efficiencies of composite solar cell efficiencies. This and other issues, such as electrode materials and development of conjugated polymers with increased hole mobilities, remain open to investigation and improvement.

14.3. QUANTUM DOT SENSITIZED PHOTOELECTROCHEMICAL SOLAR CELLS

14.3.1. Introduction

The quantum dot sensitized photoelectrochemical (PEC) solar cell concept is based on the dye-sensitized (DS) PEC structure. The photoelectrochemical effect was reported in 1839 by Becquerel [52]. This phenomenon can be observed when a semiconductor material is used as a photoelectrode, another dissimilar electrode, usually a metal is used as its counterpart, and a redox electrolyte is used as the medium to bridge the two electrodes. The structure is analogous to the well-known Schottky barrier structure in semiconductor device physics. When the incoming photon energy is higher than the bandgap of the semiconductor material used as the photoelectrode, electron–hole pairs are generated. The electrons in the conduction band are driven by the potential difference and transported through the electrolyte to the other electrode. Over the years, much progress has been made to improve this structure by using a wide bandgap semiconductor as the photoelectrode substrate and adsorbing dye molecules on to the substrate surfaces to harvest incoming photons. Liquid electrolytes have been replaced by solid matrix materials, such as polymers. The resulting improved DS-PEC structures for solar cell application is commonly referred to as dye-sensitized solar cells (DSSC).

The major breakthrough in energy conversion efficiency of DSSC structure is attributed to Grätzel and his coworkers at the Swiss Federal Institute of Technology [53–58]. This group has successfully employed a nanocrystalline TiO_2 thin film as the electrode and used Ru bipyridyl complex as a photosensitizer to reach a solar energy conversion efficiency of $>10\%$ under AM1.5 irradiation. The enhancement of solar

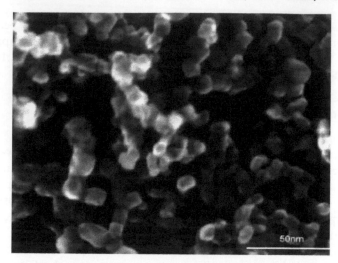

Figure 14.3. A SEM image of TiO$_2$ nanocrystalline film that may be used for DSSC development (From M. Grätzel, Photoelectrochemical cells, *Nature* **414**, 338 (2001). With permission.)

energy conversion efficiency has largely been attributed to the huge surface area of the porous TiO$_2$ network. A typical scanning electron microscopy (SEM) image of a TiO$_2$ nanocrystalline film is shown in Figure 14.3. This can increase the photon-harvesting efficiency and decrease photogenerated carrier tunneling distance, which leads to high efficiency. Metal oxides, such as TiO$_2$ and ZnO, have good chemical stability under solar irradiation, are nontoxic, and can be cheaply produced. The mechanism behind conversion of solar energy to current in a DSSC resembles the photosynthetic process, in which chlorophyll acts as the photoabsorber or photo-sensitizer and a membrane serves as the charge transporter. In DSSCs, the dye molecules, such as Ru complex, function as photosensitizers, while the electrolyte- or hole-conducting matrix and the skeletal metal oxides, such as TiO$_2$ nanocrystal network serves as the photogenerated charge carriers. This has been reviewed in greater detail by Hara and Arakawa [59].

Recently, DSSCs have enjoyed much attention and progress. However, it has become clear that the use of Ru complex or other dye molecules has significant limitations. These include low threshold of radiation resistance of dyes compared to inorganic materials and the limitation of photon to charge carrier conversion efficiency of 100%. Quantum dots may be superior in both areas. Therefore, the use of quantum dots rather than dye molecules has been proposed by Nozik [16,18,60,61]. Based on analysis by Ross and Nozik, quantum dot based solar cells are able to boost the solar energy conversion efficiency to ~66% by utilizing the hot photogenerated carriers to produce either higher photovoltages or higher photocurrents. Quantum dot sensitized solar cells (QDSSCs), in principle, will be able to produce high photocurrents through impact ionization process, which leads to the quantum yields >1 [60]. Thus far, several efforts have been made to use quantum dots as substitutes for dye molecules in solar cells [18,62–65]. The conversion efficiency has been very low and the quantum yield is far less than 1. As discussed later, the low solar conversion efficiency is likely due to the difficulties of fabricating perfect quantum dots, which is believed to be achievable in the near future.

14.3.2. Issues Concerning Materials Used as QDSSCs

QDSSC materials consist of three major components in the *active layer* of the photovoltaic structure. They are the photoelectrode, the quantum dot photosensitizers, and the hole-conducting polymers that replace the electrolytes.

In the past decades, the most commonly used photoelectrode material is nanocrystalline TiO_2. Other metal oxides, such as SnO_2, ZnO, In_2O_3, NiO, Nb_2O_5, and $SrTiO_3$, have also been investigated [66–77]. Research efforts have also been directed towards combinations of metal oxides, such as SnO_2/ZnO and SnO_2/MgO nanostructures [78–81]. Thus far, devices incorporating TiO_2 have achieved the best performance. The fundamental mechanism behind the superiority of TiO_2 remains unclear. It may be due, in part, to fewer studies on other systems. ZnO, for example, has very similar band structure to TiO_2, but ZnO nanostructure-based solar cells have much lower efficiency than TiO_2 based solar cells. Figure 14.4 illustrates a set of commonly used metal oxide and other semiconducting materials. It is clear that the band structure of both ZnO and TiO_2 is almost identical. As it will be discussed in the following section, the use of ZnO nanowire and nanorod network as a new type of photoelectrode may have a great potential for high-efficiency energy conversion [82].

Many quantum dot materials with different physical sizes and shapes have been employed for harvesting photons. InP, PbS, CdSe, CdS, and CdTe quantum dots, which have bandgaps that optimize absorption of solar radiation, have been used as

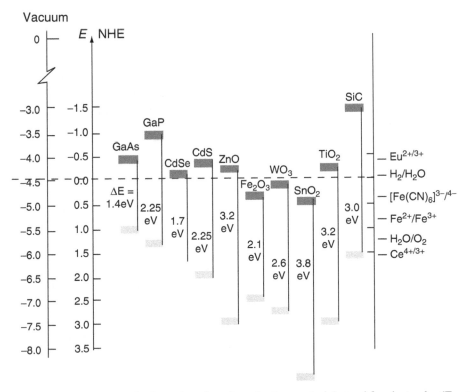

Figure 14.4. Bandgaps of a few selected semiconducting materials used for electrodes (From M. Grätzel, Photoelectrochemical cells, *Nature* **414**, 338 (2001). With permission.)

the photosensitizers. It has been demonstrated that as-synthesized quantum dots can have very high photon-to-exciton quantum efficiency in solution. However, it is difficult to transfer quantum dots from solution to the surface of the photoelectrode, such as a TiO_2 nanocrystalline-based network. Problems encountered thus far include: limited surface coverage of the electrode; aggregation of the quantum dots in voids or on the electrode surface; difficulties in removing capping agents necessary for surface passivation and quantum dot stability to facilitate charge transfer. Efforts have made recently to functionalize quantum dots and the electrode surfaces so that quantum dots can effectively attach to the electrode surface and prevent aggregation among the quantum dots in solution. With this approach, it is possible to address two important issues related to the energy conversion efficiency: increasing the rate of photogenerated charge carriers transfer to the electrode networks and building chemical barriers to enhance the stability against the degradation of the quantum dots.

The replacement of an electrolyte with solid-state matrix has very clear advantage over the conventional DS-PEC solar cell, although it has so far resulted in much lower efficiency. One of the greatest advantages of solid-state QDSSCs is their adaptability, flexibility, and applicability to harsh environmental conditions in space and on Earth. They can thus be employed on rigid device surfaces, clothing and tents, for use in civilian, military, and space applications. The commonly used solid matrices are the hole-conducting polymers, such as poly(phenylenevinylene) (PPV), MEH-PPV, poly[9,9-dioctylfluorenyl-2,7-diyl] (PFO), poly[9,9-dihexylfluorenyl-2,7-diyl)-co-(1,4-benzo-[2,1',3]-thiadiazole)] (PFOG), poly(3-octylthiophene-2,5-diyl) (P3OT), (cis-Ru(H_2dcbpy)dnbpy)(NCS)$_2$ (Z-907), and P3HT [83–87].

The use of the aforementioned organic–inorganic three-component hybrid structure as the active layer in QDSSCs has arguably taken the full advantage of the properties of both organic and inorganic components while limiting the effect of their disadvantages. For organic solar cells, a few property constraints still exist that limit the energy conversion efficiency. First, the binding energy of excitons in organic materials is usually one to two orders of magnitude larger than those of inorganic materials. Thus, it requires a strong driving force to separate electron–hole pairs. Second, charge mobility is usually poor, with charge transport dictated by hopping mechanisms. Third, the absorption bands of organic dyes are relatively narrow, resulting in low coverage of the solar spectrum. Finally, dyes are prone to thermal stability and degradation issues. These problems can be addressed with devices using quantum dots, which possess low exciton binding energies, strong and broad absorption bands, as photosensitizers and metal oxide electrodes to increase charge mobility and thermal stability. The current state-of-the-art conversion efficiency of a few percent suggests that there is much room for improvement.

14.3.3. Challenges in Materials and Device Development

As discussed by Hegedus and Luque [88], future photovoltaics will become a huge global high-tech industry. It is authors' personal belief that the basic QDSSC structure has the potential to outpace other designs due to its three key components, which have addressed issues that other structures are unable to completely cover. However, optimization of the materials and structural designs will continue to impose challenges.

High-quality quantum dots synthesized currently have very high photon–exciton quantum efficiency. However, charge separation and transport processes are the major obstacles responsible for low-energy conversion efficiency. As discussed in

Figure 14.5. A SEM image of ZnO nanoribbons that my be used for QDSSC applications (From R. Mu, M. Wu, and A. Ueda, unpublished results.)

Section 1, the effectiveness of quantum dot surface passivation to minimize the number of charge traps on the quantum dot surface and to prevent quantum dot quality degradation, such as oxidation, is critical to obtain good device performance. Although nanocrystalline TiO_2 electrodes have shown good potential, electrons are transported through TiO_2 electrodes via a hopping mechanism. It has been proposed recently [82] that TiO_2, ZnO, and other metal oxide single-crystal nanowires or nanorods may be superior electrodes, since electrons can be easily transported to external electrodes along the one-dimensional nanowires, nanoribbons, or nanorods, as shown in Figure 14.5 [89]. Certainly, the creation of new types of hole-conducting polymers is equally important.

14.4. CONCLUSION

Semiconductor quantum dot or rod–organic polymer nanocomposite solar cells are a promising alternative to conventional inorganic semiconductor cells. These cells are designed to exploit the semiconductor-like optical properties of quantum dots while maintaining the processing and mechanical advantages of polymers. Laboratory prototypes of both quantum dot or rod–polymer mixture and quantum dot sensitized photoelectrochemical solar cells have been produced. While recently reported power conversion efficiencies are lower than those of fullerene or dye-sensitized cells, research into these nanocomposite structures is still in its infancy. Significant improvements can still be made in cell designs, fabrication of quantum dots without defects or potential barriers to charge transfer on the surfaces, and the spatial uniformity in the deposition of quantum dots or rods. Development of narrow bandgap conducting polymers may also significantly improve absorption of solar photons and fully leverage the tunable bandgaps of these quantum-confined materials. Meeting these challenges will lead to a bright future for quantum dot based nanocomposite solar cells.

QDSSC structure is arguably the best design to achieve high solar energy conversion efficiency with limited sacrifice of the good values. Because the quantum dots can be attached directly on to the electrode surface, charge separation issues are effectively resolved. When the electrodes are replaced by nanowires that are directly connected to one side of the electrode, charge transport can be very effective. On the other hand, attachment of the quantum dots on to the chosen electrode materials will create new interface between the quantum dots and the electrode. The nature of the new interface, in turn, will have direct effects on charge transport from quantum dots to electrode materials. More research is required to illustrate and understand these related issues.

ACKNOWLEDGMENTS

Authors are grateful to National Renewable Energy Lab of US Department of Energy, US Army Research Office, US National Science Foundation, and US National Aeronautic and Space Administration agencies for their financial supports. R. Mu would also like to thank Sam Sun for his tireless assistance over the full course of the manuscript preparation and patience.

REFERENCES

1. L. E Brus, Electron–electron and electron–hole interactions in small semiconductor crystallites: the size dependence of the lowest excited electronic state, *J. Chem. Phys.* **80**, 4403–4409 (1984).
2. A. I. Ekimov, Al. Efros, and A. A. Onushchenko, Quantum size effect in semiconductor microcrystals, *Solid State Commun.* **56**, 921–924 (1985).
3. J. Phillips, P. Bhattacharya, S. W. Kennerly, D. W. Beekman, and M. Dutta, Self-assembled InAs–GaAs quantum dot intersubband detectors, *IEEE J. Quantum Electron.* **35**, 936–943 (1999).
4. Y. Arakawa and H. Sakaki, Multidimensional quantum well laser and temperature dependence of its threshold current, *Appl. Phys. Lett.*, **40**, 939–941 (1982).
5. I. L. Medintz, A. R. Clapp, H. Mattoussi, E. R. Goldman, B. Fisher, and J. M. Mauro, Self-assembled nanoscale biosensors based on quantum dot FRET donors, *Nat. Mater.* **2**, 575–576 (2003).
6. K. Leung and K. B. Whaley, Electron–hole interactions in silicon nanocrystals, *Phys. Rev. B* **56**, 7455–7468 (1997).
7. D. J. Norris and M. G. Bawendi, Measurement and assignment of the size-dependent optical spectrum in CdSe quantum dots, *Phys. Rev. B* **53**, 16338–16346 (1996).
8. C. Garcia, B. Garrido, P. Pellegrino, R. Ferre, A. Moreno, J. R. Morante, L. Pavesi, and M. Cazzanelli, Size dependence of lifetime and absorption cross section of Si nanocrystals embedded in SiO_2, *Appl. Phys. Lett.* **82**, 1595–1597 (2003).
9. W. Shockley and H. J. Queisser, Detailed balance limit of efficiency of p–n junction solar cells, *J. Appl. Phys.* **32**, 510–519 (1961).
10. U. Bockelmann and G. Bastard, *Phys. Rev. B* **24**, 8947 (1990).
11. J. Urayama, T. B. Norris, J. Singh, and P.Bhattacharya, Observation of phonon bottleneck in quantum dot electronic relaxation, *Phys. Rev. Lett.* **86**, 4930–4933 (2001).
12. F. Quochi, M. Dinu, N. H. Bonadeo, J. Shah, L. N. Pfeiffer, K. W. West, and P. M. Platzman, Ultrafast carrier dynamics of resonantly excited 1.3-μm InAs/GaAs self-assembled quantum dots, *Physica B* **314**, 263–267 (2002).

13. B. L. Wehrenberg, C. Wang, and P. Guyot-Sionnest, Interband and intraband optical studies of PbSe colloidal quantum dots, *J. Phys. Chem. B* **106**, 10634–10640 (2002).

14. V. Klimov, D. McBranch, A. Mihkailovsky, C. Leatherdale, and M. Bawendi, Mechanisms for intraband energy relaxation in semiconductor quantum dots: the role of electron–hole interactions, *Phys. Rev. B* **61**, R13349–R13352 (2000).

15. D. F. Schroeter, D. J. Griffiths, and P. C. Sercel, Defect assisted relaxation in quantum dots at low temperature, *Phys. Rev. B* **54**, 1486–1489 (1996).

16. A. J. Nozik, Spectroscopy and hot electron relaxation dynamics in semiconductor quantum wells and quantum dots, *Annu. Rev. Phys. Chem.* **52**, 193–231 (2001).

17. S. E. Shaheen, R. Radspinner, N. Peyghambarian, G. E. Jabbour, Fabrication of bulk heterojunction plastic solar cells by screen printing, *Appl. Phys. Lett.* **79**, 2996–2998 (2001).

18. A. Zaban, O. I. Micic, B. A. Gregg, and A. J. Nozik, Photosensitization of nanoporous TiO_2 electrodes with InP quantum dots, *Langmuir* **14**, 3153 (1998).

19. W. U. Huynh, X. Peng, A. P. Alivisatos, CdSe nanocrystal rods/poly(3-hexothiophene) composite photovoltaic devices, *Adv. Mater.* **1999**, 923–927 (1999).

20. W. C. Chan and S. Nie, Quantum dot bioconjugates for ultrasensitive nonisotopic detection, *Science* **281**, 216–218 (1998).

21. W. U. Huynh, J. J. Dittmer, W. C. Libby, G. L. Whiting, A. P. Alivisatos, Controlling the morphology of nanocrystal–polymer composites for solar cells, *Adv. Func. Mater.* **13**, 73–79 (2003).

22. See, for example, Evident Technology website, *www.evidenttech.com*, Nanophase Technologies Corporation website, *www.nanophase.com*, and Nanostructured and Amorphous Materials website, *www.nanoamor.com*.

23. A. A. Guzelian, U. Banin, A. V. Kadavanich, X. Peng, and A. P. Alivisatos, Colloidal chemical synthesis and characterization of InAs nanocrystal quantum dots, *Appl. Phys. Lett.* **69**, 1432–1434 (1996).

24. A. A. Guzelian, J. E. B. Katari, A. V. Kadavanich, U. Banin, K. Hamad, E. Juban, A. P. Alivisatos, R. H. Wolters, C. C. Arnold, and J. R. Heath, Synthesis of size-selected, surface-passivated InP nanocrystals, *J. Phys. Chem.* **100**, 7212–7219 (1996).

25. University of Illinois, Office of Technology Management.

26. I. Kang and F. W. Wise, Electronic structure and optical properties of PbS and PbSe quantum dots, *J. Opt. Soc. Am. B* **14**, 1632–1646 (1997).

27. G. Yu, J. Gao, J. C. Hummelen, F. Wudl, and A. J. Heeger, Polymer photovoltaic cells: enhanced efficiencies via a network of internal donor–acceptor heterojunction, *Science* **270**, 1789 (1995).

28. N. S. Sariciftci, Polymeric photovoltaic materials, *Curr. Opin. Solid State Mater. Sci.* **4**, 373–378 (1999).

29. W. U. Huynh, J. J. Dittmer, and A. P. Alivisatos, Hybrid nanorod–polymer solar cells, *Science* **295**, 2425–2427 (2002).

30. W. U. Huynh, J. J. Dittmer, N. Teclemariam, D. J. Milliron, and A. P. Alivisatos, Charge transport in hybrid nanorod–polymer composite photovoltaic cells, *Phys. Rev. B* **67**, 115326-1–115326-12 (2003).

31. E. Arici, N. S. Sariciftci, and D. Meissner, Hybrid solar cells based on nanoparticles of $CuInS_2$ in organic matrices, *Adv. Func. Mater.* **13**, 165–171 (2003).

32. N. C. Greenham, X. Peng, and A. P. Alivisatos, Charge separation and transport in conjugated-polymer/semiconductor–nanocrystal composites studied by photoluminescence quenching and photoconductivity, *Phys. Rev. B* **54**, 17628–17637 (1996).

33. H. J. Muller, *Semiconductors for Solar Cells*, Artech House, Boston, 1993.

34. S. V. Gaponenko, *Optical Properties of Semiconductor Nanocrystals*, Cambridge University Press, Cambridge, 1998.

35. J. Rodriguez-Viejo, K. F. Jensen, H. Mattoussi, J. Michel, B. O. Dabbousi, and M. G. Bawendi, Cathodoluminescence and photoluminescence of highly luminescent CdSe/ZnS quantum dot composites, *Appl. Phys. Lett.* **70**, 2132–2134 (1997).

36. X. Gao, W. C. W. Chan, and S. Nie, Quantum-dot nanocrystals for ultrasensitive biological labeling and multicolor optical encoding, *J. Biomed. Opt.* **7**, 532–537 (2002).
37. J. P. Wilcoxon, G. A. Samara, and P. N. Provencio, Optical and electronic properties of Si nanoclusters synthesized in inverse micelles, *Phys. Rev. B* **60**, 2704–2714 (1999).
38. J. P. Wilcoxon, P. P. Provencio, and G. A. Samara, Synthesis and optical properties of colloidal germanium nanocrystals, *Phys. Rev. B* **64**, 035417-1–035417-9 (2001).
39. O. I. Micic, S. P. Ahrenkiel, and A. J. Nozik, Synthesis of extremely small InP quantum dots and electronic coupling in their disordered solid films, *Appl. Phys. Lett.* **78**, 4022–4024 (2001).
40. D. V. Talapin, A. L. Rogach, I. Mekis, S. Haubold, A. Kornowski, M. Haase, and H. Weller, Synthesis and surface modification of amino-stabilized CdSe, CdTe and InP nanocrystals, *Coll. Surf. A* **202**, 145–154 (2002).
41. M. H. Wu, R. Mu, A. Ueda, and D. O. Henderson, Production of III–V nanocrystals by picosecond pulsed laser ablation, in *Mater. Res. Soc. Symp. Proc.*, Vol. 780, edited by M. Dinescu, Material Research Society, Boston, MA, 2003, pp. Y3.2.1–Y3.2.3.
42. D. Leonard, S. Fafard, K. Pond, Y. H. Zhang, J. L. Merz, and P. M. Petroff, Structural and optical properties of self-assembled InGaAs quantum dots, *J. Vac. Sci. and Tech. B* **12**, 2516–2520 (1994).
43. V. I. Klimov, D. W. McBranch, C. A. Leatherdale, and M. G. Bawendi, Electron and hole relaxation pathways in semiconductor quantum dots, *Phys. Rev. B* **60**, 13740–13749 (1999).
44. J. S. Salafsky, Exciton disassociation, charge transport, and recombination in ultrathin, conjugated polymer–TiO$_2$ nanocrystal intermixed composites, *Phys. Rev. B* **59**, 10885–10894 (1999).
45. D. S. Ginger and N. C. Greenham, Photoinduced electron transfer from conjugated polymers to CdSe nanocrystals, *Phys. Rev. B* **59**, 10622–10629 (1999).
46. C. A. Leatherdale, C. R. Kagan, N. Y. Morgan, S. A. Empedocles, M. A. Kastner, and M. G. Bawendi, Photoconductivity in CdSe quantum dot solids, *Phys. Rev. B* **62**, 2669–2680 (2000).
47. A. Cravino, M. A. Loi, M. C. Sharber, C. Winder, H. Neugebauer, P. Denk, H. Meng, Y. Chen, F. Wudl, and N. S. Sariciftci, Spectroscopic properties of PEDOTEHIITN, a novel soluble low band-gap conjugated polymer, *Synth. Met.* **137**, 1435–1436 (2003).
48. C. L. Yang, J. N. Wang. W. K. Ge, S. H. Wang, J. X. Cheng, X. Y. Li, Y. J. Yan, and S. H. Yang, Significant enhancement of photoconductivity on truly two-component and chemically hybridized CdS-poly (*N*-vinylcarbazole) nanocomposites, *Appl. Phys. Lett.* **78**, 760–762 (2001).
49. S. E. Shaheen, C. J. Brabec, N. S. Sariciftci, F. Padinger, T. Fromherz, and J. C. Hummelen, 2.5% Efficient organic plastic solar cells, *Appl. Phys. Lett.* **76**, 841–843 (2001).
50. F. Padinger, R. S. Rittberger, and and N. S. Sariciftci, Effects of postproduction treatment on plastic solar cells, *Adv. Func. Mater.* **13**, 85–88 (2003).
51. L.-S. Li, J. Hu, W. Yang, and A. P. Alivisatos, Band gap variation of size- and shape-controlled colloidal CdSe quantum rods, *Nano Lett.* **1**, 349–351 (2001).
52. A. E. Becquerel, Memoire sur les effets electriques produits sous l'influence des rayons solaires, *C.R. Acad. Sci Paris* **9**, 561–567 (1839).
53. B. O'Regan and M. Grätzel, A low-cost, high-efficiency solar cell based on dye-sensitized colloidal TiO$_2$ films *Nature* **353**, 737 (1991).
54. A. Hagfeldt and M. Grätzel, Light-induced redox reactions in nanocrystalline systems, *Chem. Rev.* **95**, 95 (1995).
55. A. Kay and M. Grätzel, Low cost photovoltaic modules based on dye sensitized nanocrystalline titanium dioxide and carbon powder, *Sol. Energy Mater. Sol. Cells* **44**, 99–117 (1996).
56. (a) M. Grätzel, Mesoporous oxide junctions and nanostructured solar cells, *Curr. Opin. Coll.. Interface Sci.* **4**, 314 (1999); (b) M. Grätzel, Photoelectrochemical cells, *Nature* **414**, 338 (2001).
57. A. Hagfeldt and M. Grätzel, Molecular photovoltaics, *Acc. Chem. Res.* **33**, 269 (2000).

Color Figure 1.8. Solar panels cover the rooftops of a Bremen, Germany housing complex [1].

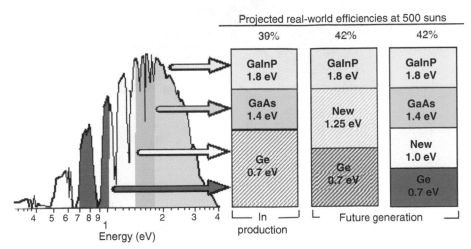

Color Figure 2.4. Diagram showing the potential photoconversion of sunlight using multi-junction III–V solar cells (courtesy of Spectrolab, Inc.).

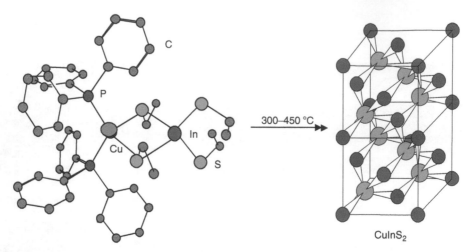

Color Figure 2.5. Pyrolysis of a single-source precursor [{PPh$_3$}$_2$Cu(SEt)$_2$In(SEt)$_2$] to produce a semiconductor material, CuInS$_2$.

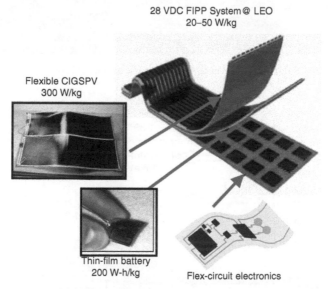

Color Figure 2.9. Flexible integrated power pack (FIPP) (courtesy of ITN Energy Systems).

Color Figure 2.12. Solar arrays of the International Space Station (graphic courtesy of NASA).

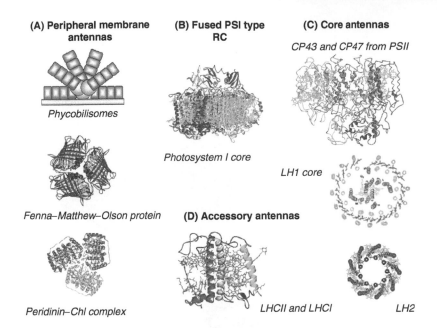

(A) Peripheral membrane antennas

Phycobilisomes

Fenna–Matthew–Olson protein

Peridinin–Chl complex

(B) Fused PSI type RC

Photosystem I core

(C) Core antennas

CP43 and CP47 from PSII

LH1 core

(D) Accessory antennas

LHCII and LHCI

LH2

Color Figure 3.4. Examples of photosynthetic light-harvesting antenna complexes. (A) Peripheral antennas: phycobilisomes from cyanobacteria and red algae (schematic); Fenna–Matthew–Olson protein from a green sulfur bacterium *Prosthecochloris aestuarii* (pdb code 4BCL, [9]); peridinin–Chl complex from a dinoflagellate *Amphidinium carterae* (pdb code 1PPR, [10]). (B) Fused PSI type RC from a cyanobacterium *Synechococcus elongatus* (pdb code 1JB0, [8]). (C) Core antennas: CP43 and CP47 from the PSII of cyanobacterium *Synechococcus elongatus* (pdb code 1FE1, [11]) and LH1 core from purple bacterium *Rhodopseudomonas palustris* (1PYH, [12]). (D) Peripheral antennas: LHCI and LHCII from algae and higher plants (pdb code 1RWT for LHCII, [13]) and LH2 from *Rhodopseudomonas acidophila* (pdb code 1KZU, [14]). Molecular graphics rendered using Web Lab Viewer from Molecular Simulations, Inc. (Figure courtesy of Dr. Alexander Melkozernov).

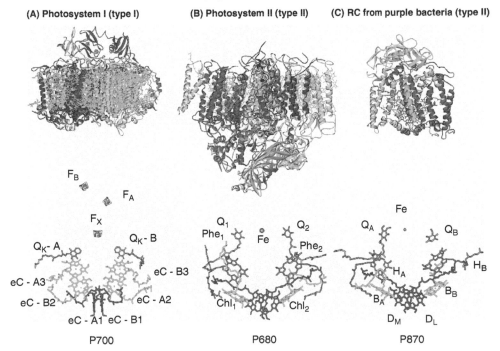

(A) Photosystem I (type I)

(B) Photosystem II (type II)

(C) RC from purple bacteria (type II)

P700

P680

P870

Color Figure 3.5. Structure of the photosynthetic reaction centers of Photosystem I (A), Photosystem II (B) and RC from purple bacteria (C). Upper panel: Side views of the pigment–protein complexes in the reaction centers. Lower panel: cofactors of electron transfer in the reaction centers. Molecular graphics rendered using Web Lab Viewer from Molecular Simulations, Inc. using atomic coordinates of the molecules in Protein Data Bank (code 1JB0 [8] for PSI from the cyanobacterium *Thermosynechococcus elongatus*, code 1S5L [9] for PSII from *T. elongatus*, and code 1M3X [15] for the RC from the purple bacterium *Rhodobacter sphaeroides*). Both models are shown at the same scale. (Figure courtesy of Dr. Alexander Melkozernov).

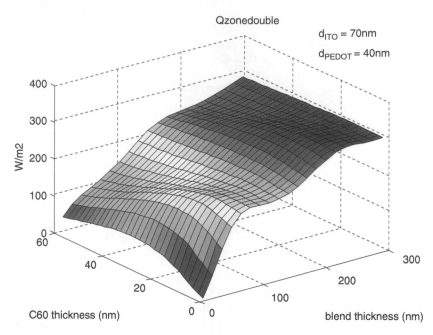

Color Figure 5.16. Absorbed power per unit area, i.e., $\int Q(z)dz$ as a function of $d_{blend} \in$ [5,290] nm and $d_{C_{60}} \in [0,58]$ nm for a zone comprising all of the blend layer, and extending $\min(d_{C_{60}}, 10)$ nm into the C_{60} layer. d_{ITO} and $d_{PEDOT-PSS}$ are fixed 70 and 40 nm, respectively.

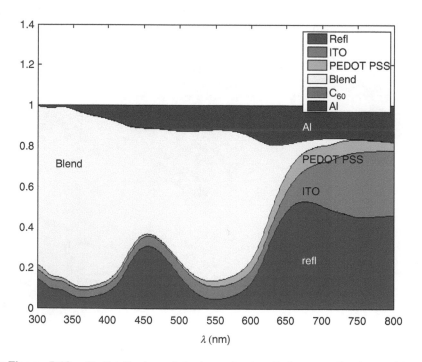

Color Figure 5.19. Redistribution of the incoming irradiation on reflection and layerwise absorption for structure 3. The structure is $(d_{ITO}, d_{PEDOT-PSS}, d_{blend}, d_{C_{60}}) = (100, 40, 210, 1)$ nm. As the C_{60} is very thin, it is not seen in the diagram. The curve is free from differences due to different number of irradiated photons.

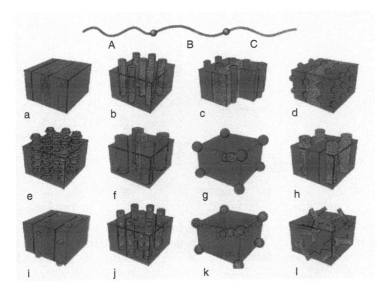

Color Figure 8.5. Representative self-assembled supramolecular nanostructures from a tri-block copolymer (From N. Hadjichristidis, S. Pispas, and G. Floudas, eds., *Block Copolymers: Synthetic Strategies, Physical Properties, and Applications*, John Wiley & Sons, New York, 2003. With permission.)

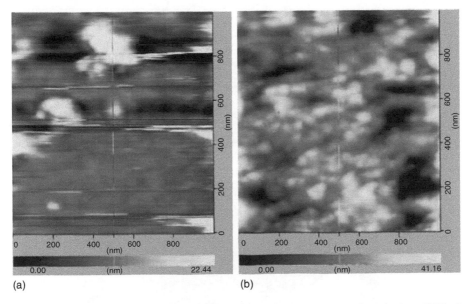

Color Figure 10.15. AFM images of H_2Pc:Me-PTC (1:1) films co-deposited on ITO glass substrates at 25°C (a) and −167°C (b). The thickness of the films was 500 nm. (From M. Hiramoto, K. Suemori, and M. Yokoyama, *Jpn. J. Appl. Phys.* **41**, 2763 (2002). With permission.)

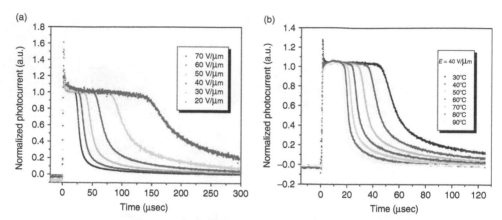

Color Figure 11.8. (a) Field dependence of the transient photocurrents measured in a guest–host sample of 4,4′-bis(phenyl-*m*-tolylamino)biphenyl (TPD) doped into polystyrene (1:1 wt.%). (b) Temperature dependence of the transient photocurrents measured in the same sample.

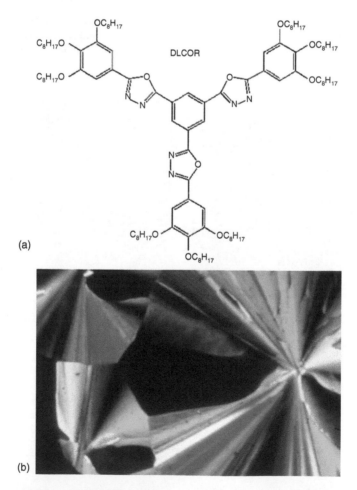

Color Figure 11.14. (a) Structure of the oxadiazole derivative DLCOR which forms a discotic columnar mesophase and (b) photograph showing the optical texture of an DLCOR sample taken at 35°C during the warming of the sample.

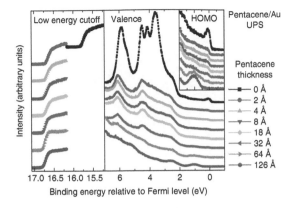

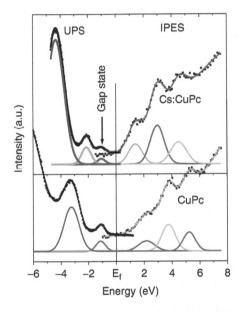

Color Figure 19.1. UPS cutoff and UPS HOMO evolution as a function of pentacene deposition onto gold. A total shift of 1 eV away from the Fermi level was seen in the UPS cutoff after the deposition of 18 Å of pentacene. Only changes in intensity of features were seen in the valence structure. (From N. J. Watkins, L. Yan, and Y. Gao, *Appl. Phys. Lett.* **80**, 4384 (2002). Copyright American Institute of Physics, 2002. With permission.)

Color Figure 19.4. Comparison of the composite of the UPS and IPES spectra of pristine and Cs-doped CuPc films. After doping, the HOMO and LUMO shift about same amount and the LUMO is close to Fermi level. (From L. Yan, N. J. Watkins, S, Zorba, Y. Gao, and C. W. Tang, *Appl. Phys. Lett.* **79**, 4148 (2001). Copyright American Institute of Physics, 2002. With permission.)

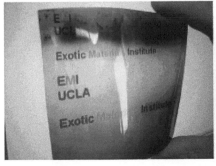

Color Figure 22.9. Semitransparent conducting films of PEDOT on a plastic substrate (left) and a glass slide (right), prepared by in situ solid-state polymerization of vacuum-deposited **22**. (From Meng, H.; Perepichka, D. F.; Bendikov, M.; Wudl, F.; Pan, G. Z.; Yu, W.; Dong, W., and Brown, S., *J. Am. Chem. Soc.* **2003**; *125*, 15151–15162.. Copyright 2003, American Chemical Society Publications. With permission.)

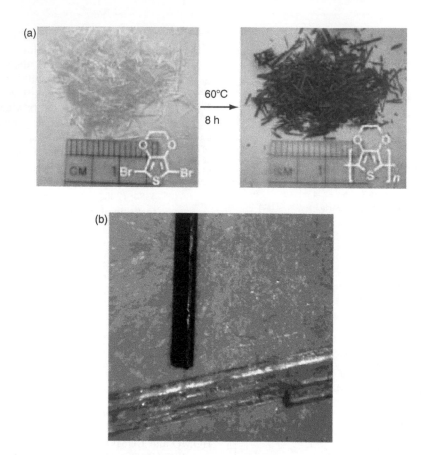

Color Figure 22.10. (a) Retention of crystal morphology of **22** even after polymerization. (From Meng, H.; Perepichka, D. F., and Wudl, F., *Angew, Chem. Int. Ed.* **2003**, *42*, 658–661. Copyright 2003, Wiley-VCH Verlag GmbH & Co. KGaA, Weinheim. With permission.) (b) Optical microscopy image of PEDOT (black, top) and **22** (colorless crystal, bottom). (From Meng, H.; Perepichaka, D. F.; Bendikov, M.; Wudl, F.; Pan, G. Z.; Yu, W.; Dong, W. and Braon, S., *J. Am. Chem. Soc.* **2003**, *125*, 15151–15162. With permission.)

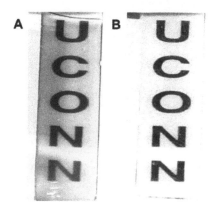

Color Figure 22.19. A 0.8 μm thick poly(T34bT) film coated on an ITO-coated glass slide in the (A) reduced state at −0.8 V and in the (B) oxidized semiconducting state at 0.4 V vs. Ag/Ag$^+$ reference electrode (0.47 V vs. NHE). The counter electrode was a platinum plate. (From Sotzing, G. A. and Lee, K., *Macromolecules* **2002**, *35*, 7281–7286. Copyright 2002, American Chemical Society Publications. With permission.)

58. R. Huber, J. Moser, M. Grätzel, and J. Wachtveitl, *J. Phys. Chem. B* **106**, 6494, (2002).

59. K. Hara and H. Arakawa, Dye-sensitized solar cells, *Handbook of Photovoltaic Science and Engineering*, edited by A. Luque and S. Hegedus, John Wiley & Sons (2003).

60. A. J. Nozik, Quantum dot solar cells, *Physica E*, **14**, 115 (2002).

61. A. J. Nozik, Quantum dot solar cells, *Proc. Electrochem. Soc.* **2001–10**, 61 (2001).

62. R. Vogel and H. Weller, Quantum-sized PbS, CdS, Ag$_2$S, Sb$_2$S$_3$, and Bi$_2$S$_3$ particles as sensitizers for various nanoporous wide-bandgap semiconductors, *J. Phys. Chem.* **98**, 3183 (1994).

63. P. Hoyer and H. Weller, Potential-dependent electron injection in nanoporous colloidal ZnO films, *J. Phys. Chem.* **99**, 14096 (1995).

64. D. Liu and P. V. Kamat, Photoelectrochemical behavior of thin cadmium selenide and coupled titania/cadmium selenide semiconductor films, *J. Phys. Chem.* **97**, 10769 (1993).

65. P. Hoyer and R. Konenkamp, Photoconduction in porous TiO$_2$ sensitized by PbS quantum dots, *Appl. Phys. Lett.* **66**, 349 (1995).

66. K. Sayama, H. Sugihara, and H. Karakawa, Photoelectrochemical properties of a porousNb$_2$O$_5$ electrode sensitized by a ruthenium dye, *Chem. Mater.* **10**, 3825 (1998).

67. S. Nasr, S. Hotsandani, and P. V. Kamat, Role of iodide in photoelectrochemical solar cells. Electron transfer between iodide ions and ruthenium polypyridyl complex anchored on nanocrystalline SiO$_2$ and SnO$_2$ Films *J. Phys. Chem. B* **102**, 4944 (1998).

68. K. Hara, T. Horiguchi, T. Kinoshita, K. Sayama, H. Sugihara, and H. Arakawa, Highly efficient photon-to-electron conversion with mercurochrome-sensitized nanoporous oxide semiconductor solar cells, *Sol. Energy Mater. Sol. Cells* **64**, 115 (2000).

69. K. Equchi, H. Koga, K. Sekizawa, and K. Sasaki, *J. Ceram. Soc. Jpn.* **108**, 1067 (2000).

70. S. Nasr, P. Kamat, and S. Hotchandani, Photoelectrochemistry of composite semiconductor thin films. Photosensitization of the SnO$_2$/TiO$_2$ coupled system with a ruthenium polypyridyl complex *J. Phys. Chem. B* **102**, 10047 (1998).

71. G. Redmond, D. Fitzmaurice, and M. Grätzel, Visible light sensitization by *cis*-bis(thiocyanato)bis(2,2′-bipyridyl-4,4′-dicarboxylato)ruthenium(II) of a transparent nanocrystalline ZnO film prepared by sol–gel techniques, *Chem. Mater.* **6**, 686 (1994).

72. H. Rensmo, K. Keis, H. Lindström, S. Södergren, A. Solbrand, A. Hagfeldt, S.-E. Lindquist, L. N. Wang, and M. Muhammed, High light-to-energy conversion efficiencies for solar cells based on nanostructured ZnO electrodes, *J. Phys. Chem.* **101**, 2598 (1997).

73. T. Rao and L. Bahadur, Band-edge movements of semiconducting diamond in aqueous electrolyte induced by anodic surface treatment, *J. Electrochem. Soc.* **144**, 179 (1997).

74. K. Keis, J. Lindgren, S. Lindquist, and A. Hagfeldt, Studies of the adsorption process of Ru complexes in nanoporous ZnO electrodes, *Langmuir* **16**, 4688 (2000).

75. J. He, H. Lindstrom, A. Hagfeldt, and S. Lindquist, Dye-sensitized nanostructured p-type nickel oxide film as a photocathode for a solar cell, *J. Phys. Chem. B* **103**, 8940 (1999).

76. S. Burnside, J. E. Moser, K. Brooks, M. Grätzel, and D. Cahen, Nanocrystalline mesoporous strontium titanate as photoelectrode material for photosensitized solar devices: increasing photovoltage through flatband potential engineering, *J. Phys. Chem. B* **103**, 9328 (1999).

77. S. Pelet, J. E. Moser, and M. Gratzel, Cooperative effect of adsorbed cations and iodide on the interception of back electron transfer in the dye sensitization of nanocrystalline TiO$_2$, *J. Phys. Chem. B* **104**, 1791(2000).

78. K. Tennakone et al. *Physica E* **14**, 190 (2002).

79. K. Tennakone, P. K. M. Bandaranayake, P. V. V. Jayaweera, A. Konno, and G. R. R. A. Kumara, Fabrication and characterization of mesoporous SnO$_2$/ZnO composite electrodes for efficient dye solar cells, *J. Mater. Chem.* **14**, 385–390 (2004).

80. G. K. R. Senadeera, K. Nakamura, T. Kimamura, Y. Wada, and S. Yanagida, Fabrication of highly efficient polythiophene-sensitized metal oxide photovoltaic cells, *Appl. Phys. Lett.* **83**, 5470–5472 (2003).

81. A. Kay and M. Grätzel, Synthesis, characterization, and adsorption studies of nanocrystalline aluminum oxide and a bimetallic nanocrystalline aluminum oxide/magnesium oxide, *Chem. Mater.* **14**, 2930 (2002).

82. R. Mu, M. Wu, and E. W. Collins, Development of Quantum Dot-Sensitized ZnO and TiO$_2$ Nanorod Arrays Solar Cells, NREL supported project (2004).

83. R. Czerw, H. S. Woo, D. L. Carroll, J. Ballato, and P. M. Ajayan, *Proceedings of the SPIE 4590* (BioMEMS and Smart Nanostructures) **4590**, 153 (2001).

84. J. Yang and J. Shen, Effects of discrete trap levels on organic light emitting diodes, *J. Appl. Phys.* **85**, 2699 (1999).

85. H. S. Woo, R. Czerw, S. Webster, D. L. Carroll, J. Ballato, A. E. Strevens, and W. J. Blau, Hole blocking in carbon nanotube–polymer composite organic light-emitting diodes based on poly(*m*-phenylene vinylene-*co*-2,5-dioctoxy-*p*-phenylene vinylene), *Appl. Phys. Lett.* **77**, 1393 (2000).

86. C. Y. Kwong, A. B. Djurisic, P. C. Chui, K. W. Cheng, and W. K. Chan, Influence of solvent on film morphology and device performance of poly(3-hexylthiophene):TiO$_2$ nanocomposite solar cells, *Chem. Phys. Lett.* 384, 372–375 (2004).

87. P. Wang, S.M. Zakeeruddin, R. Humphry-Baker, J. E. Moser, and M. Gratzel, Molecular-scale interface engineering of TiO$_2$ nanocrystals: improving the efficiency and stability of dye-sensitized solar cells, *Adv. Mater.* 15, 2101–2104 (2003).

88. S. S. Hegedus and A. Luque, Status, trend, challenges and the bright future of solar electricity from photovoltaics, in *Handbook of Photovoltaic Science and Engineering*, edited by A. Luque and S. Hegedus, John Wiley & Sons (2003).

89. R. Mu, M. Wu, and A. Ueda, unpublished results.

15

Solar Cells Based on Composites of Donor Conjugated Polymers and Carbon Nanotubes

Emmanuel Kymakis[1,2] **and Gehan A. J. Amaratunga**[1]

[1] *Engineering Department, Cambridge University, Cambridge, UK*
[2] *Photovoltaic Park, Technological Education Institute, Estavromenos, Heraklion, Crete, Greece*

Contents

Abstract   In this chapter, the photovoltaic properties occurring in single-wall carbon nanotubes (SWNTs)–donor conjugated polymer composites are reported. Photovoltaic devices based on the dispersed heterojunction concept, containing a blend of SWNTs and soluble poly(3-octylthiophene) (P3OT) were studied. The nanotubes not only act as electron acceptors, but also allow the electrons to transport efficiently along their length. The diodes (Al/polymer–nanotube composite/ITO) with low nanotube concentration (1%) show photovoltaic behavior, with an open circuit voltage of 0.75 V. The short circuit current is increased by two orders of magnitude compared with the pristine polymer diodes and the fill factor (ff) also increases from 0.3 to 0.4 for the nanotube–polymer cells. It is proposed that the photovoltaic response of these devices is based on the introduction of internal polymer–nanotube junctions within the polymer matrix, which due to a photoinduced electron transfer from the polymer to the nanotube contribute to enhanced charge separation and collection.

Keywords   polymer, nanotubes, composite, photovoltaic, dispersed, carbon, single-wall

15.1. INTRODUCTION

In the previous chapters it was seen that the photovoltaic performance of single layer devices was limited by the low electron mobility and the poor charge separation and transport. Here, we describe a novel way to tackle these problems — the formation of an interpenetrating blend of polymer electron donors and carbon nanotubes electron acceptors. Having a mixture of donor and acceptor material in these devices, charge separation is achieved due to a band offset at the interface and collection because of the existence of a bi-continuous network along which electrons and holes can travel through the electron acceptor and the electron donor, respectively, towards their respective contacts. In this way, the blend can be considered as a network of donor and acceptor heterojunctions that allows efficient exciton dissociation and balanced bipolar transport throughout its entire volume.

Since the discovery of photoinduced charge transfer between conjugated polymers (as donors) and buckminsterfullerene C_{60} and its derivatives (as acceptors), several efficient photovoltaic systems based on the donor–acceptor principle using a combination of polymer and fullerenes have been fabricated [1–3]. This charge transfer mechanism in a polymer matrix containing fullerene provides the motivation for investigating the use of the longest fullerene molecule, the carbon nanotubes as the electron transport material. The nanotubes consist of one or more sheets of graphite wrapped around each other in concentric cylinders, as shown in Figure 15.1. Individually, they could be metallic or semiconducting depending on their chirality (spiral conformation) and diameter, making them ideal reinforcing fillers in composite materials.

Firstly, Ago et al. [4] investigated the use of multiwalled carbon nanotubes (MWNTs) as the cathode electrode in poly(p-phenylene vinylene) (PPV) photodiodes. The polymer–nanotube composite was prepared by spin-coating a highly concentrated MWNT dispersion. An atomic force image showed that the PPV covers the surface of MWNTs and forms a well-mixed composite. The quantum efficiency of the device was significantly larger than the device having ITO as the electrode, suggesting that MWNT materials can act as a good cathode electrode due to the formation of a complex interpenetrating network with polymer chains. The polymer–nanotube interface quenches the radiative recombination, and hence enhances the photocurrent.

In this chapter, solar cells based on composites of donor conjugated polymers and carbon nanotubes are investigated. Poly(3-octylthiophene) (P3OT), acting as the photoexcited electron donor, is blended with SWNTs, which act as the electron acceptor. Efficient exciton dissociation is expected at the polymer–nanotube interface. Charge transfer would then follow by the transport of electrons through the nanotube length to the electron contacting contact (Al), and holes through the polymer to the hole collecting contact (ITO). This scenario is illustrated in Figure 15.2.

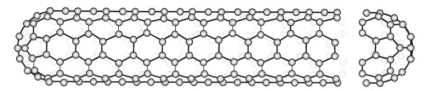

Figure 15.1.   Structural formula of a carbon nanotube.

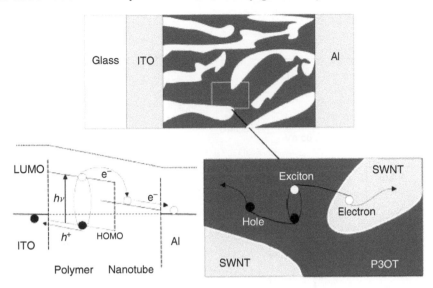

Figure 15.2. Schematic diagram of the separation and subsequent transport of excitons in a dispersed polymer–nanotube blend solar cell. The left diagram shows the dissociation process of a photogenerated exciton on polymer by electron transfer to the nanotube.

In order to efficiently analyze the exciton dissociation process at the polymer–nanotube interface, a p-semiconductor–metal junction is considered. The polymer P3OT is assumed to behave as a p-type semiconductor, since its hole mobility is much higher than its electron mobility. While the carbon nanotubes are assumed to behave as metals, due to the fact that they are either metallic, or semiconducting with a very small bandgap of 0.1–0.2 eV. When the polymer (p-semiconductor) comes into contact with the nanotube (metal), the difference in the electrochemical potential of the polymer and the contacting phase of the nanotube results in charge equilibration at the heterojunction. The electrochemical potential of the polymer is set by the position of its Fermi level (E_F) (energy level where the probability of finding an electron is one-half), while for the nanotube, the electrochemical potential of the contacting phase ($E_{F,n}$) is equal to its work function (energy required to remove an electron from the Fermi level to vacuum), as illustrated in Figure 15.3a. When the two phases are brought into contact (Figure 15.3b), electrons flow from the phase with more negative initial electrochemical potential to the other, in this case from the nanotube to the polymer, until the electrochemical potentials of both phases are in equilibrium. As a result of this charge-transfer process, both the polymer and the nanotube lose their original charge neutrality. Excess holes are produced in the nanotube while an excess of electrons appears in the polymer. This results in a positive charge on the metal (nanotube) and a negative charge accumulation on the semiconductor (polymer). Since the number of available states per unit energy in the metal (nanotube) far exceeds the number in a polymer, the accepted electrons do not change the position of the $E_{F,n}$. On the other hand, the Fermi level of the polymer becomes more negative, and the equilibrium position of the Fermi level for both phases is essentially equal to the initial value of the $E_{F,n}$.

The charge-transfer equilibration (separation of charge) causes an electric field (electric potential gradient) to rise, which is essential for effective exciton dissociation at the polymer–nanotube interface. This local electric field can split up the excitons if

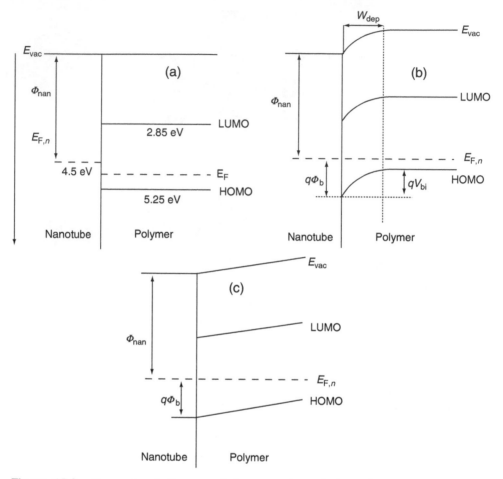

Figure 15.3. Energy band diagram of the nanotube and the polymer considered as a p-semiconductor–metal system (a) separate from each other and (b) in intimate contact. The Fermi level is the same across the whole junction. The parameter V_{bi} is defined as the amount of band bending and is called the built-in potential. (c) The case in which the depletion width is much larger than the diode thickness; an insulator–metal junction.

the exciton's binding energy (Coulomb attraction between electron and hole) is smaller than the built-in field ($q\Phi_{bi}$) of the junction. Otherwise charge transfer may occur but the excitons do not split up into their constituent charges and recombine eventually at the interface [5].

The thickness of the depletion layer width (W) at thermal equilibrium for a one-sided abrupt junction can be expressed as [6]:

$$W = 2L_D[2(\beta V_0 - 2)]^{1/2} \tag{1}$$

where $\beta = q/\kappa T$, e is the electronic charge, κ is the Boltzmann constant, T is the temperature, V_0 is the interface bending ($V = 4.5 - 2.85 = 1.65\,\text{V}$), and L_D is the Debye length, which is a characteristic length for semiconductors and is defined as:

$$L_D = \left[(\varepsilon\varepsilon_0\kappa T/q^2 N)\right]^{1/2} \tag{2}$$

where ε is the dielectric constant of the P3OT, ε_0 is the free-space dielectric permittivity, and N is the carrier concentration. In thermal equilibrium the depletion layer width of an abrupt junction at room temperature is about $8L_D$ (with $V_0 = 1.65$ V). The Debye length varies as $L_D \propto N^{-1/2}$ this results in a change of a factor of 2.82 per decade of the doping concentration. For a doping concentration of 10^{17} cm^{-3} and a dielectric constant ε of about 2.7, the Debye length L_D in P3OT is about 62 nm. The width of the depletion layer is then 500 nm. The polymer–nanotube devices have a typical thickness of about 100–200 nm. Consequently, the band bending and the corresponding depletion layer zone are smaller than the device thickness, and thus, a situation resembling that of a charged capacitor occurs [7]. The polymer–nanotube junction now treated as an insulator–metal junction is shown in the diagram of Figure 15.3c.

In this case the barrier height (Φ_b) of the interface, defined as the difference between the equilibrium Fermi level and the energy of the valence band of the polymer, equals to the built-in voltage (Φ_{bi}). Φ_{bi} can be calculated as [8]:

$$\Phi_b = (\Phi_{nan} - \mathrm{HOMO})/q \tag{3}$$

The highest occupied molecular orbital (HOMO) level of the P3OT is at 5.25 eV [9], while the nanotube work function is equal to 4.5 eV. So, the barrier height (Φ_b) at the interface is equal to 0.75 eV, which is larger than the exciton binding energy in P3OT (0.4 eV [10]). Therefore, interfacial photoinduced charge transfer that could contribute to efficient exciton dissociation can be anticipated [11].

Furthermore, in a dispersed heterojunction device, both charge separation and transport are very sensitive to the morphology of the blend. Charge separation requires uniform blending on the scale of the exciton diffusion length while transport requires bi-continuous paths from interface to contacts. In polymer–nanotube blends, the concentration of nanotubes is sufficient for charge percolation. Ideally (to ensure efficient exciton dissociation), the nanotubes should be within an exciton diffusion length from any point in the polymer. In principle, an electron can tunnel out of a common potential state when an external field is applied:

$$F_{junc} > E_{exc}/qr_{exc} \tag{4}$$

where E_{exc} is the exciton binding energy (0.4 eV) for P3OT and r_{exc} the exciton radius normally on the order of 10 nm. Thus, a field of at least 4×10^5 V/cm is required for the dissociation process. At the polymer–nanotube junction, exciton dissociation occurs in a radius r away from the junction, in which $E(r) = qV_{bi}/r = q0.75\text{V}/d > 4 \times 10^5$ V/cm. Hence, an adequate field for the excitons to dissociate will extend 19 nm away from the junction, which is within the boundary for exciton diffusion (10 nm). However, the nanotube should be treated as a cylindrical geometry rather than a planar arrangement (Figure 15.4). For an applied voltage V, the electric field amounts $E(r) = -V/r \ln(R_1/R_2)$, where R_1 (0.7 nm) and R_2 (1.4 nm) are the radii of the nanotube and the polymer coating (measured using a transmission electron micrograph [12]), as compared to $-V/d$ for two planar electrodes with an electrode distance d. Practically, this means that the field is enhanced by 1.44 times, therefore the exciton diffusion distance is about 27 nm. Therefore, all the excitons that are generated in a 27 nm radius away from the junction can be theoretically dissociated.

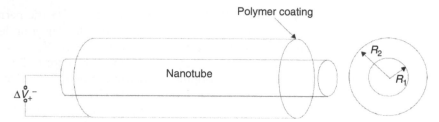

Figure 15.4. Schematic representation of the junction with the nanotube of radius R_1 coated by the polymer of radius R_2.

15.2. ABSORPTION SPECTROSCOPY

Figure 15.5 shows the unnormalized absorption spectra of P3OT and P3OT/SWNTs (1%) composite films at room temperature. Both films were 60 nm thick. The absorption spectra of the P3OT shows no significant change upon adding 1% of nanotubes by weight. This implies that in the blend, no significant ground state interaction is taking place between the two materials, and hence no charge transfer occurs in the ground state. In the photoexcited state, however, charge transfer from the photoexcited P3OT to the nanotubes is expected to occur [13].

15.3. CURRENT–VOLTAGE CHARACTERISTICS

The current–voltage characteristics of the ITO/P3OT/Al and ITO/P3OT–SWNTs/Al devices in dark and under white light illumination (AM1.5, $100\,\mathrm{mW/cm^2}$) from the

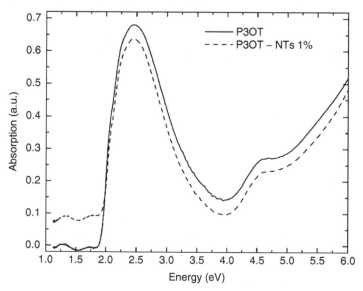

Figure 15.5. Absorption spectra of a 60 nm thick layer of P3OT and of a P3OT/SWNTs composite (1 wt.% NTs) on a quartz substrate. (From E. Kymakis and G. A. J. Amaratunga, *Appl. Phys. Lett.* **80**, 112–115 (2002). With permission.) The optical bandgap of the composite is at 2.4 eV.

ITO side, are depicted in Figure 15.6a and b, respectively. Forward bias is defined as positive voltage applied to the ITO electrode. The dark current is considerably higher in forward bias than in reverse bias, indicating a distinct diode behavior. The rectification ratio as a function of applied bias for this device is shown in Figure 15.7a. The rectification ratio reaches a maximum of 10^{-4} around 2 V. Under illumination, the blend device shows short circuit photocurrent density (I_{sc}) of 0.12 mA/cm^2 and an open circuit voltage (V_{oc}) of 0.75 V (Figure 15.7b), while the pristine device shows $I_{sc} = 0.7\,\mu$A/cm^2 and $V_{oc} = 0.35$ V. The $I–V$ characteristics for the ITO/blend/Al device (Figure 15.7b) shows an anomalous behaviour in the range -2.5 to 0 V, indicating a negative resistance (current decrease with increasing voltage). The origin of this kind of behavior is possible due to defects present in the polymer, or electron tunneling through the dispersed polymer–nanotube junctions.

The power efficiency of the blend device is dramatically increased from 2.5 × 10^{-5} to 0.04% with respect to the pristine one. Thus, a considerable improvement of photovoltaic effect is observed with the P3OT–SWNTs blend structure. The I_{sc} in the blend device is larger than that in the pristine device by about two orders of magnitude. Moreover, the V_{oc} and the fill factor in the blend device are also significantly larger than those in the pristine diode. It is proposed that the enhancement in the photovoltaic properties of the blend device is due to the introduction of internal

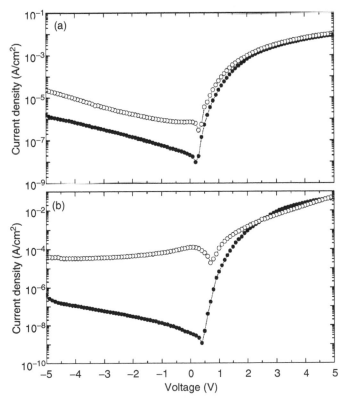

Figure 15.6. (a) $I–V$ characteristics of an ITO/P3OT/Al device in dark (filled circles) and under illumination (open circles). (b) The same data for an ITO/P3OT–SWNTs/Al device. (From E. Kymakis and G. A. J. Amaratunga, *Appl. Phys. Lett.* **80**, 112–115 (2002). With permission.)

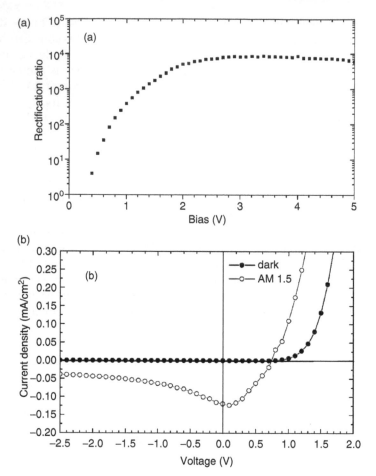

Figure 15.7. (a) Rectification ratio of an ITO/P3OT–SWNTs/Al device (b) $I-V$ characteristics of the blend device measured in the dark (filled circles) and under illumination (open circles) as a linear plot. (From E. Kymakis and G. A. J. Amaratunga, *Appl. Phys. Lett.* **80**, 112–115 (2002). With permission.)

polymer–nanotube junctions within the polymer matrix. These junctions act as dissociation centers, which are able to split up the excitons and also create a continuous pathway for the electrons to be efficiently transported to the negative electrode. This results in an increase in the electron mobility and hence balances the charge carrier transport to the electrodes.

By comparing the dark $I-V$ characteristics of the two devices, a decrease by a factor of 10 is observed at the leakage current (0 V) of the composite. This is believed to be due to the fact that the introduction of nanotubes produces a better polymer structure, the defects in the polymer chain are reduced and in effect, the conjugation length becomes longer.

15.3.1. Light Intensity Dependence Measurements

To further characterize the device ff, V_{oc}, and I_{sc} were determined as a function of the white light intensity. Figure 15.8a shows the linear variation of the I_{sc} with the light intensity. The photocurrent in a photovoltaic device varies as:

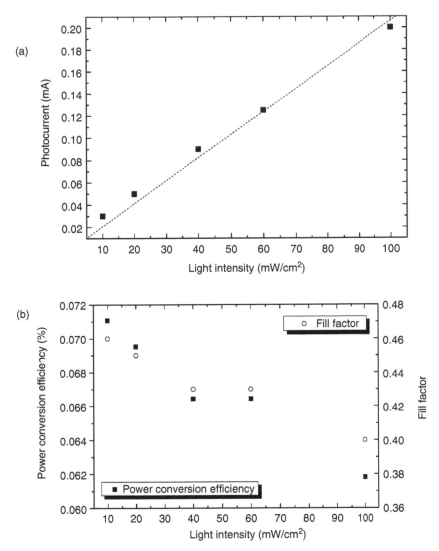

Figure 15.8. Light intensity dependence of (a) short circuit photocurrent, (b) power conversion efficiency and fill factor. (From E. Kymakis, I. Alexandrou, and G. A. J. Amaratunga, *J. Appl. Phys.* **93**, 1764–1768 (2003). With permission.)

$$I_{sc} \propto I_L^n, \quad n = f(N_f/N_t) \tag{5}$$

where I_L is the light intensity and N_f and N_t are the densities of free and trapped charges above the Fermi level. The value of n depends on the influence of traps in the photogeneration process and the conduction and recombination process in the active layer. If bimolecular recombination (electrons and holes) of charge carriers dominates, n approaches 0.5, whereas monomolecular (electrons or holes) recombination at a fixed number of sites gives rise to a linear intensity dependence ($n = 1$).

I_{sc} is found to scale linearly with the light intensity (Figure 15.8a), which is similar to the fullerene–polymer cells [15]. Also, larger photocurrents are measured when the light intensity is increased. This is due to higher rates of exciton dissociation, when the photon energy is increased. Hence, the linear dependence indicates the occurrence of a monomolecular recombination. A model, proposed by Greenham et al. [16] to explain the linear intensity dependence of the photocurrent fabricated from MEH-PPV sensitized blends, is adapted. Electrons hopping between the nanotube percolation paths and towards the aluminum electrode may reach "dead-ends" in the network, where they may be trapped before eventually recombing with holes in the polymer. An electron trapped in such a spur will repel other charge carriers from the following or the same pathway, and thus the spurs will act as recombination centers with single occupancy, giving rise to an intensity independent recombination rate.

On the other hand, the V_{oc} is constant at different intensities, its behavior will be discussed in the following section. From these results, the fill factor and the energy conversion efficiency (η) were extracted and plotted in Figure 15.8b as a function of the light intensity. The fill factor decreases from 0.47 at the lowest intensity to 0.4 at the highest level of illumination. This must indicate a light intensity dependent recombination. As a result, the power conversion efficiency decreases with increasing light intensities.

15.3.2. Open Circuit Voltage

The P3OT–SWNT solar cells show high V_{oc} values, which cannot be explained by the metal–insulator–metal (MIM) model, introduced by Parker for light emitting diodes [17]. In this model, the upper limit of the V_{oc} can be calculated from an estimate of the difference in the work functions between the two electrodes (Φ). From $\Phi_{ITO} = 4.7$ eV and $\Phi_{Al} = 4.3$ eV, we calculate $V_{oc} = 0.4$ V, which is in good agreement with the measured $V_{oc} = 0.35$ V for single P3OT devices but not for composite devices.

The effect of the difference in the work function between the two electrodes ($\Phi_{ITO} - \Phi_{Metal}$) on the V_{oc} can be ascertained by studying P3OT/SWNT-based devices and varying the metal of the negative electrode. In this case, four different metal electrodes were used: gold ($\Phi_{Au} = 5.1$ eV), aluminum ($\Phi_{Al} = 4.3$ eV), chromium ($\Phi_{Cr} = 4.5$ eV), and copper ($\Phi_{Cu} = 4.65$ eV), thus varying the $\Phi_{ITO} - \Phi_{Metal}$ from -0.4 to 0.4 eV, while keeping the rest of the device structure the same.

Figure 15.9 shows the variation of the open circuit voltage ($\Phi_{ITO} - \Phi_{Metal}$) for the four different metals utilized. V_{oc} ranges from 0.65 to 0.75 V, a total variation of 0.1 V for a 0.8 eV variation of the Φ_{Metal}. This result indicates that the work function of the metal has no significant effect on the V_{oc}, opposite to what would be if V_{oc} was determined only by the difference in the contact metal work functions according to the MIM model. The SWNTs in the active layer should also be considered. If the SWNTs form ohmic contacts with the metal electrode then the nanotube–polymer contacts could play an important role. Since the nanotubes are a single graphitic layer and the polymer coating is very thin (as seen by HREM in Ref. [12]), the electric field resulting from the built-in potential will be very strong. This field is large enough to cause exciton dissociation. The nanotubes, therefore act as the source of the dissociation field and also as the contacting channels that transfer the electrons to the negative electrode.

Therefore, the choice of the negative electrode metal would have a very little influence on V_{oc}, as is seen in Figure 15.9. In the case of gold ($\Phi_{ITO} - \Phi_{Au} = -0.4$ eV), a change of the polarity of the V_{oc} would be expected, in a similar fashion as it happens for a single polymer layer device. Interestingly, in our case, the polarity remains the same and the V_{oc} observed is lower than for the rest of the metals. Similar

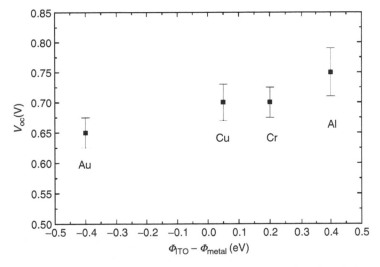

Figure 15.9. Open circuit voltage versus metal and ITO work function difference. (From E. Kymakis, I. Alexandrou, and G. A. J. Amaratunga, *J. Appl. Phys.* **93**, 1764–1768 (2003). With permission.)).

investigations have also been done by Brabec et al. [18] for polymer–fullerene based solar cells. The V_{oc} was found to be highly dependent on the lowest unoccupied molecular orbital (LUMO) level of fullerene. These results suggest that the negative electrode makes an ohmic contact to the fullerene by Fermi level pinning [19] to its LUMO level. Hence, the upper limit of the V_{oc} (built-in field of the device) can be determined by the difference of the fullerene LUMO and the polymer HOMO energy levels, and not by the difference in the electrode work functions.

Figure 15.10a shows the band diagram of a P3OT–SWNT dispersed heterojunction under flat band conditions. Also shown are the LUMO and the HOMO for P3OT relative to vacuum level and the work function levels of the SWNTs, Al, and ITO, assuming that the SWNTs are metallic. If we assume that the Brabec model applies in our case, the negative electrode will form ohmic contacts to the nanotube percolation paths and the positive electrode to the polymer (Figure 15.10b). In this way, the built-in potential of the device is the built-in potential of the polymer–nanotube junction. Therefore, the upper limit of the V_{oc} (built-in potential V_{bi} of the polymer–nanotube junction), must be equal to the difference in the P3OT HOMO and the SWNTs work function, which is 0.75 eV and very close to the measured V_{oc}. Therefore, the postulation of the pinning mechanism between the negative electrode and the SWNTs seems justified, like in the fullerene case.

15.4. NANOTUBE CONCENTRATION DEPENDENCE

According to the polymer–nanotube composite percolation characteristics [20], the composite keeps its insulating properties only for weight fractions between 0.1% and 5%. Photodiodes were constructed by varying the nanotube concentration for 0.25%, 0.5%, 1%, 2%, 3%, and 5%. The active layer thickness was the same in each case. The short circuit current was found to be significantly dependent of the nanotube concentration. Figure 15.11 shows the dependence of I_{sc} and V_{oc} on the weight

(a)   (b)

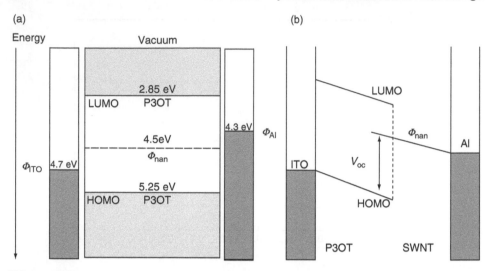

Figure 15.10. Potential energy diagrams relative to vacuum level of a P3OT/SWNT bulk heterojunction under (a) flat band conditions and (b) under short circuit conditions, assuming no interfacial layer at the metal contacts and pinning of the Al and ITO to the energy states of the polymer and SWNT, respectively.

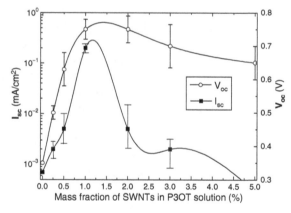

Figure 15.11. Short circuit current and open circuit voltage dependence on carbon nanotubes concentration. Error bars represent the scatter of at least four measurements for each nanotube composition.

percentage of nanotubes in the blend. Four or more samples were measured in each case using identical preparation procedures. The error bars represent the scatter of the experimental values and obtained from at least four samples. The wide range of the error bars suggest that the distribution of the junctions through the thickness of the film is not consistent in the four samples, since it is impossible to efficiently control the nanotube orientation within the polymer matrix. Therefore, the internal heterojunctions are expected to be randomly distributed, resulting in a partial discontinuity of the donor or acceptor phases. Furthermore, the likely increased energy level disorder in both phases may result in an increase of the charge trap density and in a reduction of the electron and hole mobilities. Therefore, a degradation of the overall device performance should be expected.

The photocurrent increases with increasing nanotube concentration of up to 1% and then decays. On the other hand, the open circuit voltage increases until 1%

and then tends to saturate at higher concentrations. The maximum efficiency was obtained from the composite containing 1% of carbon nanotubes. For higher concentrations, the photocurrent is believed to be limited due to a lower photogeneration rate, since the exciton generation takes place only in the polymer.

The maximum intensity of the emitted solar energy occurs at a wavelength of about 555 nm (2.2 eV), which falls within the band of green light. Upon increase in nanotube concentration, the distance between individual nanotubes becomes smaller than 555 nm, which results in a significant decrease in the absorption, and thus in the photogeneration rate. It can be concluded that there are sufficient interfaces to ensure efficient exciton dissociation and that there are bi-continuous conducting paths to provide percolation of electrons and holes to the appropriate electrode. However, as the nanotube content further increases, the photocurrent decreases, confirming the proportion of collected photons decreases, which indicates that the nanotubes do not contribute to the photocurrent.

15.5. FUTURE DIRECTIONS

Although the efficiencies of the polymer solar cells are significantly increased upon the introduction of nanotubes, they are, as yet, far from rivaling those of fullerene-based devices. The low photocurrent is mainly caused by the mismatch of the optical absorption of the polymer and the solar spectrum, and the fact that the nanotubes do not contribute to the photogeneration process, as the fullerenes do. This problem can be overcome by the incorporation of an organic dye with high absorption coefficient at the polymer–nanotube junctions, a process known as sensitization. By choosing dyes which absorb radiation over a large range, the area of photoexcitation of the diode can be broadened beyond the usual range of the polymer and cover a high proportion of the available solar energy and at the same time introduce photoconductivity properties to the nanotubes.

A molecular dispersed heterojunction solar cell consisting of a polymer and dye-coated carbon nanotubes blend sandwiched between metal electrodes has been recently developed by our group [21]. The dye naphthalocyanine (NaPc) is used as the main sensitizer, while nanotubes and P3OT act as the electron acceptor and the electron donor, respectively. In this case, sensitization at the polymer–nanotube interface is expected to take place via charge transfer:

$$D + h\nu \rightarrow D^*$$

$$D^* + SWNT \rightarrow D^+ + SWNT$$

$$D^+ + P3OT \rightarrow D + P3OT^+$$

The photoexcited dye transfers an electron to the nanotube, resulting in a positively charged dye and a negatively charged nanotube. Subsequently, a hole is transferred to the higher lying HOMO of P3OT. The charge transport to the electrodes occurs via the polymer for the holes and via the nanotubes for the electrons. In other words, the dye acts as the sensitizer, transferring electrons to the nanotube and holes to the polymer.

The incorporation of the NaPc in the P3OT–SWNT composite is found to dramatically increase the layer absorption resulting in a much higher photocurrent. The photocurrent under UV irradiation is much larger for the dye-coated cell than for the pure cell, and the photogeneration mainly occurs at the dye. However, the observed open circuit voltage and short circuit current under white light illumination for the dye-

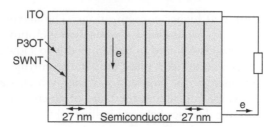

Figure 15.12.   Schematic diagram of an aligned SWNTs–polymer solar cell.

coated nanotube based cells, were significantly lower than the pure one. This attribute is believed to be due to the occurrence of a partial bimolecular recombination of the device and that the coated device is not optimized as the pure one.

Also, a significant limiting factor is the random distribution of the nanotubes in the polymer matrix. As it was seen in Section 1, for an exciton to be dissociated, it should be within a radius of 27 nm of the junction. Therefore, the ideal architecture of the solar cell is a structure in which the junctions are uniformly dispersed in a way that a junction exists every 27 nm. In order to accomplish this, an aligned nanotube configuration can be designed (Figure 15.12), consisting of a matrix of individual carbon nanotubes separated by 27 nm with each other. In this way, all the excitons generated in the P3OT can be theoretically dissociated at the polymer–nanotube junctions.

15.6. CONCLUSIONS

In this chapter, we review the prospect of using carbon nanotubes as the electron acceptor materials in polymer-based solar cells. A photovoltaic device based on SWNTs and a conjugated polymer, P3OT, is demonstrated. The operating principle of this device is that the interaction of the carbon nanotubes with the polymer, allows charge separation of the photogenerated excitons in the polymer and efficient electron transport to the electrode through the nanotubes. SWNTs doping was found to dramatically improve the photovoltaic performance of P3OT devices revealing a photocurrent larger than two orders of magnitude compared to that of the pristine diodes, and a doubling of the open circuit voltage. The I_{sc} was found to increase with light intensity, while a constant high V_{oc} was measured. This high V_{oc} was largely dependent on the negative electrode work function. These results suggest that the metal negative electrode forms ohmic contacts to the nanotube percolation paths. It is believed that the V_{oc} can be estimated from the energy differences between the work function of the nanotubes and the HOMO of the polymer.

The results show that the conjugated polymer–SWNTs composite represents an alternative class of organic semiconducting material that can be used to manufacture organic photovoltaic cells with improved performance. Further improvements in device performance are expected with more controlled film preparation and polymer doping.

REFERENCES

1. N. S. Sariciftci, L. Smilowitz, A. J. Heeger, and F. Wudl, Photoinduced electron transfer from a conducting polymer to buckminsterfullerene, *Science* **258**, 1474–1478 (1992).

2. S. Morita, A. A. Zakhidov, and K. Yoshimo, Doping effect of buckminsterfullerene in conducting polymer — change of absorption spectrum and quenching of luminescence, *Solid. State Commun.* **82**, 249 (1992).

3. J. J. M. Halls, K. Pichler, R. H. Friend, S. C. Moratti, and A. B. Holmes, Exciton diffusion and dissociation in poly(p-phenylene vinylene)/C_{60} heterojunction photovoltaic cell, *Appl. Phys. Lett.* **68**, 3120–3122 (1996).

4. H. Ago, K. Petrisch, M. S. P. Shaffer, A. H. Windle, and R. H. Friend, Composites of carbon nanotubes and conjugated polymers for photovoltaic devices, *Adv. Mater.* **11**, 1281–1285 (1999).

5. H. Becker, A. Lux, A. B. Holmes, and R. H. Friend, PL and EL quenching due to thin metal films in conjugated polymers and polymer LEDs, *Synth. Met.* **85**, 1289–1296 (1997).

6. S. M. Sze, *The Physics of Semiconductor Devices*, 2nd ed., John Wiley & Sons, New York, 1981.

7. W. R. Salaneck, S. Stafstrom, and J. L. Bredas, *Conjugated Polymer Surfaces and Interfaces*, Cambridge University Press, Cambridge, 1996.

8. E. H. Rhoderick and R. H. Williams, *Metal–Semiconductor Contacts*, Oxford University Press, New York, 1988.

9. K. Yoshino, M. Onoda, Y. Manda, and M. Yokoyama, Electronic states in polythiophene and poly(4-methylthiophene) studied by photoelectron-spectroscopy in air, *Jpn. J. Appl. Phys.* **27**, 1606–1608 (1988).

10. E. M. Conwell and H. A. Mizes, Photogeneration of polaron pairs in conducting polymers, *Phys. Rev. B* **51**, 6953–6958 (1995).

11. S. A. Jenekhe and J. A. Osaheni, Excimers and exciplexes of conjugated polymers, *Science* **265**, 765–768 (1994).

12. E. Kymakis, I. Alexandrou, and G. A. J. Amaratunga, High open-circuit voltage photovoltaic devices based on polymer–nanotube composites, *J. Appl. Phys.* **93**, 1764–1768 (2003).

13. H. Ago, M. Shaffer, D. S. Ginger, A. H. Windle, and R. H. Friend, Electronic interaction between photoexcited poly(p-phenylene vinylene) and carbon nanotubes, *Phys. Rev. B* **61**, 2286–2290 (2000).

14. E. Kymakis and G. A. J. Amaratunga, Single-wall-carbon nanotube/conjugated polymer photovoltaic devices, *Appl. Phys. Lett.* **80**, 112–115 (2002).

15. E. A. Katz, D. Faiman, S. M. Tuladhar, J. M. Kroon, M. M. Wienk, T. Fromherz, F. Padinger, C. Brabec, and N. S. Sariciftci, Temperature dependence for the photovoltaic device parameters of polymer–fullerene solar cells under operating conditions, *J. Appl. Phys.* **90**, 5343–5530 (2001).

16. N. C. Greenham, X. G. Peng, and A. P. Alivisatos, Charge separation and transport in conjugated-polymer/semiconductor-nanocrystal composites studies by photoluminescence quenching and photoconductivity, *Phys. Rev. B* **54**, 17628–17637 (1996).

17. I. D. Parker, Carrier tunneling and device characteristics in polymer light-emitting diodes, *J. Appl. Phys.* **75**, 1656–1966 (1994).

18. C. J. Brabec, A. Cravino, D. Meissner, N. S. Sariciftci, T. Fromherz, M. T. Rispens, L. Sanchez, and J. C. Hummelen, Origin of the open circuit voltage of plastic solar cells, *Adv. Funct. Mater.* **11**, 374–380 (2001).

19. L. J. Brillson, The structure and properties of metal–semiconductor interfaces, *Surf. Science Rep.* **2**, 123–326 (1982).

20. E. Kymakis, I. Alexandrou, and G. A. J. Amaratunga, Single-walled carbon nanotube polymer composites: electrical, optical and structural investigation, *Synth. Met.* **127**, 59–62 (2002).

21. E. Kymakis and G. A. J. Amaratunga, Solar cells based on dye-sensitisation of single-wall carbon nanotubes in a polymer matrix, *Sol. Energy Mater. Sol. Cells* **80**, 465–472 (2003).

16

Photovoltaic Devices Based on Polythiophene/C$_{60}$

L. S. Roman
Department of Physics, Federal University of Paraná, Curitiba-PR, Brazil

Contents

Abstract The solar spectrum matching of the low bandgap polythiophenes is attractive for device applications in photoelectric energy conversion. Polythiophenes can be chemically synthesized as soluble polymers, or deposited by electrochemical methods. They also can present higher mobility values, which is important for improving the efficiency of photovoltaic devices. The use of polythiophenes in combination with bulkminsterfullerene C$_{60}$ is promising; the photoinduced electron transfer from the excited state of a conducting polymer onto C$_{60}$ opened the possibility of constructing efficient photovoltaic devices with new approach for the development of plastic photodetectors and solar cells. As a single polymer layer device presents low efficiency due to the mechanism of charge generation and transport, the use of a C$_{60}$ molecule, which has high electron affinity value, sublimed onto the polymer in a heterojunction (bilayer) or mixed in the polymer film (blend) increased drastically the efficiency of the photovoltaic devices. In this chapter, photovoltaic devices based on several polythiophene derivatives in combination with C$_{60}$ and its derivatives will be discussed. Electrical, optical, and morphological studies in bilayers or blend active layers will be presented. The photoelectric conversion enhancement was achieved (i) by means of energy transfer between polymers in a blend active layer and (ii) by improving absorption in the active layers by light trapping.

Keywords   polythiophenes, C_{60}, bulkminsterfullerene, photovoltaic devices, organic photovoltaics, organic photodetectors, organic solar cells, polymer diodes, light trapping, photoconversion energy.

16.1.   INTRODUCTION

The research involving polymer (donor) and fullerene (acceptor) heterojunctions became intense in the area of photophysics and device physics [1,2] after the evidence of photoinduced electron transfer from the excited state of a conducting polymer onto bulkminsterfullerene (C_{60}) [3]. This discovery opened a new stream for the development of plastic photodetectors and solar cells with the possibility of constructing photovoltaic devices using this property [4,5]. As a single polymer layer device presents low efficiency due to the mechanism of charge generation and transport, the use of a C_{60} molecule, which has high electron affinity value, sublimed onto the polymer (donor) in a bilayer heterojunction or mixed in the polymer film (blend) in a bulk heterojunction increased drastically the efficiency of the photovoltaic devices [6–11]. The transfer of a photoexcited electron from the polymer onto the fullerene occurs in a timescale of 45 fsec and it is faster than any competing process [12]. The long life separated charges can be transported to the electrodes driven by the electric field thus producing the photocurrent [13,14]. Due to the low mobility values of the organic materials, the charge transport processes set the limit for the photocurrent, thus the performance of these "plastic" photovoltaic devices.

In this chapter, photovoltaic devices based on several polythiophenes derivatives in combination with C_{60} molecule (derivatized or not) will be discussed. Figure 16.1 shows the chemical structures of the materials used in the photovoltaic devices discussed here. The electrical, optical, and morphological aspects of such devices as in bilayers or blend active layers will be covered. The use of polythiophenes in combination with C_{60} is promising. The solar spectrum matching of the low bandgap polythiophenes is interesting for their applications as photoconversion energy devices because of their enhanced photochemical stability [15]. The polythiophene family of polymers can be designed to present bandgaps varying from 1.7 to 3 eV via simple substitution in the main chain and they can be chemically synthesized to be soluble or deposited by electrochemical methods [16,17]. They also can present higher mobility values, which is important for improving the efficiency of photovoltaic devices [18–21].

16.2.   ORGANIC PHOTOVOLTAIC DEVICES

The use of semiconducting organic materials as the active layer in optoelectronic devices offers some advantages in processing and new possibilities of device fabrication such as large-area devices. The mechanism of charge generation upon light absorption in organic materials is a secondary process. While in many inorganic materials the photon absorption produces free charges directly, in molecular materials the light absorption creates excited states, bound electron–hole pairs called excitons. These excitons must then be dissociated and the charges transported to the electrodes to produce photocurrent. The exciton dissociation is not a bulk process; excitons are dissociated at strong electric fields normally found at the interfaces of materials with different electron affinities and ionization potentials (in

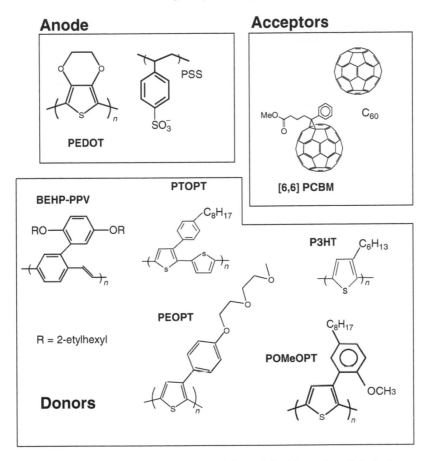

Figure 16.1. Chemical structures of the materials used in this work and their short names.

our case, polythiophenes [donor] and C$_{60}$ [acceptor]). One of the limiting aspects in these devices is the short lifetime (approximately nanoseconds) and diffusion length of excited states ($\approx$10 nm), therefore it is necessary to move the excited states to a site for charge separation within their lifetime in distance and in time. One approach to improve the exciton dissociation involves distributing the sites for photoseparation by forming a composite of two phase segregated materials with different electron affinities providing the spatially distributed interfaces necessary for exciton ionization by using a blending of donor and acceptor materials in the active layer [6]. An important issue in these devices is the phase structure formed by the mixture of the different components; the efficiency is ruled by the active layer film morphology [7,9,22–24]. To obtain good charge photogeneration and transport, one would need an intimate mixing of the components. An approach to this is the covalent linking of electron accepting and conducting moieties to a hole-transporting conjugated polymer backbone, the so called "double cable" macromolecules [25–29]. The synthesis of a novel bithiophene with a pendant fullerene substituent and its electrochemical polymerization are reported; light-induced electron spin resonance (ESR) measurements on the electrodeposited films reveal a photoinduced electron transfer from the donor cable (polythiophene) to the pendant acceptor cable (fullerene moieties). A less complex photovoltaic device is formed by a heterojunction of donor and acceptor layer on top of each other, a bilayer device. In this device, the processes occurring for the photocurrent formation can be investigated more efficiently, as it is a simpler device. By changing the thickness of the layers, it is possible to control the transport and optical properties of the devices improving the efficiencies. With the knowledge from simpler devices, it is possible to engender novel photovoltaic cells.

When comparing and characterizing photovoltaic devices, several parameters are important to set their performance. Some of them are described here, having as an example the single polymer layer device between indium tin oxide (ITO) as the anode and aluminum as the cathode (Figure 16.2). In this device under illumination, after the exciton dissociation, the charges are transported to the electrodes following the internal field raising the energy of the metal and lowering the energy of ITO, almost flattening the bands, and creating the potential difference called open circuit voltage. The open circuit voltage (V_{oc}) can be obtained from the current–voltage characteristics of an illuminated device when the current is zero or it can be measured with a high impedance voltmeter. At low temperature, the illumination can bring the flat band condition and the open circuit voltage is equal to the built-in potential (V_{bi}). At room temperature, the flat band condition is not completely achieved and a small correction term must be added to the V_{oc} value to obtain V_{bi}. Because of the charge generation and separation mechanism in organic devices, the efficiency limiting factors are therefore distinct from those in conventional cells (silicon p–n solar cells). While the maximum photovoltage (V_{oc}) achievable in silicon cells are limited to less than the magnitude of the band bending (ϕ_{bi}—built-in potential), it is common to observe experimentally that V_{oc} is greater than ϕ_{bi} in exciton-based photovoltaic devices [30]. The short circuit current density (J_{sc}) is the photocurrent value under zero applied bias. It is a result of the internal field created by the electrodes and the charge transport properties of the material. The J_{sc} dependence on light intensity (L_0) generally follows a power law, $J_{sc} = L_0^{\alpha}$ where α is a function of the densities of free n and trapped n_t charges above the Fermi level.

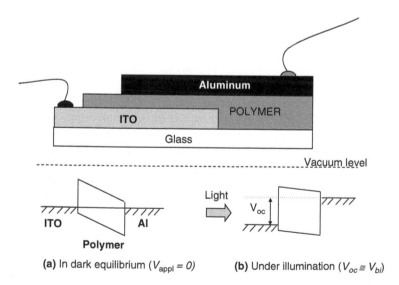

Figure 16.2. Photovoltaic device sandwich structure and the energy band schematic diagram in dark with the alignment of the Fermi levels and under illumination, the open circuit voltage at room temperature is somewhat lower than the built-in potential.

For $\alpha = 1$, there is linear dependence and the recombination is monomolecular with fixed sites, and bimolecular recombination occurs if α approaches 0.5.

The spectral response, or action spectrum, is obtained by measuring the electric response of the device upon monochromatic illumination of a wide range of wavelengths. It is given by means of the external quantum efficiency (EQE) or the incident photon to current efficiency (IPCE) of the device, that is, the ratio of the measured photocurrent by the intensity of incoming monochromatic light

$$\text{IPCE} = \frac{1240 J_{sc}}{\lambda I_\lambda} \tag{1}$$

where J_{sc} is the photocurrent density (A/cm^2), λ is the wavelength (nm), and I_λ is the light intensity (W/m^2). The action spectra can give valuable information about the mechanism of charge generation. The internal quantum efficiency (IQE) or quantum yield can be obtained from EQE by considering only the light absorbed in the device. The sum of total absorptance (A), transmittance (T), and reflectance (R) must be unity, $A + T + R = 1$, and in the case of thick devices, the transmittance is zero. Therefore the IQE is $\eta_{IQE} = (\eta/(1 - R)) = \eta/A$. In the case of thin devices, the transmittance may not be zero and the concept of IQE becomes less well-defined. Most of the polymer photovoltaic cells are quite thin and only the photons absorbed in the active regions of the device can contribute to the photocurrent. The useful absorbance in the cells is given by, $A_{active} = \sum_i A_i$ where i represents the photoactive layers in the device, where absorption can generate a charge carrier. It is possible to define quantum efficiency by describing the true efficiency of the material in the photoactive layers and the efficiency of exciton-to-charge generation (QEC), $\eta_{QEC} = \eta/A_{active}$ [31]. For solar cells, our interest is the power that can be extracted from the device and the quantity of interest is the power conversion efficiency (PCE), that is, the ratio of output electrical power (P_{out}) and input optical power (P_{in}):

$$\eta_{PCE} = \frac{P_{out}}{P_{in}} = \text{FF} \frac{J_{sc} V_{oc}}{L_0} \tag{2}$$

where the useful convention is the fill factor (FF), which is related to the quantity of electrical energy extracted from the solar cell. The fill factor for devices is described as

$$\text{FF} = \frac{(JV)_{max}}{J_{sc} V_{oc}} \tag{3}$$

where J and V are the values for the current density and voltage, respectively, for maximizing their product.

16.3. BILAYERS OF POLYTHIOPHENES/C$_{60}$

Bilayer devices can be quite efficient in photoconversion; they have an active layer formed by a heterojunction of a spin-coated donor polymer layer with C$_{60}$ sublimed on top. The exciton dissociation in a bilayer device occurs at the interface of the photoactive materials due to the difference in the electron affinity values. Bilayer photodiodes are organized systems that can be quite useful for the understanding of the processes occurring in the photocurrent formation. A polythiophene/C$_{60}$ device extensively studied is based on a donor layer of poly 3-(4'(1'',4''),7'' trioxaoctylphenylthiophene) (PEOPT) with C$_{60}$ (technocarbo, 99%) layer as the acceptor [8,32]. The spin-coated PEOPT polymer film with C_{60} molecule sublimed on top heterojunction presented a sharp interface, as can be seen from ellipsometry data, making possible to elaborate modeling for this system. Devices were fabricated using different thicknesses for the molecular layer ($L_n = 31$, 72, and 87 nm) keeping the polymer thickness constant ($L_p = 40$ nm), and the reverse way, keeping the fullerene thickness constant ($L_n = 34$ nm) and changing the polymer thickness ($L_p = 30$ and 40 nm). Optical and electrical modeling was done for these devices with different polymer and C_{60} thicknesses. These models are summarized below.

16.3.1. Optical Modeling of PEOPT/C_{60} Devices

In sandwich type devices, the active layers (multilayers) are found between two conducting electrodes, one transparent to the light and the other normally a mirror (aluminum). The stationary optical wave created inside the layers due to interference of the incoming wave with the wave reflected from the aluminum electrode strongly depends on the optical constants of the materials and principally their thicknesses. The photocurrent strongly depends on the resulting optical field distribution inside the device. In this approach, the optical electric field near the dissociation region must be enhanced, i.e., the light absorption in the active parts of the device should be increased. In a PEOPT/C_{60} heterojunction device where the exciton dissociation occurs at the interface, the light distribution could be controlled and maximized [8,32]. The metallic electrode mirror imposes a boundary condition that was used to maximize the optical electric field $|E^2|$ at one wavelength at this interface by changing the thickness of C_{60} layer, as can be seen in Figure 16.3. One can notice that the light is maximized at the polymer–C_{60} interface when the C_{60} thickness was 31 nm and minimal when it was 72 and 87 nm. This has a direct influence on the photocurrent magnitude. For modeling purposes, we assume that: (i) layers in the device are homogeneous and isotropic; (ii) the interfaces are flat and parallel compared to the wavelength of the light; (iii) the incident light are plane waves; (iv) the exciton diffusion obeys the diffusion equation

$$\frac{\partial n}{\partial t} = D\left(\frac{\partial^2 n}{\partial x^2}\right) - \frac{n}{\tau} + \frac{\theta_1}{h\nu} Q(x)$$

where n is the exciton density, D is the diffusion constant, τ is the mean lifetime of the exciton, θ_1 is the quantum efficiency of the exciton generation, $h\nu$ is the excitation energy of the incident light, and $Q(x)$ is the energy dissipated inside the multilayered device considering the distribution of light intensity due to interference of the incident wave with the one reflected from aluminum electrode. In this equation, the first term on the right represents excitons moving away by diffusion, the second term is the recombination term, and the third represents the generation rate of excitons (photogeneration). The distribution of the optical electric field is directly related to the distribution of the excited states in the active layer describing the excitation profile in the layered device; (v) excitons that contribute to the photocurrent dissociate into charge carriers at interfaces, which act as dissociation sites; (vi) the diffusion range of excitons is not dependent on excitation energy (wavelength); and (vii) all generated charges are contributing to the steady state photocurrent, i.e., no trapping of charges occur inside the device. Details about this modeling of the optical mode structure and the exciton distribution in thin-film organic structures can be found in Ref. [32]. From this model, it was found that the generated photocurrent is directly proportional to the incident light intensity generating photoexcited species at

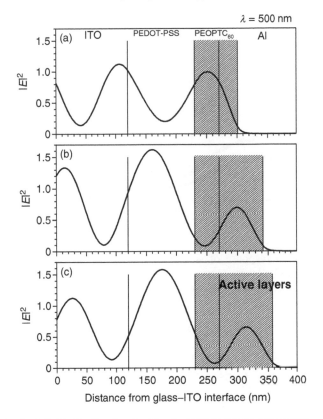

Figure 16.3. Optical field distribution inside the ITO/PEDOT/PEOPT/C$_{60}$/Al device. The polymer thickness was 40 nm and the C$_{60}$ thickness was 31, 72, and 87 nm. The dashed part refers to the PEOPT/C$_{60}$ active layers.

particular position in the film structure. This model could predict the photocurrent in devices where the molecule thickness was kept constant and the polymer thickness was changed; it was noticed that the polymer layer thickness influences the charge transport and the intensity of the light arriving at the interface but not the light distribution profile [8].

When the thickness of C$_{60}$ was varied, the model's prediction of the amount of light arriving at the PEOPT/C$_{60}$ interface due to different distributions have not reflected in linear production of photocurrent (see Figure 16.4). To model the experimental short circuit photocurrent action spectra, contributions were taken into account not only from the polymer absorption, but also from the C$_{60}$ layer to the photocurrent. For this reason, both interfaces, PEOPT/C$_{60}$ and C$_{60}$–Al, were considered as sites of exciton dissociation. The calculated action spectra were fitted to the experimental data with the Gauss–Newton algorithm by varying the thickness of the C$_{60}$ layer and diffusion ranges of the PEOPT and C$_{60}$ to get the best fit to the multiple sets of experimental data. It was possible to obtain good agreement of the model with the experimental data, resulting in diffusion ranges of 4.7 nm for PEOPT and 7.7 nm for C$_{60}$ [26]. Such bilayers were studied by photoluminescence with a view to extract the degree of photoluminescence quenching induced by a thin acceptor layer on the polymer [33]. The short exciton diffusion length in PEOPT is 5 nm, consistent with both PL quenching and photodiode performance. The modeling

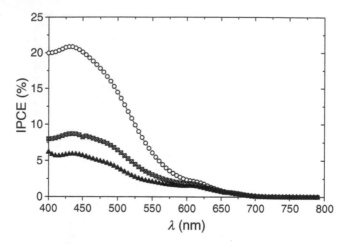

Figure 16.4. Spectral response of similar photodiodes ITO/PEDOT/PEOPT/C$_{60}$/Al with different thickness of C$_{60}$ layer, 31 nm (circles), 72 nm (squares), and 87 nm (triangles). The spectra were taken from these devices at short circuit condition.

suggests that a major part of the photocurrent must therefore be generated from the C$_{60}$ layer, and to help understand the dependence of the photocurrent on thickness of C$_{60}$, the transport of photogenerated charges in the active layers should also be taken into account. The charge transport in organic materials is much dependent on the morphology of the films. The polymer film formation can depend on the solvent, the deposition method, regioregularity, and the molecular weight, affecting mobilities of the carriers inside the material, thus the values of the charge mobilities. In C$_{60}$ film, it is not different. The mobility in C$_{60}$ is dependent on the purity, deposition conditions, and also the exposure to atmosphere (oxygen) [34]. The electrical transport model was developed in two steps: the transport of charges in these diodes in dark [35] and under illumination [36], taking into account the geometry of the devices.

16.3.2. Electrical Transport in PEOPT/C$_{60}$ Diodes in Dark

The investigations on the transport properties of these organic photovoltaic devices were done by measuring their current–voltage characteristics and developing a simple analytical electrical modeling. It was found that the forward dark current is limited by spatial charge of holes in the polymer layer and electrons in the molecule layer. The effective mobility of holes in the polymer and electrons in the C$_{60}$ are dependent on the different thicknesses of the polymer and C$_{60}$. The transport of charges in these PEOPT/C$_{60}$ bilayer devices can be described satisfactorily using this model under the following assumptions [34]:

(i)   The mobility of electrons in PEOPT is very low when compared with its mobility for holes.

(ii) The mobility of holes in C$_{60}$ is very low if compared with its mobility for electrons.

(iii) The intrinsic mobilities of holes μ_p (in the polymer layer) and the electrons μ_n (in the C$_{60}$ layer) are weakly dependent on the electric field E.

(iv) The transport of charge carriers for both materials are space charge limited.

(v) Both electrode contacts with PEOPT and the C$_{60}$ are ohmic.

(vi) PEOPT and C$_{60}$ have different electron affinity values in order to promote exciton dissociation. The interface barrier of PEOPT/C$_{60}$ is high enough to prevent the passage of charges through it, forming high charge density in each layer with a maximal recombination zone near the interface, which results in a monopolar current density of hole inside the polymer layer and electrons inside the molecule layer.

The current density and the Poisson equation for the polymer (J_p) and molecular (J_n) regions can be written as:

$$J_{p(n)} = \mu_{p(n)} \rho_{p(n)}(x) E(x) \quad \text{and} \quad dE(x)/dx = \pm \rho_{p(n)}/\varepsilon_{p(n)}$$

where $\rho_{p(n)}$ is the charge spatial density of holes (electrons) and E is the electric field. The current is $J = J_p = J_n$. The anode is located at $x = 0$ and the cathode at $x = L = L_p + L_n$. The electric field at the ohmic electrodes are null $E(0) = E(L) = 0$ and the voltage across the device is given by $\int_0^L E(x)dx = V - V_{bi}$, which is the sum of the voltage across L_n and L_p so $V - V_{bi} = V_p + V_n$, where V_{bi} is the built-in voltage and V the applied voltage across the whole device. Using the boundary conditions for E and solving the Poisson equation for the current, it is possible to find a Mott–Gurney law in each material region [37]:

$$J = 9/8 \varepsilon_{p}(n) \mu_{p(n),\text{eff}} (V - V_{bi})^2 / L_{p,n}^3. \tag{4}$$

where

$$\mu_{p(n),\text{eff}} = \mu_p/(1 + \gamma)^2 \quad \text{and} \quad \mu_{n,\text{eff}} = \mu_n/(1 + \gamma)^2 \tag{5}$$

with

$$\gamma = \left(\varepsilon_p \mu_p L_p^3 / \varepsilon_n \mu_n L_n^3 \right)^{1/2} \tag{6}$$

The space charge limited current flowing in PEOPT and C$_{60}$ regions induce an effective thickness dependent mobilities of the charges in the respective layers of the device. Using this model, it is possible to estimate the mobility of holes in PEOPT layer $\mu = 1 \times 10^{-6}$ cm^2/V s and the mobility of electrons in the fullerene layer $\mu = 3 \times 10^{-6}$ cm^2/V s, comparable to results in similar materials reported earlier [38,39]; the fitting of the curves is presented in Figure 16.5.

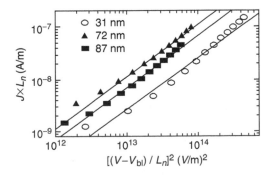

Figure 16.5. Logarithm plot of the experimental values for the current density times the C$_{60}$ thickness versus the square of the voltage across the whole device divided by the C$_{60}$ thickness values. The linear fittings have coefficients equal one, characteristic of the Mott–Gurney law.

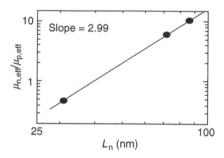

Figure 16.6. The ratio of the effective mobility values for electrons in C_{60} and holes in the polymer versus C_{60} thickness. The value of the slope is 2.99 as the linear fitting represented by the line.

The ratio between the effective mobilities is given by $\mu_{p,eff}/\mu_{n,\,eff} = L_p^3/L_n^3$, independent of the intrinsic mobility values of holes in PEOPT and electrons in C_{60}. It can be confirmed by plotting in the log–log scale the mobilities ratio versus the C_{60} thickness value in which the angular coefficient value is equal to three, as can be seen in Figure 16.6. The understanding of the transport of charges for these devices in dark conditions takes us to the next step, which is the modeling of the electrical transport under illumination.

16.3.3. Electrical Transport in PEOPT/C$_{60}$ Photodiodes under Illumination

As seen earlier in the optical model, the linear proportionality of the generated photocurrent to the light intensity distribution inside the PEOPT/C_{60} device explains very well the modification of the efficiency value for different polymer thicknesses when the C_{60} layer thickness was kept constant. The polymer acts like a filter for the active region. When the C_{60} thickness was varied, the distribution of optical electric field changed drastically and the measured photocurrent was no longer proportional to the light at the PEOPT/C_{60} interface . The bulk of the molecular layer contributes to a large part of the measured photocurrent. The electrical transport modeling for these photodiodes is an analytical description for the measured I–V characteristics under illumination of monochromatic light with the wavelength of $\lambda = 400, 460, 500$, and, 600 nm [32]. In this case also, the model is based on the basic assumptions described earlier for the transport model in dark and only the majority carriers (holes) are photoexcited in the polymer layer through exciton dissociation at the interface. It is considered that the fullerene layer behaves like a photoconductor under illumination with an electrical conductivity σ_m that comprises the contributions of the electrons generated at the heterojunction and in the bulk of C_{60}. The devices are fabricated with the active layers between two electrodes with different work function values to create an intrinsic (built-in) potential for charge collection. For a small enough applied voltage, the voltage drop across the devices is given by: $E_p L_p + E_m L_m = V_{bi} - V$, where $E_p(E_m)$ is the electric field in the polymer (molecule) layer, $L_p(L_m)$ is the thickness of the polymer (molecule) layer, V_{bi} is the built-in potential, and V is the applied voltage. The charge generation in the polymer occurs in a narrow region near the heterojunction, creating a superficial charge density Q at the polymer–fullerene interface [40]. From the discontinuity of the electric field at the interface, this charge $Q = \varepsilon_p E_p - \varepsilon_m E_m$, where $\varepsilon_p(\varepsilon_m)$ is the polymer (molecule) dielectric constants.

The total photocurrent is given by the sum of the current in the polymer and molecule layers $j_{ph} = j_p + j_m$. The photocurrent density in the polymer layer j_p is derived from the drift-diffusion equation for holes created at the interfaces and the photocurrent density in the C$_{60}$ layer that acts like a ohmic photoconductor ($j_m = \sigma_m E_m$), giving:

$$j_{ph} = j_{sc}\gamma - \frac{j_{sc}\gamma V}{V_{bi} + QL_m/\varepsilon_m} + \sigma_m\left[E_m - \gamma\frac{E_m^0 E_p}{E_p^0}\right] \tag{7}$$

where j_{sc} is the short current density ($V = 0$), the indices zero for the electric field means $V = 0$:

$$\gamma(V) = \left[\frac{1 - e^{-\beta\nu_p^0}}{1 - e^{\beta\nu_p}}\right]\left[\frac{1 - e^{\beta(V - V^*)/\alpha_p}}{1 - e^{-\beta V^*/\alpha_p}}\right], \quad \alpha_{p(m)} = 1 + [\varepsilon_{p(m)}/\varepsilon_{m(p)}][L_{m(p)}/L_{p(m)}]$$

where V^* is the applied voltage. When $V < 0$, $\gamma(V)$ tends to be independent of V for increasing reverse bias. The I–V characteristic is then ohmic and controlled by the electrical conductivity of C$_{60}$. In order to compare the experimental I–V characteristics with Equation (7), the experimental j_{sc} was taken and V_{bi} was set as 0.41 eV. The Q and σ_m were adjustable parameters to fit the curve for the device with $L_m = 31$ nm and under illumination of $\lambda = 400$ nm; for all the other measurements, the theoretical curves were calculated using the variation of Q with L_m. For $L_m = 31$ nm, $Q = 3 \times 10^{-3}$ C/m^2, for $L_m = 72$ nm, $Q = 2 \times 10^{-3}$ C/m^2, and for $L_m = 87$ nm, $Q = 1.9 \times 10^{-3}$ C/m^2. This model reproduces the main features of the experimental data, including the variation of the open circuit voltage with C$_{60}$ thickness, quite well; it is presented in Figure 16.7 for two wavelengths. From the fitting, it was also possible to obtain the electrical conductivity value for C$_{60}$ layer

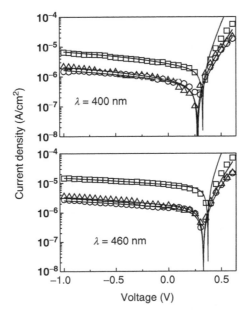

Figure 16.7. J–V characteristics for PEOPT/C$_{60}$ (40 nm) devices for three C$_{60}$ thickness values 31 nm (squares), 72 nm (triangles), and 87 nm (circles), under monochromatic illumination $\lambda = 400$ and 460 nm. The curves were plotted from Equation (7).

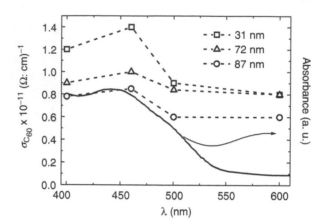

Figure 16.8. Variation of the electrical conductivity of C_{60} layer obtained from the fitting of the experimental J–V characteristics. The continuous line is the theoretical σ_e for $\mu = \mu_0$ and $\varepsilon = 3\varepsilon_0$.

depending on thickness and the wavelength of the light impinging on the device, as can be seen in Figure 16.8. It is possible to compare these conductivity values with the C_{60} electro-optical properties by applying the Maxwell's equations for waves propagating in a conducting medium. The zero-frequency electrical conductivity medium and absorption coefficient are related by

$$\sigma_e = \omega_e \sqrt{\left[\frac{2}{\mu\varepsilon}\left(\frac{\alpha}{\omega}\right)^2 + 1\right]^2 - 1}$$

where μ is the magnetic permeability of the medium, α the absorption coefficient, $\omega = 2\pi(c/\lambda)$, and c is the light speed in vacuum. Using the experimental α values for C_{60} reported in Ref. [32] and the above expression, it is possible to plot σ_e; this reflects the variation of the density of free charges in the bulk of C_{60} excited by a electromagnetic radiation of frequency ω. Comparing σ_e with σ_m, it is possible to have an idea of the fraction of the free charges created from exciton dissociation at the heterojunction device. The spectral profile σ_m follows the behavior of σ_e, even though the values are higher than σ_e, mainly for the device having a C_{60} thickness of 31 nm. In this device, the light intensity at the interface is high, the dissociation of the excitons creates high concentration of free charges in fullerene layer. For the other two devices, the light intensity at the interface is small and the σ_e is comparable to σ_m. In these devices, the optical and the transport models arrive at the same conclusion, i.e., most part of the photogenerated charges came from the fullerene layer. The deviation of σ_e values from σ_m values in the wavelength of 600 nm is expected since the absorption of the active layers in this wavelength is low and there is considerable charge transfer in C_{60}–Al interface when the device is illuminated with this energy.

16.4. BLENDS OF POLYTHIOPHENES/C$_{60}$

Photovoltaic devices based on blends of polythiophenes and C_{60} were extensively studied [5]. Blending two materials with different electron affinity values causes a distribution of the interfaces for exciton dissociation in the bulk of the organic layer. After the dissociation, the charges are transported to the electrodes using the respective material phase (donor–acceptor) driven by the electric field. The main issue when discussing devices based on mixtures of materials is their miscibility leading to different film morphologies, and the efficiency of blend devices depends on the morphology. A frequent result for solid-state molecular blends is the phase separation of the compounds; it can have structured patterns that can be on a micrometer or a nanometer scale. The structured patterns will change the area of internal interfaces drastically. Considering the diffusion length of excitons in conjugated polymers around 10 nm, this would also be the ideal phase domain size for improved exciton dissociation. Besides improving the exciton dissociation, both materials (donor and acceptor) must present continuous phase to the electrode in order to collect the charges, contributing to the photocurrent. Figure 16.9 shows the sketch of the blend active layer device with two donor–acceptor phases where the exciton in the donor phase is transferring an electron to the acceptor phase in the interface of the materials. The morphology of blended film is influenced by the deposition method, stoichiometry, the solvent properties, and the evaporation rate of the material solutions. In a three-component blend of two donor polythiophenes, (poly(3-(2′methoxy-5′octylphenyl) thiophene)) [POMeOPT] and (3-(4-octylphenyl)-2,2′-bithiophene) [PTOPT], and the acceptor C$_{60}$, the film morphology defined the efficiency of the photodiode [8]. The spin-coated films from POMeOPT:PTOPT:C$_{60}$) (1:1:2) solution presented good quality with roughness of 3 nm with no phase separation on the scale of 100 nm as can be seen in Figure 16.10, while the POMeOPT:PTOPT (1:1) spin-coated films presented phase separation in microscopic level. The three-blend diodes ITO/(POMeOPT:PTOPT:C$_{60}$) (1:1:2)/Al were significantly better than a mixture of one of the polymers with C$_{60}$.

In a different approach, a broadening the action spectra can be attempted with a mixture of several polymers absorbing in different parts of the spectrum. In a bilayer device when the donor layer was formed by a polymer blend, and the acceptor was C$_{60}$ sublimed on top, an enhancement in photoconversion was achieved by means of energy transfer, where one polymer with high absorption coefficient (PPV derivative), used as an antenna polymer to collect light, transfers the excitons to another (polythiophene derivative) with better transport properties as well as ionization of excitons [41]. The polymer blend in this case had intimate mixing in order to allow Förster energy transfer from the one polymer to the other. Besides this, the polythiophene absorption spectrum matched the emission spectrum from the PPV. Measuring the photoluminescence spectrum of the blend film, it was found to be similar to that of the neat PT-derivative film, suggesting that the energy absorbed in

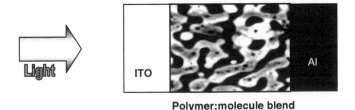

Polymer:molecule blend

Figure 16.9. Sketch of a blend active layer where the excitons are dissociated at the interface of the materials distributed in the bulk. After the dissociation the separated charges can be transported in each materials phase to the electrodes.

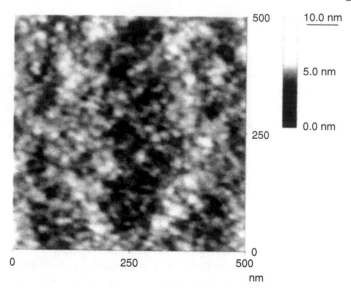

Figure 16.10. Scanning force microcopy picture of the surface of the POMeOPT: PTOPT:C_{60} (1:1:2) blend film spin-coated from dichlorobenzene solvent onto ITO/glass substrate.

PPV polymer was transferred to polythiophene. The blend donor layer device had always better performance than neat PPV/C_{60} or neat PT/C_{60} bilayer devices, even for three different polythiophenes used [39].

After achieving the bulk heterojunction interfaces with desired nanostructure for improving the exciton dissociation, the setback of these photovoltaic devices is the low mobility of the charges inside the matrix. It is known that polythiophenes can highly increase the mobility of holes if the polymer film became crystalline [42]. The enhanced crystallization of the polythiophenes can be achieved by annealing the polymer film to a temperature higher than its glass-transition temperature [43]. In this way Padinger et al. [11] proposed a postproduction treatment in photovoltaic cells based on poly-3-hexylthiophene:[6,6]-phenylC_{61}-butyric acid methyl ester, P3HT:PCBM blends as active layers, together with an external applied voltage. During the heat treatment, the chains are more mobile and the external applied potential higher than the open circuit voltage (V_{oc}) will inject charges into the matrix, supporting an orientation of the polymer chains inside the layer in the direction of the electric field. This treatment enhanced the conductivity of the charges across the polymer matrix, improving the $I–V$ characteristics of the devices and increasing the IPCE efficiency from 25% at wavelength of 420 nm to 60%, shifting the peak of the IPCE spectrum to $\lambda = 500$ nm (70% IPCE) and broadening up to 640 nm, which is even better for solar spectra match. The absorption of the posttreatment films enhanced only 10% compared to untreated devices, this result supports the idea that the enhancement of the efficiency in these devices is due to the better charge mobility because of the crystallization and orientation of the polymer chains in the matrix. The PCE of these devices is around 3.5%, equal to the state-of-the-art value for polythiophenes/C_{60} based devices.

In a device based on donor–acceptor blend as active layer, the use of asymmetric contacts of different work function gives a built-in electric field to extract a photocurrent, but this field depends also on the materials forming the active layer. In order to prevent anode–cathode contact through one and the same polymer or molecule phase, a vertical segregation was proposed enhancing the photocurrent, photovoltage, thus the fill factor in these devices [44]. The enhancement is due to the morphology of the diffuse donor–acceptor interface, where the acceptor is distributed throughout the donor layer. The vertical stratification came from the method of preparation of the active layer, by spin-coating the acceptor on top of the previously spin-coated donor layer, forming a diffusing junction due to the partial dissolution of the bottom layer and intermixing during the spin-coating of the upper layer. By making thicker polymer layers, to collect more of the light by absorption, we also decrease the field and reduce the collection efficiency, forcing us to compromise on the photocurrent generated. It is therefore desirable to make very thin devices, but to find ways of enhancing the absorption in these polymer layers. One way of reconciling these different requirements is to trap light into the polymer layers by diffraction into guided modes in the thin polymer films. This approach has been used to enhance light trapping and absorption in silicon solar cells, particularly in the energy range where optical absorption in silicon is low, thus increasing the conversion efficiency. While optical absorption in the conjugated polymers is high, it is possible to use a similar approach in order to trap light in the thin polymer films [45]. Using an elastomeric mold to transfer a submicron grating pattern from a commercial available grating template to the active polymer layer in a photovoltaic device, where the active layer was a blend of PTOPT:PCBM (1:2) was between thin gold/PEDOT–PSS as the anode and aluminum as the cathode (see Figure 16.11). The topography of the polymer layer became a grating, with a period and shape defined by the template, which in this case was a triangular-shaped grating with 2400 lines per millimeter giving a period of 416 nm. The grating profile on the polymer surface imaged by scanning force microscopy (SFM) (the upper part of Figure 16.12) confirms the shape and period of the master; the depth of the embossed grating was 15 nm, defined by processing conditions. The aluminum cathode was evaporated onto the blend following the topography of the polymer surface. We were able to find proper coupling conditions for the grating device, thus enhancing the absorption in the active layer and bringing up its efficiency. In the spectral response of the grating

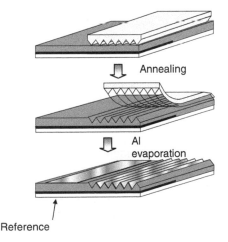

Figure 16.11. The patterned rubber stamp was left in conformal contact with half part the polymer active layer. The assembly was annealed and small pressure is applied. The stamp could be peeled, leaving the pattern on the polymer surface. The cathode was then evaporated through a shadow mask obtaining two contacts.

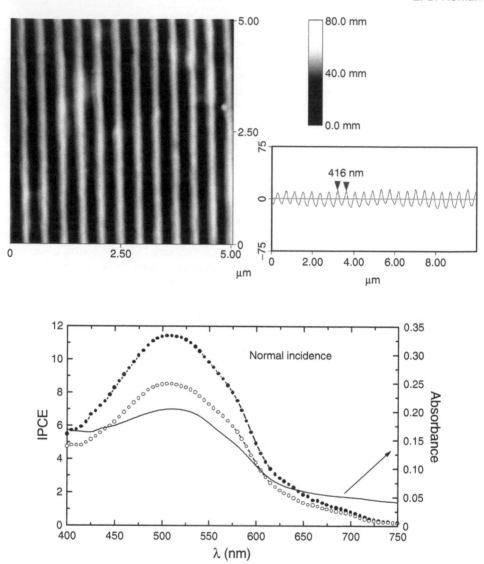

Figure 16.12. Action spectra of Au/PEDOT(PSS)/PTOPT:[6,6] PCBM (1:2)/Al device with and without grating. The SFM picture is taken from pattern blend polymeric layer reproducing the master grating period, 416 nm, as shown in the cross section of the SFM picture.

and reference diodes under normal incidence (bottom part of Figure 16.12) we notice the enhancement in efficiency all over the spectrum with the grating photodiode. At maximum peak position, the enhancement of the photon converted to electron efficiency (IPCE) was around 26%. The amount of enhancement and the peak position are qualities of the grating defined by the geometrical properties of the grating and device geometry. Among the parameters involved is the index of refraction, the thickness of the different layers, and the absorption coefficient of every layer in the device, the grating shape, period, and depth. To analyze the effects of the grating pattern, we study the spectral response using incident light with two states of polarization, with respect to the grating lines. Figure 16.13 shows the action spectra

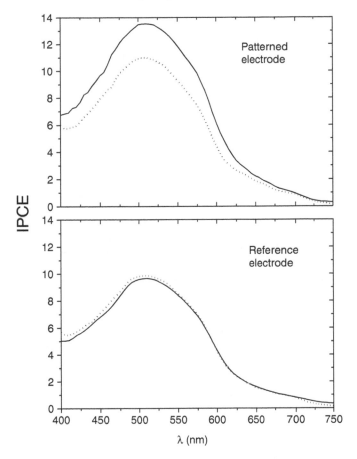

Figure 16.13. Measurements of action spectra done for Au/PEDOT(PSS)/ PTOPT:[6,6]PCBM (1:2)/Al devices in the patterned (up) and flat (down) electrodes under two orthogonal states of light polarization and normal incidence.

of the grating and reference diodes under normal light incidence in transverse and parallel light polarization. The transverse mode excites the grating coupling, trapping light, and thus improving photoconversion. In parallel mode we observe no strong influence of the grating, but we observe that the grating diode still exhibits higher photocurrent (mainly above 450 nm), because of the decrease in the polymer thickness at the valleys of the grating pattern, increasing the light penetration and the charge collection. The increase of the area of the polymer–metal interface due to the grating is less than 1% compared to the ordinary area. The reference diodes in both devices presented equal action spectra under both states of polarization.

16.5. CONCLUSIONS

The process of charge generation upon light absorption in organic photodiodes can be improved by: (i) modifying the light distribution inside the device (bilayer) and (ii) increasing the interfaces for exciton dissociation (blend) by controlling its morphology. By doing so, the charge generation in these organic devices is no longer the limiting aspect for improving efficiency. Such limitation is then transferred to the transport properties resulting from the low charge mobility found in organic materials. The low mobility of charges restricts the thickness of devices and causes losses since much of the photogenerated current is not delivered to the electrodes. An alternative for improving charge mobility in polythiophenes is the annealing of the polymer above its glass-transition temperature. The polymer crystallizes becoming more organized material with higher charge mobility. Devices based on polythiophenes/C_{60} derivative have presented higher efficiency ($\eta = 3.5\%$) by improving the organization of the blend. Applying an electric field across the sample through the electrodes at the same time as the heating treatment was done induced this organization. In these devices also the thickness is limited by the transport of charges and thus the amount of the absorbed light. To improve the absorption of light in these thin layers, light trapping in the active layer can be an alternative. It was presented as a simple method where the light trapping was done by soft embossing a reflection grating at the rear part of the blend-based device (polythiophene/C_{60} derivative).

The understanding of the optical and transport properties in these organic devices is indeed important to fabricate better devices. The light distribution inside these multilayered sandwich devices with a metallic electrode can be quite important, defining the external efficiency. It is necessary to identify how the light behaves inside a device with many interfaces of different materials. The optical modeling of a simple bilayer device (polythiophene/C_{60}) shows that depending on the fullerene thickness the molecule was responsible for a major part in the measured photocurrent. The transport of charge carriers in these devices was modeled in dark and under illumination with good experimental agreement, as well as with the optical model.

ACKNOWLEDGMENTS

The studies reported here have been performed by many researchers and I would like to acknowledge Olle Inganäs, Leif A. A. Pettersson, Thomas Granlund, Tobias Nyberg, Marcos Gomes E. da Luz, Jan C. Hummelen, Mats R. Andersson and Marlus Koehler. Funding through TFR, NFR, and SSF is gratefully acknowledged. I also acknowledge CNPq (PROFIX) for research stipendium.

REFERENCES

1. N. S. Sariciftci and A. J. Heeger, Reversible, metastable, ultrafast photoinduced electron transfer from semiconducting polymers to buckminsterfullerene and in the corresponding donor/acceptor heterojunctions, *International Journal of Modern Physics B*, **8**, 237–274 (1994).
2. C. J. Brabec, V. Dyakonov, J. Parisi, and N. S. Sariciftci (Eds.), *Organic Photovoltaics: Concepts and Realization*, Springer-Verlag, Heidelberg, ISBN 3-540-00405-x.
3. N. Sariciftci, L. Smilowitz, A. Heeger, and F. Wuld, Photoinduced electron transfer from a conduting polymer to bulkminsterfullerene, *Science*, **258**, 1474–1476 (1992).
4. N. S. Sariciftci and A. J. Heeger, Conjugated Polymer-acceptor Heterojunctions, Diodes, Photodiodes, and Photovoltaic Cells, U.S. Patent 5331183, August 17, 1992.
5. N. S. Sariciftci, D. Braun, C. Zhang, V. I. Srdanov, A. J. Heeger, G. Stucky, and F. Wudl, Semiconducting polymer–buckminsterfullerene heterojunctions: diodes, photodiodes, and photovoltaic cells, *Applied Physics Letters*, **62**, 585–587 (1993).

6. G. Yu, J. Gao, J. C. Hummelen, F. Wuld, and A. J. Heeger, Polymer photovoltaic cells: enhanced efficiencies via a network of internal donor–acceptor heterojunctions, *Science*, **270**, 1789–1791 (1995).

7. L. S. Roman, M. R. Andersson, T. Yohannes, and O. Inganäs, Photodiodes performance and nanostructure of polythiophene/C$_{60}$ blends, *Advanced Materials*, **9**, 1164–1168 (1997).

8. L. S. Roman, W. Mammo, L. A. A. Pettersson, M. Andersson, and O.Inganäs, High quantum efficiency polythiophene/C$_{60}$ photodiodes, *Advanced Materials*, **10**, 774–777 (1998).

9. S. E. Shaheen, C. J. Brabec, N. S. Sariciftci, F. Padinger, T. Fromherz, and J. C. Hummelen, 2.5% Efficient organic plastic solar cells, *Applied Physics Letters*, **78**, 841–843 (2001).

10. P. Schlinsky, C. Waldauf, and C. J. Brabec, Recombination and loss analysis in polythiophene based bulk heterojunction photodetectors, *Applied Physics Letters*, **81**, 3885–3887 (2002).

11. F. Padinger, R. S. Rittberger, and N. S. Sariciftci, Effects of postproduction treatment on plastic solar cells, *Advanced Functional Materials*, **13**, 85–88 (2003).

12. C. J. Brabec, G. Zerza, G. Cerullo, S. De Silvestris, S. Luzatti, J. C. Hummelen, and N. S. Sariciftci, Tracing photoinduced electron transfer process in conjugated polymer/fullerene bulk heterojunctions in real time, *Chemistry and Physics Letters*, **340**, 232 (2001).

13. M. C. Sharber, N. A. Shultz, and N. S. Sariciftci, Optical and photocurrent detected magnetic resonance studies on conjugated polymer/fulerene composites, *Physical Review B*, **67**, 085202 (2003).

14. C. Winder, C. Lungenschmied, G. Matt, N. S. Sariciftci, A. F. Nogueira, I. Montanari, J. R. Durrant, C. Arndt, U. Zhokhavets, and G. Gobsch, Excited state spectroscopy in polymer fullerene photovoltaic devices under operation conditions, *Synthetic Metals*, **139**, 577–580 (2003).

15. M. S. A. Abdou and S. Holdcroft, Photochemistry of electronically conducting poly(3-alkylthiophenes) containing FeCl$_4$ counterions, *Chemistry and Materials*, **6**, 962 (1994).

16. M. Berggren, G. Gustafsson, O. Inganäs, M. R. Andersson, T. Hjertberg, and O. Wernerström, Thermal control of near-infrared and visible electroluminescence in alkyl–phenyl substituted polythiophenes, *Applied Physics Letters*, **65**, 1489 (1994).

17. M. R. Andersson, M. Berggren, O. Inganäs, G. Gustafsson-Carlberg, D. Selse, T. Hjertberg, and O. Wennerström, Electroluminescence from substituted poly(thiophenes)—from blue to near infrared, *Macromolecules*, **28**, 7525–7529 (1995).

18. Z. Bao, A. Dodabalapour, and A. J. Lovinger, Soluble and processable regioregular poly(3-hexylthiophene) for thin film field effect transistor applications with high mobility, *Applied Physics Letters*, **69**, 4108–4110 (1996).

19. R. Valaski, E. Silveira, L. Micaroni, and I. A. Hümmelgen, Photovoltaic devices based on electrodeposited poly(3-methylthiophene) with tin oxide as the transparent electrode, *Journal of Solid State Electrochemical*, **5**, 261–264 (2001).

20. R. Valaski, L. S. Roman, L. Micaroni, and I. A. Hümmelgen, Photovoltaic devices based on electrodeposited poly(3-methylthiophene) with tin-oxide as transparent electrode. *Journal at Solid State Electrochemistry*, **5**, 261 (2001).

21. R. J. Kline, M. D. McGehee, E. N. Kadnikova, J. Liu, and M. J. Fréchet, Controlling the field effect mobility of regioregular polythiophene by changing the molecular weight, *Advanced Materials*, **12**, 507–510 (2003).

22. J. J. Halls, C. A. Walsh, N. C. Greenham, E. A. Marseglia, R. H. Friend, S. C. Moratti, and A. B. Holmes, Efficient photodiodes from interpenetrating polymer networks, *Nature*, **376**, 498–500 (1995).

23. C. Y. Yang and A. J. Heeger, Morphology of composites of semiconducting polymers mixed with C$_{60}$, *Synthetic Metals*, **83**, 85–88 (1996).

24. M. T. Rispens, A. Meetsma, R. Rittberger, C. J. Brabec, N. S. Sariciftci, and J. C. Hummelen, Influence of the solvent on the crystal structure of PCBM and the efficiency

of MDMO-PPV: PCBM 'plastic' solar cells, *Chemical Communications*, **17**, 2116–2118 (2003).

25. A. Cravino, G. Zerza, M. Maggini, S. Bucella, M. Svensson, M. R. Andersson, H. Neugebauer, and N. S. Sariciftci, A novel polythiophene with pendant fullerenes: toward donor/acceptor double-cable polymers, *Chemical Communications*, **24**, 2487–2488 (2000).
26. F. L. Zhang, M. Svensson, M. R. Andersson, M. Maggini, S. Bucella, E. Menna, and O. Inganäs, Soluble polythiophenes with pendant fullerene groups as double cable materials for photodiodes, *Advanced Materials*, **13**, 1871–1874 (2001).
27. A. Cravino and N. S. Sariciftci, Double-cable polymers for fullerene based organic optoelectronic applications, *Journal of Materials Chemistry*, **12**, 1931–1943 (2002).
28. S. Luzzati, M. Scharber, M. Catellani, N. Lupsac, F. Giacalone, J. L. Segura, N. Martin, H. Neugebauer, and N. S. Sariciftci, Tuning of the photoinduced charge transfer process in donor–acceptor double-cable copolymers, *Synthetic Metals*, **139**, 731–733, (2003).
29. A. Cravino, G. Zerza, M. Maggini, S. Bucella, M. Svensson, M. R. Andersson, H. Neugebauer, C. J. Brabec, and N. S. Sariciftci, A soluble donor–acceptor double-cable polymer: polythiophene with pendant fullerenes, *Monatshefte fur Chemie*, **134**, 519–527 (2003).
30. B. A. Gregg, Excitonic solar cells, *Journal of Physical Chemistry B*, **107**, 4688–4698 (2003).
31. L. A. A. Pettersson, L. S. Roman, and O. Inganäs, Quantum efficiency of exciton-to-charge generation in organic photovoltaic devices, *Journal of Applied Physics*, **89**, 5564–5569 (2001).
32. L. A. A. Pettersson, L. S. Roman, and O. Inganäs, Modeling photocurrent action spectra of photovoltaic devices based on organic thin films, *Journal of Applied Physics*, **86**, 487–496 (1999).
33. M. Theander, A. Yartsev, D. Zigmantas, V. Sundström, W. Mammo, M. R. Andersson, and O. Inganäs, Photoluminescence quenching at a polythiophene/C_{60} heterojunction, *Physical Review B*, **61**, 12957 (2000).
34. R. Könenkamp, G. Priebe, and B. Pietzak, Carrier mobilities and influence of oxygen in C_{60} films, *Physical Review B*, **60**, 11804–11808 (1999).
35. M. Koehler, L. S. Roman, O. Inganäs, and M. G. E. da Luz, Space charge limited bipolar currents in polymer/C_{60} diodes, *Journal of Applied Physics*, **92**, 5575–5577 (2002).
36. M. Koehler, L. S. Roman, O. Inganäs, and M. G. E. da Luz, Modeling bi-layer polymer/C_{60} photovoltaic devices, *Journal of Applied Physics*, **96**, 40–43 (2004).
37. M. Lampert and P. Mark, *Current Injection in Solids*, Academic Press, New York, 1970.
38. R. C. Haddon, A. S. Perel, R. C. Morris, T. T. M. Palstra, A. F. Hebard, and R. M. Fleming, C_{60} thin film transistor, *Applied Physics Letters*, **67**, 121–123 (1995).
39. L. S. Roman and O. Inganäs, Charge carrier mobility in polythiophene based diodes, *Synthetic Metals*, **125**, 419–421 (2002).
40. S. Berleb, W. Brütting, and G. Paasch, Interfacial charges and electric field distribution in organic hetero-layer light-emitting devices, *Organic Electronics*, **1**, 41 (2000).
41. L. Chen, L. S. Roman, M. D. Yohansson, M. Svensson, M. R. Andersson, R. A. J. Janssen, and O. Inganäs, Excitation transfer in polymer photodiodes for enhanced quantum efficiency, *Advanced Materials*, **12**, 1110–1115 (2000).
42. J. J. Dittmer, E. A. Marseglia, and R. H. Friend, Electron trapping in dye/polymer blend photovoltaic cells, *Advanced Materials*, **12**, 1270–1272 (2000).
43. Y. Zhao, G. X. Yuan, P. Roche, and M. Leclerc, A calorimetric study of the phase transitions in poly(3-hexylthiophene), *Polymer*, **36**, 2211–2214 (1995).
44. L. Chen, D. Godovsky, O. Inganäs, J. C. Hummelen, R. A. J. Janssen, M. Svensson, and M. R. Andersson, Polymer photovoltaic devices from stratified multilayers of donor–acceptor blends, *Advanced Materials*, **12**, 1367–1370 (2000).
45. L. S. Roman, O. Inganäs, T. Granlund, T. Nyberg, M. Svensson, M. R. Andersson, and J. C. Hummelen, Trapping light in polymer photodiodes with soft embossed gratings, *Advanced Materials*, **12**, 189–195 (2000).

17

Alternating Fluorene Copolymer–Fullerene Blend Solar Cells

Olle Inganäs[†], Fengling Zhang[†], Xiangjun Wang[†], Abay Gadisa[†], Nils-Krister Persson[†],
Mattias Svensson[*], Erik Perzon[†], Wendimagegn Mammo[*], and Mats R. Andersson[*]
[†]Biomolecular and Organic Electronics, IFM, Linköping University,
Linköping, Sweden
[*]Department of Materials and Surface Chemistry–Polymer Technology, Chalmers
University of Technology, Gothenburg, Sweden

Contents

17.1. INTRODUCTION

The extraction of energy from solar light through optical processes in conjugated organic molecules is the basis of energy supply on Earth through photosynthesis. The conversion of optical energy to electrical energy in photovoltaic conversion technology is a much less advanced technology, but can deliver high conversion efficiencies with inorganic semiconductor crystals, which are expensive. The development of plastic solar cells has the goal of delivering acceptable to moderate energy conversion efficiencies, with materials of low cost and with deposition and patterning techniques compatible with the large area application of solar cells. A great number of conditions have to be fulfilled by these materials. It is therefore important to develop families of polymers in which these conditions can be implemented, and to do that under conditions where the processability of plastics is retained.

Conditions necessary for high-efficiency photovoltaic conversion in organic polymers are not generally understood at present, but a number of aspects can easily be identified. The charge transport of photogenerated charge carriers must be easy enough to accommodate the photocurrents generated; the optical absorption must be strong enough to absorb the light impinging on a thin film; the spectral match between the optical absorption of the photovoltaic material or device and the solar spectrum should maximize photocurrent; and the photovoltage generated should be at a maximum to ensure that the energy conversion loss is minimized. There are trade-offs between several of these conditions, in particular for the spectral matching and the photovoltage extracted, and the optical absorbance in thicker devices versus transport without excessive losses. In the absence of a developed theory for the operation of plastic solar cells, no calculation is yet available for the upper limit to energy conversion efficiency in plastic solar cells. Indeed, this limit is bounded by that derived for general photovoltaic devices, however daunting that limit may look at present, where the best energy conversion efficiencies are reported at 3–3.5% in plastic solar cells. It is of interest to note that this is lower than the performance of the first silicon solar cells reported by Chapin in 1954 (6%) [1], and that the record performance today is of the order of 25% in homojunction cells and 30% in a multiple junction cell [2]. Upper limits to energy conversion, which follow the principles of thermodynamics, are found much above these numbers, and are still limited by materials and devices used.

Our choice of materials is driven by the use of the highly efficient photoinduced charge separation at closely located conjugated polymer and fullerene molecules upon optical excitation. This is a preferred method for charge generation simply because of its efficiency and because such interfaces can be generated by methods of synthetic chemistry and by processing materials as blends. We are therefore using fullerenes, such as C_{60} and their derivatives, as the acceptor compound in blends and composites with polymers. The polymer family is the polyfluorenes [3–5], chosen because of their flexibility in synthesis, propensity to generate higher mobility charge transport, possibility to design optical absorption, and their suitability for meso-scopic ordering in liquid crystalline phases [6]. They can also be produced as electron or hole conductors [7]. The high stability of polyfluorenes motivates the use of these materials in present day light-emitting diodes [8], and it is hoped that many of the integration, assembly, and stability aspects of these materials are evaluated for this device application.

The high bandgap found in the rigid polyfluorenes cause poor spectral overlap with the solar spectrum. Reports of polymer–polymer solar cells based on polyfluor-enes report moderate performance [9–17]. To compensate for the poor spectral match with the solar spectrum, it is necessary to reduce the bandgap of the polyfluorene, and to generate optical transitions in the red and near-IR part of the spectrum. Our approach is based on the inclusion of suitable comonomers in alternating polyfluor-ene copolymers (APFOs) (Figure 17.1).

The extension of the main absorption peak in the homopolymer polyfluorene down to 800 nm/1.5 eV in our copolymers is clear evidence that spectral matching is possible with these polymers also. We have been able to tune absorption peaks to energies between 1.5 and 2.1 eV, thus enabling a moderate spectral matching.

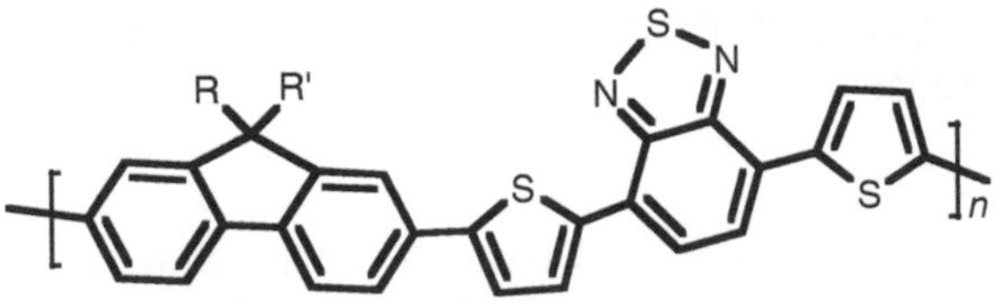

APFO-1, R = R' = C$_6$H$_{13}$
APFO-2, R = C$_6$H$_{13}$, R' = CH$_2$CH(C$_2$H$_5$)C$_4$H$_9$
APFO-3, R = R' = C$_8$H$_{17}$
APFO-4, R = R' = C$_{12}$H$_{25}$

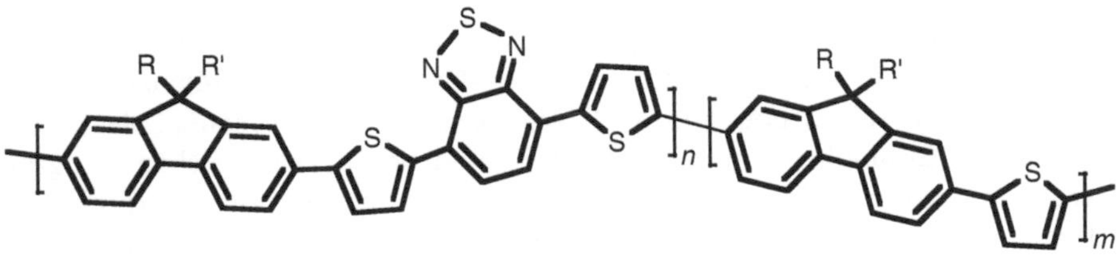

APFO-5, R = C$_6$H$_{13}$, R' = CH$_2$CH(C$_2$H$_5$)C$_4$H$_9$, n=50% m=50%

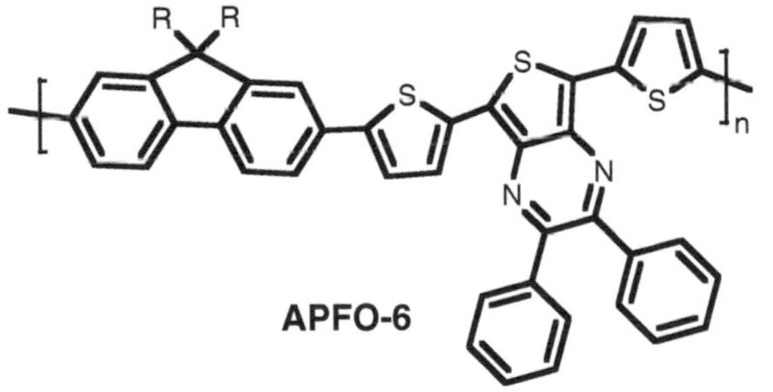

APFO-6

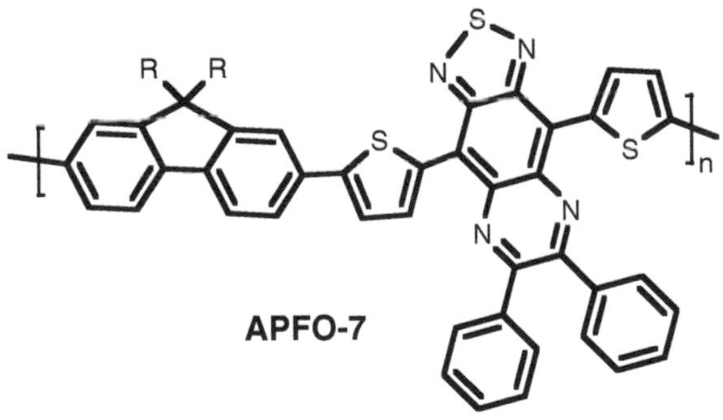

APFO-7

Figure 17.1. The structure of the prepared alternating polyfluorene polymers.

The high photovoltages recorded in the polyfluorene–fullerene blend plastic solar cells (often around 1 V) and a high quantum efficiency (typically around 50% over much of the absorption spectrum) helps to generate solar cell efficiencies in the range of 2% to 3%. In this chapter, we report current status of materials and devices, in conjunction with device physics and spectroscopy.

17.2. ALTERNATING POLYFLUORENE COPOLYMERS: SYNTHESIS AND CHARACTERIZATION

17.2.1. Chemical Synthesis

The structures of the prepared polymers is shown in Figure 17.1. The synthesis of the monomers and polymers (APFO1–APFO4) [18] with different side chains were prepared following the procedure described by Svensson et al. [19] (Figure 17.2) and references therein. Fluorene was lithiated with n-BuLi and alkylated with the corresponding alkyl bromide, followed by bromination with Br_2 in DMF to provide 2,7-dibromo-9,9′-dialkylfluorene. Both reactions gave high yields and the products were easy to purify. 2,7-Dibromo-9,9′-dialkylfluorene was dilithiated followed by the addition of 2-isopropoxy-4,4,5,5-tetramethyl-1,3,2-dioxaborolane, to obtain **1**. The boronic ester derivatives are highly stable, can be purified by chromatography, and can be stored in a normal atmosphere. 4,7-Dithien-2′-yl-2,1,3-benzothiadiazole was prepared by Suzuki coupling of 4,7-dibromo-2,1,3-benzothiadiazole with two equivalents of 2-thiopheneboronic acid and could be purified by chromatography or recrystallisation. The subsequent NBS bromination of 4,7-dithien-2′-yl-2,1,3-benzothiadiazole in DMF gave **2**, which also could be purified by recrystallization.

The polymerizations were carried out using the Suzuki coupling reaction with tetrakis(triphenylphosphine)Pd(0) as the catalyst and tetraethylammonium

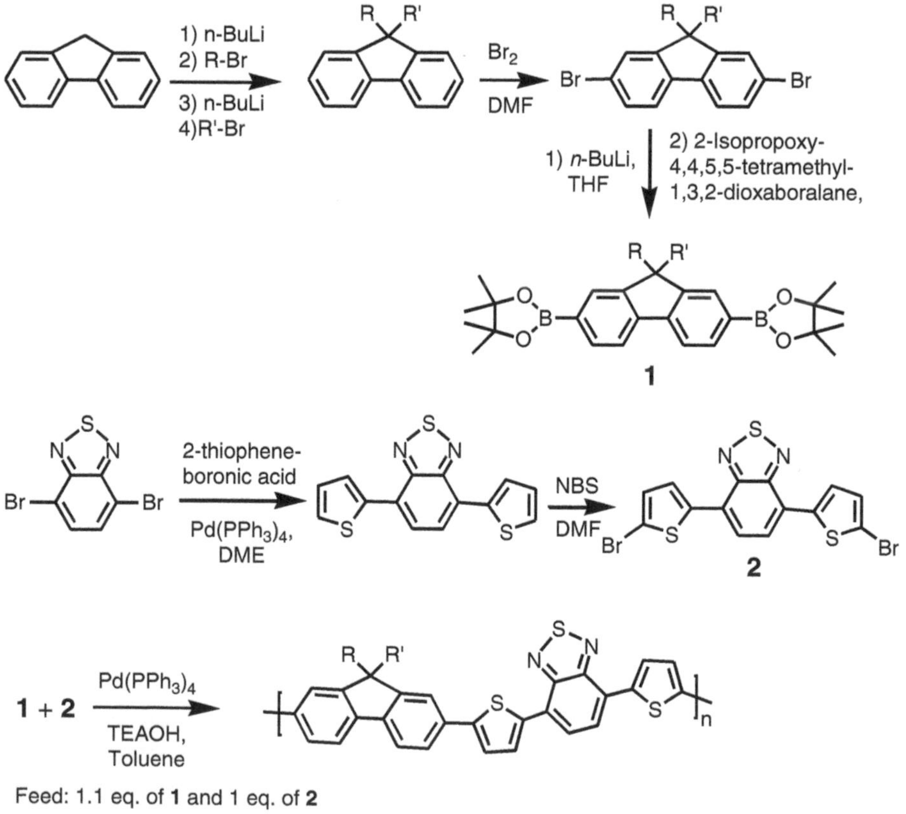

Figure 17.2. The synthetic route to polymers APFO1–APFO4.

hydroxide as the base. With tetraethylammonium hydroxide as the base, instead of K_2CO_3 or $NaHCO_3$ as for standard Suzuki conditions, the reaction is considerably faster and gives a higher molecular weight [20]. A high degree of polymerization is reached after 2–4 h with this procedure, in comparison with 24–48 h in the standard procedure [21]. For the polymers presented here, it was necessary to make the diborolane substitution on the fluorene monomer and the dihalogenation on the thiophene rings. Using borolane-substituted thiophene monomers and dihalogenated fluorene for polymerization with the tetraethylammonium hydroxide route, leads to a considerable amount of proto-deboronation and, thus, a low degree of polymerization. The polymers presented in Figure 17.1 (except APFO5–APFO7) contain relatively long segments without any solubilizing side-groups, and hence a 10% excess of the fluorene monomer was used to decrease the molecular weight so that high yields of $CHCl_3$-soluble polymers were obtained. The polymers were purified by Soxhlet extraction with diethyl ether to remove oligomers and unreacted monomers and $CHCl_3$ to extract the $CHCl_3$-soluble part. Polymer APFO5 was synthesized following a similar procedure [22], but APFO6 and APFO7 are still not fully characterized and the synthesis will be published elsewhere.

Size exclusion chromatography (SEC) was used to determine the molecular weights. The analyses were performed on a Waters 150 CV equipped with a refractive index detector using 1,2,4-trichlorobenzene as solvent at 135°C. Calibration was performed with narrow molecular weight polystyrene standards. The appearance of the liquid crystalline phase and the melting point of each polymer were measured with a Mettler FP90 hot plate coupled with an Olympus BH2 optical microscope at 10°C/min with plane-polarized light filter.

The molecular weight of these polymers are not high (Table 17.1), but sufficient to form high-quality films. These films can be heated to convert the polymer into a nematic liquid crystalline state at temperatures around 200°C (Table 17.1).

17.3. ALTERNATING POLYFLUORENE COPOLYMERS: ELECTRONIC STRUCTURE AND OPTICAL ABSORPTION

The purpose of the design of copolymers of fluorene and thiophene–benzothiazole–thiophene structures has been to generate optical absorption at lower energies by reducing the bandgap of the homopolymer polyfluorene. The homopolymer will have absorption mainly in the UV and blue part of the visible spectrum, and therefore has a particularly poor match to the solar spectrum. It is necessary to

Table 17.1. Properties of prepared polymers

	APFO1	APFO2	APFO3	APFO4	APFO5
M_n	3500	4900	4900	12000	11500
M_w	8500	12500	11800	31000	32000
LC (°C)	174	180	175	180	
Melting point (°C)	233	235	205	215	
Photoluminescence quantum yield in film (%)	18	7	20	22	25
PL emission max (nm)	702	707	692	695	682

include absorption at the green and red or near-IR part of the spectrum to generate a decent match with the solar spectrum. The philosophy therefore is to generate internal charge transfer within the main chain of the polymer, thereby reducing the dimerization and the bandgap [23–26].

17.3.1. Optical Transitions and Electronic Structure

Polymers produced with this philosophy have delivered the desired enhancement of optical absorption at low energies. The optical transmission spectra of a series of copolymers of polyfluorene (Figure 17.3) demonstrate that the optical absorption bands can be extensively modified towards the red range with the chemical approach used. The nature of the two distinct peaks found in the optical transmission spectra of all these polymers are theoretically ascertained with ZINDO semiempirical quantum chemistry calculations [27] on polymer APFO3. They suggest an interpretation of the two-peak spectrum of the copolymers as due to optical transitions at UV–blue wavelengths, between delocalized π-electron states, and a second optical absorption peak found at lower energies due to transitions localized at the comonomer TBT, and TB'T, where the B' stands for the different forms of modified benzothiazole systems found in polymers APFO6 and APFO7. These results are broadly consistent with recent modeling of closely related alternating copolymers based on fluorene [28]. These calculations are now extended to other methods to deduce the electronic structure.

Polymers APFO6, APFO7, which include more electron withdrawing groups in the T–B'–T unit, show that optical absorption bands with maxima as far into the near-IR as 800 nm can be generated. The characterization of these polymers is not yet completed, but important is that the strategy of extending optical absorption into the near-IR region is working, and also allows the formation of photodiodes. Infrared emission is possible in light-emitting diodes based on such polymers [29].

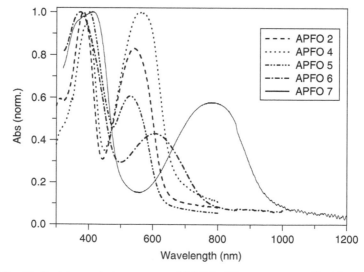

Figure 17.3.   Optical transmission spectra of APFOs. The spectra are taken from spin-coated thin polymer films on glass substrates, and have been arbitrarily normalized.

17.3.2. Dielectric Functions of APFO Films

The optical properties of spin-coated APFO2, APFO3 films are characterized by a marked optical anisotropy, both documented by birefringence and dichroism [30], as also observed in other PFOs [31]; these studies have also used structural methods to verify the expectation that polymer chains are aligned with the polymer film surface [31]. We have extracted optical anisotropy data for some of the APFOs by using spectroscopic ellipsometry to measure the dielectric function between 1 and 5 eV [30], and reproduce examples elsewhere (Chapter 5, Persson et al. in this book). We fully expect that this anisotropy is due to chain alignment within the polymer film, with chains parallel to the confining surfaces. In blends of APFOs and fullerene derivative [6,6]-phenylC_{61}-butyric acid methyl ester (PCBM), the anisotropy of spin-coated films is less pronounced, but is still very clearly expressed [32].

The high absorption coefficients of the polymer and of polymer–PCBM allows the use of thin films in photovoltaic devices to harvest the photon flow. At 300 nm thickness of the active layer, with a metallic electrode–reflector on top, the optical absorption is already close to the maximum possible.

17.3.3. Photoluminescence Processes in APFO materials

The APFOs investigated so far have a considerable quantum yield of photoluminescence. In polymer films, we have found a photoluminescence yield between 0.07 and 0.27. Appropriate to the low bandgap, they give emission in the red region of the solar spectrum, extending into the infrared region [29]. The photoluminescence processes in photodiodes are mainly evaluated for determining the degree of photoluminescence quenching in blends.

17.4. ELECTRONIC TRANSPORT IN APFOs

The polyfluorene derivatives studied here are characterized by hole transport in the majority of devices studied so far; in diodes through space charge limited currents and through time-of-flight measurements, and in field effect transistors. The picture emerging from a comparative study of a series of APFOs, where the main difference is that of the length of alkyl side chains, is that the transport can be well described within models of charge transport in disordered solids, which have been extensively and recently used by Paul Blom and others in studies of polymers and fullerenes [33,34]. Our studies of diodes constructed with anodes of PEDOT–PSS and cathodes of evaporated Al show very typical current–voltage characteristics, which can be well fitted by the assumption of a field-dependent mobility of holes [35]. By recording data at temperatures between 170 and 300 K, it is also possible to analyze the behavior within the model to deduce parameters describing the charge hopping between localized states in terms of characteristic distance a and width σ of the energy dispersion of the states. Assuming that hopping occurs between states characterized by positional and energetic disorder, with a density of states described with a Gaussian, a mobility of the form

$$\mu = \mu_\infty \exp\left[-\left(\frac{3\sigma}{5k_BT}\right) + 0.78\left(\frac{\sigma}{k_BT}\right)^{3/2} - 2\sqrt{\frac{eaE}{\sigma}}\right]$$

has been proposed. We use this theory to connect the microscopic parameters of transport to the molecular structure.

We can fit the temperature- and field-dependent $J–V$ curves using numerical fits (Figure 17.4), and deduce distances $a \approx 1.5–1.9$ nm and $\sigma \approx 0.05–0.06$ eV, depending on the polymer [35]. This is quite a narrow width, and more narrow than what has been found in modeling of transport within a PPV polymer (OC1C10-PPV) (0.11 eV) and polyalkylthiophene (P3HT) (0.1 eV) [36]. Depending on field and hopping distance, this can be helpful in obtaining high mobility.

The comparison of these data to that of other polymers should help in clarifying whether these polymers offer sufficient transport parameters. The comparative study of field-dependent mobilities in diodes and field effect transistors, for a poly(p-phenylene vinylene) and for a polythiophene has recently been reported [36]. For 0C1C10-PPV values of $\mu_0 = 5 \times 10^{-7}$ cm^2/Vs at room temperature and zero field; for P3HT values are $\mu_0 = 3 \times 10^{-5}$ cm^2/Vs, and we find values varying between 0.1 and 3×10^{-5} cm^2/Vs in APFO1, APFO3, and APFO4, depending on the choice of polymer. The field dependence of the mobility is stronger with lower mobility. Because of the low width of the Gaussian density of states, translated into a low energetic disorder, we consider that this class of materials may have the potential to make good transport photovoltaic materials. The possibility to further reduce the disorder, by transition into the liquid crystalline phase, has not yet been studied in transport measurements, but is expected to introduce higher ordering and, possibly, smaller positional and energetic disorder. On the other hand, when blending the polymer with an acceptor such as PCBM, completely new properties may be expected.

For the electron transport, no data are yet available, but the results from quantum chemistry arguing that the lowest unoccupied molecular orbital (LUMO) is localized on the comonomer segment in the APFOs would suggest that electron localization is quite strong and could suppress electron mobility. For transport in photovoltaic devices based on blends with an acceptor, this may be a small problem as long as the electron is residing on the network of acceptor molecules.

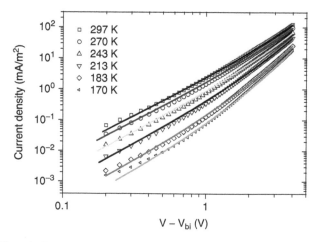

Figure 17.4. Electrical transport-fitting of diode space charge limited current to model of field-dependent mobility in layers of APFO3. (From Gadisa, D. Sharma, F. Zhang, M. Svenssonc, M. R. Andersson, and O. Inganas, submitted for publication.)

17.5. POLYFLUORENE–FULLERENE BLENDS: MORPHOLOGY AND OPTICAL PROPERTIES

The use of APFOs in photovoltaic devices normally requires an electron-carrying acceptor phase, which we supply by blending the polymer with a fullerene derivative in solvents. The choice of solvents for this process has a marked influence on the behavior of the organic thin films in devices. An important measure of the degree of interpenetration between donor and acceptor in such blends is the level of photoluminescence quenching. This is the first observation taken and it is complemented with scanning force microscopy (SFM) images of the polymer blends. With the APFOs studied here, little evidence for phase separation has been gained by these methods, which indicate an efficient photoluminescence quenching and an almost total lack of evidence for phase separation between donor and acceptor phase in the SFM images. Typical examples are shown in Figure 17.5, where a sequence of images of a polymer blend of different stoichiometries is shown.

It is to be noted that the absence of visible phase separation in amplitude and phase mode of tapping mode SFM is not a sufficient evidence for the level of interpenetration, as this is an extremely local probe; it is also always possible that the surface does not show the internal structure of the material by the formation of vertical segregation. The suppression of photoluminescence from donor and from acceptor is actually a more global and also a more sensitive probe of phase separation. However, in none of these modes is phase separation between APFOs and PCBM well documented.

To conclude that phase separation is not contributing to the formation of materials is however an overstatement. We know that thermal processing of polymer blends can drive phase separation, leading to the formation of crystalline PCBM from a blend. Therefore the balance between complete miscibility and phase separation may be a sensitive function of the processing history; the rate of solvent removal from the polymer and PCBM blend in solution, which is strongly influenced by the choice of solvents; and the thermal history of the blend, where high-temperature excursions can in some cases lead to major phase separation; the interaction between solid surfaces and the liquid polymer or molecule or solvents and atmosphere, where conditions for capillary flow may lead to reflow in the polymer layer [37]. Therefore what we obtain may be a mere kinetic stabilization of a system eventually going into phase separation. These issues need a considerable amount of studies for safe conclusions.

The optical properties of APFO–fullerene blends have been modeled through investigations by spectroscopic ellipsometry, and reveal a material that clearly has similarities to APFO and fullerene, but the dielectric function of the blend is not possible to be constructed as a combination of two materials through a effective medium model [32] (see also Persson et al., Chapter 5, this book). When blending polymer and molecule, there is a change of optical properties of the polymer, which may be caused by direct interactions with fullerene. This further strengthens the interpretation that phase separation between polymer and fullerene is small or nonexistent, as no such graininess is possible to be inferred from the modeling of the dielectric function, is not visible in SFM, and is not present according to photoluminescence quenching. A major cause of poor charge generation is therefore removed, and this is a necessary condition for high performance in photovoltaic devices.

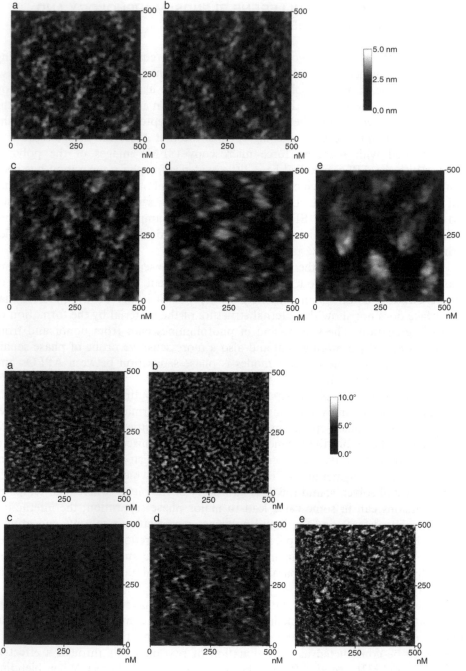

Figure 17.5. Tapping-mode AFM height (top) and phase (bottom) images (500 × 500 nm) of the thin films of APFO3–PCBM blends with different blend ratios: (a) pure APFO3, (b) APFO3:PCBM (1:1), (c) APFO3:PCBM (1:2), (d) APFO3:PCBM (1:3), and (e) APFO3:PCBM (1:4). The thin films were spin-coated from chloroform solutions at 1500 rpm on pretreated silicon substrates. To obtain approximately the same film thickness, the total concentration of the solution was adjusted (6, 12, 18, 24, and 30 mg/ml, respectively). These images have been taken by Cecilia Björström and Ellen Moons at Karlstad University, Sweden. (From Björström and E. Moons, in preparation.)

17.6. POLYFLUORENE–FULLERENE BLENDS IN DEVICES: PERFORMANCE

The performance of the APFO–PCBM blends in solar cell is often attractive [19] as shown in Figure 17.6. External quantum efficiencies in the range of 40–50% can be obtained over a considerable range of wavelengths (400–650 nm), depending on system and stoichiometry (Figure 17.7). Photovoltages V_{oc} above 1 V, fill factors of 0.5–0.6, and photocurrent densities J_{sc} of 4.7 mA/cm^2 can be obtained under AM1.5 simulated solar illumination. The first study of APFO2–PCBM gave a power conversion efficiency (PCE) of 2.2% under simulated AM1.5 solar irradiation.

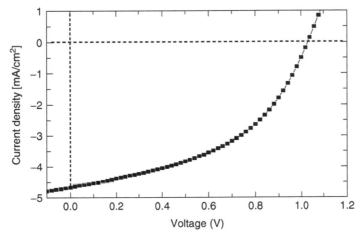

Figure 17.6. Photodiode performance for a APFO2–PCBM (1:4) device. (From M. Svensson, F. Zhang, S. C. Veenstra, W. J. H. Verhees, J. C. Hummelen, J. M. Kroon, O. Inganäs, and M. R. Andersson, *Advanced Materials* **15**, 988–991 (2003). Reproduced with permission.)

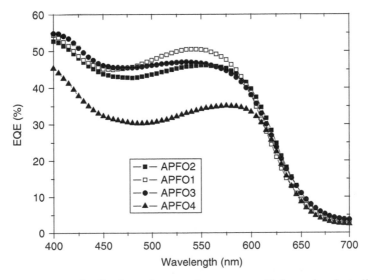

Figure 17.7. Spectral distribution of external quantum efficiency in photodiodes from APFO1–APFO4, blended with PCBM in ratio 1:4. (From Inganas, M. Svensson, F. Zhang, N. K. Persson, X. Wang, and M. R. Andersson, *Applied Physics A*. in press.)

This should be compared with other systems which have been developed and extensively studied; the poly(p-phenylene vinylene) (PPV) main chain system has delivered PCEs of 2.5% [38] and the polythiophene system has been reported to give PCE of 3.5% [39]. Our recent development of APFO–PCBM blends have generated a great number of systems where the PCE is found in the range between 2% and 3%, quantum efficiencies are around 50%, and fill factors are between 0.5 and 0.6. These are still under optimization and are increasing with the experience of device construction.

The transport of photogenerated charge carriers is often a limiting step in the function of plastic solar cells, and high mobilities are suitable to make good photovoltaic parameters. The choice of APFO–fullerene stoichiometry and of organic solid thickness has a large impact on device performance. Thicker layers generate more optical absorption in the device, but also a larger voltage drop along the path of charge carrier transport. This compromise is very relevant to PPV–PCBM devices and also to APFO–PCBM devices [22]. A study of APFO5–PCBM blends located the optimal thickness in the devices between 200 and 300 nm (Figure 17.8). This compares in a favourable way with other polymers studied.

17.6.1. Temperature Dependence of Photovoltaic Device Performance

The temperature dependence of photovoltaic action is available through studies of V_{oc}, J_{sc}, and fill factor dependence on temperature. Data from such studies are shown in Figure 17.9, where the polymer APFO3 was blended with PCBM in a 1:4 stoichiometry, and sandwiched between PEDOT–PSS and Al. It is very clear that increased temperature in the range 20°C to 90°C increases the photocurrent by a factor of 1.2, while smaller effects are found in the photovoltage and the fill factor. An almost linear decrease of photovoltage and a weak increase of fill factor with temperature is observed, and the maximum PCE is found close to 60°C. The increase

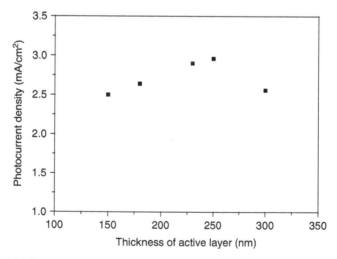

Figure 17.8. Thickness dependence of photocurrent density, under illumination in a solar simulator at 100 W/cm², AM1.5. The data derives from a study of APFO5–PCBM in stoichiometry (1:4) [22]. (Reproduced by permission from T. Yohannes, F. Zhang, M. Svensson, J. C. Hummelen, M. R. Andersson, and O. Inganas, *Thin Solid Films* **449,** 152–157 (2004)).

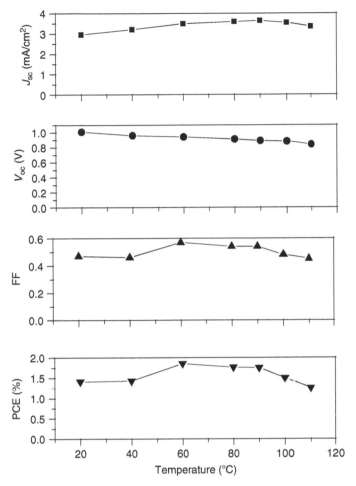

Figure 17.9. Temperature dependence of photovoltaic parameters for a APFO3–PCBM (1:4) device. Lines are added only to help the eye.

of PCE is considerable, i.e., 1.3 times larger at the temperature of maximum performance compared to that at room temperature.

Similar behavior is also observed with other APFOs studied so far, and indicate that the temperature dependence of device performance operates through effects on the photocurrent and photovoltage [40]. The decrease of photovoltage with temperature does indicate a change of injection conditions, which is also visible in the dark current–voltage characteristics of devices. In this case, injection currents may increase by two orders of magnitude in the temperature range studied. This is the injection current necessary to balance the flow of photogenerated charge carriers, under open circuit conditions, and therefore causes a decrease of the photovoltage. The increase of dark current with increasing temperature in the blend, in the space charge limited regime, is not identical to the behavior of the neat polymer, but it has a local maximum found at a temperature close to that of maximum photocurrent. A complex interplay of transport phenomena of both holes and electrons combine both in dark current and in photocurrent behavior, which is seen in the fill factor of photodiodes. This fill factor has a surprisingly weak dependence of temperature.

Summarising these results, we note that injection conditions may be dominant in the photovoltage behavior, and that the photocurrent does not simply follow the behavior of the hole conductor material.

17.7. SUMMARY

We present a new class of alternating fluorene copolymers, which can be combined with a fullerene acceptor, to make polymer blends suitable for photovoltaic energy conversion. By choice of comonomers in the polymer, it is possible to engineer the optical absorption spectrum and to cover the wavelength range down to 900 nm. The transport properties of the polymers investigated so far are competitive with other polymers used in polymer solar cells and the mixing of polymers with acceptors in the form of fullerenes is extensive. These polymers are therefore of interest in the future developments of high-performance polymer solar cells.

ACKNOWLEDGMENTS

This work has been performed with funding from the Science Research Council, Sweden, and the Strategic Research Foundation (SSF), Sweden, through the Center of Organic Electronics at Linköping University. Abay Gadisa and Wendimagegn Mammo, from Addis Ababa University, are funded by fellowships from the International Science Program at Uppsala University, and Xiangjun Wang through a National Graduate School in materials science.

REFERENCES

1. D. M. Chapin, C. S. Ruller, and G. L. Pearson, A new silicon p–n junction photocell for converting solar radiation into electrical power, *Journal of Applied Physics* **25**, 676 (1954).
2. M. A. Green, K. Emery, D. L. King, S. Igari, and W. Warta, Solar cell efficiency tables (version 21), *Progress in Photovoltaics: Research and Applications* **11**, 39–45 (2003).
3. M. Fukuda, K. Sawada, and K. Yoshino, Synthesis of fusible and soluble conducting polyfluorene derivatives and their characteristics, *Journal of Polymer Science: Part a — Polymer Chemistry* **31**, 2465–2471 (1993).
4. Q. B. Pei and Y. Yang, Efficient photoluminescence and electroluminescence from a soluble polyfluorene, *Journal of the American Chemical Society* **118**, 7416–7417 (1996).
5. M. Leclerc, Polyfluorenes: twenty years of progress, *Journal of Polymer Science: Part a — Polymer Chemistry* **39**, 2867–2873 (2001).
6. H. Sirringhaus, R. J. Wilson, R. H. Friend, M. Inbasekaran, W. Wu, E. P. Woo, M. Grell, and D. D. C. Bradley, Mobility enhancement in conjugated polymer field-effect transistors through chain alignment in a liquid-crystalline phase, *Applied Physics Letters* **77**, 406–408 (2000).
7. A. J. Campbell, D. D. C. Bradley, and H. Antoniadis, Dispersive electron transport in an electroluminescent polyfluorene copolymer measured by the current integration time-of-flight method, *Applied Physics Letters* **79**, 2133–2135 (2001).
8. M. T. Bernius, M. Inbasekaran, J. O'Brien, and W. S. Wu, Progress with light-emitting polymers, *Advanced Materials* **12**, 1737–1750 (2000).
9. J. J. M. Halls, A. C. Arias, J. D. MacKenzie, W. S. Wu, M. Inbasekaran, E. P. Woo, and R. H. Friend, Photodiodes based on polyfluorene composites: influence of morphology, *Advanced Materials* **12**, 498 (2000).

10. C. M. Ramsdale, J. A. Barker, A. C. Arias, J. D. MacKenzie, R. H. Friend, and N. C. Greenham, The origin of the open-circuit voltage in polyfluorene-based photovoltaic devices, *Journal of Applied Physics* **92**, 4266–4270 (2002).

11. C. M. Ramsdale, I. C. Bache, J. D. MacKenzie, D. S. Thomas, A. C. Arias, A. M. Donald, R. H. Friend, and N. C. Greenham, ESEM imaging of polyfluorene blend cross-sections for organic devices, *Physica E* **14**, 268–271 (2002).

12. D. M. Russell, A. C. Arias, R. H. Friend, C. Silva, C. Ego, A. C. Grimsdale, and K. Mullen, Efficient light harvesting in a photovoltaic diode composed of a semiconductor conjugated copolymer blend, *Applied Physics Letters* **80**, 2204–2206 (2002).

13. H. J. Snaith, A. C. Arias, A. C. Morteani, C. Silva, and R. H. Friend, Charge generation kinetics and transport mechanisms in blended polyfluorene photovoltaic devices, *Nano Letters* **2**, 1353–1357 (2002).

14. R. Stevenson, A. C. Arias, C. Ramsdale, J. D. MacKenzie, and D. Richards, Raman microscopy determination of phase composition in polyfluorene composites, *Applied Physics Letters* **79**, 2178–2180 (2001).

15. R. Stevenson, R. G. Milner, D. Richards, A. C. Arias, J. D. Mackenzie, J. J. M. Halls, R. H. Friend, D. J. Kang, and M. Blamire, Fluorescence scanning near-field optical microscopy of polyfluorene composites, *Journal of Microscopy — Oxford* **202**, 433–438 (2001).

16. A. C. Arias, J. D. MacKenzie, R. Stevenson, J. J. M. Halls, M. Inbasekaran, E. P. Woo, D. Richards, and R. H. Friend, Photovoltaic performance and morphology of polyfluorene blends: a combined microscopic and photovoltaic investigation, *Macromolecules* **34**, 6005–6013 (2001).

17. A. C. Arias, N. Corcoran, M. Banach, R. H. Friend, J. D. MacKenzie, and W. T. S. Huck, Vertically segregated polymer-blend photovoltaic thin-film structures through surface-mediated solution processing, *Applied Physics Letters* **80**, 1695–1697 (2002).

18. M. Inbasekaran, W. Wu, E. P. Woo, and M. T. Bernius, WO Patent 0046321 (2000).

19. M. Svensson, F. Zhang, S. C. Veenstra, W. J. H. Verhees, J. C. Hummelen, J. M. Kroon, O. Inganäs, and M. R. Andersson, High performance polymer solar cells of an alternating polyfluorene copolymer and a fullerene derivative, *Advanced Materials* **15**, 988–991 (2003).

20. C. R. Towns and R. O. Dell, WO Patent 0053656 (2000).

21. M. Ranger, D. Rondeau, and M. Leclerc, *Macromolecules* **30**, 7686–7691 (1997).

22. T. Yohannes, F. Zhang, M. Svensson, J. C. Hummelen, M. R. Andersson, and O. Inganas, Polyfluorene copolymer based bulk heterojunction solar cells, *Thin Solid Films* **449**, 152–157 (2004).

23. C. Kitamura, S. Tanaka, and Y. Yamashita, Design of narrow-bandgap polymers. Syntheses and properties of monomers and polymers containing aromatic-donor and o-quinoid-acceptor units, *Chemistry of Materials* **8**, 570–578 (1996).

24. A. Dhanabalan, J. K. J. van Duren, P. A. van Hal, J. L. J. van Dongen, and R. A. J. Janssen, Synthesis and characterization of a low bandgap conjugated polymer for bulk heterojunction photovoltaic cells, *Advanced Functional Materials* **11**, 255–262 (2001).

25. M. Jayakannan, P. A. Van Hal, and R. A. J. Janssen, Synthesis, optical, and electrochemical properties of novel copolymers on the basis of benzothiadiazole and electron-rich arene units, *Journal of Polymer Science: Part a — Polymer Chemistry* **40**, 2360–2372 (2002).

26. H. A. M. van Mullekom, J. J. J. M. Vekemans, E. E. Havinga, and E. J. Meijer, Developments in the chemistry and band gap engineering of donor–acceptor substituted conjugated polymers, *Materials Science and Engineering* **32**, 1–40 (2001).

27. K. Jespersen, G. W. Beenken, T. Pullerits, Y. Zausjitsyn, A. Yartsev, M. R. Andersson, and V. Sundström, The electronic states of a low band gap polyfluorene co-polymer with alternating acceptor units, submitted for publication (2004).

28. J. Cornil, I. Gueli, A. Dkhissi, J. C. Sancho-Garcia, E. Hennebicq, J. P. Calbert, V. Lemaur, D. Beljonne, and J. L. Bredas, Electronic and optical properties of

polyfluorene and fluorene-based copolymers: a quantum-chemical characterization, *Journal of Chemical Physics* **118**, 6615–6623 (2003).

29. M. Chen, E. Perzon, M. R. Andersson, S. Marcinkevicius, K. M. S. Jönsson, M. Fahlman, and M. Berggren, 1 Micron wavelength photo- and electroluminescence from a conjugated polymer, Applied Physics Letters **84**, 3570–3572 (2004).

30. N. K. Persson, M. Schubert, and O. Inganäs, Optical modelling of a layered photovoltaic device with a polyfluorene derivative/fullerene as the active layer, *Solar Energy Materials & Solar Cells* **83**, 169–186 (2004).

31. M. Knaapila, B. P. Lyons, K. Kisko, J. P. Foreman, U. Vainio, M. Mihaylova, O. H. Seeck, L. O. Palsson, R. Serimaa, M. Torkkeli, and A. P. Monkman, X-ray diffraction studies of multiple orientation in poly(9,9-bis(2-ethylhexyl)fluorene-2,7-diyl) thin films, *Journal of Physical Chemistry B* **107**, 12425–12430 (2003).

32. N. K. Persson, H. Arwin, S. Lucic, and O. Inganäs, Optical optimization of polyfluorene–fullerene blend photodiodes, submitted for publication (2004).

33. V. D. Mihailetchi, J. K. J. van Duren, P. W. M. Blom, J. C. Hummelen, R. A. J. Janssen, J. M. Kroon, M. T. Rispens, W. J. H. Verhees, and M. M. Wienk, Electron transport in a methanofullerene, *Advanced Functional Materials* **13**, 43–46 (2003).

34. J. K. J. van Duren, V. D. Mihailetchi, P. W. M. Blom, T. van Woudenbergh, J. C. Hummelen, M. T. Rispens, R. A. J. Janssen, and M. M. Wienk, Injection-limited electron current in a methanofullerene, *Journal of Applied Physics* **94**, 4477–4479 (2003).

35. A. Gadisa, D. Sharma, F. Zhang, M. Svenssonc, M. R. Andersson, and O. Inganas, Side chain length influence microscopic hole transport parameters in polyfluorene copolymers, submitted for publication (2004).

36. C. Tanase, E. J. Meijer, P. W. M. Blom, and D. M. De Leeuw, Unification of the hole transport in polymeric field effect transistors and light emitting diodes, *Physical Review Letters* **91**, 216601–216605 (2003).

37. X. Wang, K. Tvingstedt, and O. Inganäs, Surface energy induced polymer flow in film on PDMS modified surface, submitted for publication (2004).

38. S. E. Shaheen, C. J. Brabec, N. S. Sariciftci, F. Padinger, T. Fromherz, and J. C. Hummelen, *Applied Physics Letters* **78**, 841 (2001).

39. F. Padinger, R. S. Rittberger, and N. S. Sariciftci, Effects of postproduction treatment on plastic solar cells, *Advanced Functional Materials* **13**, 85–88 (2003).

40. F. Zhang, M. Svensson, M. R. Andersson, and O. Inganas, in preparation (2004).

41. C. Björström and E. Moons, in preparation (2004).

42. O. Inganas, M. Svensson, F. Zhang, N. K. Persson, X. Wang, and M. R. Andersson, Low bandgap alternating polyfluorene copolymers in plastic photodiodes and solar cells, *Applied Physics A* **79**, 31–35 (2004).

18

Solar Cells Based on Diblock Copolymers: A PPV Donor Block and a Fullerene Derivatized Acceptor Block

Rachel A. Segalman[†], Cyril Brochon, and Georges Hadziioannou*

Laboratoire d'Ingénierie des Polymères pour les Hautes Technologies (LIPHT) FRE 2711 CNRS, Université Louis Pasteur/ECPM, Strasbourg, France

Contents

Abstract Block copolymers made from electron-donating and electron-accepting blocks are interesting for applications in photovoltaics where self-assembly on the size scale of an exciton diffusion length ($\sim$10 nm) is advantageous. Continuous, nanometer-scale interpenetrating phases of electron donor and acceptor components would permit exciton dissociation to occur throughout the active layer and allow the separated charges to be easily transported to the proper electrode. Block copolymers self-assemble into nanometer-scale domains at equilibrium and are hence easily thermodynamically controlled and produce reproducible patterns. This chapter reviews the concepts of materials necessary for the production of

---

[†] Current Address: Dept of Chemical Engineering, University of California, Berkeley.
* To whom correspondence should be addressed: hadzii@ecpm.u-strasbg.fr

efficient photovoltaic device structures, the synthesis of donor–acceptor block copolymers, and the aggregation and device performance of these materials. The potential of semiconducting block copolymers lies in the promise of a solution processed, self-assembled active layer which is optimized to promote exciton dissociation and can be made into large area, mechanically flexible, inexpensive devices.

Keywords energy level alignment, photovoltaic cells, luminescent conjugated polymers, rod–coil copolymers, diblock copolymers, poly(*p*-phenylene vinylene), interchain interactions, heterojunctions, self-assembled active layers.

18.1. INTRODUCTION

The primary advantage of polymers over traditional semiconductor materials in photovoltaic applications is the ease and flexibility of processing. As discussed elsewhere in this book, the ability to deposit thin films over macroscopically large areas are particularly attractive. Impressive progress has been made to make the parent conjugated polymers tractable without degrading electronic properties [1,2]. Furthermore, it is now apparent that an appropriate, well-defined chemical structure as well as an understanding of the effects of processing history are essential to the control of the ultimate properties.

A relatively untapped aspect of polymeric materials in electronic applications is the capability of self-assembly. Specifically tailored polymers can self-organize into mesoscopic structures with features on the nanometer scale. This self-assembly provides access to a size scale previously untapped, but potentially extremely exploitable in the optimization of light emitting and solar cell devices. In understanding how mesoscale structuring can be advantageous in a photovoltaic device, it is useful to reemphasize the operation of such a device within the donor–acceptor framework [3,4]. The first requirement of a photovoltaic material is the generation of charge upon illumination. These charges must then drift towards electrodes for collection. Due to the low dielectric constant in organics, this photoexcitation does not directly create free charge carriers. Rather, excitons with binding energies on the order of several tenths of an electron volt are created [2]. Exciton dissociation by electron transfer is facilitated at the interface between an electron-donating and an electron-accepting material. Though it is not obvious what the nature of this interface should be in terms of scale and geometry, it is clear that the optimization of device performance will rely heavily on the tuning of this architecture. From the nature of the exciton dissociation process, we can gain several basic requirements for this idealized architecture. The donor–acceptor concept implies that the interface should be large in area and easily accessible for all excitons generated. Second, while the exciton may be created anywhere within the 100 nm light penetration depth of the light-absorbing phase, it has a finite lifetime prior to radiative decay and must reach a donor–acceptor interface within this time to contribute to energy conversion. To increase the probability of electron transfer, the interface between the light-absorbing (electron-donating phase) and the accepting phase should have a correlation length on the order of the diffusion length of the exciton, 10 nm. Following dissociation, the charges must be transferred to the appropriate electrode without recombination. This requires that both the donating and accepting phases have continuous pathways to the appropriate electrodes.

To date, this donor–acceptor concept has been implemented by using interpenetrating polymer blends of donor–acceptor homopolymers [4,5]. This interpenetrating network of donor–acceptor heterojunctions facilitates both the dissociation of excitons and charge transport. The performance of this type of device is, however, very sensitive to the morphology of the blend. These phase-separated blends, however, have relatively poorly controlled, undefined architectures which may degrade performance. Geometrical imperfections such as fully dispersed domains and disclination-type discontinuities in the phases prevent efficient charge transport. The extended interfacial area in these systems is also accompanied by an enlarged phase-boundary volume in which donor and acceptor species are molecularly mixed. This results in an increased energy level disorder, causing an increase in the charge trap density and a decrease in electron and hole mobilities.

By contrast, block copolymers are known to microphase separate into nanometer-scale domains whose geometry depends on the molecular design of the material. Since these domains are the result of a self-assembly process, there is relatively little molecular scale mixing within either domain. Finally, since many schemes exist for directing the self-assembly of standard (nonconducting polymers), there is potential for directing the formation of particular nanometer architectures within the active layer. In this chapter we explore the use of block copolymers for photovoltaic applications.

18.1.1 Microphase Separation of Block Copolymers

Block copolymers consist of chemically distinct polymer chains, A and B, joined by a covalent bond. In the melt, they are driven to segregate into a variety of ordered structures by the repulsion of the two immiscible blocks, much as in the case of a blend of two immiscible homopolymers. In this far simpler case of a blend of homopolymers, the phase behavior may be controlled by three experimental parameters: the degree of polymerization, N; the composition, f; and the A–B Flory Huggins interaction parameter, χ. In a homopolymer blend, these parameters affect the free energy through the Flory Huggins equation:

$$\frac{\Delta G_{\mathrm{mix}}}{k_{\mathrm{B}} T} = \frac{1}{N_{\mathrm{A}}} \ln (f_{\mathrm{A}}) + \frac{1}{N_{\mathrm{B}}} \ln (f_{\mathrm{B}}) + f_{\mathrm{A}} f_{\mathrm{B}} \chi \tag{1}$$

The first two terms correspond to the configurational entropy of the system, and are regulated through polymerization stoichiometry. In the third term of (1), χ is associated with the enthalpic penalty of A B monomer contacts and is a function of both the chemistry of the molecules and temperature. In general:

$$\chi = \frac{a}{T} + b \tag{2}$$

where a and b are experimentally obtained constants for a given composition of a particular polymer pair. Experimentally, χ can be controlled through temperature. Unlike macrophase separation in blends, the connectivity of the blocks in block copolymers prevents complete separation and instead the chains organize to put the A and B portions on opposite sides of an interface. The equilibrium nanodomain structure must minimize unfavorable A–B contact without overstretching the blocks [6]. In unconjugated polymers (which generally have a coil shape), the strength of segregation of the two blocks is proportional to χN. When the volume fraction of block A, f_{A}, is quite small, it forms spheres in a body-centered cubic (BCC) lattice surrounded by a matrix of B. As f_{A} is increased towards 0.5, the minority nanodomains will form first cylinders in a hexagonal lattice, then a bicontinuous double gyroid structure, and finally lamellae, as shown in Figure 18.1. The size and periodicity of the nanodomains are also functions of the polymer size (N) and the segmental interactions (χ) [6,7]. The

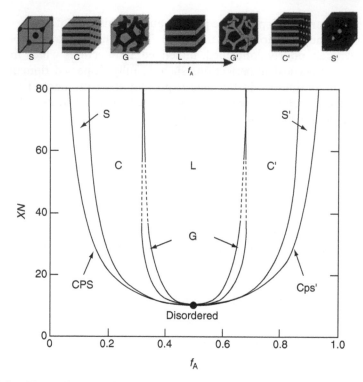

Figure 18.1. Phase diagram of a unconjugated block copolymer demonstrating the occurrence of disordered spheres (CPS), a body-centered cubic lattice of spheres (BCC), a hexagonal array of cylinders (C), double gyroid bicontinuous (G), and lamellar (L) phases (From Bates, F.S. and G.H. Fredrickson, *Annual Review of Physical Chemistry*, 41, 525–557 (1990). With permission.)

assembly of regular, well-controlled nanodomain architecture is reliant on the regularity of the molecular structure of the block copolymer molecules. It is particularly important that the molecular weight of the polymer be relatively monodisperse.

When the structure of the blocks does not form the characteristic coil structure of a traditional polymer, the microphase structures also change dramatically. When the stiffness of one of the blocks is significantly increased, a new class of intermolecular interactions is introduced, in particular the display of liquid crystalline phases such as nematic or layered smectic structures with molecules arranged with their long axes nearly parallel to each other. While the complete phase diagram of these types of molecules is not yet known, the chain rigidity of the stiff block is expected to greatly affect the details of the molecular packing and thus the nature of thermodynamically stable microphases. In rod–coil polymers constructed from nonconjugated materials, the high stiffness contrast between the blocks has been observed to drive microphase separation even at very low molecular weights as compared to flexible block copolymers. The supramolecular structures of rod–coil polymers arise from a combination of organizing forces including the repulsion of the dissimilar blocks and the packing constraints imposed by the connectivity of the blocks, and the tendency of the rod block to form orientational order. It is not surprising that many unusual symmetries such as zigzag and arrowhead phases have been observed [8,9].

Several groups have observed the self-assembly of block copolymers formed from a conjugated rod block covalently bonded to nonconjugated coil block. Yu et al.

[10–12] reported lamellar organization of spun-cast thin films of a rod–coil block copolymer containing oligo(phenylene vinylene) (PPV) derivatives as the rod blocks and polyisoprene or poly(ethylene oxide) as the coil block. Rod–coil triblocks containing PPV synthesized by Stupp [13] have been observed to self-assemble into mushroom shapes in solution. Jeneke and Chen [14] reported the self-assembly of poly(phenylquinoline) as the rod block and polystyrene as the coil block to form hollow spheres when cast from specific solvent mixtures. Lazzaroni et al. [15] showed that rod–coil polymers containing poly(p-phenylene) tend to form ribbon-like fibrils when coated on a substrate. Hempenius et al. [16] observed the formation of nonspherical micelles when using a block copolymer containing oligothiophene.

18.1.2. Block Copolymers for Photovoltaic Applications

The combination of a PPV-type polymer or oligomer as the donor material with C_{60} as the acceptor has proven promising, despite the processing challenge presented by the highly insoluble C_{60} component. By covalently incorporating C_{60} into a polymer, we can combine the conducting properties of C_{60} with the processibility of the polymer. This allows the incorporation of higher concentrations of the acceptor into the device than would be possible with pure C_{60}. By combining both the donor and acceptor onto a single molecule, we will not only bring the donor and acceptor components close together for the purposes of exciton dissociation, but also gain an extra degree of control over the architecture of the active layer and the pathways of the donating and accepting materials to the electrodes. The subsequent sections of this chapter will describe the design, synthesis, characteristics, and optoelectronic response of a material that combines the donor–acceptor concept of a PPV–C_{60} blend with the self-assembling properties of a block copolymer.

18.2. SYNTHESIS OF PPV-BASED DONOR–ACCEPTOR DIBLOCK COPOLYMER

Donor–acceptor diblock copolymers for self-assembled photovoltaic materials must have a specific and controlled design, as outlined above. One of the two blocks must be an electron donor and hole transporter that can also absorb light from the sun spectra. Conjugated polymers, such as PPV, generally meet these requirements. The second block must be an electron acceptor and conductor. The two blocks must also be chemically dissimilar to facilitate microphase separation. The conjugated block can be considered a rigid chain, since these polymers are less flexible than classical polymers (such as the nonconjugated chain polystyrene). As discussed above, any block copolymer incorporating such a component will deviate from the standard phase behavior of coil–coil block copolymers. We choose a flexible block functionalized with accepting groups like C_{60} as the second component of our block copolymer. The targeted diblock copolymer can be regarded as a rod–coil copolymer with a donating PPV rod and electron-accepting coil.

18.2.1. Synthesis of Diblock Rod–Coil Copolymers: General Aspects

Due to large difference in the chemical nature of the two blocks, the synthesis of a conjugated rod–coil copolymer is not as simple as that of a classical coil–coil block copolymer. Most conjugated rod–coil copolymers are obtained either by quenching a living coil polymer, synthesized by anionic polymerization, with an end-functionalized conjugated oligomer [12,17,18] or by a simple condensation reaction between an end-functionalized coil polymer and the rod block [16,17,19–21]. In either case, the rod block is synthesized by a polycondensation reaction and must have a reactive function at only one chain end to prevent coil–rod–coil triblock copolymer formation.

Anionic living polymerization is known to result in well-defined homopolymers and block copolymers [22]. A simple and elegant route is towards a conjugated–coil block copolymer is to first polymerize two coil blocks by living anionic polymerization and then convert one of the two blocks into a conjugated structure [23]. This method, however, is very restrictive with regard to the choice of the rod block and the existence of permanent defects in conjugation.

The use of a fully conjugated macromolecular initiator (macroinitiatior) will bypass this difficulty. Indeed, it is possible to convert the rod block into a macroinitiatior and then polymerize the second block directly onto it. This route has been used in conjunction with anionic living polymerization by Marszitzky et al. [24] to polymerize ethylene oxide with a polyfluorene macroinitiator as the conjugated rod block. Anionic polymerization, which is a popular route towards block copolymers because it yields nearly monodisperse molecular weight distributions, requires drastic reaction conditions (very high sensitivity to impurities and functional groups) and cannot be used with a wide variety of monomers, particularly those with functional groups. In fact, controlled/'living' radical polymerization techniques are more versatile than anionic polymerization. Also, a donor–acceptor rod–coil block copolymer requires the use of functional monomers for the grafting of C_{60} or other electron-accepting groups. The two main controlled/'living' radical polymerization techniques are the nitroxide-mediated radical polymerization (NMRP) [25] and the atom transfer radical polymerization (ATRP) [26]. These are compatible with a wide range of monomers (acrylates, styrene, etc.) including functionalized derivatives and result in narrow polydispersities molecular weight control, which are necessary for controlled self-assembly. Furthermore, (macro) initiators for NMRP or ATRP are very stable and it is possible to purify and characterize these compounds before (re)initiating the radical polymerization. Various rod–coil block copolymers have been synthesized by polymerization of styrene or methacrylate derivatives either by ATRP with polyfluorene as macroinitiatior [27] or by NMRP with PPV [28,29]. The use of conjugated macroinitiators for controlled/'living' radical polymerization is actually one of the best routes for the synthesis of the targeted rod–coil diblock copolymer. The general synthesis strategy is displayed in Figure 18.2.

18.2.2. Synthesis of PPV Macroinitiator

Good optoelectronic properties of the conjugated rod block are not the only requirement in building a useful block copolymer. The macroinitiator must satisfy three additional requirements: solubility in wide range of solvents (for ease of processing), low polydispersity (to improve microphase separation), and a reactive function at only one of its chain ends. PPV derivatives are known for their ability to be fine-tuned over a wide range by inclusion of functional side-chains [1,30]. Furthermore, PPV derivatives with low polydispersity and controlled molecular weight can be obtained by using the polycondensation technique developed by Meier and Kretzchmann based on the Siegrest condensation [31]. With this technique, PPVs with *trans*-configured vinyl double bonds (defect-free conjugated block) and exactly one terminal aldehyde group can be obtained.

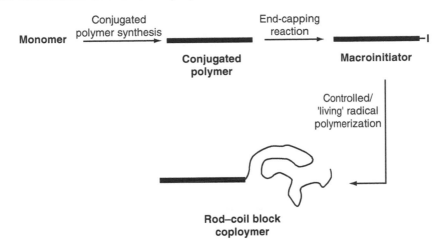

Figure 18.2. Macroinitiator approach to the controlled radical polymerization of a rod–coil polymer.

This approach has been used for the synthesis of two different kinds of PPV (with alkyl side-chains) macroinitiators for NMRP [28,29,32]. This step of the synthesis is shown in Figure 18.3. Macroinitiators (**7**) and (**11**) are made from two different stable radicals, 2,2,6,6-tetramethylpiperidine-1-oxyl (**5**, TEMPO) and 2,2,5-trimethyl-4-phenyl-3-nitroxide (**9**, TIPNO), respectively. These PPV macroinitiators were characterized primarily with ^{1}H NMR and UV–VIS spectrometry. Well-defined monofunctional oligomers (**7**) and (**11**) were obtained with degree of polymerization, N, tunable by manipulating on reaction conditions such as temperature and time.

18.2.3. Polymerization with the PPV Macroinitiator and Functionalization of the Coil Block

TEMPO is primarily used for the NMRP of styrene derivatives, while TIPNO can also be used for the polymerization of acrylate derivatives [25]. Accordingly, macro-initiators (**7**) and (**11**) have been used respectively for the polymerization of styrene and butylacrylate [29]. TEMPO is easier to obtain (a commercial product) and to graft. Styrene is also a very versatile monomer with many functional derivatives and polystyrene is a well-studied coil polymer. For these reasons, the first donor–acceptor rod–coil copolymers were synthesized using the PPV–TEMPO macroinitiator (**7**) and styrene derivatives [28,32]. The general synthesis route is shown in Figure 18.4. Elaboration of coil block from PPV–TEMPO (**7**) is a two-step process: first a statistical polymerization at 130°C of a mixture of styrene (S) and chloromethylstyrene (CMS), which leads to the PPV-*b*-P(S-*stat*-CMS) diblock copolymer (**12**); and then its functionalization with C_{60}, which leads to the PPV-*b*-P(S-*stat*-C_{60}MS) diblock copolymer (**13**).

CMS is a very useful functional monomer, which can be copolymerized by NMRP. Due to its high reactivity in radical polymerization (close to that of styrene), copolymerization of styrene and CMS leads to a random copolymer: CMS units are statistically distributed along the chain. The molecular weight of a block copolymer (**12**) initiated with a PPV–TEMPO initiator (**7**) and which incorporates both styrene and CMS monomers, increases with time and conversion, as followed by gel permeation chromatography (GPC). ^{1}H NMR confirms that the CMS composition of the

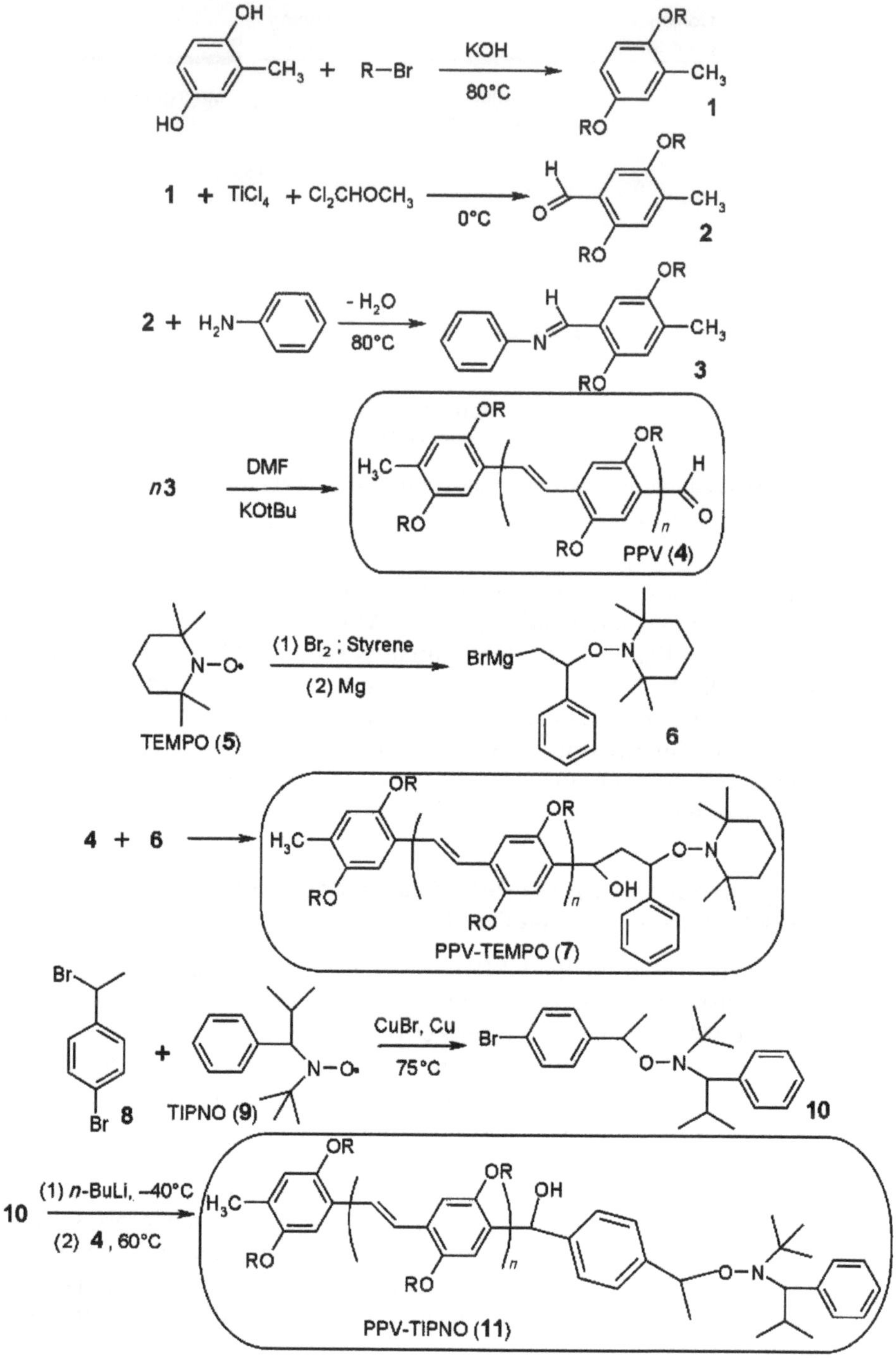

Figure 18.3. The synthesis of a PPV macroinitiator for NMRP.

final diblock copolymer (**13**) corresponds well to the feed ratio. Atom transfer radical addition (ATRA) [33] can then be used as a one-step reaction to graft C$_{60}$ directly on to CMS units. The amount of C$_{60}$ is controlled by the amount of CMS in the coil block of PPV-*b*-P(S-*stat*-CMS) diblock copolymer (**12**) [28]. The C$_{60}$ wt.% of the final

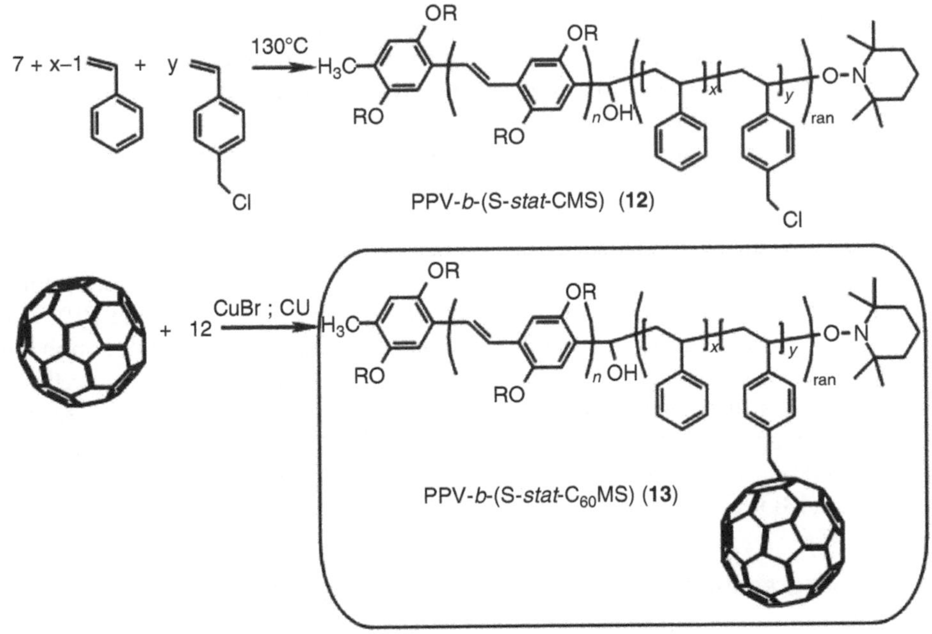

Figure 18.4. Synthesis of a PPV-*b*-PS or PPV-*b*-P(S-*stat*-CMS) polymer with a TEMPO-based macroinitiator.

product as calculated from thermogravimetric analysis (TGA) (measurement of residue weight) and UV–VIS indicates that coil blocks containing between 14 and 52 wt.% C_{60} are typically obtained.

It has been shown that it is possible to design well-defined conjugated rod–coil diblock copolymers based on the donor–acceptor concept. This new class of copolymers has a good tunability regarding the length of each block and the amount of C_{60}. These parameters of the design are a good basis for the study of self-assembly and optoelectronic properties of these new materials.

To optimize the structure of the material, sufficient C_{60} should be incorporated into the polymer so as to form a continuous electron-conducting phase between the electrodes. It is yet unclear precisely what weight fraction this corresponds to. However, it is known that a large amount of C_{60} (corresponding to CMS/S ratio below 2) results in several additions of CMS onto a single C_{60} molecule. This means that the real amount of grafted C_{60} is much lower than the theoretical amount given by CMS/S ratio. The multiple reactions of CMS on C_{60} also may result in cross-linking reactions by intermolecular grafting. These cross-linking reactions may result in an insoluble, intractable polymer. The ATRA reaction to attach C_{60} to PPV-*b*-P (S-*stat*-CMS) requires more work to carefully balance the incorporation of sufficient functionality to ensure conductivity with the cross-linking reaction, which occurs at high concentrations of CMS.

18.3. PHASE BEHAVIOR OF THE PPV-BASED DIBLOCK COPOLYMERS

One of the great attractions of PPV-based photovoltaic devices is the ability to perform solution-phase film formation, which is both inexpensive and extremely versatile. The self-assembly of diblock copolymers, however, depends strongly on solution-phase behavior and solidification process. For instance, poor solubility or aggregation of copolymers in solution can greatly impact film morphology and therefore device properties. Accordingly, a preliminary understanding of rod–coil diblock copolymer solution behavior is essential to the eventual fabrication of devices.

The specific nature of these rod–coil copolymers leads to particular solution-phase behavior. Solutions of PPV derivatives (with different alkyl side groups) in various solvents has been extensively studied [34–38]. Aggregation occurs and can dramatically affect (depending on temperature and solvent quality) the photoluminescence (PL) in solution and also the optoelectronic properties of the PPV film [36,37,39,40]. The contrast in solubility of the two blocks of a diblock copolymer can encourage specific aggregation behavior, leading to the formation of micelles or more complex structures [41]. Furthermore, the rod block might form, in specific solvents, lyotropic nematic domains. Due to the extremely low solubility of C_{60} in most organic solvents, the solution-phase behavior of the donor–acceptor block copolymer (**13**) is also greatly affected by the amount and method of C_{60} grafting.

18.3.1. Aggregation of PPV-*b*-PS Diblock Copolymers in Solution

Solutions of a model poly((2,5-dioctyloxy)-1,4-phenylene vinylene)-*block*-polystyrene) (DOO–PPV–PS) diblock copolymer have been studied by spectroscopic investigations (UV–VIS absorption and PL) to underscore aggregation phenomena, which can affect the film formation [32].

As shown in Figure 18.5, in a good solvent for each block of the PPV-*b*-PS, such as chloroform, absorption and fluorescence spectra do not change significantly with temperature (between $-20°C$ and $20°C$). The slight redshift is probably due to a small increase in conjugation length and a reduction of thermal motion. In similar experiments in carbon disulfide (CS_2), aggregation can occur at low temperature: the aggregated state emission band was identified at 630 nm (see Figure 18.5). Indeed, CS_2 still remains a good solvent for PS but becomes, at low temperatures, a poor

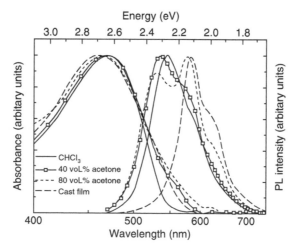

Figure 18.5. Absorption and fluorescence spectra of PPV-*b*-PS. (From de Boer, B., U. Stalmach, P. F. van Hutten, C. Melzer, V. V. Krasnikov, and G. Hadziioannou, *Polymer*, 42(21), 9097–9109 (2001). With permission.)

solvent of PPV. This selective solution behavior of each block is very favorable to the formation of aggregates.

The quality of the media can also vary by addition of a poor solvent, such as acetone, in a good solvent like chloroform. Solutions of PPV-*b*-PS have been also studied in acetone–chloroform mixture of varying solvent ratios. PL spectra, confirmed by small-angle neutron scattering (SANS) measurements, indicate the formation of micelles (probably by PPV aggregations) in dynamic equilibrium with unimers (single molecules). SANS measurements have also shown a great dependence of size and structure of these micelles with the solvent composition.

18.3.2. Self Organization of PPV-based Block Copolymers in Thin Films

As previously discussed, the thermodynamics of the solution can result in complex and relatively unexplored aggregation behavior. When this solution is cast into a film, not only is some of this aggregation behavior transferred onto the solid state, but also the rapid evaporation and consequent cooling of the film induce further mesomorphic structuring. Specific solution environments give rise to solid-state morphologies on multiple length scales, directly affecting the optoelectronic efficiencies.

The stark contrast in solution-phase behavior of PPV-*b*-PS in chloroform or *o*-dichlorobenzene versus CS_2 directly translates to the morphology of the thin film. When the block copolymer is cast from chloroform, the film has some nanometer-scale ordering [32]. Scanning force microscopy (SFM) imaging in tapping mode of a spun-cast film reveals elongated, needle-like domains with a uniform width of approximately 15 nm, as shown in Figure 18.6. This size scale suggests that the needles are formed from a double layer of PPV blocks. When C_{60} is incorporated, as shown in the bottom of Figure 18.6, the persistence length of the needles lengthens, but the structure is otherwise unchanged. While the PPV portion of this block copolymer exhibits a liquid crystalline phase between 55°C and 185°C, the SFM images are not reminiscent of either the familiar microphase separation of coil–coil block copolymers or the interesting rod–coil block structures observed by Chen et al. [8,9]. While it is clear that the presence of the side-chain mesogens strongly influences the formation of microdomains [42], the full microphase separation of this type of block copolymer is still relatively unstudied.

Guiding a maximum amount of light into the solar cell for maximum energy gain is a challenge universal to both inorganic and organic devices. Embossing has been successfully used to roughen the top surface of a photovoltaic cell on a micron-length scale to diffracted and guide incident light into the device, improving efficiency by as much as 25% [43]. This technique, however, requires an additional processing step (and associated costs). These polymer systems present a self-assembled route to the same sorts of patterns. When a polymer film is drop-cast from some organic solvents in the presence of humidity on a solid substrate, water condenses on the cooling surface as the solvent evaporates, creating a hexagonal pattern. When the solvent has completely evaporated, the water droplets are immobilized in the polymer film. As the film warms to room temperature (from temperatures as low as −6°C when the solvent was CS_2), water evaporates creating a microporous honeycomb structures with a characteristic micron-length scale (three orders of magnitude larger than the characteristic size of the polymer, as seen in Figure 18.7). This type of pattern has been seen in numerous types of polymers cast from many solvents (for example, Ref. [14,44–48]), including PPV-*b*-P

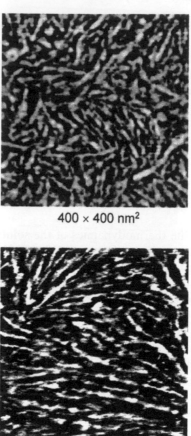

400 × 400 nm²

400 × 400 nm²

Figure 18.6. Scanning force microscopy (SFM) images of the thin-film morphology of PPV-*b*-PS (top) and PPV-*b*-P(S-*stat*-C$_{60}$MS) (bottom) films spun-cast from solutions of *o*-dichlorobenzene. (From de Boer, B., U. Stalmach, P.F. van Hutten, C. Melzer, V.V. Krasnikov, and G. Hadziioannou, *Polymer*, 42(21), 9097–9109 (2001). With permission.)

(S-*stat*-C$_{60}$MS). In the case of this donor–acceptor block copolymer, the micronscale surface roughening could be utilized in trapping solar light in the polymer layer. Since films cast in the above manner spontaneously form similar surface structures, it is possible that through a careful choice of solvents and casting conditions, the embossing step could be eliminated.

PPV-*b*-PS diblock copolymers in solution form large aggregates ranging from micelles to more complex structures. By changing the temperature or solvent quality, they reorganize from single chains to aggregated clusters of polymer chains. The design of diblock copolymers may assist in tuning the solution behavior of these macromolecules. For this purpose, the two envisaged tunable parameters are the nature of the side-chain of the PPV rod block and the nature of the coil polymer. Diblock copolymers functionalized with C$_{60}$ (**13**) show poor solubility when the amount of grafted C$_{60}$ is high. Upon casting into a thin film, the aggregate structures of both PPV-*b*-PS and PPV-*b*-(S-*stat*-C$_{60}$MS) are visible as 15 nm needle-like microphases. When the casting solvent is chosen correctly, however, the surface structure is patterned on a micronscale into a honeycomb structure, which may be useful in light trapping.

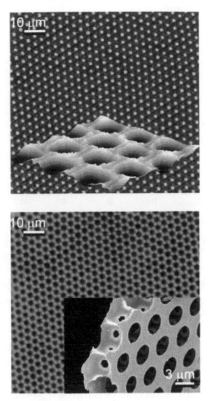

Figure 18.7. Luminescence micrographs of honeycomb structured films of PPV-*b*-PS (top) and PPV-*b*-P(S-*stat*-C_{60}MS) (bottom) cast from a CS_2 solution. (From de Boer, B., U. Stalmach, P. F. van Hutten, C. Melzer, V. V. Krasnikov, and G. Hadziioannou, *Polymer*, 42(21), 9097–9109 (2001). With permission.)

18.4. PHOTOVOLTAIC RESPONSE OF THE DONOR–ACCEPTOR BLOCK COPOLYMER

In the previous sections, we have outlined the design, synthesis, solution state, and solid-state properties of a polymer optimized for photovoltaic applications. The synthesis and use of this polymer, consisting of a PPV hole-conducting block and a C_{60} functionalized electron-conducting block, poses many challenges two of which will be outlined below:

1. The ultimate nanometer-scale morphology consisting of bicontinuous, interpenetrated network structure throughout its volume requires not only a well-defined block copolymer, but also a degree of control over microphase separation not yet developed.
2. A sufficient amount of C_{60} must be incorporated into the copolymer to provide both a donor–acceptor interface at which exciton dissociation can occur and a continuous pathway for electrons without destroying the solubility (and therefore processibility) of the material.

Since neither of these design goals has yet been optimized, the preliminary device performance, as measured by de Boer et al. [32] is only a hint of the potential of these materials. The first indication that a donor–acceptor interface had been

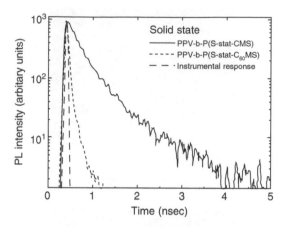

Figure 18.8. Photoluminescence (PL) decay spectra of films of PPV-*b*-PS (top) and PPV-*b*-P(S-*stat*-C_{60}MS). The PL decay time of the donor–acceptor polymer is close to the time resolution of the setup (50 ps). This reduced lifetime and yield indicate that the donor–acceptor interface is accessible to the excitons created in the PPV microphase. (From de Boer, B., U. Stalmach, P. F. van Hutten, C. Melzer, V. V. Krasnikov, and G. Hadziioannou, *Polymer*, 42(21), 9097–9109 (2001). With permission.)

formed was that PPV-*b*-(S-*stat*-C_{60}MS) has a greatly reduced (three orders of magnitude) PL lifetime relative to PPV-*b*-PS (see Figure 18.8) despite their similar morphologies (as shown previously in Figure 18.6).

A second indication of the success of nanometer-scale separation of the donor–acceptor interfaces is a direct comparison of the photovoltaic behavior of the block copolymer versus a blend of the donor and acceptor homopolymers, which comprise the blocks. This comparison was performed by fabricating cells which contained identical amounts of C_{60} and PPV, but in one case the two moieties were contained within a block copolymer and in the other the PPV homopolymers was blended with a statistical copolymer of P(S-*stat*-C_{60}MS). These cells were fabricated by spin-casting a 100 nm thick active layer from a 1 wt.% solution of polymer onto a 70 nm thick layer of poly(3,4-ethylenedioxythiophene)/poly(styrene sulphonate) (PEDOT:PSS) on an ITO-covered glass substrate. A 100 nm thick aluminum electrode was then thermally evaporated on top of the active layer to complete the sandwich-structured cell. The superior response of the block copolymer can be seen in Figure 18.9 through the measurement of current versus voltage (*I–V*) on devices under monochromatic illumination of 1 mW/cm² at 458 nm. This response is probably due to a combination of an increased donor–acceptor interfacial area and a higher continuity of each phase with the electrodes in the block copolymer film.

While these results validate the block copolymer strategy toward self-assembling materials for photovoltaic applications, there is much room for optimization. In particular, though both the blend and the block copolymer demonstrate nearly complete quenching of the fluorescence in the solid state, the collection efficiencies are quite small. This is due in part to molecular scale mixing that may lead to increased energy level disorder in the phases. This effect is probably aggravated by the presence of a small amount of homopolymer PPV (which was not functionalized with TEMPO in the early synthesis steps) and free C_{60} (which was not incorporated into the P(S-*stat*-CMS) block during the ATRA reaction). Such disorder will cause an increase in the charge trap density with a significant increase in the space charge field both of which lead to a reduction in electron and hole mobilities. Additionally,

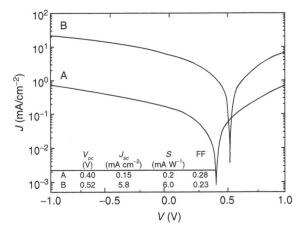

Figure 18.9. Current versus voltage curves of photovoltaic devices measured under mono-chromatic illumination of 1 mW/cm^2 at 458 nm. The two devices shown were fabricated from a 1:1 molar ratio blend of PPV and a statistical copolymer P(S-*stat*-C$_{60}$MS) (curve A) and the block copolymer PPV-*b*-P(S-*stat*-C$_{60}$MS) (curve B) such that the blend and block copolymer contain the same amounts of C$_{60}$ and PPV. Characteristic parameters such as the open circuit voltage (V_{oc}), short circuit current (J_{sc}), and the photovoltaic sensitivity (S) are much improved in the block copolymer device. (From de Boer, B., U. Stalmach, P. F. van Hutten, C. Melzer, V. V. Krasnikov, and G. Hadziioannou, *Polymer*, 42(21), 9097–9109 (2001). With permission.)

a fundamental problem of energy transfer upon photoexcitation may compete with exciton dissociation leading to a decrease in efficiency for any organic device.

18.5. CONCLUSIONS AND OUTLOOK

The attractiveness of polymeric materials to make low cost, lightweight, flexible solar cells is now obvious and the potential of low molecular weight organic semiconduct-ors to reach high power conversion efficiencies has been demonstrated by processing materials with high purity in terms of contaminants and structure. The records in power conversion efficiency, however, were achieved by optimizing the structure of donor and acceptor homopolymers blends by trial and error in casting of the films [49]. There is undeniable potential in optimizing and controlling the nanostructure of the active layer in a more systematic manner. Nanometer-scale structuring via the microstructuring of liquid crystalline materials has also led to some of the best reported photovoltaic quantum efficiencies, 34% at 490 nm [50]. In terms of process-ing efficiency and elegance, however, nanostructuring can easily be done by taking advantage of the self-assembling nature of polymers. The ability of block copolymer to be chemically tuned to microphase separate on the proper scale and into the proper structure so as to possess the desired chemical, physical, and electronic properties to optimize the overall device is in our opinion an untapped potential.

To this end, we would like to outline a few directions and opportunities necessary for the eventual development of commercial block copolymer devices. At the present time, many unexplored design parameters as to the nature of each block must be considered for the improvement of the phase behavior and optoelectronic properties of these rod–coil copolymers. In the case of the accepting block, it would be preferable to not only choose a new electron acceptor that is more soluble in higher concentrations, but also to rework the attachment mechanism to the back-bone so as to prevent cross-linking. On the donor side, this block copolymer technique is extremely flexible and as further studies suggest donors with absorbance in broader ranges of the solar spectrum, the block copolymers can be similarly tuned.

Previously, we also suggested that self-organization of a favorable microstructure is key to the overall performance of a block copolymer device. Not only is much fundamental study necessary in understanding the nanostructure of conjugated rod–coil block copolymers, but also schemes must be developed to control the orientation of these structures. This will involve much more detailed studies of the interactions and morphologies of each block at the electrode surfaces. On a polymer level, this will lead to an understanding of the self-assembly of the material on the surface. From a broader level, however, this information is necessary to our eventual understanding and control of charge transport. The formation of dipole layers, covalent bonds at the interface, or doping of the interfacial region will greatly affect the barrier height between the Fermi energy levels of the electrode and the orbital levels of the organic layers.

The ability to produce block copolymers possessing the correct electronic functions and having targeted photovoltaic characteristics is a first, encouraging step towards the bottom-up design of optoelectronic materials. The future of this technologies lies in melding together knowledge of synthetic chemistry, solid-state physics, and block copolymer architectural design to produce optimized, well-controlled nanometer-scale structures.

REFERENCES

1. Kraft, A., A. C. Grimsdale, and A. B. Holmes, *Angewandte Chemie — International Edition*, **37**(4), 402–428 (1998).
2. Heeger, A. J., *Reviews of Modern Physics*, **73**(3), 681–700 (2001).
3. Sariciftci, N. S., L. Smilowitz, A. J. Heeger, and F. Wudl, *Science*, **258**(5087), 1474–1476 (1992).
4. Yu, G., J. Gao, J. C. Hummelen, F. Wudl, and A. J. Heeger, *Science*, **270**(5243), 1789–1791 (1995).
5. Halls, J. J. M., C. A. Walsh, N. C. Greenham, E. A. Marseglia, R. H. Friend, S. C. Moratti, and A. B. Holmes, *Nature*, **376**(6540), 498–500 (1995).
6. Bates, F. S. and G. H. Fredrickson, *Annual Review of Physical Chemistry*, **41**, 525–557 (1990).
7. Bates, F. S. and G. H. Fredrickson, *Physics Today*, **52**(2), 32–38 (1999).
8. Chen, J. T., E. L. Thomas, C. K. Ober, and S. S. Hwang, *Macromolecules*, **28**(5), 1688–1697 (1995).
9. Chen, J. T., E. L. Thomas, C. K. Ober, and G. P. Mao, *Science*, **273**(5273), 343–346 (1996).
10. Yu, L. P., *Abstracts of Papers of the American Chemical Society*, **221**, 424-IEC (2001).
11. Wang, H. B., L. P. Yu, and H. H. Wang, *Abstracts of Papers of the American Chemical Society*, **218**, 426-POLY (1999).
12. Li, W. J., H. B. Wang, L. P. Yu, T. L. Morkved, and H. M. Jaeger, *Macromolecules*, **32**(9), 3034–3044 (1999).
13. Stupp, S. I., *Abstracts of Papers of the American Chemical Society*, **213**, 67-COLL (1997).
14. Jenekhe, S. A. and X. L. Chen, *Science*, **283**(5400), 372–375 (1999).
15. Lazzaroni, R., P. Leclere, A. Couturiaux, V. Parente, B. Francois, and J. L. Bredas, *Synthetic Metals*, **102**(1–3), 1279–1282 (1999).
16. Hempenius, M. A., B. M. W. Langeveld-Voss, J. van Haare, R. A. J. Janssen, S. S. Sheiko, J. P. Spatz, M. Moller, and E. W. Meijer, *Journal of the American Chemical Society*, **120**(12), 2798–2804 (1998).

17. Marsitzky, D., T. Brand, Y. Geerts, M. Klapper, and K. Mullen, *Macromolecular Rapid Communications*, **19**(7), 385–389 (1998).
18. Tew, G. N., M. U. Pralle, and S. I. Stupp, *Journal of the American Chemical Society*, **121**(42), 9852–9866 (1999).
19. Francke, V., H. J. Rader, Y. Geerts, and K. Mullen, *Macromolecular Rapid Communications*, **19**(6), 275–281 (1998).
20. Kukula, H., U. Ziener, M. Schops, and A. Godt, *Macromolecules*, **31**(15), 5160–5163 (1998).
21. Genzer, J., E. Sivaniah, E. J. Kramer, J. G. Wang, H. Korner, K. Char, C. K. Ober, B. M. DeKoven, R. A. Bubeck, D. A. Fischer, and S. Sambasivan, *Langmuir*, **16**(4), 1993–1997 (2000).
22. Szwarc, M., *Journal of Polymer Science: Part a — Polymer Chemistry*, **36**(1), IX–XV (1998).
23. Leclere, P., V. Parente, J. L. Bredas, B. Francois, and R. Lazzaroni, *Chemistry of Materials*, **10**(12), 4010–4014 (1998).
24. Marsitzky, D., M. Klapper, and K. Mullen, *Macromolecules*, **32**(25), 8685–8688 (1999).
25. Hawker, C. J., A. W. Bosman, and E. Harth, *Chemical Reviews*, **101**(12), 3661–3688 (2001).
26. Matyjaszewski, K. and J. H. Xia, *Chemical Reviews*, **101**(9), 2921–2990 (2001).
27. Lu, S., Q. L. Fan, S. J. Chua, and W. Huang, *Macromolecules*, **36**(2), 304–310 (2003).
28. Stalmach, U., B. de Boer, C. Videlot, P. F. van Hutten, and G. Hadziioannou, *Journal of the American Chemical Society*, **122**(23), 5464–5472 (2000).
29. Stalmach, U., B. de Boer, A. D. Post, P. F. van Hutten, and G. Hadziioannou, *Angewandte Chemie — International Edition*, **40**(2), 428–430 (2001).
30. Segura, J. L., *Acta Polymerica*, **49**(7), 319–344 (1998).
31. Kretzschmann, H. and H. Meier, *Tetrahedron Letters*, **32**(18), 5059–5062 (1991).
32. de Boer, B., U. Stalmach, P. F. van Hutten, C. Melzer, V. V. Krasnikov, and G. Hadziioannou, *Polymer*, **42**(21), 9097–9109 (2001).
33. Iqbal, J., B. Bhatia, and N. K. Nayyar, *Chemical Reviews*, **94**(2), 519–564 (1994).
34. Shi, Y., J. Liu, and Y. Yang, *Journal of Applied Physics*, **87**(9), 4254–4263 (2000).
35. Chang, R., J. H. Hsu, W. S. Fann, K. K. Liang, C. H. Chiang, M. Hayashi, J. Yu, S. H. Lin, E. C. Chang, K. R. Chuang, and S. A. Chen, *Chemical Physics Letters*, **317**(1–2), 142–152 (2000).
36. Hsu, J. H., W. S. Fann, P. H. Tsao, K. R. Chuang, and S. A. Chen, *Journal of Physical Chemistry A*, **103**(14), 2375–2380 (1999).
37. Nguyen, T. Q., V. Doan, and B. J. Schwartz, *Journal of Chemical Physics*, **110**(8), 4068–4078 (1999).
38. Zheng, M., G. L. Bai, and D. B. Zhu, *Journal of Photochemistry and Photobiology a—Chemistry*, **116**(2), 143–145 (1998).
39. Nguyen, T. Q., I. B. Martini, J. Liu, and B. J. Schwartz, *Journal of Physical Chemistry B*, **104**(2), 237–255 (2000).
40. Dufresne, G., J. Bouchard, M. Belletete, G. Durocher, and M. Leclerc, *Macromolecules*, **33**(22), 8252–8257 (2000).
41. Wang, Q. A., Q. L. Yan, P. F. Nealey, and J. J. de Pablo, *Journal of Chemical Physics*, **112**(1), 450–464 (2000).
42. Leclere, P., A. Calderone, D. Marsitzky, V. Francke, Y. Geerts, K. Mullen, J. L. Bredas, and R. Lazzaroni, *Advanced Materials*, **12**(14), 1042–1046 (2000).
43. Roman, L. S., O. Inganas, T. Granlund, T. Nyberg, M. Svensson, M. R. Andersson, and J. C. Hummelen, *Advanced Materials*, **12**(3), 189–195 (2000).
44. Widawski, G., M. Rawiso, and B. Francois, *Nature*, **369**(6479), 387–389 (1994).
45. Francois, B., O. Pitois, and J. Francois, *Advanced Materials*, **7**(12), 1041– (1995).
46. de Boer, B., U. Stalmach, H. Nijland, and G. Hadziioannou, *Advanced Materials*, **12**(21), 1581–1583 (2000).
47. Karthaus, O., N. Maruyama, X. Cieren, M. Shimomura, H. Hasegawa, and T. Hashimoto, *Langmuir*, **16**(15), 6071–6076 (2000).

48. Nishikawa, T., M. Nonomura, K. Arai, J. Hayashi, T. Sawadaishi, Y. Nishiura, M. Hara, and M. Shimomura, *Langmuir*, **19**(15), 6193–6201 (2003).
49. Brabec, C. J., N. S. Sariciftci, and J. C. Hummelen, *Advanced Functional Materials*, **11**(1), 15–26 (2001).
50. Schmidt-Mende, L., A. Fechtenkotter, K. Mullen, E. Moons, R. H. Friend, and J. D. MacKenzie, *Science*, **293**(5532), 1119–1122 (2001).

19

Interface Electronic Structure and Organic Photovoltaic Devices

Yongli Gao
Department of Physics and Astronomy, University of Rochester, Rochester, NY, USA

Contents

Abstract   In all organic semiconductor (OSC) devices, the charge transport process across interfaces of dissimilar materials is important for optimum device operation. For organic photovoltaic (OPV) cells, the most fundamentally important process is the charge transfer (CT), or interfacial dissociation of excitons at the donor–acceptor interface. It is critical to have efficient charge transport across the interfaces, whether between the organic or inorganic multiplayer structures, or with the electrodes. This chapter is intended to introduce the reader to the field by providing some insights through examples of different aspects of the interface formation, mostly taken from the author's own experience using surface analytical tools for the investigations. Examples include the mechanisms of interface dipole formation for the explanation of asymmetry in *I*–*V* characteristics for interfaces of the same metal and organic material, but formed differently by either metal deposition on organic or organic on metal, the treatment of indium tin oxide (ITO) and other forms of interface engineering, the interface formation and energy level alignment of organic–organic interfaces, and the charge transfer dynamics between donor and acceptor layers.

Keywords organic photovoltaic cell; interface formation; interface dipole, interface engineering, metal–organic interface; organic–organic interface; charge transfer dynamics

19.1. INTRODUCTION

Organic photovoltaic (OPV) cells have attracted substantial attention since the heterojunction OPV device reported by Tang in 1986 [1]. The potential advantages include low-cost solar energy conversion, flexible solar panels, and a large selection of materials for optimized device performance [2,3]. However, the efficiency is the main bottleneck — 2.55% in solution processed polymer blend cells [4,5] and 3.6% in vapor deposited thin film cells [6] under standard AM1.5G $100\,mW/cm^2$ solar illumination. It is still below the lower limit value of 5% viable for commercial applications, and far behind the 24% efficiency of crystalline Si [7].

In all organic semiconductor (OSC) devices, the charge transport process across interfaces of dissimilar materials is important for optimum device operation. As proven in over five decades of research, the understanding of interfaces of inorganic semiconductors (ISCs) with metals, semiconductors, and insulators, has had a tremendous impact on semiconductor device technology. Similar situation is also found in OSC devices [2,3]. Furthermore, the thickness of the active organic layer is typically only a few hundred angstroms in OSC devices, which further blurs the distinction between bulk and interface. An OPV device generally consists of, in sequence, an indium tin oxide (ITO)-covered glass as the anode, a thin layer (or multiple layers) of OSC materials, and an evaporated metal film as the cathode.

There are significant differences between OPV cells and conventional inorganic photovoltaic (IPV) solar cells, especially the relative importance of interfacial processes. ISCs have occupied (valence) and unoccupied (conduction) energy bands and the electron wave functions extend over many unit cells. The interaction between charge carriers and the lattice is generally weak, and the transport of the charge carriers can be adequately described as delocalized Bloch waves in the bands. In OSCs, the occupied and unoccupied energy levels for OSCs are formed from planar structures of sp^2 bonds as well as π-bonds, and the π-bonds between carbon atoms in organic molecules form the highest occupied molecular orbital (HOMO) and lowest unoccupied molecular orbital (LUMO) in most OSCs [8,9]. The interactions between the molecules are van der Waals in nature and the electron wave function overlap between the molecules is small. The charge carriers are localized and surrounded by significant nuclear relaxation, and are better described together with the surrounding nuclear deformation as polarons instead of electrons or holes. As a result, the transport from one molecule to another is typically described by hopping of polarons [8,9].

In IPV cells, electron–hole pairs are generated immediately upon light absorption under normal conditions throughout the bulk according to the exponential decrease of the incident light intensity. The exciton binding energy is in the range of milli electron volts, comparable to thermal energy at room temperature, and the electron–hole pairs can be easily separated by a built-in potential either through p–n junction or asymmetric cathode and anode [10]. In contrast, light absorption in organic materials almost always results in the production of a mobile excited state rather than a free electron–hole pair. Because of the non-covalent electronic interactions between organic molecules, a tightly bound electron–hole pair (Frenkel exciton or mobile excited state) is the usual product of light absorption in OSCs. Furthermore, the dielectric constant of an organic material is usually low compared to ISCs, and the attractive Coulomb potential is less screened than in ISCs. The exciton binding energy, or the correlation energy between the localized electron and holes, is substantially larger than the thermal energy at room temperature, and can be as high as 1.6 eV, resulting in virtually no probability of thermal separation of the carriers [9].

The most fundamentally important process in most OPV devices is the charge transfer (CT), or interfacial dissociation of excitons at donor–acceptor (DA) interface [2,3,11,12]. At such an interface, a donor material with a low ionization potential forms a heterojunction with an acceptor material with a high electron affinity. Depending on the alignment of the energy levels of the donor and acceptor materials, the dissociation of the strongly bound excitons can become energetically favorable, leading to a free electron polaron in the acceptor material, and a free hole polaron in the donor. The thermodynamic requirements for interfacial exciton dissociation are clear, and the energy level offset must be greater than the exciton binding energy. The CT process can occur in timescales (τ_{CT}) of a few hundred femtoseconds or less [13,14], resulting in CT efficiency approaching 100% since τ_{CT} is much shorter than any other competing process. Important processes for achieving a high efficiency include the photogenerated excitons towards the D–A interface prior to, and charge transport towards the electrodes after the CT process. Nearly complete transport of excitons to the D–A interface can be achieved by establishing an intimate contact between the donor and acceptor materials by blending [4,11], laminating [12], codeposition [15], or chemical linking [16]. However, the charge transport towards the electrodes is compromised in these approaches because of the decrease in carrier mobilities and increase in charge trap densities. Another approach is using vacuum deposited organic multiplayer structures. In these well-defined systems, exciton rather than charge transport is the limiting factor, and the power conversion efficiencies are comparable or superior to solution-processed OPV cells. The layers can be optimized for different functionalities, and the more successful ones include exciton blocking, light trapping, and optical interference optimization layers [2,6]. In all these OPV devices, it is critical to have efficient charge transport across the interfaces, whether between the organic or inorganic multiplayer structures, or with the electrodes.

Given the fundamental differences in charge separation and transport, it has become obvious that OPVs require a different description than IPVs [3]. A substantial photovoltaic (PV) effect could be achieved in an electrically symmetrical cell in which the equilibrium electrical potential energy difference $\Phi_{bi} = 0$, in sharp contrast to the requirements of a built-in equilibrium electrical potential energy difference Φ_{bi} across the cell in IPVs [17]. The PV at $\Phi_{bi} = 0$ in an OPV cell is explained by a kinetic model based on the asymmetric dissociation of excitons [3,18]. Another kinetic model for polymer PV cells has also been developed that included the spatially confined carrier generation profiles [19]. It is shown that open circuit voltage (V_{oc}) in polymer solar cells was not entirely controlled by Φ_{bi} [20], which was independently confirmed by a model that included the effect of the asymmetric interfacial dissociation of excitons [21]. More recently, it is suggested that the two types of carriers are already separated across the interface upon photogeneration in OPV cells, giving rise to a powerful chemical potential energy gradient that promotes the PV effect [3].

Understanding the interface processes in OPV devices is critical for their further advance in performance. Surface and interface analytical studies have generated critical insight into the fundamental processes at interfaces involving OSCs [22–25]. For example, it is established now from surface analytical studies that the interface energy level alignment is not from a common vacuum level as previously believed. Instead, it depends on the detailed interface interactions including wave function hyperdization, charge transfer, chemical reaction, and intermixing, etc. Understanding issues regarding metal–organic interfaces, such as the formation of the interface dipole, the injection barrier, the diffusiveness of the interface and origin of ionized species, is beginning to take shape. This chapter is intended to introduce the reader to the field by providing some insights through examples of different aspects of the interface formation, but by no means through an exhaustive study of all the research work and techniques employed in the study of interface formation in OSC devices. Most of the work has been done in understanding interfaces in organic light-emitting diode (OLED) or organic thin-film transistors (OTFT), but the physical principles should be just as well applicable to OPVs. In writing this chapter, I have naturally chosen to draw on my expertise of investigating interface electronic structures with photoemission spectroscopy and related techniques, including x-ray photoemission spectroscopy (XPS), ultraviolet photoemission spectroscopy (UPS), inverse photoemission spectroscopy (IPES), and time-resolved photoemission spectroscopy (TRPES). The UPS and IPES measurements probe directly the occupied and unoccupied electronic structure in the interface region, respectively. Core level analysis of XPS determines quantitatively the interface chemistry and morphology. TRPES provides information on the dynamics of CT at and charge transport towards the interface. However, this in no way implies that these tools are the only ones used, or are better than others in the study of OSC devices.

19.2.   SYMMETRY OF METAL–ORGANIC INTERFACES: PENTACENE WITH Au, Ag, AND Ca

Metal–organic interface is the focus of both device engineering and basic science, since it is a key factor in nearly all important aspects of device performances, including operation voltages, degradation, and efficiency. The complexity of metal–organic interfaces is also intriguing in basic organic condensed matter physics. The energy level alignment and charge injection at the interface are among the most concerned fundamental issues. This is the area in which significant efforts in research are made and impressive improvement obtained in the past few years [22–25]. There has been an outstanding puzzle in OSC devices that there can be asymmetry in I–V characteristics of interfaces of the same metal and organic material, but formed differently by either metal deposition on organic layer or organic on metal. The order of the deposition is one of the most important, yet easily overlooked, aspects in the interface formation. The major factors include the thermal energy of the material evaporated, the topology of the substrate, the condensation energy on the surface, and the growth mode of the deposited film. For the process of organic deposited on metal (denoted as organic–metal), the diffusion–disruption between the two materials is low except the most reactive metals, since the thermal energy of the evaporated organic molecules is generally low due to the low evaporating temperature. The organic material normally arrives at the surface in the molecular form, and later condenses to form thin films. The condensation energy of the organic material on the surface is low due to the nature of van der Waals interaction that bonds the organic molecules. Compared to organic–metal interfaces, there are significantly more diffusion and disruption at the metal–organic interfaces (deposit metal onto organic film), because the hot metal atoms have larger kinetic energies, which facilitates easy penetration of metal atoms into the organic film. Also, the metal arrives as isolated atoms or atom clusters that will give a high condensation energy, resulting in possible damage or reaction with the organic molecule. The relative sizes of the organic molecule and metal atom are very different. The smaller metal atoms may easily fall into the "valleys" of the organic molecules. This fact requires one to excise caution when interpreting the energy level changes at the initial metallization process. It is therefore imperative to investigate and understand the symmetry of interface formation in the cases of metal deposited onto organic and organic deposited onto metal.

The changes in the UPS spectra caused by deposition of pentacene onto Au are shown in Figure 19.1, taken from Ref. [26]. There is an initial shift of the vacuum level 0.8 eV closer to the Fermi level after the deposition of 2 Å of pentacene, reflected in the UPS cutoff shift. The valence structure of Au, the Au 5d peaks, and a surface state are immediately attenuated by deposition of 2 Å of pentacene. The Au surface state disappears completely and the Au 5d features change from sharp peaks to broad features with the adjacent peaks centered at a binding energy of −4 eV coalescing into a single broad feature. The shift of the vacuum level increases to a total of 1 eV after the deposition of 18 Å of pentacene. While these changes are observed in the cutoff position, no apparent change in position of any other feature within the UPS spectra is observed. A further decrease in the Au UPS features and a gradual increase in the pentacene UPS spectral features is observed as pentacene is deposited on the surface. So the changes that are seen in the UPS features are attributed solely to the pentacene deposition onto the Au surface.

Figure 19.2 presents a summary of the results for the interfaces between pentacene and metals Au, Ag, and Ca. In the figure, the sequence of deposition is

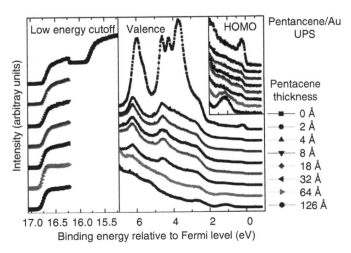

Figure 19.1. **(Color figure follows page 348).** UPS cutoff and UPS HOMO evolution as a function of pentacene deposition onto gold. A total shift of 1 eV away from the Fermi level was seen in the UPS cutoff after the deposition of 18 Å of pentacene. Only changes in intensity of features were seen in the valence structure. (From N. J. Watkins, L. Yan, and Y. Gao, *Appl. Phys. Lett.* **80**, 4384 (2002). Copyright American Institute of Physics, 2002. With permission.)

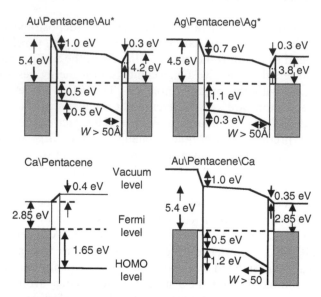

Figure 19.2. Energy diagrams for the various interfaces between pentacene and different metals (Au, Ag, and Ca), with the sequence of deposition being from left to right. When pentacene is deposited onto Au, Ag, or Ca an interface dipole results. Neither Ag nor Au achieved their expected work functions upon deposition onto the pentacene, even after 30+ Å metal deposition. On the other hand, Ca, exhibiting both intermixing and an interface dipole of 0.35 eV, did finally exhibit the features of a Ca metal film. (From N. J. Watkins, L. Yan, and Y. Gao, *Appl. Phys. Lett.* **80**, 4384 (2002). Copyright American Institute of Physics, 2002. With permission.)

from the left to right. Both the Ag and the Au interface exhibit a measurable Fermi level after ∼8 Å of metal deposition, indicating a metallic surface. Yet it can be clearly seen that neither metallic surface gives the work function of the pure metal, even after 30+ Å of metal deposition, nor the valence features of the clean metal surface. Au on pentacene results in a work function of 4.3 eV, about 1 eV lower than that of pure Au. Ag on pentacene gives a work function of 3.8 eV, which is 0.7 eV lower than that of pure Ag. It is interpreted as indication that as these metals are deposited onto the pentacene surface the metal either first penetrates the surface, thereby doping the upper layers of the pentacene, or diffuses into the pentacene [26–28]. The attenuation of the C 1s XPS peak supports metal diffusion or clustering of Au and Ag. For the deposition rates used in these experiments, <0.1 Å/sec, it is more likely that the metal has diffused into the pentacene [29]. The subsequent doping region width, $w > 50$ Å, is estimated from XPS core level analysis as described in Ref. [30]. Au* and Ag* are used in Figure 19.2 to represent the metallic mixtures at these interfaces. The difference between metal on organic and organic on metal results in asymmetric electronic properties, for example the hole injection barrier is 0.5 eV for pentacene on Au but 1 eV for Au on pentacene.

Ca deposition on pentacene does not exhibit this behavior. With the deposition of 8 Å of Ca onto the pentacene, the HOMO position relative to the Fermi level has shifted to where it was observed for the pentacene–Ca interface and the observed work function has almost reached that of a clean Ca surface, which is achieved by 64 Å of Ca deposition. Unlike Au or Ag, Ca forms a pure metal film on the surface and exhibits symmetric behavior with regard to whether Ca is deposited onto pentacene or vice versa. This type of behavior has also been observed in organic and reactive metal interfaces by Kahn et al. [31,32]. This difference in behavior of Ca and noble metals is consistent with the observation that reactive metals penetrate much less than the noble ones when deposited on organic materials [30].

19.3. MECHANISMS OF INTERFACE DIPOLE FORMATION

The energy level alignment at the interface involves a number of physical and chemical processes. In these processes, the most important ones related to the device performances include possible chemical reaction, interface dipole, diffusion, CT, and possible band bending. The vacuum level alignment (VLA) model is the earliest proposed model. In this model, the vacuum level of metal cathode and ITO anode is simply aligned with that of the organic layer. As a result, the LUMO and HOMO shifts inside the organic layer accordingly. Although used widely in the early days of metal–organic studies with various successes, this simple model has been disapproved by many studies, most persuasively by UPS studies [23]. The most important difference to VLA model is that the interface dipole model assumes that a dipole layer exists at the metal–organic and ITO–organic interfaces. The interface dipole causes an abrupt shift of potentials across the dipole layer. The interface dipole layers at cathode and anode modify the LUMO and HOMO shifts inside organic layer. This model has gained its support mainly by various UPS and Kelvin probe works from the change of work function or vacuum level across the interface [23,33].

Several possible mechanisms may contribute to the interface dipole. In their review article, Ishii et al. [23] listed six possible causes, including (a) CT across the interface, (b) image potential induced polarization of the organic material, (c) pushing back of the electron cloud tail out of the metal surface by the organic material, (d) chemical reactions, (e) formation of interface state, and (f) alignment of the permanent dipole of the organic material. Among these factors, (a)–(c) are more general in metal–organic interfaces, while (d)–(f) are specific to the individual metal–organic pair. These factors provide a plausible foundation to explain qualitatively the interface dipole formation. XPS analysis of the core level evolution indicates that (b) is unlikely to be a major factor contributing to the interface dipole formation. Should the polarization inside the organic molecule be significant, one would expect that core level peaks, especially that of the most common carbon atoms, would be broadened due to the intramolecular polarization. This is not observed for Ca/Alq$_3$ or in other organic–metal interfaces [30,34,35]. In fact, the lack of peak width change indicates that the dipole is predominantly confined to the interface between the metal surface and organic molecules.

Further quantitative understanding of the interface dipole formation can be obtained by realizing that the metal and organic material must reach thermodynamic equilibrium when in contact. Tung devised a model of ISC–metal interface based on CT and thermodynamic equilibrium between the metal and semiconductor [36]. According to his model, the interface dipole is given by

$$V_{\text{dipole}} = \frac{-ed_{\text{MS}}N_{\text{B}}}{\varepsilon_{\text{it}}} \frac{\Phi_{\text{M}} - (\text{IP} - E_{\text{g}}/2)}{E_{\text{g}} + \kappa} \tag{1}$$

where d_{MS} is the distance between metal and semiconductor atoms at the interface, N_{B} the density of bonds through which CT takes place, ε_{it} the dielectric constant in the interface region, and κ the sum of all the Coulombic interactions for charges at different sites. The model suggests that CT is a primary mechanism of Schottky barrier formation, and a good agreement with the experimental results is found for polycrystalline ISCs [36]. It should be emphasized that the model is based on thermodynamic equilibrium, facilitated by the interface bonds, of electrons across the interface between the metal and semiconductor. It therefore does not depend on the details of the interface reactions, so long as the physical properties of the semiconductor, such as IP and E_g still remain intact at the interface. The model does not apply to interfaces where strong chemical reactions result in the domination of the interface by newly reacted species.

Figure 19.3 summarizes our experimental results of the interface dipole measurements of metals with Alq$_3$, copper phthalocyanine (CuPc), pentacene, and perylene [37]. The positive dipole is defined as pointing away from the metal surface. Shown in Fig. 19.3(a) is the interface dipole at metal and Alq$_3$ interfaces as a function of the metal work function. The data show approximately linear behavior for both modes of deposition: metal deposited on organic (Na/Alq$_3$, Ca/Alq$_3$, Al/Alq$_3$) or organic deposited on metal (Alq$_3$/Ca, Alq$_3$/Mg, Alq$_3$/Au). The slope of the straight line is −0.8. In Figure 19.3(b), the dipole is plotted as a function of the difference of the metal work function Φ_M, and the energy at the mid-gap of the organic (IP − E_g/2), where IP and E_g are the ionization energy and bandgap of the organic, respectively. Despite different IP and E_g all dipoles seem to fall into straight lines with different slopes from −0.6 for CuPc to −0.4 for perylene. Similar linear dependence on metal work function has also been observed by several groups [23,37,38].

The linear dependence in Figure 19.3 is an indication that the thermal equilibrium driven CT model depicted in Equation (1) is applicable to the interface systems, and that the d_{MS} and N_B are almost identical for these systems. The latter statement is justified by assuming that $d_{MS} \sim 5$ Å and N_B is estimated to be on the order of 10^{14}/cm^2 with the observed slope (~ 0.8 for Alq$_3$). This is the same order of magnitude as the surface density of typical organic molecules, indicating that statistically almost all the molecules at the interface participated in the charge transfer.

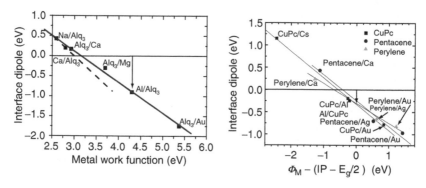

Figure 19.3. Linear relation between interface dipole and the metal work function. (a) Metal–Alq$_3$, the arrow represents the interface dipole of about −0.9 eV at the mid-gap of Alq$_3$. (b) The difference of metal work function Φ_M and the mid-gap (IP − E_g/2) versus the interface dipole. The interface dipole at the mid-gap is about −0.3 eV. (From L. Yan, N. J. Watkins, S. Zorba, Y. Gao, and C. W. Tang, *Appl. Phys. Lett.* **81**, 2752 (2002). Copyright American Institute of Physics, 2002. With permission.)

The CT is not the only factor in determining the interface dipole. A closer inspection of Equation (1) reveals that the thermodynamic equilibrium model implies that the Fermi level is shifted towards the mid-gap of the semiconductor by the interface dipole, and the interface dipole changes sign as the metal work function goes across the mid-gap energy. However, from the fitted data in Figure 19.3(a), this sign change occurs at $\Phi_M = 3.2$ eV, 1.1 eV lower than the Alq_3 mid-gap energy of 4.3 eV. In fact, an interface dipole of -0.9 eV is seen at the mid-gap energy indicated by the vertical arrow in Figure 19.3(a). On the other hand, the three organic materials in Figure 19.3(b) show a very consistent interface dipole of about -0.3 eV at their mid-gap. This apparent deviation from Equation (1) is attributed to the combination of pushing back of the electron cloud by the organic material and the permanent dipole moment of the organic [37]. On a metal surface, the electron cloud extends out of the surface defined by the positive ion cores of the metal. The relative displacement of the charge distribution produces a dipole layer pointing away from the surface, and a negative dipole potential, V_{tail}, seen by the electrons. Theoretical values of V_{tail} varies for different metals and facets but is usually about a few tenths of an electronvolt, the same order of magnitude as observed [39]. When the electron cloud tail is pushed back by the organic material, its contribution to the metal work function is also eliminated. As a result, the sign change in the metal–organic interface dipole shifts from the organic mid-gap energy towards a lower energy. This could explain the consistent discrepancy of -0.3 eV for pentacene, CuPc, and perylene as shown in Figure 19.3(b), because the three molecules are all flat molecules with small permanent dipole moments. The same contribution is expected to be in other metal–organic systems since electron tailing is universal to all metal surfaces.

19.4. DOPING AND ENERGY LEVEL SHIFT: Cs IN CuPc

Doping of organic materials often causes significant changes in these materials and it is of both fundamental and practical interest. Theoretically, it is representative of low-dimensional guest–host systems with strong correlation and anisotropic interactions. Practically, doping is widely used to improve OSC devices analogous to that in the ISC industry. It was a major breakthrough in OSC to show that doping of conjugated polymers can increase the conductivity by many orders of magnitude [40]. Although doping in OSC has proven to be very useful, the relationship between doping and the electronic structure is not well established. Often formulas derived for ISCs are adopted without verification of applicability.

In Figure 19.4, the comparison of the electronic structure and energy level alignment of pristine and Cs-doped CuPc is shown [41]. The bottom part is the IPES spectrum of a pristine CuPc film about 1000 Å thick prepared by thermal evaporation. The upper part is another CuPc film doped with Cs. The Cs doping concentration is determined to be about $R_s = 0.85$ by XPS measurement. The UPS spectrum was taken first and then IPES spectrum was recorded on the same film. The origin in Figure 19.4 is the Fermi level, which is calibrated by UPS and IPES on the same Au–Si substrate. The peaks next to Fermi level are HOMO and LUMO in UPS and IPES, respectively. After background subtraction [42], the IPES spectrum can be decomposed into three Gaussian peaks shown in Figure 19.4. The LUMO peak position is at about 2.2 eV with full width at half magnitude (FWHM) of about 1.3 eV. Deconvolution of the LUMO is performed using the IPES resolution of

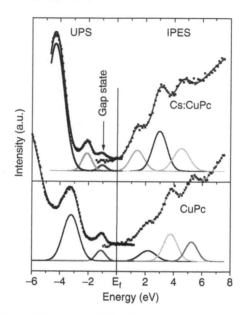

Figure 19.4. **(Color figure follows page 348).** Comparison of the composite of the UPS and IPES spectra of pristine and Cs-doped CuPc films. After doping, the HOMO and LUMO shift about same amount and the LUMO is close to Fermi level. (From L. Yan, N. J. Watkins, S, Zorba, Y. Gao, and C. W. Tang, *Appl. Phys. Lett.* **79**, 4148 (2001). Copyright American Institute of Physics, 2002. With permission.)

0.6 eV to eliminate the instrument-induced broadening, which reduces the FWHM of the LUMO slightly to ~1.2 eV. Similarly the UPS spectrum is fitted with Gaussian peaks after the removal of background [42,43], also shown in Figure 19.4 with the HOMO peak at about −1.2 eV and FWHM of about 0.5 eV. The effect of instrumental broadening is negligible in UPS. Since the HOMO is closer to the Fermi level than the LUMO, it shows the CuPc film is a weak p-type OSC, consistent with the previous knowledge [44]. As clearly shown in Figure 19.4, with the Cs doping, the LUMO shifts toward Fermi level when HOMO away from it. Using the same fitting method and background subtraction described earlier, the IPES can be decomposed to three peaks again. The LUMO peak is at about 1.3 eV with a FWHM (after deconvolution of instrument broadening) of ~1.1 eV. Similarly, the HOMO peak is at −2.1 eV with FWHM of ~ 0.7 eV. It should be noticed that a new gap-state is found to be about −1.0 eV.

UPS process involves the removal of an electron from the sample. The photo-induced hole with surrounding medium in the solid then undergoes different relaxation processes. The electronic (involving the electron distribution around the hole) and vibronic (involving geometric configuration) components are fast processes, which contribute to the final energy of the photoelectrons. The observed UPS spectrum is the representation of the filled states of a molecular cation resulting from the photoelectric process modified by the relaxation (polarization). The width of the peaks is presumably due to inhomogeneity of the film since the organic film is normally amorphous with random disorder. Therefore, the center of the HOMO peak corresponds to the HOMO energy of the most populous molecular cation, with the negative charge (photoelectron) at infinity (beyond vacuum level). Conceptually similar analysis applies to the IPES spectrum. Induced by the injected electron, the relaxation (polarization) of the surrounding medium makes the IPES spectrum the representation of the relaxed anion instead of the neutral state of the molecule. Thus the energy separation of the HOMO–LUMO peaks obtained by UPS and IPES is the energy difference of relaxed positive and negative polarons separated at infinity, sometimes referred as polaron energy gap (E_t) of the organic material. However, the charge injection into solid does not necessarily occur at the most populous or average species. Rather, when the injection is contact limited (usually the case in OLED devices), the charge injects to the molecules with the lowest energy difference to the Fermi level. For example, the hole will most likely reside on those molecules with HOMO closest to the Fermi level, resulting a cation with lowest energy that is represented by the onset of the HOMO peak close to the Fermi Level. The onset is defined as the extrapolation of the leading edge (closer to the Fermi level) in spectrum. Similarly, the barrier for the electron injection will be from the Fermi level to the onset of LUMO. In most organic molecular devices, the cathode and anode is separated far enough that the injection of hole and electron occurs at different molecules and beyond the size of the polarons. It is a general practice to define the onset of HOMO peak as the HOMO position of the CuPc film, at $-0.7\,eV$, as shown in the bottom part of Figure 19.4. Similar term is used for LUMO position, at about $1.0\,eV$ for the pristine CuPc film. The result is an injection energy gap (E_g) of about $1.7\,eV$, interestingly very close to the CuPc optical bandgap E_{opt} (~$1.7\,eV$). E_{opt} is widely used to estimate the LUMO position of organic films from UPS data, but in reality it is from the formation of a Frenkel exciton with the hole and electron on the same molecule. For the Cs-doped CuPc film, the onset of the LUMO peak is less than $0.2\,eV$ from the Fermi level. Given the width of the LUMO peak, the Fermi level is virtually aligned with the LUMO of CuPc. It can similarly be determined that the HOMO position is about $-1.5\,eV$. The experiment gives an injection energy gap of ~$1.7\,eV$, the same as in the pristine film which is a further confirmation that no chemical reaction occurs when CuPc is doped by Cs.

The energy level shifts are summarized in Figure 19.5 [45]. The core level shifts were deduced by fitting the spectra with Gaussian curves, one each for Cu $2p_{3/2}$ and N 1s, and three for C 1s. The vacuum level cutoff was obtained at the maximum of the differential curve of the UPS spectrum. Note that R_{Cs} is in logarithmic scale to emphasize the initial shift. The most remarkable feature is the nearly linear dependence of all the energy level shifts on the logarithmic of the Cs concentration for $R_{Cs} <$ 0.15. At higher R_{Cs}, the shift of vacuum cutoff, the HOMO, and the aromatic C 1s(A) continue the linear behavior approximately, approaching $1.0\,eV$ at $R_{Cs} = 4.0$, while those of the Cu $2p_{3/2}$, N 1s, and pyrrole C 1s(B) and (C) saturate or even reverse slightly.

A linear dependence is generally observed and expected for nondegenerate crystalline ISCs [10]. However, statistical physics dictates that in those cases, the slope in the semilogarithmic plot has to be the thermal energy k_BT. The slope in Figure 19.5 is more than four times of k_BT, a clear indication that the classical result for crystalline ISCs does not apply to CuPc. However, if the electron energy distribution is broad and Lorentzian, $D(\varepsilon) = N_0 \exp\left[-(\varepsilon - \varepsilon_0)/\Gamma\right]/\Gamma$, with $\Gamma >> k_BT$, the electron concentration will be dominated by the Lorentzian and will be given by

$$n \cong N_0 \exp\left[-(\varepsilon_0 - \varepsilon_F)/\Gamma\right] \qquad (2)$$

which will result in a linear dependence and a slope of Γ in the semilogarithmic plot as shown in Figure 19.5 [45]. There can be two possibilities for a broad Lorentzian distribution: that of coupling to the metal and to the molecular vibration modes. The

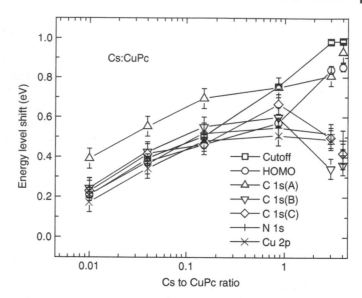

Figure 19.5. The energy level shifts as a function of doping concentration. Note that R_{Cs} is in logarithmic scale to emphasize the initial shift. The most remarkable feature is the linear dependence of the energy level shift on the logarithmic of the Cs concentration in the initial stage, and deviation from the linear dependence for Cu $2p_{3/2}$, N 1s, and pyrrole C. (From Y. Gao and L. Yan, *Chem. Phys. Lett.* **380**, 451 (2003). Copyright Elsevier B. V., 2002. With permission.)

first requires a detailed quantum chemistry calculation, for example a value of $\Gamma = 0.4\,eV$ was obtained for gold and phenyl dithiol [46]. The second effect can be observed experimentally in the low energy tail of the optical absorption known as the Urbach tail [47]. Toyozawa proposed that the Urbach tail in organic molecular semiconductors can be explained by electron–phonon coupling including the quadratic term [48], resulting in $\Gamma_p = k_B T/\sigma$ where σ is a measure of the Urbach tail width that depends on the phonon energy and temperature. An Urbach tail width $k_B T/\sigma$ of $0.132 \pm 0.006\,eV$ can be deduced by linear fitting the cutoff shift shown in Figure 19.5. Direct comparison of this value to the optical data is unavailable for CuPc, but the result does compare favorably with that of $0.11\,eV$ for Langmuir–Blodgett films of a CuPc derivative, copper amphiphilic phthalocyanine (CuAmPcII) reported by Ray et al. [49]. Presently it is unclear what is the fraction of each of the two contributions, (i) the coupling to the metal and (ii) the molecular vibration, to the slope in Figure 19.5, and a temperature dependence study may be useful for distinguishing the two.

At high doping densities, the change of electronic structure due to higher doping can no longer be ignored. This explains why the shift of the Cu $2p_{3/2}$, N 1s, and pyrrole C 1s(A) and (B) deviate from the semilogarithmic linear behavior for $R_{Cs} \geq 0.9$. The shift saturation or reversal of these core levels may result from the concentration of the doped electron distribution in the center of CuPc, including pyrrole and Cu, which screens the Coulomb attraction of the nuclei in that portion of the molecule.

19.5. INTERFACE ENGINEERING IN OSCs: LiF

A large amount of work that has been done with OSCs has focused on improving the carrier injection into the OSC in order to increase the current density and thereby the brightness of these devices. One of the successful approaches is to place a thin film of LiF, an ionic insulator, between the conducting contact and the OSC. This approach has been successful when used in devices that use various designs such as with Al/LiF/ Alq$_3$ [50,51], PEDOT/LiF/ITO [52], and Al/LiF/PFO interfaces [53]. There have been a number of studies of the interactions and energy level alignments of the Al/LiF/ Alq$_3$ interface that showed a chemical interaction, interface dipole formation, and some indication of doping of the organic by free Li [34,54,55]. The explanation for the success of this technique is a lowering of the carrier injection barrier height caused by either doping of the organic by Li that dissociates from the LiF or by formation of interface dipoles at the interfaces with LiF, but there is still controversy about the issue. Some of the results of experiments examining the insertion of LiF into devices show a dependence of the effectiveness of this technique on the thickness of the LiF, suggesting that different thicknesses of LiF result in different Fermi level positions within the bandgap of the OSC [56,57]. Schlaf et al. [58] also investigated the interfaces of LiF deposited onto Al and Pt and they concluded that there was an interface dipole at the interface, followed by band bending within the LiF that causes the thickness dependence of the Fermi level position. The ability to use LiF to design an interface that would allow the position of the Fermi level within an OSC bandgap would be useful in making efficient OSC devices.

As shown in Figure 19.6, the presence of a thin LiF layer on the Alq$_3$ surface does not substantially modify the HOMO position of Alq$_3$ although it does induce significant broadening [34]. The deposition of Al induces new states above the HOMO in the energy gap, and results in a shift of the HOMO position to higher binding energy. This is very similar to what is observed when Ca, Mg, K, or Li are deposited on Alq$_3$ and [35,59], together with the N 1s spectra, are thought to be signatures of formation of the Alq$_3$ anion [59,60]. These states are fundamentally

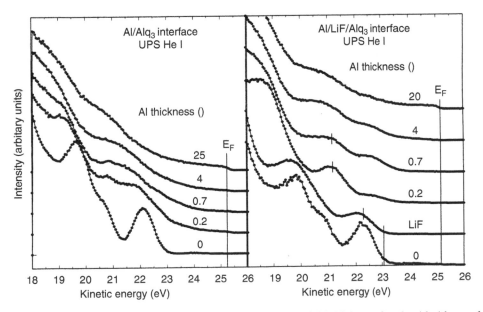

Figure 19.6. Evolution of the UPS spectra as a function of Al thickness for the Al–Alq$_3$ and Al–LiF–Alq$_3$ interfaces. (From Q. T. Le, L. Yan, Y. Gao, M. G. Mason, D. J. Giesen, and C. W. Tang, *J. Appl. Phys.* **87**, 375 (2000). Copyright American Institute of Physics, 2000. With permission.)

different from those formed in the absence of LiF. The molecular orbital structure of Alq_3 is preserved in the presence of LiF. They are shifted to higher binding energy and the gap-state formed is well-defined. In the absence of LiF, the molecular orbital structure of Alq_3 is broadened beyond recognition even at very low coverages of Al. The states that are formed in the gap appear to result simply from an extreme line broadening as a result of the Alq_3 decomposition process. The presence of the LiF layer not only decreases the extent of the reaction with Al, but also decreases the thickness of the reacted layer. The Alq_3 valence band is still identifiable at Al thickness $\Theta Al = 4 \, \text{Å}$, and the Fermi level is observed at an Al coverage as low as $2 \, \text{Å}$. Finally, it is worth noting that the HOMO position cannot be accurately measured due to the early decay of the UPS spectrum upon Al deposition. However, if one assumes that the HOMO shifts by the same energy as the core levels and that the energy gap is taken to be equal to the optical gap [61], i.e., 2.8 eV, the barrier height for electron injection are about 0.55 and 0.2 eV for Al/Alq_3 and $Al/LiF/Alq_3$ interfaces, respectively. The presence of the LiF layer at the interface reduces substantially the electron injection barrier, which represents a nearly ohmic contact character for this interface. Hence, the improvement in $I-V$ characteristics and electroluminescent (EL) efficiencies of the devices with Al/LiF cathode is aided by the reduction of the barrier height for electron injection at the cathode–Alq_3 interface [50,62], and to the formation of a stable interface. In the absence of the LiF, these states appear to result from decomposition of the Alq_3 whereas, with LiF well-defined gap-states are formed which appear to be very similar to Alq_3 anion states observed with several low work function metals. It can be concluded from the UPS and XPS data that the presence of a thin LiF film between Al and Alq_3 layers reduces the barrier height for electron injection at the cathode–Alq_3 contact, and leads to the formation of a stable interface. The reduction is attributed to the results of decomposition of LiF by Al deposition, resulting in doping of the underlying Alq_3 by Li, which lowers the LUMO towards the Fermi level of Al as observed in the UPS spectra [63].

While metal deposition on LiF seems to dope the underlying organic, organic deposition on LiF shows vacuum level alignment. Figure 19.7(a) shows the UPS spectra of the LiF-covered Au as a function of the thickness of the LiF layer on a gradual LiF thickness sample [63]. The growth of the UPS feature at ~9 eV binding energy, attributed to the F 2p orbital [64], with increasing thickness of LiF is what would be expected from earlier measurements of LiF deposition onto Al and Pt [58]. The vacuum level offset is observed to depend on the LiF thickness, from the Au vacuum level ranging from $\sim -1.2 \, \text{eV}$ for $5 \, \text{Å}$ of LiF on Au $\sim -2.3 \, \text{eV}$ for $40 \, \text{Å}$ of LiF on Au. No presence of any oxygen after deposition of LiF is detectable, showing that neither oxygen nor water was at the interface between Au and LiF. The simplest explanation for the changes in the vacuum level and F 1s peak would be a combination of interface dipole formation and energy level bending as was suggested for the interfaces with Al and Pt [58]. Changing the UV intensity by ~30% causes no observable change in the cutoff position of the UPS spectra for $128 \, \text{Å}$ of pentacene on LiF. This suggests that there is no photocurrent charging of the LiF surface at any of the thicknesses investigated here. It can therefore be safe to state that there is an observed interface dipole ranging from $\sim -1.2 \, \text{eV}$ for $5 \, \text{Å}$ of LiF on Au to $\sim -2.3 \, \text{eV}$ for $40 \, \text{Å}$ of LiF on Au.

To investigate the behavior of pentacene deposited onto LiF, we deposited $8 \, \text{Å}$ of pentacene onto the gradual thickness sample and then measured the UPS spectra. This was repeated for 32 and $128 \, \text{Å}$ of pentacene deposition. The UPS spectra of $32 \, \text{Å}$

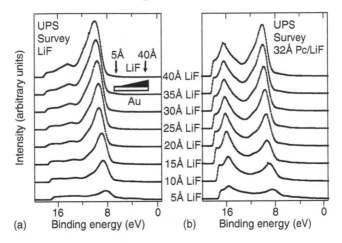

Figure 19.7. (a) UPS spectra of various thicknesses of LiF deposited onto one Au substrate. The inset shows the sample configuration of multiple LiF thicknesses. (b) UPS spectra of 32 Å of pentacene deposited onto the multiple LiF thicknesses. (From N. J. Watkins and Y. Gao, *J. Appl. Phys.* **94**, 1289 (2003). Copyright American Institute of Physics, 2003. With permission.)

of pentacene deposited onto the gradual thickness substrate are shown in Figure 19.7(b). The major change is the appearance of a feature at a binding energy of ~16 eV. This feature is roughly identical in magnitude for LiF thicknesses between 15 and 40 Å. This feature is less intense for 10 Å of LiF and even less intense for 5 Å of LiF. An important observation is that no significant shift in the vacuum level position occurs upon deposition of pentacene onto any thickness of LiF on the substrate. This indicates that there is VLA at the pentacene–LiF interface.

Of more interest in device design is the position of the Fermi level within the bandgap of the pentacene. Figure 19.8 shows the vacuum level at the top of the columns, the measured HOMO at the bottom of the empty rectangles, and the estimated LUMO, using a bandgap of 2.2 eV, deduced from combined UPS and IPES measurements, at the top of the empty rectangles. As shown in Figure 19.8, the HOMO of pentacene deposited onto a clean Au surface is located 0.5 eV below the Fermi level [26,65]. With a LiF thickness of 40 Å, the pentacene HOMO is ~2.1 eV from the Fermi level while the LUMO, which started ~1.7 eV above the Fermi level, is a mere 0.1 eV above the Fermi level. By the insertion of various thicknesses of LiF, the Fermi level position within the pentacene bandgap can be tuned to be within 0.5 eV of the HOMO or within 0.1 eV of the LUMO. Attempting to take advantage of the energy level alignment in device design requires the the fundamental insulating nature of LiF to be taken into account and the necessity of electron tunneling through the LiF be considered.

19.6. ITO SURFACE TREATMENT AND ITS INTERFACE WITH ORGANIC: NPB/ITO

ITO is one of the few metal oxides that combines many technologically interesting properties such as high transparency in the visible region of the solar spectrum, good electrical conductivity, and excellent substrate adherence. Organic optoelectronic

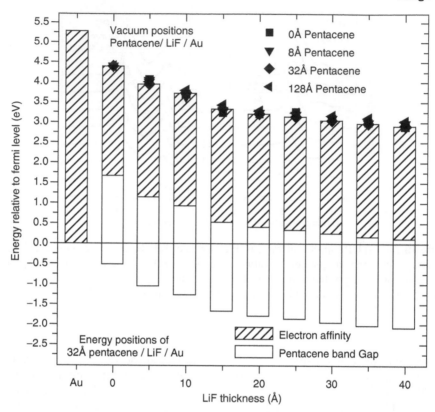

Figure 19.8. The energy levels of pentacene deposited onto LiF coated Au as a function of the LiF thickness. The vacuum level is the top of the filled rectangles. The bandgap is represented by the hollow rectangles, the bottom representing the pentacene HOMO and the top representing the calculated pentacene LUMO. The values for the pentacene/Au structure are from a separate experiment. (From N. J. Watkins and Y. Gao, *J. Appl. Phys.* **94**, 1289 (2003). Copyright American Institute of Physics, 2003. With permission.)

device performance has been further enhanced by a variety of ITO surface treatment technologies, including depositing a thin carbon layer [66], polyanilane (PANI) [67,68], and poly(3,4-ethylene dioxythiophene) (PEDOT) [68,69] by oxygen plasma [69–71]. For small molecule devices, the device stability is dramatically increased by interposing a CuPc layer on the ITO anode [72]. It had been suggested that the CuPc layer lowers the drive voltage by reducing the effective barrier between the ITO and the NPB hole-transporting layer [73,74]. Conversely, other researchers have shown that relatively thick, 15 nm CuPc layers will, in fact, reduce the rate of hole injection from the ITO anode, leading to a better balance with the electron current arriving from the cathode to the recombination zone [75]. In the latter work, it was also demonstrated that an excess of holes in the Alq$_3$ layer would create an unstable cation population, leading to a rapid degradation in device performance. For PPV, the device properties can also be improved by plasma treatments and surface morphology [69,76]. As with the cathode–organic interface, the hole injection efficiency from the ITO anode to the hole-transporting organic layer is tuned through interface engineering and selecting organic molecules with energy alignments that favor ohmic injection from the ITO anode [77]. Le et al. [78] have recently shown that significant work function shifts can be obtained simply by dipping the ITO substrate in an acid or base solution, and the modification on device performance is consistent with the work function shift [79].

Figure 19.9(a) shows a typical He I UPS spectrum for ITO in kinetic energy scale [80]. The spectrum was taken for $-3.000\,V$ sample bias so that the sample inelastic cutoff could be distinguished from that due to the spectrometer which is observable in Figure 19.9(a) as a small peak near $2\,eV$. This figure also illustrates the relation between the width of the spectrum, and sample work function ϕ and photon energy. The spectrum width is determined by the distance between the sample inelastic cutoff and the Fermi edge as illustrated in Figure 19.9(b) and (c), respectively. Ideally, these edges should be infinitely abrupt but due to the limited spectrometer resolution and thermal effects, they are broadened. The actual positions are determined as the center of the slopes, indicated by vertical lines. The estimated error in energy position was less than $0.02\,eV$. The determination of the position of the inelastic cutoff was unambiguous for all samples, while the Fermi edge did not readily show up for some samples. However, this did not pose a problem because the Fermi edge is the reference point of energy scale in photoelectron spectroscopy and should not change from sample to sample as long as they are grounded. Nevertheless we checked the position of Fermi edge using Au films electrically connected and positioned next to the ITO sample.

The most profound effects are found in ITO surfaces treated with phosphoric acid (H_3PO_4) and tetrabutylammonium hydroxide [$N(C_4H_9)_4OH$], which result in the highest work function shifts with respect to that of the standard ITO, $+0.7$ and $-0.6\,eV$, respectively. It is found that the position of the HOMO of the deposited organic material depends on the work function of the ITO substrate and that the acid-treated ITO gave the lowest hole injection barrier. No significant reactions are observed between standard ITO and N,N'-bis-(1-naphthyl)-N,N'-diphenyl-1,1'-biphenyl-4,4'-diamine (NPB), but acid-treated ITO presented interesting effects of the protons in the adsorbed acid layer. Figure 19.10 summarizes the HOMO position for NPB relative to the Fermi level, as a function of NPB thickness, for ITO substrates of different work functions. The positions of the HOMO for NPB relative to the Fermi level, as a function of NPB thickness for ITO substrates of various work functions are shown in Figure 19.10(a). The HOMO peaks for NPB deposited on a standard ITO as a function of NPB thickness are shown in Figure 19.10(b). The HOMO position is determined by the intersection between the signal and the background of the UPS spectrum. The HOMO is attributed to the central biphenyl unit with the contribution of the nitrogen lone pairs and the nitrogen-bound phenyl units [81]. For standard ITO, and in particular for the acid-treated ITO, the values corresponding to low NPB coverage are not available because the signal from NPB is weak, which prevents an accurate determination of the HOMO position. The HOMO binding energy increases slightly with the NPB thickness. The result clearly indicates that the position of the HOMO of NPB depends on the work function of the ITO substrate onto which the organic layer is deposited. The higher the ITO work function, the smaller the distance from the HOMO to the Fermi level, which corresponds to a lower barrier to hole injection. This is consistent with the current–voltage characteristics obtained from N,N'-(3-methyl phenyl)-1,1'-biphenyl-4,4'-diamine (TPD) single-layer devices based on ITO of various work functions, where the turn-on voltage of devices appears to depend qualitatively on the work function of the ITO anode [82]. The turn-on voltage increases in the following order: acid-, standard, and base-treated ITO. Note that, in this context, the current–voltage characteristics obtained from TPD- and NPB-based devices are comparable since the transport properties

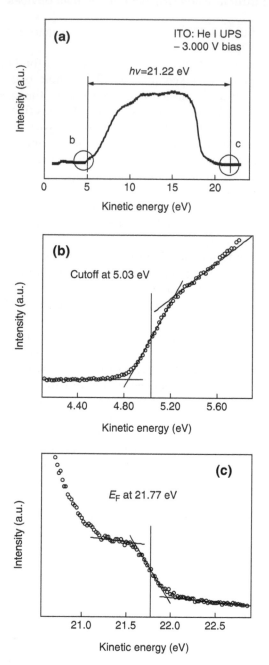

Figure 19.9. (a) A typical He I ($h\nu = 21.22\,\text{eV}$) UPS spectrum of ITO taken with $-3.000\,\text{V}$ bias applied to the sample. Also shown is inelastic cutoff (circle b) and Fermi edge (circle c). The relationship between spectrum width, $h\nu$, and work function f is illustrated. (b) Shows a detailed spectrum of inelastic cutoff region. It also shows the cutoff energy with a vertical bar. (c) Is similar to (b), but shows Fermi edge region. (From Y. Park, V. Choong, Y. Gao, B. R. Hsieh, and C.W. Tang, *Appl. Phys. Lett.* **68**, 2699 (1996). Copyright American Institute of Physics, 2003. With permission.)

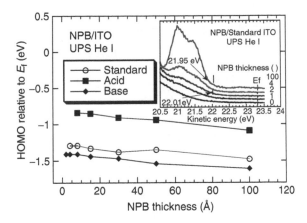

Figure 19.10. Position of the highest occupied molecular orbital (HOMO) relative to the Fermi level as a function of NPB thickness for ITO substrates treated in aqueous solutions of phosphoric acid or tetrabutylammonium hydroxide. (From Q.T. Le, F. Nuesch, L.J. Rothberg, E.W. Forsythe, and Y. Gao, *Appl. Phys. Lett.* **75**, 1357 (1999). Copyright American Institute of Physics, 2003. With permission.)

and chemical characteristics, in terms of chemical reactivity, of both molecules are very similar.

The diagrams in Figure 19.11 summarize the energy alignment at the ITO–NPB interface derived from our UPS measurements on the various ITO treatments [80]. The barrier to hole injection from the ITO anode is taken as the energy difference between the Fermi level of ITO and the HOMO of NPB. The barrier height decreases significantly from 1.61 to 1.09 eV for base- and acid-treated ITO, respectively, with the standard ITO having an intermediate value (1.35 eV). The microscopic mechanism that governs the improvement in charge injection can be extremely complex, but it is clear in the present case that the increase in the anode work function contributes to the improvement in turn-on voltage of OLEDs.

19.7. ORGANIC–ORGANIC INTERFACE: NPB AND Alq$_3$

The energy offset between two dissimilar organic films is vitally important to efficient OPV operation. Among the important information obtained from photoemission studies is the molecular energy level alignment at the interface of multilayer films. In case of doped or mixed layer OPVs, the energy level remains important as this dictates charge separation mechanism. The interfaces between organic molecules have some similar characteristics observed at the metal–organic interfaces. The fundamental distinction between metal–organic and organic–organic interfaces is that to date, researchers have not observed significant chemical interactions between organic molecular films. This is fundamentally related to the fact that the wave functions for the molecules do not extend much beyond a few nearest neighbors. However, many other features are observed, depending on the system under investigation. Interface dipoles have been observed for tetracyanoquinodimethane (TCNQ) on tetrathianaphthacence (TTN), from vacuum level shifts [83]. Similarly,

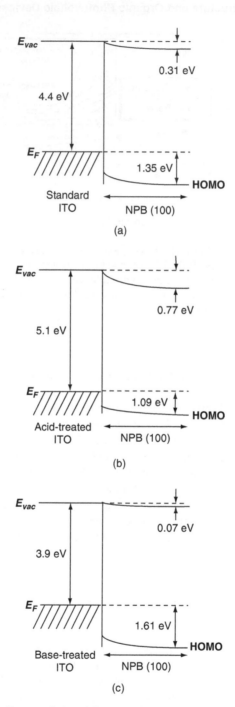

Figure 19.11. Energy diagram deduced from UPS measurements for (a) NPB/standard ITO, (b) NPB/acid-treated ITO, and (c) base-treated ITO. (From Q. T. Le, F. Nuesch, L. J. Rothberg, E. W. Forsythe, and Y. Gao, *Appl. Phys. Lett.* **75**, 1357 (1999). Copyright American Institute of Physics, 2003. With permission.)

studies of perylenetetracarboxylic dianhydride (PTCDA) on phthalocynanines (InCl and Zn, ClInPc, and ZnPc, respectively) have demonstrated significant vacuum level shifts attributed to dipole formations at the interfaces [84]. In addition, UPS results in combination with XPS measurements reveal that band bending occurs at the PTCDA–Pc interfaces [85]. Conversely, Hill and co-workers show that other donor–acceptor type molecular solid interfaces do not form appreciable interface dipoles [86]. There remains the question why some molecular solid interfaces form dipoles, while others do not. The answer may be related to the differences in the ionization potential between the molecules, though more work is required.

The UPS spectra of NPB and Alq_3 are shown in Figure 19.12 [87]. The symbols A and B mark features unique to Alq_3 and NPB, respectively. Their shifts as the thickness Θ increases again indicate the interface energy level bending at the interface. More detailed information can be obtained by spectrum subtraction. As shown in Figure 19.13, the NPB contribution in the spectrum of $\Theta = 5\,\mathring{A}$ is isolated by first scaling the Alq_3 spectrum ($\Theta = 0\,\mathring{A}$) to the ratio of the A feature in the two spectra, followed by aligning the features and subtracting the scaled spectra. The resulting difference spectrum is nearly identical to that of NPB of $\Theta = 55\,\mathring{A}$ except a rigid shift to the lower BE. Such spectrum subtraction allows us to distinguish the HOMO of NPB, which increases from -1.0 to $-1.4\,eV$ as a function of increasing Θ. The HOMO is defined as the intersection of the leading edge and the baseline of the spectrum as shown in the inset. A similar procedure was also conducted for the Alq_3 contribution UPS spectrum with $\Theta = 2\,\mathring{A}$. The resulting Alq_3 UPS spectrum is

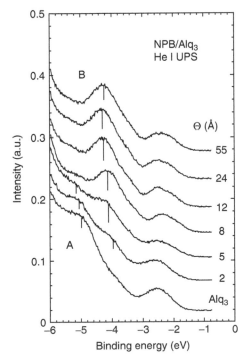

Figure 19.12. Ultraviolet photoemission spectroscopy of NPB/Alq_3 films as a function of NPB coverage, Θ. (From E.W. Forsythe, V.-E. Choong, C.W. Tang, and Y. Gao, *J. Vac. Sci. Technol.* A**17**, 3429 (1999). Copyright American Institute of Physics, 1999. With permission.)

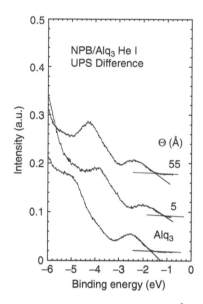

Figure 19.13. A plot of the Alq_3 UPS spectrum, the $\Theta = 5\,\text{Å}$ UPS difference spectrum, and $\Theta = 55\,\text{Å}$. The intersection of the straight lines are the HOMO level positions at the different coverages. (From E. W. Forsythe, V.-E. Choong, C. W. Tang, and Y. Gao, *J. Vac. Sci. Technol.* **A17**, 3429 (1999). Copyright American Institute of Physics, 1999. With permission.)

identical to ($\Theta = 0\,\text{Å}$) and a HOMO level increase of 0.1 eV to higher binding energy after the first NPB coverage. For $\Theta \geq 5\,\text{Å}$, the Alq_3 contribution to the UPS spectrum could not be isolated from the NPB contribution. These gradual shifts of occupied molecular orbital features provide a clear picture of energy level bending in both NPB and Alq_3. Another remarkable feature is the lack of interface states, indicating little wave function overlap or CT between the two OSCs.

The HOMO of NPB and Alq_3 are obtained by difference spectra exemplified in Figure 19.13 [87]. The Alq_3 HOMO level shifts from -1.7 eV for $\Theta = 0\,\text{Å}$ to -1.8 eV for $\Theta = 2\,\text{Å}$, similar to that of the Al 2p and O 1s core level shifts at the same coverage. For higher NPB coverages, the Alq_3 feature in the UPS spectra can hardly be distinguished because of the much shorter electron mean free path in UPS ($\leq 10\,\text{Å}$) than in XPS ($\sim 20\,\text{Å}$). The total binding energy shift in Alq_3 is between 0.4 and 0.5 eV as determined from the XPS core level shift at higher coverages. The NPB HOMO at Θ of $5\,\text{Å}$ is -1.0 eV and shifts to -1.4 eV between 12 and $24\,\text{Å}$, approximately one to two monolayers, after which it remains constant at -1.4 eV and gives a total shift of 0.4 eV downward. Such rigidity of spectra features of a given species characterizes the existence of interface level bending and the lack of chemical reactions or interface states. As seen from Figure 19.14, the saturation of the NPB shift occurs between $\Theta = 12$ and $24\,\text{Å}$, in sharp contrast to the fast saturation of work function change at $\Theta = 8\,\text{Å}$, which is approximately one monolayer of NPB. The work function has a total shift of 0.2 eV, where the work function is the difference between the Fermi level and the vacuum level, obtained from the inflection point of the secondary electron cutoff of the UPS spectra. At $\Theta = 8\,\text{Å}$, the work function, mark B in Figure 19.14, and the O 1s core level all increase by less than 0.05 eV. The origin of this increase is under investigation. The IP for Alq_3 and NPB are 5.7 and 5.2 eV, respectively. The bulk HOMO levels for Alq_3 and NPB are -1.7 and -1.4 eV, respectively.

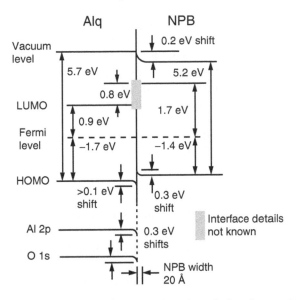

Figure 19.14. The diagram includes the HOMO and work function position from the UPS data and core level positions from XPS. The LUMO level positions relative to the HOMO position are measured from the optical absorption data. (From E. W. Forsythe, V.-E. Choong, C. W. Tang, and Y. Gao, *J. Vac. Sci. Technol.* A17, 3429 (1999). Copyright American Institute of Physics, 1999. With permission.)

The HOMO level alignment is crucial to the high efficiency observed in this NPB and Alq$_3$ heterostructure device.

The schematics of the interface energy diagram of NPB and Alq$_3$ is given in Figure 19.14 [87]. Taking an optical bandgap of 2.8 and 3.1 eV for Alq$_3$ and NPB, respectively, the LUMO energy level position for Alq$_3$ is 0.8 eV below the NPB LUMO, while the HOMO levels for Alq$_3$ and NPB are offset in the bulk by 0.3 eV. In addition to the bulk level offsets, the peculiar downward energy level shift of Alq$_3$ and upward shift of NPB towards the interface enhance the HOMO discontinuity from the bulk value of 0.3 to 0.7 eV. For Alq$_3$, the interface-bending region is determined from the shift in the core levels as determined from XPS, which is approximately 0.4 eV. The UPS difference spectra cannot distinguish Alq$_3$ contributions beyond the first NPB coverage. This additional energy level bending at the interface may contribute traps and additional barriers for the carrier transport across the interface. We cannot obtain detailed information about the LUMO interface formation region, as these levels are determined from the optical data. From the LUMO level alignment, a substantial barrier exists for electrons to be injected from the Alq$_3$ side of the heterostructure to the NPB with a relatively smaller barrier for holes to tunnel into the Alq$_3$. This energy level alignment favors electron confinement in the emissive Alq$_3$ layer of the heterostructure.

19.8. DYNAMICS OF INTERFACE CHARGE TRANSFER: TPD AND DPEP

The interest in electron transfer (ET) between the excited state manifolds in OSCs is prompted by its many uses in photoreceptor and imaging applications. Dye-sensitized

ISC surfaces have been demonstrated as potential structures for the next generation of solar panels [88,89], and the dye sensitization of silver bromide surfaces
provides the basis for color photography. From the technological point of view,
it is essential that one is able to identify and prepare a suitable class of materials
that exhibit the desired properties needed for the application. To do this, understanding the dominant mechanism and the key factors determining the CT is
required.

Earlier observations of ultrafast ET in solid systems have been based almost
solely on the measurements of fully optical transients. In time-resolved fluorescence
and transient absorption studies of oxaxine-covered SnS_2 surfaces, the time constant
for the ET from the dye to the semiconductor was determined to be 40 fsec
[90,91]. Similar transient absorption measurements on ruthenium dye and perylene
chromophores anchored onto a TiO_2 film have deduced reaction times between
25 (ruthenium) and 190 fsec (perylene) [92,93] for ET from the LUMO of the dye
to the conduction band of the semiconductor. In all of these experiments, the
LUMO of the dye molecule was found to be well above the conduction band
minimum of the semiconductor and therefore the electron transfer was considered
barrierless. More recently, ultrafast ET has also been reported for a fully organic
solid heterostructure of polymer–C_{60} composites [94,95]. In these studies, the
transfer rates for ET from the polymer to C_{60} were found to be a few hundred
femtoseconds.

The results of time-dependent photoemission spectroscopy on an organic heterostructure of *N,N′*-diphenethyl-3,4,9,10-perylenetetracarboxylic-diimide (DPEP)
and TPD [96] are presented below. DPEP is an organic photoconductor material,
and it is well known for its charge generator properties, and TPD is a widely used
organic hole transport material in many photoreceptor applications. Together, these
two materials form a photoreceptor structure where either of the molecules can act as
a donor, depending on the excitation energy. In conjunction with the TRPES
measurements of the bilayer structures, we measured the UPS spectra on the films
after each TPD evaporation. The progression of the UPS spectra of a TPD and
DPEP film is shown in Figure 19.15 as a function of the kinetic energy observed in
the energy analyzer [96]. The effect of the added TPD layer can be observed as a shift
of the HOMO level starting at 4 Å coverages. The HOMO cutoff of the pristine
DPEP film is at 29.45 eV whereas at 24 Å the level has shifted to 29.60 eV. At the
same time, the vacuum level cutoff of the film with 24 Å TPD is shifted down from
the pristine value of 13.85 to 13.70 eV. This means that the spectral width of the
bilayer film is 0.3 eV wider than the value for the pristine DPEP film. Consequently,
the ionization potential of the bilayer film is 5.3 eV, which is 0.3 eV lower than the
ionization potential of the pristine film, 5.6 eV. Furthermore, the shift of the vacuum
levels in the bilayer films indicates the presence of vacuum level offset between the
two constituent films. The observed vacuum level shift of 0.15 eV can be considered
small, and it compares well with shifts reported for other organic–organic interfaces
[83–86]. The HOMO level of the TPD and DPEP structure is 0.15 eV higher in
energy than the DPEP HOMO level. This will affect the kinetic energy spectrum
measured in the TRPES and will be discussed later. Figure 19.15 also includes a
UPS spectrum of a pure TPD film. The ionization potential of the 400 Å film was
found to be 5.2 eV, which compares relatively well with the value of 5.3 eV for the
bilayer structure. There are obvious similarities between the pristine TPD spectrum
and the high TPD coverage bilayer spectra. It turns out that the bilayer spectra
can mostly be described as a superposition of the pristine DPEP and TPD spectra,

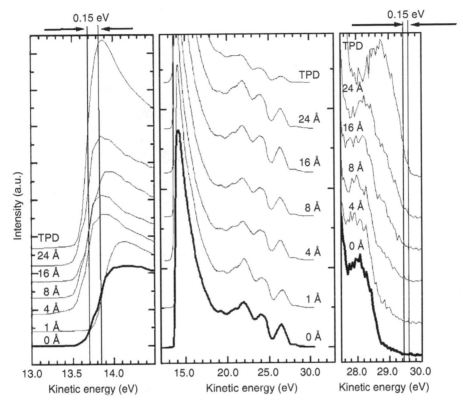

Figure 19.15. UPS spectra of DPEP-II films with increasing TPD coverage. The right panel shows the band alignment at the film interface determined from the spectra. (From A. J. Mäkinen, S. Schoemann, Y. Gao, M. G. Mason, A. A. Muenter, and A. R. Melnyk, in *Conjugated Polymer and Molecular Interfaces: Science and Technology for Photonic and Optoelectronic Applications*, W. R. Salaneck, K. Seki, A. Kahn, and J. J. Pireaux, eds. (Marcel Dekker, New York, 2001). Copyright Marcel Dekker, 2001. With permission.)

which implies that at least energetically the composite layers preserve their own characteristics.

As the TPD coverage on the DPEP films was increased, the TRPES spectra showed longer overall decay times, i.e., the delay spectra appeared broader. It was found that a single exponential function convolved with the instrument response function did not properly describe the time delay spectrum of the composite films any more, and it becomes necessary to include *two* exponential decay terms:

$$I(t) = R(t) \otimes [A_1 \exp(-|t|/\tau_1) + A_2 \exp(-|t|/\tau_2)] \tag{3}$$

where the amplitudes A_1 and A_2 were allowed to vary during the fitting process. Interestingly, the two lifetime components included in the fits are separated by more than one order of magnitude at high TPD coverages, as can be seen in Figure 19.16, where the deconvoluted lifetimes are plotted as a function of energy above the DPEP HOMO level. The fast components range from 40 to 80 fsec, and the long components are from 200 to 800 fsec in the TPD and DPEP film. The short components in bilayers are also consistently longer than the corresponding single exponential lifetimes in pristine DPEP films.

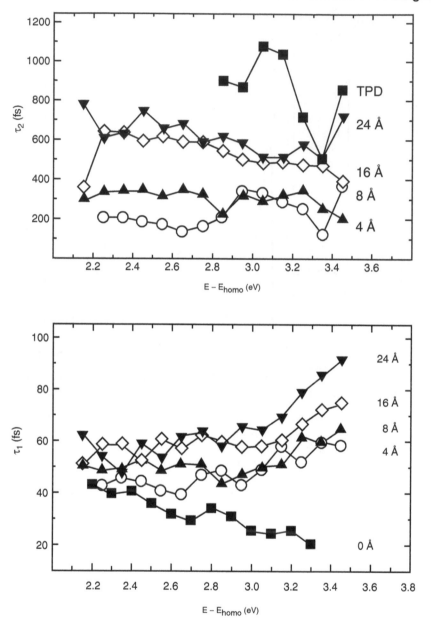

Figure 19.16. Slow and fast lifetime components in TPD and DPEP-I film. (From A.J. Mäkinen, S. Schoemann, Y. Gao, M. G. Mason, A. A. Muenter, and A. R. Melnyk, in *Conjugated Polymer and Molecular Interfaces: Science and Technology for Photonic and Optoelectronic Applications*, W. R. Salaneck, K. Seki, A. Kahn, and J. J. Pireaux, eds. (Marcel Dekker, New York, 2001). Copyright Marcel Dekker, 2001. With permission.)

The lifetime versus energy plots reveal another remarkable effect of the added TPD layer. The energy dependence of the lifetimes is different in bilayers when compared with the trend found in pristine DPEP films where lifetimes decrease monotonically with increasing energy. In bilayers, the short components seem to be almost energy independent up to 3.0 eV above the DPEP HOMO. Above 3.0 eV, however, the short components increase with increasing energy, which is opposite to what was observed in pristine DPEP films [97]. The most dramatic change observed, as the thickness of the TPD layer is increased, is the sudden growth in the absolute signal intensity. The appearance of two exponential components in the bilayer decay dynamics indicates the presence of two distinct electron populations with different decay mechanisms. It is clear from the pronounced dependence of the lifetime components on the TPD deposition thickness that the decay dynamics observed in TRPES experiments is interface and adlayer driven. Therefore, the first step in understanding the observed decay dynamics is to identify the possible electron populations formed in the bilayer structure upon photoexcitation.

An obvious question is whether the two lifetime components are just intrinsic decay times of photoexcitations in the two materials, DPEP and TPD. The lifetimes (0.8–1.0 psec) measured in the TPD layer deposited on graphite are comparable to the slow components found in the DPEP-based bilayers with a 4 Å or thicker TPD overlayer. However, there were no measurable transients present below the approximate LUMO cutoff at 3.0 eV in the TPD films deposited on graphite whereas the long components were found to persist at energies below the TPD LUMO in DPEP-based bilayer structures. This observation and the TPD thickness dependence of the long component in bilayer structures indicate that the observed long component depends on the interface, i.e., the material onto which TPD is deposited which eliminates the possibility that the long component could be just assigned to electron relaxation within the TPD layer.

As noted earlier, the short component was absent in the transients recorded in the TPD layers on graphite. However, it is found in all of the decay spectra measured on bilayer structures, and it is present at every level above DPEP HOMO accessible in the TRPES experiments. Although, the short components measured in the bilayer structures share the same order of magnitude with the single exponential lifetimes found for pristine DPEP films they are significantly longer (60–80 fsec compared to 20–50 fsec) and show very different energy dependence. Maybe the most dramatic change observed in the short component is the three orders of magnitude increase in the intensity of the component as the thickness of the TPD layer is increased. These three observations about the fast decay components in the bilayers indicate that the fast transients are not only due to electron relaxation within the DPEP film, but also additional mechanism is likely to influence the rapid relaxation dynamics in bilayer structures.

In the model depicted in Figure 19.17, after photoexciting a TPD molecule the electron is quickly transferred from the TPD molecule across the interface into the DPEP film [96]. Once in the DPEP the electron loses rapidly its energy relaxing to the lowest level of the excited anion. Therefore, the thickness dependence of the lifetime components seems to indicate that the excitation decay measured in the TRPES experiments is injection limited, i.e., the lifetime components actually reflect the injection rates of electrons into DPEP.

The considerations above of the three-population kinetics can be related to bilayer kinetics by assigning the decay rates and populations to the ones observed in the TRPES experiments. It seems that there are two parent populations $n_1(t)$ and $n_2(t)$ with fast k_1 and slow k_2 injection rates, respectively, created in the TPD layer upon photoexcitation. These rates are slower than the inherent relaxation of the electron in DPEP described by rate k_0 after its injection from TPD. Therefore, the two exponential terms describing the decay of the pump–probe spectra correspond to the fast and slow injection of electrons from the TPD layer into DPEP.

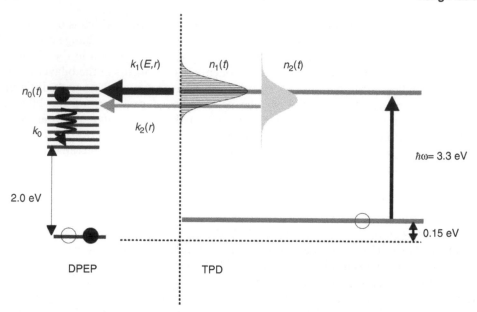

Figure 19.17. Electron transfer from TPD to DPEP. (From A. J. Mäkinen, S. Schoemann, Y. Gao, M. G. Mason, A. A. Muenter, and A. R. Melnyk, in *Conjugated Polymer and Molecular Interfaces: Science and Technology for Photonic and Optoelectronic Applications*, W. R. Salaneck, K. Seki, A. Kahn, and J. J. Pireaux, eds. (Marcel Dekker, New York, 2001). Copyright Marcel Dekker, 2001. With permission.)

19.9. FINAL REMARKS

The above is to introduce important known facts and understanding of interfaces involving OSCs, and it is not intended to be an exhaustive review at all the possible uses of surface analytical tools for investigations of OSC devices. It only means to illustrate that the electronic properties of the materials and interfaces are the key to the understanding of their PV and other characteristics under various bulk and interface conditions. It should also be mentioned that as investigations on systems of OLED and OTFT are quite extensive, those on OPV systems are still relatively limited. A large number of questions remain open for interface processes in OPVs, including the kinetic requirements and the role of interface polarizability, exciton transport rates, interfacial electronic states, the crucial process of interfacial charge-carrier recombination, the relationship of CT across, and exciton transport towards the interface, etc. One may expect that more systematic studies on OPV interfaces and more tools that probe the electronic structure of materials will be applied in the future, and a more complete picture of the interface formation in OSC devices can be expected to emerge as a result.

ACKNOWLEDGMENTS

The author gratefully acknowledges contributions by many scientists with whom he has the pleasure to work together on the topics presented in this chapter, especially Dr. C. W. Tang, Dr. M. G. Mason, Dr. Y. Park, Dr. E.W. Forsythe, Dr. Q. T. Le, Dr. V. E. Choong, Dr. A. J. Makinen, Dr. L. Yan, and Dr. N. J. Watkins. The work was supported in part by the National Science Foundation under Grant No. DMR-0305111.

REFERENCES

1. C. W. Tang, *Appl. Phys. Lett.* **48**, 183 (1986).
2. P. Peumans, A. Yakimov, and S. R. Forrest, *J. Appl. Phys.* **93**, 3693 (2003).
3. B. A. Gregg and M. C. Hanna, *J. Appl. Phys.* **93**, 3605 (2003).
4. S. F. Shaheen, C. J. Brabec, N. S. Sariciftci, F. Padiner, T. Fromherz, and J. C. Hummelen, *Appl. Phys. Lett.* **78**, 841 (2001).
5. J. M. Kroon, M. M. Wienk, W. J. J. Verhees, and J. C. Hummelen, *Thin Solid Films* **403–404**, 223 (2002).
6. P. Pewmans and S. R. Forrest, *Appl. Phys. Lett.* **79**, 126 (2001).
7. K. Bucher and S. Kunzelmann, Proc. 26th Photovoltaic Conf. 1193 (1997).
8. E. A. Silinsh and V. Capek, *Organic Molecular Crystals* (AIP Press, New York, 1994).
9. M. Pope and C. E. Swenberg, *Electronic Processes in Organic Crystals* (Oxford University Press, New York, 1982).
10. S. M. Sze, *Physics of Semiconductor Devices* (John Wiley, New York, 1981).
11. G. Yu, J. Gao, J. Hummelen, F. Wudl, and A. J. Heeger, *Science* **270**, 1789 (1995).
12. J. J. M. Halls, C. A. Walsh, N. C. Greenham, E. A. Marseglia, R. H. Friend, S. C. Moratti, and A. B. Holmes, *Nature* **376**, 498 (1995).
13. P. A. van Hall, R. A. J. Janssen, G. Cerullo, M. Zavelani-Rossi, and S. D. Silvestri, *Chem. Phys. Lett.* **345**, 33 (2001).
14. B. C. Zerza, G. Cerullo, S. D. Silvestri, and N. S. Sariciftci, *Synth. Met.* **119**, 637 (2001).
15. T. Tsuzuki, Y. Shirota, J. Rostalski, and D.Meissner, *Sol. Energy Mater. Sol. Cells* **61**, 1 (2000).
16. A. Dhanabalan, J. Knol, J. C. Hummelen, and R. A. J. Janssen, *Synth. Met.* 119, 519 (2001).
17. B. A. Gregg, M. A. Fox, and A. J. Bard, *J. Phys. Chem.* **94**, 1586 (1990).
18. B. A. Gregg and Y. I. Kim, *J. Phys. Chem.* **98**, 2412 (1994).
19. G. G. Malliaras, J. R. Salem, P. J. Brock, and J. C. Scott, *J. Phys.* **84**, 1583 (1998).
20. C. J. Brabeck, A. Cravino, D. Meissner, N. S. Sariciftci, T. Fromherz, M. T. Rispens, L. Sanchez, and J. C. Hummelen, *Adv. Funct. Mater.* **11**, 374 (2001).
21. C. M. Ramsdale, J. A. Barker, A. C. Arias, J. D. MacKenzie, R. H. Friends, and N. C. Greenham, *J. Appl. Phys.* **92**, 4266 (2002).
22. W. R. Salaneck, S. Stafstrom, and J. L. Bredas, *Conjugated Polymer Surfaces and Interfaces* (Cambridge University Press, Cambridge, 1996).
23. H. Ishii, K. Kiyoshi, E. Ito, and K. Seki, *Adv. Mater.* **11**, 605 (1999).
24. Y. Gao, Surface analytical studies of interface formation in organic light emitting devices, *Acc. Chem. Res.* **32**, 247 (1999).
25. E. W. Forsythe and Y. Gao, in *Handbook of Surfaces and Interfaces of Materials*, H. S. Nalwa, ed., p. 286 (Academic Press, New York, 2001).
26. N. J. Watkins, L. Yan, and Y. Gao, *Appl. Phys. Lett.* **80**, 4384 (2002).
27. M. Probst and R. Haight, *Appl. Phys. Lett.* **70**, 1420 (1997).
28. M. Stoessel, G. Wittmann, J. Staudigel, F. Steuber, J. Blässing, W. Roth, H. Klausmann, W. Rogler, J. Simmerer, A. Winnacker, M. Inbasekaran, and E. P. Woo, *J. Appl. Phys.* **87**, 4467 (2000).
29. F. Faupel, R. Willecke, and A. Thran, *Mater. Sci. Eng.* **R22**, 1 (1998).
30. L. Yan, M. G. Mason, C. W. Tang, and Y. Gao, *Appl. Surf. Sci.* **175–176**, 412 (2001).
31. A. Rajagopal, and A. Kahn, *J. Appl. Phys.* **84**, 355 (1998).
32. C. Shen, A. Kahn, and J. Schwartz, *J. Appl. Phys.* **89**, 449 (2001).
33. H. Ishii, K. Sugiyama, E. Ito, and K. Seki, *Adv. Mater.* **11**, 605 (1999).

34. Q. T. Le, L. Yan, Y. Gao, M. G. Mason, D. J. Giesen, and C. W. Tang, *J. Appl. Phys.* **87**, 375 (2000).
35. M. G. Mason, C. W. Tang, L. S. Hung, P. Raychaudhuri, J. Madathil, D. J. Giesen, L. Yan, Q. T. Le, Y. Gao, S. T. Lee, L. S. Liao, L. F. Cheng, W. R. Salaneck, D. A. dos Santos, and J. L. Bredas, *J. Appl. Phys.* **89**, 2756 (2001).
36. R. T. Tung, *Phys. Rev. Lett.* **84**, 6078 (2000).
37. L. Yan, N. J. Watkins, S. Zorba, Y. Gao, and C. W. Tang, *Appl. Phys. Lett.* **81**, 2752 (2002).
38. I. G. Hill, A. Rajagopal, and A. Kahn, *J. Appl. Phys.* **84**, 3236 (1998).
39. N. D. Lang, *Phys. Rev. Lett.* **46**, 842 (1981).
40. A. J. Heeger, J. R. Schriefer, and W. P. Su, *Rev. Mod. Phys.* **40**, 3439 (1988).
41. L. Yan, N. J. Watkins, S, Zorba, Y. Gao, and C. W. Tang, *Appl. Phys. Lett.* **79**, 4148 (2001).
42. I. G. Hill, A. Kahn, Z. G. Soos, and R. A. Pascal, *Chem. Phys. Lett.* **327**, 181 (2000).
43. I. Kojima and M. Kurahashi, *J. Electron Spectrosc. Relat. Phenom.* **42**, 177 (1987).
44. E. Orti and J. L. Bredas, *J. Am. Chem. Soc.* **114**, 8669 (1992).
45. Y. Gao and L. Yan, *Chem. Phys. Lett.* **380**, 451 (2003).
46. W. Tian, S. Datta, S. Hong, R. Reifenberger, J. I. Henderson, and C. P. Kubiak, *J. Chem. Phys.* **109**, 2874 (1998).
47. F. Urbach, *Phys. Rev.* **92**, 1324 (1953).
48. Y. Toyozawa, *Suppl. Progr. Theo. Phys.* **12**, 111 (1959).
49. A. K. Ray, S. Mukhopadhyay, and M. J. Cook, *Phys. Stat. Sol.* A**134**, K73 (1992).
50. L. S. Hung, C. W. Tang, and M. G. Mason, *Appl. Phys. Lett.* **70**, 152 (1997).
51. L. S. Hung, C. W. Tang, M. G. Mason, P. Raychaudhuri, and J. Madathil, *Appl. Phys. Lett.* **78**, 544 (2001).
52. F. R. Zhu, B. L. Low, K. R. Zhang, and S. J. Chua, *Appl. Phys. Lett.* **79**, 1205 (2001).
53. G. Greczynski, M. Fahlman, and W. R. Salaneck, *J. Chem. Phys.* **113**, 2407 (2000).
54. M. G. Mason, C. W. Tang, L. S. Hung, P. Raychaudhuri, J. Madathil, D. J. Giesen, L. Yan, Q. T. Le, Y. Gao, S. T. Lee, L. S. Liao, L. F. Cheng, W. R. Salaneck, D. A. dos Santos, and J. L. Bredas, *J. Appl. Phys.* **89**, 2756 (2001).
55. L. S. Hung, R. Q. Zhang, P. He, and G. Mason, *J. Phys. D* **35**, 103 (2002).
56. T. M. Brown, I. S. Millard, D. J. Lacey, J. H. Burroughes, R. H. Friend, and F. Cacialli, *Synth. Met.* **124**, 15 (2001).
57. M. Stöβel, J. Staudigel, F. Steuber, J. Blässing, J. Simmerer, and A. Winnacker, *Appl. Phys. Lett.* **76**, 115 (2000).
58. R. Schlaf, B. A. Parkinson, P. A. Lee, K. W. Nebesny, G. Jabbour, B. Kippelen, N. Peyghambarian, and N. R. Armstrong, *J. Appl. Phys.* **84**, 6729 (1998).
59. V.-E. Choong, M. G. Mason, C. W. Tang, and Y. Gao, *Appl. Phys. Lett.* **72**, 2689 (1998).
60. A. Curioni, W. Andreoni, R. Treusch, F. J. Himpsel, E. Haskal, P. Seidler, C. Heske, S. Kakar, T. van Buuren, and L. J. Terminello, *Appl. Phys. Lett.* **72**, 1575 (1998).
61. N. Johansson, T. Osada, V. Parenté, J. L. Brédas, and W. R. Salaneck, *J. Chem. Phys.* **111**, 2157 (1999).
62. G. E. Jabbour, B. Kippelen, N. R. Armstrong, and N. Peyghambarian, *Appl. Phys. Lett.* **73**, 1185 (1998).
63. N. J. Watkins and Y. Gao, *J. Appl. Phys.* **94**, 1289 (2003).
64. D. Ochs, M. Brause, P. Stracke, S. Krischok, F. Wiegershaus, W. Maus-Friedrichs, V. Kempter, V. E. Puchin, and A. L. Shluger, *Surf. Sci.* **383**, 162 (1997).
65. P. G. Schroeder, C. B. France, J. B. Park, and B. A. Parkinson, *J. Appl. Phys.* **91**, 3010 (2002).
66. A. Gyoutpku, S. Hara, T. Komatsu, M. Shirinashihama, H. Iwanaga, and K. Sakanoue, *Synth. Met.* **91**, 73 (1997).
67. A. J. Heeger, I. D. Parker, and Y. Yang, *Synth. Met.* **67**, 2160 (1994).
68. S. A. Carter, M. Angelopoulos, S. Karg, P. J. Brock, and J. C. Scott, *Appl. Phys. Lett.* **70**, 2067 (1997).

69. J. S. Kim, M. Granström, R. H. Friend, N. Johansson, W. R. Salaneck, R. Daik, W. J. Feast, and F. Cacialli, *J. Appl. Phys.* **84**, 6859 (1998).
70. C. W. Tang and S. A. VanSlyke, *Appl. Phys. Lett.* **913**, 51 (1987).
71. S. K. So, W. K. Choi, L. M. Leung, and C. F. Kwong, *Appl. Phys.* A**68**, 447 (1999).
72. S. A. Van Slyke, C. H. Chen, and C. W. Tang, *Appl. Phys. Lett.* **69**, 2160 (1996).
73. I. G. Hill and A. Kahn, *J. Appl. Phys.* **84**, 5583 (1998).
74. S. T. Lee, Y. M. Wang, and X. Y. Hou, *Appl. Phys. Lett.* **74**, 670 (1999).
75. H. Aziz, Z. D. Popovic, N. X. Hu, A. M. Hor, and G. Xu, *Science* **283**, 1900 (1999).
76. T. Osada, Th. Kugler, P. Bröms, and W. R. Salaneck, *Synth. Met.* **96**, 77 (1998).
77. C. Giebeler, H. Antoniadis, D. D. C. Bradley, and Y. Shirota, *J. Appl. Phys.* 85, 608 (1999).
78. Q. T. Le, F. Nuesch, L. J. Rothberg, E. W. Forsythe, and Y. Gao, *Appl. Phys. Lett.* **75**, 1357 (1999).
79. F. Nuesch, L. J. Rothberg, E. W. Forsythe, Q. T. Le, and Y. Gao, *Appl. Phys. Lett.* **74**, 880 (1999).
80. Y. Park, V. Choong, Y. Gao, B. R. Hsieh, and C. W. Tang, *Appl. Phys. Lett.* **68**, 2699 (1996).
81. T. Kugler and W. R. Salaneck, private communication.
82. K. G. Neoh, E. T. Kang, and K. L. Tan, *Polymer* **34**, 1630 (1993).
83. Y. Hamima, K. Yamashita, H. Ishii, and K. Seki, *Thin Solid Films* **366**, 237 (2000).
84. R. S. Deshpande, V. Bulovic, and S. R. Forrest, *Appl. Phys. Lett.* **74**, 888 (1999).
85. R. Schlaf, B. A. Parkinso, P. A. Lee, K. W. Nebesny, and N. R. Armstrong, *J. Phys. Chem.* **103**, 2984 (1999).
86. I. G. Hill and A. Kahn, *J. Appl. Phys.* **86**, 4515 (1999).
87. E. W. Forsythe, V.-E. Choong, C. W. Tang, and Y. Gao, *J. Vac. Sci. Technol.* A**17**, 3429 (1999).
88. B. O'Regan and M. Grätzel, *Nature* **353**, 737 (1991).
89. A. Hagfeldt and M. Grätzel, *Chem. Rev.* **95**, 49 (1995).
90. J. M. Lanzafame, S. Palase, D. Wang, R. J. D. Miller, and A. A. Muenter, *J. Phys. Chem.* **98**, 11020 (1994).
91. J. M. Lanzafame, L. Min, R. J. D. Miller, A. A. Muenter, and B. A. Parkinson, *Mol. Cryst. Liq. Cryst.* **194**, 287 (1991).
92. T. Hannappel, B. Burfeindt, W. Storck, and F. Willig, *J. Phys. Chem.* B**101**, 6799 (1997).
93. B. Burfeindt, T. Hannappel, W. Storck, and F. Willig, *J. Phys. Chem.* **100**, 16463 (1996).
94. D. Vacar, E. S. Maniloff, D. W. MBranch, and A. J. Heeger, *Phys. Rev. B* **56**, 4573 (1997).
95. E. S. Maniloff, V. I. Klimov, and D. W. MBranch, *Phys. Rev. B* **56**, 1876 (1997).
96. A. J. Mäkinen, S. Schoemann, Y. Gao, M. G. Mason, A. A. Muenter, and A. R. Melnyk, in *Conjugated Polymer and Molecular Interfaces: Science and Technology for Photonic and Optoelectronic Applications*, W. R. Salaneck, K. Seki, A. Kahn, and J. J. Pireaux, eds. (Marcel Dekker, New York, 2001).
97. A. J. Makinen, S. Xu, Z. Zhang, S. J. Diol, Y. Gao, M. G. Mason, A. A. Muenter, D. A. Mantell, and A. R. Melnyk, *Appl. Phys. Lett.* **74**, 1296 (1999).

20

The Influence of the Electrode Choice on the Performance of Organic Solar Cells

Aleksandra B. Djurišić[1] and Chung Yin Kwong[2]

[1] *Department of Physics, University of Hong Kong, Pokfulam Road, Hong Kong, China*

[2] *Department of Electrical and Electronic Engineering, University of Hong Kong, Pokfulam Road, Hong Kong, China*

Contents

Abstract This chapter gives an overview of the relevant issues on choice of electrodes for different types of solar cells, and mechanisms relevant to electrode choice that affect solar cell performance. The issues for the electrode choice are discussed for Schottky barrier, heterojunction, and dye-sensitized solar cells. For each of the device types, recommendations on the electrode choice are given, and an overview of commonly used electrode materials is provided. Special attention is devoted to indium tin oxide (ITO) since it is the most commonly used electrode in organic optoelectronic devices. Influence of the various surface treatments of ITO on its properties and the solar cell performance are discussed in detail. Various additional layers which can be introduced into the structure to modify the anode or cathode contact, such as PEDOT:PSS, BCP, and LiF, are also discussed. The performance of the solar cell is typically strongly dependent on the thickness of any layers introduced into the structure to modify the electrodes, so that the thicknesses of these layers should be optimized to avoid deterioration of the performance due to increased series resistance of the solar cell.

Keywords organic photovoltaic devices, indium tin oxide, surface modifications

20.1. INTRODUCTION

Organic materials are attracting much interest in photovoltaic devices due to their potential for applications in low cost, large area devices. In recent years, power conversion efficiencies higher than 1% have finally been achieved (for review, see Ref. [1–3]). Even though these efficiencies are still below the level needed for commercial applications, it is likely that further improvements by optimizing the materials, fabrication conditions, and the device structure will be made in the near future. A number of different organic solar cell architectures have been reported in the literature. Among the different solar cell architectures, Schottky barrier-type cells have the lowest efficiency, typically of the order of 10^{-3}% [4,5]. Among the highest reported efficiencies in a Schottky barrier cell is ~0.6% at 78 mW/cm^2 illumination obtained for a device consisting of merocyanine dye between Ag and Al electrodes [6]. Significant improvement in the efficiency can be achieved using heterojunction cells. The first bilayer heterojunction cell, reported by Tang [7] in 1986, achieved 1% power conversion efficiency under AM2 illumination. Since then, a number of different bilayer heterojunction cells were reported [8,9], such as indium tin oxide (ITO)/copper phthalocyanine (CuPc)/C$_{60}$/Al device with efficiency of 0.2% [8] and ITO/poly(p-phenylene vinylene) (PPV)/poly (benzimidazobenzophenanthroline ladder)/Al with efficiency of 1.4% [9].

Further improvements can be achieved if more than two layers are employed. In double heterostructure devices [10,11], power conversion efficiencies as high as 1.1% (or 2.4% with light trapping) [10] and 3.6% [11] were obtained. Multiple heterojunction cells with good performance (efficiency of 2.5% under AM1.5 illumination) were also reported [12]. Another common type of organic solar cells is a bulk heterojunction device [13–15]. Bulk heterojunctions represent blends of donor and acceptor materials, in which exciton dissociation efficiency is significantly increased. For this type of devices, efficiency of 1.04% was achieved for small molecules [13]. For polymer bulk heterojunction cells, efficiencies as high as 2.5% [14] and 3.1% [15] were reported. The fabrication conditions have been found to affect the bulk heterojunction device performance. One percent efficiency (corrected for transmitted light) was obtained when the bulk heterojunction between metal-free phthalocyanine (H$_2$Pc) and perylene derivative (Me-PTC) was fabricated at $-167°C$, with five times higher short circuit current compared to devices fabricated at 25°C [16]. In addition to these common architectures, some more exotic devices were also reported, such as solar cells based on self-organized discotic liquid crystals with external quantum efficiency (EQE) of 34% [17] and hybrid nanorod–polymer cells with EQE over 54% and power conversion efficiency of 1.7% under AM1.5 illumination [18]. Among different types of solar cells using organic materials, the highest efficiency up to date was reported for dye-sensitized solar cells (7–8% for simulated solar light, 12% in diffuse daylight) [19]. However, the efficiency of these type of devices is lower when liquid electrolyte is not used. For solid-state dye-sensitized solar cell, efficiency of 2.56% under AM1.5 illumination was reported [20].

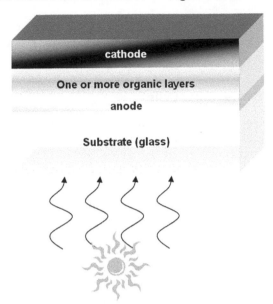

Figure 20.1.   Illustration of an organic solar cell.

All these different device architectures have one thing in common: the organic layers are placed between two electrodes, as shown in Figure 20.1. The solar cell performance is affected by the choice of the electrodes, as well as the choice of organic materials and device architecture. There are a number of issues relevant for the electrode choice, such as whether the contact is ohmic or not, the energy level alignment at organic–inorganic interface, the transparency of the electrode, etc. In this chapter, we will give an overview of the relevant issues on the choice of electrodes for different types of solar cells, and mechanisms relevant to electrode choice that affect solar cell performance. Special attention will be devoted to ITO since it is the most commonly used electrode in organic optoelectronic devices.

20.2.   CRITERIA FOR ELECTRODE CHOICE

There are several factors related to the electrode choice which influence the solar cell performance. Some of these factors are specific for certain types of solar cells (i.e., different device architectures have different requirements on electrical contacts for optimal performance), while the others, such as energy level alignment and the interface quality, are relevant for all types of solar cells. Formation of inter-facial dipoles and lack of vacuum level alignment are common phenomena at organic–metal interfaces [21–24]. The dipole, which can have different origins such as charge transfer, redistribution of the electron cloud or other types of charge rearrangement, and interfacial chemical reactions, leads to abrupt shift of the vacuum level at the interface [21]. It should also be noted that there are differences between metal–organic and organic–metal interfaces [21]. Due to different processes involved in device preparation (spin-coating or evaporation of organic layers on metal surface, followed by evaporating or sputtering of the second electrode), the properties of top and bottom contacts can be different even when the same metal is used [21,24]. When the top metal electrode is deposited, metal atoms may diffuse into the organic layer, or chemical reaction with high-reactivity hot metal atoms may occur at the interface [21].

20.2.1. Schottky Barrier Solar Cells

For optimal performance, organic Schottky barrier solar cells require one ohmic and one barrier contact [25]. Calculations predict that with appropriate electrode choice (i.e., one ohmic and on Schottky contact), two orders of magnitude improvement in the efficiency can be achieved [25]. When a Schottky barrier cell is illuminated from the rectifying contact, good correlation between the absorption spectrum and the short circuit current action spectrum can be observed, which is not the case when the cell is illuminated through the ohmic contact [4]. However, it should be pointed out that this theory (i.e., excitons dissociated due to built-in electrical field in depletion region) cannot explain observed photovoltaic effect in a cell consisting of a liquid crystal porphyrin between the two ITO electrodes [26,27]. Since the device was symmetrical (the space between two ITO electrodes was filled with zinc octakis (octylohexylethyl)porphyrin in liquid phase and cooled into solid state), there was no built-in electrical field [26,27]. The observed photoeffect was explained by asymmetric exciton dissociation at illuminated electrode, and it was demonstrated that the cell performance could be improved by carrier-selective contacts [26]. Even though the efficiency of Schottky barrier solar cells can be improved by electrode modifications [26], this type of devices is not likely to be able to compete with more complex device architectures [1,2] but they may be useful for evaluation purposes due to their simple structure [2,5].

20.2.2. Heterojunction Solar Cells

In donor–acceptor heterojunction solar cells, the excitons dissociate at the interface between donor and acceptor materials [26]. There are different heterojunction geometries, such as bilayer and bulk heterojunctions. In bilayer heterojunctions, only a small region near the interface contributes to the photocurrent due to short exciton diffusion length [28]. This problem can be overcome by using bulk heterojunctions (see an example of a multilayer device with a bulk heterojunction layer in Figure 20.2). This is because the number of exciton dissociation sites is significantly increased in the bulk heterojunction, while the separated electrons and holes are

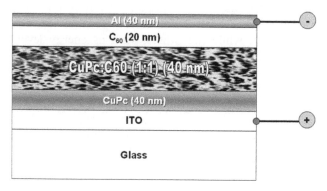

Figure 20.2. Bulk heterojunction organic solar cell.

transported through donor and acceptor materials to the electrodes. Therefore, the photocurrent can be significantly increased. In both bilayer and bulk heterojunction architectures, contact electrodes can influence the device efficiency. For example, bilayer cell efficiency is different for different sequences of deposition of two dyes [4]. Typically, cells with "n-type" material in contact with ITO have higher efficiency than those with "p-type" material in contact with ITO. For all kinds of donor–acceptor heterojunctions, the electrode choice should be carrier selective, i.e., one good contact for the hole collection (on the "p-type" material side in bilayer or multilayer cells) and one good contact for electron collection (on the "n-type" material side in bilayer or multilayer cells) should be used. Since exciton dissociation is caused by the band offset at donor–acceptor interface and the exciton binding energy is larger than the internal electric potential drop across the exciton, the work function difference between two electrodes is not an important factor [26]. It was shown that the open circuit voltage in donor–acceptor heterojunction cells shows only weak dependence on the work function of the electrodes [29–31].

It addition to purely organic devices, in recent years, different combinations of organic and inorganic materials have been explored for photovoltaic cells. There are two main types of such hybrid cells: cells with composite layers of polymer–inorganic nanostructures and dye-sensitized solar cells. Due to their different device architecture, dye-sensitized cells will be discussed separately in the next subsection. Excellent performance with EQE over 54% and power conversion efficiency of 1.7% under AM1.5 illumination was reported for a hybrid poly(3-hexyl thiophene):CdSe nanorod solar cell [18]. Transport properties and the influence of morphology were studied for this promising material combination [32,33]. Other material combinations have been explored [34–36], but reported efficiencies have been relatively low: 0.42% for poly(3-hexyl thiophene):TiO_2 cells [36] or even below 0.1% for some other material combinations. [34,35]. Carbon nanotube–polymer composites have also been proposed for application in photovoltaic cells [37–40]. Two types of devices have been reported: composites of poly(3-octylthiophene) (P3OT) and single-wall carbon nanotubes (SWNTs) [37–39] and multiwall carbon nanotube (MWNT)/poly(p-phenylene vinylene) devices where MWNT layer was used as an electrode instead of the conventional ITO. Low efficiency below 0.1% was reported for P3OT:SWNT nanocomposites [37,38]. It was also shown that the photocurrent in P3OT:SWNT devices can be increased by the addition of naphthalocyanine dye [39], but further improvements are still necessary. The device using MWNT layer as an electrode also showed relatively low efficiency of 0.08% [40]. Nevertheless, such hybrid nanostructure/polymer devices are of great interest since they offer a promise of combining good properties of inorganic nanostructures with inexpensive fabrication inherent to organic technology, and it is likely that their performance will improve in the near future. Up to date, the works on the influence of electrode choice on the performance of these devices have been scarce [38]. Recent work on the open circuit voltage of P3OT:SWNT devices found only small variation of the open circuit voltage with the change of the metal electrode [38], similar to the conventional organic donor–acceptor heterojunction devices [29–31]. Therefore, it is likely that similar guidelines, i.e., carrier-selective ohmic contacts, would also apply for the optimal performance of these devices.

20.2.2. Dye-Sensitized Solar Cells

Dye-sensitized solar cells represent the most efficient type of organic photovoltaic devices, with efficiencies as high as 7–8% for simulated solar light and 12% in diffuse daylight [19]. These devices consist of a nanoporous electrode (typically TiO_2) and an organic dye. The cells can contain either liquid electrolyte or a solid-state medium whose function is to regenerate the dye molecules. The efficiency of solid-state devices [20] is typically lower. The comparison of charge transport in electrolyte and solid-state devices can be found in Ref. [41]. The most commonly used material for the nanoporous electrode is TiO_2 [19,20,41–43]. Other metal oxides, such as SnO_2–ZnO [43,44] and NiO [45], have also been investigated for solar cell applications. Both n-type (TiO_2, SnO_2–ZnO) and p-type (NiO) materials can be used. For the same device structure, the performance of SnO_2–ZnO cell was inferior compared to the TiO_2 cell [43]. In spite of the inferior efficiency of some of the reported alternative nanoporous electrodes, reliability concerns due to photocatalytic activity of TiO_2 under UV illumination represent strong motivation for research in alternative porous electrodes for dye-sensitized solar cells. Core-shell electrode structures such as $TiO_2/SrTiO_3$ have also been proposed [46]. In addition to various organic sensitizers, quantum dot sensitization has also been proposed [47,48]. While different sensitizers and different metal oxides have been explored for photovoltaic applications, majority of the dye-sensitized solar cells with liquid electrolyte use Pt-coated counter-electrode. A carbon counter-electrode has been proposed recently [49]. Several types of carbon and different electrode thicknesses were used to optimize the carbon electrode and finally achieve better performance compared to the sputtered Pt counter-electrode [49].

Influence of the variations of the front contact on the performance of dye-sensitized solar cells was investigated [50,51]. It was shown that the open circuit voltage is determined by photoinduced chemical gradients, and that it does not depend on the work function of the contact to TiO_2 [51]. Only a small variation of the open circuit voltage was obtained for different substrates whose work functions covered 1.4 eV range [51]. Lack of the significant influence of the front contact on the open circuit voltage was confirmed by a more recent study [50]. However, it was found that the fill factor was affected by the contact to the TiO_2 layer [50]. Therefore, overall device performance will be affected by the contact choice. The best performance was obtained for a conventional SnO_2:F contact [50]. The degradation of the fill factor for large barrier height between the contact and TiO_2 was explained with a secondary diode model to account for TiO_2–contact (SnO_2:F,Al, ZnO:Al, Au) interface [50]. This indicates that even though open circuit voltage is not sensitive to the contact choice, it is advisable to choose carrier-selective contacts for all types of cells, except the Schottky barrier-type which requires one Schottky and one ohmic contact for the best performance.

20.3. COMMONLY USED ANODES

There are several possible choices for a bottom electrode (anode) in organic solar cells. Most commonly used bottom electrode is ITO. Since ITO/glass substrates are commercially available and widely used in organic light-emitting diodes (OLEDs) fabrication, it is very convenient to use them for organic solar cells as well. In addition to ITO and ITO/conductive polymer electrodes, semitransparent metal can also be used as a bottom electrode.

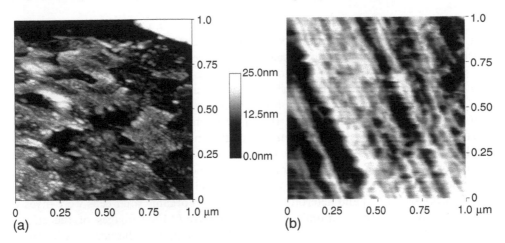

Figure 20.3. Comparison between SPM images of ITO surface from two different suppliers. (From A. B. Djurišić, C. Y. Kwong, P. C. Chui, and W. K. Chan, *J. Appl. Phys.* **93**, 5472–5479 (2003). With permission.)

20.3.1. Indium Tin Oxide

ITO is a highly degenerate n-type semiconductor with low resistance, high work function, and high transmittance in the visible spectral region. In addition, ITO glass substrates are widely available, so that ITO is commonly used as a substrate for both OLEDs and organic solar cells. However, the variations of ITO parameters, such as work function, surface morphology, sheet resistance, etc. are common not only for ITO from different manufacturers but also for different batches from the same manufacturer [52,53]. Figure 20.3 shows the surface probe microscopy (SPM) images of the ITO surface for ITO substrates obtained from China Southern Glass Holding Co. Ltd., Shenzhen, China (Figure 20.3a) and Varitronix Limited, Hong Kong (Figure 20.3b) [53]. The difference in the surface morphology is obvious.

The variation in the properties of the ITO will influence the device performance. The effects of various surface treatments (plasma, chemical, UV ozone, etc.) on the ITO properties and the OLED performance were extensively studied [54–62]. Increase in the work function of ITO with oxidative treatments (oxygen plasma, UV ozone) has been attributed to carbon removal from the ITO surface [59,60], larger number of states created close to and possibly below the edge of the conduction band [54], or the shift of the Fermi level [63]. Another possible explanation for the work function shift is the formation of surface dipoles [56–58]. However, none of these hypotheses is fully consistent with the experimental data. For example, if carbon contamination played a major role, then an increase in the work function should be obtained with Ar plasma treatment which significantly reduces carbon content but does not result in the increase of the work function. The Fermi level shift hypothesis is not consistent with carrier concentration changes. The surface dipoles hypothesis [56–58,63] is not in direct contradiction with the experimental data, but it explains only the work function change and not the change of the other properties, so that very likely other phenomena also play a role in changes of the ITO properties.

While the ITO work function change is significant for OLED performance, it is not very relevant for the performance of donor–acceptor organic solar cells since the open circuit voltage shows only a weak dependence on the electrode work function [29–31]. ITO sheet resistance also does not appear to be a significant factor [5]. However, other factors such as carbon contamination, surface defects and traps, surface dipoles, and surface morphology can have significant influence on the solar cell performance. The influence of the ITO parameters on the organic solar cell performance was not studied as much as the influence of ITO parameters on OLED performance. However, it was found that the ITO surface treatments can significantly affect the solar cell performance [5,53,64,65]. The influence of UV ozone treatment and the combination of UV ozone and acid treatments on the performance of Schottky barrier [5] and heterojunction [53,64,65] organic solar cells were investigated. Figure 20.4 shows atomic force microscopy (AFM) images of the effects of surface treatments (UV ozone, UV ozone + HCl, and mechanical + UV ozone + HCl) on the ITO surface morphology, while the ITO parameters are given in Table 20.1 [53]. It can be observed that the surface treatments have large impact on ITO surface roughness and morphology, as well as work function, carrier concentration, and mobility. UV ozone treatment and mechanical treatment followed by UV ozone and HCl result in the reduction of surface roughness, while UV ozone followed by HCl results in significant increase of the surface roughness compared to the untreated ITO. For ITO from different suppliers, the trends are the same but the

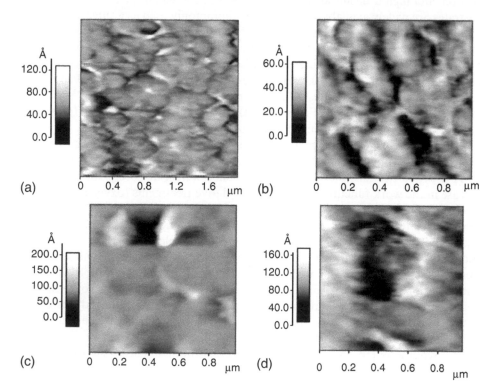

Figure 20.4. AFM images of ITO substrates: (a) untreated; (b) UV ozone treated; (c) UV ozone and HCl treated; (d) mechanical and UV ozone and HCl treated. (From A. B. Djurišić, C. Y. Kwong, P. C. Chui, and W. K. Chan, *J. Appl. Phys.* **93**, 5472–5479 (2003). With permission.)

Table 20.1. ITO parameters for different surface treatments

ITO treatment	$\Delta\Phi$ (eV)	N_s (10^{17} cm^{-2})	μ (cm^2/V s)	R_s (Ω/() Hall	Rms roughness (nm)
Untreated	—	0.54	14.3	8.02	1.28
UV ozone	0.75	1.48	6.0	7.03	1.10
UV ozone + HCl	0.50	0.73	11.0	7.76	2.63
Mechanical + UV ozone + HCl	0.19	0.85	10.0	7.37	1.06

Source: From A. B. Djurišić, C. Y. Kwong, P. C. Chui, and W. K. Chan, Indium-tin-oxide surface treatments: influence on the performance of CuPc/C$_{60}$ solar cells, *J. Appl. Phys.* **93**, 5472–5479 (2003). With permission.

magnitude of decrease (or increase) of the surface roughness, as well as change in the other properties, will depend on the initial properties of ITO.

In Schottky barrier cells, the ITO surface treatment affects both the open circuit voltage and the short circuit current, while the fill factor is independent of the surface treatments [5]. Treating the surface with HCl alone resulted in the reduction of the open circuit voltage, as well as significant worsening of the rectification ratio and the increase in the dark current, illustrating that the quality of the ITO–organic interface has strong influence on the current–voltage characteristics of the cell [5]. However, in a heterojunction solar cell, the open circuit voltage is not significantly affected by the surface treatment because the open circuit voltage in this case is determined by the difference in energy levels of the "p-type" and "n-type" materials [29]. The cell parameter which shows the strongest dependence on the ITO surface treatments is the short circuit current [53,64,65]. *I–V* characteristics of a heterojunction ITO/CuPc(40 nm)/C$_{60}$ (40 nm)/Al cell is shown in Figure 20.5, while the cell parameters are listed in Table 20.2 [53]. The fill factor does not show strong dependence on the ITO surface treatments, similar to Schottky barrier solar cells. The open circuit voltage is somewhat lower for the cell fabricated on untreated ITO ($V_{oc} = 0.22$ V) compared to UV ozone ($V_{oc} = 0.44$ V) and treatments involving HCl

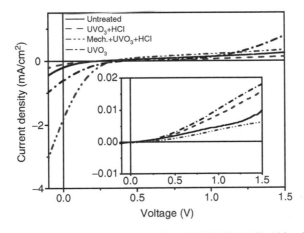

Figure 20.5. The current–voltage characteristics of ITO/CuPc/C$_{60}$/Al cells with different ITO surface treatments under AM1 illumination. The inset shows the dark current. (From A. B. Djurišić, C. Y. Kwong, P. C. Chui, and W. K. Chan, *J. Appl. Phys.* **93**, 5472–5479 (2003). With permission.)

Table 20.2. Comparison of the ITO/CuPc/C_{60}/Al solar cell parameters for different ITO treatments for AM1 98 mW/cm^2 excitation

Treatment/parameter	Untreated	UV ozone	UV ozone + HCl	Mechanical + UV ozone + HCl
I_{sc} (mA/cm^2)	0.20	0.60	0.10	1.81
V_{oc} (V)	0.22	0.44	0.34	0.34
FF	0.17	0.13	0.13	0.15
η (%)	0.008	0.035	0.004	0.093

Source: From A. B. Djurišić, C. Y. Kwong, P. C. Chui, and W. K. Chan, Indium-tin-oxide surface treatments: influence on the performance of CuPc/C_{60} solar cells, *J. Appl. Phys.* **93**, 5472–5479 (2003). With permission.

($V_{oc} = 0.34$ V), which is likely due to carbon contamination on the ITO surface. Carbon contamination on the untreated ITO surface is common, and it was shown that it can be significantly reduced with UV ozone treatment [60]. The short circuit current is strongly affected by the surface treatment and it increases from 0.2 mA/cm^2 for the cell on untreated ITO to 1.8 mA/cm^2 for cell on ITO treated with mechanical + UV ozone + HCl treatment, resulting in the increase in power conversion efficiency from 0.008% to 0.09% [53]. The reasons for the increase in the short circuit current and the power conversion efficiency are not fully clear. The only distinguishing factor which can be identified from the ITO properties is the surface roughness and morphology. Another possible factor is the concentration of defects and traps on ITO–organic interface.

In order to further investigate the influence of the ITO surface treatments on the solar cell performance, cells with different organic materials in contact with ITO have been fabricated [64,65]. We fabricated the following two-layer structures ITO/C_{60} (40 nm)/CuPc (40 nm)/Cu, ITO/CuPc (25 nm)/perylene tetracarboxylic acid diimide (PTCDI) (20 nm)/Al, ITO/PTCDI (20 nm)/CuPc (25 nm)/Cu, as well as a three-layer cell ITO/CuPc(20 nm)/CuPc:C_{60} (1:1) (40 nm)/C_{60} (20 nm)/Cu. In case when the "p-type" material is used in contact with the ITO, low work function metal is used for the other contact to ensure good collection of electrons. When an "n-type" material is used in contact with ITO, higher work function metal (Cu or Au) should be used for the other electrode to ensure good collection of holes. Figure 20.6 shows the I–V characteristics of ITO/C_{60}/CuPc/Cu cells with different ITO surface treatments. In this case, NaOH was used instead of HCl, since base treatment lowers the work function of ITO, while acid treatment increases the work function [56,57]. The AM1 power conversion efficiency η for the cell fabricated on untreated substrate is 0.092%, for UV ozone treated substrate, $\eta = 0.15$%, for UV ozone + NaOH-treated cell $\eta = 0.19$%, while $\eta = 0.25$% for mechanical + UV ozone + NaOH-treated cell [64,65]. In this case, the short circuit current again represents the parameter which is most strongly affected by the surface treatment. However, this type of cells are less sensitive to the surface treatments compared to the cells with CuPc in contact with ITO.

Similar trend can be observed in CuPc/PTCDI and PTCDI/CuPc cells. I–V characteristics of ITO/CuPc/PTCDI/Al solar cell is shown in Figure 20.7. The obtained AM1 power conversion efficiencies are 0.0004% for untreated substrate, 0.07% for UV ozone, 0.001% for UV ozone + HCl, 0.02% for UV ozone + NaOH, and 0.003% for mechanical + UV ozone + HCl treated substrate [64]. Similar to

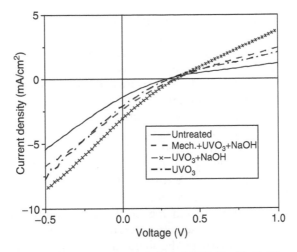

Figure 20.6. The current–voltage characteristics of ITO/C$_{60}$/CuPc/Cu cells with different ITO surface treatments under AM1 illumination. (From (a) C. Y. Kwong, A. B. Djurišić, P. C. Chui, and W. K. Chan, in *Proc. of the SPIE*, Vol. 5215 "Organic Photovoltaics IV," edited by Z. Kafafi and P. A. Lane, SPIE, Bellingham, WA, 2004, pp. 153–160. With permission; (b) C. Y. Kwong, A. B. Djurišić, P. C. Chui, and W. K. Chan, *Jpn. J. Appl. Phys.* **43**, 1305–1311 (2004).).

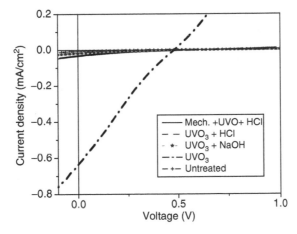

Figure 20.7. The current–voltage characteristics of ITO/CuPc/PTCDI/Al cells with different ITO surface treatments under AM1 illumination. (From C. Y. Kwong, A. B. Djurišić, P. C. Chui, and W. K. Chan, in *Proc. of the SPIE*, Vol. 5215 "Organic Photovoltaics IV," edited by Z. Kafafi and P. A. Lane, SPIE, Bellingham, WA, 2004, pp. 153–160. With permission.)

ITO/CuPc/C$_{60}$/Al cells, the cell performance is strongly dependent on the ITO surface treatment. However, optimal treatment is different for ITO/CuPc/PTCDI/Al and ITO/CuPc/C$_{60}$/Al cells. In case of OLEDs, different performance was also obtained for the same treatment when different hole transport material was used [54]. This can possibly be explained by different molecules affected by the substrate roughness and morphology in a different manner. The surface treatment can affect the growth of the organic layer and its surface morphology. It has been shown that the ITO surface treatments can change the growth mode of N,N'-bis-(1-naphtyl)-N,N'-diphenyl-1,1'-biphenyl-4,4'-diamine (NPB) from island growth on an untreated substrate to layer-by-layer growth on a treated substrate [66,67]. Therefore, the surface treatment affects the interface between the two organic materials where the exciton dissociation takes place. For two different "n-type" materials, it is possible that the interface morphologies resulting in best exciton dissociation would be different. Also, the presence and nature of the interfacial chemical reactions is also determined by the chemical structure of the molecules. I–V characteristics of the ITO/PTCDI/CuPc/Cu solar cell under AM1 illumination is shown in Figure 20.8. AM1 power conversion efficiency is 0.13% for untreated substrate, 0.41% for mechanical + UV + NaOH, 0.42% for UV ozone + NaOH, and 0.42% for UV ozone [64]. Similar to CuPc/C_{60}-based cells, inversion of layer order results in reduced dependence on ITO surface treatments and higher efficiency compared to the cells with CuPc in contact with ITO. Increased efficiency for devices with "n-type" material in contact with ITO was reported in the literature [68–70]. The increased efficiency when reversing the order of layer deposition was explained by Förster energy transfer when the shorter wavelength absorber is placed in contact with the transparent electrode surface [69,70]. However, this hypothesis does not explain lower sensitivity of the cell performance to ITO surface treatments when an "n-type" material is in contact with the ITO.

In order to investigate in more detail the influence of the surface treatments on the solar cell performance, we fabricated ITO/C_{60} (20 nm)/C_{60}:CuPc (1:1) (40 nm)/CuPc (20 nm)/Cu solar cells with six different surface treatments. The obtained results are shown in Figure 20.9 and summarized in Table 20.3. Similar trends can be observed for a three-layer cell and a two-layer cell with C_{60} in contact with ITO. Low sensitivity to surface treatments can be observed (power conversion efficiency in the range from 0.37% to 0.39%), with the exception of NaOH + UV ozone and HCl + UV ozone treatments, which results in reduced open circuit voltage and reduced short circuit current, as well as less stable performance [64,65]. Since these two treatments cause opposite changes in the work function of ITO (base treatment

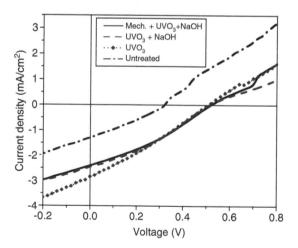

Figure 20.8. The current–voltage characteristics of ITO/PTCDI/CuPc/Cu cells with different ITO surface treatments under AM1 illumination. (From C. Y. Kwong, A. B. Djurišić, P. C. Chui, and W. K. Chan, in *Proc. of the SPIE*, Vol. 5215 "Organic Photovoltaics IV," edited by Z. Kafafi and P. A. Lane, SPIE, Bellingham, WA, 2004, pp. 153–160. With permission.)

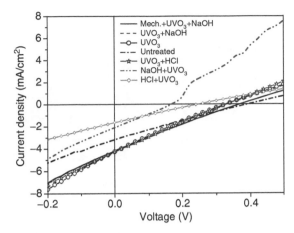

Figure 20.9. The current–voltage characteristics of ITO/C_{60}/CuPc:C_{60}/CuPc/Cu cells with different ITO surface treatments under AM1 illumination. (From C. Y. Kwong, A. B. Djurišić, P. C. Chui, and W. K. Chan, in *Proc. of the SPIE*, Vol. 5215 "Organic Photovoltaics IV," edited by Z. Kafafi and P. A. Lane, SPIE, Bellingham, WA, 2004, pp. 153–160. With permission; (b) C. Y. Kwong, A. B. Djurišić, P. C. Chui, and W. K. Chan, *Jpn. J. Appl. Phys.* **43**, 1305–1311 (2004).)

Table 20.3. Comparison of the ITO/C_{60}/CuPc:C_{60}/CuPc/Cu solar cells parameters for different ITO treatments for AM1 98 mW/cm^2 excitation

Treatment/ parameter	Untreated	UV ozone	UV ozone + NaOH	NaOH + UV ozone	Mechanical + UV ozone + NaOH	UV ozone + HCl	HCl + UV ozone
I_{sc} (mA/cm^2)	3.18	4.24	4.37	2.09	4.18	4.29	1.64
V_{oc} (V)	0.38	0.32	0.32	0.16	0.36	0.32	0.24
FF	0.24	0.24	0.26	0.27	0.24	0.25	0.25
η (%)	0.24	0.35	0.38	0.1	0.39	0.37	0.1

Source: From (a) C. Y. Kwong, A. B. Djurišić, P. C. Chui, and W. K. Chan, *Proc. of the SPIE*, Vol. 5215 "Organic Photovoltaics IV," edited by Z. Kafafi and P. A. Lane, SPIE, Bellingham, WA, 2004, pp. 153–160; (b) C. Y. Kwong, A. B. Djurišić, P. C. Chui, and W. K. Chan, *Jpn. J. Appl. Phys.* **43**, 1305–1311 (2004).

decreases while acid treatment increases the work function [56,57]), the observed reduction in V_{oc} is not related to the ITO work function change. One possible explanation is that if the carbon contamination (consisting mainly of organic hydrocarbons) is not removed first, it will react with the chemical and result in an interfacial layer containing traps and recombination centers which lower the carrier collection efficiency.

In order to investigate the difference of the influence of the surface treatments on cell performance for cells with CuPc and C_{60} in contact with ITO, we performed AFM measurements on CuPc and C_{60} films deposited on ITO treated with different surface treatments. The obtained results are shown in Figure 20.10 and Figure 20.11. It can be observed that the surface morphology and roughness of C_{60} films are not very sensitive to ITO surface treatments, while the morphology and roughness of CuPc films are dependent on the morphology and roughness of the ITO substrate. Therefore, the difference in sensitivity to ITO surface treatments can be attributed to

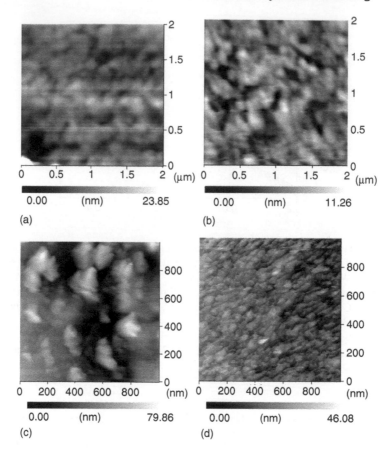

Figure 20.10. AFM images of the CuPc films deposited on ITO substrates treated with different surface treatments: (a) untreated; (b) UV ozone; (c) UV ozone + HCl; (d) mechanical + UV ozone + HCl. (From C. Y. Kwong, A. B. Djurišić, P. C. Chui, and W. K. Chan, *Jpn. J. Appl. Phys.* **43**, 1305–1311 (2004).).

the sensitivity of CuPc film morphology to the morphology of the underlying substrate. Another factor, independent of the type of material in contact with ITO, which influences the cell performance is the surface contamination, and this should be removed to obtain optimal performance. Based on our results [64,65], as well as results obtained in the literature [68–70], it appears that cells with the "n-type" material in contact with ITO have higher efficiency compared to those where "p-type" material is in contact with ITO. In addition to proposed Förster energy transfer [69,70], another possible reason is that ITO acts as a more efficient electron collector than hole collector, which would be expected from a n-type semiconductor. However, further studies of the interface chemistry for ITO–CuPc and ITO–C_{60} interfaces with different surface treatments are necessary to fully explain the observed phenomena.

The influence of the UV ozone and oxygen plasma treatments of the ITO anode on the performance of multilayer organic photodetectors was also studied [71]. It was found that the treatment of the ITO resulted in the decrease of the EQE of the photodetectors, which was attributed to the increased density of the defect states at the ITO–organic interface [71]. However, it should be noted that only a fraction of the

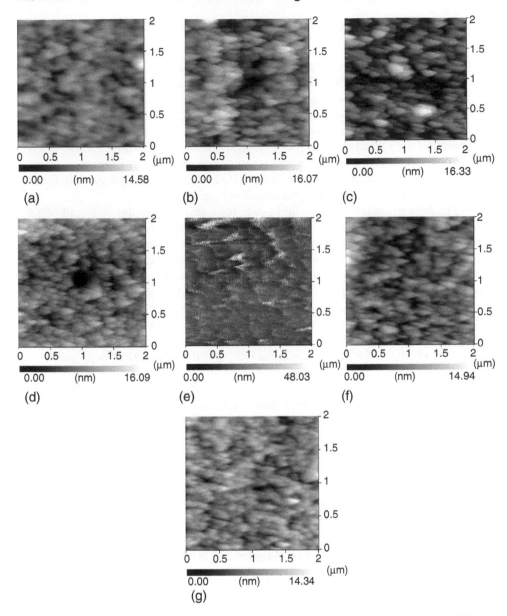

Figure 20.11. AFM images of C_{60} films deposited on ITO substrates treated with different surface treatments: (a) untreated; (b) UV ozone; (c) UV ozone + NaOH; (d) NaOH+UV ozone; (e) mechanical + UV ozone + NaOH; (f) UV ozone + HCl; (g) HCl + UV ozone. (From C. Y. Kwong, A. B. Djurišić, P. C. Chui, and W. K. Chan, *Jpn. J. Appl. Phys.* **43**, 1305–1311 (2004).)

ITO surface (at most 40–50%) has electronically active sites [72]. Therefore, ITO–organic interface is of large importance to the device performance, even though the work function of the ITO does not have significant effect on the open circuit voltage. In addition to surface treatments, ITO surface can be modified with conductive polymers or self-assembled monolayers, which will be discussed in the following subsections.

20.3.2. Poly(Ethylene Dioxythiophene):Polystyrene Sulfonic Acid (PEDOT:PSS)/ITO

Poly(ethylene dioxythiophene):polystyrene sulfonic acid (PEDOT:PSS)/ITO is a commonly used anode in OLEDs [73,74] and organic solar cells [3,15,75–78]. The work function of PEDOT:PSS is independent of the work function of the substrate [74], so that the use of PEDOT:PSS is promising for reproducible fabrication of OLEDs with good performance. In organic solar cells, PEDOT:PSS layer ensures improved interface between the active layer and the electrode [75]. Use of PEDOT:PSS also changes the shape of the EQE curve, which was attributed to the doping of the active layer by PSS [76,77]. It was found that the use of PEDOT:PSS increased the yield and, in the case of fullerene-based cells, the open circuit voltage [3].

The best efficiency reported up to date for a small molecular solar cell of 3.6% was obtained for a device structure using PEDOT:PSS [11]. However, it was also reported that the use of PEDOT:PSS causes an increase in the series resistance of the solar cell, which results in a lower fill factor [15]. The obtained AM1.5 power conversion efficiencies in a poly(2-methoxy-5-(3′,7′-dimethyloctyloxy)-1,4-phenylene vinylene) (MDMO-PPV):(6,6)-phenyl-C_{61}–butyric acid methyl ester (PCBM) bulk heterojunction cells with and without PEDOT:PSS were 2.3% and 3.2%, respectively [15]. The contradictions between different works on the influence of PEDOT:PSS on the solar cell performance may be due to the fact that the device performance is dependent on the PEDOT:PSS thickness [78], and that PEDOT:PSS properties are dependent on the thermal treatments [79].

PEDOT:PSS layer thickness should be optimized since very thin layers yield more pin holes, while thicker layers cause larger increase in the series resistance [78]. The conductivity and roughness of PEDOT:PSS films are dependent on the annealing temperature and atmosphere [79], so that the thermal treatment of PEDOT:PSS layer after spin-coating needs to be optimized as well. It should also be noted that there are some indications that PEDOT:PSS interface is not stable and that indium diffuses into the PEDOT:PSS film during annealing [80]. The stability of the interface can be improved by modifying the ITO surface with a monolayer of alkylsilohexanes before spin-coating PEDOT:PSS [81].

20.3.3. Polyaniline/ITO and Self-Assembled Monolayers on ITO

In OLEDs, conductive polymer electrodes other than PEDOT:PSS, such as polyaniline (PANI), are also explored [82–84]. It was shown that the use of PANI anodes in OLEDs improves the device performance [82–84]. However, the conductivity of PANI may be lower than that of PEDOT:PSS [82], which may have detrimental effect on the solar cell performance. Therefore, if PANI anodes are used in solar cells, it is necessary to compromise between the improvement in the interface stability and anode roughness which can be obtained by coating a conductive polymer layer on ITO and the increase in the series resistance of the cell. PANI layer can also prevent interfacial oxidation of the active polymer layer, since ITO acts as a source of oxygen [84].

Another promising method to modify the ITO surface is the use of self-assembled monolayers (SAMs). Unlike spin-coated conductive polymer layers, SAMs are very thin and hence they are not likely to significantly change the cell resistance. It was shown that the efficiency of polymer light-emitting diodes can be increased using SAMs on ITO [85]. SAMs were also used to tune the Schottky barrier height between the metal and the polymer over a range of 1 eV [86]. Enhancement of photocurrent generation by ITO electrodes modified with SAMs of linked donor–acceptor molecules was also demonstrated [87]. An improvement in the solar cell performance by modifying the ITO surface with electroactive small molecules (ferrocene dicarboxylic acid and 3-thiophene acetic acid) was reported recently [72]. The improvement was attributed to the improvement of the hole capture and electrical contact between the ITO and organic layer [72]. Instead of a SAM or a conductive polymer film, a layer of a commonly used hole transport material, such as N,N'-diphenyl-N,N'-ditolyl benzidine (TPD), can be used to modify ITO surface and enhance carrier generation [88]. Use of thin layers of redox polymers to modify the ITO surface and control the polarity and efficiency of the exciton dissociation was also demonstrated [27].

20.3.4. Semitransparent Metals

Instead of an ITO anode, a thin, semitransparent metal can also be used. Two metal electrodes were commonly used in early works on organic Schottky barrier solar cells [4,6]. The majority of recent studies use ITO or ITO/PEDOT:PSS electrodes (for review, see Ref. [3]), although there are some exceptions. For example, Au/PEDOT:PSS electrode was used in a polymer heterojunction cell with EQE of 29% [89]. The general rule for the choice of semitransparent metal is that a high work function metal should be used if good contact for hole collection is desired, while for good contact for electron collection, low work function metal should be chosen. For example, gold (work function $\phi = 5.1\,eV$), nickel ($\phi = 5.15\,eV$), platinum ($\phi = 5.65\,eV$), or copper ($\phi = 4.65\,eV$) [90] can be used for hole-collecting contact, while aluminum ($\phi = 4.28\,eV$), silver ($\phi = 4.26\,eV$), or indium ($\phi = 4.12\,eV$) [90] can be used for electron-collecting contact. While group II elements have lower work functions (for Mg $\phi = 3.66\,eV$, while for Ca $\phi = 2.87\,eV$ [90]), they are also very reactive. For this reason, when group II metals are used as a cathode, their deposition is typically followed by the deposition of a protective layer, usually Ag.

20.4. COMMONLY USED CATHODES

The top contact of an organic solar cell typically consists of a single metal layer or a two-layer structure. The choice of the metal depends on the type of carriers to be collected. In the majority of devices reported in the literature, the top electrode is usually used for electron collection.

20.4.1. Single Layer Metal Cathode

Aluminum is a very commonly used cathode in organic solar cells [3,4]. It should be pointed out that photocorrosion of the aluminum electrode was proposed as origin of the photocurrent in porphyrin–Al single layer solar cells [91]. However, this may not be general phenomenon since the initial transmittance of Al electrode was relatively high (over 30%) [91], indicating very thin metal layer. A number of other metal contacts have also been reported in the literature, such as Ag, Au, Cu, In, Cr, Ca, etc. The choice of metal depends on its work function — low work function metals for electron collection, high work function metals for hole collection.

Reversal of the photocurrent was observed in Au/H_2-phthalocyanine/C_{60}/metal solar cells when metal was changed from Pt to In and Cu, while for Au as the top electrode the sign of the photocurrent was dependent on the C_{60} layer thickness [92]. Studies on other material systems, however, found no photocurrent reversal and only a weak dependence of the open circuit voltage on the metal work function [30,31]. The photocurrent reversal was explained by competition between p–n junction of H_2-phthalocyanine/C_{60} and Schottky junctions at the metal–organic contacts. However, the reported photocurrent reversal [92] is not easily explained considering the highest occupied molecular orbital (HOMO) and lowest unoccupied molecular orbital (LUMO) levels of H_2-phthalocyanine and C_{60} and the charge transport properties of these materials, since the current reversal would require C_{60} to transport holes and H_2-phthalocyanine to transport electrons.

There are other factors which need to be taken into account when considering metal cathode, not only its work function. It is known that the dipole layers would form at the C_{60}–metal interface [93], which is not surprising since the formation of dipoles at organic–metal interfaces is a common phenomenon [21–23]. Due to the interfacial dipole, Schottky–Mott model of common vacuum levels is not valid at the interface [21]. It was demonstrated that the large component of the open circuit voltage is not dependent on the work function of the electrodes, both for materials which do not exhibit alignment of the vacuum levels [30] and for materials which show vacuum level alignment at polymer–metal interface [31]. Therefore, the appropriate criterion for the choice of the top metal electrode is that it should form a selective contact for the desired type of carriers [26].

20.4.2. LiF/Aluminum

The LiF/Al cathode is commonly used in OLEDs to enhance electron injection [94]. The LiF layer is typically very thin (<1 nm), since thicker layers are found to be detrimental to electron injection [94]. Recent ion mass spectroscopy and x-ray photoelectron spectroscopy study has disproved the hypothesis of the performance improvement due to formation of AlF_3 and doping of Li into the organic layer, while LiF was found to prevent a reaction between Al and PCBM [94]. The formation of a dipole in LiF layer and consequent lowering of the cathode work function was also proposed [95,96]. Since LiF/Al cathode is likely to have lower work function, this cathode has been used in solar cells as well [14], with obtained power conversion efficiency of 2.5%. However, it should be taken into account that the cathode work function does not have significant influence on the open circuit voltage [30]. The use of an insulating layer such as LiF and SiO_2 resulted in lower short circuit current compared to Al electrode in Schottky barrier solar cells [5]. Similar result was obtained in a three-layer photovoltaic cell with device structure ITO/CuPc/chlorotricarbonyl rhenium(I) bis(phenylimino)acenaphthene (Re-DIAN):C_{60}/C_{60}/Al, as shown in Figure 20.12, where the insertion of a thin LiF layer resulted in the reduction of AM1 power conversion efficiency from 0.48% to 0.39% [97]. The obtained results are likely to be strongly dependent on LiF layer thickness. It was found that the insertion of 5 Å LiF improved the performance of CuPc/C_{60}/bathocuproine (BCP) solar cells [98].

20.4.3. BCP/Ag and BCP/Al

Use of BCP was proposed to prevent the damage of the organic layer during metal deposition [3]. The cell performance is dependent on the BCP thickness [3]. The BCP layer can also be used to optimize the EQE via optical interference effects, but large

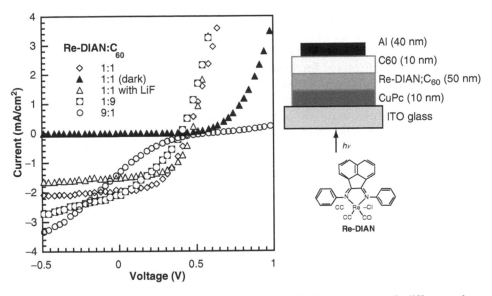

Figure 20.12. Current–voltage characteristics and device structure of different three layer photovoltaic cells in the dark or under illumination with AM1.5 simulated solar radiation. (From H. L. Wong, L. S. M. Lam, K. W. Cheng, W. K. Chan, C. Y. Kwong, and A. B. Djurišić, *Appl. Phys. Lett.* **84**, 2557–2559 (2004).)

BCP thickness was found to be detrimental to carrier collection efficiency [3]. The highest efficiency of 3.6% reported up to date in a small molecule organic photovoltaic device was obtained for a ITO/PEDOT:PSS/CuPc/C_{60}/BCP/Al solar cell [3,11]. Improvement with the use of BCP in phthalocyanine/C_{60} solar cells was confirmed by others [78,98], although the obtained efficiencies were lower. The main improvement with the BCP insertion is increase of the short circuit current and the fill factor [78].

Based on the reported results, the insertion of a BCP layer is beneficial to the cell performance, though the thickness of the BCP layer requires optimization. It would probably be worthwhile to explore the influence of other wide bandgap electron transport materials commonly used in OLEDs. It is expected that a material with higher electron mobility should enable more flexibility in thickness optimization and hence yield better improvement in the device performance.

20.5. CONCLUSIONS

The choice of electrodes is important for achieving good performance in organic solar cells. Heterojunction devices require carrier-selective contacts, so that one electrode should provide ohmic contact for electrons, while the other should provide ohmic contact for the holes. In order to achieve improved carrier collection, surface treatment of ITO electrode is beneficial. In all cases, the removal of carbon contamination by UV ozone (oxygen plasma can be used for the same purpose) improves the cell performance. However, optimal treatment is dependent on the material. When "n-type" material is in contact with ITO, the solar cell performance exhibits weaker dependence on the ITO surface properties. When "p-type" material is in contact with ITO, the use of interfacial layer such as PEDOT:PSS can enable improved device performance, provided that the thickness and thermal treatment of the PEDOT:PSS layer is optimized. The use of self-assembled monolayers to improve the stability and adjust the energy levels at the ITO–organic interface is promising, but up to date it has been largely unexplored for solar cell applications.

The choice of the top electrode depends on the type of carriers which should be collected. In the cells where electrons are collected at the top electrode, the insertion of BCP layer prior to metal deposition significantly improves the device performance. The BCP layer acts as a buffer layer to prevent damage of the organic material during the metal deposition. For a suitably chosen thickness of BCP layer, both the fill factor and the short circuit current of the device are enhanced. Other modifications of the cathode include insertion of a thin LiF layer. However, this may result in the decrease of the current, so that the thickness of the LiF layer needs to be carefully adjusted. The performance of the solar cell is typically strongly dependent on the thickness of any layers introduced into the structure to modify the electrodes, such as PEDOT:PSS, BCP, and LiF, so that the thicknesses of these layers should be optimized to avoid deterioration of the performance due to increased series resistance of the solar cell.

REFERENCES

1. J. Nelson, Organic photovoltaic films, *Curr. Opin. Solid State Mater. Sci.* **6**, 87–95 (2002).
2. J.-M. Nunzi, Organic photovoltaic materials and devices, *C. R. Physique* **3**, 523–542, (2002).
3. P. Peumans, A. Yakimov, and S. R. Forrest, Small molecular weight organic thin-film photodetectors and solar cells, *J. Appl. Phys.* **93**, 3693–3723 (2003).
4. D. Wöhrle, L. Kreienhoop, and D. Schlettwein, Phthalocyanines and related macrocycles in organic photovoltaic junctions, in *Phthalocyacyanines, Properties and Applications.* C. C. Leznoff and A. B. P. Lever (Eds.), VCH Publishers Inc., New York, 1996, pp. 219–284.
5. C. Y. Kwong, A. B. Djurišić, P. C. Chui, L. S. M. Lam, and W. K. Chan, Improvement in the efficiency of phthalocyanine organic Schottky solar cells with ITO electrode treatment, *Appl. Phys. A* **77**, 555–560 (2003).
6. D. L. Morel, A. K. Ghosh, T. Feng, E. L. Stogryn, P. E. Purwin, R. F. Shaw, and C. Fishman, High efficiency organic solar cells, *Appl. Phys. Lett.* **32**, 495–497 (1978).
7. C. W. Tang, Two-layer organic photovoltaic cell, *Appl. Phys. Lett.* **48**, 183–185 (1986).
8. T. Stübinger and W. Brütting, Exciton diffusion and optical interference in organic donor–acceptor photovoltaic cells, *J. Appl. Phys.* **90**, 3632–3641 (2001).
9. S. A. Jenekhe and S. Yi, Efficient photovoltaic cells from semiconducting polymer heterojunctions, *Appl. Phys. Lett.* **77**, 2635–2637 (2000).
10. P. Peumans, V. Bulovic, and S. R. Forrest, Efficient photon harvesting at high optical intensities in ultrathin organic double-heterostructure photovoltaic diodes, *Appl. Phys. Lett.* **76**, 2650–2652 (2000).
11. P. Peumans and S. R. Forrest, Very high-efficiency double-heterostructure copper phthalocyanine/C_{60} photovoltaic cells, *Appl. Phys. Lett.* **79**, 126–128 (2001).
12. A. Yakimov and S. R. Forrest, High photovoltage multiple-heterojunction organic solar cells incorporating interfacial metallic nanoclusters, *Appl. Phys. Lett.* **80**, 1667–1669 (2002).
13. D. Gebeyehu, B. Maennig, J. Drechsel, K. Leo, and M. Pfeiffer, Bulk-heterojunction photovoltaic devices based on donor-acceptor organic small molecule blends, *Sol. Energy Mater. Sol. Cells* **79**, 81–92 (2003).
14. S. E. Shaheen, C. J. Brabec, N. S. Sariciftci, F. Padinger, T. Fromherz, and J. C. Hummelen, 2.5% Efficient organic plastic solar cells, *Appl. Phys. Lett.* **78**, 841–843 (2001).

15. T. Aernouts, W. Geens, J. Poortmans, P. Heremans, S. Borghs, and R. Mertens, Extraction of bulk and contact components of the series resistance in organic bulk donor–acceptor heterojunctions, *Thin Solid Films* **403–404**, 297–301 (2002).

16. M. Hiramoto, K. Suemori, and M. Yokoyama, Photovoltaic properties of ultramicrostructure-controlled organic co-deposited films, *Jpn. J. Appl. Phys.* **41**, 2763–2766 (2002).

17. L. Schmidt-Mende, A. Fechtenkötter, K. Müllen, E. Moons, R. H. Friend, and J. D. MacKenzie, Self-organized discotic liquid crystals for high-efficiency organic photovoltaics, *Science* **293**, 1119–1122 (2001).

18. W. U. Huynh, J. J. Dittmer, and A. P. Alivisatos, Hybrid nanorod–polymer solar cells, *Science* **295**, 2425–2427 (2002).

19. B. O'Regan and M Grätzel, A low-cost, high-efficiency solar-cell based on dye-sensitized colloidal TiO_2 films, *Nature* **353**, 737–740 (1991).

20. J. Krüger, R. Plass, L. Cevey, M. Piccirelli, M. Grätzel, and U. Bach, High efficiency solid-state photovoltaic device due to inhibition of interface charge recombination, *Appl. Phys. Lett.* **79**, 2085–2087 (2001).

21. H. Ishii, K. Sugiyama, E. Ito, and K. Seki, Energy level alignment and interfacial electronic structures at organic/metal and organic/organic interfaces, *Adv. Mater.* **11**, 605–625 (1999).

22. K. Seki, N. Hayashi, H. Oji, E. Ito, Y. Ouchi, and H. Ishii, Electronic structure of organic/metal interfaces, *Thin Solid Films* **393**, 298–303 (2001).

23. I. G. Hill, D. Milliron, J. Schwartz, and A. Kahn, Organic semiconductor interfaces: electronic structure and transport properties, *Appl. Surf. Sci.* **166**, 354–362 (2000).

24. J. Campbell Scott, Metal–organic interface and charge injection in organic electronic devices, *J. Vac. Sci. Technol. A* **21**, 521–531 (2003).

25. P. H. Fang, Analysis of conversion efficiency of organic–semiconductor solar cells, *J. Appl. Phys.* **45**, 4672–4673 (1974).

26. B. A. Gregg, Excitonic solar cells, *J. Phys. Chem. B* **107**, 4688–4698 (2003).

27. B. A. Gregg and Y. I. Kim, Redox polymer contacts to molecular semiconductor films: toward kinetic control of interfacial exciton dissociation and electron-transfer processes, *J. Phys. Chem.* **98**, 2412–2417 (1994).

28. J. Rostalski and D. Meissner, Photocurrent spectroscopy for the investigation of charge carrier generation and transport mechanisms in organic p/n junction solar cells, *Sol. Energy Mater. Sol. Cells* **63**, 37–47 (2000).

29. C. J. Brabec, A. Cravino, D. Meissner, N. S. Sariciftci, M. T. Rispens, L. Sanchez, J. C. Hummelen, and T. Fromherz, The influence of materials work function on the open circuit voltage of plastic solar cells, *Thin Solid Films* **403–404**, 368–372 (2002).

30. C. J. Brabec, A. Cravino, D. Meissner, N. S. Sariciftci, T. Fromherz, M. T. Rispens, L. Sanchez, and J. C. Hummelen, Origin of the open circuit voltage in plastic solar cells, *Adv. Funct. Mater.* **11**, 374–380 (2001).

31. C. M. Ramsdale, J. A. Barker, A. C. Arias, J. D. MacKenzie, R. H. Friend, and N. C. Greenham, The origin of the open circuit voltage in polyfluorene-based photovoltaic devices, *J. Appl. Phys.* **92**, 4266–4270 (2002).

32. W. U. Huynh, J. J. Dittmer, N. Teclemariam, D. J. Milliron, A. P. Alivisatos, and K. W. Barnham, Charge transport in hybrid nanorod-polymer composite photovoltaic cells, *Phys. Rev. B* **67**, 115326-1–115326-12 (2003).

33. W. U. Huynh, J. J. Dittmer, W. C. Libby, G. L. Whiting, and A. P. Alivisatos, Controlling the morphology of nanocrystal-polymer composites for solar cells, *Adv. Funct. Mater.* **13**, 73–79 (2003).

34. C. L. Huisman, A. Goossens, and J. Schoonman, Preparation of a nanostructured composite of titanium dioxide and polythiophene: a new route towards 3D heterojunction solar cells, *Synth. Met.* **138**, 237–241 (2003).

35. E. Arici, N. S. Sariciftci, and D. Meissner, Hybrid solar cells based on nanoparticles of $CuInS_2$ in organic matrices, *Adv. Funct. Mater.* **13**, 165–171 (2003).

36. C. Y. Kwong, A. B. Djurišić, P. C. Chui, K. W. Cheng, and W. K. Chan, Influence of solvent on film morphology and device performance of poly(3-hexylthiophene): TiO$_2$ nanocomposite solar cells, *Chem. Phys. Lett.* **384**, 372–375 (2004).

37. E. Kymakis and G. A. J. Amaratunga, Single-wall carbon nanotube/conjugated polymer photovoltaic devices, *Appl. Phys. Lett.* **80**, 112–114 (2002).

38. E. Kymakis, I. Alexandrou, and G. A. J. Amaratunga, High open-circuit voltage photovoltaic devices from carbon-nanotube–polymer composites, *J. Appl. Phys.* **93**, 1764–1768 (2003).

39. E. Kymakis and G. A. J. Amaratunga, Photovoltaic cells based on dye-sensitization of single-wall carbon nanotubes in a polymer matrix, *Sol. Energy Mater. Sol. Cells* **80**, 465–472 (2003).

40. H. Ago, K. Petritsch, M. S. P. Shaffer, A. H. Windle, and R. H. Friend, Composite of carbon nanotubes and conjugated polymers for photovoltaic devices, *Adv. Mater.* **11**, 1281–1285 (1999).

41. G. Kron, T. Egerter, J. H. Werner, and U. Rau, Electronic transport in dye-sensitized nanoporous TiO$_2$ solar cells — comparison of electrolyte and solid-state devices, *J. Phys. Chem. B* **107**, 3556–3564 (2003).

42. S. Uchida, M. Tomiha, N. Masaki, A. Miyazawa, and H. Takizawa, Preparation of TiO$_2$ nanocrystalline electrode for dye-sensitized solar cells by 28 GHz microwave irradiation, *Sol. Energy Mater. Sol. Cells* **81**, 135–139 (2004).

43. G. K. R. Senadeera, K. Nakamura, T. Kitamura, Y. Wada, and S. Yanagida, Fabrication of highly efficient polythiophene-sensitized metal oxide photovoltaic cells, *Appl. Phys. Lett.* **83**, 5470–5472 (2003).

44. S. Ito, Y. Makari, T. Kitamura, Y. Wada, and S. Yanagida, Fabricaterization of mesoporous SnO$_2$/ZnO-composite electrodes for efficient dye solar cells, *J. Mater. Chem.* **14**, 385–390 (2004).

45. J. He, H. Lindström, A. Hagfeldt, and S.-E. Lindquist, Dye-sensitized nanostructured p-type nickel oxide film as a photocathode for a solar cell, *J. Phys. Chem. B* **103**, 8940–8943 (1999).

46. Y. Diamant, S. G. Chen, O. Melamed, and A. Zaban, Core-shell nanoporous electrode for dye-sensitized solar cells: the effect of the SrTiO$_3$ shell on the electronic properties of the TiO$_2$ core, *J. Phys. Chem. B* **107**, 1977–1981 (2003).

47. K. Imoto, K. Takahashi, T. Yamaguchi, T. Komura, J.-I. Nakamura, and K. Murata, High-performance carbon counter electrode for dye-sensitized solar cells, *Sol. Energy Mater. Sol. Cells* **79**, 459–469 (2003).

48. R. Plass, S. Pelet, J. Krueger, M. Gratzel, and U. Bach, Quantum dot sensitization of organic–inorganic hybrid solar cells, *J. Phys. Chem. B* **106**, 7578–7580 (2002).

49. L. M. Peter, K. G. U. Wijayantha, D. J. Riley, and J. P. Waggett, Band-edge tuning in self-assembled layers of Bi$_2$S$_3$ nanoparticles used to photosensitize nanocrystalline TiO$_2$, *J. Phys. Chem. B* **107**, 8378–8381 (2003).

50. G. Kron, U. Rau, and J. H. Werner, Influence of the built-in voltage on the fill factor of dye-sensitized solar cells, *J. Phys. Chem. B* **107**, 13258–13261 (2003).

51. F. Pichot and B. Gregg, The photovoltage-determining mechanism in dye-sensitized solar cells, *J. Phys. Chem. B* **104**, 6–10 (2000).

52. M. Fahlman and W. R. Salaneck, Surfaces and interfaces in polymer based electronics, *Surf. Sci.* **500**, 904–922 (2002).

53. A. B. Djurišić, C. Y. Kwong, P. C. Chui, and W. K. Chan, Indium-tin-oxide surface treatments: influence on the performance of CuPc/C$_{60}$ solar cells, *J. Appl. Phys.* **93**, 5472–5479 (2003).

54. J. S. Kim, M. Granström, R. H. Friend, N. Johanson, W. R. Salaneck, and F. Cacialli, Indium-tin oxide treatments for single- and double-layer polymeric light-emitting diodes: the relation between the anode physical, chemical, and morphological properties and the device performance, *J. Appl. Phys.* **84**, 6859–6870 (1998).

55. J. S. Kim, F. Cacialli, A. Cola, G. Gigli, and R. Cingolani, Increase of charge carriers density and reduction of Hall mobilities in oxygen-plasma treated indium-tin-oxide anodes, *Appl. Phys. Lett.* **75**, 19–21 (1999).

56. F. Nüesch, E. W. Forsythe, Q. T. Le, Y. Gao, and J. Rothberg, Importance of indium tin oxide surface acido basicity for charge injection into organic materials based light emitting diodes, *J. Appl. Phys.* **87**, 7973–7980 (2000).

57. F. Nüesch, J. Rothberg, E. W. Forsythe, Q. T. Le, and Y. Gao, A photoelectron spectroscopy study on the indium tin oxide treatment by acids and bases, *Appl. Phys. Lett.* **74**, 880–882 (1999).

58. F. Nüesch, K. Kamaras, and L. Zuppiroli, Protonated metal–oxide electrodes for organic light emitting diodes, *Chem. Phys. Lett.* **283**, 194–200 (1998).

59. K. Sugiyama, H. Ishii, Y. Ouchi, and K. Seki, Dependence of indium-tin-oxide work function on surface cleaning method as studied by ultraviolet and x-ray photoemission spectroscopies, *J. Appl. Phys.* **87**, 295–298 (2000).

60. S. K. So, W. K. Choi, C. H. Cheng, L. M. Leung, and C. F. Kwong, Surface preparation and characterization of indium tin oxide substrates for organic electroluminescent devices, *Appl. Phys. A* **68**, 447–450 (1999).

61. F. Li, H. Tang, J. Shinar, O. Resto, and S. Z. Weisz, Effects of aquaregia treatment of indium-tin-oxide substrates on the behavior of double layered organic light-emitting diodes, *Appl. Phys. Lett.* **70**, 2741–2743 (1997).

62. A. B. Djurišić, T. W. Lau, C. Y. Kwong, W. L. Guo, Y. Bai, E. H. Li, and W. K. Chan, Surface treatments of indium tin oxide substrates: comprehensive investigation of mechanical, chemical, thermal, and plasma treatments, in *Proc. of the SPIE*, Vol. 4464 "Organic Light-Emitting Materials and Devices V," edited by Z. Kafafi, SPIE, Bellingham, WA, 2002, pp. 273–280.

63. H. Y. Yu, X. D. Feng, D. Grozea, Z. H. Lu, R. N. Sodhi, A.-M. Hor, and H. Aziz, Surface electronic structure of plasma-treated indium tin oxides, *Appl. Phys. Lett.* **78**, 2595–2597 (2001).

64. C. Y. Kwong, A. B. Djurišić, P. C. Chui, and W. K. Chan, Influence of the device architecture to the ITO surface treatment effects on organic solar cell performance, in *Proc. of the SPIE*, Vol. 5215 "Organic Photovoltaics IV," edited by Z. Kafafi and P. A. Lane, SPIE, Bellingham, WA, 2004, pp. 153–160.

65. C. Y. Kwong, A. B. Djurišić, P. C. Chui, and W. K. Chan, CuPc/C$_{60}$ solar cells — influence of the ITO substrate and device architecture on the solar cell performance, *Jpn. J. Appl. Phys.* **43**, 1305–1311 (2004).

66. Q. T. Le, E. W. Forsythe, F. Nüesch, L. J. Rothberg, L. Yan, and Y. Gao, Interface formation between NPB and processed indium tin oxide, *Thin Solid Films* **363**, 42–46 (2000).

67. Q. T. Le, F. Nüesch, L. J. Rothberg, E. W. Forsythe, and Y. Gao, Photoemission study of the interface between phenyl diamine and treated indium-tin-oxide, *Appl. Phys. Lett.* **75**, 1357–1359 (1999).

68. A. J. Breeze, A. Salomon, D. S. Ginsley, B. A. Gregg, H. Tollmann, and H.-H. Hörhold, Polymer–perylene diimide heterojunction solar cells, *Appl. Phys. Lett.* **81**, 3085–3087 (2002).

69. P. Panayotatos, G. Bird, R. Sauers, A. Piechowski, and S. Husain, An approach to the optimal-design of p–n heterojunction solar-cells using thin-film organic semiconductors, *Solar Cells* **21**, 301–311 (1987).

70. J. B. Whitlock, P. Panayotatos, G. D. Sharma, M. D. Cox, R. R. Sauers, and G. R. Bird, Investigations of materials and device structures for organic semiconductor solar-cells, *Opt. Eng.* **32**, 1921–1934 (1993).

71. J. Xue and S. R. Forrest, Carrier transport in multilayer organic photodetectors: II. Effects of anode preparation, *J. Appl. Phys.* **95**, 1869–1877 (2004).

72. N. R. Armstrong, C. Carter, C. Donley, A. Simmonds, P. Lee, M. Brumbach, B. Kippelen, B. Domercq, and S. Yoo, Interface modification of ITO thin films: organic photovoltaic cells, *Thin Solid Films* **445**, 342–352 (2003).

73. A. Elschner, F. Bruder, H.-W. Heuer, F. Jonas, A. Karbach, S. Kirchmeyer, S. Thurm, and F. Wehrmann, PEDT/PSS for efficient hole-injection in hybrid organic light-emitting diodes, *Synth. Met.* **111–112**, 139–143 (2000).

74. G. Greczynski, Th. Kugler, W. Osikowicz, M. Fahlman, and W. R. Salaneck, Photo-electron spectroscopy of thin films of PEDOT–PSS conjugated polymer blend: a mini-review and some new results, *J. Electron Spectrosc. Related Phenom.* **121**, 1–17 (2001).

75. J. C. Hummelen, J. Knol, and L. Sánchez, Stability issues of conjugated polymer/fullerene solar cells from a chemical viewpoint, in *Proc. of the SPIE*, Vol. 4108 "Organic Photovoltaics," edited by Z. Kafafi, SPIE, Bellingham, WA, 2001, pp. 76–84.

76. A. C. Arias, M. Granström, K. Petritsch, and R. H. Friend, Organic photodiodes using polymeric anodes, *Synth. Met.* **102**, 953–954 (1999).

77. A. C. Arias, M. Granström, D. S. Thomas, K. Petritsch, and R. H. Friend, Doped conducting-polymer–semiconducting-polymer interfaces: their use in organic photovoltaic devices, *Phys. Rev. B* **60**, 1854–1860 (1999).

78. K. Fostiropoulos, M. Vogel, B. Mertesacker, and A. Weidinger, Preparation and investigation of phthalocyanine/C_{60} solar cells, in *Proc. of the SPIE*, Vol. 4801 "Organic Photovoltaics III," edited by Z. Kafafi, SPIE, Bellingham, WA, 2003, pp. 1–6.

79. J. Huang, P. F. Miller, J. C. de Mello, A. J. de Mello, and D. D. C. Bradley, Influence of thermal treatment on the conductivity and morphology of PEDOT:PSS films, *Synth. Met.* **139**, 569–572 (2003).

80. M. P. de Jong, L. J. van Ijzendoorn, and M. J. A. de Voigt, Stability of the interface between indium-tin-oxide and poly(3,4,ethylenedioxythiophene)/poly (styrene sulfonate) in polymer light emitting diodes, *Appl. Phys. Lett.* **77**, 2255–2257 (2000).

81. K. W. Wong, H. L. Yip, Y. Luo, K. Y. Wong, W. M. Lau, K. H. Low, H. F. Chow, Z. Q. Gao, W. L. Yeung, and C. C. Chang, Blocking reactions between indium-tin oxide and poly(3,4,ethylenedioxythiophene):poly(styrene sulphonate) with a self-assembly monolayer, *Appl. Phys. Lett.* **80**, 2722–2724 (2002).

82. B. Werner, J. Posdorfer, B. Wessling, S. Heun, H. Becker, H. Vestweber, and T. Hassenkam, PANI as hole injection layer for OLEDs and PLEDs, in *Proc. of the SPIE*, Vol. 4800 "Organic Light-Emitting Materials and Devices VI," edited by Z. Kafafi, SPIE, Bellingham, WA, 2003, pp. 115–122.

83. S. Karg, J. C. Scott, J. R. Salem, and M. Angelopoulos, Increased brightness and lifetime of polymer light-emitting diodes with polyaniline anodes, *Synth. Met.* **80**, 111–117 (1996).

84. J. C. Scott, S. A. Carter, S. Karg, and M. Angelopoulos, Polymeric anodes for organic light-emitting diodes, *Synth. Met.* **85**, 1197–1200 (1997).

85. P. K. H. Ho, M. Granström, R. H. Friend, and N. C. Greenham, Ultrathin self-assembled layers at the ITO interface to control charge injection and electroluminescence efficiency in polymer light emitting diodes, *Adv. Mater.* **10**, 769–774 (1998).

86. I. H. Campbell, S. Rubin, T. A. Zawodzinski, J. D. Kress, R. L. Martin, D. L. Smith, N. N. Baraskov, and J. P. Ferraris, Controlling Schottky energy barriers in organic electronic devices using self-assembled monolayers, *Phys. Rev. B* **54**, 14321–14324 (1996).

87. H. Yamada, H. Imahori, Y. Nishimura, I. Yamazaki, and S. Fukuzumi, Enhancement of photocurrent generation by ITO electrodes modified chemically with self-assembled monolayers of porphyrin-fullerene dyads, *Adv. Mater.* **14**, 892–895 (2002).

88. B. A. Gregg, Photovoltaic properties of a molecular semiconductor modulated by an exciton-dissociating film, *Appl. Phys. Lett.* **67**, 1271–1273 (1995).

89. M. Granström, K. Petritsch, A. C. Arias, A. Lux, M. Anderson, and R. H. Friend, Laminated fabrication of polymeric photovoltaic diodes, *Nature* **395**, 257–260 (1998).

90. H. B. Michaelson, The work function of elements and its periodicity, *J. Appl. Phys.* **48**, 4729–4733 (1977).

91. K. Murata, S. Ito, K. Takahashi, and B. M. Hoffman, Photocurrent from photo-corrosion of aluminum electrode in porphyrin/Al Schottky barrier cells, *Appl. Phys. Lett.* **71**, 674–676 (1997).

92. I. Hiromitsu, M. Kitano, R. Shinto, and T. Ito, Photocurrent characteristics of phthalo-cyanine/C_{60} p–n heterojunctions controlled by the work functions of metal electrodes, *Solid State Commun.* **113**, 165–169 (2000).

93. C. Melzer, V. V. Krasnikov, and G. Hadziioannou, Organic donor/acceptor photovol-taics: the role of C_{60}/metal interfaces, *Appl. Phys. Lett.* **82**, 3101–3103 (2003).

94. W. J. H. van Gennip, J. K. J. van Duren, P. C. Thüne, R. A. J. Janssen, and J. W. Niemantsverdriet, The interfaces of poly(p-phenylene vinylene) and fullerene derivatives with Al, LiF, Al/LiF studied by secondary ion mass spectroscopy: formation of AlF_3 disproved, *J. Chem. Phys.* **117**, 5031–5035 (2002).

95. G. Greczynski, M. Fahlman, and W. R. Salaneck, An experimental study of poly(9,9-dioctyl-fluorene) and its interface with Li and LiF, *Appl. Surf. Sci.* **166**, 380–386 (2000).

96. B. Masenelli, E. Tutis, M. N. Bussac, and L. Zuppiroli, Numerical model for injection and transport in multilayers OLEDs, *Synth. Met.* **122**, 141–144 (2001).

97. H. L. Wong, L. S. M. Lam, K. W. Cheng, W. K. Chan, C. Y. Kwong, and A. B. Djurišić, Low-band-gap, sublimable rhenium(I) diimine complex for efficient bulk heterojunction photovoltaic devices. *Appl. Phys. Lett.* **84**, 2557–2559 (2004).

98. Y. J. Ahn, G. W. Kang, and C. H. Lee, Photovoltaic properties of multilayer hetero-junction organic solar cells, *Mol. Cryst. Liq. Cryst.* **377**, 301–304 (2002).

21

Conducting and Transparent Polymer Electrodes

Fengling Zhang and Olle Inganäs

Biomolecular and Organic Electronics, Department of Physics, Linköping University, Linköping, Sweden

Contents

Abstract Polymer electronics shows the possibility to build flexible devices. Conducting polymer poly(3,4-ethylenedioxythiophene) doped with polystyrene sulfonic acid (PEDOT–PSS) was widely used as interfacial layer or polymer electrode in polymer electronic devices, such as light-emitting diodes (LEDs) and photovoltaic devices (PVDs). In this chapter, the applications of PEDOT–PSS used as buffer layer and anode in polymer devices were briefly reviewed. The important properties for its application in polymer PVDs, such as conductivity, the optical properties, and work function were discussed. The applications of PEDOT–PSS in polymer PVDs as a buffer layer or transparent electrode were demonstrated.

Keywords polymer, photovoltaic devices, PEDOT–PSS, polymer electrode, flexible solar cells.

21.1. INTRODUCTION

Semiconducting polymers have emerged as novel electronic materials after the invention of light-emitting diode (LED) using poly(p-phenylene vinylene) (PPV) as an emitter in 1990 [1]. This created a new scientific field named polymer electronics or plastic electronics. Polymer electronics developed into a hot scientific research area, which includes polymer-based LEDs [2], transistors [3], solar cells [4,5], etc. and has potential application in developing flexible devices. Displays based on electroluminescence in organic molecular LEDs have already been used in commercial products, and polymer LEDs are approaching the market. In organic photovoltaic devices (PVDs), the energy conversion efficiency (ECE) up to 3% under illumination of AM 1.5 has now been reached, in devices based on organic small molecule solar cells [6,7] and in polymer-based systems such as poly(3-hexylthiophene) (P3HT) and poly (2-methoxy, 5-(3′,7′-dimethyl-octyloxy))-p-phenylene vinylene) (MDMO-PPV) combined with a fullerene derivative [6,6]-phenyl-C_{61}-butyric acid methyl ester (PCBM) [5,8]. The merits of polymer electronics compared to conventional inorganic semiconductor devices are easy processing, flexibility, low weight, large area deposition possibility, and low cost.

Polymer PVDs are thin film cells with a thickness of only a few hundred nanometers (nm), in which a polymer film deposited by solution processing is the active layer. The film is sandwiched between two electrodes with different work functions, one of which is transparent for incident light. The requirements for electrodes in such devices are that they give (transparent) solid film with high conductivity and desired work function. Indium tin oxide (ITO) coated on glass substrates were commonly used as transparent anodes and metals like Ca, Al were used as cathodes. Polymers have not been routinely used as electrodes due to their low conductivity compared to their inorganic counterpart ITO. The desire to build flexible electronic devices, which can be rolled up or pasted on curved surfaces, like windows, motivate efforts to exploit flexible polymer electrodes for plastic electronics. Even though ITO coated on flexible substrates is an option for flexible electrodes [9], its flexibility is limited. Gustafsson et al. made a fully flexible PLED by using poly(ethylene terephthalate) (PET) as the substrate, soluble polyaniline (PANI) as the hole injection electrode and polymer poly(2-methoxy,5-(2′-ethyl-hexoxy)-1,4-phenylene vinylene) (MEH-PPV) as emitter. This can be sharply bent and curled without failure [10–12]. Their results demonstrated the possibility to use conducting polymer PANI as an electrode for polymer electronics. The optical absorption of polyaniline is found in the visible wavelengths and compromises the transmission to the active layer. The use of a doped conjugated polymer as an injection layer in polymer devices was therefore extended to the polythiophene family of polymers. Early work demonstrated that poly(3,4-ethylenedioxythiophene) doped with polystyrene sulfonic acid (PSS) could be prepared by electropolymerization, and used as the anode in nano LEDs and flexible nano LEDs [13–15]. A related material PEDOT–PSS, an aqueous dispersion prepared by chemical polymerization of EDOT was later used, demonstrating dramatic improvement of performance and stability of LEDs [16–18]. Flexible organic small molecular LEDs, using commercial PEDOT–PSS conducting film coated on flexible substrate PET as an anode, were fabricated and reached maximum illumination $2760 \, cd/m^2$, and maximum external quantum efficiency (EQE) of 1.4%, which is close to the performance of similar organic light-emitting diodes (OLEDs) using ITO as electrode, and with better flexibility [19].

Polypyrrole, polyaniline, and other conducting polymers are now widely used as electrodes in capacitors and batteries, etc. [20,21], but PEDOT–PSS alone or combined with ITO is the most common conducting polymer used as transparent polymer electrode in organic and polymer LEDs and polymer PVDs. The conductivity can be modified by several methods [22]. Therefore, in following sections, we will focus only on the use of PEDOT–PSS as polymer anode in polymer PVDs. First, we summarize the properties of PEDOT–PSS, which enable its use as a flexible polymer electrode, and then give a general view on using it as buffer layer on ITO to improve the performance of diodes, and as flexible polymer anodes to make flexible devices.

21.2. PROPERTIES OF PEDOT–PSS

PEDOT was first developed at Bayer AG, and later much developed by many groups worldwide [23–25]. PEDOT–PSS, a polyelectrolyte complex between poly(3,4-ethylenedioxythiophene) and poly(styrenesulfonate) was originally used as antistatic layer in photographic films. The molecular structure of PEDOT–PSS is shown in Figure 21.1. It became an extremely attractive material when it was found that the performance and stability of polymer LEDs could be improved by inserting this material between the polymer active layer and ITO as a buffer layer [16–18]. PEDOT–PSS is available in an aqueous dispersion, which can form uniform, transparent, and conductive film by spin-coating. The most important physical properties for its application in devices are the high work function and the smooth surface.

To be used as an electrode in polymer PVDs, the materials need to be highly conducting, transparent in solid film, stable, with easy processing and patterning characteristics. It should also have desirable work function to decrease the energy barrier at the interface between electrode and active layer, and to facilitate charge carrier collection. In this section, we will cover all these aspects, which are crucial properties for PEDOT–PSS to be used as an electrode in polymer electronics, especially polymer PVDs.

PEDOT–PSS

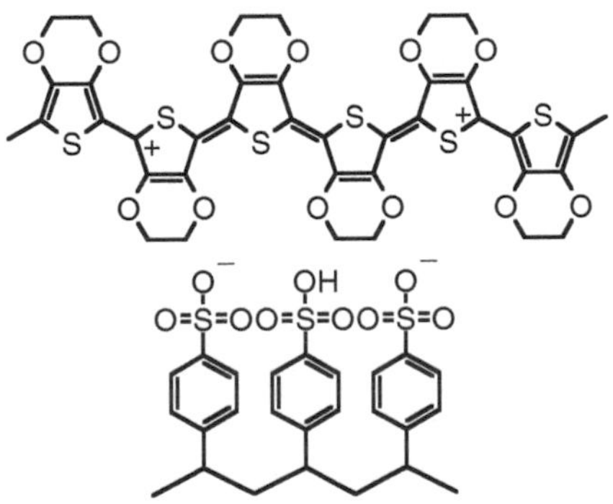

Figure 21.1. The molecular structure of PEDOT–PSS.

21.2.1. Conductivity of PEDOT–PSS

To be used as an electrode, high conductivity is necessary. The conductivity of PEDOT–PSS produced by electropolymerization is 80 S/cm and produced by chemical polymerization is 0.03 S/cm, which is due to different compositions [26]. There are several methods to modify the conductivity of PEDOT–PSS. Conductivity of PEDOT–PSS not only can vary some orders of magnitudes by using different polymerization processes, but also by doping and annealing [27]. Secondary doping, first employed to increase the conductivity of PANI [28], can also be used to increase the conductivity of PEDOT–PSS. It was demonstrated that gels formed from PEDOT–PSS could be dramatically influenced by the choice of solvents and polymer additives, which strongly affect the conductivity [29,30]. Similar results can be found in films. The conductivity of PEDOT–PSS can be increased as much as two orders of magnitude by thermal processing [31]. Sheet resistance of PEDOT–PSS as low as 350–500 Ohm/□ was reported [32]. Kim and coworkers observed the enhancement of conductivity of the PEDOT–PSS free standing film by adding different organic solvents: dimethyl sulfoxide (DMSO), N,N-dimethyl formamide (DMF), and tetrahydrofuran (THF). A 100-fold increase of conductivity was achieved in the film doped with DMSO [33]. By adding glycerol or sorbitol into PEDOT–PSS, aqueous dispersion, and annealing the films made from mixed solutions, the conductivity of PEDOT–PSS film can be increased about two orders of magnitude [34]. Recently, an increase of the conductivity of PEDOT–PSS by a factor of about 600 through solvent and heat treatment was reported [35]. Louwet et al. [36] achieved an increase of three orders of magnitude of conductivity in PEDOT–PSS film by adding organic solvent N-methyl pyrrolidone (NMP).

The nanostructure of the films have been reported to be crucial for effective conductivity. In brief, the PEDOT–PSS particles dressed in a shell of insulating PSS is modified during chemical and thermal treatment to remove some of the limiting insulating layer and make better particle–particle electrical contact. This is also visible as a change of the surface composition of films of PEDOT–PSS due to the treatment [35].

21.2.2. The Optical Properties of PEDOT–PSS Film

The thickness, color, and conductivity of PEDOT–PSS film vary with process condition. For samples produced by spin-coating, high speeds give a thinner film with high transmission, and larger surface resistance, whereas low speeds give a thicker film with decreasing transmission and lower surface resistance. The optical properties of early forms of PEDOT have been studied in detail by Pettersson et al. [37], who reported that the material was optically anisotropic. This anisotropy is also observed in spin-coated films of PEDOT–PSS [38] as well as in thermally annealed films. From these data, we observe a marked minimum of optical absorbance at wavelengths close to 400 nm, and increasing almost linearly with increasing wavelengths (see Figure 21.2 and Figure 21.3). The full dielectric function of the layer is also reported, which allows the optical absorption and reflection properties of this layer, in combination with other layers in organic PVDs, to be calculated [39]. Instructive plots of the amount of energy deposited in the different layers in PVDs are found therein, which demonstrate that above the bandgap absorption of the polymers modeled, the optical absorption of PEDOT–PSS is very minute, but considerable loss term has to be added to account for absorption below the bandgap. Therefore the use of PEDOT anodes is advantageous for purposes of light detection

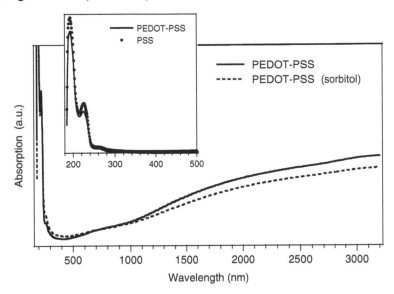

Figure 21.2. Absorbance of normal incidence of thin films of PEDOT–PSS and a PEDOT–PSS (sorbitol) of equal thickness (110 nm) in the spectral range from 185 to 3200 nm. In the inset, absorbance of PEDOT–PSS and PSS in the spectral range from 185 to 500 nm. (From L. A. A. Pettersson, S. Ghosh, and O. Inganäs, *Organic Electronics*, **3**, 143–148 (2002). With permission.)

and photovoltaic conversion, where the optical loss of the PEDOT layer only sets in at wavelengths where little absorption can be found in the active layer.

The anode of polymer PVDs requires a solid film with both high transparency and conductivity, which are in conflict and hard to satisfy with untreated PEDOT–PSS, such as PEDOT–PSS (Baytron P). When used as a buffer layer on top of ITO, it is usually a very thin film (less than 100 nm) to guarantee the high transmission because the conductivity of PEDOT–PSS is not important in that case. When used as a current carrying anode, it needs to be a thicker film (150–200 nm) to keep its conductivity high enough for lateral conduction of photocurrent. This results in loss of optical transmission. Different modified high conductivity forms of PEDOT–PSS contribute to solve this conflict. With high conductivity form of PEDOT–PSS, the film retained its original optical transparency over 75% to 80% in the visible range and kept its surface resistance low enough to be an electrode in organic molecular LEDs or polymer PVDs [36,40].

21.2.3. Processing and Patterning

Conventional semiconductors usually need multiple etching steps, often in complex equipment, which make the processing complicated and also make the devices expensive. However, polymer electronic devices can be fabricated by solution processes, which use simple, cheap equipments, consumes short time, and results in inexpensive and possibly disposable devices. This big advantage makes polymer electronics attractive.

The deposition of solids from liquids by spin-coating is inherently wasteful, but dip-coating, doctor blading, and similar methods use the active material efficiently, via processes of deposition from liquids, which may be batch-like or based on reel-to-reel processing, thin layer electrodes can be deposited.

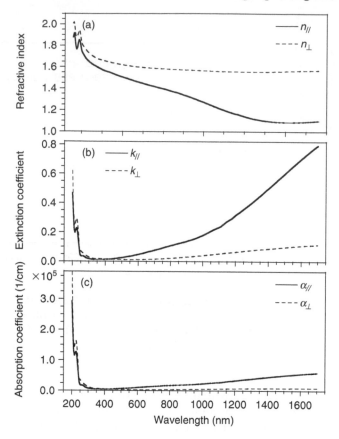

Figure 21.3. Uniaxial anisotropic complex indices of refraction $\tilde{n}_i = n_i + ik_i$, ($i = \parallel$, $\perp$) and the absorption coefficient $\alpha_i = 4\pi k_i/\lambda$ parallel (ordinary) and perpendicular (extraordinary) to the surface of thin films of PEDOT–PSS in the spectral range 200–1700 nm. (a) The refractive index, (b) the extraction coefficient, and (c) the absorption coefficient. (From L. A. A. Pettersson, S. Ghosh, and O. Inganäs, *Organic Electronics*, **3**, 143–148 (2002). With permission.)

In polymer electronics, patterning the materials to form electronic devices is necessary. There are many methods, from classical lithography to soft lithography, to pattern polymer films. Rogers et al. [41] demonstrated a microcapillary method (MIMIC) with PANI to pattern source–drain electrodes for organic transistors and liquid embossing to produce relief features in the film of PPV and used the structures in PLEDs [42]. These methods were also used for patterning PEDOT–PSS and polymer active layer into nanowires and nanodots [43], which has potential application to enhance photocurrent by trapping more light in polymer PVDs [44]. Touwslager et al. [45] reported their work on using process based on waterborne solution only and making use of negatively acting, highly sensitive, lithography with UV light of 365 nm (I-line) to structure PEDOT–PSS. Minimum feature size of 1 μm spacing and 2.5 μm lines with the sheet resistance of about 1 kOhm/□ was achieved. A new method namely line patterning used in polypyrrole and PEDOT was developed by Hohnholz and MacDiarmid [46]. Flexible and transparent all-polymer field-effect transistors (FETs) were fabricated by using this technique and PEDOT–PSS as a channel materials without using photolithography, vacuum deposition, or printing of conducting inks [47]. All these examples show that there are many options to pattern devices of plastic electronics and they are all simple processes compared with inorganic semiconductors.

21.2.4. Work Function

The work function of electrodes in plastic electronics plays an important role in organic or polymer LEDs and PVDs because it determines the distribution of internal electric field and the height of the energy barrier at the interface between electrode and polymer active layers. This barrier influences the charge carrier injection from electrodes into active layer in LEDs, or oppositely, dominates charge carrier collection from active layer into electrodes in PVDs [5,16]. To facilitate the hole injection or collection, high work function materials are preferred as an anode. The work function of ITO is about 4.7 eV [48] and PEDOT–PSS is around 5.2 eV [49]. This increase of work function by 0.5 eV from ITO to PEDOT–PSS is one of the reasons for the significant improvement of the performance of polymer LEDs with PEDOT–PSS as buffer layer between ITO and active layers. PEDOT–PSS alone as the anode gives comparable performance in organic and polymer LEDs and PVDs, even though the surface resistance of ITO (10–20 Ohm/□) is at least one order of magnitude lower than that of PEDOT–PSS (>350 Ohm/□) partly due to its high work function compensating its high resistance.

21.3. APPLICATION IN PVDs

Arias et al. [50] fabricated a polymer PVD by using PPV as the active layer, PEDOT–PSS as an anode and Al as a cathode. They obtained a broader spectrum of EQE in the diode with polymer anode than when using ITO as an anode. We have investigated different forms of PEDOT–PSS as polymer anodes in polymer solar cells using MEH-PPV and PCBM as active layers, under both low-intensity monochromatic light illumination ($1.2 \, W/m^2$) and high-intensity white light illumination AM1.5 ($780 \, W/m^2$). All solar cells with three different PEDOT–PSS anodes gave better EQE than ITO anode under low-intensity monochromatic light illumination, even though their surface resistances are several orders higher than ITO. These are small devices with size of 5–7 mm^2. The surface resistance limits the performance of polymer solar cells only in the diode with the PEDOT–PSS (Baytron P) without any treatment as an anode under high-intensity white light illumination AM1.5 ($780 \, W/m^2$). For the diodes with high conductive PEDOT–PSS doped sorbitol, the EQE is very close to the diode with ITO anode and ITO with PEDOT–PSS buffer layer on the top of it [34].

21.3.1. PEDOT–PSS Used as a Buffer Layer

21.3.1.1. Increase the Rectification and Photovoltage of Polymer PVDs

As the conductivity of polymer electrode materials is much lower than its inorganic counterparts, such as ITO, it produces a significant voltage drop. Thicker polymer film can decrease the surface resistance, but results in a lower transmission. However, combining a polymer electrode with ITO will overcome the drawback of polymer electrodes with low conductivity, and utilize its high work function.

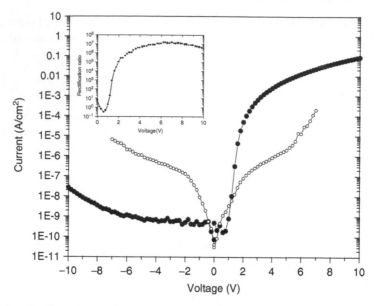

Figure 21.4. Semilog plot of the current–voltage characteristics of two similar MEH-PPV based diodes using different anodes Cu (open circles) and Cu/PEDOT–PSS (solid circles). The inset graph shows a semilog plot of the rectification ratio versus voltage for the Cu/PEDOT–PSS anode diode. The diodes used were fabricated in sandwich structure Cu/PEDOT–PSS/MEH-PPV/Al. (Reprinted with permission from L. S. Roman, M. Berggren, and O. Inganäs, *Applied Physics Letters*, **75**, 3557–3559 (1999). Copyright American Institute of Physics, 1999. With permission).

Roman et al. [51] investigated the function of PEDOT–PSS (Baytron P) layer in diodes using semiconducting polymer MEH-PPV in sandwich geometry between two low work function metal electrodes Cu and Al, where Cu was covered with PEDOT–PSS. The *I–V* characteristics of diodes with and without PEDOT–PSS layer are shown in Figure 21.4. The results indicated that PEDOT–PSS significantly increased rectification in polymer diodes with Cu as anode, by reaching seven orders of magnitude under $\pm$ 10 V. The change of injection conditions due to the PEDOT–PSS layer accounts for this behavior.

Alem et al. [52] systematically investigated the influence of interfacial layer of PEDOT–PSS on performance of polymer PVDs and concluded that inserting a thin layer (30 nm) of PEDOT–PSS between active layer (MEH-PPV:PCBM) and ITO did not change the value of short circuit current density (J_{sc}), but increased the open circuit voltage (V_{oc}) (from 0.4 to 0.66 V) and fill factor (FF), and resulted in an improved ECE [52]. However, Aernouts et al. [53] reported the lower ECE in PVDs based on MDMO-PPV:PCBM with PEDOT–PSS (2.3%) than that without PEDOT–PSS (3.1%) besides the enhanced V_{oc}. Their results showed significant decrease of J_{sc} from 7.2 to 5.7 mA/cm² upon insertion of PEDOT–PSS layer with a thickness of 140 nm. It demonstrated that only thin layer of PEDOT–PSS should be used as buffer layer to avoid decrease transparency of anode.

21.3.1.2.   Tuning the Photovoltage of Polymer PVDs

PEDOT–PSS as buffer layer can increase the rectification of diodes and enhance the performance of polymer PVDs [51,52]. There are different kinds of PEDOT–PSS used in polymer PVDs, such as PEDOT–PSS (Baytron P) and PEDOT–PSS (EL grade). Do they have same effects on the performance of polymer PVDs? We fabricated solar cells by using three different forms of PEDOT–PSS as buffer layers on top of ITO and polyfluorene copolymer poly(2,7-(9,9-dioctyl-fluorene)-*alt*-5,5-(4′,7′-di-2-thienyl-2′,1′,3′-benzothiadiazole)) (DiO-PFDTBT) blended with PCBM as the active layer [54]. The different PEDOT–PSS polymers used in this study are PEDOT–PSS (EL), PEDOT–PSS (P), and PEDOT–PSS:Sorbitol (PEDOT–PSS:S). PEDOT–PSS (EL) and PEDOT–PSS (P) are commercial products, PEDOT–PSS:S has the same composition as that in Ref. [34] and was prepared in our laboratory. The EQEs of these three diodes are similar, as the active layers are the same, but the *J–V* characteristics of these diodes both in dark and under illumination are quite sensitive to the buffer layers (Figure 21.5 and Figure 21.6). First, the rectification of *J–V* curves in dark increases

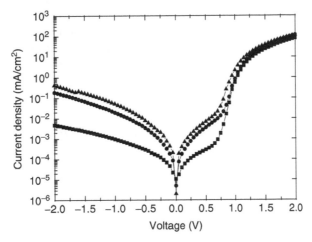

Figure 21.5. The *I–V* characteristics of PVDs fabricated with different buffer layers in the dark, where the line with squares is for PEDOT–PSS(EL), with circles for PEDOT–PSS(P), and with triangles for PEDOT–PSS(P):S. (Reprinted with permission from *Appl, Phys. Lett.*, **84**, 3906–3908 (2004). Copyright 2004, American Institute of Physics)

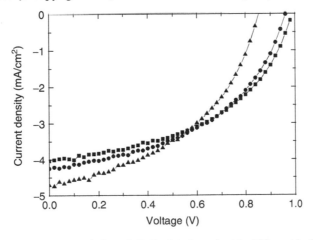

Figure 21.6. The *I–V* characteristics of PVDs fabricated with different buffer layers under illumination of AM1.5, where the line with squares is for PEDOT–PSS(EL), with circles for PEDOT–PSS(P), and with triangles for PEDOT–PSS(P):S. (Reprinted with permission from *Appl, Phys. Lett.*, **84**, 3906–3908 (2004). Copyright 2004, American Institute of Physics)

two orders of magnitude in the diodes with buffer layer of PEDOT–PSS (EL) compared to the other two buffer layers, PEDOT–PSS (P) and PEDOT–PSS:S under $\pm$ 2 V. Second, the J–V curves of the diodes under illumination of AM1.5 (1000 W/m^2) from solar simulator shows different behavior; J_{sc}, V_{oc}, and FF are all influenced by buffer layers. The highest J_{sc} was achieved in the diode with buffer layer PEDOT–PSS:S, but V_{oc} was lowest. Conversely, highest V_{oc} was achieved in the diode of PEDOT–PSS (EL), but J_{sc} was lowest.

There are different views on the origin of V_{oc}, which depends on the device structure, work functions of electrodes, the HOMO, LUMO of donor and acceptor, etc. [55–58]. Mihailetchi and coworkers [59] demonstrated that for nonohmic contacts the V_{oc} is determined by the work function difference of the electrodes and for ohmic contacts the V_{oc} is governed by the LUMO and HOMO level of the acceptor and donor [59]. The V_{oc} is the voltage where photocurrent is balanced by injection current from electrodes, so the injection condition also influences V_{oc}. Lower injection current in diode with PEDOT–PSS (EL) results in higher V_{oc} when other conditions are same, such as same active layer and same cathode. Our result shows that different forms of PEDOT–PSS not only can be used as polymer anodes, but also adjust the hole injection from anode to active layer and tune the photovoltage by more than 100 mV in polymer PVDs. Detailed discussion on this issue can be found in Ref. [60].

21.3.2.   PEDOT–PSS as an Electrode in Polymer PVDs

To avoid a large voltage drop across the polymer electrode, which depends on resistivity, geometry, and the current density flow through the electrode, the surface resistance of polymer electrode should be smaller than 100 Ohm/□ to be used as electrode. However, so far the photocurrent density in polymer PVDs is not more than 10 mA/cm^2, therefore the conductivity of modified PEDOT–PSS is high enough to be used as an electrode for diodes with low current density.

We fabricated diodes by using the high conductivity form of PEDOT–PSS, that is, PEDOT–PSS doped with sorbitol (PEDOT–PSS:S) as an anode on glass substrate and obtain a surface resistance $\approx$1 kOhm [34]. We used same active layer as in Section 3.1.2. The EQE of this diode and that of the diode made with ITO alone are similar (Figure 21.7), which shows that comparable photocurrent was achieved in diode using PEDOT–PSS:S as an anode under low-intensity monochromatic illumination ($\sim$1 W/m^2). The J–V curves in the dark shown in Figure 21.8 indicate smaller rectification in the diode with polymer anode than that with ITO anode. The J–V curves of the diodes under high-intensity illumination of AM1.5 (1000 W/m^2) (Figure 21.9) present a smaller short circuit current density (J_{sc}) and FF in the diode with polymer anode than in another diode with ITO anode, which reflects the transport limitation in the diode with polymer anode due to its low conductivity. However, larger open circuit voltage (V_{oc}) due to higher work function of PEDOT–PSS was obtained in the diode with polymer anode, which compensates the smaller J_{sc} and FF than in the diode with ITO, which results in the similar ECE in two diodes. The comparison of diodes with two different anodes is listed in Table 21.1, which demonstrates that highly conductive PEDOT–PSS can be used as anode for polymer PVDs when J_{sc} is in the order of 5 mA/cm^2.

Besides the polymer PVDs fabricated on hard substrates, such as glass, we also built the polymer PVDs on flexible substrates like normal transparencies. The photos of such flexible PVD and its photovoltage and photocurrent under room light are shown in Figure 21.10a and b, which demonstrate the possibility of using polymer anode for flexible PVDs.

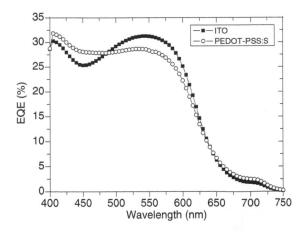

Figure 21.7. EQEs of polymer PVDs fabricated on ITO and polymer anode glass/PDOT–PSS:Sorbitol.

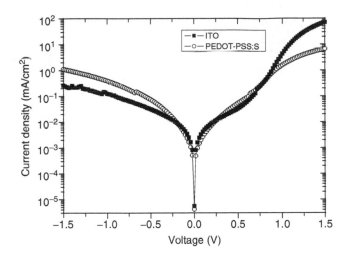

Figure 21.8. The I–V characteristics of same polymer PVDs as in Figure 21.7 in the dark.

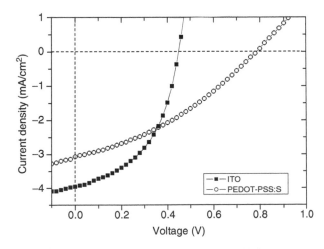

Figure 21.9. The I–V characteristics of same polymer PVDs under illumination of AM 1.5.

Table 21.1. Comparison of diodes with polymer anode and ITO anode

Anode	J_{sc} (mA/cm^2)	V_{oc} (V)	FF	ECE (%)
PEDOT–PSS:Sorbitol	3.07	0.78	0.35	0.85
ITO	3.96	0.45	0.47	0.85

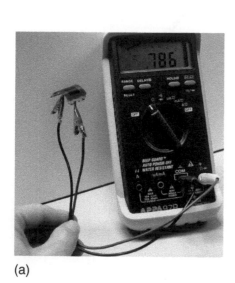 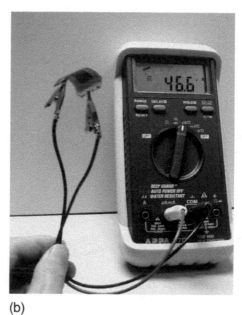

(a)   (b)

Figure 21.10. Photos of a solar cell fabricated on normal transparency with a polymer anode. The voltage display on multimeter is the V_{oc} (a) and current display on multimeter is the photocurrent (b) of the flexible PVD with area of 1.4 cm^2 under room light ($\sim$1 W/m^2).

21.4. OUTLOOK

The results presented indicate that PEDOT–PSS alone used as an electrode can be an alternative to ITO, but not to ITO combined with PEDOT–PSS, which gave much better performance than either the PEDOT–PSS or the ITO electrode. Therefore, polymer electrode of PEDOT–PSS may be a replacement for ITO in case of flexible devices, but cannot totally replace ITO in plastic electronics at the moment. However, the progress in this field demonstrates the possibility of using polymer electrodes in plastic electronics in the near future. Of course, it will always be necessary to include current collectors from high-conductivity metals when making solar cell panels. At what level this current collector is introduced may be a simple consequence of the conductivity of polymers to come.

ACKNOWLEDGMENT

We acknowledge funding from the Swedish Research Council for Engineering Sciences (TFR) and the Göran Gustafsson foundation.

REFERENCES

1. J. H. Burroughes, D. D. C. Bradley, A. R. Brown, R. N. Marks, K. Mackay, R. H. Friend, P. L. Burns, and A. B. Holmes, Light-emitting-diodes based on conjugated polymers, *Nature*, **347**, 539–541 (1990).

2. X. Gong, D. Moses, A. J. Heeger, S. Liu, and A. K. Y. Jen, High-performance polymer light-emitting diodes fabricated with a polymer hole injection, *Applied Physics Letters*, **83**, 183–185 (2003).

3. N. Stutzmann, R. H. Friend, and H. Sirringhaus, Self-aligned, vertical-channel, polymer field-effect transistors, *Science*, **299**, 1881–1884 (2003).

4. D. M. Russell, A. C. Arias, R. H. Friend, C. Silva, C. Ego, A. C. Grimsdale, and K. Mullen, Efficient light harvesting in a photovoltaic diode composed of a semiconductor conjugated copolymer blend, *Applied Physics Letters*, **80**, 2204–2206 (2002).

5. F. Padinger, R. S. Rittberger, and N. S. Sariciftci, Effects of postproduction treatment on plastic solar cells, *Advanced Functional Materials*, **13**, 85–88 (2003).

6. P. Peumans and S. R. Forrest, Very-high-efficiency double-heterostructure copper phthalocyanine/C-60 photovoltaic cells, *Applied Physics Letters*, **79**, 126–128 (2001).

7. P. Peumans, A. Yakimov, and S. R. Forrest, Small molecular weight organic thin-film photodetectors and solar cells. *Journal of Applied Physics*, **93**, 3693–3723 (2003).

8. C. J. Brabec, N. S. Sariciftci, and J. C. Hummelen, Plastic solar cells, *Advanced Functional Materials*, **11**, 15–26 (2001).

9. G. Gu, P. E. Burrows, S. Venkatesh, S. R. Forrest, and M. E. Thompson, Vacuum-deposited, nonpolymeric flexible organic light-emitting devices, *Optics Letters*, **22**, 172–174 (1997).

10. G. Gustafsson, Y. Cao, G. M. Treacy, F. Klavetter, N. Colaneri, and A. J. Heeger, Flexible light-emitting-diodes made from soluble conducting polymers, *Nature*, **357**, 477–479 (1992).

11. G. Gustafsson, G. M. Treacy, Y. Cao, F. Klavetter, N. Colaneri, and A. J. Heeger, The plastic led — a flexible light-emitting device using a polyaniline transparent electrode, *Synthetic Metals*, **57**, 4123–4127 (1993).

12. Y. Yang and A. J. Heeger, Polyaniline as a transparent electrode for polymer light-emitting-diodes — lower operating voltage and higher efficiency, *Applied Physics Letters*, **64**, 1245–1247 (1994).

13. M. Granström and O. Inganäs, Flexible arrays of submicrometer-sized polymeric light emitting diodes, *Advanced Materials*, **7**, 1012–1014 (1995).

14. M. Granström, M. Berggren, and O. Inganäs, Micro and nanometer size light sources using polymer light emitting diodes, *Science*, **267**, 1479–1481 (1995).

15. M. Granström and O. Inganäs, White light emission from a polymer blend light emitting diode, *Applied Physics Letters*, **68**, 147–149 (1996).

16. Y. Cao, G. Yu, C. Zhang, R. Menon, and A. J. Heeger, Polymer light-emitting diodes with polyethylene dioxythiophene-polystyrene sulfonate as the transparent anode, *Synthetic Metals*, **87**, 171–174 (1997).

17. S. A. Carter, M. Angelopoulos, S. Karg, P. J. Brock, and J. C. Scott, Polymeric anodes for improved polymer light-emitting diode performance, *Applied Physics Letters*, **70**, 2067–2069 (1997).

18. J. C. Scott, S. A. Carter, S. Karg, and M. Angelopoulos, Polymeric anodes for organic light-emitting diodes, *Synthetic Metals*, **85**, 1197–1200 (1997).

19. Y. Qiu, L. Duan, and L. D. Wang, Flexible organic light-emitting diodes with poly-3,4-ethylenedioxythiophene as transparent anode, *Chinese Science Bulletin*, **47**, 1979–1982 (2002).

20. K. S. Ryu, K. M. Kim, N. G. Park, Y. J. Park, and S. H. Chang, Symmetric redox supercapacitor with conducting polyaniline electrodes, *Journal of Power Sources*, **103**, 305–309 (2002).

21. K. Henkel, A. Oprea, I. Paloumpa, G. Appel, D. Schmeisser, and P. Kamieth, Selective polypyrrole electrodes for quartz microbalances: NO_2 and gas flux sensitivities, *Sensors and Actuators B — Chemical*, **76**, 124–129 (2001).

22. W. H. Kim, G. P. Kushto, H. Kim, and Z. H. Kafafi, Effect of annealing on the electrical properties and morphology of a conducting polymer used as an anode in organic light-emitting devices, *Journal of Polymer Science Part B — Polymer Physics*, **41**, 2522–2528 (2003).

23. Q. B. Pei, G. Zuccarello, M. Ahlskog, and O. Inganäs, Electrochromic and highly stable poly(3,4-ethylenedioxythiophene) gives blue-black and transparent sky blue colors, *Polymer*, **35**, 1347–1351 (1994).

24. Bayer AG, European patent 339 340 (1988).

25. B. L. Groenendaal, F. Jonas, D. Freitag, H. Pielartzik, and J. R. Reynolds, Poly(3,4-ethylenedioxythiophene) and its derivatives: past, present, and future, *Advanced Materials*, **12**, 481–494 (2000).

26. X. Crispin, S. Marciniak, W. Osikowicz, G. Zotti, A. W. D. Van der Gon, F. Louwet, M. Fahlman, L. Groenendaal, F. De Schryver, and W. R. Salaneck, Conductivity, morphology, interfacial chemistry, and stability of poly(3,4-ethylene dioxythiophene)–poly(styrene sulfonate): a photoelectron spectroscopy study, *Journal of Polymer Science Part B — Polymer Physics*, **41**, 2561–2583 (2003).

27. J. Huang, P. F. Miller, J. C. de Mello, A. J. de Mello, and D. D. C. Bradley, Influence of thermal treatment on the conductivity and morphology of PEDOT/PSS films, *Synthetic Metals*, **139**, 569–572 (2003).

28. A. G. MacDiarmid and A. J. Epstein, The concept of secondary doping as applied to polyaniline, *Synthetic Metals*, **65**, 103–116 (1994).

29. S. Ghosh, J. Rasmusson, and O. Inganäs, Conductivity in conjugated polymer blends: Ionic crosslinking in blends of poly(3,4-ethylenedioxythiophene) poly(styrenesulfonate) and poly(vinylpyrrolidone), *Advanced Materials*, **10**, 1097–1099 (1998).

30. S. Ghosh and O. Inganäs, Self-assembly of a conducting polymer nanostructure by physical crosslinking: applications to conducting blends and modified electrodes, *Synthetic Metals*, **101**, 413–416 (1999).

31. T. Granlund, L. A. A. Pettersson, and O. Inganäs, Determination of the emission zone in a single-layer polymer light-emitting diode through optical measurements, *Journal of Applied Physics*, **89**, 5897–5902 (2001).

32. A. J. Mäkinen, I. G. Hill, R. Shashidhar, N. Nikolov, and Z. H. Kafafi, Hole injection barriers at polymer anode/small molecule interfaces, *Applied Physics Letters*, **79**, 557–559 (2001).

33. J. Y. Kim, J. H. Jung, D. E. Lee, and J. Joo, Enhancement of electrical conductivity of poly(3,4-ethylenedioxythiophene)/poly(4-styrenesulfonate) by a change of solvents, *Synthetic Metals*, **126**, 311–316 (2002).

34. F. L. Zhang, M. Johansson, M. R. Andersson, J. C. Hummelen, and O. Inganäs, Polymer photovoltaic cells with conducting polymer anodes, *Advanced Materials*, **14**, 662–665 (2002).

35. S. K. M. Jönsson, J. Birgerson, X. Crispin, G. Greczynski, W. Osikowicz, A. W. D. van der Gon, W. R. Salaneck, and M. Fahlman, The effects of solvents on the morphology and sheet resistance in poly(3,4-ethylenedioxythiophene)–polystyrenesulfonic acid (PEDOT–PSS) films, *Synthetic Metals*, **139**, 1–10 (2003).

36. F. Louwet, L. Groenendaal, J. Dhaen, J. Manca, J. Van Luppen, E. Verdonck, and L. Leenders, PEDOT/PSS: synthesis, characterization, properties and applications, *Synthetic Metals*, **135**, 115–117 (2003).

37. L. A. A. Pettersson, F. Carlsson, O. Inganäs, and H. Arwin, Spectroscopic ellipsometry studies of the optical properties of doped poly(3,4-ethylenedioxythiophene): an aniso-tropic metal, *Thin Solid Films*, **313**, 356–361 (1998).

38. L. A. A. Pettersson, S. Ghosh, and O. Inganäs, Optical anisotropy in thin films of poly(3,4-ethylenedioxythiophene)–poly(4-styrenesulfonate), *Organic Electronics*, **3**, 143–148 (2002).

39. N. K. Persson, Chapter 5: Simulations of optical processes in organic photovoltaic devices, in *Organic Photovoltaics: Mechanisms, Materials, and Devices*, eds. Sun and Sariciftci, CRC Press, Boca Raton, FL, 2005.

40. W. H. Kim, A. J. Mäkinen, N. Nikolov, R. Shashidhar, H. Kim, and Z. H. Kafafi, Molecular organic light-emitting diodes using highly conducting polymers as anodes, *Applied Physics Letters*, **80**, 3844–3846 (2002).

41. J. A. Rogers, Z. N. Bao, and V. R. Raju, Nonphotolithographic fabrication of organic transistors with micron feature sizes, *Applied Physics Letters*, **72**, 2716–2718 (1998).

42. J. A. Rogers, Z. N. Bao, and L. Dhar, Fabrication of patterned electroluminescent polymers that emit in geometries with feature sizes into the submicron range, *Applied Physics Letters*, **73**, 294–296 (1998).

43. F. L. Zhang, T. Nyberg, and O. Inganäs, Conducting polymer nanowires and nanodots made with soft lithography, *Nano Letters*, **2**, 1373–1377 (2002).

44. L. S. Roman, O. Inganäs, T. Granlund, T. Nyberg, M. Svensson, M. R. Andersson, and J. C. Hummelen, Trapping light in polymer photodiodes with soft embossed gratings, *Advanced Materials*, **12**, 189–195 (2000).

45. F. J. Touwslager, N. P. Willard, and D. M. de Leeuw, Water based I-line projection lithography of PEDOT, *Synthetic Metals*, **135**, 53–54 (2003).

46. D. Hohnholz and A. G. MacDiarmid, Line patterning of conducting polymers: new horizons for inexpensive, disposable electronic devices, *Synthetic Metals*, **121**, 1327–1328 (2001).

47. H. Okuzaki, M. Ishihara, and S. Ashizawa, Characteristics of conducting polymer transistors prepared by line patterning, *Synthetic Metals*, **137**, 947–948 (2003).

48. G. Yu, C. Zhang, and A. J. Heeger, Dual-function semiconducting polymer devices: light-emitting and photodetecting diodes, *Applied Physics Letters*, **64**, 1540–1542 (1994).

49. T. M. Brown, J. S. Kim, R. H. Friend, F. Cacialli, R. Daik, and W. J. Feast, Built-in field electroabsorption spectroscopy of polymer light-emitting diodes incorporating a doped poly(3,4-ethylene dioxythiophene) hole injection layer, *Applied Physics Letters*, **75**, 1679–1681 (1999).

50. A. C. Arias, M. Granström, K. Petritsch, and R. H. Friend, Organic photodiodes using polymeric anodes, *Synthetic Metals*, **102**, 953–954 (1999).

51. L. S. Roman, M. Berggren, and O. Inganäs, Polymer diodes with high rectification, *Applied Physics Letters*, **75**, 3557–3559 (1999).

52. S. Alem, R. de Bettignies, J. M. Nunzi, and M. Cariou, Efficient polymer-based interpenetrated network photovoltaic cells, *Applied Physics Letters*, **84**, 2178–2180 (2004).

53. T. Aernouts, W. Geens, J. Poortmans, P. Heremans, S. Borghs, and R. Mertens, Extraction of bulk and contact components of the series resistance in organic bulk donor–acceptor-heterojunctions, *Thin Solid Films*, **403**, 297–301 (2002).

54. O. Inganäs, M. Svensson, F. Zhang, A. Gadisa, N. K. Persson, X. Wang, and M. R. Andersson, Low bandgap alternating polyfluorene copolymers in plastic photodiodes and solar cells, *Applied Physics A: Materials Science & Processing*, **79**, 31–35 (2004).

55. C. J. Brabec, A. Cravino, D. Meissner, N. S. Sariciftci, T. Fromherz, M. T. Rispens, L. Sanchez, and J. C. Hummelen, Origin of the open circuit voltage of plastic solar cells, *Advances in Functional Materials*, **11**, 374–380 (2001).

56. C. J. Brabec, A. Cravino, D. Meissner, N. S. Sariciftci, M. T. Rispens, L. Sanchez, J. C. Hummelen, and T. Fromherz, The influence of materials work function on the open circuit voltage of plastic solar cells, *Thin Solid Films*, **403**, 368–372 (2002).

57. B. A. Gregg, Excitonic solar cells, *Journal of Physical Chemistry B*, **107**, 4688–4698 (2003).

58. C. M. Ramsdale, J. A. Barker, A. C. Arias, J. D. MacKenzie, R. H. Friend, and N. C. Greenham, The origin of the open-circuit voltage in polyfluorene-based photovoltaic devices. *Journal of Applied Physics*, **92**, 4266–4270 (2002).

59. V. D. Mihailetchi, P. W. M. Blom, J. C. Hummelen, and M. T. Rispens, Cathode dependence of the open-circuit voltage of polymer: fullerene bulk heterojunction solar cells, *Journal of Applied Physics*, **94**, 6849–6854 (2003).
60. F. L. Zhang, A. Gadisa, O. Inganäs, M. Svensson, and M. R. Andersson, Influence of buffer layers on the performance of polymer solar cells, *Applied Physics Letters*, **84**, 3906–3908 (2004).

22

Progress in Optically Transparent Conducting Polymers

Venkataramanan Seshadri and Gregory A. Sotzing
Institute of Materials Science, U-136, University of Connecticut, Storss, CT, USA

Contents

Abstract All polymeric devices that are lightweight, easy to fabricate, and inexpensive have been the focus of many research groups for several years and the discovery of intrinsically conducting polymers (ICPs) has made this dream a reality. Till date, there are many applications of ICPs: antistatic coatings, electrochromics, organic light-emitting diodes (LEDs), batteries, photovoltaics, sensors, supercapacitors, etc. to name a few. The preparation of newer conjugated polymers with better properties has paved way to extend the applications of these specialty materials. Increasing energy demands has given the impetus for researching renewable energy sources such as the solar energy. Polymer-based photovoltaics have been considered as a cost-effective solution compared to the inorganic photovoltaics, and ICPs have played a key role in this regard. ICPs have been utilized as the transparent electrode as well as the light-harvesting dye in photovoltaics with promising results. With the wealth of knowledge now available towards tuning the optoelectronic properties of these conjugated polymers, it is possible to tailor the ICPs to make photovoltaics with higher efficiency. In this chapter, optically transparent ICPs, which have been synthesized using different strategies for tuning the bandgap, are reviewed.

Keywords optically transparent, conjugated, low bandgap, conductive, photovoltaics, polythiophenes.

22.1. INTRODUCTION

Ever since the discovery of highly conductive polyacetylene [1] by iodine doping, there has been a pursuit to make highly conductive polymers. Several conjugated polymers with different basic structures have been made, e.g., anilines [2], benzenes [3], thiophenes [4], pyrroles [5], and other heterocycles. A tremendous amount of knowledge has been gained from this research on the structure–property relationship of molecules to make highly conductive and transparent intrinsically conducting polymers (ICPs). Along with material development, there has been an enormous increase in the application areas of these conjugated polymers such as in antistatics batteries [6], sensors [7], organic light-emitting diodes (OLEDs) [8], photovoltaics [9], supercapacitors [10], molecular electronics [11], etc. Research on applications of conducting polymer like flexible display devices, OLEDs, and photovoltaics has created a huge demand for conjugated polymers that are transparent, stable, and highly conducting. ICPs have been used as hole-injection layers (HILs), as emitting layers, and even as the anode. The idea of making cheap and easily processible materials has been the impetus for such research. While use of poly(3,4-ethylenediox-ythiophene)–poly(styrenesulfonate) (PEDOT–PSS) as replacement for indium-doped tin oxide (ITO) in inorganic luminescence has been known for some time, Inganas et al. [12], Reynolds et al. [13], and Lee et al. [14] have recently demonstrated the potential use of ICPs as one of the electrodes in the areas of photovoltaics, smart windows, and flexible organic film speakers. Although there are a very limited number of current publications wherein ICPs have been used as the optically trans-parent electrode (OTE), some recent publications on making transparent and highly conducting polymers gives hope that prospects for highly efficient all-polymeric devices are not mere fiction. Thus, it would be a constant endeavor to make an organic equivalent of the present inorganic materials used as OTEs so that processing could be simplified and the cost of devices reduced such that flexible devices could be made a reality. There are innumerable reports on photovoltaics (PVs) and LEDs, wherein ICPs have been used as hole-injection layers. For an ICP to be used as a hole-injection layer in PVs, it is necessary that they be transparent in the oxidized form. Generally, the absorption of conjugated polymers is redshifted when oxidized. If the absorption maximum for a neutral conjugated polymer is very high (>600 nm), there is a very strong possibility that its spectrum in the oxidized form will be redshifted to longer wavelengths, leading to low absorbance in the visible region, thus making it almost transparent. Apart from finding use as hole-injection layers, ICPs are also used as donor materials in PVs made using donor–acceptor heterojunctions. Of the total solar emission spectrum, 40% of the sunlight reaching us is above 600 nm with a maximum solar flux at 600–800 nm (Figure 22.1) [15].

Therefore there is a strong need to match the absorption spectrum of semicon-ducting conjugated polymers to that of the solar emission spectrum to attain max-imum absorption in this region. It would be our endeavor to present low bandgap conducting polymers, which could fit into the criteria described earlier. Roncali [16] Patil et al. [17], Yamamoto and Hayashida [18], Scherf and Mullen [19], Ajayaghosh [20], and others have previously reviewed the tuning of bandgap of conjugated polymers. In this chapter, we would like to present those polymers which have absorption maxima greater than 600 nm in the neutral form and also polymers that can show extreme change in color upon oxidation, possibly attaining a highly transmissive state.

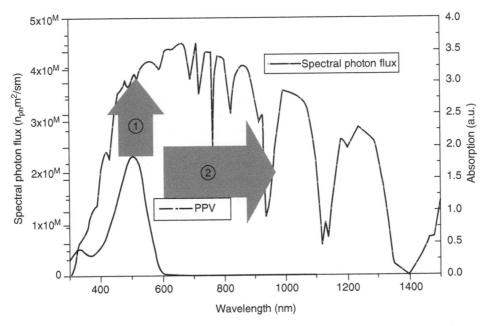

Figure 22.1. Comparison of solar emission spectrum and absorption spectrum of most conjugated polymers. (From Wormser, P. and Gaudiana, R., *NCPV and Solar Prog. Rev. Meeting Proc.* **2003**, 307–310. Copyright 2003, NREL and Konarka Technologies.)

22.2. TOWARDS TRANSPARENT ICPs

Conducting polymers can be classified as semiconducting and the bandgap of these polymers has been defined as the energy gap between the highest occupied molecular orbital (HOMO) and lowest unoccupied molecular orbital (LUMO). This gap is usually obtained from the higher wavelength absorption edge of the neutral polymer (which is attributed to the π to π^* transition) and it is a common practice to report the λ_{max} to show the breadth of the bandgap. Alternatively this gap is also obtained from the cyclovoltammogram of the polymer itself. Practically, it is obtained by calculating the difference in the onset of oxidation and reduction waves seen in the cyclovoltammogram. In this chapter, we would characterize a transparent polymer as the one showing very low or no absorbance in the visible region (400–800 nm). Thus ICPs with a λ_{max} greater than 600 nm in the neutral state have a very strong chance to exhibit very low absorbance within the visible spectrum in the oxidized form as thin films. Thus several strategies have been adopted to increase the conjugation like introducing electron-donating or -withdrawing groups, spacer groups to reduce steric interactions between adjacent rings, polymers containing donor–acceptor type repeat units, which can exhibit internal charge transfer, repeat units containing fused aromatic backbone, and rigidified ladder type polymers.

22.2.1. Polymers with Electron-Rich Moieties

Of all the heterocycles, thiophene-based conjugated polymers have been shown to be environmentally stable and robust in nature. The conjugated polymer from parent thiophene has a bandgap of about ~2.1 eV and reduction of the bandgap of these stable polythiophenes have been of great interest to create newer stable conducting polymers. Raising the HOMO of the polythiophene by electron-donating substituents has evolved as a key strategy in this regard with emphasis on 3,4-alkylenedioxythiophene (**1–16**), owing to its high transmissivity in the oxidized form [21]. PEDOT is one of the most studied polymers in terms of structure–property relationships, and is presently commercially sold under the tradename BAYTRON-P (Bayer) as a processible aqueous dispersion. Polystyrenesulfonate is the charge compensating counter-ion to the positively charged PEDOT that renders BAYTRON-P dispersible in water. Since the β-positions in the molecule have been blocked by the alkylenedioxy unit the polymerization proceeds exclusively through the α-positions thereby minimizing defects in the polymeric structure. Electrochemically generated PEDOT has a band edge gap of 1.6–1.7 eV with a λ_{max} at 580–610 nm and PEDOT is over 75% transmissive throughout the visible region for a 300-nm thick film (Figure 22.2) [13].

The high transparency of PEDOT has attracted several research groups to study the effect of different substituents on the repeat unit on its optical properties. Dietrich et al. [22], Reynolds et al. [23], Ahonen et al. [24], Czardybon et al. [25], and Krishnamoorthy et al. [26] have reported the synthesis of different alkylenedioxy derivatives of thiophenes and by electrochemical polymerization of these monomers (Figure 22.3).

Following the work of Jonas and Schrader on ProDOT (**3**) [21], Reynolds and Kumar [23] have varied the bulkiness of the side-groups in order to study the contrast in transmittance in the oxidized form and neutral forms of the polymer. Reynolds and co-workers [23d] report a 76% contrast for dimethyl-substituted PProDOT (**9**) and Kumar and co-workers [26] recently reported 89% contrast for dibenzyl substitutions on PrODOT (**11**). During the spectroelectrochemistry of alkyl-substituted PProDOT, poly(**9**) and poly(**11**) were reported to exhibit a peak at

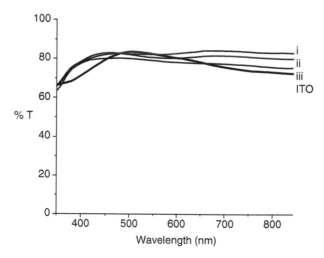

Figure 22.2. Percent transmittance (%T) of the PEDOT–PSS-coated transparent film electrodes in the visible region with (i) one layer, (ii) two layers, and (iii) three layers. Electrodes with three layers yield a surface resistivity of 600 Ω/square with an average %T value of 77% through the visible spectrum. %T spectrum of an ITO electrode (bold line) is also shown for comparison. (From Argun, A. A.; Cirpan, A., and Reynolds, J. R., *Adv. Mater.* **2003**, *15*, 1338–1341. Copyright 2003, Wiley-VCH Verlag GmbH & Co. KGaA, Weinheim.)

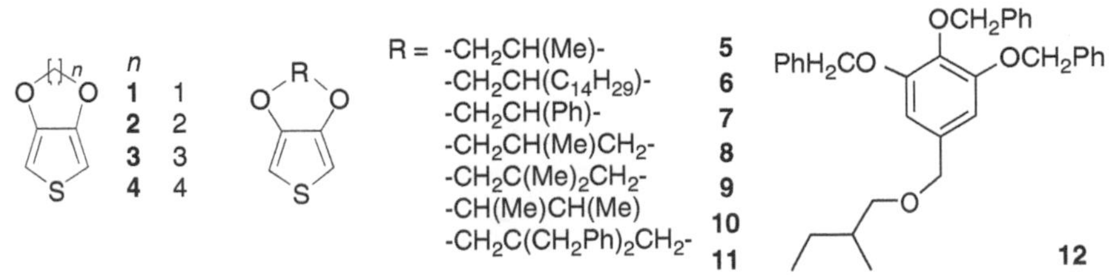

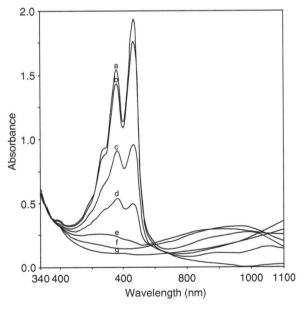

n
1 1
2 2
3 3
4 4
R = -CH2CH(Me)- 5
-CH2CH(C14H29)- 6
-CH2CH(Ph)- 7
-CH2CH(Me)CH2- 8
-CH2C(Me)2CH2- 9
-CH(Me)CH(Me) 10
-CH2C(CH2Ph)2CH2- 11
OCH2Ph
PhH2CO
OCH2Ph
12

Figure 22.3. The different substituted and unsubstituted 3,4-alkylenedioxythiophenes prepared by several groups.

1.25 eV, which was found to increase in intensity during the initial oxidation and with further increasing in the potential, this peak disappears (Figure 22.4).

This has been the key reason for the high optical transmissivity and contrast of these polymers in the oxidized state. Groenendaal et al. [27] have published exhaustive reviews on the electrochemical polymerization of different 3,4-alkylenedioxythiophenes as well as the different derivatives of EDOT that have been prepared so far. The high transparency of this polymer has prompted the synthesis of water-soluble derivatives of the polymer by hanging ionizable sulfonate groups or nonionic poly (ethylene glycol) derivatives (Figure 22.5) [28].

This gives rise to a versatile approach to make pure polymers without external contaminants in the form of dopants. With so much of bandgap tunability, replacing the oxygens in EDOT with sulfur looks attractive. Blanchard et al. [29] studied the sulfur analogs of EDOT by substituting either one or both the oxygen atoms with sulfur atom and electrochemically polymerized (Figure 22.6).

2.0
1.5
1.0
0.5
0.0
Absorbance
a
b
c
d
e
f
g
340 400
400
800
1000 1100
Wavelength (nm)

Figure 22.4. Optoelectrochemical spectra of the poly(**11**) (charge density 22.5 mC/cm^2) as a function of applied potential in 0.1 *M* tetrabutylammonium tetrafluoroborate–acetonitrile: (a) −1 V, (b) −0.4 V, (c) −0.2 V, (d) 0.1 V, (e) 0.3 V, (f) 0.5 V, and (g) 1.0 V. (From Krishnamoorthy, K.; Ambade, A. V.; Kanungo, M.; Contractor, A. Q., and Kumar, A., *J. Mater. Chem.* **2001**, *11*, 2909–2911. Copyright 2001, Royal Society of Chemistry, UK. With permission.)

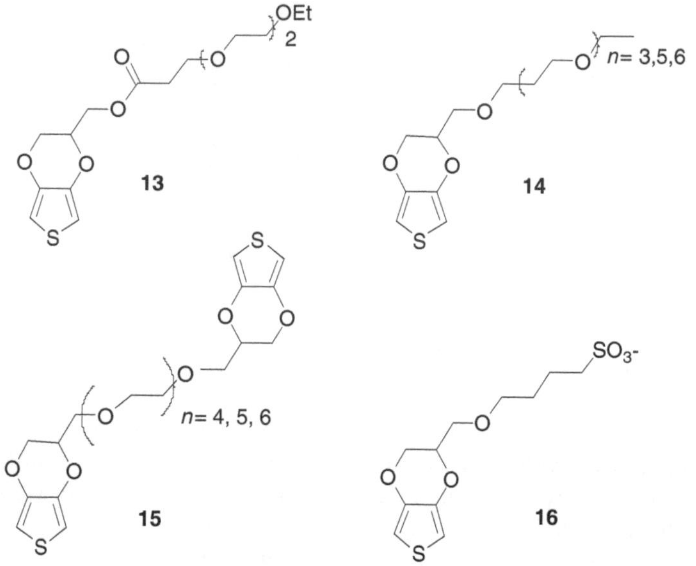

Figure 22.5. Some water-soluble derivatives of **2** that have been prepared and polymerized.

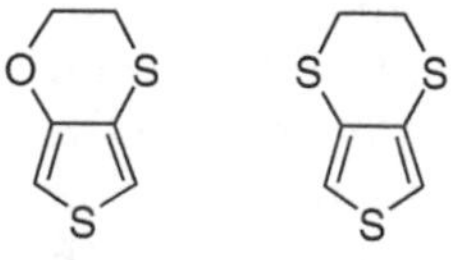

Figure 22.6. Sulfur-containing analogs of 3,4-ethylenedioxythiophene.

In the 3,4-ethylenethiathiophenes, steric interaction plays an important role, causing the rings to go out of the plane and resulting in higher bandgap. The λ_{max} for these two polymers (**17** and **18**) have been measured to be 532 and 438 nm, indicating a strong steric interaction between adjacent rings. Research on low bandgap polymers has been extended to other heteroaromatic systems like pyrrole- and selenophone-based heteroaromatics and their electron-rich counterparts. Reynolds et al. [30] and Cava et al. [31] have prepared the pyrroles and selenophene-based electron-rich monomers similar to alkylenedioxythiophenes and polymerized these electrochemically (Figure 22.7).

Reynolds and co-workers studied a systematic variation of 3,4-alkylenedioxypyrrole (**16**) analogs of PEDOT/ProDOT and reported these to have a higher bandgap than PEDOT [30]. Both the HOMO and LUMO of the poly(XDOP) was found to have shifted to higher values compared to PEDOT, leading to a higher E_g value. Of all the

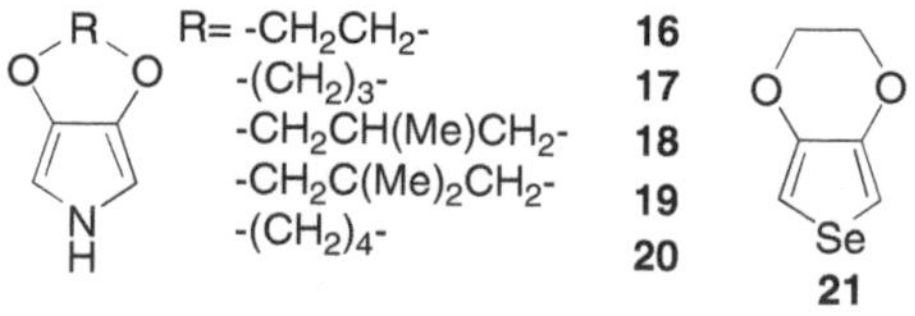

Figure 22.7. Repeat unit structures for the PEDOT analogs: 3,4-alkylenedioxy derivatives of pyrrole and 3,4-ethylenedioxyselenophene.

alkylenedixoypyrroles studied, PProDOP, poly(**17**), was reported to show very high contrast at 534 nm. It was reported to be orange in the neutral form and transmissive sky blue in the oxidized form. Such extreme color changes are very rare in conducting polymers and more so the maximum contrast was observed at 534 nm, a value close to the wavelength (550 nm) at which the human eye is very sensitive (Figure 22.8).

Thin films of PProDOP (thickness 0.1 μm) were found to exhibit a transmittance of about 70% throughout the visible region. Poly(**21**), PEDOS, or the selenium-containing analog of PEDOT prepared by Aqad et al. was found to exhibit a λ_{max} at 594 nm in the neutral state (deep blue color) and was sky blue in the oxidized form.

PEDOT is one of the commercially available conducting polymers in a water-processible form. Bayer first introduced PEDOT in the form of aqueous dispersions in the presence of polystyrene sulfonic acids, BAYTRON-P, for antistatic coatings for photographic films [32]. These dispersions typically contain 1.3 wt.% of the

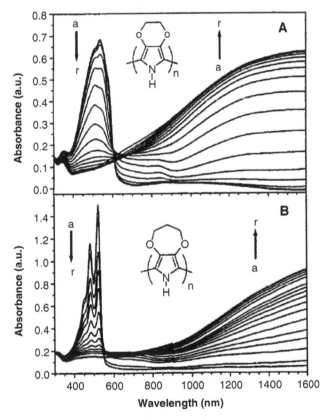

Figure 22.8. (A) Spectroelectrochemistry of poly(**16**) in 0.1 M LiClO$_4$/propylene carbonate at applied potentials of (a) -1.33 V, (b) -1.23 V, (c) -1.13 V, (d) -1.03 V, (e) -0.98 V, (f) -0.93 V, (g) -0.88 V, (h) -0.83 V, (i) -0.78 V, (j) -0.73 V, (k) -0.68 V, (l) -0.63 V, (m) -0.53 V, (n) -0.43 V, (o) -0.33 V, (p) -0.23 V, (q) -0.13 V, and (r) $+0.07$ V vs. Fc/Fc$^+$. (B) Spectroelectrochemistry of poly(**17**) in 0.1 M LiClO$_4$/PC at applied potentials of (a) -1.13 V, (b) -1.03 V, (c) -0.93 V, (d) -0.88 V, (e) -0.83 V, (f) -0.73 V, (g) -0.62 V, (h) -0.58 V, (i) -0.53 V, (j) -0.48 V, (k) -0.43 V, (l) -0.38 V, (m) -0.33 V, (n) -0.23 V, (o) -0.13 V, (p) -0.03 V, (q) $+0.07$ V, and (r) $+0.17$ V vs. Fc/Fc$^+$. (From Schottland, P.; Zong, K.; Gaupp, C. L.; Thompson, B. C.; Thomas, C. A.; Giurgiu, I.; Hickman, R.; Abboud, K. A., and Reynolds J. R., *Macromolecules* **2000**, *33*, 7051–7061. Copyright 2000, American Chemical Society Publications. With permission.)

combined polymers with a PEDOT to PSS ratio of approximately 1:2.5 by weight and several grades with conductivity values ranging from 10^{-5} to $10\,S\,cm^{-1}$ have been made available. These dispersions can be coated onto glass or plastic substrates by conventional techniques such as spin-coating, dip-coating, or drop-casting to thin-film transparent films. AGFA introduced a modified form of these dispersions (ORGACON) [32] with conductivities one order higher than that of BAYTRON-P and has reduced solids in its dispersions, typically 0.9 wt.%. Recently Jönsson et al. [33] reported highly conducting PEDOT–PSS coatings prepared by adding sorbitol to the PEDOT–PSS dispersions and preparing thin films from them. In their report, they claim that pristine PEDOT–PSS films with conductivities as low as 0.05–0.075 S cm^{-1} can be made highly conducting ($\sigma = 12$–$19\,S\,cm^{-1}$) by addition of sorbitol, a reduced glucose. Further heating of these samples at 200°C for 4 min on a hot plate increases the conductivity by another 2.5 times (i.e., $\sigma = 30$–$48\,S\,cm^{-1}$).

While tremendous efforts in understanding the structure–property relationship of 3,4-alkylenedioxy thiophenes and improving its properties have been undertaken, novel processing techniques to make transparent thin films with high conductivity have also been carried out.

Meng et al. [34] report an unexpected and unprecedented solid-state polymerization technique to make transparent PEDOT films onto glass or plastic substrates (Figure 22.9) using 2,5-dibromo-3,4-ethylenedioxythiophene (**22**). Compound **22**, on prolonged storage at ambient conditions or gentle heating at 50–80°C, produced PEDOT with very high conductivity (Scheme 1).

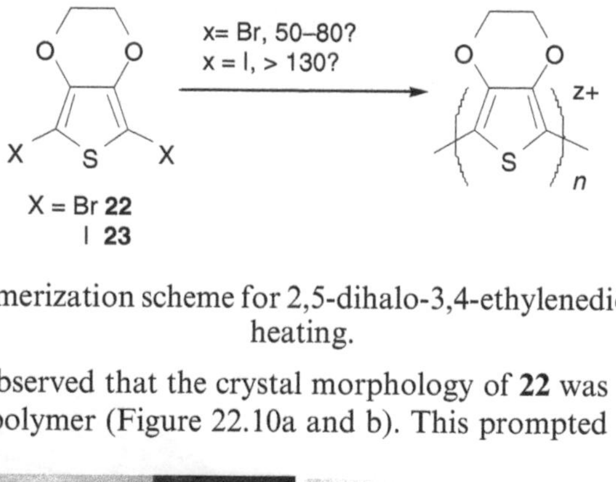

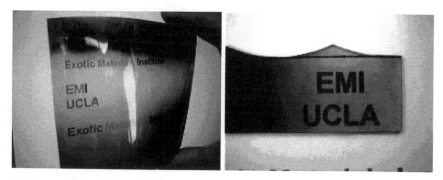

Scheme 1.   Polymerization scheme for 2,5-dihalo-3,4-ethylenedioxythiophene by heating.

Also it was observed that the crystal morphology of **22** was not changed upon conversion to the polymer (Figure 22.10a and b). This prompted them to form thin

Figure 22.9.   (**Color figure follows page 348**). Semitransparent conducting films of PEDOT on a plastic substrate (left) and a glass slide (right), prepared by in situ solid-state polymerization of vacuum-deposited **22**. (From Meng, H.; Perepichka, D. F.; Bendikov, M.; Wudl, F.; Pan, G. Z.; Yu, W.; Dong, W., and Brown, S., *J. Am. Chem. Soc.* **2003**; *125*, 15151–15162. Copyright 2003, American Chemical Society Publications. With permission.)

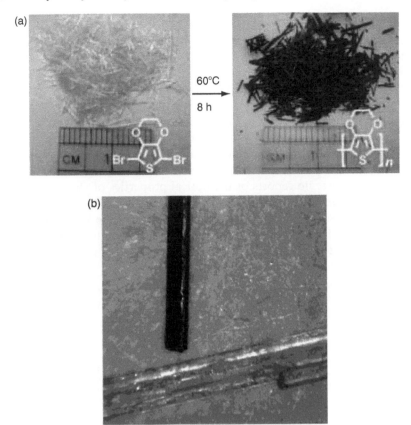

Figure 22.10. **(Color figure follows page 348).** (a) Retention of crystal morphology of **22** even after polymerization. (From Meng, H.; Perepichka, D. F., and Wudl, F., *Angew, Chem. Int. Ed.* **2003**, *42*, 658–661. Copyright 2003, Wiley-VCH Verlag GmbH & Co. KGaA, Weinheim. With permission.) (b) Optical microscopy image of PEDOT (black, top) and 22 (colorless crystal, bottom). (From Meng, H.; Perepichaka, D. F.; Bendikov, M.; Wudl, F.; Pan, G. Z.; Yu, W.; Dong, W. and Braon, S., *J. Am. Chem. Soc.* **2003**, *125*, 15151–15162. With permission.)

films of PEDOT onto glass as well as flexible substrates by subliming **22** by gentle heating and subsequent in situ polymerization of the dibromo derivative. In a recent article from the same group, they reported the attempts to polymerize the dichloro and diiodo derivatives. While polymerization of **23** occurred at relatively high temperatures (>130°C), the dichloro derivative was reported not to polymerize under these conditions.

Skabara and co-workers [35] have proved that this polymerization is not only limited to PEDOT but also could be extended to other thiophene-based systems. They have reported the extension of the solid-state polymerization technique to 3,4-alkylenedithiathiophenes to produce conjugated polymers.

Other derivatives of PEDOT involving vinylene spacers and copolymers with heteroaromatic compounds will be discussed in the later sections of this chapter.

22.2.2. Polymers Containing Electron-Withdrawing Repeat Units

One other strategy to reduce the gap between the HOMO and LUMO has been to reduce the LUMO of the polymer. Studies in this direction, although limited, have been useful in tuning the bandgap of conjugated polymers. This has been accomplished by using electron-withdrawing groups like carbonyl, cyano, or imino functionalities on the conjugated backbone. Ferraris et al. [36] first reported the use of electron-withdrawing groups in bithienyls (Figure 22.11), 4*H*-cyclopenta[2,1-*b*;3,4-*b'*]dithiophen-4-one (**24**) and its dicyanomethylene (**25**) derivative prepared by the Knoevanagel condensation of the ketone with malononitrile.

These monomers were modeled based on the strategy that reduced aromaticity leads to decreased HOMO–LUMO gap and can be persistent in the polymers. While the polymer from the ketone was found to exhibit a bandgap of $\leq 1.2\,\text{eV}$, the polymer from the dicyanomethylene derivative was found to exhibit an even lower bandgap of $0.67\,\text{eV}$. But, there are no reports of the optical properties of such polymers in the oxidized form (Figure 22.12).

Following these reports, several other derivatives of the ketone were prepared and polymerized, which include dioxolane (**26**) and its sulfur analogs (**27**), cyano (nonafluorobutylsulfonyl) (**28**), 1,3-dithiole derivatives (**29**), etc. Loganathan et al.

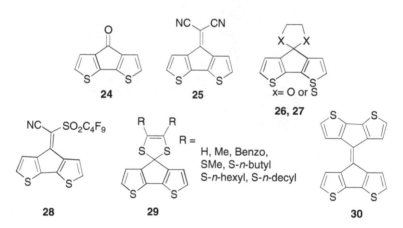

Figure 22.11.   Bithienyl-based monomers used for polymerization.

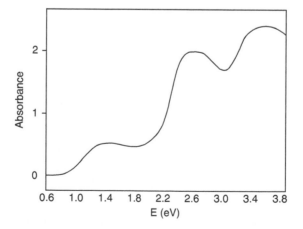

Figure 22.12.   Electronic absorption spectrum of neutral poly(**25**) coated on ITO. [From Ferraris, J. P. and Lambert, T. L., *J. Chem. Soc., Chem. Commun.* **1991**, 1268–1270. Copyright 1991, Royal Society of Chemistry, UK. With permission.]

[37] have reported a conjugated bridged monomer based on the cyclopentathiophene described earlier. He reported electrochemical polymerization of the monomer (**30**) to yield a polymer with a bandgap of 0.5 eV. The optical spectrum of the polymeric film shows an onset for π to π^* transition of 0.5 eV, with a peak absorption at around 420 nm (820 nm, shoulder). The low gap was also confirmed by electrochemistry of the polymer and further with in situ conductivity measurements. As it can be seen, coupling through all the α-positions will lead to cross-conjugation through the bridge. More recently, copolymers of these with electron-rich monomers have been synthesized to further reduce the bandgap. Such copolymers will be discussed in a later section dealing with donor–acceptor type polymers.

22.2.3. Fused Aromatics as Repeat Units

Wudl et al. [38] first prepared a conjugated polymer using isothianapthene (ITN, **31**), a fused aromatic compound. This has served as a model for several other ICPs based on the concept of fused aromatic backbone. It originated with the idea that if the polymerization proceeds through coupling of one of the rings, then the stabilization of the quinonoid form, which in most polymers is lower in energy, is possible though the aromatization of the second ring. The optical spectrum of PITN indeed indicated a lower bandgap of 1.0 eV (Figure 22.13).

This polymer was reported to be dark blue-black in the neutral form ($\lambda_{max} = 1.2$ eV) and upon oxidation was a transparent greenish-yellow in color. With the attractive low bandgap properties of PITN, several reports of different routes to PITN and processible PITN have been put forth [39]. Following the successful reduction of the bandgap by the use of fused aromatics, several other fused aromatic compounds have been investigated to make low bandgap π-conjugated polymers. Initial reports consisted of polymerization of thienothiophenes (Figure 22.14) and dithieno[3,4-*b*;3′,4′-*d*]thiophene (**34**) was the only molecule that was found to yield a low bandgap polymer (e.g., ~1.1 to 1.2 eV) [40].

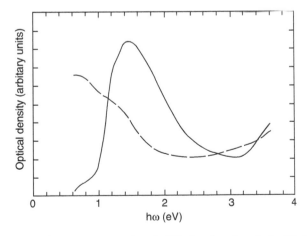

Figure 22.13. Absorption spectrum of poly(**31**) in the doped (solid line) and neutral (dashed) state. (From Kobayashi, M.; Colaneri, N.; Boysel, M.; Wudl, F., and Heeger, A. J., *J. Chem. Phys.* **1985**, *82*, 5717–5723. Copyright 1985, American Institute of Physics Publications. With permission.)

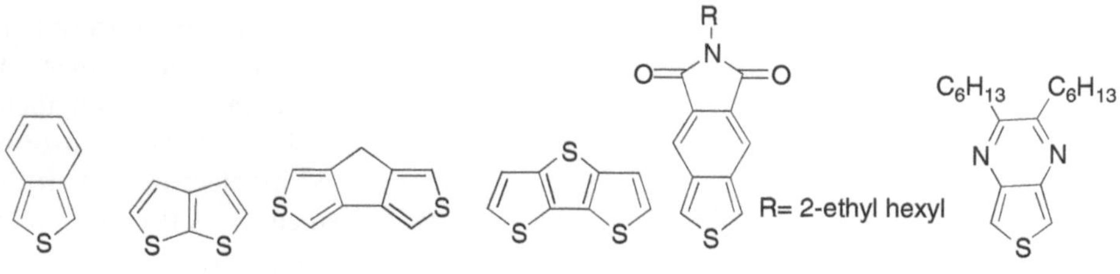

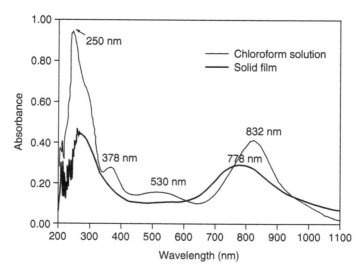

Figure 22.14. Fused thiophenes monomers used to produce low bandgap polymers.

It was claimed that the polymer from this dithienothiophene was transparent in the doped form. Recently Meng and Wudl [41] have reported the synthesis of a new monomer based on **31**. They introduced an alkyl-substituted imide ring on the benzene ring of **31** to alter the HOMO–LUMO levels of the molecule. The monomer was prepared by a seven-step synthetic scheme and the monomer was further polymerized in chloroform using ferric chloride. Poly(**35**) thus obtained was purified using acetone and THF to get rid off the low molecular weight oligomers and polymers. Finally the high molecular weight chloroform-soluble fraction was characterized by UV–vis–NIR spectroscopy. While the solution shows low-energy absorption at 832 nm (λ_{max}), the solid film is blueshifted to 778 nm (Figure 22.15). The band edge gap of this polymer was found to be 1.24 eV, comparable to the parent PITN and this polymer was reported to be n-dopable.

In 1992 Pomerantz et al. [42] reported the synthesis and polymerization of thieno[3,4-*b*]pyrazines. Thieno[3,4-*b*]pyrazine (**36**) closely resembles **31** except that the carbons next to the β-positions of the thiophene are replaced by nitrogen atoms. They also reported soluble alkyl derivatives of the polypyrazine. They had calculated that the quinonoid form of the polymer is more stable than the aromatic form and claimed that even the aromatic form of the polymer is planar. These

Figure 22.15. UV–vis–NIR spectrum of poly(**35**) on a glass slide as a solid film (bold line) and as a solution in chloroform (thin line). (From Meng, H. and Wudl, F., *Macromolecules* **2001**, *34*, 1810–1816. Copyright 2001, American Chemical Society Publications. With permission. *The solution spectrum was digitally enhanced using Adobe photoshop 6.0 for clarity.*)

prompted the synthesis of the poly(thieno[3,4-*b*]pyrazines) by chemical methods as the resultant polymers were expected to be soluble. Poly(2,3-dihexylthieno[3,4-*b*]pyrazine) was shown to exhibit a band edge gap of 1.14 eV with a λ_{max} at 875 nm in solution. The polymer film showed a λ_{max} at 287 and 915 nm with a band edge gap of 0.86–1.02 eV. They also reported that upon doping the dark blue-black polymer solution in chloroform with nitrosonium tetrafluoroborate the solution turned light yellow. Schrof et al. [43] prepared a number of copolymers using a tool kit of bis(trimethylstannyl) thiophenes and dihalo thiophenes with fused pyrazine rings. All of these polymers were reported to show an absorption edge >900 nm and also a λ_{max} > 600 nm. The different copolymers (**37–43**) made by this group are shown in Figure 22.16 with the corresponding peak absorption for each of the copolymers.

Karikomi et al. [44] have prepared several polymers from terthiophene structures with benzobis(1,2,5-thiadiazole) as the central ring. Polymers electrochemically generated from the monomers **44** and **45** exhibited very low bandgap of ≈0.5 eV (band edge) with a λ_{max} of 702 and 618 nm.

Fused thiophenes have generated a lot of synthetic and theoretical interest (Figure 22.17). Of all the thienothiophenes reported, thieno[3,4-*b*]thiophene

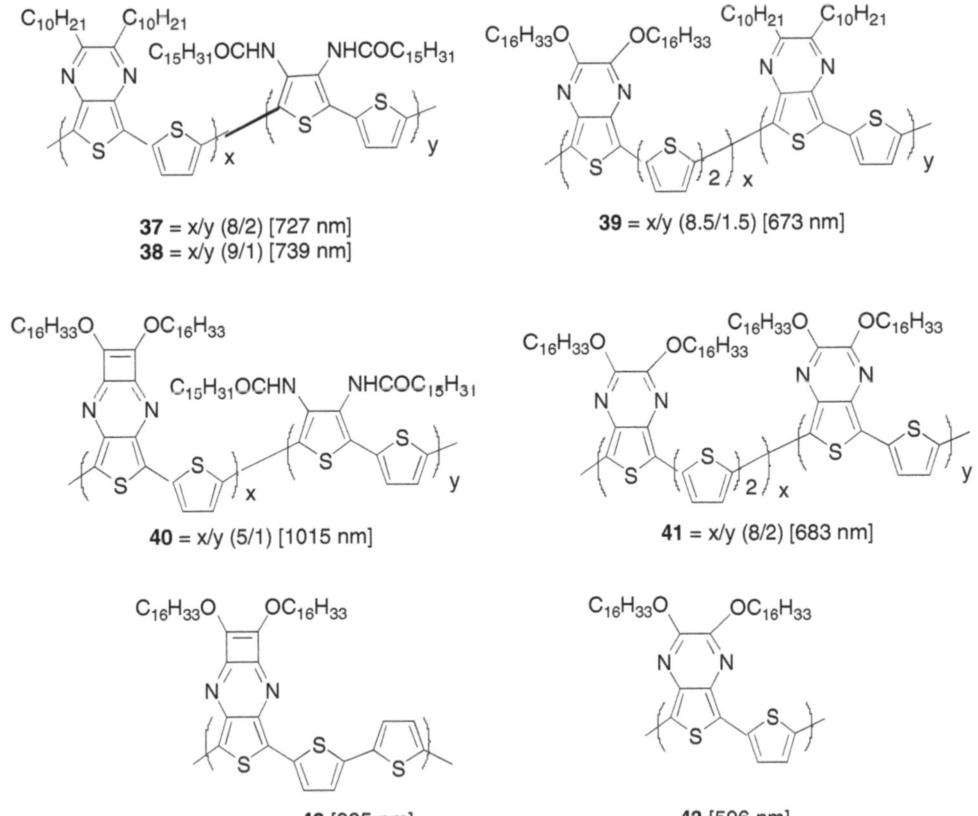

37 = x/y (8/2) [727 nm]
38 = x/y (9/1) [739 nm]

39 = x/y (8.5/1.5) [673 nm]

40 = x/y (5/1) [1015 nm]

41 = x/y (8/2) [683 nm]

42 [985 nm]

43 [596 nm]

Figure 22.16. Different copolymers prepared by Schrof et al. using thieno[3,4-*b*]pyrazine derivatives and thiophenes. The absorption maxima for each polymer are reported in square bracket.

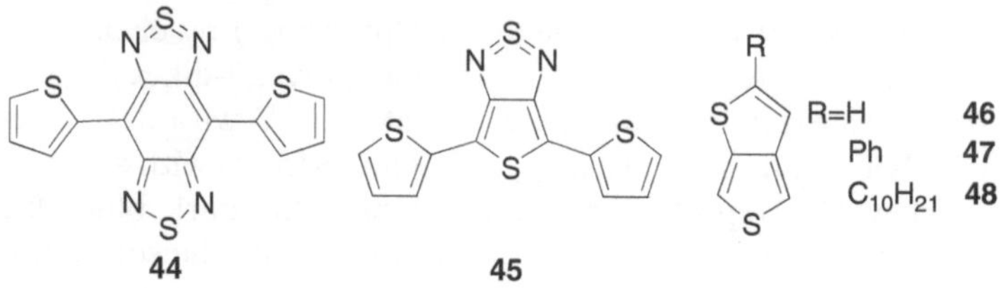

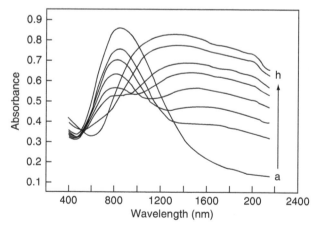

Figure 22.17. Other fused thiophenes that have yielded low bandgap polymers.

(T34bT) based polymers have shown high promise in terms of much reduced bandgaps. Wynberg and Zwanenberg [45] first synthesized T34bT (**46**) and till 1999 there was no report of any polymer from this monomer.

It must be noted that T34bT is an unsymmetrical and planar molecule, giving rise to the possibility of obtaining regioregular polymers. Neef et al. [46] first reported the electrochemically generated low band PT34bT ($E_g. = 0.85\,eV$, $\lambda_{max} = 1.30\,eV$) with a 2-phenyl-substituted monomer (**47**). Following this, Pomerantz et al. [47] reported an alkyl chain derivatized soluble PT34bT with a band edge gap of $0.98\,eV$ and λ_{max} at 925 nm. The polymer was obtained by the chemical polymerization (using ferric chloride) of 2-decyl T34bT (**48**), and hence will lead to a regioirregular structure. In the same year we had reported the synthesis of T34bT by a slightly modified route reported by Brandsma and Verkuijsse [48] and electropolymerized the unsubstituted T34bT (**46**) using tetra(*n*-butyl)ammonium perchlorate as the dopant electrolyte.

The PT34bT generated electrochemically was found to have a band edge gap of $0.85\,eV$ (Figure 22.18) and a λ_{max} of $1.46\,eV$ (846 nm). [49] We had also reported the

Figure 22.18. UV–vis–NIR spectrum of a 0.1 mm thick poly(**46**) film on ITO glass at different potentials. The film was first electrochemically reduced at −0.8 V, dipped into a 0.1 *M* TBAP/acetonitrile containing 0.2 vol.% hydrazine for full reduction, and then placed into a 0.1 *M* TBAP/acetonitrile, and the UV–vis–NIR spectrum was taken (a). Poly(T34bT) was then sequentially oxidized to (b) −0.4 V, (c) −0.3 V, (d) −0.2 V, (e) −0.1 V, (f) 0.0 V, (g) 0.15 V, and (h) 0.4 V vs. Ag/Ag+ reference electrode (0.47 V vs. NHE). (From Sotzing, G. A. and Lee, K., *Macromolecules* **2002**, *35*, 7281–7286. Copyright 2002, American Chemical Society Publications. With permission.)

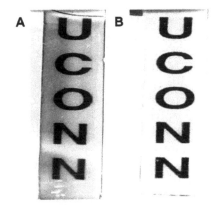

Figure 22.19. **(Color figure follows page 348).** A 0.8 μm thick poly(T34bT) film coated on an ITO-coated glass slide in the (A) reduced state at −0.8 V and in the (B) oxidized semiconducting state at 0.4 V vs. Ag/Ag⁺ reference electrode (0.47 V vs. NHE). The counter electrode was a platinum plate. (From Sotzing, G. A. and Lee, K., *Macromolecules* **2002**, *35*, 7281–7286. Copyright 2002, American Chemical Society Publications. With permission.)

high optical transparency of PT34bT in the oxidized form and Figure 22.19 shows an 800 nm thick film of PT34bT in the oxidized and reduced forms.

Poly(thieno[3,4-*b*]thiophene) was sky blue in the neutral form and transparent in the oxidized form (for a thickness of 800 nm). The regioregular PT34bT with coupling through the 4 and 6 positions is yet to be made as none of the three reported polymers can form regioregular structures. While the first two reports by Ferraris et al. [47] and Pomerantz et al. [46] ruled out the possibility of coupling through the 2 position due to blockage of these positions, the third report by Sotzing and Lee [49] does not rule out the possibility of branched or networked structures. A closer inspection at the hypothetical branched or networked structure will indicate that coupling through all the α-positions in fact will produce cross-conjugation (Scheme 2).

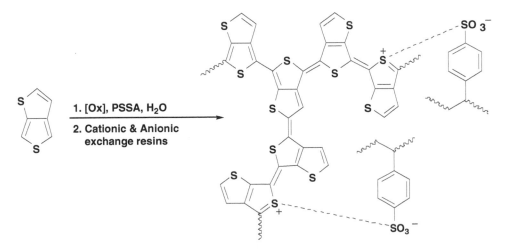

Scheme 2.   Polymerization of **46** to yield a hypothetical networked poly(**46**), showing cross-conjugation.

Moreover the α-positions are far removed from each other and should not lead to any steric hindrance that might lead to twisted conformations of the rings. Lee and Sotzing [50] reported the preparation of processible forms of T34bT much like PEDOT. These dispersions were found to be highly transmissive as they are in the oxidized forms. We also copolymerized T34bT and EDOT together electrochemically from a solution containing equimolar concentration of the two monomers [51]. The copolymer was characterized and confirmed by vis–NIR spectroscopy, which showed only one peak for the copolymers, while the spectrum of a layered structure of the two polymers showed two distinct peaks (Figure 22.20).

The band edge gap of the copolymers prepared using tetrabutylammonium perchlorate and tetrabutylammonium hexafluorophosphate was 1.06 and 1.19 eV, respectively. The peak absorption for the two neutral copolymers was 750 and 650 nm when perchlorate and hexafluorophosphate were used as the dopant ions.

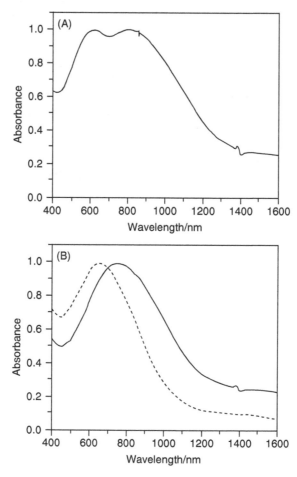

Figure 22.20. Vis–NIR spectrum of (A) neutral PEDOT deposited over PT34bT and (B) neutral copolymer containing T34bT and EDOT obtained from electrochemical polymerization in (solid line) TBAP and (---) TBAPF$_6$. (From Seshadri, V.; Wu, L., and Sotzing, G. A., *Langmuir* **2003**, *19*, 9479–9485. Copyright 2003, American Chemical Society Publications. With permission.)

Furthermore, we had calculated the copolymer composition from elemental analysis and the ion transport values obtained using electrochemical quartz crystal microbalance (EQCM). The copolymer prepared using perchlorate dopant ion was found to contain 79.9% of T34bT and 20.1% EDOT, and the copolymer obtained using hexafluorophosphate as the dopant ion consisted of 67.7% T34bT and 32.3% EDOT. The copolymer obtained using PF_6^- was also found to exhibit extraordinary stability to n-doping. This monomer therefore has enormous potential to produce polymers with even lower bandgaps by already known techniques used for bandgap reduction.

22.2.4. Arylvinylenes and Arylmethines

Many polymers exhibit higher bandgap or absorb at higher wavelength due to steric interactions of the adjacent rings. These interactions could arise from the substituents on the rings or from the ring itself. Reducing the twist between rings by spatially separating the rings with less sterically hindered groups has been utilized deftly to lower the bandgap of conjugated polymers. A second advantage of vinylene–ethylenic bond is the restricted rotation due to the defined configuration of the double bond, i.e., *cis* or *trans*. This was first utilized to make poly(phenylenevinylene) (PPV) [52] and later extended to other heteroaromatic polymers for lowering the bandgap of the polymers by reducing the steric interactions. Thus PThV (**49**), a thiophene-based system, shows a bandgap of 1.7–1.8 eV about 0.4 eV lesser than the parent PTh [53]. In an attempt to further reduce the bandgap of poly(thiophenevinylene), substituted thiophenes such as 3,4-dimethoxy (**50**) and the 3,4-dibutoxy derivatives (**51**) have been synthesized. Improved thermal elimination routes to poly(**51**) gave rise to further redshift in the spectrum of the neutral polymer (Figure 22.21) [54].

The polymer was estimated to have a bandgap of 1.2 eV with a λ_{max} at 700 nm. Kim et al. [55] have reported a convenient synthesis of polyheteroarylvinylene based on pyrrole (**52–54**). They prepared the 1-alkyl-2,5-pyrrolevinylene using Mannich

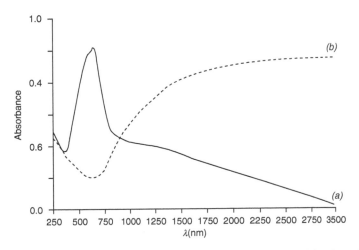

Figure 22.21. Electronic absorption spectra of poly(3,4-dibutoxythienylenevinylene) on ITO: (a) neutral and (b) doped. (From Jen, K-Y.; Maxfield, M. R.; Shacklette, L. W., and Elsenbaumer, R. L., *J. Chem. Soc., Chem. Commun.* **1987**, 309–311. Copyright 1995, Royal Society of Chemistry, UK. With permission.)

reaction. The synthetic procedure adopted by them is shown in Scheme 3. They report an optical bandgap of 1.65 eV for the onset with a λ_{max} at 563 nm for the N-dodecyl derivative.

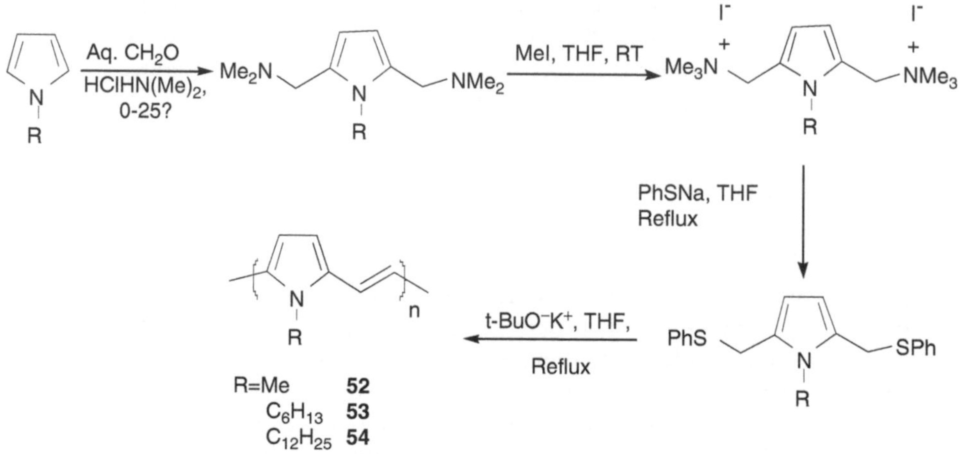

Scheme 3. Synthetic scheme for the preparation of poly(pyrrolevinylene).

Arylethylenes is an excellent model for arylvinylenes and offer the advantage of being electrochemically polymerizable due to the presence of terminal electroactive functionalities (Figure 22.22). The first to be tried in this series were the thiophene-based polymers (**55**), which under optimized electrochemical conditions produced polymeric films with a bandgap of 1.8 eV [56]. Later, alkyl-substituted thiopheny-lethylenes (**55–58**) were prepared by Stille coupling of 2-iodo-3-alkyl thiophene and bis(tributyl stannyl)ethylene, leading to a rigidified structure with a bandgap of 1.7 eV. This idea of olefinic spacer with terminal electroactive unit was extended to EDOT as well and the resultant monomer (**59**) was electrochemically polymerized to give a polymer with a band edge gap of 1.4 eV [57].

The heteroarylidenemethines form a special class of nonclassical π-conjugated polymers. Based on the idea that the quinonoid form of the conjugated polymer has a lower bandgap than the aromatic form, some parts of the π-backbone is forced to be in the quinonoid geometry. Jenekhe et al. [58] first proposed these methine-containing structures and synthesized a thiophene with methine linkage. This polymer was

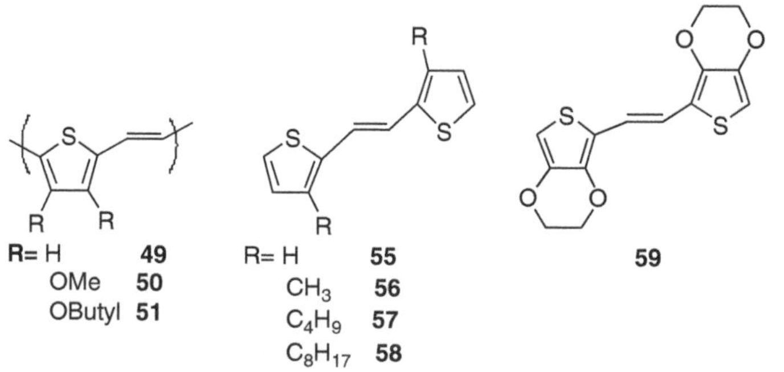

Figure 22.22. Some poly(thiophenevinylenes) and poly(thiophene ethylenes).

claimed to have a low bandgap of 0.75 eV. Following this, several low bandgap polymers with similar architectures were prepared and characterized.

Figure 22.23 shows some of the low bandgap heteroarylidenemethine structures that have been synthesized and studied. ITN has been shown to be a very low bandgap polymer, and Kurti et al. [59] calculated that heteroarylmethines containing ITN would lead to further reduction in the bandgap of the parent PITN. According to his calculations, poly(bis-isothianaphthene–methine) (**PBITNM-62**) will have a bandgap of 0.7 eV and the convergence was observed for just 11 units. Following the procedure of Okuda et al. [60], the first PBITN [61] with substituted methine was synthesized by condensation of 1,3-bis(*t*-butyldimethylsilyl)isothianaphthene with 4-hexyloxy benzaldehyde in the presence of POCl$_3$. This polymer was found to exhibit a low bandgap of 1.2 eV and was found to have a very low oxidation potential of 375 mV against silver wire. The utility of POCl$_3$ towards the polycondensation was first carried out by Goto et al. [62] to make soluble heteroarylmethines of thiophenes and pyrroles. This procedure was found to give better yields of the polymer and also devoid of a separate dehydrogenation step involved in the synthesis by Chen et al. [63]. Recently Kiebooms et al. [64] prepared a series of side-chain liquid crystalline polymers with a π-conjugated backbone based on PBITNM (**63–67**). Several polymers were synthesized by varying the liquid crystalline side-chains and all of the polymers exhibited a band edge gap of $\leq$1.3 eV. Upon doping with iodine, the π to π* transition of these polymers was redshifted into the near-infrared region as shown in Figure 22.24.

Along with the redshift of the long wavelength absorption, a significant absorption increase in the region of 300–400 nm was observed, which has been attributed to Cotton effect, due to the presence of optically active side-groups. More recently,

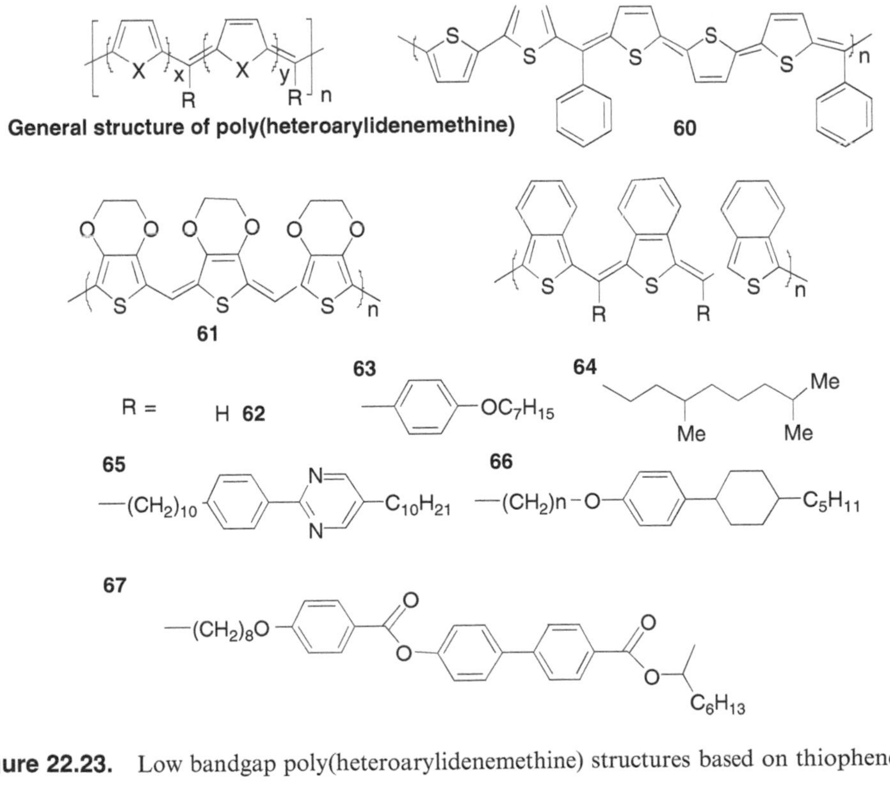

Figure 22.23. Low bandgap poly(heteroarylidenemethine) structures based on thiophenes.

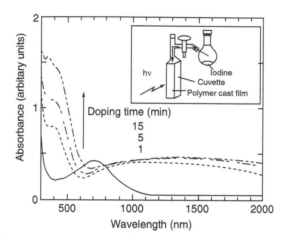

Figure 22.24. Vis–NIR spectrum of **65** during in situ iodine doping. (From Kiebooms, R. H. L.; Goto, H., and Akagi, K., *Macromolecules* **2001**, *34*, 7989–7998. Copyright 2001, American Chemical Society Publications. With permission.)

Zotti and co-workers [65] have reported electrochemical polymerization of bis(3,4-ethylenedioxy thiophene)methane. The polymerization proceeds as a result of removal of at least one of the methine hydrogens. They claim that the polymer has a very low bandgap of 1.0 eV based on the onset of oxidation and reduction waves in the cyclo-voltammogram, but the absorption edge for the neutral polymer is observed around 800–900 nm with a peak at ca. 500 nm (Figure 22.25).

The polymer containing methine was reported to be transmissive blue in the oxidized form and gray in the neutral form. Thus it can be seen that this nonclassical method of making polyarylenemethine could indeed by a useful tool in tuning bandgap of polymers.

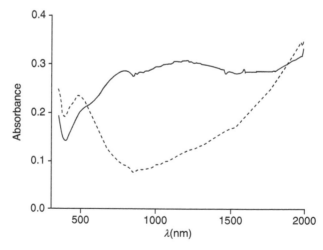

Figure 22.25. Vis–NIR spectrum of doped and neutral forms of nonclassical conjugated polymer based on EDOT. (From Benincori, T.; Rizzo, S.; Sannicolo, F.; Schiavon, G.; Zecchin, G.; Zecchin, S., and Zotti, G., *Macromolecules* **2003**, *36*, 5114–5118. Copyright 2003, American Chemical Society Publications. With permission.)

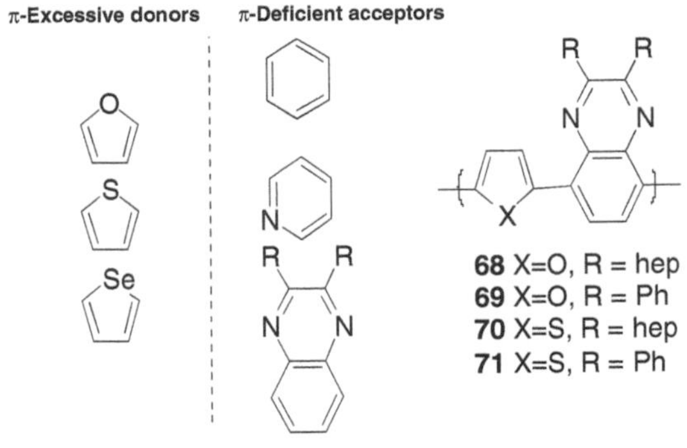

π-Excessive donors   π-Deficient acceptors

Figure 22.26. π-Excessive and π-deficient monomers (left side) used by Yamamoto et al. and some low bandgap polymers generated using them (right side).

22.2.5. Donor–Acceptor Type Polymers

The donor–acceptor strategy for making low bandgap polymers has been utilized by several groups in modeling new monomers as well as synthesizing and characterizing new monomers and their polymers. Both regular alternating donor–acceptor type polymers as well as random polymeric structures obtained by mixed monomer polymerization have been reported. Several research groups have reported π-conjugated polymers with internal charge transfer to yield low bandgap polymers. In the following section, we will discuss some of these polymers, which have a $\lambda_{max} \geq 600$ nm. Yamamoto et al. [66] had investigated several polymers synthesized chemically, consisting of alternating structures containing a π-excessive unit and a π-deficient unit (Figure 22.26).

The electron-rich units consisted of thiophenes, furans, selenophenes, among others, while the electron-deficient units used for their study were pyridine and 2,3-disubsituted quioxlines (thieno[3,4-b]pyrazines). These alternating donor–acceptor type polymers were prepared using nickel or palladium catalyzed coupling of aryl dihalides and distannyl aryl compounds. Of all the polymers synthesized, alternating structures comprised of thiophene or furan as the electron-rich unit and diphenyl-substituted quinoxaline (**68–71**) as the electron-deficient unit exhibited λ_{max} at higher wavelengths, 603 and 692 nm (as thin films). But there is no comment on the transparency of the oxidiation state of these polymers.

This strong intramolecular charge transfer resulting in donor–acceptor type architectures was also shown to be effective in reducing the bandgaps of polythiophenes. Tour and co-workers [67] synthesized several BOC-protected 3,4-diamino-thiophene, and polymerized with 3,4-dinitro-2,5-dihalo-thiophene intitially by Stille coupling reaction to produce polymers of alternating donor–acceptor structures (Scheme 4).

BOC protection was removed after the polymer formation to yield unprotected amino functionalities to the donating unit. They also reported other polymers with groups capable of internal charge transfer using alkoxy substitutions for the donors and carbonyl or imide functionality on the aromatic backbone for the acceptors. Only **72** and **73** showed a huge redshift of the spectrum. This polymer was found to be soluble in polar solvents like MeOH, owing to its zwitterionic character. The optical spectrum of the polymer in MeOH exhibited a λ_{max} at 662 nm with an absorption edge at ca. 900 nm and thin films exhibited a λ_{max} at 768 nm with an absorption edge at

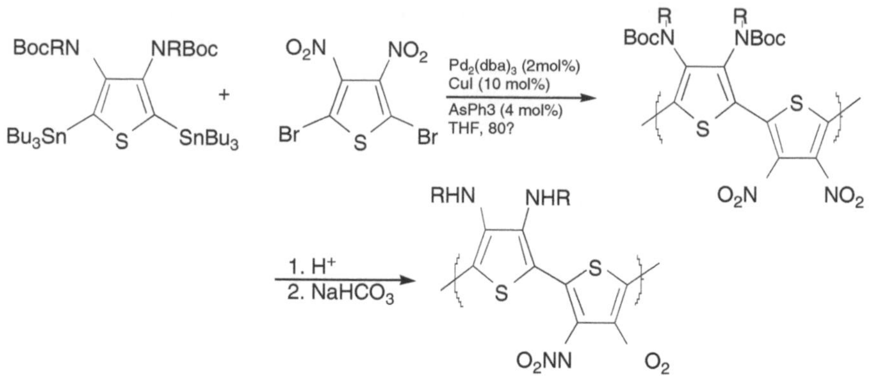

Scheme 4.   Synthesis of donor–acceptor type zwitterionic poly(thiophenes) prepared by Tour and co-workers.

about 1000–1100 nm. Recently Meng et al. [68] reported the synthesis of an alternating copolymer consisting of EDOT (2) (with relatively high HOMO level) and ben-zo[c]thiophene-N-2′-ethylhexyl-4,5-dicarboxylic imide (EHI-ITN, **35**) (with relatively low LUMO level) using Stille coupling from the di(tributyl stannyl) derivative of EDOT and the dihalide of EHI-ITN. The neutral polymer was found to be blue in color with a λ_{max} at 807 nm and an absorption edge around 1240 nm (1.0 eV). The polymer was reported to be highly soluble due to the alkyl chains attached to the imide ring and could be easily spun-coated to form thin films from common organic solvents.

   In an earlier section, we had discussed the effect of vinylene spacers between aromatic rings. An extension of this is placing electron-withdrawing groups like cyano on one of the olefinic carbon to give a push–pull or donor–acceptor system. It was first shown that cyano-substituted poly(phenylenevinylene) [69] had a higher electron affinity due to the electron-withdrawing effects of the cyano substitution. This push–pull type conjugation could be utilized to stabilize the quinonoid form of the conjugated molecule and thereby reduce the bandgap. Monomeric structures incorporating these cyanovinylenic spacers can be easily synthesized by Knoevanagel condensation of an aromatic aldehyde with an aryl acetonitrile in the presence of a base (Scheme 5).

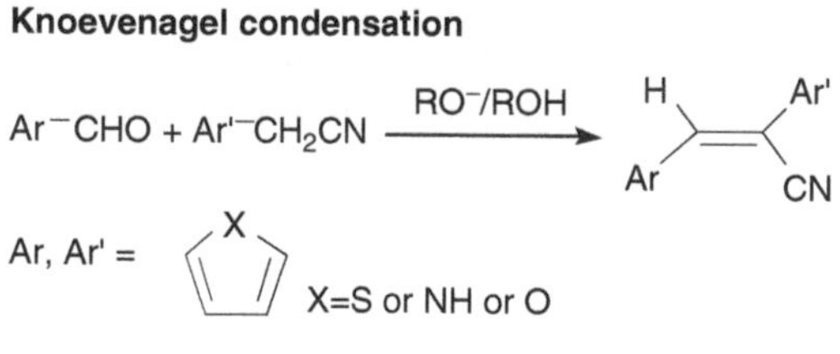

Knoevenagel condensation

Scheme 5.   General procedure for Knoevenagel condensation reaction.

   Several such structures with push–pull architecture have been designed and some of the monomers, which were polymerized to yield low bandgap polymers, are shown in Figure 22.27.

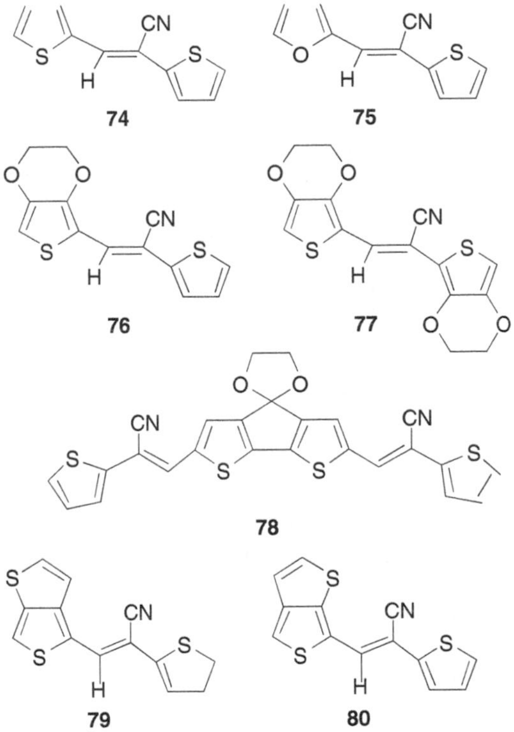

Figure 22.27. π-Conjugated polymer building blocks containing cyanovinylene spacers.

Poly(**75**) [70] has been reported to show greater stability compared to the unsusbstituted poly(bisfurylvinylene). The presence of thiophene also helps in reducing the overoxidation of the polymer, thereby giving rise to a stable polymer. Poly(**74**) was found to have a very low band edge gap of 0.5 eV [70]. Sotzing et al. [71] reported the cyanovinylenes prepared using EDOT aldehyde and the 2-acetonitrile of EDOT and thiophene. The in situ spectroelectrochemistry of electrochemically generated polymers from Th–CNV–EDOT and BEDOTCNV (**76** and **77**) showed a band edge gap of 1.1 and 1.3 eV, respectively. The in situ spectroelectrochemistry of the two polymers were reported to be very similar and the λ_{max} of Th–CNV–EDOT was reported to be 1.8 eV. Figure 22.28 shows the in situ spectroelectrochemistry of Th–CNV–EDOT in 0.1 M TBAP/ACN.

Compound **78**, although has an extended π-conjugation in the monomeric state, was successfully electropolymerized to yield a low bandgap polymer ($E_g = 1.5$ eV) [72]. Recently our group has made two cyanovinylenes based on **46** by the Knoevenagel condensation of the 4- and 6-formyl **46** with 2-thiopheneacetonitrile. The lowest energy absorption for the two monomers (**79** and **80**) was 397 and 387 nm. Electrochemical polymerization of the two monomers was carried out to yield two low bandgap polymers and in situ spectroelectrochemistry results show that the E_g is about ca. 1.2 eV for the onset for both the polymers and the peak absorption is 1.72 and 1.85 eV for poly(**79**) and poly(**80**), respectively (V. Seshadri et al., personal communication) [73].

Gallazzi et al. [74] had used this donor–acceptor alternance strategy incorporating the cyanovinylene as a side-chain rather than in the main chain. Figure 22.29

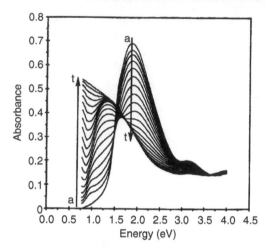

Figure 22.28. Optoelectrochemical analysis of poly(Th–CNV–EDOT) in 0.1 *M* TBAP/ACN at applied potentials of (a) −0.6 V, (b) 0.10 V, (c) 0.15 V, (d) 0.20 V, (e) 0.25 V, (f) 0.30 V, (g) 0.35 V, (h) 0.40 V, (i) 0.45 V, (j) 0.50 V, (k) 0.55 V, (l) 0.60 V, (m) 0.65 V, (n) 0.70 V, (o) 0.75 V, (p) 0.80 V, (q) 0.85 V, (r) 0.90 V, (s) 0.95 V, and (t) 1.00 V vs. Ag/Ag⁺. (From Sotzing, G. A.; Thomas, C. A., and Reynolds, J. R., Low bandgap cyanovinylene polymers based on ethylenedioxythiophene, *Macromolecules* **1998**, *31*, 3750–3752. Copyright 1998, American Chemical Society Publications. With permission.)

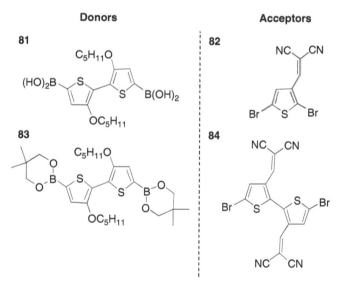

Figure 22.29. Donor and acceptor monomer kit utilized to produce low bandgap polymers by Gallazzi et al.

shows the monomer-kit used by them to produce these alternating donor–acceptor type polymers.

Suzuki coupling of the dihalide and the diboronic acid in ethanol or THF gave the polymers, which was further purified by Soxhlet extraction with methanol, ethyl ether, and chloroform, in that order. The optical studies reported were for the chloroform extract. With these monomers, they could obtain polymers with a band edge gap as low as 1.2 eV and $\lambda_{max} > 600$ nm. Interestingly, the donor monomers,

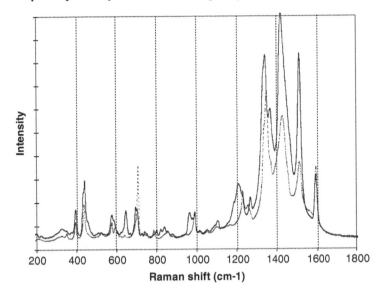

Figure 22.30. Raman spectrum of a copolymer film prepared at 1.34 V (solid, A from Figure 22.1) and a simulated spectrum of a blend of poly-**1** and poly-**2** (dashed, average of the poly(**2**) and poly(**25**) Raman spectra. [From Huang, Hs. and Pickup, P. G., *Chem. Mater.* **1998**, *10*, 2212–2216. Copyright 1998, American Chemical Society Publications.]

though head-to-head coupled, were found to be planar (**81** and **83**), while the bithiophene acceptor (**84**) was highly twisted (x-ray crystallography reveals a *cis* conformation for the dimer with torsional angle of 60°). Hence copolymers containing (**84**) were found to exhibit higher bandgap.

In another work by Huang and Pickup [75], donor–acceptor copolymers have been obtained using electrochemical copolymerization of an acceptor and a donor monomer. The donor used was EDOT while the acceptor was 4-dicyanomethylene-4*H*-cyclopenta[2,1-*b*;3,4-*b'*]dithiophene (**25**). The oxidation potential of both the monomers are very close, 1.43 and 1.44 V vs. SCE for EDOT and dithiophene, respectively. The formation of copolymers at various potentials was confirmed by Raman spectra (Figure 22.30).

Figure 22.30 shows the differences in the spectra of a copolymer and that of a simulated spectrum of a blend of the homopolymers as reported by the authors. Furthermore, the copolymer compositions were calculated from the cyclovoltammo-grams of the polymer in acetonitrile containing 0.1 *M* tetrabutylammonium hexa-fluorophosphate.

In a recent work by Reynolds et al. [76], an alternating donor–acceptor type copolymer was synthesized using EDOT, with a high HOMO level, as donor and a silole with a low-lying LUMO as the acceptor. The polymers were electrochemically generated from the terminal polymerizable bis-EDOT silole derivative (**86**) synthesized by the Stille coupling of dihalo silole (**85**) and the trimethyl stannyl EDOT (Scheme 6).

The polymer was reported to be blue in color in the neutral form and transmissive yellow-green in the oxidized form. The band edge was observed at 1.3–1.4 eV with a broad absorption between 1.6 and 1.8 eV. Figure 22.31 shows the spectro-electrochemistry of the above polymer and as it can be observed that in the oxidized form (at 0.17 V vs. Fc/Fc⁺), the polymer exhibits a broad transition starting at 830 nm and extending into the mid-infrared region.

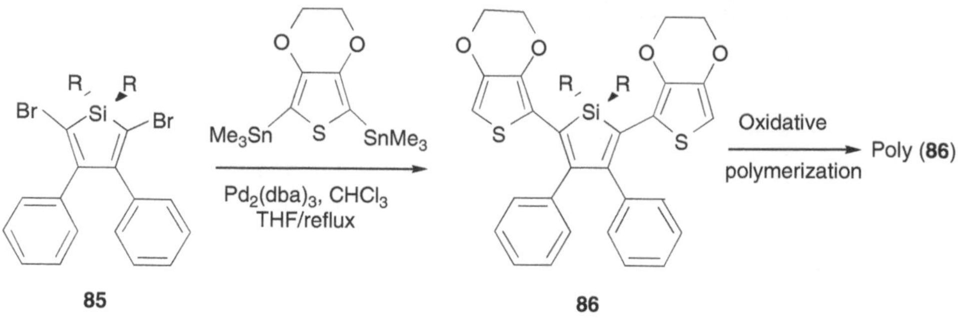

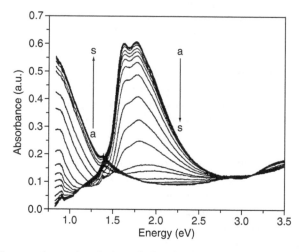

Scheme 6. Synthesis of alternating donor–acceptor type polymers consisting of EDOT (donor) and a silole (acceptor).

Figure 22.31. Spectroelectrochemical analysis of poly(BEDOT–silole) in 0.1 M LiClO$_4$/PC at applied potentials of (a) to (s): –0.43 V to 0.17 V vs. Fc/Fc$^+$ in 20 mV potentials steps. (From Lee, Y.; Sadki, S.; Tsuie, B.; Reynolds, J. R., *Chem. Mater.* **2001**, *13*, 2234–2236. Copyright 2001, American Chemical Society Publications. With permission.)

Squarine-based polymers deserve a special mention in this category of donor–acceptor type polymers. Squarine-based small molecular dyes are known to exhibit intense colors from the visible region into the near-infrared region. This is due to the strong zwitterionic character of these molecules consisting of a donor molecule (like pyrrole, benzothiazole, phenol, azulene, *N,N*-dialkyl aniline, etc.) and the acceptor, 3,4-dihydroxy-3-cyclobutene-1,2-dione (squaric acid). The extension of these to polymers based on squarines to yield low bandgap material had been reported as early as 1965 by Triebs and Jacobs [77], but the intractability had prevented further characterization. Havinga et al. [78] have studied the synthesis and properties of several polysquarines and polycroconaines (a higher homolog of squaric acid). The polysquarines can be classified as low bandgap materials since the bandgap ranges anywhere from 0.5 to 1.0 eV. Ajayaghosh [20] recently reviewed these polysquarine-based donor–acceptor type polymers. He also had reported the synthesis of low bandgap polymers based on squarine acceptors with π-extended donors made up of 1,4-dialkoxy–divinylene bridged bispyrroles [79]. Several polymers with different alkyl chains were prepared and the general synthetic scheme for the preparation of

R	R'	
CH_3	C_4H_9	**87**
CH_3	C_8H_{17}	**88**
CH_3	$C_{12}H_{25}$	**89**
$C_{12}H_{25}$	CH_3	**90**
$C_{12}H_{25}$	C_4H_9	**91**
$C_{12}H_{25}$	C_8H_{17}	**92**
$C_{12}H_{25}$	$C_{12}H_{25}$	**93**

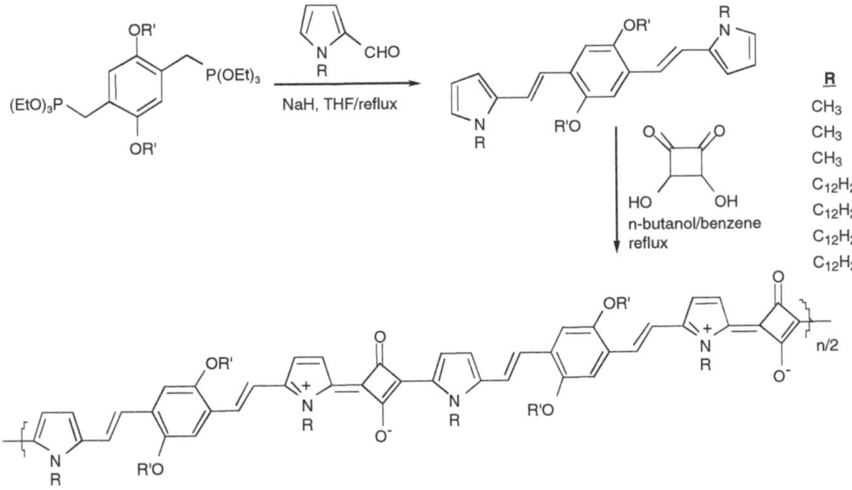

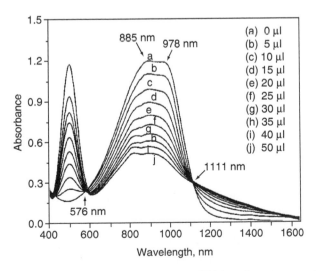

Scheme 7. Synthesis of soluble polysquarine based on a π-extended conjugation system by chemical means.

these polymers is shown in Scheme 7. These polymers were characterized as very low bandgap polymers, E_g of 0.79 to 1.02 eV, for the onset in the solid state with peak absorption at 860–910 nm. The polymers prepared were reported to have a metallic luster and exhibited conductivity as high as 10^{-2} S cm^{-1} for the doped state, while the conductivities in the undoped state were measured to be as low as 10^{-4} to 10^{-7} S cm^{-1}. There is no comment on the optical transparency in the oxidized state, though changes in the absorption as a function of iodine doping of the polymer solution has been reported.

The absorption spectrum (Figure 22.32) indicates a reduction of the peak absorption (at 885 nm) as a result of oxidation with a simultaneous increase in

Figure 22.32. Changes in the absorption spectrum of **91** in toluene upon addition of iodine solution (0.07 *M* in toluene). (From Eldo, J. and Ajayghosh, A., *Chem. Mater.* **2002**, *14*, 410–418. Copyright 2002, American Chemical Society Publications. With permission.)

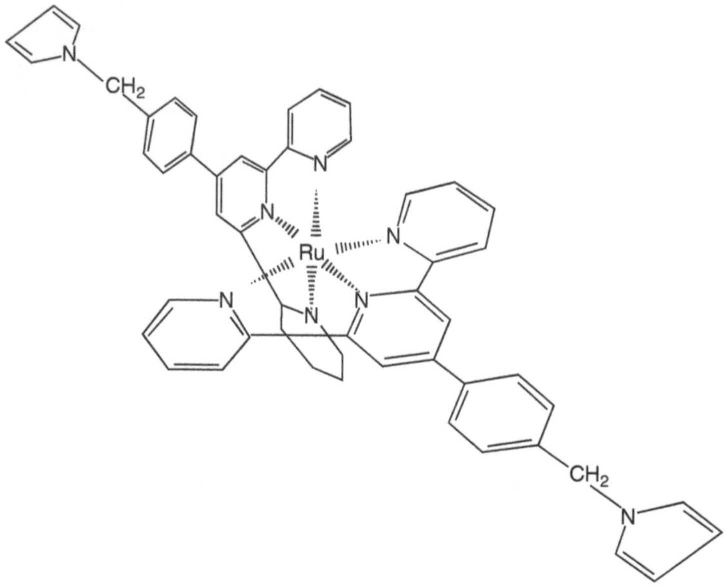

Figure 22.33. Schematic representation of [Ru(ttpy-pyrrole)2]$^{2+}$. (From Jiang, B.; Yang, S. W.; Bailey, S. L.; Hermans, L. G.; Niver, R. A.; Bolcar, M. A., and Jones, Jr., W. E., *Coord. Chem. Rev.* **1998**, *171*, 365–386. Copyright 1998, Elsevier Science S. A. With permission.)

absorption at 500 nm, which the authors attribute to residual iodine. With these results, it can be seen that polysquarine-based conjugated polymers could be a useful tool to modulate the bandgap of conjugated polymers.

We would like to briefly mention that π-conjugated donor–acceptor type polymers have also been prepared using porphyrin systems incorporated into the main chain conjugation by using ruthenium complexes containing terminal electropolymerizable pyrrole. In a review by Jiang et al. [80] on transparent conducting wires using transition metal derivatized conducting polymers, there is a communication on electrochemically prepared transparent conducting copolymers from [Ru(ttpy-pyrrole)$_2$](PF$_6$)$_2$ (**94**) (Figure 22.33) and isothianaphthene (**31**).

This polymer was found to be blue in the neutral form and transparent in the oxidized form. The electrosynthesis utilized 10:1 molar excess of ITN in acetonitrile containing 0.1 *M* TBAPF$_6$. There is no report on the polymer composition, or any other characterization of the polymer.

22.3. CONCLUSIONS

Over the past decade, tremendous strides have been made in the development of conjugated and conductive polymers having high optical transparency. Without doubt, with all of the tremendously brilliant scientists working in this area, there will be further advances within the next 10 years that will place the conductivity of these optically transparent conductive polymers equivalent to or beyond that of ITO. As a prime example of the tremendous development over the past 10 years, PEDOT has advanced in conductivity from approximately 10^{-1} S cm^{-1} in the early 1990s for processible PEDOT–PSS to approximately 100 S cm^{-1} in the year 2002, thus demonstrating a three orders of magnitude increase in conductivity. Is it to be believed that this is the limit in light of this extraordinary development? Surely, within the next 10 years, advancement will be made with processible transparent conductive polymers, whether it be through the development of different chemical repeat units or through different formulations, that will place the conductivities of processible optically transparent conductive polymers at higher than 1000 S cm^{-1}. Most definitely, this is a very exciting time to be involved in the area of intrinsically conductive polymers, whether it is from an academic or industrial perspective.

REFERENCES

1. (a) Ito, T.; Shirakawa, H., and Ikeda, S., Simultaneous polymerization and formation of polyacetylene film on the surface of a concentrated soluble Ziegler-type catalyst solution, *J. Polym. Sci. Chem. Ed.* **1974**, *12*, 11–20; (b) Chiang, C. K.; Park, Y. W.; Heeger, A. J.; Shirakawa, H.; Louis, E. J., and MacDiarmid, A. G., Electrical conductivity in doped polyacetylene, *Phys. Rev. Lett.* **1977**, *39*, 1098–1101.

2. Diaz, A. F. and Logan, J. A., Electroactive polyaniline films, *J. Electroanal. Chem.* **1980**, *111*, 111–114.

3. Grimsdale, A. C. and Mullen, K., 1-, 2-, and 3-Dimensional polyphenylenes — from molecular wires to functionalised nanoparticles, *The Chemical Record* **2001**, *1*, 243–257.

4. (a) Tourillon, G. and Garnier, F., New electrochemically generated organic conducting polymers, *J. Electroanal. Chem.* **1982**, *135*, 173–178; (b) Kobayashi, M.; Chen, J.; Chung, T. C.; Moraes, F.; Heeger, A. J., and Wudl, F., Synthesis and properties of chemically coupled poly(thiophene), *Synth. Met.* **1984**, *9*, 77–86.

5. (a) Dall'Olio, A.; Dascola, G.; Varacco, V., and Bocchi, V., Electron paramagnetic resonance and conductivity of an electrolytic oxypyrrole [(pyrrole polymer)] black, *C. R. S. Acad. Sci. Ser, C,* **1968**, *267*, 433–435; (b) Gardini, G. P., Oxidation of mono-cyclic pyrroles, *Adv. Heterocyc. Chem.* **1973**, *15*, 67–98.

6. Novak, P.; Muller, K.; Santhanam, K. S. V., and Haas, O., Electrochemically active polymers for rechargeable batteries, *Chem. Rev.* **1997**, *97*, 207–281.

7. Collins, G. E. and Buckley, L. J., Conductive polymer-coated fabrics for chemical sensing, *Synth. Met.* **1996**, *78*, 93–101.

8. Burroughes, J. H.; Bradley, D. D. C.; Brown, A. R.; Marks, R. N.; Mackay, K.; Friend, R. H.; Burn, P. L., and Holmes, A. B., Light-emitting diodes based on conjugated polymers, *Nature* **1990**, *347*, 539–541.

9. Marks, R. N.; Halls, J. J. M.; Bradley, D. D. C.; Friend, R. H., and Holmes, A. B., The photovoltaic response in poly(*p*-phenylene vinylene) thin-film devices, *J. Phys.: Condens. Matter* **1994**, *6*, 1379–1394.

10. Rudge, A.; Davey, J.; Raistrick, I.; Gottesfield, S., and Ferraris, J. P., Conducting polymers as active materials in electrochemical capacitors, *J. Power Sources* **1994**, *47*, 89–107.

11. Donhauser, Z. J.; Mantooth, B. A.; Kelly, K. F.; Bumm, L. A.; Monnell, J. D.; Stapleton, J. J.; Price, D. W., Jr.; Rawlerr, A. M.; Allara, D. L.; Tour, J. M., and Wiess, P. S., Conductance switching in single molecules through conformational changes, *Science* **2001**, *292*, 2303–2307.

12. Zhang, F.; Johansson, M.; Andersson, M. R.; Hummelen, J. C., and Inganäs O., Polymer photovoltaic cells with conducting polymer anodes, *Adv. Mater.* **2002**, *14*, 662–665.

13. Argun, A. A.; Cirpan, A., and Reynolds, J. R., The first truly all-polymer electrochromic devices, *Adv. Mater.* **2003**, *15*, 1338–1341.

14. Lee, C. S.; Kim, J. Y.; Lee, D. E.; Joo, J.; Wagh, B. G.; Han, S.; Beag, Y. W., and Koh, S. K., Flexible and transparent organic film speaker by using highly conducting PEDOT/PSS as electrode, *Synth. Met.* **2003**, *139*, 457–461.

15. Wormser, P. and Gaudiana, R., Polymer photovoltaics — challenges and opportunities, *NCPV and Solar Prog. Rev. Meeting Proc.* **2003**, 307–310.

16. Roncali, J., Synthetic principles for bandgap control in linear π-conjugated systems, *Chem. Rev.* **1997**, *97*, 173–205.

17. Patil, A. O.; Heeger, A. J., and Wudl, F., Optical properties of conducting polymers, *Chem. Rev.* **1988**, *88*, 183–200.

18. Yamamoto, T. and Hayashida, N., π-Conjugated polymers bearing electronic and optical functionalities. Preparation, properties and their applications, *React. Func. Polym.* **1998**, *37*, 1–17.

19. Scherf, U. and Mullen, K., Design and synthesis of extended π-systems: monomers, oligomers, polymers, *Synthesis* **1992**, 23–38.

20. Ajayaghosh, A. donor–acceptor type low band gap polymers: polysquaraines and related systems, *Chem. Soc. Rev.* **2003**, *32*, 181–191.

21. Jonas, F. and Schrader, L., Conductive modifications of polymers with polypyrroles and polythiophenes, *Synth. Met.* **1991**, *41*, 831–836.

22. Dietrich, M.; Heinze, J.; Heywang, G., and Jonas, F., Electrochemical and spectroscopic characterization of polyalkylenedioxythiophenes, *J. Electroanal. Chem.* **1994**, *369*, 87–92.

23. (a) Kumar, A. and Reynolds, J. R., Soluble alkyl-substituted poly(ethylenedioxythiophenes) as electrochromic materials, *Macromolecules* **1996**, *29*, 7629–7630; (b) Kumar, A.; Welsh, D. M.; Morvant, M. C.; Piroux, F.; Abboud, K. A., and Reynolds, J. R., Conducting poly(3,4-alkylenedioxythiophene) derivatives as fast electrochromics with high-contrast ratios, *Chem. Mater.* **1998,** *10*, 896–902; (c) Zong, K.; Madrigal, L.; Groenendaal, L. B., and Reynolds, J. R., 3, 4-Alkylenedioxy ring formation via double efficient route for the synthesis of 3,4-ethylenedioxythiophene (EDOT) and 3,4-propylenedioxythiophene (Pro-DOT) derivatives as monomers for electron-rich conducting polymers, *J. Chem. Soc., Chem. Commun.* **2002,** 2498–2499; (d) Welsh, D. M.; Kumar, A.; Meijer E. W., and Reynolds, J. R., Enhanced contrast ratios and rapid switching in electrochromics based on poly(3,4-propylenedioxythiophene) derivatives, *Adv. Mater.*, **1999**, *11*, 1379–1382.

24. Ahonen, H. J.; Kankare, J.; Lukkari, J., and Pasanen, P., Electrochemical synthesis and spectroscopic study of poly(3,4-methylenedioxythiophene), *Synth. Met.* **1997,** *84*, 215–216.

25. Czardybon, A.; Domagala, W., and Lapkowski, M., Synthesis and electropolymerisation of 3,4-alkylenedioxythiophenes, *Synth. Met.* **2003**, *135–136*, 27–28.

26. (a) Krishnamoorthy, K.; Ambade, A. V.; Kanungo, M.; Contractor, A. Q., and Kumar, A., Rational design of an electrochromic polymer with high contrast in the visible region: dibenzyl substituted poly(3,4-propylenedioxythiophene), *J. Mater. Chem.* **2001**, *11*, 2909–2911; (b) Krishnamoorthy, K.; Ambade, A. V.; Mishra, S. P.; Kanungo, M.; Contractor, A. Q., and Kumar, A., Dendronized electrochromic polymer based on poly(3,4-ethylenedioxythiophene), *Polymer* **2002**, *43*, 6465–6470.

27. (a) Groenendaal, L. B.; Jonas, F.; Frietag, D.; Pielartzik, H., and Reynolds, J. R., Poly(3,4-ethylenedioxythiophene) and its derivatives: past, present, and future, *Adv. Mater.* **2000**, *12*, 481–494; (b) Groenendaal, L. B.; Zotti, G.; Aubert, P-H.; Waybright, S. M., and Reynolds, J. R., Electrochemistry of poly(3,4-alkylenedioxythiophene) derivatives, *Adv. Mater.* **2003**, *15*, 855–879.

28. Perepichka, I. F.; Besbes, M.; Levillain, E.; Salle, M., and Roncali, J., Hydrophilic oligo(oxyethylene)-derivatized poly(3,4-ethylenedioxythiophenes): cation-responsive optoelectroelectrochemical properties and solid-state chromism, *Chem. Mater.* **2002**, *14*, 449–457.

29. (a) Blanchard, P.; Cappon, A.; Levillain, E.; Nicolas, Y.; Frere, P., and Roncali, J., Thieno[3,4-*b*]-1,4-oxathiane: an unsymmetrical sulfur analogue of 3,4-ethylenedioxythiophene (EDOT) as a building block for linear π-conjugated systems, *Org. Lett.* **2002**, *4*, 607–609; (b) Wang, C.; Schindler, J. L.; Kannewurf, C. R., and Kanatzidis, M. G., Poly(3,4-ethylenedithiathiophene). A new soluble conductive polythiophene derivative, *Chem. Mater.* **1995**, *7*, 58–68.

30. Schottland, P.; Zong, K.; Gaupp, C. L.; Thompson, B. C.; Thomas, C. A.; Giurgiu, I.; Hickman, R.; Abboud, K. A., and Reynolds J. R., Poly(3,4-alkylenedioxypyrrole)s:

highly stable electronically conducting and electrochromic polymers, *Macromolecules* **2000**, *33*, 7051–7061.

31. Aqad, E.; Lakshmikantham, M. V., and Cava, M. P., Synthesis of 3,4-ethylenedioxy-selenophene (EDOS): A novel building block for electron-rich π-conjugated polymers, *Org. Lett.* **2001**, *3*, 4283–4285.

32. (a) www.hcstarck.de/index.php?page_id = 307. (b) www.agfa.com/sfc/polymer/.

33. Jönsson, S. K. M.; Birgerson, J.; Crispin, X.; Greczynski, G.; Osikowicz, W.; Denier van der Gon, A. W.; Salaneck, W. R., and Fahlman, M., The effects of solvents on the morphology and sheet resistance in poly(3,4-ethylenedioxythiophene)–polystyrenesulfonic acid (PEDOT–PSS) films, *Synth. Met.* **2003**, *139*, 1–10.

34. (a) Meng, H.; Perepichka, D. F., and Wudl, F., Facile solid-state synthesis of highly conducting poly(ethylenedioxythiophene), *Angew, Chem. Int. Ed.* **2003**, *42*, 658–661; (b) Meng, H.; Perepichka, D. F.; Bendikov, M.; Wudl, F.; Pan, G. Z.; Yu, W.; Dong, W., and Brown, S., Solid-state synthesis of a conducting polythiophene via an unprecedented heterocyclic coupling reaction, *J. Am. Chem. Soc.* **2003**; *125*, 15151–15162.

35. Spencer, H. J.; Berridge, R.; Crouch, D. J.; Wright, S. P.; Giles, M.; McCulloch, I.; Coles, S. J.; Hursthouse, M. B., and Skabara, P. J., Further evidence for spontaneous solid-state polymerisation reactions in 2,5-dibromothiophene derivatives, *J. Mater. Chem.* **2003**, *13*, 2075–2077.

36. Ferraris, J. P. and Lambert, T. L., Narrow band gap polymers: poly(cyclopenta [2,1-b;3,4-b']dithiophen-4-one), *J. Chem. Soc., Chem. Commun.* **1991**, 1268–1270.

37. Loganathan, K.; Cammisa, E. G.; Myron, B. D., and Pickup, P. G., $\Delta^{4,4'}$-Dicyclopenta[2,1-*b*,3,4-*b'*]dithiophene. A conjugated bridging unit for low band gap conducting, *Chem. Mater.* **2003**, *15*, 1918–1923.

38. (a) Wudl, F.; Kobayashi, M., and Heeger, A. J., Poly(isothinaphthene), *J. Org. Chem.* **1984**, *49*, 3382–3384; (b) Kobayashi, M.; Colaneri, N.; Boysel, M.; Wudl, F., and Heeger, A. J., The electronic and electrochemical properties of poly(isothianaphthene), *J. Chem. Phys.* **1985**, *82*, 5717–5723.

39. Chen, S.-A. and Lee, C-C., Processable low band gap π-conjugated polymer, poly (isothianaphthene), *Polymer* **1996**, *37*, 519–522.

40. Bolognesi, A.; Catellani, M.; Destri, S.; Zamboni, R., and Taliani, C., Poly(dithieno[3,4-*b*:3',4'-*d*]thiophene): a new transparent conducting polymer, *J. Chem. Soc., Chem. Commun.* **1988**, 246–247.

41. Meng, H. and Wudl, F., A robust low band gap processable n-type conducting polymer based on poly(isothianaphthene), *Macromolecules* **2001**, *34*, 1810–1816.

42. (a) Pomerantz, M.; Chaloner-Gill, B.; Harding, L. O.; Tseng, J. J., and Pomerantz, W. J., Poly(2,3-dihcxylthieno[3,4-*b*]pyrazine). A new processable low band-gap polyheterocycle, *J. Chem. Soc., Chem. Commun.* **1992**, 1672–1673; (b) Pomerantz, M.; Chaloner-Gill, B.; Harding, L. O.; Tseng, J. J., and Pomerantz, W. J., New processable low band-gap, conjugated polyheterocycles, *Synth. Met.* **1993**, *55–57*, 960–965.

43. Shrof, W.; Rozouvan, S.;. Hartmann, T., and Mohwald, H., Nonlinear optical properties of novel low-bandgap polythiophenes, *J. Opt. Soc. Am. B*, **1998**, *15*, 889–894.

44. Karikomi, M.; Kitamura, C.; Tanaka, S., and Yamashita, Y., New narrow-bandgap polymer composed of benzobis(1,2,5-thiadiazole) and thiophenes, *J. Am. Chem. Soc.* **1995**, *117*, 6791–6792.

45. Wynberg, H. and Zwanenberg, D. J., Thieno[3,4-*b*]thiophene. The third thiophthene, *Tetrahedron Lett.* **1967**, *8*, 761–764.

46. Neef, C. J.; Brotherston, I. D., and Ferraris, J. P., Synthesis and electronic properties of poly(2-phenylthieno[3,4-*b*]thiophene): a new low band gap polymer, *Chem. Mater.* **1999**, *11*, 1957–1958.

47. Pomerantz, M.; Gu, X., and Zhang, S. X., Poly(2-decylthieno[3,4-*b*]thiophene-4,6-diyl). A new low band gap conducting polymer, *Macromolecules* **2001**, *34*, 1817–1822.

48. Brandsma, L. and Verkruijsse, H. D., An alternative synthesis of thieno[3,4-*b*]thiophene, *Synth. Commun.* **1990**, *20*, 2275–2277.

49. (a) Lee, K. and Sotzing, G. A., Poly(thieno[3,4-*b*]thiophene). A new stable low band gap conducting polymer, *Macromolecules* **2001**, *34*, 5746–5747; (b) Sotzing, G. A. and Lee, K., Poly(thieno[3,4-*b*]thiophene): A p- and n-dopable polythiophene exhibiting high optical transparency in the semiconducting state, *Macromolecules* **2002**, *35*, 7281–7286.

50. Lee, B. and Sotzing, G. A., Aqueous phase polymerization of thieno[3,4-*b*]thiophene, *Polym. Prep.* **2002**, *43*, 568–569.

51. Seshadri, V.; Wu, L., and Sotzing, G. A., Conjugated polymers via electrochemical polymerization of thieno[3,4-*b*]thiophene (T34bT) and 3,4-ethylenedioxythiophene (EDOT), *Langmuir* **2003**, *19*, 9479–9485.

52. (a) Gilch, H. G.; Wheelwright, W. L., Polymerization of α-halogenated *p*-xylenes with base, *J Polym Sci: A-1* **1966**, *4*, 1337–1349; (b) Wessling, R. A., The polymerization of xylylene bisdialkylsulfonium salts, *J. Polym. Sci. Polym. Symp.* **1985**, *72*, 55–66.

53. (a) Jen, K-Y.; Maxfield, M. R.; Shacklette, L. W., and Elsenbaumer, R. L., Highly-conducting, poly(2,5-thienylene vinylene) prepared via a soluble precursor polymer, *J. Chem. Soc., Chem. Commun.* **1987**, 309–311; (b) Barker, J., An electrochemical investigation of the doping processes in poly(thienylenevinylene), *Synth. Met.* **1989**, *32*, 43–50.

54. Cheng, H. and Elsenbaumer, R. L., New precursors and polymerization route for the preparation of high molecular mass poly(3,4-dialkoxy-2,5-thienylenevinylene)s: low band gap conductive polymers, *J. Chem. Soc., Chem. Commun.***1995**, 1451–1452.

55. Kim, I. T. and Elsenbaumer, R. L., Solution processible poly(1-alkyl-2,5-pyrrolenevinylenes): new low band gap conductive polymers, *J. Chem. Soc., Chem. Commun.* **1998**, 327–328.

56. (a) Martinez, M.; Reynolds, J. R.; Basak, S.; Black, D. A.; Marynick, D. S., and Pomerantz, M., Electrochemical synthesis and optical analysis of poly[(2,2′-dithienyl)-5,5′-diylvinylene], *J. Polym. Sci. B*, **1988**, *26*, 911–920; (b) Roncali, J.; Thobie-Gautier, C.; Elandaloussi, E., and Frere, P., Control of the bandgap of conducting polymers by rigidification of the π-conjugated system, *J. Chem. Soc., Chem. Commun.* **1994**, 2249–2250.

57. Sotzing, G. A. and Reynolds, J. R. Poly[trans-bis(3,4-ethylenedioxythiophene)vinylene]: a low band-gap polymer with rapid redox switching capabilities between conducting transmissive and insulating absorptive states, *J. Chem. Soc., Chem. Commun.* **1995**, 703–704.

58. (a) Jenekhe, S. A., A class of narrow-band-gap semiconducting polymers, *Nature* **1986**, *322*, 345–347; (b) Jenekhe, S. A., Synthesis of conjugated polymers with alternating aromatic and quinonoid sequences via elimination on precursors, *Macromolecules* **1986**, *19*, 2663–2664.

59. Kuerti, J.; Surjan, P. R., and Kertesz, M., Electronic structure and optical absorption of poly(biisothianaphthene–methine) and poly(isonaphthothiophene–thiophene): two low-band-gap polymers, *J. Am. Chem. Soc.* **1991**, *113*, 9865–9867.

60. Okuda, Y.; Lakshmikantham, M. V., and Cava, M. P., A new route to 1,3-disubstituted benzo[*c*]thiophenes, *J. Org. Chem.* **1991**, *56*, 6024–6026.

61. Kiebooms, R. and Wudl, F., Synthesis and characterisation of poly(isothianaphthene methine), *Synth. Met.* **1999**, *101*, 40–43.

62. Goto, H.; Akagi, K., and Shirakawa, H., Syntheses and properties of small-bandgap liquid crystalline conjugated polymers, *Synth. Met.* **1997**, *84*, 385–386.

63. (a) Chen, W. C. and Jenekhe, S. A., Small-bandgap conducting polymers based on conjugated poly(heteroarylene methines). 1. Precursor poly(heteroarylene methylenes), *Macromolecules* **1995**, *28*, 454–464. (b) Chen, W. C. and Jenekhe, S. A., Small-bandgap conducting polymers based on conjugated poly(heteroarylene methines). 2. Synthesis, structure, and properties, *Macromolecules* **1995**, *28*, 465–480.

64. Kiebooms, R. H. L.; Goto, H., and Akagi, K., Synthesis of a new class of low-band-gap polymers with liquid crystalline substituents, *Macromolecules* **2001**, *34*, 7989–7998.

65. Benincori, T.; Rizzo, S.; Sannicolo, F.; Schiavon, G.; Zecchin, G.; Zecchin, S., and Zotti, G., An electrochemically prepared small-bandgap poly(biheteroarylidenemethine): poly{bi[(3,4-ethylenedioxy)thienylene]methine}, *Macromolecules* **2003**, *36*, 5114–5118.

66. Yamamoto, T.; Zhou, Z.; Kanbara, T.; Shimura, M.; Kizu, K.; Maruyama, T.; Nakamura, Y.; Fukuda, T.; Lee, B-L.; Ooba, N.; Tomaru, S.; Kurihara, T.; Kaino, T.; Kubota, K., and Sasaki, S., π-Conjugated donor–acceptor copolymers constituted of π-excessive and π-deficient arylene units. Optical and electrochemical properties in relation to CT structure of the polymer, *J. Am. Chem. Soc.* **1996**, *118*, 10389–10399.

67. (a) Brockmann, T. W. and Tour, J, M., Synthesis and properties of low-bandgap zwitterionic and planar conjugated pyrrole-derived polymeric sensors. Reversible optical absorption maxima from the UV to the near-IR, *J. Am. Chem. Soc.* **1995**, *117*, 4437–4447; (b) Zhang, Q. T. and Tour, J. M., Low optical bandgap polythiophenes by an alternating donor/acceptor repeat unit strategy, *J. Am. Chem. Soc.* **1997**, *119*, 5065–5066; (c) Zhang, Q. T. and Tour, J. M., Alternating donor/acceptor repeat units in poly-thiophenes. Intramolecular charge transfer for reducing band gaps in fully substituted conjugated polymers, *J. Am. Chem. Soc.* **1998**, *120*, 5355–5362.

68. Meng, H.; Tucker, D.; Chaffins, S.; Chen, Y.; Helgeson, R.; Dunn, B., and Wudl, F., An unusual electrochromic device based on a new low-bandgap conjugated polymer, *Adv. Mater.* **2003**, *15*, 146–149.

69. (a) Lenz, R. W. and Handlovits, C. E., Thermally stable hydrocarbon polymers: poly-terephthalylidenes, *J. Org. Chem.* **1960**, *25*, 813–817; (b) Hoerold, H. H. and Opferman, J., Poly-*p*-xylylidenes. Synthesis and relation between structure and electrophysical properties, *Makromol. Chem.* **1970**, *131*, 105–132; (c) Hoerold, H. H. and Helbig, M., Poly(phenylenevinylenes) — synthesis and redox chemistry of electroactive polymers, *Makromol. Chem. Makromol. Symp.* **1987**, *12*, 229–258.

70. Ho, H. A.; Brisset, H.; Freìre, P., and Roncali, J., Electrogenerated small bandgap π-conjugated polymers derived from substituted dithienylethylenes, *J. Chem. Soc., Chem. Commun.* **1995**, 2309–2310.

71. Sotzing, G. A.; Thomas, C. A., and Reynolds, J. R., Low band gap cyanovinylene polymers based on ethylenedioxythiophene, *Macromolecules* **1998**, *31*, 3750–3752.

72. Ho, H. A.; Brisset, H.; Elandaloussi, E.; Fre're, P., and Roncali, J., Thiophene-based conjugated oligomers and polymers with high electron affinity, *Adv. Mater.* **1996**, *8*, 990–994.

73. Seshadri, V. and Sotzing, G. A., Low band-gap poly(thieno[3,4-*b*]thiophene) consisting of cyanovinylene spacer units, *Polym. Prep.* **2003**, *44*, 398–399.

74. Gallazzi, M. C.; Toscano, F.; Paganuzzi, D.; Bertarelli, C.; Farina, A., and Zotti, G., Polythiophenes with unusual electrical and optical properties based on donor acceptor alternance strategy, *Macromol. Chem. Phys.* **2001**, *202*, 2074–2085.

75. Huang, Hs. and Pickup, P. G., A donor–acceptor conducting copolymer with a very low band gap and high intrinsic conductivity, *Chem. Mater.* **1998**, *10*, 2212–2216.

76. Lee, Y.; Sadki, S.; Tsuie, B., and Reynolds, J. R., A new silole containing low band gap electroactive polymer, *Chem. Mater.* **2001**, *13*, 2234–2236.

77. Treibs, A. and Jacob, K., Cyclotrimethine dyes derived from quadratic acid [1,2-dihydroxycyclobutenedione], *Angew. Chem.* **1965**, *77*, 680–681.

78. (a) Havinga, E. E.; ten Hoeve, W., and Wynberg, H., A new class of small band gap organic polymer conductors, *Polym. Bull.* **1992**, *29*, 119–126; (b) Havinga, E. E.; ten Hoeve, W., and Wynberg, H., Alternate donor–acceptor small-band-gap semiconducting polymers; polysquaraines and polycroconaines, *Synth. Met.* **1993**, *55*, 299–306.

79. (a) Ajayaghosh, A. and Eldo, J., A novel approach toward low optical band gap poly-squaraines, *Org. Lett.* **2001**, *3*, 2595–2598; (b) Eldo, J. and Ajayghosh, A., New low band gap polymers: control of optical and electronic properties in near infrared absorbing π-conjugated polysquaraines, *Chem. Mater.* **2002**, *14*, 410–418.

80. Jiang, B.; Yang, S. W.; Bailey, S. L.; Hermans, L. G.; Niver, R. A.; Bolcar, M. A., and Jones, Jr., W. E., Towards transparent molecular wires: electron and energy transfer in transition metal derivatized conducting polymers, *Coord. Chem. Rev.* **1998**, *171*, 365–386.

23

Optoelectronic Properties of Conjugated Polymer/Fullerene Binary Pairs with Variety of LUMO Level Differences

Steffi Sensfuss and Maher Al-Ibrahim
Department of Functional Polymer Systems, Thuringian Institute for Textile and Plastics Research, Rudolstadt, Germany

Contents

Abstract It is the objective of this chapter to show the correlation between optoelectronic properties of different donor polymer–fullerene acceptor combinations (acceptors: C_{60}, [6,6]-phenyl-C_{61}-butyric acid methyl ester (PCBM), and bismorpholino-C_{60}) as well as their influence on the parameters of corresponding photovoltaic devices. Therefore the following optoelectronic properties have been considered: absorption coefficients, detection of photoinduced donor–acceptor charge transfer (CT) (by photoluminescence (PL) quenching experiments and light-induced electron spin resonance spectroscopy), and the determination of highest occupied molecular orbital (HOMO)/lowest unoccupied molecular orbital (LUMO) levels of donors and acceptors by means of cyclic voltammetry. Regioregular poly(3-alkylthiophene)s (P3AT) (poly(3-hexylthiophene), poly(3-ocylthiophene), poly(3-dodecylthiophene)) were investigated concerning the influence of the alkyl side chain length on the optoelectronic properties and on the photovoltaic device parameters in P3AT/PCBM (1:3 wt.%) composites. Novel phenylene ethynylene/phenylene vinylene (PPE–PPV) hybrid polymers DE69 and DE21 were compared with the state-of-the-art material poly[2-methoxy-5-(3′,7′-dimethyloctyloxy)-1,4-phenylene vinylene] (MDMO-PPV). The presence of $-C{\equiv}C-$ within the polymer backbone leads to a higher oxidation potential (helpful to improve the intrinsic material stability), energetically lower HOMO–LUMO levels and higher optical bandgaps for the hybrid polymers. Photovoltaic devices were prepared on ambient conditions on flexible PET–ITO foils with MDMO-PPV/PCBM, giving $\eta_{AM1.5} = 3\%$ and the new polymers DE69 or DE21/PCBM with $\eta_{AM1.5} = 1.75\%$ and 0.5% ($A = 0.25$ cm^2, $P_{IN} = 100$ mW/cm^2).

Keywords optoelectronic properties, conjugated polymers, fullerene, absorption coefficient, HOMO–LUMO levels, photoinduced electron transfer, photoluminescence, light-induced electron spin resonance (LESR), donor–acceptor polymer solar cells

23.1. INTRODUCTION

Conjugated polymers combine both the electronic properties of common inorganic semiconductors and the advantages of plastic processing. The discovery of ultrafast charge transfer in p-conducting polymer–fullerene composites by Sariciftci in 1992 [1] represents a decisive milestone in the development of polymer solar cells. It has to be pointed out that especially the high asymmetry between forward and back polymer–fullerene electron transfer rate that covers nine orders of magnitude [2] allows blending of donor and acceptor components to prepare solar cell architectures possessing only one photoactive composite layer instead of donor monolayers or donor–acceptor bilayers. At present, the typical device structure of a polymer solar cell consists of a semiconducting polymer–organic acceptor composite layer sandwiched between two asymmetrical work function metal electrodes, typically indium tin oxide (ITO) and aluminum (Al), calcium (Ca), or magnesium (Mg) (Figure 23.1). State-of-the-art is an external AM1.5 power conversion efficiency of 3.5% using a regioregular poly(3-hexylthiophene) (P3HT/([6,6]-phenyl-C_{61}-butyric acid methyl ester) (PCBM) composite (active area: 5–8 mm^2, $P_{IN} = 80$ mW/cm^2, $\eta = 3.5\%$) [3]. It is the objective of this chapter to show the correlation between chemical structure and resulting optoelectronic properties of different donor polymer–fullerene acceptor combinations as well as their influence on the parameters of corresponding photovoltaic devices. Therefore, the following optoelectronic properties have been considered.

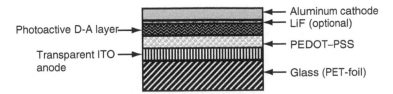

Figure 23.1. Cross-section of typical device configuration of polymer solar cells (PET, polyester; ITO, indium tin oxide; PEDOT–PSS, poly(3,4-ethylene dioxythiophene) doped with poly(styrene sulphonate)).

Absorption coefficients: Compared to inorganic semiconductors, the significantly lower charge carrier mobility in polymeric donor (D)–acceptor (A) composites ($\mu_{hole} \sim 10^{-4}\,cm^2\,V^{-1}\,s^{-1}$ for poly(p-phenylene-vinylene) (PPV) [4], $\mu_e \sim 2\text{--}4.5 \times 10^{-3}\,cm^2\,V^{-1}\,s^{-1}$ for PCBM [5]) currently restricts the applicable thickness of the photoactive layer to about 100–300 nm [6]. In order to achieve high optical densities, absorption coefficients as high as $10^5\,cm^{-1}$ are required. This is generally attainable using organic materials. Additionally, most of the semiconducting polymers currently available possess bandgaps higher than 2.0 eV ($\sim$600 nm), which will restrict the amount of light absorbed from the incident radiation to about 30% only [7] (Figure 23.2).

Detection of donor–acceptor charge transfer: The working principle of donor–acceptor polymer solar cells is based on three fundamental steps [8]:

(a) Excitation of the light absorber transferring an electron from its HOMO to its LUMO with formation of excitons

(b) Formation of spatially separated charge carriers at the donor–acceptor interface including several intermediate steps (e.g., diffusion of photoexcited excitons towards the D–A interface, formation of ion radical pairs, exciton dissociation)

(c) Selective transport of the charge carriers through the composite layer of the device towards the corresponding electrodes because each component is able to ensure the transport of one type of charge carriers only (the

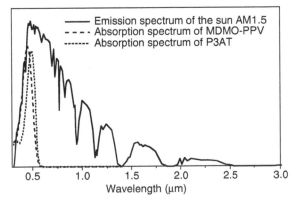

Figure 23.2. Absorption behavior of typical donor polymers (MDMO-PPV, poly(3-alkylthiophene)s applied in polymer solar cells in comparison with the emission spectrum of sun (AM1.5).

donor transports the holes, the acceptor transports the electrons). For both the donor as well as the acceptor continuous pathways are required.

The comparison of the photoluminescence (PL) of the donor with that of the donor–acceptor composite provides an important, but also simple method to detect the charge transfer (CT), which is indicated by the quenching effect of the composite [1]. Radical cations (polarons) and anions are generated during the charge transfer at the donor polymer and the acceptor, respectively. For studying the paramagnetic species formed and to understand their kinetic behavior, light-induced electron spin resonance spectroscopy (ESR) was found to be a useful tool. The understanding of the correlation of the ESR kinetics of donor–acceptor-composites with their parameters in photovoltaic devices could be a key issue to evaluate potential photovoltaic materials.

HOMO–LUMO levels of donors and acceptors: The relative position of the HOMO–LUMO levels (HOMO, highest occupied molecular orbital; LUMO, lowest unoccupied molecular orbital) and the choice of the electrode materials that have to possess suitable work functions are basic requirements to make a solar cell work successfully (Figure 23.3). Thus the HOMO of the donor should be higher in energy than the HOMO of the acceptor. Additionally, the energy level of the LUMO$_{acceptor}$ has to be below that of the LUMO$_{donor}$. Typically, an organic acceptor possesses a reduction potential (determining the LUMO) lower than its oxidation potential (corresponds with the HOMO). If the LUMO$_{donor}$ − LUMO$_{acceptor}$ difference is relatively high as it is for the actual standard material combination MDMO-PPV/PCBM (0.9 eV, MDMO-PPV, poly[2-methoxy-5-(3′,7′-dimethyloctyloxy)-1,4-phenylene vinylene]), the photoinduced electron transfer occurs within picoseconds, so that no back transfer or other disturbing processes can take place. This is additionally due to the extremely high electron affinity of fullerenes and reflects in an 100% internal quantum efficiency for the PPV/PCBM systems. However, the required minimum value for the LUMO$_{donor}$ − LUMO$_{acceptor}$ difference is still uncertain and will probably depend on the material combination applied. On the other hand, it should only be as high as necessary to just achieve a directed donor–acceptor charge transfer since the LUMO$_{acceptor}$ − HOMO$_{donor}$ difference determines the open circuit voltage of the device [9]. The presented chapter shall give a contribution to this discussion.

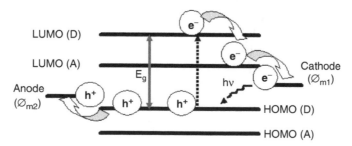

Figure 23.3. Working principle of the charge transfer in a donor–acceptor polymer solar cell.

23.2. OPTOELECTRONIC PROPERTIES OF DONOR–ACCEPTOR MATERIALS AND THEIR INFLUENCE ON THE PHOTOVOLTAIC PARAMETERS

23.2.1. Fullerene and Fullerene Derivatives as Electron Acceptors

Concerning charge separation and electron transport properties, fullerenes and their soluble derivatives have proven to be the most efficient materials for solar cell applications among the currently available acceptors. Figure 23.4 shows the chemical structures of C_{60}-fullerene, its most frequently used derivative PCBM as well as that of bismorpholino-C_{60}-fullerene (BM-C_{60}). PCBM and BM-C_{60} exhibit a much better solubility compared to unsubstituted C_{60}. As one can learn from cyclic voltammetric measurements, all three compounds have distinctively different reduction potentials. Compared with PCBM (E_p^{red} : −0.62 V), C_{60} (E_p^{red}: −0.54 V) appears to be a more effective electron acceptor, whereas BM-C_{60} shows a clearly lower electron affinity (E_p^{red}: −0.88 V) in comparison to PCBM and C_{60}. This may be explained by the nucleophilic character of the attached morpholine moieties.

The LUMO of C_{60} (−3.83 eV) is lower in energy than that of PCBM (−3.75 eV) and of BM-C_{60} (−3.51 eV), which means that the acceptor strength of C_{60} is higher than that of PCBM and BM-C_{60}. According to the empirical equation $E^{LUMO} = [-e (E_{1/2(vs.Ag/AgCl)}^{red} - E_{onset(Fc/Fc+ \ vs. \ Ag/AgCl)})] - 4.8$ eV [10,11] including the ferrocene value −4.8 eV (with respect to the vacuum level, which is defined as zero), the LUMO levels can be estimated from the $E_{1/2}^{red}$ value, that has been obtained from measurements in solution (Table 23.1).

23.2.2. Regioregular P3ATs as Electron Donors

P3ATs have been found to be an unusual class of polymers with good solubility (increasing with the side chain length), processability, environmental stability, elec-

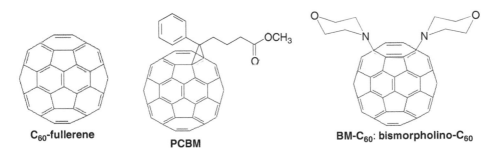

C$_{60}$-fullerene　　　　**PCBM**　　　　**BM-C$_{60}$: bismorpholino-C$_{60}$**

Figure 23.4. Chemical structures of C_{60}-fullerene and its derivatives [6,6]-phenyl-C_{61}-butyric acid methyl ester (PCBM) and bismorpholino-C_{60}-fullerene (BM-C_{60}).

Table 23.1. Cyclic voltammetry (CV) data and corresponding LUMO levels of three different fullerenes measured in solution (o-dichlorobenzene/acetonitrile 4:1) [12]

Material	E_p^{ox} vs. Ag/AgCl (V)	$E_{1/2}^{red}$ vs. Ag/AgCl (V)	LUMO (eV)
C_{60}	−0.54	−0.50	−3.83
PCBM	−0.62	−0.58	−3.75
BM-C_{60}	−0.88	−0.82	−3.51

troactivity, and other interesting properties [13,14]. The used regioregular P3ATs have the incorporated 3-alkylthiophene monomer into the polymer chain only in one type of arrangement (>98.5% head to tail, the previous 3-alkylthiophene monomer is coupled with the next via 2–5 position) whereas regiorandom P3ATs contain different arrangements (head to tail, head to head, tail to tail) statistically distributed. Consequently, regioregular P3ATs are superior to regiorandom P3ATs concerning the degree of intermolecular order and the resulting electronic properties [13,15]. The chemical structures of regioregular P3HT, poly(3-octylthiophene) (P3OT), and poly(3-dodecylthiophene) (P3DDT) are given in Figure 23.5.

23.2.2.1. *Electrochemical Properties of Poly(3-alkylthiophene)s*

The electrochemical data can give valuable information and allow the estimation of the relative position of HOMO–LUMO levels of investigated materials. The knowledge of these values is required for finding suitable donor–acceptor pairs. Table 23.2 contains the onset and the peak values for oxidation and reduction potentials of P3HT, P3OT, and P3DDT determined by means of cyclic voltammetry vs. Ag/AgCl. The values of E_p^{ox} and E_p^{red} for the P3ATs slightly depend on the length of the alkyl side chain. As it can be recognized from Table 23.2, the oxidation potential is directly proportional to the chain length whereas the reduction potential shows an inverse dependence. This is due to the fact that the density of π-electrons and charges is increasing as the length of the side chain decreases, caused by better π-stacking and improved interchain coupling [16], which enhances the donor character and favors oxidation. Additionally, it is conceivable that longer alkyl chains generate a type of shielding effect hampering the electron donation.

For the performed film measurements, the E_{onset}^{ox} and E_{onset}^{red} values serve to the calculation of the corresponding HOMO and LUMO levels according to the empirical equation as given earlier (Section 23.2.1, $E^{HOMO/LUMO} = [-e\,(E_{onset(vs.\ Ag/AgCl)} - E_{onset(Fc/Fc+\ vs.\ Ag/AgCl)})] - 4.8\,eV$). As determined from the absorption edges of UV–vis spectra of polymer films, the electrochemical gap is distinctively lower than the optical gap for all investigated polymers (see Table 23.2 and Table 23.3). Al-

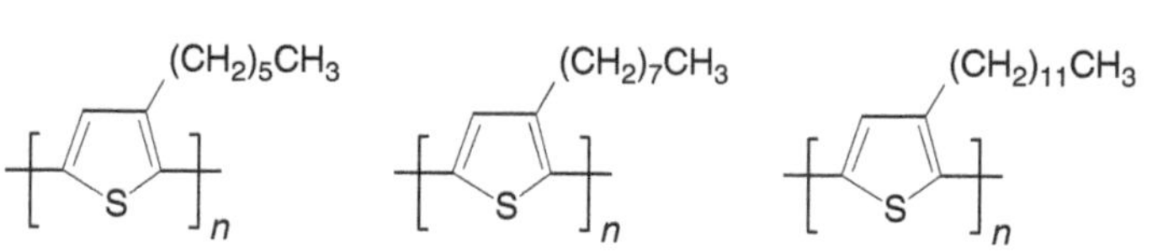

Figure 23.5. Chemical structures of the investigated regioregular poly(3-hexylthiophene) P3HT, poly(3-octylthiophene) P3OT, and poly(3-dodecylthiophene) P3DDT.

Table 23.2. Cyclic voltammetry (CV) data, HOMO and LUMO levels, and E_g^{CV} of P3HT, P3OT, and P3DDT measured in form of films [12]

Polymer	E_p^{ox} vs. Ag/AgCl (V)	E_{onset}^{ox} vs. Ag/AgCl (V)	HOMO (eV)	E_p^{red} vs. Ag/AgCl (V)	E_{onset}^{red} vs. Ag/AgCl (V)	LUMO (eV)	E_g^{CV} (eV)
P3HT	+1.12	+0.82	−5.20	−1.15	−0.85	−3.53	1.67
P3OT	+1.25	+0.87	−5.25	−1.10	−0.83	−3.55	1.70
P3DDT	+1.30	+0.91	−5.29	−1.04	−0.83	−3.55	1.74

though the effect of discrepancy between E_g^{opt} and E_g^{CV} has already been reported in several references [9,17,18], a comprehensive discussion is still lacking. The HOMO of P3HT is slightly higher in energy than that the HOMOs of P3OT and of P3DDT. These values are in accordance with the literature for P3OT [19], for P3HT [20,21], and for P3DDT [21]. The LUMOs of the three polythiophenes are almost at the same energetic level.

23.2.2.2. Optical Properties of Poly(3-alkylthiophene)s

Figure 23.6 shows the absorption spectra of films made from the three poly (3-alkylthiophene)s. The absorption maximum (λ_{max}) of P3HT, P3OT, and P3DDT can be observed at 500 nm (2.48 eV), 512 nm (2.42 eV), and 515 nm (2.41 eV), respectively. These values for λ_{max} correlating with the energy of $\pi-\pi^*$ interband transition indicate a better planarity of the backbone chains for P3DDT than for P3HT [22]. This is due to the fact that longer side groups restrict the number of possible conformations and that two adjacent thiophene rings can form by rotating around their shared C$-$C bond. With the decreasing length of the side chain, the absorption coefficient increases for the three investigated polymers in the following graduation: P3DDT < P3OT < P3HT (Table 23.3). In accordance with the literature [19,23–25], the optical gap was found to be approximately 1.92 eV for all three polythiophenes and does not depend on the side chain length. When measuring the absorption spectra of P3HT/PCBM, P3OT/PCBM, and P3DDT/PCBM the resulting graphs represent simple superpositions of the spectra of the individual components, but without ground state doping of the polythiophenes initiated by PCBM [26,27].

23.2.2.3. Photoinduced Electron Transfer in P3AT/Fullerene Systems

Figure 23.7 exhibits the energy diagram of P3ATs and fullerenes in relation to the work function of ITO and Al, which are usually applied as electrodes in polymer

Table 23.3. Optical data of regioregular poly(3-alkylthiophene)s determined from ellipsometric data [12]

Material	λ_{max} [nm (eV)]	Absorption coefficient at λ_{max} (1/cm)	E_g^{opt} (eV)
P3HT	500 (2.48)	1.75×10^5	1.92
P3OT	512 (2.42)	1.38×10^5	1.92
P3DDT	515 (2.41)	1.12×10^5	1.93

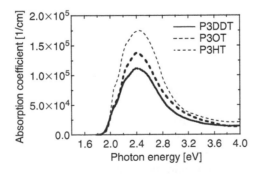

Figure 23.6. Absorption spectra for P3HT, P3OT, and P3DDT thin films.

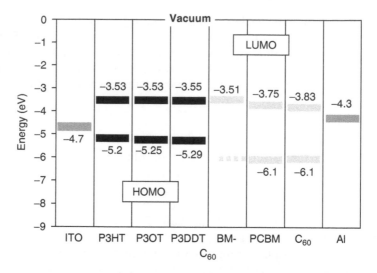

Figure 23.7. Energy band diagram with HOMO and LUMO levels of donor polymers P3HT, P3OT, and P3DDT and fullerene acceptors estimated from cyclic voltammogramic data in relation to the work function of the common electrode materials ITO and Al (HOMO values of fullerenes are literature data [1]).

solar cells. The HOMO of all three polymers is distinctively higher in energy than that of C_{60} and PCBM. In the case of BM-C_{60}, the value is unknown. Nevertheless, the relative positions of the donor LUMO and the acceptor LUMO are important for the intended charge transfer. As shown in Figure 23.7, the differences between the LUMOs of P3HT, P3OT, or P3DDT and the LUMOs of C_{60} or PCBM are in the range of 0.2–0.3 eV, respectively. It seems that these differences are sufficiently high in order to enable an unrestricted and directed charge transfer, as it will be described in the following. Irrespective of this, we found the position of LUMO of BM-C_{60} at nearly the same level like that of P3HT, P3OT, and P3DDT; consequently a selective charge transfer into this fullerene material becomes uncertain or impossible. Although $\Delta LUMO_{D-A}$ data obtained from cyclic voltammetric measurements do not represent absolute values, they should be suitable to compare materials if applying identical experimental conditions.

Within all investigated systems, such as P3AT/PCBM and P3AT/C_{60}, the charge transfer can be considered to be the most preferred process that can occur. This assumption has been supported by both, the strong PL of P3AT polymers accompanied by its quenching in the presence of PCBM or C_{60} and the detection of two types of LESR signals (one of polarons $P^{+\bullet}$ on the polymer chain and one of radical anions of PCBM or C_{60}) as demonstrated later and in Ref. [26,28,29]. Films made from composites of P3HT/PCBM (1:3 wt.%) (Figure 23.8) show a distinctive PL quenching effect. The same behavior has been found for P3OT/PCBM (1:2 wt.%) (Figure 23.9). Figure 23.10 contains the photoluminescence spectra of P3DDT and blends of P3DDT with three different fullerenes. In contradiction to the blends consisting of P3DDT/C_{60} and P3DDT/PCBM (1:1 wt.%), the photoluminescence quenching of P3DDT/BM-C_{60} (1:1 wt.%) composite film is significantly reduced. This could be due to an energetic problem caused by the relative LUMO$_{P3DDT}$/LUMO$_{BM-C60}$ position and would be in agreement with the cyclic voltammetry data for this D–A combination (Figure 23.7).

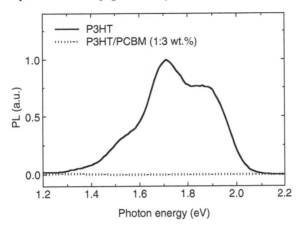

Figure 23.8. Film photoluminescence spectra of P3HT and a P3HT/PCBM blend (1:3 wt.%).

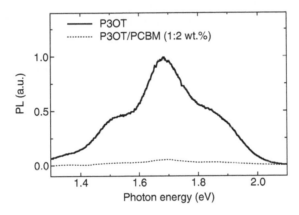

Figure 23.9. Film photoluminescence spectra of P3OT and a P3OT/PCBM blend (1:2 wt.%).

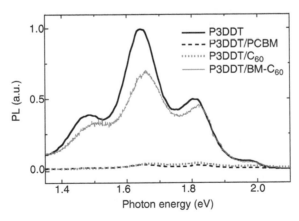

Figure 23.10. Film photoluminescence spectra of P3DDT and blends of P3DDT with different fullerenes (1:1 wt.%).

Light-induced spin resonance (LESR) represents the most efficient quantitative measurement technique to detect radicals exclusively generated by charge separation processes via light excitation. In addition to the detection of cation and anion radicals formed by light excitation, the analysis of spectroscopic g-factor components provides information on the symmetry of the paramagnetic centers and the position of localized polarons along the polymer backbone. Analyzing the LESR signal line width, information on the spin–lattice and spin–spin relaxations, the spin exchange and dipole–dipole interactions can be obtained. Detailed conclusions regarding the charge separation and recombination kinetics may be withdrawn from studying the signal amplitudes measured when switching the radiation source off or on (separation) and on or off (recombination), respectively. Nevertheless, even quantitative estimations of the charge carrier concentrations are conceivable.

The effective charge separation in the systems P3HT/PCBM (1:2 wt.%), P3OT/PCBM (1:2 wt.%), and P3DDT/PCBM (1:2 wt.%) has been established by two LESR signals. They can be assigned to positive polarons ($P^{+ \bullet}$) on the polymer chain with a g-factor $g_{iso} \approx 2.0017$ and to radical anions of PCBM with $g_{iso} \approx 1.9995$ [20,28,30] (Figure 23.11–Figure 23.13). P3HT/PCBM (1:2 wt.%) and P3OT/PCBM (1:2 wt.%) composite films show a strong $P^{+ \bullet}$ ESR signal without light excitation (Figure 23.11 and Figure 23.12) [12], which may be caused by traces of impurities or the effect of entrapped oxygen. Despite the fact that previous investigations using P3DDT/C$_{60}$ [26,28] with another batch of P3DDT could prove the existence of a dark signal, P3DDT/PCBM (1:2 wt.%) does not yield any signal without exposure to radiation (Figure 23.13).

Dark ESR and LESR spectra (Ar$^+$ laser operating at $\lambda = 488$ nm) for films of P3HT/PCBM and P3OT/PCBM are depicted in Figure 23.11 and Figure 23.12, respectively. After starting light exposition, the amplitude of the polaron signal increases as the time proceeds and a signal to be assigned to the high-field component (PCBM radical anion) starts to appear additionally. Hence, the distinctive spectroscopic features between spectra taken during radiation and in the dark, respectively, clearly indicate the charge separation process of the investigated composites (Figure 23.11 and Figure 23.12). Applying the same conditions, the signal amplitude ratio of P3DDT/PCBM (1:2 wt.%) films appears to be changed compared to Figure 23.11.

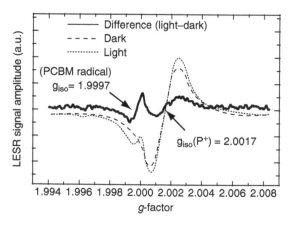

Figure 23.11. Dark, light, and difference ESR spectra of P3HT/PCBM (1:2 wt.%, measured in form of a film, Ar$^+$ laser with $\lambda = 488$ nm, $P_{\mu W} = 6$ mW, $I = 30$ mW/cm^2, $T = 77$ K).

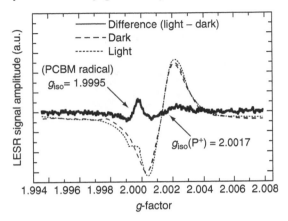

Figure 23.12. Dark, light, and difference ESR spectra of P3OT/PCBM (1:2 wt.%, measured in form of a film, Ar$^+$ laser with $\lambda = 488$ nm, $P_{\mu W} = 0.6$ mW, $I = 30$ mW/cm^2, $T = 77$ K).

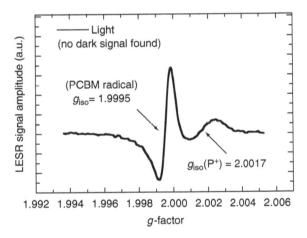

Figure 23.13. Light-induced ESR spectrum of P3DDT/PCBM (1:2 wt.%, measured in form of a film, Ar$^+$ laser with $\lambda = 488$ nm, $P_{\mu W} = 6$ mW, $I = 30$ mW/cm^2, $T = 77$ K).

and Figure 23.12 (the PCBM$^{\cdot-}$ signal amplitude is distinctively higher in intensity than the polaron signal). Contrary to the P3ATs/PCBM composites, no ESR signal could be detected in P3DDT/BM-C$_{60}$ films. This is in accordance with the results from photoluminescence measurements (Figure 23.10) and HOMO/LUMO determinations (Figure 23.7).

23.2.2.4. *Photovoltaic Devices Based on P3ATs/PCBM*

The photovoltaic devices consist of four layers as shown in Figure 23.14. A flexible polyester foil coated with ITO is used as the substrate (the device area amounts 0.25 cm^2). All cells were prepared and measured on ambient conditions illuminated with white light of a Steuernagel solar simulator AM1.5 ($P_{IN} = 100$ mW/cm^2). Figure 23.15–Figure 23.17 show the current–voltage characteristics of the solar cell based on P3HT/PCBM (1:3 wt.%), P3OT/PCBM (1:3 wt.%), and P3DDT/PCBM (1:3 wt.%), respectively, in the dark and on AM1.5 conditions with light intensity of 100 mW/cm^2

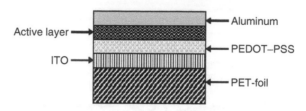

Figure 23.14. Cross-section of typical device configuration of prepared flexible polymer solar cells.

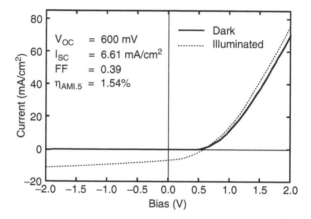

Figure 23.15. $I-V$ characteristics of a PET foil/ITO/PEDOT–PSS/P3HT:PCBM (1:3 wt.%)/ Al cell ($A = 0.25\,cm^2$) in dark and under illumination with white light ($P_{IN} = 100\,mW/cm^2$).

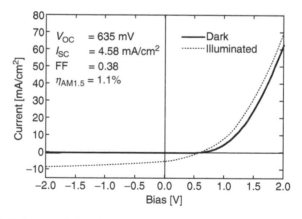

Figure 23.16. $I-V$ characteristics of a PET foil/ITO/PEDOT–PSS/P3OT:PCBM (1:3 wt.%)/ Al cell ($A = 0.25\,cm^2$) in dark and under illumination with white light ($P_{IN} = 100\,mW/cm^2$).

at room temperature. The P3HT/PCBM cell has an open circuit voltage (V_{oc}) of 600 mV, a short circuit current (I_{sc}) of 6.61 mA/cm^2, and a calculated fill factor (FF) of 0.39. The overall efficiency of this device is 1.54%. The P3OT/PCBM cell provides an open circuit voltage of 635 mV, a short circuit current of 4.58 mA/cm^2, and a fill factor of 0.38. The overall efficiency of this device reaches 1.1%. For P3DDT/PCBM cells these parameters are: open circuit voltage 650 mV, short circuit current 2.53 mA/ cm^2, and fill factor 0.36. The overall efficiency of this device is 0.59%. Table 23.4

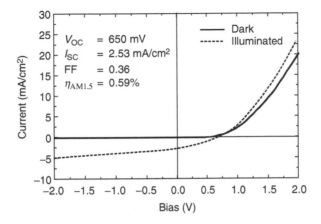

Figure 23.17. *I–V* characteristics of a PET foil/ITO/PEDOT–PSS/P3DDT:PCBM (1:3 wt.%)/Al cell ($A = 0.25\,\text{cm}^2$) in dark and under illumination with white light ($P_{IN} = 100\,\text{mW/cm}^2$).

Table 23.4. Photovoltaic performance parameters of the cells based on poly(3-alkylthiophene)/PCBM (1:3 wt.%) depicted in Figure 23.15 Figure 23.17 on AM1.5 condition

Active layer	V_{oc} (mV)	I_{sc} (mA/cm^2)	FF	$\eta_{e\,(AM1.5)}$ (%)
P3HT/PCBM	600	6.61	0.39	1.54
P3OT/PCBM	635	4.58	0.38	1.10
P3DDT/PCBM	650	2.53	0.36	0.59

summarizes the photovoltaic performance parameters of the cells depicted in Figure 23.15–Figure 23.17 on AM1.5 condition.

The open circuit voltage of the three cells mentioned above increases in the graduation P3HT (600 mV) < P3OT (635 mV) < P3DDT (650 mV). Different acceptors (fullerenes) have been investigated in combination with the same donor [9], where the authors found that the open circuit voltage depends on the acceptor strength of the fullerenes applied. This result fully does support the assumption that the open circuit voltage of a donor–acceptor bulk heterojunction cell is directly related to the energy difference between the HOMO level of the donor and the LUMO level of the acceptor component. In agreement with this result and the realizable trend comparing E_p^{ox}, E_{onset}^{ox} of P3HT, P3OT, and P3DDT (see Table 23.2), a possible explanation could be that the relatively small differences in the HOMO levels of the three polythiophenes slightly affect their donor strength and this corresponds to the energy difference between the HOMO level of the donor polymers and the LUMO level of PCBM.

The cell based on P3HT possesses a higher short circuit current (6.61 mA/cm^2) than the cells based on P3OT (4.58 mA/cm^2) and P3DDT (2.53 mA/cm^2). Regioregular head-to-tail P3HT is known for a high degree of intermolecular order leading to high charge carrier mobilities ($\mu_{hole} \sim 0.1\,\text{cm}^2\,\text{V}^{-1}\,\text{s}^{-1}$ [31,32], $\mu_{hole} \sim 0.2\,\text{cm}^2\,\text{V}^{-1}\,\text{s}^{-1}$ [33] measured in form of field effect transistor geometries). The hole mobility for P3DDT was described to be lower ($\mu_{hole} \sim 0.002$–$0.005\,\text{cm}^2\,\text{V}^{-1}\,\text{s}^{-1}$ [34] measured in form of field effect transistor geometries). Assuming the same degree of regioregularity for all three P3ATs, the hole mobility should increase as the length of the side chain decreases due to their contribution to the degree of intermolecular order and chain packaging density. Additionally, the potential barrier at P3OT/ITO is slightly higher than that at P3HT/ITO and somewhat lower than that one at P3DDT/ITO (see Figure 23.7 and Table 23.2) with a probably clearer trend in E_p^{ox} comparing P3HT–P3OT–P3DDT than it is expressed by the differences in HOMO levels based on E_{onset}^{ox}. Thus the hole injection from the HOMO of the polymers into ITO becomes less restricted in the case of P3HT if compared to the other two polythiophenes. P3HT shows a higher absorption coefficient than P3OT and P3DDT (see Figure 23.6). Consequently, compared to the other two polymers, P3HT is enabled to convert larger amounts of photons promoting an increased current within the cell.

The cells based on P3HT/PCBM and P3OT/PCBM have approximately the same fill factor (0.39 and 0.38, respectively). A disturbed or less ordered morphology of the photoactive layer may be the reason for the slightly lower fill factor (0.36) found for devices based on P3DDT/PCBM with respect to the values of P3HT and P3OT. Therefore, the external AM1.5 power conversion is found to be higher for the cell based on P3HT (1.54%) than for the P3OT (1.1%) and P3DDT (0.59%) devices.

According to the single diode model described in Ref. [35,36], the serial resistance of the cells based on P3HT amounts to 4.5 Ω cm^2, 11 Ω cm^2 for P3OT, and 39 Ω cm^2 for P3DDT device. The parallel resistance of the cell based on P3HT is 246 Ω cm^2, 303 Ω cm^2 for P3OT, and 1000 Ω cm^2 for P3DDT device. The different photovoltaic parameters of the cells based on P3HT, P3OT, and P3DDT may be explained in accordance with the values of their resistances. The cell based on P3HT has the lowest serial resistance and the lowest parallel resistance; therefore, it shows the highest current and the lowest voltage. The cell based on P3DDT has the highest serial resistance and the highest parallel resistance, consequently it gives the lowest current and the highest voltage. The cell based on P3OT is a compromise between both borderline cases. These results are in agreement with the analysis of the influence of the resistance values on the photovoltaic parameters of organic solar cells as described in Ref. [37].

23.2.2.5. Summary of P3ATs/Fullerene Systems

The HOMO and LUMO values of regioregular P3HT, P3OT, and P3DDT were estimated from cyclic voltammetry data. These values slightly depend on the length of the alkyl side chain, whereas the energy of the HOMO and the absorption coefficients of the three P3ATs are inversely proportional to this length. Induced D–A charge transfer processes in P3AT–PCBM composites were detected by photoluminescence quenching after blending with PCBM and the polaron (P$^{+\cdot}$ on polymer chain) as well as the PCBM radical anion could be assigned to signals found in LESR spectra.

According to the results obtained, it can be concluded that an energetic difference of only ~0.2 eV between the LUMO of P3HT, P3OT, or P3DDT and the LUMO of PCBM seems to be sufficient to ensure a directed donor–acceptor charge transfer. On the other hand, the position of BM-C$_{60}$ was found on nearly the same level like that of P3HT, P3OT, and P3DDT, which is in agreement with the PL quenching result (only partially) and the ESR measurements (no signal detected) for P3DDT/BM-C$_{60}$.

The large area flexible polymer solar cells based on P3HT/PCBM (1:3 wt.%), P3OT/PCBM (1:3 wt.%), and P3DDT/PCBM (1:3 wt.%) show V_{oc} values between 600 and 650 mV. V_{oc} systematically increases in the order P3HT/PCBM < P3OT/ PCBM < P3DDT/PCBM cells, which is probably due to the slightly lower HOMO levels of P3DDT and P3OT compared with P3HT. I_{sc} of the P3HT/PCBM cell (6.61 mA/cm^2) is higher than that of P3OT/PCBM (4.58 A/cm^2) and P3DDT/ PCBM device (2.53 mA/cm^2). These values are determined by

(a) Higher absorption coefficients
(b) An increased holes mobility
(c) A lower energy transition barrier for holes undergoing transfer from the HOMO level into the ITO anode, which is preferred for P3HT against P3OT and P3DDT.

All three materials do also possess fill factors having the same order of magnitude. Within the series of material compositions P3HT/PCBM < P3OT/PCBM < P3DDT/PCBM, the serial and parallel resistance as well as the open circuit voltage increases, whereas the short circuit current decreases.

23.2.3. MDMO-PPV, DE69, and DE21 as Electron Donors

Polymers with a higher oxidation potential should reduce the photooxidation of the photoactive layer in presence of oxygen or water to enhance the stability of polymer solar cells. Poly(phenylene ethynylene)s (PPEs) are known to be more stable towards photooxidation than their PPV analogs [38]. Contrary to PPVs, the PPEs have attracted much less attention despite their favorable emission characteristics and high fluorescence quantum yields [39]. From this view point, it seemed to be interesting to test PPE–PPV hybrid polymers, which have not been the subject of intensive studies up till now [40–43].

In this section, MDMO-PPV and defect-free PPE–PPV hybrid polymers, DE69, and DE21 [44–46] are described and their use as electron donors in polymer solar cells is presented. PCBM and BM-C$_{60}$ (see Figure 23.4) serve as electron acceptors. The chemical structures of the polymers MDMO-PPV, DE69, and DE21 are shown in Figure 23.18.

Despite the extensive use of PPV, PPE–PPV hybrid polymers have not been widely used in solar cell applications [47]. The synthesis of the PPE–PPV hybrid polymers, DE69, and DE21, have been described elsewhere [44–46]. Instead of applying the Sonogashira Pd-catalyzed cross-coupling reaction [43,47], a synthetic pathway based on the Horner–Wadsworth–Emmons olefination reaction of luminophoric dialdehydes with bisphosphonates was preferred. Combined with a thorough purification procedure, high molecular weight and defect-free polymers (synonym DE69 and DE21) may be obtained. Contrary to the reaction mentioned above, the Sonogashira polycondensation reaction of arene diiodes with diethynyl derivatives is known to give polymers with minor amounts of diyne ($-C\equiv C-C\equiv C-$) defect units, resulting from a partial oxidative coupling of acetylenes in the presence of CuI (serving as cocatalyst) [40,48].

23.2.3.1. Electrochemical Properties of MDMO-PPV, DE69, and DE21

Table 23.5 displays the oxidation and reduction potentials of the polymers MDMO-PPV, DE69, and DE21 determined in form of a film by means of cyclic voltammetry. The electrochemical data of MDMO-PPV are $E_{onset}^{ox} = +0.93$ V, $E_{onset}^{red} = -1.55$ V

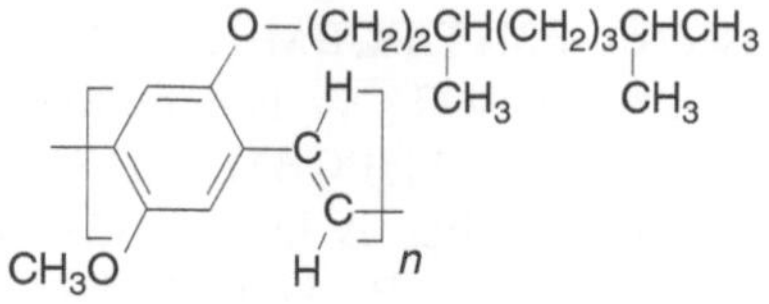

MDMO-PPV

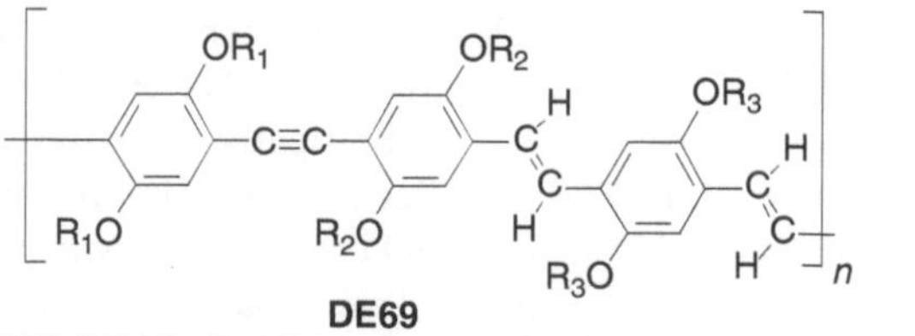

DE69

PPE-PPV:R_1,R_2:$(CH_2)_7CH_3$ R_3:$CH_2CH(C_2H_5)(CH_2)_3CH_3$

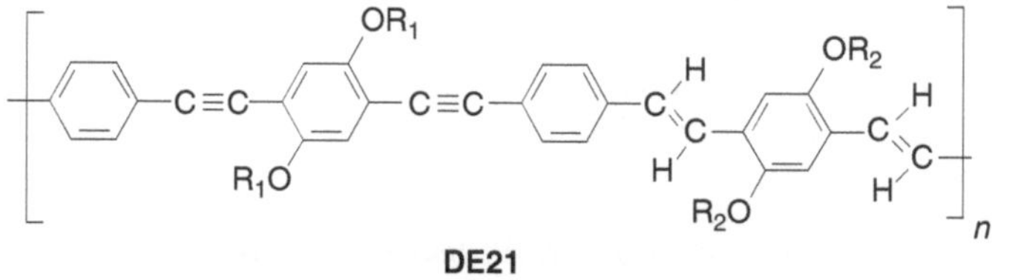

DE21

PPE-PPV: R_1: $(CH_2)_{17}CH_3$ R_2: $(CH_2)_{17}CH_3$

Figure 23.18. Chemical structures of the investigated polymers MDMO-PPV, DE69, und DE21.

Table 23.5. Electrochemical data, HOMO/LUMO levels, and E_g^{CV} of MDMO-PPV, DE69, and DE21 [26,50]

Polymer	E_p^{ox} vs. Ag/AgCl (V)	E_{onset}^{ox} vs. Ag/AgCl (V)	HOMO (eV)	E_p^{red} vs. Ag/AgCl (V)	E_{onset}^{red} vs. Ag/AgCl (V)	LUMO (eV)	E_g^{CV} (eV)
MDMO-PPV	+1.22	+0.93	−5.31	−1.72	−1.55	−2.83	2.48
DE69	+1.33	+1.08	−5.46	−1.16	−0.82	−3.56	1.90
DE21	+1.44	+1.22	−5.60	−1.06	−0.80	−3.58	2.02

measured vs. Ag/AgCl. It is probably due to the presence of triple bonds in the polymer backbones that the onset and peak values for oxidation processes of DE69 and DE21 are higher, and these values for reduction processes are lower than those of MDMO-PPV. These bonds are also responsible for the reduction of the donor strength of PPE–PPV hybrid polymers if compared to PPV homopolymers. Thus, the HOMO and LUMO levels of MDMO-PPV, DE69, and DE21 were estimated based on the reference energy level of ferrocene (4.8 eV below the vacuum level) according to the already introduced empirical equation: $E^{HOMO/LUMO} = [-e(E_{onset(vs. Ag/AgCl)} - E_{onset(Fc/Fc+ vs. Ag/AgCl)})] - 4.8$ eV [10,11].

The electrochemical bandgap of MDMO-PPV (2.48 eV) is higher than the bandgaps of DE69 (1.90 eV) and DE21 (2.02 eV). With the exception of MDMO-PPV, the electrochemical gap of the investigated polymers is clearly lower than the optical gap, which was determined from the low-energy edge of the absorption spectra of films. The same effect was observed for P3ATs (see Section 23.2.2.1) and reported by other authors [9,17,18].

23.2.3.2. Optical Properties of MDMO-PPV, DE69, and DE21

Figure 23.19 shows the optical absorption spectra of MDMO-PPV, DE69, and DE21 obtained from films deposited from a chlorobenzene solution. Compared to MDMO-PPV, the λ_{max} values of the absorption spectra of DE69 (2.64 eV/470 nm)

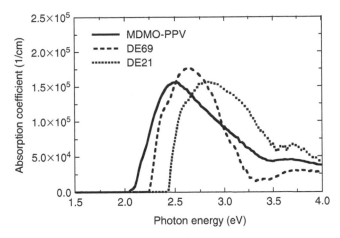

Figure 23.19. Optical absorption spectra of MDMO-PPV, DE69, und DE21 measured in form of films spin-coated from chlorobenzene solutions.

Table 23.6. Optical data of MDMO-PPV, DE69, and DE21 determined from ellipsometric data

Polymer	λ_{max} [nm (eV)]	Absorption coefficient at λ_{max} (1/cm)	E_g^{opt} (eV)
MDMO-PPV	492 (2.52)	1.57×10^5	2.12
DE69	470 (2.64)	1.76×10^5	2.24
DE21	437 (2.84)	1.57×10^5	2.44

and DE21 (2.84 eV/437 nm) undergo a distinctive blueshift (for DE69: 0.12 eV/22 nm, for DE11: 0.32 eV/55 nm). These values and the corresponding bandgap energies of the investigated polymers are shown in Table 23.6. Both polymers also exhibit slightly higher optical bandgap energies (0.12 and 0.32 eV, respectively), but lower electrochemical bandgaps in comparison to MDMO-PPV (see Section 23.2.3.1).

Regarding the absorption coefficients, one can recognize that the value of hybrid polymer DE69 is in fact somewhat higher, but still within the same order of magnitude, whereas DE21 has almost the same value as MDMO-PPV (Table 23.6). Despite identical or even longer side chains in DE69 and DE21, the higher or identical absorption coefficients are probably the result of symmetrical substitution patterns when compared to MDMO-PPV. Obviously, the resulting enhancement of the conformational order within individual or between different chains will strongly influence the aggregation behavior of the macromolecules [51]. As already reported for P3ATs, the absorption graphs of the composites consisting of MDMO-PPV, DE69, or DE21 with PCBM (1:2 wt.%) represent simple superpositions of the spectra of the individual components without any ground state doping by PCBM [26,50].

23.2.3.3. Photoinduced Electron Transfer in MDMO-PPV, DE69, or DE21–Fullerene Systems

Figure 23.20 shows the energy band diagram of MDMO-PPV, DE69, DE21, and PCBM derived from the cyclic voltammographic data in relation to the work

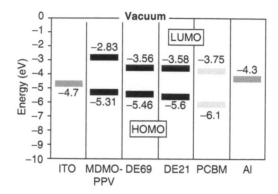

Figure 23.20. Energy band diagram of MDMO-PPV, DE69, DE21, and PCBM from the cyclic voltammogramic data in relation to the work function of the common electrode materials ITO and Al (HOMO of PCBM is a literature value [1]).

function of the common electrode materials ITO and Al. The LUMO position of BM-C_{60} is given in Figure 23.7 (−3.51 eV). Although the hybrid polymers DE69 and DE21 possess higher oxidation potentials that ensure a higher stability, it simultaneously leads to an increased hole injection barrier in combination with ITO electrodes. As already described by Brabec et al. [9], this phenomenon leads to an increased difference between the HOMO of the donor and the LUMO of the acceptor, which should promote higher open circuit voltages (V_{co}) in solar cell devices.

All three polymers described here show distinctively higher energy levels of their HOMOs in comparison to PCBM, whereas the HOMO of BM-C_{60} has obviously not been measured yet. As explained in the "Introduction", for the intended charge transfer, the relative position of donor LUMO and acceptor LUMO is decisive. Figure 23.20 clearly displays the difference of 1.08 eV between the LUMOs of MDMO-PPV and that of PCBM, which should allow an effective charge transfer. It seems that a charge transfer is still enabled in the case of DE69 and DE21 in combination with PCBM, although their LUMO–LUMO-differences are markedly lower (DE69: 0.2 eV and DE21: 0.17 eV). For MDMO-PPV/BM-C_{60} the $\Delta LUMO_{donor} - LUMO_{acceptor}$ amounts to 0.68 eV, which should be sufficient for a directed electron transfer. However, the photoluminescence quenching factor differs markedly from MDMO-PPV/PCBM composite.

Photoluminescence measurement and LESR technique were used to investigate the photoinduced charge generation and charge transfer in composites of MDMO-PPV, DE69, or DE21 with PCBM and MDMO-PPV with BM-C_{60}, respectively. Photoexcited excitons can undergo several consecutive processes: recombination within the polymer (donor) or at the donor–acceptor interface, energy transfer between the materials followed by radiative or nonradiative recombination or charge transfer (exciton splitting) at the donor–acceptor interface. Within the scope of investigated systems MDMO-PPV/PCBM, DE69/PCBM, DE21/PCBM, and MDMO-PPV/BM-C_{60}, one can consider the charge transfer as the preferred process due to:

(a)   the intensive photoluminescence of the polymers MDMO-PPV, DE69, and DE21 and its quenching in the presence of PCBM or BM-C_{60}, respectively and

(b) the detection of two types of LESR signals (one to be assigned to the polaron $P^{+\bullet}$ formed on the polymer chain and another one belonging to the radical anion of PCBM or $BM\text{-}C_{60}$).

The measurements of photoluminescence were performed on a thin polymer (donor) film and compared with the photoluminescence of the blend film of donor and acceptor. The effective quenching of photoluminescence of the polymers MDMO-PPV, DE69, and DE21 in the presence of PCBM indicates [1,28] an effective charge transfer from these polymers to the acceptor PCBM (see Figure 23.21–Figure 23.23). The quenching behavior of MDMO-PPV in the presence of $BM\text{-}C_{60}$ is depicted in Figure 23.24. With decreasing $LUMO_{donor} - LUMO_{acceptor}$ difference ($\Delta LUMO_{D-A}$), the PL quenching factor ($PL_{polymer} : PL_{composite}$) is dropping for the investigated polymer systems MDMO-PPV/PCBM, DE69/PCBM, and DE21/PCBM (see Table 23.7). This corresponds to the reduced donor strength of DE69 and DE21 compared to MDMO-PPV, and is caused by an increasing content of $C\equiv C$ triple bonds per monomer unit. This structural feature is also responsible for the lower reduction potential of these polymers compared to MDMO-PPV. A similar correlation of the PL quenching factor with the $\Delta LUMO_{D-A}$ energy is found comparing MDMO-PPV/PCBM with MDMO-PPV/$BM\text{-}C_{60}$. This is in agreement

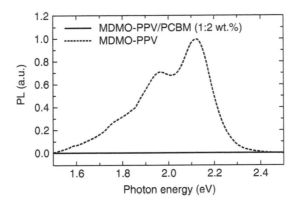

Figure 23.21. Film photoluminescence spectra of MDMO-PPV and of a MDMO-PPV/PCBM blend (1:2 wt.%).

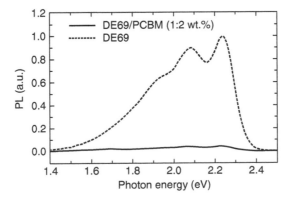

Figure 23.22. Film photoluminescence spectra of DE69 and of a DE69/PCBM blend (1:2 wt.%).

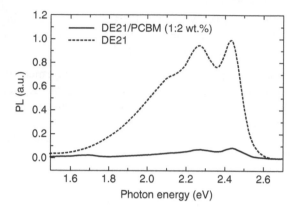

Figure 23.23. Film photoluminescence spectra of DE21 and of a DE21/PCBM blend (1:2 wt.%).

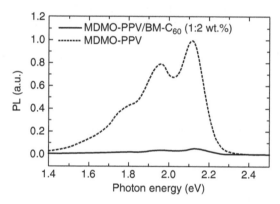

Figure 23.24. Film photoluminescence spectra of MDMO-PPV and of a MDMO-PPV/ BM-C_{60} blend (1:2 wt.%).

Table 23.7. Relationship of photoluminescence quenching factor of composites MDMO-PPV/PCBM, MDMO-PPV/BM-C_{60}, DE69/PCBM, and DE21/PCBM with the LUMO$_{donor}$ − LUMO$_{acceptor}$ difference

Composite	PL quenching factor[a]	ΔLUMO$_{D-A}$ [eV][b]
MDMO-PPV/PCBM (1:2)	495	0.92
MDMO-PPV/BM-C_{60} (1:2)	21	0.68
DE69/PCBM (1:2)	26	0.19
DE21/PCBM (1:2)	12	0.17

[a] PL$_{polymer}$:PL$_{composite}$ ratio.
[b] LUMO$_{donor}$ − LUMO$_{acceptor}$ difference (ΔLUMO$_{D-A}$).

with the lower acceptor strength of BM-C_{60} compared to PCBM, which is due to the nucleophilic morpholino groups. Summarizing the results given above and considering the charge transfer detected by LESR, there is no discrepancy in the exciton binding energies in the literature (0–0.6 eV for MEH-PPV, 0.05–1.1 eV for PPV, and 0.36 eV for alkoxy-PPV) as reported elsewhere [52] despite the small LUMO−LUMO differences for DE69/PCBM and DE21/PCBM. Such reported values may differ significantly.

The effective charge separation in the systems MDMO-PPV/PCBM, DE69/PCBM, and DE21/PCBM has been proven by two LESR signals, which were only observed under light excitation. They can be attributed to positive polarons ($P^{+\bullet}$) on the polymer chain with a g-factor $g_{iso} \approx 2.0027$ and to radical anions of PCBM molecules with a $g_{iso} \approx 1.9995$ corresponding to [20,28,30] (Figure 23.25–Figure 23.27). Polymer and PCBM radicals may show significant differences in their relaxation dynamics, causing considerable differences in the LESR signal amplitudes.

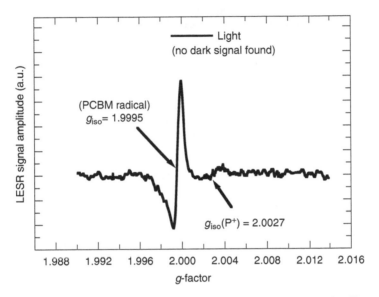

Figure 23.25. LESR spectra of MDMO-PPV/PCBM (1:2 wt.%) composite film under light illumination $I = 15\,\mathrm{mW/cm^2}$ (Ar^+ laser, λ: 488 nm) measured at a microwave power of 0.4 mW at 77 K.

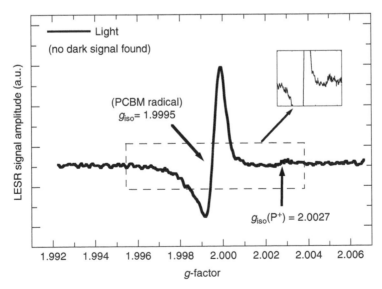

Figure 23.26. LESR spectra of DE69/PCBM (1:2 wt.%) composite film under light illumination $I = 15\,\mathrm{mW/cm^2}$ (Ar^+ laser, λ = 488 nm) measured at a microwave power of 0.4 mW at 77 K.

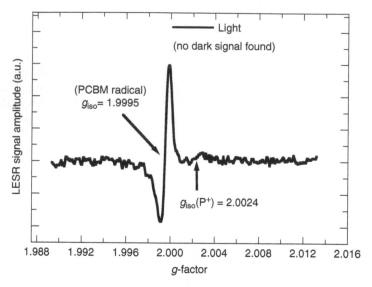

Figure 23.27. LESR spectra of DE21/PCBM (1:2 wt.%) composite film under light illumination $I = 15\,mW/cm^2$ (Ar$^+$ laser, $\lambda = 488\,nm$) measured at a microwave power of 0.4 mW at 77 K.

They are observed for the polarons on the polymer chains at a much lower saturation microwave power than for the PCBM radicals. These results and the determined *g*-factors are very similar to the obtained spectra reported in Ref. [20]. In all experiments, no dark signals of both types of radicals could be found which clearly proves the occurrence of charge separation processes after light excitation in DE21/PCBM (1:2 wt.%) and DE69/PCBM (1:2 wt.%) blends. This behavior is comparable to that of MDMO-PPV/PCBM (1:2 wt.%).

An effective charge separation was analogously found in the system MDMO-PPV/BM-C_{60} (1:2 wt.%) (Figure 23.28) perceptible from two anisotropic LESR

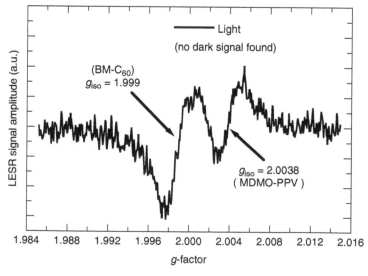

Figure 23.28. LESR spectra of MDMO-PPV/BM-C_{60} (1:2 wt.%) composite film under light illumination $I = 8\,mW/cm^2$ (Ar$^+$ laser, $\lambda = 488\,nm$) measured at a microwave power of 2 mW at 77 K.

signals. For MDMO-PPV, the polaron signal is more pronounced than for a blend of MDMO-PPV/PCBM, indicating different spin concentrations or relaxation dynamics. Nevertheless, it permits the determination of the isotropic g-factor with higher accuracy than for the weak polaron signal in Figure 23.25 (polaron: $g_{iso} \approx$ 2.0038, BM-C_{60} radical anion: $g_{iso} \approx 1.999$). The latter one is useful to explain the different g-factors assigned to the polaron signals in both spectra. No dark signal could be detected for the MDMO-PPV/BM-C_{60} composite.

23.2.3.4.   Photovoltaic Devices Based on MDMO-PPV/PCBM, DE69/PCBM, and DE21/PCBM

Photovoltaic cells using the systems MDMO-PPV/PCBM (1:3 wt.%), DE69/PCBM (1:3 wt.%), and DE21/PCBM (1:3 wt.%) in the photoactive layer (100–150 nm) sandwiched between a transparent ITO front electrode covered with a 100 nm layer of poly(3,4-ethylene-dioxythiophene) doped with poly(styrene sulfonic acid) (PEDOT–PSS) and an Al back electrode (80 nm) were prepared. The substrate is a flexible polyester foil (PET-foil, 100 nm ITO, surface resistance: 60 Ω/square) as given in Figure 23.14. The device area amounts to 0.25 cm^2. Figure 23.29 shows the I–V curves of a PET-foil/ITO/PEDOT–PSS/MDMO-PPV:PCBM (1:3 wt. ratio)/Al solar cell in the dark and under white-light illumination (100 mW/cm^2 with AM1.5 conditions). The investigated solar cell possesses the following parameters: open circuit voltage $V_{oc} = 0.76$ V, short circuit current $I_{sc} = 8.2$ mA/cm^2, fill factor, defined as $(I_{max} \times V_{max})/(I_{sc} \times V_{oc})$, with I_{max} and V_{max} corresponding to the maximum power point, FF = 0.49, and the power conversion efficiency $\eta_{AM1.5}$ [%] = $(P_{out}/P_{light}) \times$ 100% = (FF $\times V_{oc} \times I_{sc} / P_{light}) \times$ 100% = 3% [53].

In Figure 23.30 and Figure 23.31, the I–V characteristics of a PET-foil/ITO/ PEDOT–PSS/DE69:PCBM (1:3 wt.%)/Al solar cell and of a PET-foil/ITO/PEDOT– PSS/DE21:PCBM (1:3 wt. ratio)/Al solar cell, respectively, in the dark and under white-light illumination (100 mW/cm^2 with AM1.5 conditions) are plotted. The open circuit voltage V_{oc}, the short circuit current I_{sc}, the fill factor FF, and the power conversion

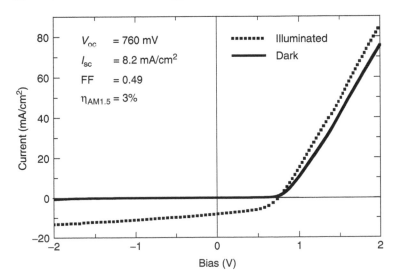

Figure 23.29. I–V characteristics of PET-foil/ITO/PEDOT–PSS/MDMO-PPV:PCBM (1:3 wt.%)/Al solar cell in the dark and under 100 mW/cm^2 white-light illumination (AM1.5), device area = 0.25 cm^2.

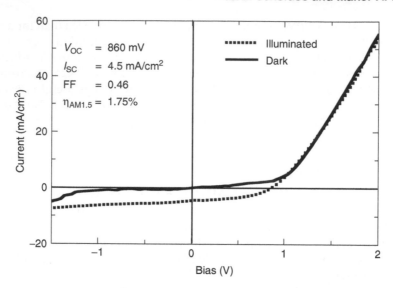

Figure 23.30. *I–V* characteristics of PET-foil/ITO/PEDOT–PSS/DE69:PCBM (1:3 wt.%)/Al solar cell in the dark and under 100 mW/cm² white-light illumination (AM1.5), device area = 0.25 cm².

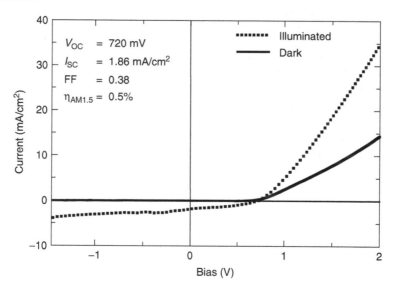

Figure 23.31. *I–V* characteristics of PET-foil/ITO/PEDOT–PSS/DE21:PCBM (1:3 wt.%)/Al solar cell in the dark and under 100 mW/cm² white-light illumination (AM1.5), device area = 0.25 cm².

efficiency $\eta_{AM1.5}$ of DE69/PCBM device under white-light illumination (100 mW/cm²) are 860 mV, 4.5 mA/cm², 0.46%, and 1.75%, respectively. The photovoltaic parameters of the cell based on DE21/PCBM (1:3 wt.%) are: V_{oc}, 720 mV; I_{sc}, 1.86 mA/cm²; FF, 0.38, and $\eta_{AM1.5}$ under white-light illumination (100 mW/cm²): 0.5%. Table 23.8 summarizes the photovoltaic parameters of cells based on these three systems.

The open circuit voltage of DE69/PCBM cell is higher than that of MDMO-PPV/PCBM, which should be attributed to the higher difference between HOMO of DE69 and LUMO of PCBM [9] (Figure 23.20). The presented performance of DE21/

Table 23.8. Photovoltaic parameters of cells based on MDMO-PPV/PCBM (1:3 wt.%), DE69/PCBM (1:3 wt.%), and DE21/PCBM (1:3 wt.%) under illumination with $100\,mW/cm^2$ of AM1.5 white light

Cell based on	V_{oc} (mV)	I_{sc} (mA/cm^2)	FF (%)	$\eta_{AM1.5}$ (%)
MDMO-PPV/PCBM	760	8.20	49	3.00
DE69/PCBM	860	4.50	46	1.75
DE21/PCBM	720	1.86	38	0.50

PCBM device is not yet on an optimized level, which is demonstrated by the relatively low open circuit voltage, and is normally expected to be higher than for DE69/PCBM cell.

Since MDMO-PPV has a lower optical bandgap than the polymers DE21 and DE69, it is able to absorb more photons, which leads to an increased number of photogenerated charge carriers. Consequently, the short circuit current of the MDMO-PPV/PCBM cell is higher compared to DE69/PCBM, but significantly higher in comparison to DE21/PCBM.

Additionally, the triple bonds in the main chain of DE69 and DE21 probably reduce the mobility of charge carriers in these polymers, which may be due to a lower conjugation compared with PPV homopolymers [43]. A lower hole mobility reduces the photocurrent of MDMO-PPV/PCBM, DE69/PCBM, and DE21/PCBM cells (85.3, 54.4, and 34.9 mA/cm^2, respectively, at +2 V). As described above, in the case of short circuit current, the serial resistance of the MDMO-PPV/PCBM device cell is lower compared to DE69/PCBM, but significantly lower in comparison to DE21/PCBM [50]. Furthermore the hole injection barrier at the HOMO$_{polymer}$–ITO interface increases in the order MDMO-PPV/PCBM < DE69/PCBM < DE21/PCBM as a result of the different HOMO levels of the investigated polymers (see Figure 23.20). Especially, the performance of the cell based on DE69/PCBM could be improved towards a higher short circuit current, for example by optimizing the thickness of the photoactive layer.

23.2.3.5. Summary of MDMO-PPV/PCBM and PPE–PPV/PCBM Systems

Novel hybrid PPE–PPV polymers, DE69 and DE21, were compared with the state-of-the-art material MDMO-PPV with regard to HOMO and LUMO levels, absorption coefficients, and charge transfer properties in composites with PCBM (photoluminescence quenching and LESR). In comparison to MDMO-PPV homopolymer, the hybrid polymers show typical features such as energetically lower HOMO and LUMO levels. Low HOMO levels will cause higher oxidation potentials, leading to an improved material stability. On the other hand, lower LUMO levels are accompanied by a diminished donor strength if PCBM is used as the acceptor. Nevertheless, photoluminescence quenching and the detected LESR signals for polarons on polymer chain and PCBM radical anion show that an effective donor–acceptor charge transfer takes place. The photoluminescence quenching factor was found to be proportional to the LUMO$_{donor}$ − LUMO$_{acceptor}$ difference when comparing MDMO-PPV/PCBM with DE69/PCBM, and DE21/PCBM and MDMO-PPV/PCBM with MDMO-PPV/BM-C$_{60}$. Large area photovoltaic devices on flexible PET–ITO foils with MDMO-PPV/PCBM (1:3 wt.%), yielding an efficiency of

$\eta_{AM1.5} = 3\%$ could be realized. An $\eta_{AM1.5}$ value of 1.75% was found for the DE69/ PCBM (1:3 wt.%) system prepared and measured on ambient conditions.

23.3. CONCLUSIONS

It was the objective of the present paper to demonstrate that a systematic preliminary investigation of the optoelectronic properties involving well-investigated, as well as novel donor and acceptor materials represents a helpful tool to evaluate potential materials for polymer solar cell applications. Although it is not possible to demonstrate the extent of the value of these methods within the scope of this paper, the importance of other methods such as photoinduced absorption (PIA) spectroscopy or incident photon-to-current efficiency (IPCE) measurements to estimate the photovoltaic potential of several donor–acceptor combinations is not in doubt.

The knowledge of HOMO and LUMO levels of the organic materials applied is a required precondition to predict the probability of photoinduced charge transfer processes of binary donor–acceptor pairs and will help to understand how a specified material property influences a special parameter of the photovoltaic device. This is the key issue to design materials with the most optimal compromise of properties (e.g., of optoelectronic and film forming properties), resulting in the maximum device performance. To find a minimum $LUMO_{donor} - LUMO_{acceptor}$ difference, which simultaneously ensures a directed charge transfer and a high open circuit voltage, represents one of these compromises. This chapter is regarded as a contribution to this discussion.

ACKNOWLEDGMENTS

We wish to thank Professor E. Klemm and Dr. D.A.M. Egbe, Friedrich-Schiller, University of Jena for the synthesis of PPE–PPV hybrid polymers, Professor P. Scharff, Technical University of Ilmenau for the preparation of bismorpholine-C_{60}, Professor G. Gobsch and U. Zhokhavets, Technical University of Ilmenau for the optical and photoluminescent measurements and our colleagues Dr. A. Konkin for the ESR experiments, Dr. M. Schroedner, Professor H.-K. Roth, and Dr. T. Schulze for the helpful discussions. Financial support from the German Ministries BMWI (project No. 347/01) and BMBF (project No. 01SF0119) is gratefully acknowledged.

REFERENCES

1. N. S. Sariciftci, L. Smilowitz, A. J. Heeger, and F. Wudl, Photoinduced electron transfer from a conducting polymer to buckminsterfullerene, *Science* **258**, 1474–1476 (1992).
2. C. J. Brabec, G. Zerza, G. Cerullo, S. De Silvestri, S. Luzzati, J. C. Hummelen, and N. S. Sariciftci, Tracing photoinduced electron transfer process in conjugated polymer/fullerene bulk heterojunctions in real time, *Chem. Phys. Lett.* **340**, 232–236 (2001).
3. F. Padinger, R. S. Rittberger, and N. S. Sariciftci, Effects of postproduction treatment on plastic solar cells, *Adv. Funct. Mater.* **11(13)**, 1–4 (2003).
4. D. Gebeyehu, C. J. Brabec, F. Padinger, T. Fromherz, J. C. Hummelen, D. Badt, H. Schindler, and N. S. Sariciftci, The interplay of efficiency and morphology in photo-

voltaic devices based on interpenetrating networks of conjugated polymers with fullerenes, *Synth. Met.* **118**, 1–9 (2001).

5. C. Waldauf, P. Schilinsky, M. Perisutti, J. Hauch, and C. J. Brabec, Solution-processed organic n-type thin-film transistors, *Adv. Mater.* **15**, 2084–2088 (2003).

6. P. Schilinsky, C. Waldauf, and C. J. Brabec, Recombination and loss analysis in polythiophene based bulk heterojunction photodetectors, *Appl. Phys. Lett.* **81**, 3885–3887 (2002).

7. J. -M. Nunzi, Organic photovoltaic materials and devices, *C.R. Physique* **3**, 523–542 (2002).

8. C. J. Brabec, Semiconductor aspects of organic bulk heterojunction solar cells [Chapter 5], in *Organic Photovoltaics*, C. J. Brabec, V. Dyakonov, J. Parisi, and N. S. Sariciftci, eds., Springer-Verlag, Heidelberg, 2003, pp. 159–248.

9. C. J. Brabec, A. Cravino, D. Meissner, N. S. Sariciftci, T. Fromherz, M. T. Rispens, L. Sanchez, and J. C. Hummelen, Origin of the open circuit voltage of plastic solar cells, *Adv. Funct. Mater.* **11(5)**, 374–380 (2001).

10. Y. Liu, M. S. Liu, and A. K.-Y. Jen, Synthesis and characterization of a novel and highly efficient light-emitting polymer, *Acta Polymerica* **50**, 105–108 (1999).

11. S.-W. Hwang and Y. Chen, Synthesis and electrochemical and optical properties of novel poly(aryl ether)s with isolated carbazole and *p*-quaterphenyl chromophores, *Macromolecules* **34**, 2981–2986 (2001).

12. M. Al-Ibrahim, H.-K. Roth, M. Schroedner, A. Konkin, U. Zhokhavets, G. Gobsch, and S. Sensfuss, The influence of the optoelectronic properties of poly(3-alkylthiophenes) on the device parameters in flexible polymer solar cells, *Organic Electronics* (submitted 2004).

13. T.-A Chen, X. Wu, and R. D. Rieke, Regiocontrolled synthesis of poly(3-alkylthiophenes) mediated by Rieke zinc: their characterization and solid-state properties, *J. Am. Chem. Soc.* **117**, 233–244 (1995).

14. H. Xie, S. O'Dwyer, J. Corish, and D. A. Morton-Blake, The thermochromism of poly (3-alkylthiophene)s: the role of the side chains, *Synth. Met.* **122**, 287–296 (2001).

15. N. Camaioni, M. Catellani, S. Luzzati, A. Martelli, and A. Migllori, Effect of regioregularity on the photoresponse of Schottky-type junctions based on poly(3-alkylthiophenes), *Synth. Met.* **125**, 313–317 (2002).

16. A. Heeger and P. Smith, The novel science and technology of highly conducting and nonlinear optically active materials, in *Conjugated Polymers*, J.L. Bredas and R. Silbey, eds., Kluwer, Dordrecht, 1991, pp. 142–210.

17. R. Cervini, X.-C. Li, G. W. Spencer, A. B. Holmes, S. C. Morath, and R. H. Friend, Electrochemical and optical studies of PPV derivatives and poly(aromatic oxadiazoles), *Synth. Met.* **84**, 359–360 (1997).

18. B. Liu, W.-L. Yu, Y.-H. Lai, and W. Huang, Blue-light-emitting fluorene-based polymers with tunable electronic properties, *Chem. Mater.* **13**, 1984–1996 (2001).

19. T. Ahn, B. Choi, S. H. Ahn, S. H. Han, and H. Lee, Electronic properties and optical studies of luminescent polythiophene derivatives, *Synth. Met.* **117**, 219–221 (2001).

20. D. Chirvase, Z. Chigvare, M. Knipper, J. Parisi, V. Dyakonov, and J. C. Hummelen, Electrical and optical design and characterisation of regioregular poly(3-hexylthiophene-2,5diyl)/fullerene-based heterojunction polymer solar cells, *Synth. Met.* **138**, 299–304 (2003).

21. K. Takahashi, K. Tsuji, K. Imoto, T. Yamaguchi, T. Komura, and K. Murata, Enhanced photocurrent by Schottky-barrier solar cell composed of regioregular polythiophene with merocyanine dye, *Synth. Met.* **130**, 177–183 (2002).

22. X. Qiao, X. Wang, and Z. Mo, The effects of different alkyl substitution on the structures and properties of poly(3-alkylthiophenes), *Synth. Met.* **118**, 89–95 (2001).

23. G. Dicker, T. J. Savenije, B.-H. Huisman, D. M. de Leeuw, M. P. de Haas, and J. M. Warman; Photoconductivity enhancement of poly(3-hexylthiophene) by increasing inter- and intra-chain order, *Synth. Met.* **137**, 863–864 (2003).

24. J. J. Apperloo, R. A. J. Janssen, M. M. Nielsen, and K. Bechgaard, Doping in solution as an order-inducing tool prior to film formation of regio-irregular polyalkylthiophenes, *Adv. Mater.* **12**, 1594–1596 (2000).

25. R. H. Friend, Conjugated polymers and related materials, in *Proceedings on the Eighty-First Nobel Symposium*, W.R. Salaneck, I. Lundström, and B. Ranby, eds., Oxford University Press, Oxford, 1993, pp. 285–323.

26. S. Sensfuss, M. Al-Ibrahim, A. Konkin, G. Nazmutdinova, U. Zhokhavets, G. Gobsch, D. A. M. Egbe, E. Klemm, and H.-K. Roth, Characterisation of potential donor acceptor pairs for polymer solar cells by ESR, optical and electrochemical investigations, in *Proc. SPIE*, Vol. 5215, "Organic Photovoltaics IV," Z. Kafafi and P. A. Lane, eds., SPIE, San Diego, CA, August 3–8, 2003, pp. 129–140.

27. M. Al-Ibrahim, H.-K. Roth, U. Zhokhavets, G. Gobsch, and S. Sensfuss, Flexible large area polymer solar cells based on poly(3-hexylthiophene)/fullerene, *Sol. Energy Mater. Sol. Cells*, in press 2004.

28. S. Sensfuss, A. Konkin, H.-K. Roth, M. Al-Ibrahim, U. Zhokhavets, G. Gobsch, V. I. Krinichnyi, G. A. Nazmutdinova, and E. Klemm, Optical and ESR studies on poly (3-alkylthiophene)/fullerene composites for solar cells, *Synth. Met.* **137/1–3**, 1433–1434 (2003).

29. K. Muramoto. N. Takeuchi, T. Ozaki, and S. Kuroda, ESR studies of photogenerated polarons in regioregular poly(3-alkylthiophenes)-fullerene composite, *Synth. Met.* **129**, 239–247 (2002).

30. V. Dyakonov, I. Riedel, C. Deibel, J. Parisi, C. J. Brabec, N. S. Sariciftci, and J. C. Hummelen, Electronic properties of polymer–fullerene solar cells, *Mater. Res. Soc. Symp. Proc.* **665**, C7.1.1.–C7.1.12 (2001).

31. H. Sirringhaus, N. Tessler, and R. H. Friend, Integrated optoelectronic devices based on conjugated polymers, *Science* **280**, 1741–1744 (1998).

32. H. Sirringhaus, P. J. Brown, R. H. Friend, M. M. Nielsen, K. Bechgaard, B. M. W. Langefeld-Voss, A. J. H. Spiering, R. A. J. Janssen, E. W. Meijer, P. Herwig, and D. M. de Leeuw, Two-dimensional charge transport in self-organised, high-mobility conjugated polymers, *Nature* **401**, 685–688 (1999).

33. A. Ullmann, J. Ficker, W. Fix, H. Rost, W. Clemens, I. McCulloch, and M. Giles, High performance organic field-effect transistors and integrated inverters, *Mater. Res. Soc. Symp. Proc.* **665**, C7.5 (2001).

34. S. Scheinert, G. Paasch, M. Schrödner, H.-K. Roth, S. Sensfua, and Th. Doll, Subthreshold characteristics of field effect transistors based on poly(3-dodecylthiophene) and an organic insulator, *J. Appl. Phys.* **92(1)**, 330–337 (2002).

35. C. J. Brabec, A. Cravino, G. Zerza, N. S. Sariciftci, R. Kiebooms, D. Vanderzande, and J. C. Hummelen, Photoactive blends of poly(*p*-phenylenevinylene) (PPV) with methano-fullerenes from a novel precursor: photophysics and device performance, *J. Phys. Chem. B* **105**, 1528–1536 (2001).

36. C. Winder, Spectral Sensitization of Low-Band-Gap Polymer Bulk Heterojunction Solar Cells, Diploma thesis, University of Linz, 2001.

37. K. Petrisch, Organic Solar Cell Architectures, PhD thesis, Technische Universität Graz, Austria, 2000.

38. B. H. Cumpston and K. F. Jensen, Photooxidative stability of substituted poly(phenylene vinylene) (PPV) and poly(phenylene acetylene) (PPA), *J. Appl. Polym. Sci.* **69**, 2451–2458 (1998).

39. U. H. F. Bunz, V. Enkelmann, L. Kloppenburg, D. Jones, K. D. Shimizu, J. B. Claridge, H.-C. zur Loye, and G. Lieser, Solid-state structures of phenyleneethynylenes: comparison of monomers and polymers, *Chem. Mater.* **11**, 1416–1424 (1999).

40. S. A. Jeglinski, A. Amir, X. Wei, Z. V. Vardeny, J. Shinar, T. Cerkvenik, W. Chen, and T. J. Barton, Symmetric light emitting devices from poly(*p*-di ethynylene phenylene) (*p*-di phenylene vinylene) derivatives, *Appl. Phys. Lett.* **67**, 3960–3962 (1995).

41. G. Brizius, N. G. Pschirer, W. Steffen, K. Stizer, H.-C. zur Loye, and U. H. F. Bunz, Alkyne metathesis with simple catalyst systems: efficient synthesis of conjugated polymers containing vinyl groups in main or side chain, *J. Am. Chem. Soc.* **122**, 12435–12440 (2000).

42. A. M. Ramos, M. T. Rispens, J. K. J. van Duren, J. C. Hummelen, and R. A. J. Janssen, A poly(*p*-phenylene ethynylene vinylene) with pendant fullerenes, *Synth. Met.* **119**, 171–172 (2001).

43. A. P. H. J. Schenning, A. C. Tsipis, S. C. J. Meskers, D. Beljonne, E. W. Meijer, and J. L. Bre'das, Electronic structure and optical properties of mixed phenylene vinylene/phenylene ethynylene conjugated oligomers, *Chem. Mater.* **14**, 1362–1368 (2002).

44. D. A. M. Egbe, H. Tillmann, E. Birckner, and E. Klemm, Synthesis and properties of novel well-defined alternating PPE/PPV copolymers, *Macromol. Chem. Phys.* **202**, 2712–2726 (2001).

45. D. A. M. Egbe, C. P. Roll, E. Birckner, U.-W. Grummt, R. Stockmann, and E. Klemm, Side chain effects in hybrid PPV/PPE polymers, *Macromolecules* **35**, 3825–3837 (2002).

46. D. A. M. Egbe, C. Bader, J. Nowotny, and E. Klemm, Synthesis and characterization of three types of alkoxy-substituted hybrid PPV–PPE polymers: potential candidates for photovoltaic applications, *in Proc. SPIE*, Vol. 5215, "Organic Photovoltaics IV," Z. Kafafi and P.A. Lane, eds., SPIE, San Diego, CA, August 3–8, 2003, pp. 79–88.

47. A. M. Ramos, M. T. Rispens, J. K. J. van Duren, J. C. Hummelen, and R. A. J. Janssen, Photoinduced electron transfer and photovoltaic devices of a conjugated polymer with pendant fullerenes, *J. Am. Chem. Soc.* **123**, 6714–6715 (2001).

48. D. L. Trumbo and C. S. Marvel, Polymerization using palladium (II) salts: homopolymers and copolymers from phenylethynyl compounds and aromatic bromides, *J. Polym. Sci., Part A: Polym. Chem.* **24**, 2311–2326 (1986).

49. D. Mühlbacher, H. Neugebauer, A. Cravino, N. S. Sariciftci, J. van Duren, A. Dhanabalan, P. van Hal, R. Janssen, and J. Hummelen, Comparison of electrochemical and spectroscopic data of the low bandgap polymer PTPTB, *Mol. Crystal. Liq. Crystal.* **385**, 85–92 (2002).

50. M. Al-Ibrahim, A. Konkin, H.-K. Roth, D. A. M. Egbe, E. Klemm, U. Zhokhavets, G. Gobsch, and S. Sensfuss, PPE–PPV copolymers: optical and electrochemical characterization, comparison with MDMO-PPV and application in flexible polymer solar cells, *Thin Solid Films* (in press, 2004).

51. M. Kemerink, J. K. J. van Duren, P. Jonkheijm, W. F. Pasveer, P. M. Koenraad, R. A. J. Janssen, H. W. M. Salemink, and J. H. Wolter, Relating substitution to single-chain conformation and aggregation in poly(*p*-phenylene vinylene) films, *Nano Letters* **3(9)**, 1191–1196 (2003).

52. M. Knupfer, Exciton binding energies in organic semiconductors, *Appl. Phys. A* **77**, 623–626 (2003).

53. M. Al-Ibrahim, H.-K. Roth, and S. Sensfuss, Efficient large area polymer solar cells on flexible substrates, *Appl. Phys. Lett.* **85(9)**, 1481–1483 (2004).

24

Polymer–Fullerene Concentration Gradient Photovoltaic Devices by Thermally Controlled Interdiffusion

Martin Drees,[a,*] **Richey M. Davis,**[b] **and Randy Heflin**[a]
[a] *Department of Physics, Virginia Polytechnic Institute and State University, Blacksburg, Virginia, USA*
[b] *Department of Chemical Engineering, Virginia Polytechnic Institute and State University, Blacksburg, Virginia, USA*

Contents

Abstract   Efficient polymer–fullerene photovoltaic devices require close proximity of the two materials to ensure photoexcited electron transfer from the semiconducting polymer to the fullerene acceptor. We review a series of studies in which thermally controlled interdiffusion is used to establish a gradient bulk heterojunction in photovoltaic devices incorporating conjugated polymers and fullerenes. First, a bilayer system consisting of spin-cast 2-methoxy-5-(2′-ethylhexyloxy)-1,4-phenylenevinylene (MEH-PPV) copolymer and sublimed C_{60} is heated above the MEH-PPV glass transition temperature (T_g) in an inert environment, inducing an interdiffusion of the polymer and the fullerene layers. With this process, a controlled, bulk gradient heterojunction is created bringing the fullerene molecules within the exciton diffusion radius of the MEH-PPV throughout

---

* Present address: Linz Institute for Organic Solar Cells (LIOS), Physical Chemistry, Johannes Kepler University Linz, Linz, Austria.

the film to achieve highly efficient charge separation. The interdiffused devices show a dramatic decrease in photoluminescence and concomitant increase in short circuit currents, demonstrating the improved interface. The dependence of the photovoltaic performance on the thickness of the MEH-PPV layer is examined, and it is found that, within the range examined, thinner layers yield enhanced efficiency as a result of improved charge transport. Cross-sectional transmission electron microscopy (TEM) images show that the interdiffused films contain large aggregates of C_{60} as a result of the relatively low miscibility of MEH-PPV and C_{60}. Further improved device performance is achieved using poly(3-octylthio-phene-2,5-diyl), which is more miscible with C_{60}. Auger spectroscopy combined with ion beam milling reveals a smooth concentration gradient from one surface of the film to the other in this latter case.

Keywords   organic photovoltaic, interdiffusion, fullerene, semiconducting polymer, concentration gradient

24.1.   INTRODUCTION

The discovery of ultrafast photoinduced charge transfer from conducting polymers to the Buckminsterfullerene C_{60} [1] has fueled research towards organic photovoltaic devices [2–5]. Upon photoexcitation of an electron–hole pair, the transfer of electrons from the polymer onto the fullerene leads to an efficient separation of charges that prevents luminescent recombination and is required in photovoltaic devices. Since the charge transfer can only occur if the photoexcitation on the polymer is within less than 10 nm of a C_{60} molecule [6–8], close proximity of polymer and fullerene is essential for efficient charge separation. One approach for achieving such close proximity of the electron donor and acceptor materials is by creating a bulk heterojunction. This means that the donor and acceptor form an interface through-out the bulk of the active layer of the photovoltaic device. In this manner, photons absorbed anywhere within the active layer yield separated charges. However, a bulk heterojunction formed by a blend of the donor and acceptor is not the ideal com-position from the point of charge transport, which is improved by having a film that is donor-rich in the vicinity of the anode and acceptor-rich in the vicinity of the cathode.

We describe here an approach to create such a bulk heterojunction with a concentration gradient of donor and acceptor material by using thermally controlled interdiffusion of polymer and fullerene layers [9]. Photoluminescence quenching with concomitant increase in photocurrent in interdiffused devices shows the improved interface between the polymer and fullerene throughout the bulk of the active layer. Furthermore, variation of the polymer layer thickness shows that, within the exam-ined thickness regime (70–110 nm), the photocurrents in the interdiffused devices increase with decreasing thickness as a result of improved charge transport out of the film. Transmission electron microscopy (TEM) studies on cross sections of the films reveal that C_{60} forms clusters of up to 30 nm diameter in the polymer bulk of the interdiffused devices. These large clusters are a constraint for interdiffusion in this pair of relatively immiscible materials and therefore hinder the creation of a bulk heterojunction. In addition, the interdiffusion was observed in situ by monitoring the photocurrents during the heat treatment. Further improvements in device perform-ance are achieved through use of poly(3-octylthiophene-2,5-diyl), which is more miscible with C_{60}.

24.2. EXPERIMENTAL

Photovoltaic devices were prepared by first spin-coating a poly(3,4-ethylenedioxythiophene):poly(styrenesulfonate) (PEDOT:PSS) complex (Bayer Corp.) onto indium tin oxide (ITO)-covered glass substrates. Subsequently, 2-methoxy-5-(2′-ethylhexyloxy)-1,4-phenylenevinylene (MEH-PPV) copolymer (M.W. ~85,000, H.W. Sands Corp.) was spin-coated from a 1% wt./vol. chlorobenzene solution. By varying the spin speed between 2000 and 4000 rpm, film thicknesses of 110, 90, and 70 nm were achieved. After annealing films under vacuum at 150°C to remove water and solvents, a 100 nm C_{60} (MER Corp.) film was sublimed onto the MEH-PPV. This polymer/fullerene bilayer structure was then heated in a nitrogen glove box at 150°C or 250°C for 5 min to induce an interdiffusion of the two layers. Finally, a 200-nm-thick Al layer was evaporated under vacuum as the cathode.

Film thicknesses were determined using the optical density values obtained with a Filmetrics F20-UV thin film spectrometer system from transmission and reflection data. The absorption coefficients used for calibration are $18 \times 10^4 \, cm^{-1}$ at 490 nm for MEH-PPV and $6 \times 10^4 \, cm^{-1}$ at 435 nm for C_{60}, as determined by interference fringes in the reflection spectra of thicker films.

The photoresponsivity spectra were recorded under argon atmosphere by illuminating the samples through the glass and ITO with light from a 300 W Xe-lamp that was passed through a CVI CM100 monochromator to select wavelengths between 300 and 700 nm and recording the photocurrents with a Keithley 485 picoammeter. A calibrated Si photodiode was used in the same setup to correct the photocurrent spectra for the power spectrum of the Xe-lamp/monochromator system. Photoluminescence was measured with an Ocean Optics S2000 fiber optics spectrometer. The sample was illuminated through the glass and ITO with a wavelength of 470 nm and an intensity of 3.8 mW/cm². The luminescence signal was collected by the fiber at an angle of 45° with respect to the incident light. $I-V$ curves were recorded with a Keithley 236 Source Measure Unit both in the dark and under 470 nm illumination. All of the above measurements were made in an argon environment to inhibit oxidation of the polymer. The illuminated area on the samples was 2 mm² in all cases.

For the in situ studies, complete devices (including Al cathode) were first prepared without the interdiffusion heating step. This way, the devices could be measured in the unheated bilayer structure before the heat treatment. Devices were then heated in an argon atmosphere while monitoring the photocurrents in the same setup used for photoresponsivity measurements.

To study the morphology of the interdiffused devices, TEM was performed on cross sections of the films. MEH-PPV/C_{60} bilayers were prepared on glass substrates and taken through the interdiffusion heating step. The films were then lifted off the substrate by submerging them in deionized water. After drying films overnight in a nitrogen glovebox, films were embedded in an epoxy and slices of 50 nm thickness revealing the cross section of the films were microtomed with a Reichert–Jung Ultracut E microtome. These specimens were studied in a Philips 420T TEM system at 100 keV.

24.3. MEH-PPV/C_{60} DEVICES

The optical absorption spectrum of a pure MEH-PPV film between 300 and 850 nm has an absorption edge at 590 nm and an absorption maximum at 490 nm in the visible portion of the spectrum. For a C_{60} film, the absorption starts a slow rise around 700 nm, followed by a peak at 435 nm and a stronger maximum at 338 nm. Correcting for reflection effects, the optical density of a bilayer system of MEH-PPV/C_{60} is the superposition of the two single layer films. The optical density spectra of an ITO/PEDOT-coated glass susbtrate before and after adding layers of MEH-PPV and MEH-PPV/C_{60} are shown in Figure 24.1. The heating of the films for interdiffusion does not lead to any measurable changes in the absorption spectrum, confirming that there is no degradation in the film constituents due to the heat treatment.

24.3.1. Thermally Controlled Interdiffusion

The photoluminescence of the devices can be used as a measure of the exciton dissociation in the films. If the photogenerated excitons are not dissociated, they can recombine radiatively, leading to photoluminescence. The well-known effect of ultrafast electron transfer from the polymer to C_{60} is used to increase the exciton dissociation efficiency in photovoltaic devices. Figure 24.2 shows the photoluminescence of a sequence of samples. The unheated MEH-PPV/C_{60} bilayer shows a strong photoluminescence signal, which is not significantly reduced compared to a MEH-PPV single layer device. This is because effective exciton dissociation is only possible for excitons that are created within ~10 nm of a fullerene molecule. Thus, the photoluminescence is quenched only within a small volume at the interface of MEH-PPV and C_{60}. In the MEH-PPV bulk, radiative recombination of the excitons remains efficient. The devices heated at 150°C and 250°C show a decrease of an order of magnitude in photoluminescence, indicating that the heat treatment induced interdiffusion of the polymer and fullerene layers. The interdiffusion of C_{60} molecules into the polymer provides electron acceptors throughout the bulk, therefore strongly

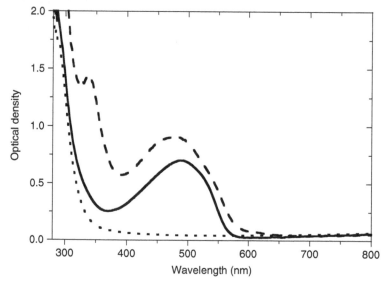

Figure 24.1. Optical density of a MEH-PPV/C_{60} bilayer device. The dotted line shows the absorption of the ITO/PEDOT-coated glass substrate. The solid line shows the absorption after spin-casting an additional layer of 90 nm MEH-PPV. The dashed line shows the absorption after subliming an additional layer of 100 nm C_{60}. The absorption of the bilayer is a superposition of the absorption of the two single layers.

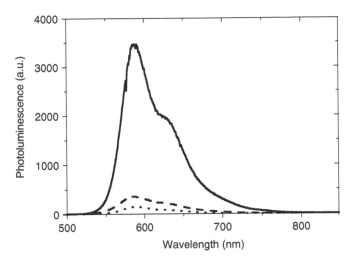

Figure 24.2. Photoluminescence of an unheated MEH-PPV/C_{60} bilayer device (solid) and two heated MEH-PPV/C_{60} devices (5 min at 150°C (dotted) and 5 min at 250°C (dashed)).

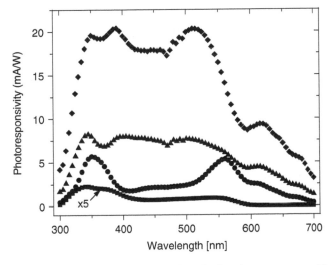

Figure 24.3. Photoresponsivity of a MEH-PPV device (squares, magnified by 5×), an unheated MEH-PPV/C_{60} bilayer device (circles), a MEH-PPV/C_{60} bilayer device that was heated at 150°C for 5 min (triangles) and a MEH-PPV/C_{60} bilayer device that was heated at 250°C for 5 min (diamonds).

quenching the photoluminescence. The number of MEH-PPV photoexcitation sites that do not undergo exciton dissociation is thus reduced by 90%.

Figure 24.3 shows the spectral photoresponsivity of various MEH-PPV single layer and MEH-PPV/C_{60} bilayer films. MEH-PPV has a sharp increase in photoresponsivity at the absorption edge and a peak at 550 nm. The addition of a C_{60} layer increases the photoresponsivity by a factor of ~20 due to the electron–hole dissociation by the fullerene at the polymer–fullerene interface. In both films, a minimum of the photocurrent is observed in the region of maximum absorption of MEH-PPV (400–550 nm). We attribute this to the filter effect [10,11]. Because of the large MEH-PPV absorbance, the majority of excitons are created near the ITO–MEH-PPV interface. For the MEH-PPV single layer device, the electrons are not easily collected at the Al electrode since they are created far from it and are not transported well through the hole-transporting MEH-PPV. In the case of the MEH-PPV/C_{60} bilayer, most excitons are created far away from the C_{60} interface where efficient exciton dissociation occurs.

The heated MEH-PPV/C_{60} devices, one heated below the T_g of MEH-PPV and one above this temperature, show several important distinct characteristics. The T_g of MEH-PPV was confirmed to be ~230°C by differential scanning calorimetry (DSC) analysis. The devices heated at 150°C for 5 min no longer have a photocurrent minimum at the maximum MEH-PPV absorption. Instead, the photoresponsivity is nearly constant over the wavelength range between 340 and 560 nm. This indicates that, even at temperatures significantly below T_g of MEH-PPV, C_{60} can partially interdiffuse into the polymer bulk and therefore reduce the filter effect observed in unheated bilayer systems. This result is supported by the reduction of photoluminescence observed in the devices. On the other hand, no significant increase in the maximum photoresponsivity is observed. Even though fullerenes have diffused into the polymer bulk and enhanced the charge separation, the charge transport out of the device is not optimized for these conditions. Heating at 250°C for 5 min increases the photoresponsivity by an order of magnitude throughout most of the visible range. This shows that heating above T_g improves the interdiffusion of the fullerenes into the polymer bulk such that charge separation and charge transport out of the devices are enhanced.

The photocurrent measurements can also be expressed in terms of the external quantum efficiency (EQE). The EQE is the ratio of electrons produced as current to photons that enter the device. At a given wavelength, the EQE can be calculated from the photoresponsivity (PR) by

$$\text{EQE}(\lambda) = \frac{hc\text{PR}(\lambda)}{e\lambda} \tag{1}$$

where h is Planck's constant, c is the speed of light in vacuum, e is the charge of an electron, and λ is the wavelength. Figure 24.4 shows the data of Figure 24.3 expressed as EQE.

The corresponding dark and illuminated $I-V$ curves for the devices described above are shown in Figure 24.5. All devices have more than two orders of magnitude rectification in the dark and more than one order of magnitude when illuminated at 470 nm. The open circuit voltage (V_{OC}) is reduced from 0.95 V in a single MEH-PPV layer device to 0.39 V in the unheated MEH-PPV/C_{60} bilayer device. Upon heating, V_{OC} improves to 0.64 V for the device heated at 150°C and 0.47 V for the device heated at 250°C.

The point of maximum power efficiency of a photovoltaic device is where the product IV is largest. The ideal maximum power would by $I_{SC}V_{OC}$, where I_{SC} is the short circuit current. The fill factor (FF) is defined as the actual maximum power over the ideal maximum power and is given by

$$\text{FF} = \frac{I_{MP}V_{MP}}{I_{SC}V_{OC}} \tag{2}$$

where I_{MP} and V_{MP} are the current and voltage at the maximum power point, respectively. Figure 24.6 shows the $I-V$ characteristics of the interdiffused devices in the fourth quadrant. FF is calculated to be 0.24 for the device interdiffused at 150°C and 0.29 for the one at 250°C. These values are quite low. For comparison, the unheated bilayer has a FF of 0.40.

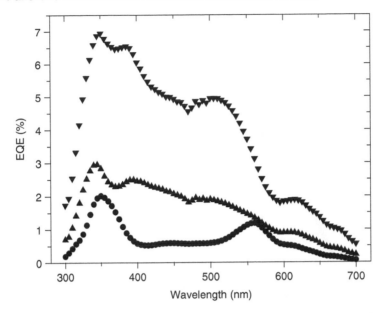

Figure 24.4. External quantum efficiencies of interdiffused devices. The graph shows the EQE spectrum of a device heated at 150°C for 5 min (up triangles) and a device heated at 250°C for 5 min (down triangles). For comparison, an unheated bilayer device is shown (circles).

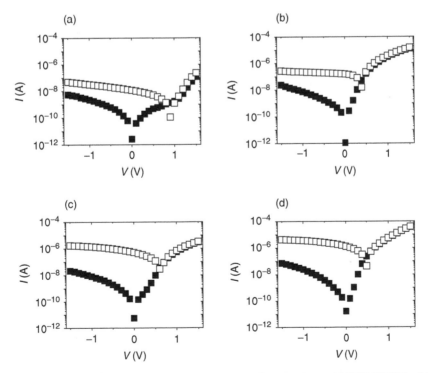

Figure 24.5. $I-V$ curves for (a) a MEH-PPV device, (b) an unheated MEH-PPV/C_{60} bilayer device, (c) a MEH-PPV/C_{60} bilayer device heated at 150°C for 5 min, and (d) a MEH-PPV/C_{60} bilayer device heated at 250°C for 5 min. The filled symbols show the dark $I-V$ curves, and the open show the illuminated $I-V$ curves (470 nm, 3.8 mW/cm^2).

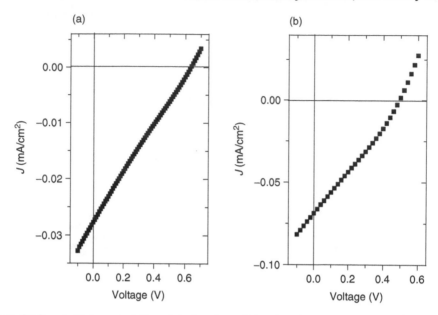

Figure 24.6. $I-V$ characteristics of devices interdiffused at (a) 150°C and (b) 250°C in the fourth quadrant. FF is 0.24 for 150°C and 0.29 for 250°C.

24.3.2. Dependence on MEH-PPV Thickness

While the fraction of incident photons absorbed increases with the thickness of the MEH-PPV, the ability to extract separated charges decreases due to the relatively low electrical conductivity. To examine these competing effects, MEH-PPV/C_{60} bilayer devices with 70, 90, and 110 nm polymer and 100 nm fullerene layer thicknesses were prepared without heating and with interdiffusion at 250°C and 150°C. The photo-responsivity spectra of the unheated bilayer devices are shown in Figure 24.7. With decreasing polymer layer thickness, the photoresponsivity of the devices increases throughout most of the spectrum. The increase in the photocurrents is due to two reasons: (1) the number of photons that can reach the polymer–fullerene interface, where efficient charge separation occurs, is increased in devices with thinner polymer films and (2) the distance that charges have to travel to the collecting electrodes is reduced, therefore reducing the series resistance of the device. In addition to the increase in photocurrent, the peak position of the photocurrents shifts towards the absorption maximum at 490 nm. The peak photocurrent moves from $\lambda = 575$ nm in the 110 nm device to $\lambda = 560$ nm and $\lambda = 540$ nm in the 90 and 70 nm devices, respectively. This behavior can be explained by a reduction in the filter effect of the MEH-PPV bulk with decreasing layer thickness. Since the MEH-PPV layer acts like an optical filter, most of the photons in wavelength regions of strong absorption are absorbed near the anode/MEH-PPV interface and do not reach the MEH-PPV/C_{60} interface where efficient charge separation occurs. Therefore, devices with a filter effect exhibit low photocurrents in regions of strong absorption, which leads to a mismatch between the absorption spectrum and photocurrent spectrum. Since the filter effect is reduced with decreasing MEH-PPV layer thickness, the peak of the photocurrent spectrum shifts towards the absorption peak. This mismatch between absorption and photocurrent spectra can thus be used to characterize the quality of

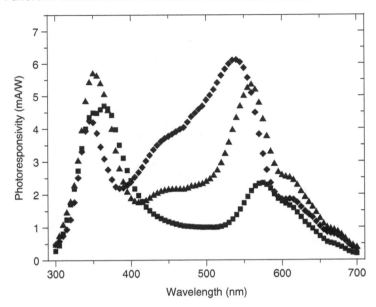

Figure 24.7. Photoresponsivity of MEH-PPV/C$_{60}$ bilayer devices with varying MEH-PPV thickness. The photocurrents increase with decreasing MEH-PPV thickness. In addition, the photocurrent peak for 110 nm MEH-PPV (squares) shifts from 575 nm to 560 and 540 nm for 90 nm MEH-PPV (triangles) and 70 nm MEH-PPV (diamonds), respectively. The C$_{60}$ film thickness was 100 nm for all devices.

the bulk heterojunction in the active layer of the device. Only devices with a bulk heterojunction will have similar absorption and photocurrent spectra.

Figure 24.8 shows the photoresponsivity for a similar set of devices heated above the T_g of MEH-PPV at 250°C for 5 min. The interdiffusion leads to an order of magnitude increase in the photocurrents throughout most of the spectrum. Again, devices with thinner polymer layers show higher photocurrents. Importantly, the shape of the photocurrent spectrum for devices with 70 nm MEH-PPV matches the absorption spectrum of the devices reasonably well, indicating that a bulk heterojunction has been achieved.

The $I–V$ characteristics, fill factors, and monochromatic power conversion efficiencies of the various devices are listed in Table 24.1. While the open circuit voltage and the short circuit current are improved by the interdiffusion, the fill factor is reduced. The overall efficiency of our unheated bilayer devices is comparable to efficiencies previously observed in MEH-PPV/C$_{60}$ bilayers [12]. The 0.3% (under monochromatic illumination) efficiency of the device with 70 nm MEH-PPV heated at 250°C is low compared to MEH-PPV–fullerene blend devices, which have obtained 2.9% monochromatic efficiency [4]. But these latter devices utilize a highly soluble C$_{60}$ derivative that is known to provide enhanced efficiency.

These results indicate that the interdiffusion of MEH-PPV and C$_{60}$ is most likely restricted by the strong phase separation of polymer and fullerene in this pair of materials. C$_{60}$ does not easily diffuse long distances into MEH-PPV. Therefore, the charge transfer is not optimized throughout the entire bulk and only the interdiffused devices with 70 nm of MEH-PPV come close to forming a bulk

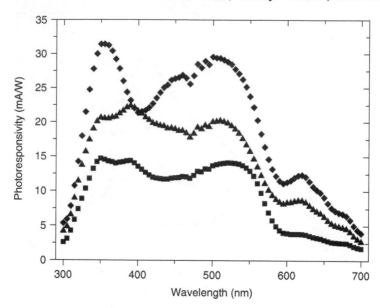

Figure 24.8. Photoresponsivity of MEH-PPV/C_{60} bilayer devices heated at 250°C with varying MEH-PPV thicknesses. A decrease in MEH-PPV thickness results in an increase of photocurrents. Devices shown have MEH-PPV thickness of 110 nm (squares), 90 nm (triangles) and 70 nm (diamonds). C_{60} thickness is 100 nm for all devices. The photocurrent spectrum of the 70 nm device is very similar in shape to the absorption spectrum which indicates that the interdiffusion resulted in a bulk heterojunction device.

Table 24.1. $I-V$ characteristics, fill factor, and monochromatic power conversion efficiency for MEH-PPV/C_{60} devices with varying thicknesses and thermal treatment under monochromatic illumination (470 nm, 3.8 mW/cm^2)

MEH-PPV thickness (nm)	Interdiffusion Temperature (°C)	V_{OC} (V)	I_{SC} (mA/cm^2)	FF	Efficiency (%)
110	Unheated	0.3	3.8×10^{-3}	0.50	0.02
90	Unheated	0.4	7.9×10^{-3}	0.40	0.03
70	Unheated	0.4	16×10^{-3}	0.39	0.07
110	150	0.6	24×10^{-3}	0.25	0.10
90	150	0.6	26×10^{-3}	0.24	0.10
70	150	0.6	28×10^{-3}	0.25	0.11
110	250	0.5	42×10^{-3}	0.25	0.14
90	250	0.5	67×10^{-3}	0.29	0.26
70	250	0.5	92×10^{-3}	0.25	0.30

heterojunction, which is indicated by a better match of photoresponsivity and absorption spectrum. The overall efficiency of the devices is low because of the difficulty of the interdiffusion process with this pair of materials.

It should also be noted here that in devices with even thinner polymer layers the photocurrents are expected to eventually decrease. Even though the series resistance is continually reduced, the percentage of absorbed photons is concomitantly reduced and this effect at some point outweighs the reduction in series resistance.

24.3.3. *In situ* Observation of Interdiffusion

In situ studies of interdiffusion during the heating process were carried out to determine the optimum conditions for the time–temperature thermal profile. In the studies presented above, the interdiffusion heating was done before the final production step of depositing the top electrode. Because of this, measurement of the photoresponse of the same device before and after the heat treatment was not possible. This was initially done to prevent damage of the devices due to cracking of the polymer bulk that was often observed when heating was carried out in thick (layer thicknesses of 120 nm or more) polymer films. We found, however, that this problem was alleviated with thinner films at 150°C, thus enabling in situ observations.

Figure 24.9 shows the photocurrent of an MEH-PPV/C_{60} bilayer with 70 nm of MEH-PPV and 110 nm of C_{60} that was monitored during the heating. The data shown are the raw photocurrent values resulting from illumination of the device with monochromatic light (470 nm, 3.8 mW/cm^2). The temperature of the hot stage is indicated for selected times. With increasing temperature, the photocurrent of the illuminated sample rapidly begins to rise. When the temperature reaches 150°C, it is maintained for 5 min before the hot plate is turned off and the temperature starts decreasing. While the temperature stays constant at 150°C, the photocurrent continually increases and even after the temperature starts decreasing, it rises further for a few minutes. The final photocurrent after cooling to room temperature is four times larger than the initial photocurrent. Since the photocurrent decreases to some extent during cooling, it is clear that not all of the increase in photocurrent during the heating is due to interdiffusion. Part of it is simply due to the elevated temperature. Since charge transport in conjugated polymers and C_{60} is typically a hopping process, which is thermally activated, the photocurrents are expected to be larger at higher temperatures because of the increased conductivity. This temperature dependence of the short circuit current has been previously observed in polymer–fullerene photovoltaic devices [13,14].

Interdiffusion could not be observed in situ at temperatures above T_g of MEH-PPV. Typically, the photocurrent signal decreased to zero at temperatures in the vicinity of T_g and did not recover even after cooling of the device. The reason for this

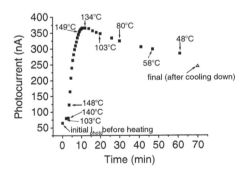

Figure 24.9. In situ photocurrent measurements. The photocurrent of a 70 nm MEH-PPV/110 nm C_{60} device was observed during the interdiffusion heating at 150°C (470 nm, 3.8 mW/cm^2 illumination). The squares show the photocurrent during the thermal treatment. The temperature at various times is indicated. The final photocurrent of the device (after complete cooling down to room temperature) is given by the open triangle.

permanent device failure could not be determined but is possibly due to delamination of the electrode from the organic film. It is apparent, though, that device failure is in the vicinity of the glass transition temperature of the polymer where the polymer chains can undergo segmental motion.

It should be mentioned here that heating the devices for more than 5 min at 150°C led to a decrease in photocurrent at the elevated temperature. The time after which this decrease began varied between devices from 5 to 10 min after reaching 150°C.

24.3.4. Morphology of the Interdiffused Devices

The TEM image of the cross-section of an unheated MEH-PPV/C_{60} bilayer is shown in Figure 24.10. By imaging the cross section of an MEH-PPV single layer, it was determined that the light phase is the polymer. The dark phase is therefore C_{60}. The gray background surrounding the film is due to the epoxy in which the film is embedded. The image shows a smooth, distinct interface between polymer and fullerene, which is expected from an unheated bilayer. A very similar image was obtained for bilayers heated at 150°C. Since the TEM image does not provide atomic

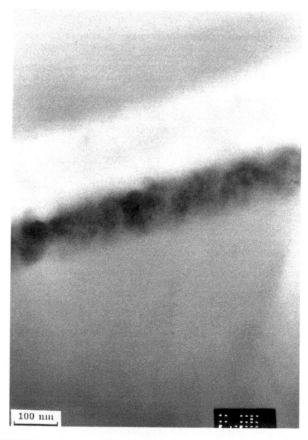

Figure 24.10.   TEM image of an unheated MEH-PPV/C_{60} bilayer. The light phase is MEH-PPV, the dark phase is C_{60}, and the gray background is from the epoxy. The image shows a distinct and smooth interface between the two materials.

resolution, it does not provide information concerning single C_{60} molecules or very small clusters of C_{60} diffused into the MEH-PPV. The reduction of the filter effect in the devices heated at 150°C indicates that some interdiffusion occurs, but the TEM image shows that no large-scale intermixing of the two materials takes place.

The TEM image (Figure 24.11) of a device heated at 250°C (above T_g of MEH-PPV) shows dramatic changes compared to the unheated bilayer. The distinct smooth interface between the two materials has vanished. Instead, clusters of C_{60} with diameters up to 30 nm have migrated into the MEH-PPV bulk. The image clearly shows that there is no continuous concentration gradient of donor and acceptor materials from one end of the film to the other. Instead, there is a polymer-rich phase at one end and a fullerene-rich phase at the other end of the cross section with a blend of large C_{60} clusters and MEH-PPV in the middle. The formation of C_{60} clusters is partly due to the strong tendency towards crystallization of C_{60} [15] and due to the tendency of MEH-PPV and C_{60} to phase separate. It is therefore not surprising that the interdiffusion is limited in these materials. In fact, in light of

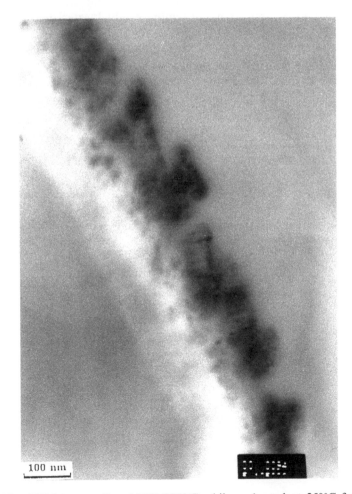

Figure 24.11. TEM image of an MEH-PPV/C_{60} bilayer heated at 250°C for 5 min. The image shows a rough interface with clusters of C_{60} with a diameter of up to 30 nm present in the MEH-PPV bulk.

the TEM images, it is remarkable that thermally induced interdiffusion is so effective in enhancing the photovoltaic performance of this material system. It might be expected that a more miscible material pair would yield further improvement, as will be described in the next section.

24.4. P3OT/C$_{60}$ DEVICES

In order to achieve improved miscibility, devices were fabricated from regioregular poly(3-octylthiophene) (P3OT, Sigma-Aldrich Corp.) and C$_{60}$. Interdiffusion of C$_{60}$ into the P3OT bulk has been reported to occur even at room temperature [16]. Furthermore, P3OT has an absorption onset at 670 nm and peak at 512 nm. The redshifted and broadened spectrum relative to MEH-PPV provide a better match to the solar spectrum. The absorption spectra of P3OT and P3OT/C$_{60}$ films are shown in Figure 24.12.

After deposition of the PEDOT film on ITO-coated glass, P3OT was spin-cast from a 1.5% wt./vol. chloroform solution. Devices were then annealed for 1 h at 120°C under vacuum (3×10^{-6} Torr) to remove residual water and solvents before a 100 nm layer of C$_{60}$ (MER Corp.) was sublimed on the P3OT layer. Finally, a 200 nm Al layer was evaporated as a top electrode. The typical film thicknesses of the P3OT and fullerene layers were ~100 nm each, as determined from absorption measurements with the Filmetrics F20-UV thin film spectrometer system.

Regioregular P3OT is a microcrystalline polymer that shows a melting transition [17]. The melting temperature T_m of P3OT was determined to be 187°C by DSC. The glass transition temperature of the polymer could not be observed but is believed to be below 100°C. To test whether the melting transition has any influence on the interdiffusion process, the temperatures chosen for the interdiffusion were 130°C, which is below T_m but above T_g, and 210°C, which is above T_m.

The photocurrent spectra of the unheated and heated devices are shown in Figure 24.13. Upon heating, the EQE improves by an order of magnitude. Little difference is observed in the EQE whether the interdiffusion heating is done below or above T_m. For the photocurrent, there is no advantage to performing

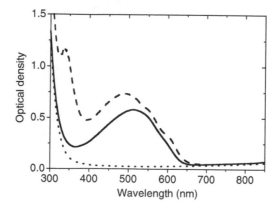

Figure 24.12. Optical density of a P3OT/C$_{60}$ bilayer device. The dotted line shows the absorption of the ITO–PEDOT substrate, the solid line of the P3OT single layer, and the dashed line of the P3OT/C$_{60}$ bilayer.

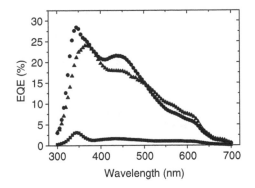

Figure 24.13. EQE spectrum of an unheated $P3OT/C_{60}$ bilayer device (squares), a $P3OT/C_{60}$ bilayer device heated at $130°C$ for 5 min (circles), and a $P3OT/C_{60}$ bilayer device heated at $210°C$ for 5 min (triangles). The EQE spectra of the two heat-treated devices are comparable and are an order of magnitude larger than the unheated bilayer device.

the interdiffusion heating above the melting transition of the P3OT. The sharp cutoff of the photocurrents below 350 nm is due to the strong absorption of the glass substrate and the ITO/PEDOT electrode, which prevents any light below 350 nm to penetrate into the active layer.

The $I–V$ characteristics of the interdiffused devices are plotted on a semilog scale in Figure 24.14. $I–V$ curves were measured under monochromatic illumination at 470 nm ($3.8\,\text{mW/cm}^2$). The devices heated below T_m have four orders of magnitude rectification in the dark and three orders of magnitude rectification under illumination. This is the same rectifying behavior that the unheated bilayer devices showed. The rectification is reduced to three orders of magnitude in the dark and two orders of magnitude under illumination when the temperature of the heat treatment is raised above the melting temperature ($210°C$). The $I–V$ characteristics of the devices in the fourth quadrant are shown in Figure 24.15. Upon heat treatment, the $I–V$

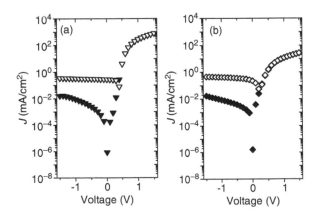

Figure 24.14. $I–V$ characteristics of heat-treated $P3OT/C_{60}$ devices on a semi-log scale. (a) The device heated at $130°C$ shows four orders of magnitude rectification in the dark (solid) and three orders of magnitude under illumination (open). (b) The device heated at $210°C$ has three orders of magnitude rectification in the dark (solid) and two orders of magnitude under illumination (open). Illumination was at 470 nm, $3.8\,\text{mW/cm}^2$.

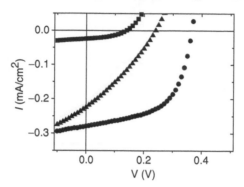

Figure 24.15. Current–voltage characteristics for an unheated P3OT/C_{60} bilayer device (squares), a P3OT/C_{60} bilayer device heated at 130°C for 5 min (circles) and a P3OT/C_{60} bilayer device heated at 210°C for 5 min (triangles).

characteristics dramatically improve. The open circuit voltage and short circuit current of devices heated at 130°C or 210°C increase compared to the unheated films. However, the fill factor only improves for devices heated below T_m and decreases for devices heated above T_m. A summary of the device performance is given in Table 24.2. The $I–V$ curves of the various devices clearly show that it is advantageous to perform interdiffusion below T_m. The monochromatic conversion efficiency calculated for devices heated at 130°C is found to be 1.5%.

To study the morphology of the interdiffused films, we used a 610 Perkin–Elmer scanning Auger spectroscopy system in combination with Ar-ion beam milling. In this system, the surface layer of a film can be tested for its atomic constituents. After the Auger scan, the surface layer is milled off with an Ar-ion beam and the new surface layer can be tested. For these experiments, P3OT films were spin-coated directly onto ITO-covered glass slides and C_{60} was subsequently sublimed. During the measurement, the ITO film was grounded to prevent charging of the films and substrate caused by the electron beam. We used the fact that the thiophene backbone contains sulfur that can be detected with Auger spectroscopy. In contrast, C_{60} is made up entirely of carbon. From the strength of the sulfur signal, we can therefore make a qualitative statement about the polymer content in the surface layer. The limitations of this technique are the penetration depth of the electron beam, which is about 5 nm, and the fact that the ion-beam milling does not remove a perfectly smooth layer. Furthermore, the technique cannot distinguish between carbon from the polymer and carbon from the C_{60}. Therefore, no determination can be made whether a layer is free of C_{60}.

The peak-to-peak Auger signal of an unheated bilayer of C_{60} and P3OT as a function of depth in the film is shown in Figure 24.16. The P3OT and C_{60} layers each

Table 24.2. $I–V$ characteristics of unheated and heated P3OT/C_{60} devices under monochromatic illumination (470 nm, 3.8 mW/cm^2)

Interdiffusion temperature (°C)	V_{OC} (V)	I_{SC} (mA/cm^2)	FF	Efficiency (%)
Unheated	0.14	16×10^{-3}	0.41	0.02
130	0.36	277×10^{-3}	0.57	1.5
210	0.23	226×10^{-3}	0.32	0.44

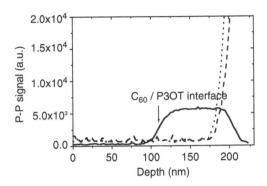

Figure 24.16. Depth profile of an unheated P3OT/C$_{60}$ bilayer device. The concentration of sulfur (solid), indium (dotted), and oxygen (dashed) was monitored. The arrow indicates the position of the P3OT–C$_{60}$ interface as determined from absorption measurements. The scan shows a rather sharp interface between P3OT and C$_{60}$.

had a thickness of 110 nm as determined from absorption measurements. The sulfur signal is zero while the tested surface is within the C$_{60}$ layer and then shows a sharp increase in the vicinity of the interface between the P3OT and C$_{60}$. In the P3OT layer, the sulfur signal remains constant, indicating a constant concentration of the polymer throughout the layer. Finally, the sulfur signal drops and the indium and oxygen signals appear when the scan hits the P3OT–ITO interface. This depth scan shows that the unheated bilayer is a rather clear two-layer system. It should be mentioned here that the Auger scan does not rule out small amounts of C$_{60}$ in the P3OT layer. As a matter of fact, photoluminescence studies indicate that small amounts of C$_{60}$ already diffuse into the P3OT bulk during sublimation of the C$_{60}$, leading to photoluminescence quenching. This is in agreement with observations by Schlebusch et al. [16].

The depth profile of a heated bilayer system is shown in Figure 24.17. In this system, the C$_{60}$ layer was 80 nm and the P3OT was 160 nm thick. The bilayer was heated at 130°C for 5 min. After etching through 35 nm of C$_{60}$ with no sign of sulfur, the sulfur signal starts increasing steadily throughout most of the film. After more than 100 nm of increasing sulfur concentration, the signal levels off and stays constant until it drops off when the etching hits the ITO layer. The jump in signal at a depth of 185 nm is due to a readjustment in the electronics of the system during the experiment. This scan shows that in fact the heat treatment has created a concentration gradient of the P3OT throughout most of the bulk of the active layer, which results in dramtically enhanced photovoltaic performance.

24.5. SUMMARY

In conclusion, we have fabricated polymer–fullerene bilayer systems in which the charge transfer and charge transport have been improved by thermally controlled interdiffusion, which results in a concentration gradient of the electron donor and acceptor species. This leads to increased proximity of C$_{60}$ molecules to optical excitations throughout the bulk of the film as demonstrated by luminescence quenching of one order of magnitude. The corresponding order of magnitude increase in photocurrent in the region of maximum absorption of the MEH-PPV

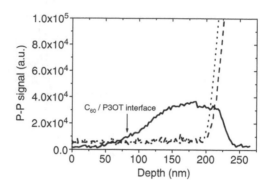

Figure 24.17. Depth profile of a P3OT/C_{60} bilayer device heated at 130°C. The concentration of sulfur (solid), indium (dotted), and oxygen (dashed) was monitored. The arrow indicates the position of the P3OT–C_{60} interface as determined from absorption measurements. The scan shows that the P3OT concentration rises slowly throughout more than 100 nm (from 35 to 150 nm) of the bulk of the device.

indicates the formation of a bulk heterojunction due to thermally controlled interdiffusion.

The photocurrent in MEH-PPV/C_{60} photovoltaic devices was also examined as a function of the polymer layer thickness. The studies were carried out on devices with and without the use of thermally induced interdiffusion of the layers to create a gradient bulk heterojunction. For MEH-PPV in combination with C_{60}, the interdiffusion process is limited due to strong phase separation between the polymer and fullerene. TEM studies show a tendency of C_{60} to form large clusters of up to 30 nm in diameter in the polymer bulk. These clusters illustrate the strong phase separation of MEH-PPV and C_{60} that limits the interdiffusion process and therefore formation of a bulk heterojunction. A bulk heterojunction is only formed by interdiffusion in devices with thin MEH-PPV layers, in our case 70 nm. In addition, in situ studies showed that the interdiffusion can be observed by monitoring the photocurrent of the device. An optimization of the time–temperature profile for the interdiffusion of the materials can therefore be easily implemented.

Further improvement is obtained by use of regioregular P3OT, which is more miscible with C_{60}. The best performance is obtained by heating at temperatures below T_m of the polymer. Heating above T_m results in lower open circuit voltages and fill factors than heating below T_m.

Auger spectroscopy in combination with ion-beam milling revealed a morphology of the film in which the unheated films have a clear bilayer structure and the heated films show a concentration gradient of the polymer throughout most of the active layer. This study shows that the heat treatment indeed results in the intended concentration gradient of donor and acceptor. For P3OT/C_{60} films heated at 130°C, a monochromatic power conversion efficiency of 1.5% was achieved.

ACKNOWLEDGMENTS

We would like to thank Steve McCartney at Virginia Tech for his help with the TEM studies and Frank Cromer for help with the Auger spectroscopy measurements.

REFERENCES

1. N. S. Sariciftci, L. Smilowitz, A. J. Heeger, and F. Wudl, *Science* **258**, 1474–1476 (1992).
2. S. E. Shaheen, C. J. Brabec, N. S. Sariciftci, F. Padinger, T. Fromherz, and J.C. Hummelen, *Appl. Phys. Lett.* **78**, 841–843 (2001).
3. M. Granstrom, K. Petritsch, A. C. Arias, A. Lux, M. R. Andersson, and R. H. Friend, *Nature* **395**, 257 (1998).
4. G. Yu, J. Gao, J. C. Hummelen, F. Wudl, and A. J. Heeger, *Science* **270**, 1789–1791 (1995).
5. L. Chen, D. Godovsky, O. Inganäs, J. C. Hummelen, R. A. J. Janssens, M. Svensson, and M.R. Andersson, *Adv. Mater.* **12**,1367 (2000).
6. D. Vacar, E. S. Maniloff, D. W. McBranch, and A. J. Heeger, *Phys. Rev. B* **56**, 4573–4577 (1997).
7. J.J.M. Halls, K. Pichler, R. H. Friend, S. C. Moratti, and A. B. Holmes, *Appl. Phys. Lett.* **68**, 3120–3122 (1996).
8. A. Haugeneder, M. Neges, C. Kallinger, W. Spirkl, U. Lemmer,, J. Feldmann, U. Scherf, E. Harth, A. Gügel, and K. Müllen, *Phys. Rev. B* **59**, 15346–15351 (1999).
9. M. Drees, K. Premaratne, W. Graupner, J.R. Heflin, R.M. Davis, D. Marciu, and M. Miller, *Appl. Phys. Lett.* **81**, 4607–4609 (2002); M. Drees, R. M. Davis, J. R. Heflin, *Phys. Rev. B* **69**, 165320:1–6 (2004).
10. C. W. Tang and A. C. Albrecht, *J. Chem. Phys.* **62**, 2139 (1975).
11. A. C. Arias, M. Granstrom, D. S. Thomas, K. Petrisch, and R. H. Friend, *Phys. Rev. B* **60**, 1854 (1999).
12. N. S. Sariciftci, L.Smilowitz, A.J. Heeger, and F. Wudl, *Synth. Met.* **59**, 333–352 (1993).
13. E. A. Katz, D. Faiman, S. M. Tuladhar, J. M. Kroon, M. M Wienk, T. Fromherz, F. Padinger, C.J. Brabec, and N.S. Sariciftci, *J. Appl. Phys.* **90**, 5343–5350 (2001).
14. I. Riedel, J. Parisi, V. Dyakonov, L. Lutsen, D. Vanderzande, and J. C. Hummelen, *Adv. Func. Mater.* **14**, 38–44 (2004).
15. C. Y. Yang and A. J. Heeger, *Synth. Met.* **83**, 85–88 (1996).
16. C. Schlebusch, B. Kessler, S. Cramm, and W. Eberhardt, *Synth. Met.* **77**, 151–154 (1996).
17. T. A. Chen, X. Wu, and R. D. Rieke, *J. Am. Chem. Soc.* **117**, 233–244 (1995).

25

Vertically Aligned Carbon Nanotubes for Organic Photovoltaic Devices

Michael H.-C. Jin[a,b] and Liming Dai[c]

[a] *Ohio Aerospace Institute, Brook Park, OH, USA*

[b] *NASA Glenn Research Center, Cleveland, OH, USA*

[c] *Department of Chemical and Materials Engineering, University of Dayton, Dayton, OH, USA*

Contents

Abstract As promising thin-film photovoltaic (PV) devices, organic solar cells offer many advantages, including low-cost, lightweight, large-area processability, and versatility for applications. Recently developed heterojunction device structure can support a highly efficient photoinduced charge transfer (exciton dissociation) at the junction between electron-donating and -accepting materials, leading to a high photocurrent quantum efficiency (QE) up to 70%. The device performance, however, is currently limited by the low light absorption, poor carrier transport, and degradation of the materials. As reviewed in this chapter, dispersed heterojunction devices containing vertically aligned carbon nanotubes (VA-CNTs) can show improved carrier transport properties, as each of constituent aligned CNT is connected directly onto an electrode to maximize the electron mobility. The hole transportation can also be improved by electrochemical or chemical vapor deposition of appropriate conjugated polymers onto the individual VA-CNT to provide a well-controlled conjugation path and phase morphology. The recent advances in the synthesis of VA-CNTs and new photovoltaic-active polymers for the fabrication of conjugated polymer/CNT dispersed heterojunction solar cells, will be summarized in this chapter to provide a blueprint for the development of high efficiency organic solar cells with the new heterojunction architecture containing VA-CNTs.

Keywords   photovoltaics; organic solar cells; plastic solar cells; carbon nanotube; conjugated polymer; heterojunction; composite

25.1. INTRODUCTION

Extensive reviews on the history and the fundamentals of organic solar cells have been given in previous chapters and the associated further reading materials [1–7]. In addition, comparison between organic and inorganic photovoltaic (PV) devices has been made by Gregg and Hanna [8] to help scientists and engineers who have mostly worked on inorganic solar cells understand some of the critical differences in their PV processes. This chapter will review the recent developments and structural evolution of organic PV devices with nanoarchitectures including carbon nanotube (CNT)-containing organic solar cells.

Since the development of the first single-junction inorganic (Si) solar cell made at Bell Laboratories in 1954 [9], tremendous success has been made during the last 50 years and a power conversion efficiency (PCE) up to about 35% was achieved for inorganic (III–V semiconductor) multijunction solar cells in a research laboratory [10]. However, these inorganic solar cells are still too expensive to compete with conventional grid electricity [11]. Meanwhile, alternative approaches using organic materials have been vastly studied because of their potential benefits over the inorganic materials, including their low cost, lightweight, flexibility, and versatility for fabrication (especially over a large area). Another driving force for the thin-film PV devices including organic solar cells is that they can offer a variety of applications for future space missions [12].

Many organic materials initially studied for their use as photoconductors in imaging applications [13] also showed PV effects. Common dyes and organic pigments including biological molecules have attracted a tremendous attention since the synthesis of the field in 1950s [14,15]. The discovery of semiconducting (conjugated) polymers [16] further boosted research in the organic PV devices. Despite many years of devoted work worldwide, the performance of organic PV devices including polymer solar cells was rather discouraging and PCE stayed well below a single-digit number. Unlike inorganic semiconductor materials, conjugated organic molecules require high exciton dissociation energy (a few meV vs. an order of 100 meV) in order to generate free electrons and holes. The simple single-layer solar cell structure (Figure 25.1) [3] used in the early organic PV solar cells did not allow an effective exciton dissociation because the electric field established by the work function difference between two electrodes or by the Schottky contact made between an organic material and an electrode was insufficient.

To overcome this limitation, the concept of a layered heterojunction organic PV cell (Figure 25.2) was introduced by Tang in 1986 [17]. The junction was made between two different layers of dyes (copper phthalocyanine (CuPc) and a perylene tetracarboxylic derivative) leading to a PCE of about 1% under simulated AM2 illumination. The efficient exciton dissociation at the junction was attributed to the strong electric field formed due to the energy level offset at the junction between two materials. Free electrons and holes generated from the exciton dissociation were transported through electron-accepting (EA) and electron-donating (ED) materials, respectively. There was less chance for carrier recombination in the layered heterojunction devices than the single-layer devices because free electrons and holes did not come across each other after the charge separation at the junction in the former case.

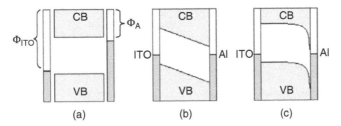

(a)	(b)	(c)

Figure 25.1. Schematic representation of single-layer organic solar cell devices. (a) The energy levels for the conduction band (CB) and the valence band (VB) are shown, along with the work function for the low work-function electrode (Al) and the high work-function electrode (indium tin oxide, ITO). (b) Assembled and short-circuited solar cell with an insulating organic material. (c) A hole-conducting polymer forms a Schottky junction at the high work-function electrode. (From H. Spanggaard and F. C. Krebs, *Sol. Energy Mater. Sol. Cells* **83**, 125–146 (2004).)

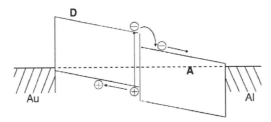

Figure 25.2. Schematic representation of a bilayer heterojunction device. The junction is made between electron-donating material (D) and electron-accepting material (A). The high work-function (Au) and the low work-function (Al) electrodes make contact with D and A, respectively. Photogenerated excitons can only be dissociated at the D–A heterojunction and thus the device is exciton diffusion limited. (From H. Hoppe and N. S. Sariciftci, *J. Mater. Res.* **19**, 1924–1945 (2004).)

Examples of ED conjugated molecules and EA materials are listed in Figure 25.3 [1]. Further improvement in PCE was made by inserting a transparent, organic exciton-blocking layer (EBL) between an EA material and a negative electrode [18,19]. The overall device structure was indium tin oxide (ITO)/CuPc/C_{60}/EBL/Ag, and bathocuproine (BCP) was used as the EBL. The use of the EBL offers advantages, including the prevention of the EA material (C_{60}) from damage during the deposition of the metal contact (Ag), parasitic exciton quenching at the interface between C_{60} and Ag, and increasing optical density at the heterojunction between CuPc and C_{60}. PCEs of 3.6 $\pm$ 0.2% [18] and 4.2 $\pm$ 0.2% [19] have been achieved under simulated 1 sun and 4–12 suns AM1.5G illumination, respectively. The layered heterojunction approach was also made for conjugated polymer PV cells. In 1993, Sariciftci et al. [20] first demonstrated the junction between C_{60} and (poly(2-methoxy-5-(2'-ethyl-hexyloxy)1,4-phenylenevinylene)) (MEH-PPV), a soluble derivative of PPV. Various polymer materials and deposition techniques were used in many subsequent studies [17,21–25]. Examples of layered heterojunctions between organic electronic materials include PPV/C_{60} [20], poly(3-alkylthiophene) (P3AT)/C_{60} [21], PPV/CNT [22], CuPc/perylene [17], PPV/perylene [23], P3AT/poly(p-pyridylvinylene) (PPyV) [24], and PPV/poly(benzimidazobenzophenanthroline ladder) (BBL) [25]. It is also notable that hybrid organic–inorganic layered heterojunction PV cells were fabricated using conjugated polymer and TiO_2 as an EA inorganic semiconductor [26].

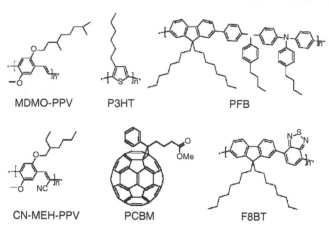

Figure 25.3. Several solution processable conjugated polymers and a fullerene derivative used in organic solar cells. Upper row: the p-type electron-donating polymers, MDMO-PPV (poly[2-methoxy-5-(3,7-dimethyl-octyloxy)]-1,4-phenylenevinylene), P3HT (poly (3-hexyl-thiophene-2,5-diyl), and PFB (poly(9,9′-dioctylfluorene-*co*-bis-*N,N*′-(4-butylphenyl)-bis-*N,N*′-phenyl-1,4-phenylenediamine). Lower row: the electron-accepting materials, CN-MEH-PPV (poly-[2-methoxy-5-(2′-ethylhexyloxy)-1,4-(1-cyanovinylene)-phenylene]), a fullerene derivative, PCBM (1-(3-methoxycarbonyl) propyl-1-phenyl[6,6]C$_{61}$), and F8TB (poly(9,9′-dioctylfluoreneco-benzothiadiazole). (From H. Hoppe and N. S. Sariciftci, *J. Mater. Res.* **19**, 1924–1945 (2004).)

Despite the remarkable improvements in device performance, the quantum efficiency of the layered heterojunction devices was still limited by a loss due to a short exciton diffusion length ($\sim$10 nm) [4] — excitons will recombine before they reach the junction by diffusion if the absorber material is thicker than the order of 10 nm. Because typical absorption length of the conjugated polymer is in the order of 100 nm [5], the zero order approximation is that only one tenth of the photons absorbed as a form of an exciton in the 100 nm thick polymer will contribute to the free carrier generation. In fact, monochromatic external quantum efficiency (MEQE) of $\sim$9% was measured for PPV/C$_{60}$ under short-circuit condition and monochromatic illumination (490 nm) with an intensity of the order of 0.1 mW/cm^2 [27]. The analysis made by Patterson et al. [28] showed MEQE was up to 25% for their poly(3-(4′-(1″,4″,7″-trioxaoctyl)phenyl)thiophene) (PEOPT)/C$_{60}$ layered heterojunction structure.

The problem associated with a short exciton diffusion length motivated the development of a new device structure, called a dispersed (or bulk) heterojunction (Figure 25.4) [1]. The main idea is that both the ED and EA materials can be randomly blended as a composite. Compared to the layered heterojunction device, this will create more interfacial area (junction) between two materials within a limited absorption length of the polymer, resulting in a device configuration with which the most efficient harvest of excitons generated by photons can occur. This scheme also requires a nanoscale blending morphology in order to maximize the interfacial area and the collection of excitons before their recombination. Remarkably improved MEQE ($\sim$70%) with a PCE of about 3.5% under white light (80 mW/cm^2) was recently demonstrated with the dispersed heterojunction between poly(3-hexylthiophene) (P3HT) and a soluble derivative of C$_{60}$, PCBM (1-(3-methoxycarbonyl) propyl-1-phenyl[6,6]C$_{61}$) after postproduction treatment (Figure 25.5) [29].

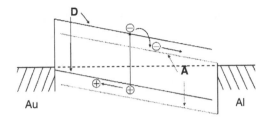

Figure 25.4. Schematic representation of a dispersed (bulk) heterojunction device. The electron-donating material (D) is blended with the electron-accepting material (A) to form interfaces at the molecular level for exciton dissociation throughout the whole film. (From H. Hoppe and N. S. Sariciftci, *J. Mater. Res.* **19**, 1924–1945 (2004).)

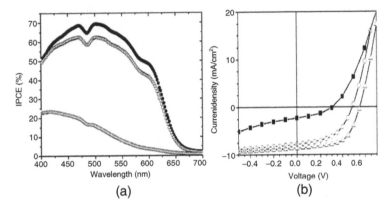

Figure 25.5. Characteristics of P3HT-PCBM dispersed heterojunction solar cells: (a) External quantum efficiency vs. photoexcitation wavelength and (b) Current–voltage curves measured under white light with an irradiation intensity of 800 W/m^2. As-produced solar cell (open triangles in (a) and filled squares in (b)), annealed solar cell (open squares in (a) and open circles in (b)), and cell simultaneously treated by annealing and applying an external voltage (filled circles in (a) and open triangles in (b)). (From F. Padinger, R. S. Rittberger, and N. S. Sariciftci, *Adv. Func. Mater.* **13**, 85–88 (2003)).

For the past few years, a variety of materials and methods have been developed to fabricate the dispersed heterojunction organic PV cells. They include mixtures between (i) two conjugated polymers [30], (ii) organic and nanoscale inorganic materials (TiO$_2$ [31], CdSe [32], and CuInSe$_2$ [33]), and (iii) polymer and CNTs [34]. More advanced nanoarchitecturing of the blend is essential to improve the performance of the dispersed heterojunction organic PV devices. The following sections will review the most recent understanding of the dispersed heterojunction with a particular focus on polymer/CNT composites proposing more advanced solar cell architecture for the composites. Different nanoengineered approaches for the near-future organic PV devices will be also discussed.

25.2. POLYMER/CARBON NANOTUBE SOLAR CELLS

This section is designed to review polymer/CNT PV research, whereas Chapter 15 authored by Kymakis and Amaratunga has covered recent results on dispersed heterojunction organic PV cells particularly made with poly(3-octylthiophene)

(P3OT) and single-wall CNTs (SWNTs). Although there have been many review articles and book chapters devoted to organic PV devices [1–8], little attention was given to the polymer/CNT PV devices.

In spite of the high MEQEs (up to 70%) obtained from the dispersed heterojunction structure and open circuit voltages (V_{oc}) (0.5–1 V) for the corresponding organic solar cells, these PV devices still suffer from low short circuit currents (I_{sc}) (5–10 mA/cm^2) and the best PCE achieved remains under 5%. To achieve better carrier transport without sacrificing EQE and V_{oc} could be a solution for a higher PCE. In the dispersed heterojunction organic solar cells, free charges are percolating through the device as drift and diffusion currents. With an assumption of 100% exciton dissociation, drift current can be maximized by having materials with the highest possible carrier mobilities, whereas diffusion current will increase until the film thickness matches with its carrier diffusion length. Unlike layered heterojunction, free carriers generated from the dissociation of the excitons at the dispersed heterojunctions see their opposite carriers during transport, leading to a loss from the high rate of carrier recombination. Furthermore, miscibility between two different solid-state materials is another inherent nature of the blending that also creates problems. Typical method for fabricating a dispersed heterojunction conjugated polymer solar cell is to spin-coat a mixture solution of soluble derivatives of both ED and EA materials onto a substrate. The evaporation of the solvent and subsequent phase segregation can be important factors determining overall material properties and device performance [30,35,36] and the soluble derivatives often contain side groups and chains that normally cause the material degradation and provide hindrance to carrier transport [2].

Recently performed time-of-flight (TOF) photocurrent measurements of the charge transport in polyfluorene/PCBM [37] and PPV/PCBM [38] composites showed that electron mobility (μ_e) (10^{-4}–10^{-3} cm^2/V·s) was higher than hole mobility (μ_h) (10^{-6} cm^2/V·s) at the composition optimized for their best cell performance with a PCBM composition of more than 75 wt.%. In literature [39,40], electron mobilities of PCBM calculated from TOF experiments and the analyses of field-effect transistors ranged between 10^{-6} and 10^{-1} cm^2/V·s. Meanwhile, the highest field-effect hole mobilities (μ_{fh}) of organic materials were measured to be between 10^{-5} and 1 cm^2/V·s).

While there are intrinsic limitations to enhance μ_h of conjugated polymers [40], replacing PCBM with an EA material like CNT, which has a higher μ_e than PCBM, is considered to be promising route to improve the solar cell performance. CNT is known to be an extremely electron-conductive semiconductor and field-effect electron mobility (μ_{fe}) up to 77,000 cm^2/V·s was recently reported for a SWNT [41]. Another benefit from the high carrier mobility is that more photons can be harvested by making thicker devices. Typical thickness of the organic solar cell is in the order of 100 nm, which is limited by optical absorption and diffusion current at a given carrier mobility [2,4]. If the device thickness becomes too thin, there is too much loss in optical absorption. In the case of 100 nm PPV/PCBM, for example, more than half of the incoming light was lost [42]. In comparison, recombination becomes competitive if the device becomes too thick. A simplified model demonstrated that composites with better transport properties (higher carrier mobilities) allowed making thicker devices to absorb more photons [43].

A number of experiments with polymer/CNT composites have been pursued in an attempt to improve current transport of organic solar cells [22,44]. The first conjugated polymer/CNT layered heterojunction as an optoelectronic device was fabricated by Romero et al. [22] in 1996. The junction was made between PPV and multiwalled CNTs (MWNTs) prepared by filtration. Although a rectifying heterojunction was realized, no photocurrent was observed. Two years later, light-emitting diodes were prepared with composites based on PPV derivatives and arc-grown MWNTs, and much better device stability in air was demonstrated compared to devices without MWNTs [44]. The PV effect from organic solar cells that contain CNTs was observed by Ago et al. [45] for the first time in 1999. These authors deposited thin films of MWNTs by spin-coating a MWNT dispersion onto a glass substrate, followed by spin-coating a PPV precursor (sulfonium salt) on top of the MWNT films and thermal treatment. Thin semitransparent Al contacts were made on top to complete the devices. The device showed encouraging MEQE (at 390–430 nm) of 1.8%.

The best effort to advance the performance of organic PV devices based on polymer/nanotube composites was more recently made by Kymakis et al. [34,46]. Compared to the previous studies with MWNTs as EA materials, SWNTs were used within P3OT matrix to promote better charge transfer and electron transport in the dispersed heterojunction architecture. The conductivity of the P3OT/SWNT composites increased five orders of magnitude as the SWNT concentration increased from 0 to 20 wt.%. The percolation threshold was about 11 wt.% and the best cell performance was achieved with only 1 wt.% SWNTs [46]. The composite was sandwiched by Al and ITO and V_{oc} of the solar cell was found to be 0.75 V, which was weakly dependent on the negative electrode work function. Fermi-level pinning between the negative electrode and the EA material was proposed to explain the observed weak dependence of V_{oc} on the electrode work function [34]. The V_{oc} was well agreed with the energy differences between the work function of the nanotubes and the HOMO (highest occupied molecular orbital) of the P3OT [34].

As an effort to realize space-qualified organic solar cells, NASA Glenn Research Center has been teamed up with the NanoPower Research Laboratories in Rochester Institute of Technology and Ohio Aerospace Institute to study P3OT/SWNT dispersed heterojunction PV devices. V_{oc} of 0.65 V was previously achieved [47] and it was further improved in a recent study in which the initial device performance with V_{oc} of 0.75 V was presented [48]. In the report, CdSe–SWNT complexes were synthesized by covalent attachment of carboxylic acid-functionalized SWNTs with CdSe–aminoethanethiol (AET) nanoparticles (Figure 25.6). Composites between P3OT and CdSe–AET–SWNTs were prepared as dispersed heterojunction solar cells that contain both nanoparticles and CNTs for the first time. A similar approach was made by incorporating MWNTs into the BBL layer in PPV/BBL bilayer heterojunction PV devices [49]. I_{sc} was doubled while V_{oc} was decreased by adding MWNTs into the BBL layer.

Vignali et al. [50] paid their attention to the material degradation of polymer/CNT composites. Unmodified conjugated conducting polymers are generally not soluble whereas their solubility can be significantly enhanced by grafting various side chains onto the conducting polymer chains. However, carrier mobility and cell efficiency can be considerably decreased within hours in the air because these side-moieties can cause photooxidation of polymers as well as lowering crystallinity and chain-to-chain conduction. CNTs possess advantages over more popular soluble C_{60} derivatives containing side groups that are vulnerable to photooxidations. In order to avoid the need of organic side chains, unsubstituted polyterthiophene (P3T) was deposited by electropolymerization in the presence of CNTs to form a dispersed heterojunction structure. So far only a Schottky diode between P3T and ITO was

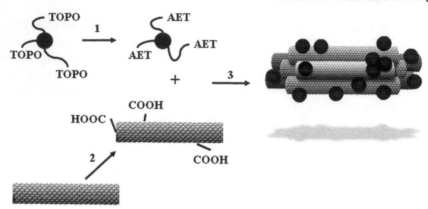

Figure 25.6. Reaction scheme for covalent attachment of CdSe–TOPO (trioctylphosphine oxide) quantum dots to single-wall CNTs (SWNTs). (1) Ligand exchange of TOPO with AET (aminoethanethiol), (2) carboxylic acid functionalization of SWNTs, and (3) coupling reaction to form an amide linkage. (From S. G. Bailey, S. L. Castro, B. J. Landi, H. J. Ruf, and R. P. Raffaelle, *19th European Photovoltaic Solar Energy Conference and Exhibition*, chaired by W. Hoffmann et. al., Paris, France, June 7–11, 2004.)

prepared and a PV effect was observed under ambient light. Conductivity measurements showed that P3T layers deposited by electropolymerization were quite tolerant to photooxidation compared to the polythiophene derivatives. Composites with high-molecular weight polyaniline (PANI) and SWNTs were also investigated for electronic devices with enhanced thermal and mechanical stabilities [51].

Despite the respectable V_{oc} (0.75 V), the best PCE of polymer/CNT solar cell is less than 1% [34]. The device performance of the polymer/CNT was limited by low photocurrent due to incomplete phase separation, the lack of light absorption, and a low hole mobility. In an attempt to increase the photocurrent of the cells, the SWNTs were coated with dye molecules [52]. Improved light absorption in the UV and red region due to the addition of the dye molecules contributed to the current, and I_{sc} increased more than five times higher than the case with pure SWNTs. Despite much effort, PV devices based on polymer/CNT composites have been less discussed in the literature compared to polymer/PCBM devices. Further analysis of charge transport in polymer/CNT composite is essential to gain a better fundamental understanding of the device performance [53]. Meanwhile, dispersed heterojunction devices made of P3HT and one-dimensional CdSe semiconductor nanorods were recently analyzed using a modified Shockley equation to include series and shunt resistance at low currents as well as space-charge effect at high currents [54]. This quantitative study provides more insights of the dispersed heterojunction device with one-dimensional EA materials.

25.3. VERTICALLY ALIGNED CARBON NANOTUBES FOR OPTOELECTRONIC APPLICATIONS

Instead of randomly mixing a polymer with CNTs in a solution to cast a film, an ideal device structure for polymer/CNT dispersed heterojunction solar cell can be made by building a VA-CNT network and combining it with a polymer (Figure 25.7). Spacing between VA-CNTs needs to be within exciton diffusion length to maximize exciton

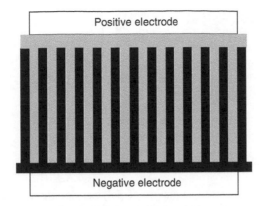

Figure 25.7. Schematic representation of polymer/vertically-aligned CNT (VA-CNT) composite. Black, VA-CNT layer; gray, conjugated polymer.

dissociation at interfaces between the polymer and the VA-CNTs [53]. The percolation threshold of VA-CNTs could be even lower than randomly mixed CNTs without having any isolated electron trapping sites. In addition, each component of the dispersed heterojunction makes contact with only one electrode, which prevents shunting between positive and negative electrodes. VA-CNTs offer additional advantages for many different applications including electron emitters and molecular membranes [55].

VA-CNTs can be prepared in different routes. Templates that have a vertical porous structure were used to synthesize CNTs inside pores [56]. They can be also directly realized on substrates without any template [57]. Postsynthesis method like self-assembling end-functionalized CNTs can provide further versatility [58]. Template-based materials synthesis was pioneered by Charles Martin from the University of Florida and tremendous progress has been made in the last decade [59]. The first report of the large-scale CVD growth of aligned CNTs with a mesoporous silica template was made by Xie and co-workers in 1996 [60] and numerous experiments using a similar template were reported afterwards [61–64]. The selection of the template is essential to produce well-aligned CNTs and anodic aluminum oxide membranes are normally used to provide pores with uniform diameters and lengths [59,63]. More recently, Kim and Jin [56] demonstrated the preparation of PPV and (poly(2,5-thienylenevinylene)) (PTV) [64] nanotubes and nanorods inside the nanopores (10–200 nm diameter) in alumina membranes by chemical vapor deposition polymerization (CVDP) (Figure 25.8). Carbonized products including CNTs were also synthesized through postdeposition thermal treatments [56] and remarkable field-emission characteristics were obtained from both CNTs and Au nanoparticle-embedded CNTs [65]. They have also synthesized CNT/PPV bilayer nanotubes with alumina templates and photoluminescence (PL) quenching due to exciton-dissociation at the interface between two phases was observed (J.-I. Jin, pers. commun., 2004). A similar bilayer structure was also fabricated for composites between TiO_2 and polypyrrole in a different study [66]. The composite was made in a 60-mm thick alumina template with 200 nm diameter pores.

Because the use of a template requires postsynthesis processes like template removal using an etchant, it is more desirable to produce VA-CNTs without a template. Mainly there are two methods to prepare VA-CNTs without a template: plasma-assisted processes [67–71] and chemical pyrolysis in CVD reactors. Ren et al. [67]

Figure 25.8. Electron microscopic images of polymer nanotubes synthesized by chemical vapor deposition polymerization and isolated from alumina membranes: (a) Scanning electron micrograph of PPV nanotubes (from K. Kim and J.-I. Jin, *Nano Lett.* **1**, 631–636 (2001)) and (b) Transmission electron micrograph of PTV (Poly(2, 5-thienylenevinylene)) nanotubes [64].

synthesized large arrays of VA-CNTs using a plasma-assisted hot-wire CVD of acetylene in the presence of ammonia gas; thin layer of nickel was deposited beforehand as a catalyst. Other plasma techniques including direct current [68], radio frequency [69], and microwave plasma [70,71] were also experimented. Pyrolysis of methane–argon under an electric field was able to produce VA-CNTs [72]. A number of researchers, including the present authors have demonstrated the synthesis of VA-CNTs by the pyrolysis of precursors that contain both metal catalysts and carbons for CNT growth [57,73–75]. An example of a pyrolysis apparatus for the generation of VA-CNTs was shown in Figure 25.9(a) [76]. A typical scanning-electron microscopic image of the CNTs, prepared from the setup by pyrolysis of ferrocene (FePc) showed fairly good uniformities in length and diameter (Figure 25.9(b)) [77]. The formation of the CNTs was explained by the catalytic effect of Fe particles produced upon the thermal decomposition of FePc [57]. The growth mechanism study indicated that small metal particles segregated on the substrate surface supported the nucleation of the nanotubes and the larger Fe particles on the top of growing CNTs provided the carbon atoms for the growth of CNTs. An attempt to realize the formation of CNTs on a variety of thin conducting layers (substrates) like Au, Pt, Cu, and ITO was successfully made by a unique transfer technique described in literatures (Figure 25.9(b)) [77]. Particularly a metal substrate can be used as a negative electrode that contacts with VA-CNTs in the configuration of the dispersed heterojunction PV devices. The thin conducting substrate also allowed electrochemical deposition to produce highly electroactive polyaniline–CNT coaxial nanowires (Figure 25.10) [77]. Alternatively, VA-CNTs can be prepared by a post-synthesis self-assembling technique. For instance, Liu and co-workers [58] have demonstrated self-assembling of end-functionalized CNTs on gold and metal oxide substrates [78]. The resulting CNTs were so stable that ultrasonication could not remove them from the substrate. Nan et al. [79] have demonstrated the region-specific self-assembly of CNTs by chemically prepatterning substrates using micro-contact printing [79].

While there have been many developments in depositing VA-CNTs, the fabrication of polymer/VA-CNT composites has been limited to the nano- or microscale [77]. Spin-coating is a relatively easy and effective method to cast a large-area thin film on top of porous or nanostructured EA materials like a layer of VA-CNTs. A recent study on composites between P3HT and well-ordered mesostructured TiO_2 found that the polymer deposited by spin-coating can be infiltrated to the bottom of

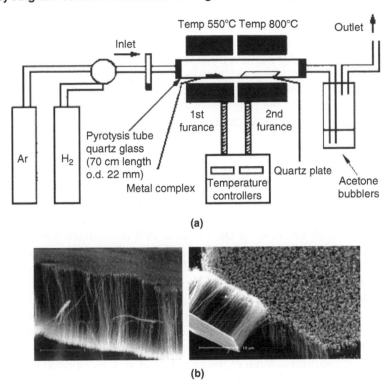

(a)

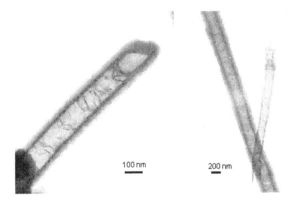

Figure 25.9. Deposition of vertically aligned CNTs (VA-CNTs): (a) Apparatus for the generation of CNTs by pyrolysis of FePc (iron(II) phthalocyanine). (From S. Huang, L. Dai, and A. W. H. Mau, *J. Phys. Chem. B* **103**, 4223–4227 (1999)) and (b) SEM images of typical VA-CNTs produced (left) and VA-CNTs after transfer onto a gold foil (a small piece of the as-synthesized VA-CNT film is included at the bottom-left corner to show the amorphous carbon layer as well) (right). (From M. Gao, S. Huang, L. Dai, G. Wallace, R. Gao, and Z. Wang, *Angew. Chem. Int. Ed.* **39**, 3664–3667 (2000).)

Figure 25.10. Typical transmission electron microscopic images of the conducting polymer–carbon nanotube coaxial nanowires formed from the cyclic voltammetric method, the images are in the tip region (left) and on the wall (right). (From M. Gao, S. Huang, L. Dai, G. Wallace, R. Gao, and Z. Wang, *Angew. Chem. Int. Ed.* **39**, 3664–3667 (2000).)

the 50–300 nm thick TiO$_2$ films [80]. This study, however, indicated polymer chains were twisted and consequently not sufficiently π-stacked due to the structure and size of pores in the TiO$_2$ film. Earlier approaches like electropolymerization [50,77] and CVDP [56] might be promising alternatives to achieve better polymer morphology, improving exciton diffusion length, and hole mobility. We anticipate tremendous technical challenges ahead but also believe much more developments will be made in a few years to advance the performance of organic solar cells.

25.4. NANO-ENGINEERING FOR THE FUTURE

As shown previously, the performance of organic solar cells has been mainly limited by I_{sc}. In order to improve EQE, the concept of a dispersed heterojunction was introduced. Although the best MEQE of about 70% (at 500 nm) was demonstrated with composites made of P3HT and PCBM [29], MEQE was typically under 30%. In previous section, VA-CNTs were introduced in order to improve electron transport. In addition, it is also possible that an appropriate selection of a polymer preparation method could result in better hole mobility at the same time; because of the regularity of VA-CNT network, a controlled deposition can produce better stacking of polymer chains. A self-assembly technique coupled with electropolymerization and CVDP can be applied to synthesize vertically stacked polymer chains between CNTs.

In general, the optimization of the morphology of the dispersed heterojunction is essential to improve photoinduced carrier transfer (exciton dissociation) and current transport [36]. Approaches to make better composite morphology include optimizing the solvent for film formation [35], the ratio of ED and EA materials [36], and new material design to promote both better PV effects and composite morphology [48,52,81–87]. Nano-engineering of the material properties suitable for solar cell applications is an interesting research field and some of the current research activities are briefly summarized below.

Applications of the new fullerene derivatives other than PCBM have been active because they can be designed to be more red-absorbing [81] or more soluble than PCBM in a wide range of organic solvents and be still effectively electron-accepting [82]. For instance, dispersed heterojunction solar cells have been recently fabricated with composites of MEH-PPV and new methanofullerene derivatives (Figure 25.11(a)) [82]. Very efficient PL quenching was observed from the composites and the PV devices made with MEH-PPV/TDC$_{60}$ composites showed V_{oc} of 0.7 V and MEQE of about 6%. More examples of fullerene derivatives can be found in literature [81,83]. Chemical modification of CNTs is another method to boost the characteristics of the solar cells [84]. In Section 2, dye-sensitized SWNTs to improve light absorption [52] and SWNTs covalently coupled with CdSe quantum dots as an example of a complex of two different EA materials have been discussed [48]. Inorganic EA materials are always attractive due to their high electron mobility and stability although they cannot be much versatile in terms of chemical modification. New development in the controlled growth of nanomaterials can open new possibilities. Recently, CdSe tetrapods were synthesized to enhance current transport in solar cells and MEQE was doubled compared to the solar cell with CdSe nanorods [85]. The benefits from the three-dimensional structure of CdSe tetrapods are comparable to what is anticipated in the case with VA-CNTs. The last category of new material will be the complex of ED and EA materials [83,86]. The complexes between polymer materials including co-polymers that contain both ED and EA units are well described in Chapters 8 and 18.

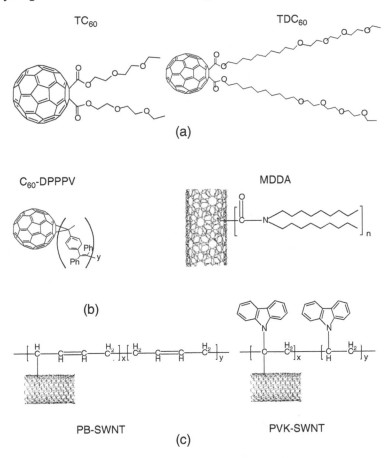

Figure 25.11. Molecular structures of nano-engineered photoactive materials for organic solar cells. (a) Mono-functionalized methanofullerene derivatives: TDC_{60} contains an inner hydrophobic part (decyl chain) and an outer hydrophilic part (triethylene glycols) and TC_{60} is C_{60} substituted by hydrophilic chains only (From J. Li, N. Sun, Z.-X. Guo, C. Li, Y. Li, L. Dai, D. Zhu, D. Sun, Y. Cao, and L. Fan, *Synth. Met.* **137**, 1527–1528 (2003).). (b) An example of C_{60}-containing conjugated polymer, poly[1,4-phenylene (1,2-diphenyl)vinylene] (DPPPV)-grafted C_{60} (From T. Lin, L. Dai, G. Wallace, A. Burrell, and D. Officer, *Polymer preprints* **43**(1), 98 (2002).). (c) Examples of surface-modified CNTs, MDDA (didecylamine-modified multiwall CNT), PB-SWNT (polybutadiene-modified single-wall CNT), and PVK-SWNT (polyvinylcarbazole-modified SWNT) (From W.Wu, J. Li, L. Liu, L. Yanga, Z.-X. Guo, L. Dai, and D. Zhu, *Chem. Phys. Lett.* **364**, 196–199 (2002). W. Wu, S. Zhang, Y. Li, J. Li, L. Liu, Y. Qin, Z.-X. Guo, L. Dai, C. Ye, and D. Zhu, *Macromolecules* **36**, 6286–6288 (2003).).

Although current transport in these complexes will vary case by case, the underlined idea is very similar to the concept of a "double-cable" approach pictured in Figure 25.12 [83]. In this configuration, loss from exciton recombination due to irregularly formed long diffusion paths can be prevented because of close proximity between two phases that are bound together very uniformly through the entire composite. Both C_{60} and CNT can be used as an EA unit and vertically aligned composite structures using these materials have great potential for the PV devices. There have been several attempts to try the double-cable configuration. For example, Lin et al. [86] fabricated solar cells with C_{60} containing conjugated polymers (Figure 25.11(b)) and Wu et al.

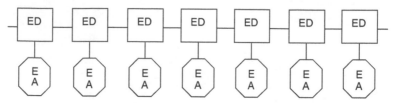

Figure 25.12. Schematic representation of "double-cable" material. ED, electron-donating conjugated polymer; EA, electron-accepting moieties.

[84,87] successfully synthesized functionalized CNTs (both SWNTs and MWNTs) and their photoinduced electron transfer behavior was studied (Figure 25.11(c)). Further research activities related to the complexes can be found in the literature [83]. Although it was not emphatically addressed in this review, synthesizing new low-bandgap organic materials that can harvest more light is one of the biggest challenges in the organic PV devices. Encouraging progress has been made in this topic over the last few years and the authors sincerely recommend the reader to consult the relevant references [88,89].

25.5. CONCLUSION

The device structure of organic solar cells has developed from a single-layer device to a layered heterojunction device and to a dispersed heterojunction device most recently. The main focus of this development was to increase EQE of the solar cell by promoting better exciton dissociation and a P3HT/PCBM dispersed hetero-junction solar cell has achieved MEQE (at 500 nm) of ~70% with a PCE of 3.5% under white light (80 mW/cm^2) [29]. Recent effort has been more oriented toward understanding carrier transport of the cells in order to identify the limiting factor for the low cell current. Studies found that both electron and hole mobilities from the solar cells were not as high as field-effect mobilities extracted from organic thin-film transistor measurements; hole mobilities in the solar cells are often five orders lower than those calculated from the characteristics of thin-film transistors [39].

In an attempt to increase carrier mobilities and overall current transport of the cells, VA-CNTs are introduced to engineer the architecture of the near-future version of the dispersed heterojunction solar cells. VA-CNTs will provide a more facile route for electron current and better light harvest is possible due to lower percolation threshold of VA-CNTs. Recent studies proved that VA-CNTs can be readily prepared on various conducting substrates by pyrolysis of precursors containing both metal catalyst and carbons for the CNT growth [57,76]. It is emphasized that the improvement of hole transport is also challenging in organic solar cells and the authors have proposed bottom-up approaches for the deposition of polymer materials into the network of nanostructured EA materials including VA-CNTs; electro-polymerization [50,77] and CVDP [56] have demonstrated the deposition of conjugated polymers and they have great potential for the VA-CNT/polymer architecture. Further research effort related to nanoengineering of the organic solar cell materials, for example, self-assembly of CNTs and polymer nanostructures, will complement the realization of the architecture.

ACKNOWLEDGMENTS

The authors would like to thank Jeremiah S. McNatt from NASA Glenn Research Center and David A. Scheiman from Ohio Aerospace Institute for their help with proofreading. LD is grateful to AFRL/ML, WBI, The Dayton Development of Colation, and University of Dayton for the Wright Brothers Institute Endowed Chair Professorship in Nanomaterials.

REFERENCES

1. H. Hoppe and N. S. Sariciftci, Organic solar cells: an overview, *J. Mater. Res.* **19**, 1924–1945 (2004).
2. N. S. Sariciftci, Plastic photovoltaic devices, *Materials Today* **7**(9), 36–40 (2004).
3. H. Spanggaard and F. C. Krebs, A brief history of the development of organic and polymeric photovoltaics, *Sol. Energy Mater. Sol. Cells* **83**, 125–146 (2004).
4. C. Brabec, V. Dyakonov, J. Parisi, and N. S. Sariciftci, *Organic Photovoltaics: Concepts and Realization*, Springer, New York, 2003.
5. J. Nelson, Organic photovoltaic films, *Mater. Today* **5**(5), 20–27 (2002).
6. J. J. M. Halls and R. H. Friend, Organic photovoltaic devices, in *Clean Electricity From Photovoltaics*, M. D. Archer, R. Hill, eds., Imperial College Press, London, 2001, pp. 377–445.
7. C. J. Brabec, N. S. Sariciftci, and J. C. Hummelen, Plastic solar cells, *Adv. Func. Mater.* **11**, 15–26 (2001).
8. B. A. Gregg and M. C. Hanna, Comparing organic to inorganic photovoltaic cells: theory, experiment, and simulation, *J. Appl. Phys.* **93**, 3605–3614 (2003).
9. D. M. Chapin, C. S. Fuller, and G. L. Pearson, A new silicon p–n junction photocell for converting solar radiation into electrical power, *J. Appl. Phys.* **25**, 676–677 (1954).
10. M. A. Green, K. Emery, D. L. King, S. Igari, and W. Warta, Solar cell efficiency tables (version 24), *Prog. Photovolt: Res. Appl.* **12**, 365–372 (2004).
11. J. Johnson, Power from the sun, *Chem. Eng. News* **82**(25), 25–28 (2004).
12. D. J. Hoffman, T. W. Kerslake, A. F. Hepp, M. K. Jacobs, and D. Ponnusamy, Thin-film photovoltaic solar array parametric assessment, in *Proc. AIAA 35th Intersociety Energy Conversion Engineering Conference*, AIAA, Las Vegas, NW, 2000, pp. 670–680.
13. P. M. Borsenberger and D. S. Weiss, *Organic Photoreceptors For Imaging Systems*, Marcel Dekker, New York, 1993.
14. G. A. Chamberlain, Organic solar cells: a review, *Solar Cells* **8**, 47–81 (1983).
15. R. H. Bube, *Photoconductivity of Solids*, John Wiley & Sons, New York, 1960.
16. H. Shirakawa, E. J. Louis, A. G. MacDiarmid, C. K. Chiang, and A. J. Heeger, Synthesis of electrically conducting organic polymers: halogen derivatives of polyacetylene, $(CH)_x$, *J. Chem. Soc. Chem. Commun.* 578–580 (1977).
17. C. W. Tang, Two-layer organic photovoltaic cell, *Appl. Phys. Lett.* **48**, 183–185 (1986).
18. P. Peumans and S. R. Forrest, Very-high-efficiency double-heterostructure copper phthalocyanine/C_{60} photovoltaic cells, *Appl. Phys. Lett.* **79**, 126–128 (2001).
19. J. Xue, S. Uchida, B. P. Rand, and S. R. Forrest, 4.2% Efficient organic photovoltaic cells with low series resistances, *Appl. Phys. Lett.* **84**, 3013–3015 (2004).
20. N. S. Sariciftci, D. Braun, C. Zhang, V. I. Srdanov, A. J. Heeger, G. Stucky, and F. Wudl, Semiconducting polymer–buckminsterfullerene heterojunctions: diodes, photodiodes, and photovoltaic cells, *Appl. Phys. Lett.* **62**, 585–587 (1993).
21. S. Morita, A. A. Zakhidov, and K. Yoshino, Wavelength dependence of junction characteristics of poly(3-alkylthiophene)/C_{60} layer, *Jpn. J. Appl. Phys.* **32**, 873–874 (1993).

22. D. B. Romero, M. Carrard, W. De Heer, and L. Zuppiroli, A carbon nanotube/organic semiconducting polymer heterojunction, *Adv. Mater.* **8**, 899–902 (1996).

23. J. J. M. Halls and R. H. Friend, The photovoltaic effect in a poly(*p*-phenylenevinylene)/ perylene heterojunction, *Synth. Met.* **85**, 1307–1308 (1997).

24. K. Tada, M. Onoda, H. Nakayama, and K. Yoshino, Photocell with heterojunction of donor/acceptor polymers, *Synth. Met.* **102**, 982–983 (1999).

25. S. A. Jenekhe and S. Yi, Efficient photovoltaic cells from semiconducting polymer heterojunctions, *Appl. Phys. Lett.* **77**, 2635–2637 (2000).

26. T. J. Savenije, J. M. Warman, and A. Goossens, Visible light sensitisation of titanium dioxide using a phenylene vinylene polymer, *Chem. Phys. Lett.* **287**, 148–153 (1998).

27. J. J. M. Halls, K. Pichler, R. H. Friend, S. C. Moratti, and A. B. Holmes, Exciton diffusion and dissociation in a poly(*p*-phenylenevinylene)/C$_{60}$ heterojunction photovoltaic cell, *Appl. Phys. Lett.* **68**, 3120–3122 (1996).

28. L. A. A. Patterson, L. S. Roman, and O. Inganas, Modeling photocurrent action spectra of photovoltaic devices based on organic thin films, *J. Appl. Phys.* **86**, 487–496 (1999).

29. F. Padinger, R. S. Rittberger, and N. S. Sariciftci, Effects of postproduction treatment on plastic solar cells, *Adv. Func. Mater.* **13**, 85–88 (2003).

30. J. J. M. Halls, A. C. Arias, J. D. MacKenzie, W. Wu, M. Inbasekaran, E. P. Woo, and R. H. Friend, Photodiodes based on polyfluorene composites: influence of morphology, *Adv. Mater.* **12**, 498–502 (2000).

31. P. Ravirajan, S. A. Haque, J. R. Durrant, D. Poplavskyy, D. D. C. Bradley, and J. Nelson, Hybrid nanocrystalline TiO$_2$ solar cells with a fluorene–thiophene copolymer as a sensitizer and hole conductor, *J. Appl. Phys.* **95**, 1473–1480 (2004).

32. W. U. Huynh, J. J. Dittmer, and A. P. Alivisatos, Hybrid nanorod–polymer solar cells, *Science* **295**, 2425–2427 (2002).

33. E. Arici, H. Hoppe, F. Schäffler, D. Meissner, M. A. Malik, and N. S. Sariciftci, Hybrid solar cells based on inorganic nanoclusters and conjugated polymers, *Thin Solid Films*, **451–452**, 612–618 (2004).

34. E. Kymakis, I. Alexandrou, and G. A. J. Amaratunga, High open-circuit voltage photovoltaic devices from carbon-nanotube-polymer composites, *J. Appl. Phys.* **93**, 1764–1768 (2003).

35. S. E. Shaheen, C. J. Brabec, N. S. Sariciftci, F. Padinger, T. Fromherz, and J. C. Hummelen, 2.5% Efficient organic plastic solar cells, *Appl. Phys. Lett.* **78**, 841–843 (2001).

36. J. K. J. van Duren, X. Yang, J. Loos, C. W. T. Bulle-Lieuwma, A. B. Sieval, J. C. Hummelen, and R. A. J. Janssen, Relating the morphology of poly(*p*-phenylene vinylene)/methanofullerene blends to solar-cell performance, *Adv. Func. Mater.* **14**, 425–434 (2004).

37. R. Pacios, J. Nelson, D. D. C. Bradley, and C. J. Brabec, Composition dependence of electron and hole transport in polyfluorene:[6,6]-phenyl C$_{61}$-butyric acid methyl ester blend films, *Appl. Phys. Lett.* **83**, 4764–4766 (2003).

38. S. A. Choulis, J. Nelson, Y. Kim, D. Poplavskyy, T. Kreouzis, and J. R. Durrant, Investigation of transport properties in polymer/fullerene blends using time-of flight photocurrent measurements, *Appl. Phys. Lett.* **83**, 3812–3814 (2003).

39. G. H. Bauer and P. Würfel, Quantum solar energy conversion in organic solar cells, in *Organic Photovoltaics: Concepts and Realization*, C. Brabec, V. Dyakonov, J. Parisi, and N. S. Sariciftci, eds., Springer, New York, 2003, pp. 150–151.

40. V. D. Mihailetchi, J. K. J. Duren, P. W. M. Blom, J. C. Hummelen, R. A. J. Janssen, J. M. Kroon, M. T. Rispens, W. J. H. Verhees, and M. M. Wienk, Electron transport in a methanofullerene, *Adv. Func. Mater.* **13**, 43–46 (2003).

41. T. Durkop, S. A. Getty, E. Cobas, and M. S. Fuhrer, Extraordinary mobility in semiconducting carbon nanotubes, *Nano Lett.* **4**, 35–39 (2004).

42. C. J. Brabec, Semiconductor aspects of organic bulk heterojunction solar cells, in *Organic Photovoltaics: Concepts and Realization*, C. Brabec, V. Dyakonov, J. Parisi, and N. S. Sariciftci, eds., Springer, New York, 2003, p. 196.

43. C. J. Brabec, Semiconductor aspects of organic bulk heterojunction solar cells, in *Organic Photovoltaics: Concepts and Realization*, C. Brabec, V. Dyakonov, J. Parisi, and N. S. Sariciftci, eds., Springer, New York, 2003, pp. 200–204.

44. S. A. Curran, P. M. Ajayan, W. J. Blau, D. L. Carroll, J. N. Coleman, A. B. Dalton, A. P. Davey, A. Drury, B. McCarthy, S. Maier, and A. Strevens, A composite from Poly(*m*-phenylenevinylene-*co*-2,5-dioctoxy-*p*-phenylenevinylene) and carbon nanotubes: a novel material for molecular optoelectronics, *Adv. Mater.* **10**, 1091–1093 (1998).

45. H. Ago, K. Petritsch, M. S. P. Shaffer, A. H. Windle, and R. H. Friend, Composites of carbon nanotubes and conjugated polymers for photovoltaic devices, *Adv. Mater.* **11**, 1281–1285 (1999).

46. E. Kymakis, I. Alexandou, and G. A. J. Amaratunga, Single-walled carbon nanotube–polymer composites: electrical, optical and structural investigation, *Synth. Met.* **127**, 59–62 (2002).

47. S. G. Bailey, B. J. Landi, T. G. Gennett, and R. P. Raffaelle, Carbon nanotube solar cells, at *Space Photovoltaic Research and Technology XVIII*, chaired by P. P. Jenkins, Cleveland, OH, September 16–18, 2003.

48. S. G. Bailey, S. L. Castro, B. J. Landi, H. J. Ruf, and R. P. Raffaelle, Nanotube/quantum dot-polymer solar cells, at *19th European Photovoltaic Solar Energy Conference and Exhibition*, chaired by W. Hoffmann et al., Paris, France, June 7–11, 2004.

49. E. E. Donaldson, M. F. Durstock, B. E. Taylor, D. W. Tomlin, L. C. Richardson, and J. W. Baur, Hybrid polymer-based photovoltaics via carbon nanotubes and electrostatic self-assembly, in *Proc. SPIE*, Vol. 4465, "Organic Photovoltaics II", Z. H. Kafafi and D. Fichou, eds., SPIE, San Diego, CA, 2001, pp. 85–93.

50. M. Vignali, R. Edwards, and V. J. Cunnane, Polythiophene layers made by electropolymerisation for a new polythiophene-nanotube photovoltaic cell concept, at *19th European Photovoltaic Solar Energy Conference and Exhibition*, chaired by W. Hoffmann et al., Paris, France, June 7–11, 2004.

51. P. C. Ramamurthy, A. M. Malshe, W. R. Harrell, R. V. Gregory, K. McGuire, and A. M. Rao, Electronic device fabricated from polyaniline/single walled carbon nanotubes composite, in *Proc. Mat. Res. Soc. Symp.*, Vol. 772 "Nanotube-based Devices", P. Bernier et al., eds., MRS, San Francisco, CA, 2003, pp. M4.3.1–M4.3.6.

52. E. Kymakis and G. A. J. Amaratunga, Photovoltaic cells based on dye-sensitisation of single-wall carbon nanotubes in a polymer matrix, *Sol. Energy Mater. Sol. Cells* **80**, 465–472 (2003).

53. B. Kannan, K. Castelino, and A. Majumdar, Design of nanostructured heterojunction polymer photovoltaic devices, *Nano Lett.* **3**, 1729–1733 (2003).

54. W. U. Huynh, J. J. Dittmer, N. Teclemariam, D. J. Milliron, and A. P. Alivisatos, Charge transport in hybrid nanorod-polymer composite photovoltaic cells, *Phys. Rev. B* **67**, 115326-1–115326-12 (2003).

55. L. Dai, A. Patil, X. Gong, Z. Guo, L. Liu, Y. Liu, and D. Zhu, Aligned nanotubes, *Chem. Phys. Chem.* **4**, 1150–1169 (2003).

56. K. Kim and J.-I. Jin, Preparation of PPV nanotubes and nanorods and carbonized products derived there from, *Nano Lett.* **1**, 631–636 (2001).

57. D.-C. Li, L. Dai, S. Huang, A. W. H. Mau, and Z. L. Wang, Structure and growth of aligned carbon nanotube films by pyrolysis, *Chem. Phys. Lett.* **316**, 349–355 (2000).

58. Z. Liu, Z. Shen, T. Zhu, S. Hou, L. Ying, Z. Shi, and Z. Gu, Organizing single-walled carbon nanotubes on gold using a wet chemical self-assembling technique, *Langmuir*, **16**, 3569–3573 (2000).

59. J. C. Hulteen and C. R. Martin, A general template-based method for the preparation of nanomaterials, *J. Mater. Chem.* **71**, 1075–1087 (1997).

60. W. Z. Li, S. S. Xie, L. X. Qian, B. H. Chang, B. S. Zou, W. Y. Zhou, R. A. Zhao, and G. Wang, Large-scale synthesis of aligned carbon nanotubes, *Science* **274**, 1701–1703 (1996).

61. J. Li, C. Papadopoulos, and J. M. Xu, Highly-ordered carbon nanotube arrays for electronics applications, *Appl. Phys. Lett.* **75**, 367–369 (1999).

62. T. Iwasaki, T. Motoi, and T. Den, Multiwalled carbon nanotubes growth in anodic alumina nanoholes, *Appl. Phys. Lett.* **75**, 2044–2046 (1999).

63. G. Che, B. B. Lakshmi, C. R. Martin, E. R. Fisher, and R. S. Ruoff, Chemical vapor deposition based synthesis of carbon nanotubes and nanofibers using a template method, *Chem. Mater.* **10**, 260–267 (1998).

64. J.-I. Jin, Polyconjugated polymers in nano shapes-preparation by chemical deposition polymerization and their properties, at Great Lakes photonics symposium, *Nano and MEMS (Micro-Systems) Conference*, chaired by J. G. Grote et al., Cleveland, OH, June 7–11, 2004.

65. K. Kim, S. H. Lee, W. Yi, J. Kim, J. W. Choi, Y. Park, and J.-I. Jin, Efficient field emission from highly aligned, graphitic nanotubes embedded with gold nanoparticles, *Adv. Mater.* **15**, 1618–1622 (2003).

66. V. M. Cepak, J. C. Hulteen, G. Che, K. B. Jirage, B. B. Lakshmi, E. R. Fisher, and C. R. Martin, Chemical strategies for template syntheses of composite micro-and nanostructures, *Chem. Mater.* **9**, 1065–1067 (1997).

67. Z. F. Ren, Z. P. Huang, J. W. Xu, J. H. Wang, P. Bush, M. P. Siegal, and P. N. Provencio, Synthesis of large arrays of well-aligned carbon nanotubes on glass, *Science* **282**, 1105–1107 (1998).

68. M. Tanemura, K. Iwata, K. Takahashi, Y. Fujimoto, F. Okuyama, H. Sugie, and V. Filip, Growth of aligned carbon nanotubes by plasma-enhanced chemical vapor deposition: optimization of growth parameters, *J. Appl. Phys.* **90**, 1529–1533 (2001).

69. Y. H. Wang, J. Lin, C. H. Huan, and G. S. Chen, Synthesis of large area aligned carbon nanotube arrays from C_2H_2–H_2 mixture by rf plasma-enhanced chemical vapor deposition, *Appl. Phys. Lett.* **79**, 680–682 (2001).

70. S. H. Tsai, C. T. Shiu, S. H. Lai, L. H. Chan, W. J. Hsieh, and H. C. Shih, *In situ* growing and etching of carbon nanotubes on silicon under microwave plasma, *J. Mater. Sci. Lett.* **21**, 1709–1711 (2002).

71. J. Y. Lee and B. S. Lee, Nitrogen induced structure control of vertically aligned carbon nanotubes synthesized by microwave plasma enhanced chemical vapor deposition, *Thin Solid Films* **418**, 85–88 (2002).

72. Y. Avigal and R. Kalish, Growth of aligned carbon nanotubes by biasing during growth, *Appl. Phys. Lett.* **78**, 2291–2293 (2001).

73. C. N. R. Rao and R. Sen, Large aligned-nanotube bundles from ferrocene pyrolysis, *Chem. Commun.* 1525–1526 (1998).

74. B. J. Landi, H. Ruf, C. Schauerman, R. P. Raffaelle, J. D. Harris, and A. F. Hepp, Injection CVD grown MWNTs for PEM fuel cells, at AIAA symposium *2nd International Energy Conversion Engineering Conference*, chaired by D. M. Allen, Providence, RI, August 16–19, 2004.

75. X. B. Wang, Y. Q. Liu, and D. B. Zhu, Two-and three-dimensional alignment and patterning of carbon nanotubes, *Adv. Mater.* **14**, 165–167 (2002).

76. S. Huang, L. Dai, and A. W. H. Mau, Patterned growth and contact transfer of well-aligned carbon nanotube films, *J. Phys. Chem. B* **103**, 4223–4227 (1999).

77. M. Gao, S. Huang, L. Dai, G. Wallace, R. Gao, and Z. Wang, Aligned coaxial nanowires of carbon nanotubes sheathed with conducting polymers, *Angew. Chem. Int. Ed.* **39**, 3664–3667 (2000).

78. B. Wu, J. Zhang, Z. Wei, S. Cai, and Z. Liu, Chemical alignment of oxidatively shortened single-walled carbon nanotubes on silver surface, *J. Phys. Chem. B* **105**, 5075–5078 (2001).

79. X. Nan, Z. Gu, and Z. Liu, Immobilizing shortened single-walled carbon nanotubes (SWNTs) on gold using a surface condensation method, *J. Col. Inter. Sci.* **245**, 311–318 (2002).

80. K. M. Coakley, Y. Liu, M. D. McGehee, K. L. Frindell, and G. D. Stucky, Infiltrating semiconducting polymers into self-assembled mesoporous titania films for photovoltaic applications, *Adv. Func. Mater.* **13**, 301–306 (2003).

81. M. M. Wienk, J. M. Kroon, W. J. H. Verhees, J. Knol, J. C. Hummelen, P. A. van Hal, and R. A. J. Janssen, Efficient methano[70]fullerene/MDMO-PPV bulk heterojunction photovoltaic cells, *Angew. Chem. Int. Ed.* **42**, 3371–3375 (2003).

82. J. Li, N. Sun, Z.-X. Guo, C. Li, Y. Li, L. Dai, D. Zhu, D. Sun, Y. Cao, and L. Fan, Photovoltaic properties of MEH-PPV doped with new methanofullerene derivatives, *Synth. Met.* **137**, 1527–1528 (2003).

83. M. T. Rispens and J. C. Hummelen, Photovoltaic applications, in *Fullerenes: From Synthesis to Optoelectronic Properties*, D. M. Guldi, N. Martin, eds., Kluwer, Boston, 2002, pp. 387–435.

84. W. Wu, J. Li, L. Liu, L. Yanga, Z.-X. Guo, L. Dai, and D. Zhu, The photoconductivity of PVK–carbon nanotube blends, *Chem. Phys. Lett.* **364**, 196–199 (2002).

85. B. Sun, E. Marx, and N. C. Greenham, Photovoltaic devices using blends of branched CdSe nanoparticles and conjugated polymers, *Nano Lett.* **3**, 961–963 (2003).

86. T. Lin, L. Dai, G. Wallace, A. Burrell, and D. Officer, C_{60}-containing conjugated polymers and carbon nanotubes as optoelectronic nanomaterials, *Polym. Preprints* **43**(1), 98 (2002).

87. W. Wu, S. Zhang, Y. Li, J. Li, L. Liu, Y. Qin, Z.-X. Guo, L. Dai, C. Ye, and D. Zhu, PVK-modified single-walled carbon nanotubes with effective photoinduced electron transfer, *Macromolecules* **36**, 6286–6288 (2003).

88. C. J. Brabec, C. Windcr, N. S. Sariciftci, J. C. Hummelen, A. Dhanabalan, P. A. van Hal, and R. A. J. Janssen, A low-bandgap semiconducting polymers for photovoltaic devices and infrared emitting diodes, *Adv. Func. Mater.* **12**, 709–712 (2002).

89. C. Winder and N. S. Sariciftci, Low bandgap polymers for photon harvesting in bulk heterojunction solar cells, *J. Mater. Chem.* **14**, 1077–1086 (2004).

Index